CHRONIC INFLAMMATION

Molecular Pathophysiology, Nutritional and Therapeutic Interventions

CHRONIC INFLAMMATION

Molecular Pathophysiology, Nutritional and Therapeutic Interventions

Edited by
SASHWATI ROY
DEBASIS BAGCHI
SIBA P. RAYCHAUDHURI

CRC Press is an imprint of the
Taylor & Francis Group, an **informa** business

CRC Press
Taylor & Francis Group
6000 Broken Sound Parkway NW, Suite 300
Boca Raton, FL 33487-2742

First issued in paperback 2016

Version Date: 20120711

ISBN 13: 978-1-138-19955-2 (pbk)
ISBN 13: 978-1-4398-7211-6 (hbk)

Visit the Taylor & Francis Web site at
http://www.taylorandfrancis.com

and the CRC Press Web site at
http://www.crcpress.com

To my mentors in science, beloved parents, and my family and friends
Sashwati Roy

To my beloved father, the late Sri Tarak Chandra Bagchi
Debasis Bagchi

To Smriti Kana Raychaudhuri and to my mother Bilwabasani Raychaudhuri
Siba P. Raychaudhuri

Contents

SECTION II Pathologies Associated with Inflammation

SECTION III Nutrition & Therapeutics for Inflammatory Diseases

Preface

While acute inflammation is a healthy physiological response aimed at wound healing, chronic inflammation has been directly implicated in a wide range of degenerative human health disorders. These pathologies encompass almost all of present-day noncommunicable diseases such as obesity, diabetes, atherosclerosis, and high blood pressure, as well as cancer. Thus, the February 23, 2004, cover of *TIME* magazine featured inflammation as "The Secret Killer." To preserve good health, it is important that we are capable of rapidly mounting an inflammatory response to tissue injury. It is also equally important that such inflammation be resolved in a timely manner such that the state of chronic inflammation is averted. Lifestyle factors such as diet, stress, tobacco, obesity, infection, and pollutants are known to contribute to chronic inflammation. Prudent nutritional and exercise habits are powerful tools to fight chronic inflammation.

In this volume, the following three sections provide cutting-edge and comprehensive treatment of the process and factors that influence chronic inflammation.

(I) Systems Biology of Inflammation and Regulatory Mechanisms (Chapters 1–10)
(II) Pathologies Associated with Inflammation (Chapters 11–18)
(III) Nutrition & Therapeutics for Inflammatory Diseases (Chapters 19–28)

Section I addresses the understanding of the process of chronic inflammation including initiation, progression, and resolution. Section II includes a rigorous and critical treatment of specific human health disorders where chronic inflammation plays a major role. Section III discusses countermeasures for protection including nutritional and other interventions. Protective abilities of structurally diverse antioxidants, phytochemicals, anti-inflammatory diets, omega-3 fatty acids, NSAIDs, disease-modifying antirheumatic drugs, and novel regimens have been extensively discussed by authoritative experts in the discipline.

We extend our special thanks to Randy Brehm and Jill Jurgensen at Taylor & Francis for their continued support and help for putting this volume together. It has been a rewarding experience to interact with the generous authors who were enthusiastic and willing to contribute to this volume.
We hope you enjoy this volume as much as we have enjoyed putting it together.

Sashwati Roy, PhD
The Wexner Medical Center at The Ohio State University

Debasis Bagchi, PhD, MACN, CNS, MAIChE
University of Houston College of Pharmacy

Siba P. Raychaudhuri, MD, FACR
School of Medicine, University of California, Davis

Editors

Sashwati Roy, PhD, is an associate professor of surgery at the Ohio State University, Columbus. She received her PhD in 1994 in physiology and environmental sciences. She completed her postdoctoral training at the University of California, Berkeley. Her research interests include wound inflammation, mechanisms of resolution of diabetic wound inflammation, and the role of miRNA in tissue repair processes.

Dr. Roy has more than 150 peer-reviewed publications. She is an expert in significance of inflammation in chronic wounds and has delivered dozens of lectures in international and national level meetings. Dr. Roy's research is currently funded by the National Institutes of Health to investigate the role of inflammation in diabetic wounds.

Debasis Bagchi, PhD, MACN, CNS, MAIChE, received his PhD in medicinal chemistry in 1982. He is a professor in the Department of Pharmacological and Pharmaceutical Sciences at the University of Houston. He is also Director of Innovation & Clinical Affairs at Iovate Health Research Sciences Inc., Oakville, Ontario. Dr. Bagchi is the immediate past president of the American College of Nutrition, Clearwater, Florida; a distinguished advisor at the Japanese Institute for Health Food Standards, Tokyo, Japan; and immediate past chairman of the Nutraceuticals and Functional Foods Division of the Institutes of Food Technologists, Chicago, Illinois. Dr. Bagchi received the Master of American College of Nutrition Award in October 2010. His research interests include free radicals, human diseases, carcinogenesis, pathophysiology, mechanistic aspects of cytoprotection by antioxidants, regulatory pathways in obesity, diabetes, and gene expression.

Dr. Bagchi has 281 papers in peer-reviewed journals, 11 books, and 15 patents. He has delivered invited lectures in various national and international scientific conferences, organized workshops, and group discussion sessions. He is a fellow of the American College of Nutrition, member of the Society of Toxicology, member of the New York Academy of Sciences, fellow of the Nutrition Research Academy, and member of the TCE stakeholder Committee of the Wright Patterson Air Force Base, Ohio. Dr. Bagchi is a member of the Study Section and Peer Review Committee of the National Institutes of Health. He is the associate editor of the *Journal of Functional Foods* and *Journal of the American College of Nutrition*, and also serves as an editorial board member of numerous peer-reviewed journals including *Antioxidants and Redox Signaling*, *Cancer Letters*, *Toxicology Mechanisms and Methods*, and other scientific and medical journals. He is also a consulting editor of CRC Press/Taylor & Francis.

Dr. Bagchi has received funding from various institutions and agencies including the U.S. Air Force Office of Scientific Research, Nebraska State Department of Health, Biomedical Research Support Grant from National Institutes of Health (NIH), National Cancer Institute (NCI), Health Future Foundation, the Procter & Gamble Company, and Abbott Laboratories.

Siba P. Raychaudhuri, MD, FACR, received his rheumatology training at Stanford University. He received his MD in 1987. In his early research career, Dr. Raychaudhuri directed one of the most successful psoriasis research programs at the Psoriasis Research Institute, Palo Alto, California, and worked on cutting-edge immune-based therapy for autoimmune diseases at the Stanford University School of Medicine. His research group works in arthritis, inflammation, human autoimmune diseases, and animal models of inflammation. The long-term goal of his research group is to explore the inflammatory cascades in inflammatory diseases and develop safe and effective therapies by targeting the critical molecular events specific for these groups of diseases.

Dr. Raychaudhuri is currently an associate professor in the Division of Rheumatology, Allergy and Clinical Immunology of the University of California, Davis, and chief of the Rheumatology Division at the VA Sacramento Medical Center. He is a fellow of the American College of Rheumatology and the American Academy of Dermatology, and is a member of the American College of Physicians.

Contributors

Amita Aggarwal
Department of Clinical Immunology
Sanjay Gandhi Postgraduate Institute of Medical Sciences
Lucknow, India

Bharat B. Aggarwal
Department of Experimental Therapeutics
The University of Texas MD Anderson Cancer Center
Houston, Texas

Reeva Aggarwal
The Dorothy M. Davis Heart and Lung Research Institute
The Wexner Medical Center at The Ohio State University
Columbus, Ohio

Myron Allukian
Department of Surgery
University of Pennsylvania School of Medicine
Philadelphia, Pennsylvania

Ali Alqahtani
Faculty of Pharmacy
University of Sydney
New South Wales, Australia

Praveen R. Arany
School of Engineering and Applied Sciences
Harvard University
Cambridge, Massachusetts

Debasis Bagchi
Department of Pharmacological and Pharmaceutical Sciences
University of Houston
Houston, Texas

Clifton A. Baile
Departments of Animal & Dairy Science and Foods & Nutrition
University of Georgia
Athens, Georgia

Jaideep Banerjee
Department of Surgery
Wexner Medical Center at The Ohio State University
Columbus, Ohio

Annadora J. Bruce-Keller
Pennington Biomedical Research Center
Louisiana State University
Baton Rouge, Louisiana

Philip C. Calder
Human Development and Health Academic Unit
University of Southampton
Southampton, United Kingdom

Kelvin Chan
Faculty of Pharmacy
University of Sydney
New South Wales, Australia
and
Centre for Complementary Medicine Research
University of Western Sydney
New South Wales, Australia

Vinod Chandran
Department of Medicine
University of Toronto
Toronto, Ontario, Canada

Shampa Chatterjee
Institute for Environmental Medicine
University of Pennsylvania Medical Center
Philadelphia, Pennsylvania

Hun-Taeg Chung
School of Biological Sciences
University of Ulsan
Ulsan, Republic of Korea

Amitava Das
Department of Surgery
The Ohio State University
Columbus, Ohio

Hiranmoy Das
The Dorothy M. Davis Heart and Lung Research Institute
The Wexner Medical Center at The Ohio State University
Columbus, Ohio

MaryAnne Della-Fera
Department of Animal & Dairy Science
University of Georgia
Athens, Georgia

Tim D. Eubank
Division of Pulmonary, Allergy, Critical Care, and Sleep Medicine
The Ohio State University
Columbus, Ohio

Ujjawal H. Gandhi
Department of Veterinary and Biomedical Sciences
The Pennsylvania State University
University Park, Pennsylvania

Alakendu Ghosh
Department of Rheumatology
Institute of Postgraduate Medical Education and Research
Kolkata, India

Gowrishankar Gnanasekaran
Department of Internal Medicine, Lipid Signaling, Lipidomics, and Vasculotoxicity Laboratory
The Dorothy M. Davis Heart and Lung Research Institute
The Ohio State University College of Medicine
Columbus, Ohio

Subash C. Gupta
Department of Experimental Therapeutics
The University of Texas MD Anderson Cancer Center
Houston, Texas

Nigil Haroon
Department of Medicine
University of Toronto
Toronto, Ontario, Canada

Kiichi Hirota
Department of Anesthesia
Kyoto University Hospital
Kyoto, Japan

George X. Huang
School of Engineering and Applied Sciences
Harvard University
Cambridge, Massachusetts

Xueting Jiang
Burnett School of Biomedical Sciences
College of Medicine
University of Central Florida
Orlando, Florida

Antony Kam
Faculty of Pharmacy
University of Sydney
New South Wales, Australia

Naveen Kaushal
Department of Veterinary and Biomedical Sciences
The Pennsylvania State University
University Park, Pennsylvania

Savita Khanna
Department of Surgery
The Wexner Medical Center at The Ohio State University
Columbus, Ohio

Ji Hye Kim
Department of Experimental Therapeutics
The University of Texas MD Anderson Cancer Center
Houston, Texas

Woo Seob Kim
School of Engineering and Applied Sciences
Harvard University
Cambridge, Massachusetts
and
College of Medicine
Chung-Ang University
Seoul, Republic of Korea

Anil Kumar Kotha
Department of Veterinary and Biomedical Sciences
The Pennsylvania State University
University Park, Pennsylvania

Sainath R. Kotha
Department of Internal Medicine, Lipid Signaling, Lipidomics, and Vasculotoxicity Laboratory
The Dorothy M. Davis Heart and Lung Research Institute
The Ohio State University College of Medicine
Columbus, Ohio

George Q. Li
Faculty of Pharmacy
University of Sydney
New South Wales, Australia

Kong M. Li
Sydney Medical School
University of Sydney
New South Wales, Australia

Kenneth W. Liechty
Department of Surgery
The University of Mississippi Medical Center
Jackson, Mississippi

Smitha Malireddy
Department of Internal Medicine, Lipid Signaling, Lipidomics, and Vasculotoxicity Laboratory
The Dorothy M. Davis Heart and Lung Research Institute
The Ohio State University College of Medicine
Columbus, Ohio

Ginger L. Milne
Division of Clinical Pharmacology
Vanderbilt University
Nashville, Tennessee

Ananya Datta Mitra
School of Medicine
University of California, Davis
Davis, California

Anupam Mitra
School of Medicine & VA Medical Center Sacramento
University of California, Davis
Davis, California

Arpita Myles
Department of Clinical Immunology
Sanjay Gandhi Postgraduate Institute of Medical Sciences
Lucknow, India

Chandrakala Aluganti Narasimhulu
Burnett School of Biomedical Sciences
College of Medicine
University of Central Florida
Orlando, Florida

Vivek Narayan
Department of Veterinary and Biomedical Sciences
The Pennsylvania State University
University Park, Pennsylvania

Shakira M. Nelson
Department of Veterinary and Biomedical Sciences
The Pennsylvania State University
University Park, Pennsylvania

John Noel
Institute for Environmental Medicine
University of Pennsylvania Medical Center
Philadelphia, Pennsylvania

Eshaifol A. Omar
Faculty of Pharmacy
University of Sydney
New South Wales, Australia

Hyun-Ock Pae
Department of Microbiology and Immunology
Wonkwang University School of Medicine
Iksan, Republic of Korea

Narasimham L. Parinandi
Department of Internal Medicine, Lipid Signaling, Lipidomics, and Vasculotoxicity Laboratory
The Dorothy M. Davis Heart and Lung Research Institute
The Ohio State University College of Medicine
Columbus, Ohio

Sampath Parthasarathy
Burnett School of Biomedical Sciences
College of Medicine
University of Central Florida
Orlando, Florida

Vincent J. Pompili
The Dorothy M. Davis Heart and Lung Research Institute
The Wexner Medical Center at The Ohio State University
Columbus, Ohio

K. Sandeep Prabhu
Department of Veterinary and Biomedical Sciences
The Pennsylvania State University
University Park, Pennsylvania

Sahdeo Prasad
Department of Experimental Therapeutics
The University of Texas MD Anderson Cancer Center
Houston, Texas

Irfan Rahman
Department of Environmental Medicine
University of Rochester Medical Center
Rochester, New York

Saravanan Rajendrasozhan
Department of Chemistry
University of Hail
Hail, Saudi Arabia

Srujana Rayalam
Department of Animal & Dairy Science
University of Georgia
Athens, Georgia

Siba P. Raychaudhuri
Department of Medicine
School of Medicine
University of California, Davis
Davis, California
and
Department of Rheumatology
Veterans Administration Medical Center Sacramento
Mather, California

Smriti K. Raychaudhuri
School of Medicine & VA Medical Center Sacramento
Division of Rheumatology, Allergy, & Clinical Immunology
University of California, Davis
Davis, California

Valentina Razmovski-Naumovski
Faculty of Pharmacy
University of Sydney
New South Wales, Australia
and
Centre for Complementary Medicine Research
University of Western Sydney
New South Wales, Australia

Flávio Reis
Medicine Faculty
Coimbra University
Coimbra, Portugal

Cameron Rink
Department of Surgery
The Wexner Medical Center at The Ohio State University
Columbus, Ohio

Julie M. Roda
Division of Pulmonary, Allergy, Critical Care, and Sleep Medicine
The Ohio State University
Columbus, Ohio

Sashwati Roy
Department of Surgery
The Dorothy M. Davis Heart and Lung Research Institute
The Wexner Medical Center at The Ohio State University
Columbus, Ohio

Jordan D. Secor
Department of Internal Medicine, Lipid Signaling, Lipidomics, and Vasculotoxicity Laboratory
The Dorothy M. Davis Heart and Lung Research Institute
The Ohio State University College of Medicine
Columbus, Ohio

Krithika Selvarajan
Burnett School of Biomedical Sciences
College of Medicine
University of Central Florida
Orlando, Florida

Chandan K. Sen
Department of Surgery
Wexner Medical Center at The Ohio State University
Columbus, Ohio

Pradyot Sinhamahapatra
Department of Rheumatology
Institute of Postgraduate Medical Education and Research
Kolkata, India

Isaac K. Sundar
Department of Environmental Medicine
University of Rochester Medical Center
Rochester, New York

Edite Teixeira de Lemos
Medicine Faculty
Coimbra University
Coimbra, Portugal
and
ESAV, Polytechnic Institute of Viseu
Viseu, Portugal

Yoram Vodovotz
Department of Surgery
University of Pittsburgh
Pittsburgh, Pennsylvania

Traci A. Wilgus
Department of Pathology
The Ohio State University
Columbus, Ohio

Ka H. Wong
Faculty of Pharmacy
University of Sydney
New South Wales, Australia

Brian C. Wulff
Department of Pathology
The Ohio State University
Columbus, Ohio

Zhaohui Yang
Burnett School of Biomedical Sciences
College of Medicine
University of Central Florida
Orlando, Florida

Xian Zhou
Faculty of Pharmacy
University of Sydney
New South Wales, Australia

Section I

Systems Biology of Inflammation and Regulatory Mechanisms

1 At the Interface between Acute and Chronic Inflammation

Insights from Computational Modeling

Yoram Vodovotz
Department of Surgery, University of Pittsburgh
Pittsburgh, Pennsylvania

CONTENTS

1.1 INFLAMMATION IN HEALTH AND DISEASE: A ROBUST, ADAPTIVE, MULTISCALE SYSTEM

Inflammatory diseases include infection/sepsis, trauma, inflammatory bowel diseases, chronic wounds, rheumatologic disorders, and asthma; many other diseases, such as cancer, diabetes, atherosclerosis, Alzheimer's, and obesity, are also associated with dysregulated inflammation. Inflammation can be thought of as acute (in settings such as sepsis, trauma, and wound healing) or chronic (in diseases such as rheumatoid arthritis, ulcerative colitis, Crohn's disease, etc.). Previously, chronic and acute inflammatory processes were thought to be driven by different causes, through the activities of different cells and inflammatory mediators, and to result in quite different ultimate outcomes. However, a more modern view suggests that these processes are interlinked (Figure 1.1) [66] and have evolved to give organisms the robustness and flexibility to deal with diverse insults as well as regulating key homeostatic processes [39, 80, 85, 109, 143]. Moreover, inflammation is linked to stress and cellular damage/dysfunction, not merely to overt tissue damage [24, 96].

The acute inflammatory response involves a cascade of events mediated by a large array of cells and molecules that locate invading pathogens or damaged tissue, alert and recruit other cells and molecules, eliminate the offending agents, and finally restore the body to equilibrium. In sepsis and

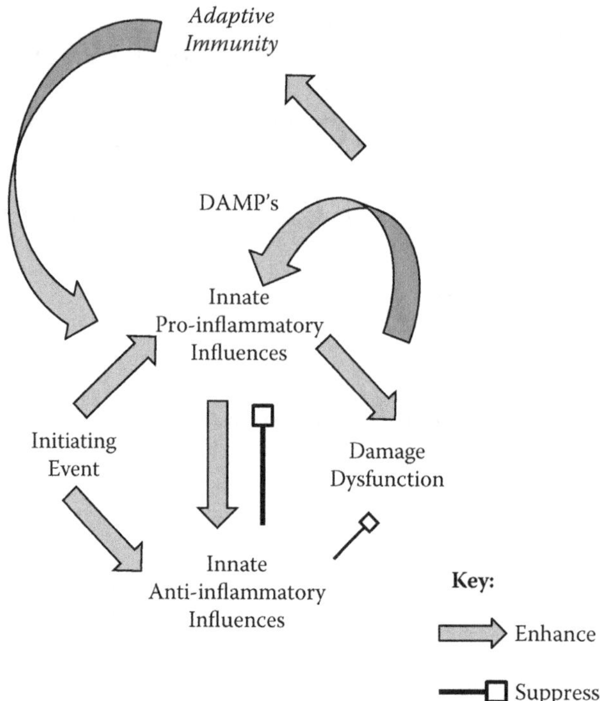

FIGURE 1.1 Interconnected processes in acute and chronic inflammation. Following an initiating event (e.g., trauma, hemorrhage, infection, or exposure to a chemical irritant or immunostimulant), both pro- and anti-inflammatory influences (e.g., chemokines, cytokines, lipid products, and free radicals) are elaborated, leading to tissue damage or dysfunction. These stressed tissues elaborate DAMPs, which further propagate innate immune mechanisms. By activating DCs, DAMPs also lead to the initiation and propagation of adaptive immune mechanisms that in turn help drive chronic inflammation.

trauma, this response is accompanied by macroscopic manifestations such as fever and elevated heart rate, which contribute to optimize the various defense mechanisms involved. In other tissues, inflammation manifests as redness, swelling, and pain. Perplexingly, the feed-forward loop of inflammation → damage/dysfunction → inflammation can lead to persistent, dysregulated inflammation that promotes organ dysfunction and death [68, 92], and yet a well-regulated inflammatory response is necessary for proper tissue healing (Figure 1.1) [113, 122, 124, 126] .

Indeed, inflammation is not in and of itself detrimental. It is in most cases a well-coordinated communication network operating at an intermediate time scale between neural and longer-term endocrine processes [109, 154]. Inflammation is necessary for the removal or reduction of challenges to the organism and subsequent restoration of homeostasis [109]. For example, plasma levels of the early pro-inflammatory cytokine tumor necrosis factor-α (TNF-α) were significantly elevated within 6 h postadmission in human trauma survivors versus nonsurvivors. Moreover, plasma TNF-α was inversely correlated with indices of organ dysfunction, both in the trauma patients taken together and in the survivor cohort. Furthermore, swine exhibiting a robust TNF-α response following two different, clinically realistic trauma/hemorrhage paradigms survived these insults, and those animals that did not mount an adequately robust TNF-α response died [106]. More recent studies suggest that well-organized dynamic networks of inflammatory mediators are invoked early in the response to experimental surgical trauma in mice, and that these networks become disorganized and less complex when hemorrhagic shock is superimposed on this minor trauma [97]. This requirement for a robust, properly interconnected early inflammatory response may represent a means of communicating the extent of injury and perhaps also of preparing the body for a possible secondary infection.

Moreover, these findings suggest that a hallmark of a healthy response to stress involves both physiological and inflammatory responsiveness commensurate with the degree of stress experienced by cells, tissues, organs, and the whole organism. Thus, physiology and inflammation appear to be linked—and in a sense "integrated"—via the release of damage-associated molecular pattern (DAMP) molecules (Figure 1.1) [98, 148]. DAMPs such as high-mobility group box-1 (HMGB1) activate macrophages [17] and neutrophils [114], and are therefore key innate immune stimulants. In addition, DAMPs activate dendritic cells [131], thereby leading to T cell–mediated responses. Thus, a large body of literature suggests that DAMPs propagate inflammation in both infectious and sterile inflammatory settings using similar signaling pathways [16, 95, 101, 153]. Insights derived from computational modeling, discussed in the next section, raise the hypothesis that the body integrates the signals communicated via inflammatory mediators via DAMPs derived from damaged or dysfunctional tissues, and thus in a direct, mechanistic sense, the dynamics of these DAMPs may reflect the tissue's or organism's health [98, 147, 148, 152, 153]. Both classical inflammatory mediators and DAMPs affect important physiological functions, which in turn lead to further production of inflammatory mediators and perpetuate a positive feedback loop [16, 60, 115, 129, 160]. Thus, unrecoverable simulated tissue damage/dysfunction may serve as a surrogate for progressive cell/tissue/organ dysfunction and eventual death, while damage/dysfunction that tends to return to baseline may be a proxy for survival.

1.2 INFLAMMATION IS A COMPLEX SYSTEM WITH CONTEXT-DEPENDENT REGULATION: INSIGHTS FROM COMPUTATIONAL MODELING

The acute inflammatory response exhibits the features of a complex system: dependence on initial conditions and relative strengths of interaction among components, nonlinear dynamics of and the key role of thresholds of activation, and the emergence of outcomes at higher levels of organization that would not be predicted simply from knowledge of the individual components [110]. The tool set for computational modeling of complex systems involves data-driven methods [67], traditional ordinary differential equation (ODE) and partial differential equation (PDE)-based approaches [148, 154], rules-based signaling modeling platforms such as BioNetGen [45], and agent-based modeling platforms such as NetLogo and SPARK [15]. Data-driven models are based on associations among variables but can be used to infer mechanistic information regarding principal drivers (through the use of methods such as principal component analysis) [67, 152]. Equation-based models describe mechanistically the sequential change in the states of the components of the system over time, in which the variables of the equations generally represent average concentrations of the various components. These systems of equations are generally most accurate in settings in which large numbers of individuals of these components are assumed to exist and to exert their effects in the aggregate [148, 154]. When the numbers become small, differential equation descriptions break down, and agent-based models (ABM; a type of cellular automata simulation) may be more accurate than equation-based models in settings in which the stochastic actions of these agents are a better approximation of biological reality as compared to the actions of these components in the aggregate [15].

Despite this extensive tool set and advances in computational modeling as applied to well-controlled biological systems in experimental settings, the use of these methods and techniques has lagged in the preclinical and clinical arena. Various groups have already created simulations of inflammation in wound healing at various scales, from cell culture studies to the whole organism [28, 37, 88, 140, 155, 156]. As a whole, however, these diverse studies have not been focused on the core problem of clinical translation of preclinical findings. Translational systems biology [13, 14, 154] involves using mechanistic computational modeling generated in basic science research to simulate higher-level behaviors at the organ and organism levels, without shying away from the realities of the preclinical and clinical settings. Simply stated, translational systems biology is a framework in which translational rather than basic insights are primary, mathematical models are designed to facilitate *in silico* clinical trials, simulations are appropriate for *in vivo* validation, and mechanistic simulations of whole-organism responses could guide therapeutic approaches. Translational systems biology spans a range of applications from drug discovery and development, animal models of disease, simulated clinical trials, and personalized medicine, to rational drug/device design [12, 13, 98, 147, 148, 152, 153, 154].

Translational systems biology was spurred on by the clinical challenge of integrating acute inflammation and organ dysfunction in critical illness, including trauma/hemorrhage, sepsis, and related processes such as necrotizing enterocolitis [14, 151, 154]. This work progressed to computational modeling of the next logical temporal step, such as healing of injuries/wounds and infections, including chronic, nonhealing, diabetic foot ulcers [99] as well as experimental injuries to the vocal folds in both animals [83] and human volunteers [82], as well as post-spinal cord injury pressure ulcers. These studies then went on to cover the spectrum of inflammation → wound healing → inflammation-associated cancer, based on insights derived from these computational studies, and also based on the long-established hypothesis in which cancer is considered to be the outcome of dysregulated inflammation and wound healing [94, 142]. In the following section, the insights into these processes derived from computational simulations are discussed.

1.2.1 Modeling Inflammation in Trauma, Hemorrhage, and Sepsis

Traumatic injury, often accompanied by hemorrhage, continues to be the most common cause of death for young people and represents a significant source of morbidity and mortality for all ages [73, 74]. Initial survivors of acute trauma are particularly susceptible to multiple organ dysfunction syndrome, a poorly understood syndrome of sequential and gradual loss of organ function [102]. Multiple organ failure is the most frequent cause of late deaths after injury, accounting for substantial morbidity and mortality [58, 59] and is thought to be due, in part, to excessive or maladaptive activation of inflammatory pathways [25]. Organs such as the liver and the gut not only become damaged or dysfunctional from trauma-induced inflammation, but in turn further perpetuate this vicious inflammatory cycle [30, 35, 117]. Furthermore, patients admitted to the intensive care unit following trauma and hemorrhage often become susceptible to infection (a so-called "second hit"), further complicating attempts at immunomodulation early in the clinical course [130].

In the initial innate immune phase of the response to trauma, hemorrhage, or infection, neutrophils and macrophages are activated directly by bacterial endotoxins or indirectly by various stimuli elicited systemically upon trauma and hemorrhage [34, 79, 121, 145]. These stimuli, including endotoxins, enter the systemic circulation quickly and activate circulating monocytes and neutrophils [116]. Activated neutrophils also reach compromised tissue by migrating along a chemoattractant gradient [23]. Once activated, macrophages and neutrophils produce and secrete effectors that activate these same cells and also other cells, such as endothelial cells. Pro-inflammatory cytokines—primary among them TNF-α [22, 31, 33, 40, 69, 76, 119]—promote immune cell activation and therefore feedback positively to promote further production of inflammatory cytokines.

The initial inflammatory response to pathogens, pathogen-derived immunostimulants, trauma, and hemorrhage is known as the innate immune response. To address the complexity of this process, a series of computational models of the acute inflammatory response were developed; several of these models have various degrees of abstraction of relevant mechanisms of inflammation and tissue damage [36, 38, 78, 125]. Alt and Lauffenburger were the first to create a spatially distributed, equation-based model of bacterial-induced chemotaxis of leukocytes, suggesting that the dependence of the dynamic behavior of the inflammatory response was due to specific parameters of acute inflammation [9]. An was the first to suggest that agent-based models could shed insight into the pathobiology of sepsis and associated multiple organ dysfunction [10]. Later, equation-based mathematical models of acute inflammation consisting of a bacterial pathogen, a single population of inflammatory cells, and a measure of global tissue damage/dysfunction interrelated the actions of pro-inflammatory cytokines and DAMPs for the first time, and described both recoverable infection and septic shock, as well as suggested different therapeutic avenues for the diverse manifestations of sepsis [78]. More complex equation-based models incorporating the positive feedback loop of inflammation → damage → inflammation were used to create simulated populations of septic patients [36] to simulate population responses to anthrax in the presence or absence of vaccination [77], and to examine the inflammatory response in necrotizing enterocolitis as well as the possible effects of probiotic treatment in this disease of premature newborn infants [19]. In parallel, an agent-based model of inflammation and organ dysfunction in sepsis was also developed and used to address both the failure of prior sepsis clinical trials and to suggest the likelihood of success of combination therapies [11]. Another equation-based model was calibrated in various inflammatory scenarios in mice [34] and was able to predict responses to combinations of insults on which it was not trained [34, 149, 150]. To increase its translational utility, this latter model was calibrated on easily accessible circulating levels of inflammatory cytokines and NO reaction products, and explicitly included measurable physiological parameters such as blood pressure along with the more abstract global tissue damage/dysfunction (a surrogate for both DAMPs and the health status of the individual; Figure 1.1) in mice [34, 79, 121]. This calibrated model contains key elements of the innate immune response, despite having been calibrated only on survivable doses of endotoxin, and was capable of predicting dose ranges of endotoxin at which death is known to occur [34]. Both equation- and agent-based models of acute inflammation were developed based on the concept of indirect-response modeling, using transcriptomic data derived from circulating leukocytes of human volunteers subjected to endotoxemia [43, 51]. These models were extended to include the reciprocal effects of inflammation and physiological variability, the effects of corticosteroids on these processes, and the role of the autonomic nervous system [50, 52, 53, 134].

Additionally, two systems biology methods, namely mathematical modeling and transcriptomic analysis, were compared and suggested that the role of initial trauma in the murine response to experimental trauma/hemorrhagic shock is central in driving the inflammatory response, both systemically and in the liver (a central organ in acute inflammation) [79]. Despite this overlap in the levels of inflammatory mediators between sham surgical cannulation and that same cannulation combined with hemorrhagic shock in this mouse model, which could be demonstrated upon multiplexed analysis of circulating cytokines [97], a mechanistic mathematical model could differentiate the inflammatory responses to sham cannulation versus hemorrhagic shock [145]. In addition, multivariate regression, principal component analysis, and dynamic network analysis all suggested major mechanistic differences between sham cannulation and hemorrhagic shock [97], and predicted that the majority of the inflammatory response to survivable trauma/hemorrhage was due mostly to the underlying tissue trauma induced by cannulation surgery [79]. The model was extended to include details of experimental trauma/hemorrhage in mice (e.g., bleeding rate and target blood pressure), and further validated using a unique, computerized platform for automated hemorrhage that was constructed specifically to test the behavior of this mathematical model [145].

The acute inflammatory response differs in aged versus young individuals, through mechanisms not yet fully elucidated [54, 55, 120, 136]. We utilized a combination of *in silico*, *in vivo*, and *in vitro* studies to gain insights into acute inflammation in response to bacterial endotoxin (LPS) as a function of age. Using data from a time course endotoxemia study in "middle-aged" (6–8 months old) C57BL/6 mice, we recalibrated a mechanistic mathematical model of acute inflammation originally calibrated for "young" (2–3 months old) C57BL/6 mice using cytokine and NO_2^-/NO_3^- data obtained following exposure to LPS, surgical trauma, or hemorrhagic shock [34]. The central hypothesis that emerged from the recalibration process was that activated macrophages from middle-aged mice would exhibit a greater degree of cell death concomitantly with producing higher levels of TNF-α and IL-6 than macrophages from young mice. This prediction was borne out when intraperitoneal macrophages from young versus "middle-aged" mice were assessed for cell numbers, cell viability, and cytokine production *in vitro*. These studies demonstrate the utility of a combined *in silico*, *in vivo*, and *in vitro* approach to the study of acute inflammation in shock states, and suggest hypotheses with regard to the changes in the cytokine milieu that accompany aging.

Following significant hemorrhage, severe trauma, and/or persistent infection, the innate immune response triggers activation of the more specific and focused adaptive immune response, which is regulated largely by dendritic cells (DC) [20, 30, 44, 64, 93, 100, 108, 133, 135, 146, 162]. A simplified cascade of events leading to the transition from innate to adaptive immune responses was used to augment the mathematical models of inflammation. In these models, "immature" DC phagocytose pathogens become activated and migrate to the lymph nodes. Here, they activate $CD8^+$ T cells and $CD4^+$ T cells by displaying pathogen-specific antigens in the context of major histocompatibility complex class II molecules. $CD4^+$ cells, in turn, activate antigen-specific B cells, which produce antibodies. $CD4^+$ cells also differentiate into T helper cells, either T_H1 or T_H2, which differ in the spectrum of cytokines they produce. $CD8^+$ T cells contribute to tissue damage by eliminating infected cells [70, 85]. Cytokines produced by these cells of the adaptive immune response will modulate both the adaptive and innate immune responses.

The immature DC differentiate, transforming into two subsets, $CD8^+$ and $CD8^-$ [89]. These differentiated DCs will drive the differentiation of naïve T cells into T_H0 cells. In turn, T_H0 cells will differentiate into T_H1 cells (pro-inflammatory in the setting of sepsis) or T_H2 cells (anti-inflammatory in the setting of sepsis). Cytokines will play a major role in determining which response, T_H1 or T_H2, dominates. The cytokine milieu may also affect whether or not a particular DC subset is produced. Another subset of T cells that is stimulated by DCs is the regulatory T cell subset (T_{reg}). This latter subset of T cells suppresses helper T cell subsets, and has been recognized to play a role in the response to trauma and sepsis [146]. An emerging consensus is that the T_H cell response is skewed toward a T_H2/T_{reg} phenotype following trauma and hemorrhage [1, 2, 21, 128, 146]. Cytokines such as transforming growth factor-β1 (TGF-β1) and IL-10 [6, 29, 81, 103] have been implicated in this process.

Building on prior mathematical modeling and experimental studies on DC/T cell interactions [87], a mathematical model encompassing the above processes reproduced the dynamics of cytokine and NO_2^-/NO_3^- production following endotoxemia, surgical trauma, or surgery + hemorrhage over 24 hours. In the simulation, both endotoxemia and trauma ± hemorrhage resulted in T_H1 and T_H2 waves. The model also predicted that in trauma ± hemorrhage, both T_H1 and T_H2 responses are reduced as compared to endotoxemia.

1.2.2 Modeling Inflammation in Wound Healing

Based on these insights into the acute inflammatory response to trauma, hemorrhage, endotoxemia, and infection, mathematical modeling approaches were utilized to gain mechanistic and potentially clinically useful insights into the wound healing response. Wounds can be classified into acute and chronic wounds. Acute wounds include burns, lacerations, and penetrating injuries and are typically observed more frequently among military personnel as opposed to the civilian population. Chronic wounds that are more common among the civilian population include diabetic foot ulcers and pressure ulcers (e.g., those arising following prolonged bed rest and inadequate innervation in spinal cord injury patients), as well as venous ulcers. Wound healing is a complex process that involves both inflammation and the resolution of the inflammatory response, which culminates in remodeling [61, 62]. The first phase of the wound healing response involves the degranulation of platelets and infiltration of inflammatory cells, followed by proliferation of fibroblasts and epithelial cells that deposit collagen and cause contraction of wounds. Pro-inflammatory cytokines such as TNF-α [123] and interferon-γ (IFN-γ) [3] inhibit wound healing both *in vitro* and *in vivo*. Interleukin-6 (IL-6), a cytokine central to inflammation [72], is also necessary for proper healing [56]. Interleukin-10, a potent anti-inflammatory cytokine, appears to suppress inflammation and induce the remodeling necessary for proper wound healing [111]. Another cytokine that is central to the wound healing cascade is TGF-β1 [127]. Recently, DAMPs, which are released from injured tissue and stimulate or augment the inflammatory response, have also been recognized as key regulators of wound healing [71, 95].

Multiple mathematical models, largely developed by the Sherratt group, were used to gain insights into the interactions among macrophages, fibroblasts, and their products in wound healing [139]. Later, the relationship between inflammation and healing in the chronic wound was modeled using an ABM of diabetic foot ulcers [99]. This ABM was calibrated using values published in the literature regarding normal skin healing, which the ABM was capable of reproducing at baseline. This simulation demonstrated delayed healing in the setting of elevated TNF-α or reduced TGF-β1 expression (both known aspects of deranged inflammation in diabetes and/or diabetic foot ulcers), highlighted both beneficial and detrimental effects of known therapies for diabetic foot ulcers (wound debridement and treatment with platelet-derived growth factor [PDGF]), and was used to suggest novel therapies [99]. In a similar vein, an equation-based model was used to study the effects of engineered skin substitutes in diabetic foot ulcers [156].

Another type of ulcer, namely pressure ulcers, can occur with high frequency in patients with spinal cord injury (SCI). An ABM of pressure ulcer formation in SCI patients was created with the goal of preventing pressure ulcer development by uncovering potentially effective clinical interventions. This tissue-realistic ABM exhibits a range of behaviors comparable to outcomes seen in a prospective cohort of SCI patients, including ulcers of various shapes. There are two types of initial injury in these simulations: (1) a circumscribed area of pressure whose intensity decreases radially, simulating the effect of a bony protrusion against soft tissue, followed by ischemia/reperfusion injury, activation of an inflammatory cascade, and production of oxygen free radicals, or (2) shear force injury, in which epithelial cells are damaged by tissue stretching. In both cases, tissue damage stimulates inflammatory cells to produce cytokines (and chemokines), and further damage caused by cytokines coupled with repeated cycles leads to ulceration. The short-term behavior of the model was compared against data on blood flow from experiments in noninjured human volunteers. The longer-term behavior of the model was validated against images of pressure ulcers from SCI patients. The shape of the simulated ulcers qualitatively match patterns seen clinically, including position and size relative to the protrusion and the development of secondary satellite ulcers that eventually coalesce with the primary ulcer.

An important aspect of mechanistic computational modeling of biological systems involves the search for appropriate model parameters, that is, the constants that govern the strengths of interactions among model variables [153]. In seeking to determine optimal therapeutic interventions, model parameters that cause the greatest change in predicted outcome (amount of epithelial damage) with the least relative change in value were determined, focusing on parameters governing production of mediators such as TNF-α, TGF-β1, and DAMPs. Studies such as these may have implications for the rational design of drugs and devices that modulate inflammation and wound healing [12] and may lead to personalized therapy [98].

Indeed, personalized medicine is a central pillar of translational systems biology [98, 157]. A proof of this concept was suggested by studies involving patient-specific ABMs for vocal fold inflammation [82]. An ABM consisting of interactions similar to those described before for the skin ABM [99] could reproduce trajectories of inflammatory mediators in laryngeal secretions of individuals subjected to experimental phonotrauma early post-injury, and predict the levels of inflammatory mediators at late time points post-injury. Subject-specific simulations also predicted the effects of behavioral treatment regimens to which subjects had not been exposed [82]. This ABM could also reproduce the trajectories of inflammatory mediators in vocal fold tissue of rats subjected to surgical injury to the vocal folds [83] and was recently extended to include additional mediators in the laryngeal secretions of human subjects [84].

1.3 FROM WOUND HEALING TO CANCER: MODELING INTERRELATIONSHIPS AMONG ACUTE INFLAMMATION, CHRONIC INFLAMMATION, AND CANCER

For many years, chronic and acute inflammatory processes were thought to be driven by different causes, through the activities of different cells and inflammatory mediators, and to result in quite different ultimate outcomes. However, a more modern view suggests that these processes are interlinked (see Figure 1.1) [85]. Moreover, in the setting of acute inflammation, well-regulated tissue healing can go awry and drive a chronic inflammatory process intertwined with fibrosis and related processes. In hepatic, pancreatic, and gastrointestinal tissues, among others, this pro-inflammatory/pro-fibrotic environment can stimulate carcinogenesis [27, 47, 65, 159], which, in turn, can lead to an altered immune/inflammatory milieu [75].

Computational modeling approaches have been used to study chronic inflammatory/fibrotic processes. One organ that has been studied in this fashion is the lung. Multiple studies using both equation- and agent-based models have examined the inflammatory and fibrotic/granulomatous response of the lung in the setting of tuberculosis, focusing on the role of TNF-α in granuloma formation [46, 91, 137, 158]. In another example, an ABM was created to examine the response of inflammatory cells and cells involved in remodeling to particulate exposure (combined with experimental smoke inhalation in mice), focusing on TNF-α, TGF-β1, collagen deposition, and tissue damage/DAMPs; this study may improve the understanding of lung diseases such as chronic obstructive pulmonary disease (COPD) [26]. Another inflammatory/fibrotic lung disease that was examined using equation-based modeling is sarcoidosis; in that study, the authors utilized a similar approach to the one used in modeling COPD, in that they varied model parameters in order to define key controlling processes in this disease [5].

As mentioned previously, a chronic inflammatory/fibrotic environment can lead to cancer. An excellent example of this type of inflammation-associated cancer is hepatocellular carcinoma (HCC) [104]. There is increasing evidence that, like many other cancers, HCC is initiated through so-called cancer stem cells, which can be extremely resistant to standard cancer therapy [7, 8, 138, 141]. Malignant cell formation in HCC is a consequence of increased hepatocyte turnover, induced by chronic liver injury and regeneration as a result of inflammation induced by viral infections such as hepatitis C virus (HCV) [104]. To study the interlinked processes of inflammation, tissue healing/fibrosis/cirrhosis, and HCC, an ABM was created that was calibrated qualitatively based on published literature data regarding the dynamics of cell populations and cytokines and their influence on inflammation and the early stages of HCC caused by chronic HCV infection, as well as published models of the effect of these inflammatory response elements on already established HCC. This model simulates the onset and progression of HCC in a patch of liver, arising from only a few cells (hypothetical cancer stem cells). The tissue environment of pro- and anti-inflammatory responses present in chronic infections, such as HCV, accelerates tumor formation. The model is initialized with a region of healthy liver cells, which includes several cancer cells. Agents are used to simulate virus, inflammatory cells (macrophages), and the tumor. Infection by viruses elicits an inflammatory response mediated by local macrophages, which are attracted to the site of injury by DAMPs (e.g., HMGB1) [57, 86]. The macrophages secrete the pro-inflammatory TNF, and the anti-inflammatory cytokine TGF-β1, which eventually suppresses the pro-inflammatory response. In the presence of multiple viral flare-ups, a chronic inflammatory state causes progressive, cumulative damage. The tumor grows larger, and tumor-associated macrophages appear in the presence of chronic inflammation [42], compared to the case when a single acute viral infection occurs and resolves. This type of modeling work is likely to integrate with an extensive suite of computational models of cancer already in existence [32, 90, 106, 112, 132, 144].

1.4 FROM MODELING TO RATIONAL REPROGRAMMING OF INFLAMMATION

As may be inferred from the previous discussion of the complexity of the inflammatory response, controlling inflammation while minimizing adverse effects and also allowing for control of infection, stimulating appropriate responses to trauma, and promoting tissue healing is a daunting task. Our long-term therapeutic goal is not to abolish inflammation *per se* but to reduce damage/dysfunction (i.e., promote healing) by modulating the inflammatory response in a rational fashion.

Our initial approach involved hemoadsorption (HA) in the setting of a rat model of bacterial sepsis. Hemoadsorption is an emerging approach to modulate sepsis-induced inflammation [118]. We sought to define the effects of HA on inflammation in *E. coli*-induced fibrin peritonitis in rats, hypothesizing that HA both reduces and reprograms inflammation in sepsis. We found that plasma cytokine levels at baseline were the same in HA and sham. Plasma TNF-α, IL-6, GRO/KC, MCP-1, and the organ damage marker AST were significantly reduced by HA versus sham animals. Principal component analysis suggested that inflammation in sepsis alone (sham) was driven by IL-6 and TNF-α, while inflammation in the presence of HA was driven primarily by TNF-α, GRO/KC, IL-10, and MCP-1. In support of this analysis, the peritoneal:plasma ratios of TNF-α, GRO/KC, and MCP-1 were higher in HA animals. Moreover, peritoneal IL-5, IL-6, IL-18, IFN-γ, and NO_2^-/NO_3^- were significantly lower—and GRO/KC and MCP-1 were significantly higher—in HA versus sham. Peritoneal bacterial counts were significantly lower in HA versus sham. Organ damage, assessed by plasma AST as well as histologically in the lung, kidney, and liver, was reduced in the HA group. These results, elucidated in part by computational analysis, suggest that HA may reduce, reprogram, and relocalize inflammation, while improving bacterial clearance and reducing organ damage in experimental sepsis.

To take the concept of inflammation reprogramming further, we have conceived of a self-regulating device for individualized regulation of inflammation. The basic concept of the proposed device is to create negative feedback proportional to the exact degree of inflammatory stimulus. More precisely, for every unit of a given inflammatory cytokine, the device would produce or release a unit of the neutralizing protein. Examples include: (1) TNF and its endogenous inhibitor, soluble TNF receptor sTNFR [4, 48, 63]; (2) IL-1 and IL-1 receptor antagonist L-1ra [42]; and (3) TGF-β1 and TGF-β1 latency-associated peptide (LAP) [18, 161]. For proof of concept, we focused on modulating the production of sTNFR in a manner dependent on the production of TNF. We have created multiple, stably transfected human hepatocyte (HepG2-derived) cell lines expressing the mouse NF-κB/sTNFR, and characterized these cell lines *in vitro*. All cell lines exhibited different base expression levels of sTNFR production but have similar stimulation response (~ 3.5 fold). We have also created a variant of HepG2 that produces sTNFR constitutively via the cytomegalovirus (CMV) promoter. In addition, we have established a rat model of bacterial endotoxin infusion in which cytokine dynamics over 6 h mimic those of true bacterial sepsis over 48 h. In this animal model, we tested the effect of the bioreactor versus sham cannulation or sham bioreactor [107].

1.5 SUMMARY: CLINICAL AND BASIC SCIENCE PERSPECTIVES

Inflammation is a complex system whose complexity is being deciphered through the use of computational modeling. Key clinical and basic science aspects of this work are summarized as follows:

- Inflammation is a robust, adaptive, and multiscale process that serves to maintain homeostasis in health.
- The positive feedback between inflammation and tissue damage/dysfunction may create a vicious, feed-forward cycle (Figure 1.1) that drives the detrimental effects of inflammation

in trauma/hemorrhage, sepsis, chronic nonhealing wounds/ulcers, chronic inflammatory diseases, and cancer.
- DAMPs are the key intermediates in the inflammation → damage/dysfunction → inflammation positive feedback loop, thereby integrating inflammation and organ (dys)function (Figure 1.1).
- DAMPs are central to innate immunity, but by also activating DCs, these cells form a gateway to the activation of adaptive (T cell) mediated, chronic inflammatory processes (Figure 1.1).
- Computational modeling of the processes described herein has had utility in sepsis, trauma/hemorrhage, wound healing, chronic inflammatory/fibrotic diseases, and cancer.

1.6 CONCLUSIONS AND FUTURE DIRECTIONS

The inflammatory response has fascinated clinicians and scientists for millennia. In both industrialized and developing nations, inflammatory diseases caused by accidents, war, infection, stress, aging, and lifestyle take a significant toll in terms of quality of life and incidence of death. And yet, inflammation is an evolutionarily conserved, multiscale framework crucial to the maintenance of homeostasis in the face of precisely these same insults. It is therefore not surprising that attempts at modulating inflammation in various diseases have at times resulted in adverse effects. Mechanistic computational modeling may offer a way out of this conundrum, and thereby help devise new approaches to rational, personalized diagnosis and therapy for inflammation-associated diseases [12, 98, 147, 148, 152, 153, 154]. Such a strategy may involve using mechanistic simulations of inflammatory diseases for drug design coupled to simulated clinical trials [12], personalized diagnostics and therapy [98], and, ultimately, self-adaptive therapies. While this vision is still off in the future, the future is approaching fast—for example, simulated clinical trials of sepsis have already informed real trials [12]. Computational modeling has suggested links among processes previously thought to be disparate, highlighting the common functional underpinnings of acute and chronic inflammation, wound healing/fibrosis, and cancer. With the increasing cost of drug development and the concomitant decreasing rate of development of new drugs [49], biomedicine can ill afford to continue to attempt to develop new therapies without involving these insights into the process.

TAKE-HOME MESSAGES

- In health, inflammation is a robust, evolutionarily conserved process that allows organisms to adapt to both rapid and chronic changes in their internal and external environments.
- Inflammation is a multiscale process in which key functions can be observed at the cellular, tissue/organ, and whole-organism levels.
- Inflammation affects essentially all diseases in some fashion, from the response to infection and trauma, through prolonged wound healing processes, to chronic diseases such as rheumatoid arthritis, obesity, and cancer.
- Inflammation is a complex system in which outcomes are likely determined by a combination of differing initial conditions as well as positive and negative feedback loops.
- It is important to think of inflammation as being regulated by functional processes that are context dependent, rather than focusing on individual cells or mediators.
- Inflammation is largely driven by a positive feedback cycle of pro-inflammatory influences → damage/dysfunction → pro-inflammatory influences, and negatively regulated by anti-inflammatory influences. These processes can be modeled computationally using both data-driven and mechanistic approaches.
- Acute inflammation is comprised of fast feedback loops that can activate adaptive immunity, which in turn is both regulated by slower internal feedback loops and affects acute inflammation through these slower feedback loops.

- Both acute and chronic inflammatory processes share and invoke key elements and functions, thereby helping explain why drugs nominally targeting acute inflammation work well in chronic diseases such as rheumatoid arthritis, while nominal elements of adaptive immunity are activated early in acute states such as sepsis and trauma.
- Rational modulation of inflammation must therefore take into account the interconnected nature of acute and chronic inflammation and the multiscale nature of inflammation that may progress differently in different compartments, as well as individual patient factors. Thus, computational modeling may be the only rational means by which to design the next generation of therapeutics for inflammatory diseases.
- Rational reprogramming of inflammation, rather than the complete abolition of the inflammatory response, may be the key to balancing the beneficial and detrimental effects of inflammation.

ACKNOWLEDGMENTS

This work was supported in part by the National Institutes of Health grants R01GM67240, P50GM53789, R33HL089082, R01HL080926, R01AI080799, R01HL76157, R01DC008290, and UO1 DK072146; National Institute on Disability and Rehabilitation Research grant H133E070024; a Shared University Research Award from IBM, Inc.; and grants from the Commonwealth of Pennsylvania, the Pittsburgh Lifesciences Greenhouse, and the Pittsburgh Tissue Engineering Initiative/Department of Defense.

REFERENCES

1. Abraham E: Effects of stress on cytokine production. *Methods Achiev Exp Pathol* 1991, 14:45–62.
2. Abraham E, Chang YH: Haemorrhage-induced alterations in function and cytokine production of T cells and T cell subpopulations. *Clin Exp Immunol* 1992, 90:497–502.
3. Adelmann-Grill BC, Hein R, Wach F, Krieg T: Inhibition of fibroblast chemotaxis by recombinant human interferon gamma and interferon alpha. *J Cell Physiol* 1987, 130:270–275.
4. Aggarwal BB, Shishodia S, Takada Y, Jackson-Bernitsas D, Ahn KS, Sethi G, Ichikawa H: TNF blockade: An inflammatory issue. *Ernst Schering Res Found Workshop* 2006,161–186.
5. Aguda BD, Marsh CB, Thacker M, Crouser ED: An *in silico* modeling approach to understanding the dynamics of sarcoidosis. *PLoS ONE* 2011, 6:e19544.
6. Ahmad S, Choudhry MA, Shankar R, Sayeed MM: Transforming growth factor-beta negatively modulates T-cell responses in sepsis. *FEBS Lett* 1997, 402:213–218.
7. Alison MR: Liver stem cells: Implications for hepatocarcinogenesis. *Stem Cell Rev* 2005, 1:253–260.
8. Alison MR, Lovell MJ: Liver cancer: The role of stem cells. *Cell Prolif* 2005, 38:407–421.
9. Alt W, Lauffenburger DA: Transient behavior of a chemotaxis system modelling certain types of tissue inflammation. *J Math Biol* 1987, 24:691–722.
10. An G: Agent-based computer simulation and SIRS: Building a bridge between basic science and clinical trials. *Shock* 2001, 16:266–273.
11. An G: In-silico experiments of existing and hypothetical cytokine-directed clinical trials using agent based modeling. *Crit Care Med* 2004, 32:2050–2060.
12. An G, Bartels J, Vodovotz Y: *In silico* augmentation of the drug development pipeline: Examples from the study of acute inflammation. *Drug Dev Res* 2010, 72:1–14.
13. An G, Faeder J, Vodovotz Y: Translational systems biology: Introduction of an engineering approach to the pathophysiology of the burn patient. *J Burn Care Res* 2008, 29:277–285.
14. An G, Hunt CA, Clermont G, Neugebauer E, Vodovotz Y: Challenges and rewards on the road to translational systems biology in acute illness: Four case reports from interdisciplinary teams. *J Crit Care* 2007, 22:169–175.
15. An G, Mi Q, Dutta-Moscato J, Solovyev A, Vodovotz Y: Agent-based models in translational systems biology. *WIRES* 2009, 1:159–171.
16. Andersson U, Tracey KJ: HMGB1 is a therapeutic target for sterile inflammation and infection. *Annu Rev Immunol* 2011, 29:139–162.

17. Andersson U, Wang H, Palmblad K, Aveberger AC, Bloom O, Erlandsson-Harris H, Janson A, Kokkola R, Zhang M, Yang H, Tracey KJ: High mobility group 1 protein (HMG-1) stimulates proinflammatory cytokine synthesis in human monocytes. *J Exp Med* 2000, 192:565–570.
18. Annes JP, Munger JS, Rifkin DB: Making sense of latent TGFbeta activation. *J Cell Sci* 2003, 116:217–224.
19. Arciero J, Rubin J, Upperman J, Vodovotz Y, Ermentrout GB: Using a mathematical model to analyze the role of probiotics and inflammation in necrotizing enterocolitis. *PLoS ONE* 2010, 5:e10066.
20. Ayala A, Chaudry IH: Immune dysfunction in murine polymicrobial sepsis: Mediators, macrophages, lymphocytes and apoptosis. *Shock* 1996, 6 Suppl 1:S27–S38.
21. Ayala A, Lehman DL, Herdon CD, Chaudry IH: Mechanism of enhanced susceptibility to sepsis following hemorrhage. Interleukin-10 suppression of T-cell response is mediated by eicosanoid-induced interleukin-4 release. *Arch Surg* 1994, 129:1172–1178.
22. Baugh JA, Bucala R: Mechanisms for modulating TNF alpha in immune and inflammatory disease. *Curr Opin Drug Discov Devel* 2001, 4:635–650.
23. Bellingan G: Inflammatory cell activation in sepsis. *Br Med Bull* 1999, 55:12–29.
24. Black PH: The inflammatory response is an integral part of the stress response: Implications for atherosclerosis, insulin resistance, type II diabetes and metabolic syndrome X. *Brain Behav Immun* 2003, 17:350–364.
25. Bone RC: Immunologic dissonance: a continuing evolution in our understanding of the systemic inflammatory response syndrome (SIRS) and the multiple organ dysfunction syndrome (MODS). *Ann Intern Med* 1996, 125:680–687.
26. Brown BN, Price IM, Toapanta FR, Dealmeida DR, Wiley CA, Ross TM, Oury TD, Vodovotz Y: An agent-based model of inflammation and fibrosis following particulate exposure in the lung. *Math Biosci* 2011, 231:186–196.
27. Budhu A, Wang XW: The role of cytokines in hepatocellular carcinoma. *J Leukoc Biol* 2006, 80:1197–1213.
28. Callaghan T, Khain E, Sander LM, Ziff RM: A stochastic model for wound healing. *J Statistical Phys* 2006, 122:909–924.
29. Cameron SB, Nawijn MC, Kum WW, Savelkoul HF, Chow AW: Regulation of helper T cell responses to staphylococcal superantigens. *Eur Cytokine Netw* 2001, 12:210–222.
30. Catania RA, Chaudry IH: Immunological consequences of trauma and shock. *Ann Acad Med Singapore* 1999, 28:120–132.
31. Cavaillon JM: Cytokines and macrophages. *Biomed Pharmacother* 1994, 48:445–453.
32. Chaplain MA, Anderson AR: Mathematical modelling, simulation and prediction of tumour-induced angiogenesis. *Invasion Metastasis* 1996, 16:222–234.
33. Chen G, Goeddel DV: TNF-R1 signaling: A beautiful pathway. *Science* 2002, 296:1634–1635.
34. Chow CC, Clermont G, Kumar R, Lagoa C, Tawadrous Z, Gallo D, Betten B, Bartels J, Constantine G, Fink MP, Billiar TR, Vodovotz Y: The acute inflammatory response in diverse shock states. *Shock* 2005, 24:74–84.
35. Clark JA, Coopersmith CM: Intestinal crosstalk: A new paradigm for understanding the gut as the "motor" of critical illness. *Shock* 2007, 28:384–393.
36. Clermont G, Bartels J, Kumar R, Constantine G, Vodovotz Y, Chow C: *In silico* design of clinical trials: A method coming of age. *Crit Care Med* 2004, 32:2061–2070.
37. Cobbold CA, Sherratt JA: Mathematical modelling of nitric oxide activity in wound healing can explain keloid and hypertrophic scarring. *J Theor Biol* 2000, 204:257–288.
38. Day J, Rubin J, Vodovotz Y, Chow CC, Reynolds A, Clermont G: A reduced mathematical model of the acute inflammatory response: II. Capturing scenarios of repeated endotoxin administration. *J Theor Biol* 2006, 242:237–256.
39. Dempsey PW, Vaidya SA, Cheng G: The art of war: Innate and adaptive immune responses. *Cell Mol Life Sci* 2003, 60:2604–2621.
40. Dinarello CA: Proinflammatory cytokines. *Chest* 2000, 118:503–508.
41. Dinarello CA: Therapeutic strategies to reduce IL-1 activity in treating local and systemic inflammation. *Curr Opin Pharmacol* 2004, 4:378–385.
42. Dirkx AE, Oude Egbrink MG, Wagstaff J, Griffioen AW: Monocyte/macrophage infiltration in tumors: Modulators of angiogenesis. *J Leukoc Biol* 2006, 80:1183–1196.
43. Dong X, Foteinou PT, Calvano SE, Lowry SF, Androulakis IP: Agent-based modeling of endotoxin-induced acute inflammatory response in human blood leukocytes. *PLoS ONE* 2010, 5:e9249.

44. Efron P, Moldawer LL: Sepsis and the dendritic cell. *Shock* 2003, 20:386–401.
45. Faeder JR, Blinov ML, Hlavacek WS: *Rule-based modeling of biochemical systems with BioNetGen. Methods in Molecular Biology: Systems Biology.* Edited by Maly IV. Totowa, NJ, Humana Press, 2008, pp. 113–168.
46. Fallahi-Sichani M, Schaller MA, Kirschner DE, Kunkel SL, Linderman JJ: Identification of key processes that control tumor necrosis factor availability in a tuberculosis granuloma. *PLoS Comput Biol* 2010, 6:e1000778.
47. Farrow B, Evers BM: Inflammation and the development of pancreatic cancer. *Surg Oncol* 2002, 10:153–169.
48. Fernandez-Botran R, Crespo FA, Sun X: Soluble cytokine receptors in biological therapy. *Expert Opin Biol Ther* 2002, 2:585–605.
49. Food and Drug Administration. Innovation or Stagnation: Challenge and Opportunity on the Critical Path to New Medical Products. 1–38. 2004. Ref Type: Report.
50. Foteinou PT, Calvano SE, Lowry SF, Androulakis IP: In silico simulation of corticosteroids effect on an NFkB- dependent physicochemical model of systemic inflammation. *PLoS ONE* 2009, 4:e4706.
51. Foteinou PT, Calvano SE, Lowry SF, Androulakis IP: Modeling endotoxin-induced systemic inflammation using an indirect response approach. *Math Biosci* 2009, 217:27–42.
52. Foteinou PT, Calvano SE, Lowry SF, Androulakis IP: Multiscale model for the assessment of autonomic dysfunction in human endotoxemia. *Physiol Genomics* 2010, 42:5–19.
53. Foteinou PT, Calvano SE, Lowry SF, Androulakis IP: A physiological model for autonomic heart rate regulation in human endotoxemia. *Shock* 2011, 35:229–239.
54. Franceschi C, Bonafe M, Valensin S: Human immunosenescence: The prevailing of innate immunity, the failing of clonotypic immunity, and the filling of immunological space. *Vaccine* 2000, 18:1717–1720.
55. Franceschi C, Bonafe M, Valensin S, Olivieri F, De LM, Ottaviani E, De BG: Inflamm-aging. An evolutionary perspective on immunosenescence. *Ann N Y Acad Sci* 2000, 908:244–254.
56. Gallucci RM, Simeonova PP, Matheson JM, Kommineni C, Guriel JL, Sugawara T, Luster MI: Impaired cutaneous wound healing in interleukin-6-deficient and immunosuppressed mice. *FASEB J* 2000, 14:2525–2531.
57. Gallucci S, Matzinger P: Danger signals: SOS to the immune system. *Curr Opin Immunol* 2001, 13:114–119.
58. Harbrecht BG, Doyle HR, Clancy KD, Townsend RN, Billiar TR, Peitzman AB: The impact of liver dysfunction on outcome in patients with multiple injuries. *Am Surg* 2001, 67:122–126.
59. Harbrecht BG, Zenati MS, Doyle HR, McMichael J, Townsend RN, Clancy KD, Peitzman AB: Hepatic dysfunction increases length of stay and risk of death after injury. *J Trauma* 2002, 53:517–523.
60. Harrison DG, Guzik TJ, Lob HE, Madhur MS, Marvar PJ, Thabet SR, Vinh A, Weyand CM: Inflammation, immunity, and hypertension. *Hypertension* 2011, 57:132–140.
61. Hart J: Inflammation. 1: Its role in the healing of acute wounds. *J Wound Care* 2002, 11:205–209.
62. Hart J: Inflammation. 2: Its role in the healing of chronic wounds. *J Wound Care* 2002, 11:245–249.
63. Hasegawa A, Takasaki W, Greene MI, Murali R: Modifying TNFalpha for therapeutic use: A perspective on the TNF receptor system. *Mini Rev Med Chem* 2001, 1:5–16.
64. Huang X, Venet F, Chung CS, Lomas-Neira J, Ayala A: Changes in dendritic cell function in the immune response to sepsis. Cell- & tissue-based therapy. *Expert Opin Biol Ther* 2007, 7:929–938.
65. Hussain SP, Harris CC: Inflammation and cancer: An ancient link with novel potentials. *Int J Cancer* 2007, 121:2373–2380.
66. Iwasaki A, Medzhitov R: Regulation of adaptive immunity by the innate immune system. *Science* 2010, 327:291–295.
67. Janes KA, Yaffe MB: Data-driven modelling of signal-transduction networks. *Nat Rev Mol Cell Biol* 2006, 7:820-828.
68. Jarrar D, Chaudry IH, Wang P: Organ dysfunction following hemorrhage and sepsis: Mechanisms and therapeutic approaches (Review). *Int J Mol Med* 1999, 4:575–583.
69. Jones AL, Selby P: Tumour necrosis factor: Clinical relevance. *Cancer Surv* 1989, 8:817–836.
70. Joyce S: Immune recognition, response, and regulation: How T lymphocytes do it. *Immunol Res* 2001, 23:215–228.
71. Kaczorowski DJ, Mollen KP, Edmonds R, Billiar TR: Early events in the recognition of danger signals after tissue injury. *J Leukoc Biol* 2008, 83:546–552.
72. Kamimura D, Ishihara K, Hirano T: IL-6 signal transduction and its physiological roles: The signal orchestration model. *Rev Physiol Biochem Pharmacol* 2003, 149:1–38.

73. Kauvar DS, Lefering R, Wade CE: Impact of hemorrhage on trauma outcome: An overview of epidemiology, clinical presentations, and therapeutic considerations. *J Trauma* 2006, 60:S3–11.
74. Kauvar DS, Wade CE: The epidemiology and modern management of traumatic hemorrhage: US and international perspectives. *Crit Care* 2005, 9 Suppl 5:S1–S9.
75. Kiessling R, Wasserman K, Horiguchi S, Kono K, Sjoberg J, Pisa P, Petersson M: Tumor-induced immune dysfunction. *Cancer Immunol Immunother* 1999, 48:353–362.
76. Kox WJ, Volk T, Kox SN, Volk HD: Immunomodulatory therapies in sepsis. *Intensive Care Med* 2000, 26 Suppl 1:S124–S128.
77. Kumar R, Chow CC, Bartels J, Clermont G, Vodovotz Y: A mathematical simulation of the inflammatory response to anthrax infection. *Shock* 2008, 29:104–111.
78. Kumar R, Clermont G, Vodovotz Y, Chow CC: The dynamics of acute inflammation. *J Theoretical Biol* 2004, 230:145–155.
79. Lagoa CE, Bartels J, Baratt A, Tseng G, Clermont G, Fink MP, Billiar TR, Vodovotz Y: The role of initial trauma in the host's response to injury and hemorrhage: Insights from a comparison of mathematical simulations and hepatic transcriptomic analysis. *Shock* 2006, 26:592–600.
80. Laroux FS: Mechanisms of inflammation: The good, the bad and the ugly. *Front Biosci* 2004, 9:3156–3162.
81. Letterio JJ, Vodovotz Y, Bogdan C: *TGF-b and IL-10: Inhibitory Cytokines Regulating Immunity and the Response to Infection. Novel Cytokine Inhibitors.* Edited by Henderson B, Higgs G. Basel, Birkhauser Verlag, 2000, pp. 217–242.
82. Li NYK, Verdolini K, Clermont G, Mi Q, Hebda PA, Vodovotz Y: A patient-specific i*n silico* model of inflammation and healing tested in acute vocal fold injury. *PLoS ONE* 2008, 3:e2789.
83. Li NYK, Vodovotz Y, Hebda PA, Verdolini K: Biosimulation of inflammation and healing in surgically injured vocal folds. *Ann Otol Rhinol Laryngol* 2010, 119:412–423.
84. Li NYK, Vodovotz Y, Kim KH, Mi Q, Hebda PA, Verdolini Abbott K: Biosimulation of acute phonotrauma: An extended model. *Laryngoscope* 2011, 121:2418–2428.
85. Lo D, Feng L, Li L, Carson MJ, Crowley M, Pauza M, Nguyen A, Reilly CR: Integrating innate and adaptive immunity in the whole animal. *Immunol Rev* 1999, 169:225–239.
86. Lotze MT, DeMarco RA: Dealing with death: HMGB1 as a novel target for cancer therapy. *Curr Opin Investig Drugs* 2003, 4:1405–1409.
87. Ludewig B, Krebs P, Junt T, Metters H, Ford NJ, Anderson RM, Bocharov G: Determining control parameters for dendritic cell-cytotoxic T lymphocyte interaction. *Eur J Immunol* 2004, 34:2407–2418.
88. MacArthur BD, Please CP, Taylor M, Oreffo RO: Mathematical modelling of skeletal repair. *Biochem Biophys Res Commun* 2004, 313:825–833.
89. Maldonado-Lopez R, Moser M: Dendritic cell subsets and the regulation of Th1/Th2 responses. *Semin Immunol* 2001, 13:275–282.
90. Mansury Y, Kimura M, Lobo J, Deisboeck TS: Emerging patterns in tumor systems: Simulating the dynamics of multicellular clusters with an agent-based spatial agglomeration model. *J Theor Biol* 2002, 219:343–370.
91. Marino S, El-Kebir M, Kirschner D: A hybrid multi-compartment model of granuloma formation and T cell priming in tuberculosis. *J Theor Biol* 2011, 280:50–62.
92. Marshall JC: Inflammation, coagulopathy, and the pathogenesis of multiple organ dysfunction syndrome. *Crit Care Med* 2001, 29:S99–106.
93. Marshall JC, Charbonney E, Gonzalez PD: The immune system in critical illness. *Clin Chest Med* 2008, 29:605–616, vii.
94. Martins-Green M, Boudreau N, Bissell MJ: Inflammation is responsible for the development of wound-induced tumors in chickens infected with Rous sarcoma virus. *Cancer Res* 1994, 54:4334–4341.
95. Matzinger P: The danger model: A renewed sense of self. *Science* 2002, 296:301–305.
96. Medzhitov R: Origin and physiological roles of inflammation. *Nature* 2008, 454:428–435.
97. Mi Q, Constantine G, Ziraldo C, Solovyev A, Torres A, Namas R, Bentley T, Billiar TR, Zamora R, Puyana JC, Vodovotz Y: A dynamic view of trauma/hemorrhage-induced inflammation in mice: Principal drivers and networks. *PLoS ONE* 2011, 6:e19424.
98. Mi Q, Li NYK, Ziraldo C, Ghuma A, Mikheev M, Squires R, Okonkwo DO, Verdolini Abbott K, Constantine G, An G, Vodovotz Y: Translational systems biology of inflammation: Potential applications to personalized medicine. *Personalized Medicine* 2010, 7:549–559.
99. Mi Q, Rivière B, Clermont G, Steed DL, Vodovotz Y: Agent-based model of inflammation and wound healing: Insights into diabetic foot ulcer pathology and the role of transforming growth factor-b1. *Wound Rep Reg* 2007, 15:617–682.

100. Miller AC, Rashid RM, Elamin EM: The "T" in trauma: The helper T-cell response and the role of immunomodulation in trauma and burn patients. *J Trauma* 2007, 63:1407–1417.
101. Mollen KP, Anand RJ, Tsung A, Prince JM, Levy RM, Billiar TR: Emerging paradigm: Toll-like receptor 4-sentinel for the detection of tissue damage. *Shock* 2006, 26:430–437.
102. Moore FA, Moore EE, Sauaia A: *Postinjury Multiple-Organ Failure. Trauma.* Edited by Mattox KL, Feliciano DV, Moore EE. New York, McGraw-Hill, 1999, pp. 1427–1459.
103. Moore KW, de Waal MR, Coffman RL, O'Garra A: Interleukin-10 and the interleukin-10 receptor. *Annu Rev Immunol* 2001, 19:683–765.
104. Moradpour D, Blum HE: Pathogenesis of hepatocellular carcinoma. *Eur J Gastroenterol Hepatol* 2005, 17:477–483.
105. Myers ER, McCrory DC, Nanda K, Bastian L, Matchar DB: Mathematical model for the natural history of human papillomavirus infection and cervical carcinogenesis. *Am J Epidemiol* 2000, 151:1158–1171.
106. Namas R, Ghuma A, Torres A, Polanco P, Gomez H, Barclay D, Gordon L, Zenker S, Kim HK, Hermus L, Zamora R, Rosengart MR, Clermont G, Peitzman A, Billiar TR, Ochoa J, Pinsky MR, Puyana JC, Vodovotz Y: An adequately robust early TNF-a response is a hallmark of survival following trauma/hemorrhage. *PLoS ONE* 2009, 4:e8406.
107. Namas R, Mikheev M, Yin J, Over P, Young M, Constantine G, et al: Biohybrid device for the systemic control of acute inflammation. *Disrupt Science & Technol* In Press.
108. Napolitano LM, Faist E, Wichmann MW, Coimbra R: Immune dysfunction in trauma. *Surg Clin North Am* 1999, 79:1385–1416.
109. Nathan C: Points of control in inflammation. *Nature* 2002, 420:846–852.
110. Oda K, Kitano H: A comprehensive map of the toll-like receptor signaling network. *Mol Syst Biol* 2006, 2:EPub.
111. Ohshima T, Sato Y: Time-dependent expression of interleukin-10 (IL-10) mRNA during the early phase of skin wound healing as a possible indicator of wound vitality. *Int J Legal Med* 1998, 111:251–255.
112. Owen MR, Sherratt JA: Pattern formation and spatiotemporal irregularity in a model for macrophage-tumour interactions. *J Theor Biol* 1997, 189:63–80.
113. Park JE, Barbul A: Understanding the role of immune regulation in wound healing. *Am J Surg* 2004, 187:11S–16S.
114. Park JS, Arcaroli J, Yum HK, Yang H, Wang H, Yang KY, Choe KH, Strassheim D, Pitts TM, Tracey KJ, Abraham E: Activation of gene expression in human neutrophils by high mobility group box 1 protein. *Am J Physiol Cell Physiol* 2003, 284:C870–C879.
115. Park S, Yoon SJ, Tae HJ, Shim CY: RAGE and cardiovascular disease. *Front Biosci* 2011, 16:486–497.
116. Parker SJ, Watkins PE: Experimental models of gram-negative sepsis. *Br J Surg* 2001, 88:22–30.
117. Peitzman AB, Billiar TR, Harbrecht BG, Kelly E, Udekwu AO, Simmons RL: Hemorrhagic shock. *Curr Probl Surg* 1995, 32:925–1002.
118. Peng ZY, Carter MJ, Kellum JA: Effects of hemoadsorption on cytokine removal and short-term survival in septic rats. *Crit Care Med* 2008, 36:1573–1577.
119. Pinsky MR: Sepsis: A pro- and anti-inflammatory disequilibrium syndrome. *Contrib Nephrol* 2001, 354–366.
120. Plackett TP, Boehmer ED, Faunce DE, Kovacs EJ: Aging and innate immune cells. *J Leukoc Biol* 2004, 76:291–299.
121. Prince JM, Levy RM, Bartels J, Baratt A, Kane JM, III, Lagoa C, Rubin J, Day J, Wei J, Fink MP, Goyert SM, Clermont G, Billiar TR, Vodovotz Y: *In silico* and *in vivo* approach to elucidate the inflammatory complexity of CD14-deficient mice. *Mol Med* 2006, 12:88–96.
122. Ramadori G, Saile B: Inflammation, damage repair, immune cells, and liver fibrosis: Specific or nonspecific, this is the question. *Gastroenterology* 2004, 127:997–1000.
123. Rapala K: The effect of tumor necrosis factor-alpha on wound healing. An experimental study. *Ann Chir Gynaecol Suppl* 1996, 211:1–53.
124. Redd MJ, Cooper L, Wood W, Stramer B, Martin P: Wound healing and inflammation: Embryos reveal the way to perfect repair. *Philos Trans R Soc Lond B Biol Sci* 2004, 359:777–784.
125. Reynolds A, Rubin J, Clermont G, Day J, Vodovotz Y, Ermentrout GB: A reduced mathematical model of the acute inflammatory response: I. Derivation of model and analysis of anti-inflammation. *J Theor Biol* 2006, 242:220–236.
126. Richardson M: Acute wounds: An overview of the physiological healing process. *Nurs Times* 2004, 100:50–53.
127. Roberts AB, Sporn MB: *Transforming growth factor-b. The Molecular and Cellular Biology of Wound Repair.* Edited by Clark RAF. New York, Plenum Press, 1996, pp. 275–308.

128. Rose S, Marzi I: Mediators in polytrauma—pathophysiological significance and clinical relevance. *Langenbecks Arch Surg* 1998, 383:199–208.
129. Rosin DL, Okusa MD: Dangers within: DAMP responses to damage and cell death in kidney disease. *J Am Soc Nephrol* 2011, 22:416–425.
130. Rotstein OD: Modeling the two-hit hypothesis for evaluating strategies to prevent organ injury after shock/resuscitation. *J Trauma* 2003, 54:S203–S206.
131. Rovere-Querini P, Capobianco A, Scaffidi P, Valentinis B, Catalanotti F, Giazzon M, Dumitriu IE, Muller S, Iannacone M, Traversari C, Bianchi ME, Manfredi AA: HMGB1 is an endogenous immune adjuvant released by necrotic cells. *EMBO Rep* 2004, 5:825–830.
132. Sanga S, Frieboes HB, Zheng X, Gatenby R, Bearer EL, Cristini V: Predictive oncology: A review of multidisciplinary, multiscale in silico modeling linking phenotype, morphology and growth. *Neuroimage* 2007, 37 Suppl 1:S120–S134.
133. Schaffer M, Barbul A: Lymphocyte function in wound healing and following injury. *Br J Surg* 1998, 85:444–460.
134. Scheff JD, Calvano SE, Lowry SF, Androulakis IP: Modeling the influence of circadian rhythms on the acute inflammatory response. *J Theor Biol* 2010, 264:1068–1076.
135. Schneider DF, Glenn CH, Faunce DE: Innate lymphocyte subsets and their immunoregulatory roles in burn injury and sepsis. *J Burn Care Res* 2007, 28:365–379.
136. Schroder AK, Rink L: Neutrophil immunity of the elderly. *Mech Ageing Dev* 2003, 124:419–425.
137. Segovia-Juarez JL, Ganguli S, Kirschner D: Identifying control mechanisms of granuloma formation during M. tuberculosis infection using an agent-based model. *J Theor Biol* 2004, 231:357–376.
138. Sell S, Leffert HL: Liver cancer stem cells. *J Clin Oncol* 2008, 26:2800–2805.
139. Sherratt JA, Dallon JC: Theoretical models of wound healing: Past successes and future challenges. *C R Biol* 2002, 325:557–564.
140. Sherratt JA, Murray JD: Models of epidermal wound healing. *Proc Biol Sci* 1990, 241:29–36.
141. Shupe T, Petersen BE: Evidence regarding a stem cell origin of hepatocellular carcinoma. *Stem Cell Rev* 2005, 1:261–264.
142. Sieweke MH, Bissell MJ: The tumor-promoting effect of wounding: A possible role for TGF-beta-induced stromal alterations. *Crit Rev Oncog* 1994, 5:297–311.
143. Stavitsky AB: The innate immune response to infection, toxins and trauma evolved into networks of interactive, defensive, reparative, regulatory, injurious and pathogenic pathways. *Mol Immunol* 2007, 44:2787–2799.
144. Tanaka G, Hirata Y, Goldenberg SL, Bruchovsky N, Aihara K: Mathematical modelling of prostate cancer growth and its application to hormone therapy. *Philos Transact A Math Phys Eng Sci* 2010, 368:5029–5044.
145. Torres A, Bentley T, Bartels J, Sarkar J, Barclay D, Namas R, Constantine G, Zamora R, Puyana JC, Vodovotz Y: Mathematical modeling of post-hemorrhage inflammation in mice: Studies using a novel, computer-controlled, closed-loop hemorrhage apparatus. *Shock* 2009, 32:172–178.
146. Venet F, Chung CS, Monneret G, Huang X, Horner B, Garber M, Ayala A: Regulatory T cell populations in sepsis and trauma. *J Leukoc Biol* 2008, 83:523–535.
147. Vodovotz Y: Translational systems biology of inflammation and healing. *Wound Repair Regen* 2010, 18:3–7.
148. Vodovotz Y, An G: *Systems biology and inflammation. Systems Biology in Drug Discovery and Development: Methods and Protocols.* Edited by Yan Q. Totowa, NJ, Springer Science & Business Media, 2009, pp. 181–201.
149. Vodovotz Y, Chow C, Bartels J, Lagoa C, Kumar R, Day J, Rubin J, Ermentrout B, Riviere B, Yotov I, Constantine G, Billiar T, Fink M, Clermont G. Mathematical simulations of sepsis and trauma. Proceedings of the 11th Congress of the European Shock Society. 2005.
150. Vodovotz Y, Chow CC, Bartels J, Lagoa C, Prince J, Levy R, Kumar R, Day J, Rubin J, Constantine G, Billiar TR, Fink MP, Clermont G: *In silico* models of acute inflammation in animals. *Shock* 2006, 26:235–244.
151. Vodovotz Y, Clermont G, Hunt CA, Lefering R, Bartels J, Seydel R, Hotchkiss J, Ta'asan S, Neugebauer E, An G: Evidence-based modeling of critical illness: An initial consensus from the Society for Complexity in Acute Illness. *J Crit Care* 2007, 22:77–84.
152. Vodovotz Y, Constantine G, Faeder J, Mi Q, Rubin J, Sarkar J, Squires R, Okonkwo DO, Gerlach J, Zamora R, Luckhart S, Ermentrout B, An G: Translational systems approaches to the biology of inflammation and healing. *Immunopharmacol Immunotoxicol* 2010, 32:181–195.
153. Vodovotz Y, Constantine G, Rubin J, Csete M, Voit EO, An G: Mechanistic simulations of inflammation: Current state and future prospects. *Math Biosci* 2009, 217:1–10.

154. Vodovotz Y, Csete M, Bartels J, Chang S, An G: Translational systems biology of inflammation. *PLoS Comput Biol* 2008, 4:1–6.
155. Walker DC, Southgate J, Hill G, Holcombe M, Hose DR, Wood SM, Mac NS, Smallwood RH: The epitheliome: Agent-based modelling of the social behaviour of cells. *Biosystems* 2004, 76:89–100.
156. Waugh HV, Sherratt JA: Modeling the effects of treating diabetic wounds with engineered skin substitutes. *Wound Repair Regen* 2007, 15:556–565.
157. Weston AD, Hood L: Systems biology, proteomics, and the future of health care: Toward predictive, preventative, and personalized medicine. *J Proteome Res* 2004, 3:179–196.
158. Wigginton JE, Kirschner D: A model to predict cell-mediated immune regulatory mechanisms during human infection with Mycobacterium tuberculosis. *J Immunol* 2001, 166:1951–1967.
159. Yang L, Pei Z: Bacteria, inflammation, and colon cancer. *World J Gastroenterol* 2006, 12:6741–6746.
160. Yirmiya R, Goshen I: Immune modulation of learning, memory, neural plasticity and neurogenesis. *Brain Behav Immun* 2011, 25:181–213.
161. Zamora R, Vodovotz Y: Transforming growth factor-b in critical illness. *Crit Care Med* 2005, 33:S478–S481.
162. Zimmer S, Pollard V, Marshall GD, Garofalo RP, Traber D, Prough D, Herndon DN: The 1996 Moyer Award. Effects of endotoxin on the Th1/Th2 response in humans. *J Burn Care Rehabil* 1996, 17:491–496.

2 The Cellular Component of Chronic Inflammation

Julie M. Roda and Tim D. Eubank
The Ohio State University
Columbus, Ohio

CONTENTS

2.1 WHAT IS CHRONIC INFLAMMATION?

In a normal inflammatory response, tissue injury triggers mast cells and resident macrophages to release pro-inflammatory mediators such as tumor necrosis factor (TNF)-α, interleukin (IL)-1β, IL-6, histamines, prostaglandins, and leukotrienes into the surrounding area, resulting in vasodilation and leaky blood vessels. Plasma proteins called *complement* are released and summon phagocytic cells, namely monocytes and neutrophils, to the area to remove necrotic tissue, invading bacteria, and debris. The final stage of inflammation is *resolution*, where neutrophils convert prostaglandins and leukotrienes to lipoxins—initiating the termination sequence while alternatively activated macrophages produce anti-inflammatory factors such as transforming growth factor (TGF)-β and IL-10. Neutrophils apoptose and reparative fibroblasts infiltrate the area to release matrix metalloproteinases (MMPs) for tissue remodeling while producing extracellular matrix (ECM) and collagen. Finally, macrophages depart from the site through lymph vessels. If these steps are strictly followed, then acute inflammation resolves without tissue injury.

Chronic inflammation involves persistent acute inflammation due to the mismanagement of the well-orchestrated inflammation resolution phase. This persistent activation can be due to a failure to remove the inflammatory stimulus, a continual procession of leukocytes that regulate inflammation through the production of pro-inflammatory cytokines and reactive oxygen species (ROS) that damage and continually remodel tissue, or a situation that maintains these leukocytes at the site of inflammation. Well-known chronic inflammatory diseases, or diseases known to initiate due to persistent inflammation, include asthma, autoimmune diseases like lupus and rheumatoid arthritis, chronic prostatitis, inflammatory bowel diseases, reperfusion injury, sarcoidosis, transplant rejection, vasculitis, chronic obstructive pulmonary disease (COPD), atherosclerosis, psoriasis, and sepsis.

There are two components of chronic inflammation: the *exudative* component and the *cellular* component. The exudative component involves the fluid portion of inflammation regulated by vascular permeability and the resultant accumulation of fluid in the tissue (*edema*). This component will not be discussed in detail in this chapter.

2.2 THE CELLULAR COMPONENT OF CHRONIC INFLAMMATION

2.2.1 Monocytes and Macrophages

Monocytes are circulating cells that comprise about 10% of the leukocyte population in peripheral blood. Originating in the bone marrow from hematopoietic stem cells, monocytes are nonproliferative cells with a half-life of about three days and arise from the same myeloid progenitor cells that produce granulocytes. Within the human monocyte population, there is a heterogeneity consisting of three subgroups defined by the presence or absence of CD14 antigen (receptor for bacterial cell membrane component lipopolysaccharide) and CD16 antigen (FcγRIIIa and FcγRIIIb receptors that bind the Fc portion of IgG antibodies for clearance of immune complexes): *classical* CD14^{++}/CD16^{-} monocytes, *intermediate* CD14^{++}/CD16^{+} monocytes, and *nonclassical* CD14^{+}/CD16^{+} monocytes. Mouse homologues to the human CD14^{++}/CD16^{-} and CD14^{+}/CD16^{+} monocyte subtypes are defined as Ly6C^{hi}/CXCR1^{lo}/CCR2^{+}/CD62L^{+} and Ly6C^{lo}/CXCR1^{hi}/CCR2^{-}/CD62L^{-} monocytes, respectively.

Not surprisingly, each subpopulation displays its own phenotype and functionality. The CD14^{++}/CD16^{-} classical monocytes make up approximately 90% of peripheral blood monocytes. Under normal physiological conditions, these classical monocytes migrate into tissue and differentiate to replenish resident tissue macrophages or dendritic cells in response to environmental stimuli. CD14^{++}/CD16^{-} monocytes specialize in phagocytosis, antigen presentation to T cells to mount an immune response, and production of pro-inflammatory cytokines TNF-α, IL-1, and IL-12.

A second subtype of monocytes is the CD14^{+}/CD16^{+} nonclassical monocytes. Typically, CD14^{+}/CD16^{+} monocytes comprise about 10% of circulating monocytes in healthy individuals. CD16 upregulation on monocytes is mediated by TGF-β expression and was confirmed using TGF-β neutralizing antibodies in mice [1]. Interestingly, CD14^{+}/CD16^{+} monocytes exhibit a phenotype resembling inflammatory tissue macrophages, specifically higher major histocompatibility complex (MHC) antigens HLA-DR, -DP, and -DQ, increased numbers of adhesion molecules like ICAM-1, and reduced expression of the anti-inflammatory cytokine IL-10 [2], suggesting that monocytes expressing CD16 are more mature than their CD16-negative relatives and supporting their role in chronic inflammatory diseases. For example, CD14^{+}/CD16^{+} blood monocytosis is observed in such inflammatory diseases as sepsis [3], rheumatoid arthritis [4], and asthma [5]. In a clinical study of patients with bacterial sepsis, CD14^{+}/CD16^{+} monocytes comprised more than 50% of the total circulating monocyte number accompanied by significantly higher levels of IL-6 compared to healthy individuals [3]. In patients with active rheumatoid arthritis, synovial infiltration of immune cells in the monocyte/macrophage lineage is suggested to be responsible for augmented inflammation as these cells are producers of pro-inflammatory mediators IL-1, TNFα, IL-8, and granulocyte/macrophage colony-stimulating factor (GM-CSF) and of tissue damaging molecules like ROS [4].

The CD14^{++}/CD16^{+} *intermediate* monocyte subgroup represents a transitory phenotype between the CD14^{++}/CD16^{-} classical and the more mature, CD14^{+}/CD16^{+} nonclassical monocytes. The importance of characterizing heterogeneous populations of immune cells with potent inflammatory potential is highlighted by the fact that circulating CD14^{++}/CD16^{+} monocytes correlated with cardiovascular outcome in patients with chronic kidney disease (CKD) [5]. A study of 119 patients with non-dialysis CKD was assessed for monocyte heterogeneity by flow cytometry, and the results support the idea that CD16-positive monocytes rather than CD16-negative monocytes are involved in human atherosclerosis [6].

Recently, a unique subset of monocytes characterized by the expression of the endothelial cell marker Tie2, known as *Tie2-expressing monocytes* (TEMs), have been described [7]. The Tie2 receptor binds the growth factor family known as the angiopoetins (ANGPT-1, -2, -3, and -4).

On endothelial cells, ANGPT-1 and ANGPT-2 have contradictory roles. ANGPT-1 functions as an anti-inflammatory cytokine by abrogating vascular permeability while ANGPT-2 serves a pro-inflammatory role by inducing vascular leakage, propagating an inflammatory response. On monocytes, Tie2 receptor expression designates TEMs as pro-inflammatory cells as the presence of TEMs correlates with increased angiogenesis in both human breast cancer [8] and a mouse model of breast cancer [9].

The main function of resident tissue macrophages is tissue homeostasis—immune surveillance, tissue remodeling, and inflammation resolution. Under normal physiological conditions, it is suggested that macrophage numbers are mostly maintained by local proliferation. However, during an inflammatory insult, chemokines like CCL2 released at the onset of tissue injury recruit circulating monocytes from peripheral blood. Interestingly, there seems to be a temporal orchestration for the recruitment of specific monocyte subpopulations. In support of this hypothesis, Nahrendorf et al. [10] observed early and specific recruitment of pro-inflammatory $Ly6C^{hi}$ monocytes with enhanced phagocytic and proteolytic activity during the first three days of insult in a resolving model of myocardial infarction in mice. After three days, $Ly6C^{lo}$ monocytes infiltrated the tissue possessing an anti-inflammatory phenotype. In another study utilizing a model of atherosclerosis, $Ly6C^{hi}$ blood monocytes were identified as the predominant monocyte subpopulation in mouse atheromas and recruited in a CCR2-dependent manner [11].

Macrophages are derived from circulating monocytes and monocytic precursors from the bone marrow and inherit the characteristics of their surrounding tissue. For example, Kupffer cells are liver-specific macrophages that develop an increased phagocytic capacity to remove the high level of toxins and other invading matter from portal circulation. During tissue injury, the polarity of macrophages present at the inflammatory site dictates the duration and magnitude of the inflammatory response. For example, interferon (IFN)-γ-priming of resident macrophages followed by subsequent exposure to bacterial lipopolysaccharide (LPS) results in a *classically activated* M1-type macrophage. M1 macrophages are cytotoxic and antiproliferative through their expression of both ROS and reactive nitrogen species. These macrophages promote inflammation through expression of inflammatory cytokines IL-1β, IL-6, and TNF-α and by recruiting neutro-phils, dendritic cells, and T lymphocytes via production of chemokines IL-8, IP-10, MIP-1α, MIP-1β, and RANTES. Classically activated macrophages contribute to tissue remodeling by expressing numerous MMPs that degrade extracellular matrix components. On the contrary, *alternatively activated* M2-type macrophages present a phenotype conducive to *resolving* inflammation by promoting wound repair, producing ECM components like fibronectin, promoting angiogenesis by expressing FGF basic, TGF-α, and VEGF, and augmenting cell proliferation by secreting IGF, TGF-β, and PDGF, as well as summoning immune cells via expression of CCL2 for monocytes, and CCL18 and CCL17 for T cells.

Asthma is a chronic inflammatory disease of the airways characterized by reversible airway obstruction. In response to an allergen, macrophages produce the pro-inflammatory cytokines TNF-α,IL-1β, and IL-6, as well as the chemokines CCL2, CXCL1, and CXCL8 for monocytes and neutro-phils. Both neutrophils and macrophages release neutrophil elastase and proteases that remodel the alveolar walls as well as induce mucus hypersecretion. Further, macrophages secrete CXCL9, CXCL10, and CXCL11 to recruit T cells, subsequently leading to IFN-γ production, epithelial cell release of TGF-β and fibroblast growth factor (FGF), fibroblast activation and connective tissue growth factor (CTGF) production, and finally fibrosis of the small airways [12]. IL-17 is another key pro-inflammatory molecule in asthmatic patients recently suggested to be produced by macrophages, and not Th17 cells as once thought, in allergic asthma. In a mouse model of OVA-induced allergic asthma, an increase in $IL\text{-}17^{(+)}$ cells in the lung were predominantly $CD11b^{(+)}F4/80^{(+)}$ macrophages and not T cells, and these $IL\text{-}17^{(+)}$ macrophages were alveolar and not invading interstitial. Depleting alveolar macrophages or neutralizing IL-17 prevented this asthma-related inflammation by inhibiting the increase of inflammatory cells and inflammatory factors in bronchoalveolar lavage fluid (BAL) in this model [13].

Sepsis, sometimes classified as systemic inflammatory response syndrome (SIRS), is induced by immune activation in response to microbes in the blood or tissues leading to organ dysfunction. Sepsis is characterized by acute inflammation present throughout the entire body and associated with leukocytosis. Macrophage dysfunction, as a component of immune suppression seen during sepsis, appears to be one of the contributing factors to morbidity and mortality in septic patients. Newton et al. suggested when using a CLP model of sepsis in wild type and transgenic mice deficient in CD86 that expression of CD86 (receptor involved in co-stimulatory signals essential for T cell activation and survival) and MHC II, but not CD40 or CD80 (other co-stimulatory cells on antigen presenting cells) was decreased on peritoneal macrophages after the onset of sepsis. Further, the production of anti-inflammatory IL-10 was reduced implicating the role of CD86 and macrophages as important regulators of sepsis [14]. Importantly, an abundance of reports implicates macrophages as contributing to all stages of atherosclerosis by regulating cholesterol trafficking and inflammation in the arterial walls [15, 16, 17].

2.2.2 Neutrophils

Neutrophils are innate immune cells and professional phagocytic cells derived from the granulocyte compartment. Under homeostatic conditions, neutrophils are the most abundant form of leukocytes in the blood stream. During an acute inflammatory response, neutrophils are the first responders on the scene as chemotaxic factors for these cells are released during tissue injury, namely IL-8, complement component C5a, and leukotriene B4. Once at the site of inflammation, neutrophils act as phagocytic cells and release antimicrobial cytotoxic factors such as lytic enzymes and ROS. In the past, neutrophils were described as short-lived effector cells, but our understanding of the role of neutrophils in inflammation is expanding. Recently, neutrophils are being recognized as long-lasting effectors of tissue inflammation and are key mediators of acute inflammation resolution through their expression of anti-inflammatory cytokines during the resolution phase. The list of cytokines produced by neutrophils is vast and includes CXC and CC chemokines, both pro- and anti-inflammatory cytokines, immunoregulatory cytokines, colony-stimulating growth factors, angiogenic and fibrogenic factors, and members of the TNF super family. For a detailed review on cytokines produced by neutrophils, see Mantovani et al. [18]. In addition, neutrophils have become recognized as important in the TH17 response as T helper lymphocytes secreting IL-17 regulates neutrophil recruitment, proliferation, and accumulation at the inflammatory site. Further, neutrophils have been shown to express nucleotide-binding oligomerization domain protin-1 (NOD1), potentially integrating the role of neutrophils with the inflammosome [19].

A major function of neutrophils is inflammation resolution, and their activity is three-fold: (1) neutrophils have recently been credited with producing pro-resolving lipid mediators at the final stage of acute inflammation by changing eicosanoid output from leukotriene B4 to lipoxins A4 [20]—inhibiting recruitment of migrating neutrophils and increasing the removal of resident apoptosing neutrophils by infiltrating macrophages; (2) neutrophils upregulate the expression of CCR3 to act as a sink for free CCL3 and CCL5 remaining in the environment [21]. CCL3, also known as macrophage inflammatory protein-1α (MIP-1α), is CC cytokine involved in the acute inflammatory state and in recruitment and activation of polymorphonuclear leukocytes. CCL5, also known as RANTES (Regulated upon Activation, Normal T-cell Expressed, and Secreted), is a chemokine for infiltrating leukocytes in the inflammatory response; and (3) neutrophils release IL-1 receptor antagonist (IL-1RA), which binds and blocks IL-1 receptor activity, as well as produces the decoy receptor, IL-1R2, which sequesters and inactivates free, pro-inflammatory IL-1 [18].

In 2001, Lefkowitz et al. enhanced their 1995 theory of *perpetuation of inflammation*, a model describing cellular "cross talk" between neutrophils, macrophages, and endothelial cells in chronic inflammatory diseases [22]. This model involved the pro- and anti-inflammatory effects through cytokine output of neutrophil myeloperoxidase (MPO) and eosinophil peroxidase

(EPO) on both macrophages and endothelial cells in chronic inflammation. Briefly, this model highlights the binding of MPO to mannose receptors on macrophages to induce the secretion of reactive oxygen intermediates (ROI) as well as TNF-α, IL-1, IL-6, IL-8, and GM-CSF. Reactive oxygen intermediates can induce tissue damage as well as enhance the secretion of more TNF-α. TNF-α, produced by various cell types in the tissue environment, is a key mediator in inflammation as it acts in both an autocrine and paracrine fashion to regulate immune cells in inflammatory situations. A primary acute phase protein, TNF-α initiates the inflammatory response and its persistent expression would hinder inflammation resolution and drive chronic inflammatory states.

Not surprisingly, neutrophils have become relevant in the development and maintenance of several chronic inflammatory diseases such as COPD, sepsis, rheumatoid arthritis, and atherosclerosis. The potent neutrophil chemoattractant proline-glycine-proline (PGP) tripeptide is implicated in the persistence of COPD. PGP, normally degraded in the lungs, accumulates in smokers due to the inactivation of aminopeptidase leukotriene A4 hydrolase (LTA4H), leading to PGP accumulation and neutrophil accumulation. Neutrophils are also being implicated in systemic inflammation induced by patients with sepsis. One study found that neutrophil apoptosis is inversely proportional to the severity of sepsis, leading one to believe that it is not the cytotoxic contents of apoptotic neutrophils perpetuating systemic inflammation in these patients [23] but functional neutrophils themselves.

Recently, a novel role for IL-33 has been reported linking neutrophil recruitment with sepsis. Using cecal ligation puncture (CLP), a well-known mouse model of sepsis, the authors showed that administration of IL-33 (an IL-1 super family member that induces various leukocytes to produce anti-inflammatory type 2 cytokines) increased neutrophil migration into the peritoneal cavity of these mice, improving bacterial clearance. The result was a reduction in the systemic, but not local, pro-inflammatory response. The mechanism of action was determined to be the reversal of Toll-like receptor 4 (TLR4)-induced reduction of the CXCR2 receptors on neutrophils by inhibiting the G protein-coupled receptor kinase-2 (GRK2), a serine-threonine protein kinase that induces internalization of chemokine receptors [24].

Another inflammatory disease implicating neutrophils is rheumatoid arthritis (RA). Neutrophil populations are increased in the synovial fluid of patients with RA. Neutrophils undergo a *respiratory burst* in inflamed joints involving the activation of NADPH oxidase, producing massive amounts of superoxide, a reactive oxygen species normally used for killing invading organisms, implicating these cells in the pathogenesis of joint destruction.

Finally, neutrophil numbers have been correlated with increased severity in mouse models of atherosclerosis [25]. $LDLR^{-/-}$ mice (mice deficient in low-density lipoprotein receptors) were subjected to a high-fat diet resulting in atherosclerotic lesions. Immunohistochemistry using Ly6C antibodies specific for neutrophils revealed increased cell numbers in the lesions and the cap and colocalization of neutrophils with myeloperoxidase, suggesting neutrophils as a potential important mediator in the development of atherosclerosis.

2.2.3 Dendritic Cells

Dendritic cells (DCs) are cells of the innate immune system that are derived from the same hematopoietic precursors as monocytes. Dendritic cells reside in areas that are exposed to the external environment like the skin (called Langerhans cells) or the lining of the nose, lungs, stomach, and intestines, but can also reside in peripheral blood. As such, DCs are professional antigen presenting cells (APCs)—phagocytizing and presenting foreign components to B cells and T cells to induce an immune response—and thus serve as the link between the innate and adaptive immune systems. DCs remain in a relatively "immature" state under normal physiological conditions but quickly mature in response to bacterial lipopolysaccharide (LPS)—augmenting their antigen processing and presentation capabilities as well as their ability to regulate T cell-mediated immunity. DCs have been implicated in the initial phases of the acute inflammatory response as LPS activation showed MIP-2, RANTES, IP-10, and MCP-1/CCL2 modulation during DC maturation as well as surface upregulation of CCR1 in a mouse model. As CCR1 binds ligands such as known chemokines MCP-3, RANTES, and MIP-1α, these data implicate the CCR1 receptor in the recruitment and maintenance of dendritic cells at an inflammatory site [26].

In chronic inflammatory diseases such as asthma, a TH2-cell-mediated disorder, DCs regulate adaptive immunity by inducing TH2-cell sensitization, as well as maintaining effector TH2-cell responses during ongoing allergic disease. In fact, DCs have been reported to be essential for generating allergen-specific effector TH2 responses in ongoing inflammation in sensitized mice. Persistent colonization of DCs in the airway lining due to maintained chemokine expression from inflamed epithelial cells leads to a failure of DCs to be removed through the lymphatic network.

In other inflammatory diseases such as sepsis, dendritic cells have been shown to be depleted in the spleens of septic patients. Spleens from 26 septic patients and 20 non-septic trauma patients were evaluated by immunohistochemistry, and a dramatic reduction in the percentage area of follicular DCs was observed [27]. This dramatic decrease in dendritic cell number in septic spleens may significantly impair B and T cell function and contribute to the immune suppression that is a hallmark of sepsis. To further support the role of dendritic cells in sepsis, CD11c+ dendritic cell diphtheria toxin-depleted mice showed reduced survival in a sepsis model compared with wild type mice, while repletion of wild type DCs in these same DC-depleted mice rescued survival [28]. These data confirm that dendritic cells are essential in the septic response and suggest that strategies to maintain DC numbers may improve outcome. Further, this supports dendritic cells as a cellular target for treating chronic inflammatory diseases.

In atherosclerosis, immune mechanisms are required to control its pathogenesis. But the role of dendritic cells essential for priming the immune response has been elusive. CCL17-producing dendritic cells have been shown to be associated with advanced forms of both human and mouse atherosclerosis, and increased numbers of DCs are found co-localizing with T cells in advanced disease [29, 30]. In atherosclerotic mice, a deficiency in CCL17, a potent chemokine for T cells, correlated with decreased atherosclerosis; this decrease was associated with a limited expansion of immune suppressor T cells (Tregs), highlighting dendritic cell-derived CCL17 as a central regulator of Tregs, characterized as powerful inhibitors of atherosclerosis, and implicating DCs in atherogenesis [31]. Further, dendritic cells phagocytize modified lipoproteins like oxidized low-density lipoprotein (oxLDL) deposited in the arterial wall and can initiate early lesion formation [32] as well as instigate immune activation by upregulating co-stimulatory molecules on dendritic cells and increasing T cell proliferation [33]. These lipid-loaded dendritic cells are capable of priming CD4+ T cells in atherogenesis [34] as it has been shown that different T cell subpopulations with a specific signature of pro- or anti-inflammatory cytokines control the atherogenic process [35, 36, 37, 38, 39, 40].

Dendritic cells have also been found in synovium and joint fluid in patients with rheumatoid arthritis and often, similar to DC accumulation in atherosclerosis, co-localize with T cells. These resident DCs can alter T cell polarity into TH1 or TH2 phenotypes depending on cytokine expression in the environment. The cytokine milieu of RA synovium induces DC differentiation and function that leads to autoantigen presentation to T cells. In such cases, dendritic cells could be central to the pathogenesis of rheumatoid arthritis. For an in-depth review of DCs in rheumatoid arthritis, see Sarkar and Fox [41].

The importance of dendritic cells in chronic inflammation is becoming more established. In fact, there are therapeutic strategies attempting to target dendritic cell function in these diseases. For example, tolerogenic dendritic cells (TolDC) have been generated ex vivo. Harry et al. is reinstating immune tolerance in mouse models of RA by vaccination with autologous dendritic cells to recover potent tolerogenic DC function [42].

2.2.4 Fibrocytes

There is a growing awareness of fibrocytes as contributing regulators to chronic inflammatory diseases. Fibrocytes are mesenchymal cells derived from monocyte precursors and a phenotype resembling both fibroblasts and macrophages. Fibrocytes can be identified by expression of hematopoietic markers such as CD45, CD11b, CD11c, CD11d, and the MHC class I and II molecules CD80 and CD86. Further, fibrocytes express CD34, which represents a discriminating factor absent from macrophages and fibroblasts. Like macrophages, CD14 (LPS receptor)-positivity on fibrocytes suggests these cells can be derived from a monocyte subpopulation. Physiologically, fibrocytes are few in number in circulation but have been shown to expand in certain pathologies where macrophages drive a sustained inflammatory state resulting in mismanaged tissue remodeling, such as autoimmune disorders, cardiovascular diseases, and airway disease. In these types of macrophage-directed inflammatory diseases, infiltrating classical macrophages produce pro-inflammatory mediators and reactive oxygen species promoting tissue injury while alternatively activated macrophages abrogate the inflammatory process by recruiting fibroblasts that contribute to the resolution of inflammation by removing survival signals and chemokine gradients, allowing infiltrating leukocytes to undergo apoptosis. Because fibrocyte numbers are augmented in the tissue of such inflammatory conditions, it is suggested that these cells display the plasticity to perform the roles of both macrophages and fibroblasts.

While increased circulating fibrocyte numbers have been reported to correlate with various chronic inflammatory diseases, the function of fibrocytes remains unknown in these pathologies. In asthma patients, persistent lung inflammation causes airway obstruction. Lung biopsies from these patients reveal increased fibrocyte numbers in the airway submucosa. The highest numbers of circulating fibrocytes reside in patients with the most severe chronic airway obstruction compared to patients presenting mild forms of asthma. Interestingly, even mild asthmatics display an increased number of fibrocytes in peripheral blood compared to normals, suggesting a role for fibrocytes in inflammatory progression [43].

In cardiovascular disease, atherosclerosis is initiated when classically activated macrophages transform into *foam cells* in response to IFN-γ and oxidized low-density lipoprotein to drive *atheroma*—the formation of plaque in blood vessel walls. In turn, these foam cells produce pro-inflammatory cytokines, MMPs, and ROS. Fibrocytes, which localize in response to TGF-β expression, have been observed in the fibrous caps of both human and mouse atheromas. For example, fibrocyte recruitment was increased in the plaques of ApoE-null mice (animal model of human atherosclerosis) when TGF-β was overexpressed, suggesting a regulatory role of fibrocytes for plaque development [44].

Finally, in autoimmune diseases such as rheumatoid arthritis, fibroblast-like synovial cells infiltrate joints to induce an inflammatory tissue environment leading to bone/cartilage loss, swollen joint capsules, and inflamed synovium. In these patients, circulating fibrocytes are not elevated, but those fibrocytes that subsist in peripheral blood show enhanced activation of such pro-inflammatory signaling pathways as the mitogen-activated protein kinase (MAPK) and signal transducer and activator of transcription 3 and 5 (STAT3 and STAT5) [45].

These studies suggest that fibrocytes play a role in maintaining an inflammatory milieu as they perform functions similar to pro-inflammatory macrophages and tissue remodeling-fibroblasts and should be further investigated as mediators of chronic inflammation.

2.2.5 T Cells

To this point, we have discussed the cellular component of chronic inflammation representing innate immunity. But, cells of the adaptive immune system can also regulate inflammation. Thymocytes, or T cells, are hematopoietic progenitor cells present in the thymus. T cells are classified by expression of the CD4 or CD8 antigens. To focus on the role of T cells in chronic inflammation, we will discuss a specialized subset of CD4$^+$ T cells called regulatory T cells, or *Tregs*. Tregs are functionally defined by their co-expression of CD25 (the IL-2 receptor α chain) and the transcription factor, Foxp3. These cells were first identified based on the observation that depletion of the CD4$^+$CD25$^+$ T cells left a population of T cells that induced a spectrum of autoimmune diseases when transferred to an immunocompromised recipient mouse, while co-transfer of the CD25$^+$ cells prevented the development of autoimmunity. This population of cells was therefore first identified based on its immunosuppressive activity. It is now known that the primary function of Tregs is to inhibit the activation and proliferation of other effector cells. Therefore, either an overabundance or a deficit of these cells can lead to the development of diseases of deregulated inflammation.

Unlike other T cell subsets, Tregs do not proliferate in response to clonal stimulation. Rather, these cells exert their suppressive functions primarily through the secretion of anti-inflammatory cytokines, such as IL-10, TGF-β, and the recently defined cytokine IL-35 (a member of the IL-12 cytokine family). IL-10 has anti-inflammatory and suppressive effects on most hematopoietic cells, and indirectly suppresses cytokine production and proliferation of antigen-specific CD4$^+$ T effector cells by inhibiting the antigen-presenting capacity of different types of professional APCs, including dendritic cells and macrophages. TGF-β induces apoptosis in autoreactive immune cells, and also inhibits NK cell function by downregulating the activating NK cell receptor, NKG2D. Tregs also express granzyme A and B and perforin and have been shown to directly lyse activated CD4$^+$ and CD8$^+$ T cells and B cells [46]. Tregs have also been shown to suppress effector T cells by directly transferring the inhibitory second messenger cAMP into target cells through gap junctions, resulting in a downregulation of target cells' effector functions [47]. Furthermore, because IL-2 is required for Treg survival and expansion, it has been suggested that IL-2 production from activated effector T cells is responsible for the maintenance of the peripheral pool of Tregs. The Tregs then compete with the activated effector T cells for the available IL-2, which serves to deplete the IL-2 from activated T cells, resulting in their apoptosis. These data suggest the presence of a feedback loop during immune responses where Tregs respond and expand via IL-2 to inflammation, and then sequester the IL-2 from activated T cells in order to limit the duration of the inflammatory response.

In addition to their effects on T cell activation, Tregs can also inhibit APC function through their expression of CTLA-4, which downregulates CD80 and CD86, the major co-stimulatory molecules on APCs. In addition, the hydrolysis of extracellular ATP by CD39 expressed on the Treg cell surface might prevent the inhibition of APC function by ATP. Altogether, these studies demonstrate that Tregs are directly capable of suppressing effector T cell functions as well as limiting the ability of other cell types to stimulate effector T cells.

Tregs actively suppress the activation and expansion of autoreactive immune cells to limit the duration and extent of inflammation. Therefore, a decrease in Treg activity can contribute to autoimmunity or inflammatory disease. For example, Tregs can inhibit the formation of atherosclerotic plaques through TGF-β signaling [48]. Administration of Tregs to mice resulted in reduced plaque size, decreased inflammatory cells within the plaque, and preservation of the fibrous cap. In murine models of systemic lupus erythematosus (SLE), adoptive transfer of Tregs slowed the progression of renal damage and decreased mortality. Adoptive transfer of Tregs also alleviates joint inflammation in a murine model of collagen-induced arthritis [49].

Predictably, decreased Treg numbers have been found in patients with numerous autoimmune diseases. Patients with atherosclerosis exhibit a reduced number of Tregs, as well as reduced Treg function [50]. Decreased Treg numbers are observed in the peripheral blood of patients experiencing active flares of SLE, with an inverse correlation between Treg number and disease severity and autoantibody levels [51]. A reduced number of peripheral Tregs is also observed in patients with rheumatoid arthritis, multiple sclerosis, and inflammatory bowel disease, while functional deficits in Tregs have been observed in patients with psoriasis and coronary heart disease. Furthermore, mutations in FOXP3, the Treg-specific transcription factor, result in decreased numbers of circulating Tregs and have been identified as the etiologic mutation in the lethal autoimmune condition

IPEX (X-linked neonatal diabetes mellitus, enteropathy, and endocrinopathy syndrome) [52]. The wide variety of human autoimmune diseases in which a defect in Treg numbers or function has been demonstrated suggests that this is a common denominator in uncontrolled immune responses to self-antigen.

Although standard therapies for autoimmune diseases were not designed to specifically target Tregs, several therapies have been shown to restore Treg numbers in affected patients. For example, statins have been shown to increase circulating Tregs, suggesting that one mechanism through which statins reverse atherosclerosis may be by inducing the development of Tregs [50]. In addition, corticosteroids have also been found to increase the number of circulating Tregs in SLE patients who responded well to therapy, suggesting that the number of circulating Tregs might be a useful marker for the evaluation of disease activity or the analysis of the therapeutic effects of steroids in patients with autoimmune disorders [53].

Solid organ transplantation has become the standard of care for end-stage organ diseases. Although rates of acute allograft rejection following transplantation have improved, chronic allograft dysfunction remains a complicating factor in the long-term outcome of transplant patients. The ultimate goal in transplantation medicine is the induction of specific tolerance to donor alloantigens and long-term graft acceptance with minimal immunosuppressive drug exposure. There is evidence that in transplantation protocols where robust peripheral tolerance was achieved, the induction and maintenance of tolerance was associated with, and dependent on, the expansion of donor-specific Tregs. Because extensive data obtained in rodents and large animal models of transplantation have revealed a role for Tregs in graft rejection and transplantation tolerance, these cells are being investigated for cell-based therapies in the transplant field.

Because Tregs represent a small proportion of the peripheral T cells, comprising approximately 10% of the $CD4^+$ T cell population, these protocols require that Tregs be expanded to obtain a peripheral Treg/T effector cell ratio that favors regulation. There are several approaches being explored in the clinic to expand Tregs in the context of transplantation. However, in each of these methods, purification is a critical issue, as contaminating effector T cells might expand as well, causing immune pathologies. One of the strategies being explored is the expansion of Tregs in the presence of IL-2 and the mTOR inhibitor rapamycin. Conventional $CD4^+$ T cells and Tregs activate different signaling pathways following IL-2 stimulation—whereas conventional T cells signal primarily through the mTOR pathway, Tregs utilize the JAK2/STAT5 pathway. This difference can therefore be exploited to preferentially expand Tregs by stimulation of a donor-derived T cell population with IL-2 in the presence of an mTOR inhibitor. Tregs could be expanded in rapamycin and IL-2 ex vivo without loss of Foxp3 expression or suppressive function, and were functional at suppressing graft versus host disease (GVHD) when transferred into recipient mice in a murine transplantation model, demonstrating that the rapamycin-expanded Tregs retained their suppressive function in vivo [54]. In a similar approach, Shin et al. induced GVHD in mice and then therapeutically administered IL-2 and rapamycin. In vivo treatment with rapamycin and IL-2 had additive effects on the expansion of donor-derived Tregs, inhibited the proliferation of effector T cells, and reduced the frequency and severity of GVHD. Therefore, treatment with rapamycin and IL-2 could be a viable therapeutic option in clinical settings where expansion of Tregs could result in clinical utility, such as in the context of autoimmunity or solid organ transplantation.

Clinical trials are currently underway to compare rapamycin versus other immunosuppressive regimens on Treg numbers in order to determine whether rapamycin more effectively induces transplantation tolerance via expansion of Tregs (NCT01014234; Principal Investigator, Antonio Dal Canton). Cellular therapy with Tregs is also being explored in the context of type 1 diabetes, in which downregulation of Tregs is associated with the induction of autoantibodies that result in the destruction of insulin-producing pancreatic β cells. Preclinical studies in nonobese diabetic mice have demonstrated that adoptive transfer of Tregs can slow diabetes progression and, in some cases, reverse new onset diabetes. The ability of autologous Tregs expanded ex vivo to alter β cell function and alleviate measures of disease severity is being studied in patients with type 1 diabetes (NCT01210664; Principal Investigator, Stephen E. Gitelman).

Another strategy being employed to increase Treg function is the use of DNA methyltransferase inhibitors (DNMTs) such as 5-aza-2'-deoxycytidine (5-Aza). Treg cells are defined by their expression of Foxp3, a member of the forkhead/winged-helix family of transcription factors that serves as a master regulator for Treg differentiation. Recent reports have demonstrated that the expression and stability of Foxp3 is epigenetically regulated [55]. The Foxp3 locus contains several conserved noncoding regions that are methylated in effector T cells but unmethylated in Tregs, allowing for Foxp3 to be expressed in Tregs but repressed in other T cell subsets. Demethylation of the Foxp3 promoter in naïve $CD4^+$ T cells with 5-Aza resulted in the re-expression of Foxp3 and the subsequent differentiation of $CD4^+$ T cells into Tregs [55]. In the same study, a different DNMT, decitabine, was shown to suppress GVHD by inducing Treg suppressor function in a murine transplantation model. The epigenetic regulation of Foxp3 could therefore be exploited to drive Treg differentiation for therapeutic purposes using clinically relevant DNMTs.

Nutritional interventions have also been explored as a strategy to improve Treg numbers and function in the context of autoimmune diseases. Dietary punicic acid, which is found in pomegranate seed oil, has recently been shown to ameliorate experimental IBD by upregulating Foxp3 expression in Tregs [56]. Similarly, supplementation of the diet with sphingomyelin (a cell membrane phospholipid) has been shown to reduce the incidence of precancerous inflammatory lesions in the colon. This effect was associated with the upregulation of Treg-associated genes, including the IL-10 receptor [57]. The epigenetic regulation of Foxp3, as described previously, can be exploited to generate Treg for therapeutic purposes. For example, epigallocatechin-3-gallate (EGCG), a component of green tea, has been shown to have anti-inflammatory properties based on its ability to inhibit DNMT activity, resulting in the reactivation of methylation-silenced genes. A recent study demonstrated that EGCG could reverse Foxp3 promoter demethylation and induce the differentiation and expansion of Tregs, both in vitro and in vivo [58]. EGCG also increased systemic Treg activity and decreased the incidence of lupus nephritis in a murine model, although it was not determined whether the induction of Tregs was dependent on the demethylating ability of EGCG or the restoration of Foxp3 transcription [59]. Although it remains to be seen whether dietary EGCG could reduce the incidence or symptoms of Treg-dependent autoimmune diseases in humans, ongoing clinical trials are investigating the ability of EGCG to reduce inflammation in relapsing-remitting multiple sclerosis (NCT00525668, Principal Investigator, Judith Bellmann-Strobl), hepatitis C (NCT01018615, Principal Investigator, Roy L. Hawke), and ulcerative colitis (NCT00718094, Principal Investigator, Gerald W. Dryden). The ability of dietary supplements to target similar mechanisms as pharmacological agents offers opportunities for sustained and long-term exposures without the associated toxicities.

TAKE-HOME MESSAGES

- Chronic inflammation is a result of persistent inflammatory stimuli and aberrant resolution resulting in sustained infiltration of invading leukocytes.
- CD16-positivity on monocytes correlates with increases in chronic inflammatory diseases such as sepsis, rheumatoid arthritis, asthma, and atherosclerosis.
- Newly described Tie2-expressing monocytes are of a pro-inflammatory phenotype and support angiogenesis.
- In mouse models of chronic inflammation, there is a temporal infiltration of specific monocyte subtypes from peripheral blood with pro-inflammatory $Ly6C^{hi}$ monocytes arriving first, followed by $Ly6C^{lo}$ monocytes to help resolve the inflammatory response.
- Macrophages promote inflammation through expression of inflammatory cytokines IL-1β, IL-6, and TNFα and by recruiting neutrophils, dendritic cells, and T lymphocytes via production of chemokines IL-8, IP-10, MIP-1α, MIP-1β, and RANTES. There are two

major macrophage phenotypes: classically activated (predominantly pro-inflammatory) and alternatively activated (anti-inflammatory).

- Neutrophils are the master orchestrators of acute inflammation resolution, but studies have shown that pathologies leading to accumulation of neutrophils in tissues result in chronic inflammatory diseases.
- Dendritic cells serve as a link between the innate and adaptive immune systems and are implicated in various chronic inflammatory diseases through their regulation of TH2-immunity.
- Regulatory T cells (Tregs) actively suppress the activation and expansion of autoreactive immune cells to limit the duration and extent of inflammation. Therefore, a decrease in Treg activity can contribute to autoimmunity or persistent inflammatory diseases.

REFERENCES

1. Allen JB, Wong HL, Guyre PM, Simon GL, Wahl SM. 1991. Association of circulating receptor Fc gamma RIII-positive monocytes in AIDS patients with elevated levels of transforming growth factor-beta. *J Clin Invest.* May; 87(5):1773–9.
2. Ziegler-Heitbrock L. 2007. The CD14+ CD16+ blood monocytes: Their role in infection and inflammation. *J Leukoc Biol.* 81(3):584–92.
3. Fingerle G, Pforte A, Passlick B, Blumenstein M, Ströbel M, Ziegler-Heitbrock HW. 1993. The novel subset of CD14+/CD16+ blood monocytes is expanded in sepsis patients. *Blood.* Nov 15; 82(10):3170–6.
4. Kawanaka N, Yamamura M, Aita T, Morita Y, Okamoto A, Kawashima M, et al. 2002. CD14+,CD16+ blood monocytes and joint inflammation in rheumatoid arthritis. *Arthritis Rheum.* Oct; 46(10):2578–86.
5. Rivier A, Pène J, Rabesandratana H, Chanez P, Bousquet J, Campbell AM. 1995. Blood monocytes of untreated asthmatics exhibit some features of tissue macrophages. *Clin Exp Immunol.* May; 100(2):314–8.
6. Rogacev KS, Seiler S, Zawada AM, Reichart B, Herath E, Roth D, et al. 2011. CD14++CD16+ monocytes and cardiovascular outcome in patients with chronic kidney disease. *Eur Heart J.* Jan; 32(1):84–92.
7. De Palma M, Venneri MA, Galli R, Sergi Sergi L, Politi LS, Sampaolesi M, et al. 2005. Tie2 identifies a hematopoietic lineage of proangiogenic monocytes required for tumor vessel formation and a mesenchymal population of pericyte progenitors. *Cancer Cell.* Sep; 8(3):211–26.
8. Sangaletti S, Tripodo C, Ratti C, Piconese S, Porcasi R, Salcedo R, et al. 2010. Oncogene-driven intrinsic inflammation induces leukocyte production of tumor necrosis factor that critically contributes to mammary carcinogenesis. *Cancer Res.* Oct 15; 70(20):7764–75.
9. De Palma M, Murdoch C, Venneri MA, Naldini L, Lewis CE. 2007. Tie2-expressing monocytes: Regulation of tumor angiogenesis and therapeutic implications. *Trends Immunol.* Dec; 28(12):519–24.
10. Nahrendorf M, Swirski FK, Aikawa E, Stangenberg L, Wurdinger T, Figueiredo JL, et al. 2007. The healing myocardium sequentially mobilizes two monocyte subsets with divergent and complementary functions. *J Exp Med.* Nov 26; 204(12):3037–47.
11. Combadière C, Potteaux S, Rodero M, Simon T, Pezard A, Esposito B, et al. 2008. Combined inhibition of CCL2, CX3CR1, and CCR5 abrogates Ly6C(hi) and Ly6C(lo) monocytosis and almost abolishes atherosclerosis in hypercholesterolemic mice. *Circulation.* Apr 1; 117(13):1649–57.
12. Barnes PJ. 2008. The cytokine network in asthma and chronic obstructive pulmonary disease. *J Clin Invest.* Nov; 118(11):3546–56.
13. Song C, Luo L, Lei Z, Li B, Liang Z, Liu G, et al. 2008. IL-17-producing alveolar macrophages mediate allergic lung inflammation related to asthma. *J Immunol.* Nov 1; 181(9):6117–24.
14. Newton S, Ding Y, Chung CS, Chen Y, Lomas-Neira JL, Ayala A. 2004. Sepsis-induced changes in macrophage co-stimulatory molecule expression: CD86 as a regulator of anti-inflammatory IL-10 response. *Surg Infect (Larchmt).* Winter; 5(4):375–83.
15. Seneviratne AN, Sivagurunathan B, Monaco C. 2011. Toll-like receptors and macrophage activation in atherosclerosis. *Clin Chim Acta.* Aug 22; 413:3–14.
16. Saito R, Matsuzaka T, Karasawa T, Sekiya M, Okada N, Igarashi M, et al. 2011. Macrophage elovl6 deficiency ameliorates foam cell formation and reduces atherosclerosis in low-density lipoprotein receptor-deficient mice. *Arterioscler Thromb Vasc Biol.* Sep; 31(9):1973–9.
17. Hirata Y, Tabata M, Kurobe H, Motoki T, Akaike M, Nishio C, et al. 2011. Coronary atherosclerosis is associated with macrophage polarization in epicardial adipose tissue. *J Am Coll Cardiol.* Jul 12; 58(3):248–55.

18. Mantovani A, Cassatella MA, Costantini C, Jaillon S. 2011. Neutrophils in the activation and regulation of innate and adaptive immunity. *Nat Rev Immunol.* Jul 25; 11(8):519–31.
19. Dagenais M, Dupaul-Chicoine J, Saleh M. 2010. Function of NOD-like receptors in immunity and disease. *Curr Opin Investig Drugs.* Nov; 11(11):1246–55.
20. Serhan CN, Chiang N, Van Dyke TE. 2008. Resolving inflammation: Dual anti-inflammatory and pro-resolution lipid mediators. *Nat Rev Immunol.* May; 8(5):349–61.
21. Viola A, Luster AD. 2008. Chemokines and their receptors: Drug targets in immunity and inflammation. *Annu Rev Pharmacol Toxicol.* 48:171–97.
22. Lefkowitz DL, Lefkowitz SS. 2001. Macrophage-neutrophil interaction: A paradigm for chronic inflammation revisited. . *Immunol Cell Biol.* Oct; 79(5):502–6.
23. Fialkow L, Fochesatto Filho L, Bozzetti MC, Milani AR, Rodrigues Filho EM, Ladniuk RM, et al. 2006. Neutrophil apoptosis: A marker of disease severity in sepsis and sepsis-induced acute respiratory distress syndrome. *Crit Care.* 10(6):R155.
24. Alves-Filho JC, Sônego F, Souto FO, Freitas A, Verri WA Jr, Auxiliadora-Martins M, et al. 2010. Interleukin-33 attenuates sepsis by enhancing neutrophil influx to the site of infection. *Nat Med.* Jun; 16(6):708–12.
25. van Leeuwen M, Gijbels MJ, Duijvestijn A, Smook M, van de Gaar MJ, Heeringa P, et al. 2008. Accumulation of myeloperoxidase-positive neutrophils in atherosclerotic lesions in LDLR-/- mice. *Arterioscler Thromb Vasc Biol.* Jan; 28(1):84–9.
26. Foti M, Granucci F, Aggujaro D, Liboi E, Luini W, Minardi S, et al. 1999. Upon dendritic cell (DC) activation chemokines and chemokine receptor expression are rapidly regulated for recruitment and maintenance of DC at the inflammatory site. *Int Immunol.* Jun; 11(6):979–86.
27. Hotchkiss RS, Tinsley KW, Swanson PE, Grayson MH, Osborne DF, Wagner TH, et al. 2002. Depletion of dendritic cells, but not macrophages, in patients with sepsis. *J Immunol.* Mar 1; 168(5):2493–500.
28. Scumpia PO, McAuliffe PF, O'Malley KA, Ungaro R, Uchida T, Matsumoto T, et al. 2005. CD11c+ dendritic cells are required for survival in murine polymicrobial sepsis. *J Immunol.* Sep 1; 175(5):3282–6.
29. Bobryshev YV. 2005. Dendritic cells in atherosclerosis: Current status of the problem and clinical relevance. *Eur Heart J.* 26(17):1700–04.
30. Yilmaz A, Lochno M, Traeg F, Cicha I, Reiss C, Stumpf C, et al. 2004. Emergence of dendritic cells in rupture-prone regions of vulnerable carotid plaques. *Atherosclerosis.* Sep; 176(1):101–10.
31. Weber C, Meiler S, Döring Y, Koch M, Drechsler M, Megens RT, et al. 2011. CCL17-expressing dendritic cells drive atherosclerosis by restraining regulatory T cell homeostasis in mice. *J Clin Invest.* Jul 1; 121(7):2898–910.
32. Paulson KE, Zhu SN, Chen M, Nurmohamed S, Jongstra-Bilen J, Cybulsky MI. 2010. Resident intimal dendritic cells accumulate lipid and contribute to the initiation of atherosclerosis. *Circ Res.* 106(2):383–90.
33. Alderman CJ, Bunyard PR, Chain BM, Foreman JC, Leake DS, Katz DR. 2002. Effects of oxidised low density lipoprotein on dendritic cells: A possible immunoregulatory component of the atherogenic microenvironment? *Cardiovasc Res.* 55(4):806–19.
34. Packard RR, Maganto-García E, Gotsman I, Tabas I, Libby P, Lichtman AH. 2008. CD11c(+) dendritic cells maintain antigen processing, presentation capabilities, and CD4(+) T-cell priming efficacy under hypercholesterolemic conditions associated with atherosclerosis. *Circ Res.* Oct 24; 103(9):965–73.
35. Kleemann R, Zadelaar S, Kooistra T. 2008. Cytokines and atherosclerosis: A comprehensive review of studies in mice. *Cardiovasc Res.* 79(3):360–76.
36. Binder CJ, et al. 2004. IL-5 links adaptive and natural immunity specific for epitopes of oxidized LDL and protects from atherosclerosis. *J Clin Invest.* 114(3):427–37.
37. Mallat Z, et al. 1999. Protective role of interleukin-10 in atherosclerosis. *Circ Res.* 85(8):e17–e24.
38. Ait-Oufella H, et al. 2006. Natural regulatory T cells control the development of atherosclerosis in mice. *Nat Med.* 12(2):178–180.
39. Taleb S, et al. 2009. Loss of SOCS3 expression in T cells reveals a regulatory role for interleukin-17 in atherosclerosis. *J Exp Med.* 206(10):2067–77.
40. van Es T, et al. 2009. Attenuated atherosclerosis upon IL-17R signaling disruption in LDLr deficient mice. *Biochem Biophys Res Commun.* 388(2):261–5.
41. Sarkar S, Fox DA. 2005. Dendritic cells in rheumatoid arthritis. *Front Biosci.* Jan 1; 10:656–65. Print 2005 Jan 1.
42. Harry RA, Anderson AE, Isaacs JD, Hilkens CM. 2010. Generation and characterisation of therapeutic tolerogenic dendritic cells for rheumatoid arthritis. *Ann Rheum Dis.* Nov; 69(11):2042–50.
43. Reilkoff RA, Bucala R, Herzog EL. 2011. Fibrocytes: Emerging effector cells in chronic inflammation. *Nat Rev Immunol.* Jun; 11(6):427–35.

44. Buday A, Orsy P, Godó M, Mózes M, Kökény G, Lacza Z, et al. 2010. Elevated systemic TGF-beta impairs aortic vasomotor function through activation of NADPH oxidase-driven superoxide production and leads to hypertension, myocardial remodeling, and increased plaque formation in apoE(-/-) mice. *Am J Physiol Heart Circ Physiol.* Aug; 299(2):H386–95.
45. Galligan CL, Siminovitch KA, Keystone EC, Bykerk V, Perez OD, Fish EN. 2010. Fibrocyte activation in rheumatoid arthritis. *Rheumatology* (Oxford). Apr; 49(4):640–51.
46. Grossman WJ, Verbsky JW, Tollefsen BL, Kemper C, Atkinson JP, Ley TJ. 2004. Differential expression of granzymes A and B in human cytotoxic lymphocyte subsets and T regulatory cells. *Blood* 104:2840–8.
47. Bopp T, Becker C, Klein M, Klein-Hessling S, Palmetshofer A, Serfling E, et al. 2007. Cyclic adenosine monophosphate is a key component of regulatory T cell-mediated suppression. *J. Exp. Med.* 204:1303–10.
48. Ait-Oufella H, Salomon BL, Potteaux S, Robertson AK, Gourdy P, Zoll J, et al. 2006. Natural regulatory T cells control the development of atherosclerosis in mice. *Nat. Med.* 12:178–180.
49. Morgan ME, Flierman R, van Duivenvoorde LM, Witteveen HJ, van Ewijk W, van Laar JM, et al. 2005. Effective treatment of collagen-induced arthritis by adoptive transfer of CD25+ regulatory T cells. *Arthritis Rheum.* 52:2212–21.
50. Mor A, Luboshits G, Planer D, Keren G, George J. 2006. Altered status of CD4(+)CD25(+) regulatory T cells in patients with acute coronary syndromes. *Eur. Heart J.* 27:2530–7.
51. Lee JH, Wang LC, Lin YT, Yang YH, Lin DT, Chiang BL. 2006. Inverse correlation between CD4+ regulatory T-cell population and autoantibody levels in paediatric patients with systemic lupus erythematosus. *Immunology* 117:280–6.
52. Wildin RS, Ramsdell F, Peake J, Faravelli F, Casanova JL, Buist N, et al. 2001. X-linked neonatal diabetes mellitus, enteropathy and endocrinopathy syndrome is the human equivalent of mouse scurfy. *Nat. Genet.* 27:18–20.
53. Cepika AM, Marinic I, Morovic-Vergles J, Soldo-Juresa D, Gagro A. 2007. Effect of steroids on the frequency of regulatory T cells and expression of FOXP3 in a patient with systemic lupus erythematosus: A two-year follow-up. *Lupus* 16:374–7.
54. Hippen KL, Merkel SC, Schirm DK, Sieben CM, Sumstad D, Kadidlo DM, et al. 2011. Massive ex vivo expansion of human natural regulatory T cells (T(regs)) with minimal loss of in vivo functional activity. *Sci. Transl. Med.* 3:83ra41.
55. Lal G, Bromberg JS. 2009. Epigenetic mechanisms of regulation of Foxp3 expression. *Blood* 114:3727–35.
56. Bassaganya-Riera J, Diguardo M, Climent M, Vives C, Carbo A, Jouni ZE, et al. 2011. Activation of PPARgamma and delta by dietary punicic acid ameliorates intestinal inflammation in mice. *Br. J. Nutr.* 106:878–86.
57. Mazzei JC, Zhou H, Brayfield BP, Hontecillas R, Bassaganya-Riera J, Schmelz EM. 2011. Suppression of intestinal inflammation and inflammation-driven colon cancer in mice by dietary sphingomyelin: Importance of peroxisome proliferator-activated receptor gamma expression. *J. Nutr. Biochem.* 22:1160–71.
58. Wong CP, Nguyen LP, Noh SK, Bray TM, Bruno RS, Ho E. 2011. Induction of regulatory T cells by green tea polyphenol EGCG. *Immunol. Lett.* 139:7–13.
59. Tsai PY, Ka SM, Chang JM, Chen HC, Shui HA, Li CY, et al. 2011. Epigallocatechin-3-gallate prevents lupus nephritis development in mice via enhancing the Nrf2 antioxidant pathway and inhibiting NLRP3 inflammasome activation. *Free Radic. Biol. Med.* 51:744–54.

3 Mast Cells in Chronic Inflammation

Traci A. Wilgus and Brian C. Wulff
The Ohio State University
Columbus, Ohio

CONTENTS

3.1 MAST CELL BIOLOGY

The study of mast cells was pioneered by Paul Ehrlich in the late 1800s (Crivellato et al. 2003). Ehrlich, in some of his first studies, and many researchers after him have noted that mast cells are frequently found in chronically inflamed tissues. He described their unique metachromatic staining with aniline dyes, which can be used to visualize their abundant intracellular granules. Mast cell granules contain a myriad of mediators that are quickly released upon activation, including histamine, which gives rise to the hallmark symptoms of allergic reactions like those seen in allergic rhinitis, atopic dermatitis, and allergic asthma (Rao and Brown 2008).

3.1.1 Development, Maturation, and Distribution

Mast cells are resident immune cells and are typically found within tissues but not in the circulation. They are bone marrow–derived cells that circulate as immature hematopoietic progenitors that then mature locally within the tissue of residence (Galli and Tsai 2008). The process by which mast cell progenitors are recruited to certain anatomical sites and subsequently undergo maturation has not been completely defined; however, stem cell factor (SCF) is known to play a critical role in this process. SCF, which is produced by stromal cells, is a critical factor for the survival and differentiation of both rodent and human mast cells (Metz et al. 2007). In fact, mice that lack the ability to respond to SCF due to the absence of a functional SCF receptor (c-kit) are deficient in mast cells, and these strains are commonly used to study mast cell function (Metz et al. 2007). In addition to SCF, interleukin (IL)-3 and IL-6 are also important for mast cell growth and survival (Kirshenbaum and Metcalfe 2006).

Mature mast cells are widely distributed throughout the body and are common inhabitants of vascularized tissues. Areas of the body that are continually exposed to the environment such as the skin (Figure 3.1), respiratory tract, and gastrointestinal tract contain high numbers of mast cells (Metz et al. 2007). At these sites, which interface with the environment, mast cells are poised to detect and respond to injury and potential pathogens. They are often found adjacent to cells or structures with which they are known to interact like blood vessels, lymphatic vessels, smooth muscle cells, and nerves (Galli and Tsai 2008). Mast cells have a relatively long lifespan within resident tissues and are capable of undergoing local proliferation when properly stimulated (Tsai et al. 2011).

3.1.2 Phenotypes

Mast cells are not a homogeneous population of cells; rather, they have different characteristics depending on their local environment. Individual mast cells or populations of mast cells can differ in the types or amount of stored mediators present within granules, how responsive they are to external stimuli, and what mediators they produce upon activation. Mast cells can also alter their characteristics after exposure to certain factors and over the progression of a biological response (Galli and Tsai 2008).

Mast cells are typically divided into subpopulations that differ based on species. In humans, mast cells are categorized by the proteases stored in their granules. Mast cells containing tryptase as the main protease (MC_T) or those with both tryptase and chymase (MC_{TC}) are most common, although mast cells that contain chymase but lack tryptase have also been described. MC_T are prevalent in mucosal tissues of the lung and gut, and MC_{TC} are more commonly found in connective tissue organs like the skin (Rao and Brown 2008). In rodents, mast cells are usually characterized as connective tissue mast cells (CTMC) or mucosal mast cells (MMC). CTMCs reside in the skin and peritoneal cavity and contain heparin, histamine, and carboxypeptidase A (Rao and Brown 2008). MMCs, on the other hand, are found mostly in intestinal and lung mucosa and contain high levels of chondroitin sulfate. These mast cell subtypes also produce different lipid mediators after stimulation, with CTMC releasing predominately prostaglandin D_2 (PGD_2), while leukotriene B_4

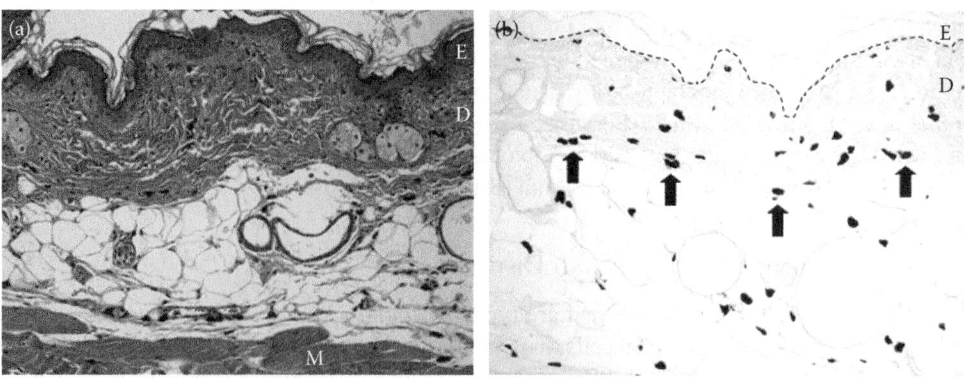

FIGURE 3.1 Resident mast cells in the skin. Mast cells are commonly found at sites that are exposed to environmental insults and potential pathogens such as the skin. (a) A hematoxylin and eosin-stained section of mouse skin is shown to illustrate the architecture of the skin (E = epidermis; D = dermis; M = muscle). (b) Abundant dermal mast cells (arrows) can be seen in sections stained with toluidine blue, a common histological stain that stains mast cell granules. The dotted line indicates the border between the epidermal and dermal skin layers.

(LTB_4) is produced at higher levels in MMC (Rao and Brown 2008). Mast cells display a certain amount of plasticity, as they can change their phenotype and take on different characteristics as a result of microenvironmental cues. For instance, mast cell protease content can be altered transcriptionally by IL-10 in mice, and treatment of human mast cells with IL-4 increases the amount of chymase stored within granules (Ghildyal et al. 1992; Toru et al. 1998). The plasticity of these cells is highlighted by studies demonstrating a change from one mast cell subtype to another under certain conditions. Bone marrow–derived cultured mast cells, which share some characteristics of MMC, can mature into CTMC depending on the location in which they are injected (Nakano et al. 1985). Furthermore, peritoneal CTMC can take on characteristics of MMC when cultured *in vitro*, and then revert to a CTMC-like phenotype when injected into the peritoneum (Kanakura et al. 1988). These studies imply that mast cell phenotype is based largely on signals present in the local microenvironment, and exposure to cytokines and growth factors like SCF, IL-3, IL-4, and IL-9 are thought to be some of the critical signals involved in determining mast cell phenotype (Galli and Tsai 2008). The phenotypic heterogeneity and plasticity of mast cells likely contribute to the multifunctional nature of these cells.

3.1.3 Activation

There are multiple mechanisms by which mast cells can be activated. The most well-studied mechanism of activation is mediated by IgE. Crosslinking of IgE bound to high-affinity surface receptors (FcεRI) by antigen promotes FcεRI aggregation and activation of downstream signaling. In addition, mast cells can be activated independently of IgE by many different stimuli (Rao and Brown 2008; Theoharides et al. 2011). They can be directly activated by pathogens (bacteria, viruses, and parasites) or pathogen products either directly by binding toll-like receptors (TLRs) or indirectly through complement receptors. Mast cells can also be activated by cytokines, neuropeptides, and chemicals, as well as physical stimuli such as heat or mechanical injury.

Mast cell activation results in the release of preformed or newly synthesized mediators, and there are several possible modes of release (Figure 3.2). Often, mediators are liberated into the surrounding tissue as the mast cells release cytoplasmic granules. Activation of the IgE receptor FcεRI, in particular, gives rise to what is known as compound exocytosis or anaphylactic degranulation. This involves microtubule-dependent translocation of the granules to the surface of the cell, and exocytosis through calcium-dependent fusion of the granule with the plasma membrane of the cell (Nishida et al. 2005). In many cases, this type of activation causes virtually all granules to be released by exocytosis such that the degranulated mast cells, so-called phantom mast cells, cannot be detected using granule-specific dyes (Claman et al. 1986). This activation pathway often results in the rapid release of high levels of histamine and other pre-stored mediators, initiating allergic and inflammatory reactions. Mast cells can also release mediators and be involved in inflammatory processes without undergoing complete degranulation (Theoharides et al. 2011). When exposed to some stimuli, mast cells can secrete individual granules, secrete mediators independently of granules, or selectively release certain mediators. For example, IL-6 can be secreted without the release of histamine in response to IL-1 stimulation (Kandere-Grzybowska et al. 2003). Selective release of specific mediators is thought to occur via secretory vesicles that are much smaller in size than granules (Theoharides et al. 2011). In some chronic inflammatory processes there is not always evidence of widespread mast cell degranulation, and some have proposed that these alternative mechanisms of mediator secretion could allow mast cells to provide a slow and steady release of mediators that help maintain a chronic inflammatory state (Kovanen 2009).

3.1.4 Mediators

Mast cells are capable of releasing a large number of biologically active mediators upon activation (Figure 3.3). Mast cells store a host of preformed mediators that can be released quickly

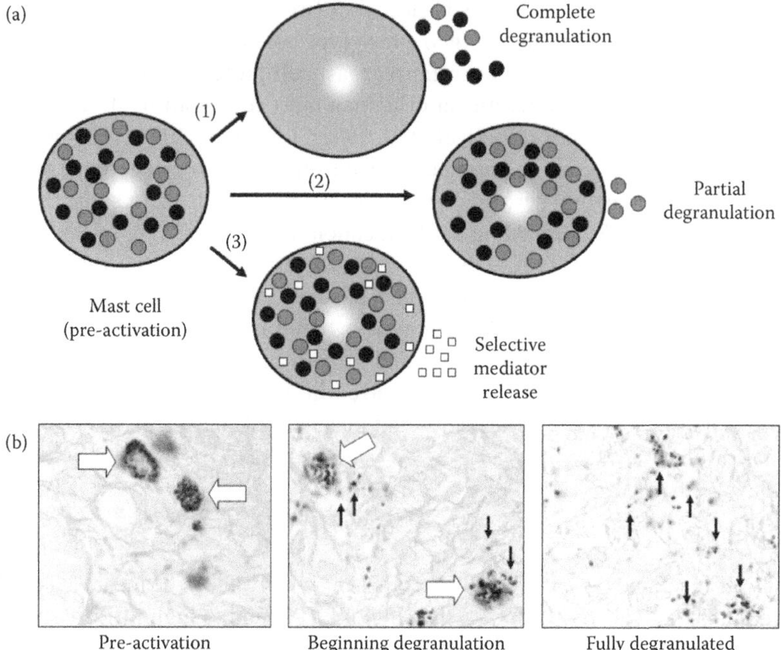

FIGURE 3.2 Potential mechanisms of mediator release by activated mast cells. Mast cells contain numerous granules that act as storage sites for preformed mediators. (a) Prior to activation, all granules are located within the cytoplasm of the mast cell. After mast cells are activated, they release preformed and/or newly synthesized mediators that can affect neighboring cells. (1) Mast cells can undergo complete degranulation, whereby all or most of the granules are released into the surrounding tissue. This is sometimes referred to as anaphylactic degranulation and is a hallmark of allergic reactions. (2) Mast cells can also release a subset of granules, undergoing partial degranulation rather than completely releasing all granules. (3) Finally, in response to certain stimuli, mast cells can selectively release specific mediators without releasing granules. (b) Toluidine blue-stained tissue sections illustrate mast cell degranulation in wounded skin. Prior to injury (pre-activation), mast cells (open arrows) contain numerous intracellular granules (left). After injury, mast cells are activated and begin to release granules (closed arrows) into the tissue (middle). Eventually, the majority of the granules are released and individual mast cells can no longer be visualized with a granule-specific stain (right).

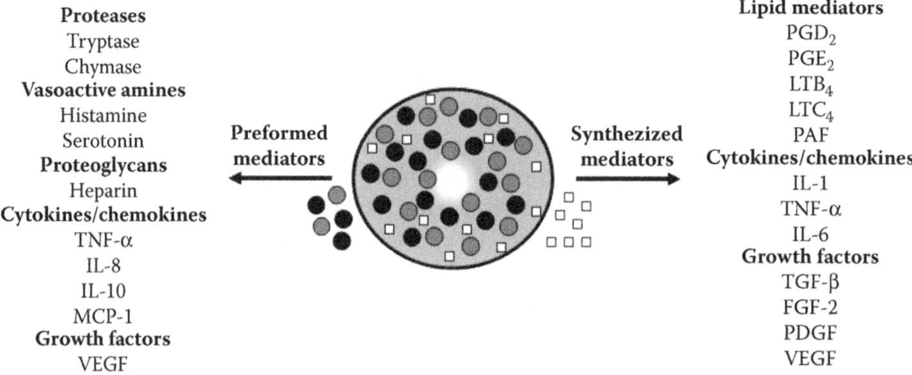

FIGURE 3.3 Mast cell mediators. The release of both preformed mediators stored within granules (left) and mediators synthesized *de novo* after activation (right) by mast cells allows them to carry out multiple biological functions. The list, which is not comprehensive, contains some of the most common mast cell mediators. While most of the mediators are considered pro-inflammatory, mast cells can also be a source of anti-inflammatory or immunosuppressive cytokines.

via degranulation, including histamine and TNF. The neutral proteases tryptase and chymase are also stored in mast cells granules and are released into the tissue upon degranulation. Mast cells also synthesize mediators *de novo* upon activation. They quickly convert arachidonic acid to pro-inflammatory lipid mediators such as prostaglandins and leukotrienes. Over a longer period of time, mast cells also synthesize and release numerous cytokines and growth factors. Many mast cell–derived mediators are pro-inflammatory and cause the hallmark features of inflammation, including vasodilation, vascular permeability, and the recruitment and activation of circulating immune cells. However, mast cells also produce anti-inflammatory and immunosuppressive cytokines, such as IL-10 and transforming growth factor-β (TGF-β). This allows mast cells to have diverse effects during innate or adaptive immune responses under different circumstances.

3.2 PHYSIOLOGICAL FUNCTIONS

The idea that mast cells are multifunctional cells with various roles beyond allergy is becoming more accepted. In fact, studies have shown that mast cells help maintain homeostasis, regulate innate and adaptive immune responses, and contribute to several aspects of wound repair.

3.2.1 Homeostasis

Mast cells appear to be important for maintaining homeostasis in organs that undergo repeated cycles of growth and remodeling. Hair follicles go through a series of changes during the hair cycle, which includes growth, regression, and resting stages. Distinct changes in the number and activation of mast cells have been shown to correlate with each stage of the hair cycle (Maurer et al. 1995), and mast cell–deficient mice display abnormalities in hair follicle cycling (Maurer et al. 1997). A role for mast cells in bone remodeling has also been demonstrated. Mast cell–deficient mice have lighter, thinner, and more fragile femurs (Cindik et al. 2000) as well as delayed bone remodeling with reduced production of new bone matrix in a model of induced bone remodeling (Silberstein et al. 1991). It is believed that factors produced by mast cells, such as histamine, IL-1, IL-6, and TGF-β, influence the recruitment and development of osteoclasts (Weller et al. 2011). Mast cells also produce osteopontin (Nagasaka et al. 2008), which can regulate bone metabolism. Mast cells are often localized near nerve endings in many different tissues, and they actively interact with one another. Several mast cell mediators, including histamine, serotonin, and tryptase, affect the activity of sensory neurons, while calcitonin gene-related peptide, substance P, and endothelin-1 are examples of mediators released by nerves that stimulate mast cells (Rao and Brown 2008). The importance of these interactions is evident in the intestine, where mast cells and neurons affect ion transport, mucous secretion, vascular permeability, and intestinal mobility to maintain homeostasis (Van Nassauw et al. 2007).

3.2.2 Innate and Adaptive Immunity

In addition to causing detrimental effects by mediating hypersensitivity reactions, mast cells can also have protective effects by guiding the course of innate and adaptive immune responses. Mast cells are optimal first responders due to their location at anatomical sites exposed to potential pathogens and their ability to quickly respond through degranulation. Two of the important initial studies demonstrating a role for mast cells in the innate response to bacteria were performed by Echtenacher et al. and Malaviya et al. One study used cecal ligation and puncture (CLP) to mimic acute septic peritonitis and showed that mast cell–deficient mice had a higher mortality rate compared to normal mice, in part due to the absence of TNF produced by mast cells (Echtenacher et al. 1996). Another study showed a higher mortality rate from enterobacteria in mast cell–deficient mice and demonstrated that the protective response was due to the recruitment of bacteria-clearing neutrophils to the infection site by mast cell–derived TNF (Malaviya et al. 1996). A direct role for mast cells in neutralizing pathogens has also been suggested by studies showing that mast cells can produce antimicrobial peptides such as cathelicidins (Di Nardo et al. 2003).

Mast cells are also involved in adaptive immunity, as they can interact directly and indirectly with adaptive immune cells. It has been suggested that mast cells are capable of processing and presenting antigen, thereby interacting directly with T cells (Tsai et al. 2011). Mast cells influence lymphocytes by secreting mediators that regulate migration, trafficking, maturation, and activation. For example, histamine, lipid mediators, and TNF secreted by mast cells can stimulate the activity of lymphocytes (Tsai et al. 2011). Mast cells have also been shown to stimulate lymphocyte activation through the release of exosomes (Skokos et al. 2001). Furthermore, mast cells are involved in the enlargement of lymph nodes early after infection, increasing the retention of lymphocytes and the chances that the appropriate antigen-specific lymphocytes will be involved in the immune response to a particular pathogen (Abraham and St John 2010).

Although mast cells can stimulate adaptive immune responses, recent evidence suggests that they can also exert immunosuppressive effects. This is likely due, at least in part, to the fact that mast cells can produce both positive as well as negative regulatory mediators. Examples of mast cell–derived products with immunosuppressive activity include histamine, IL-4, TGF-β, and IL-10 (Tsai et al. 2011). Several studies have indicated a role for mast cells in immunosuppression. Hart and colleagues showed that mast cells are involved in acute ultraviolet light-induced immunosuppression, an effect that is dependent on histamine (Hart et al. 1998). Mast cell–derived IL-10 appears to be important for immunosuppression following *Anopheles* mosquito bites and limits the immune response and subsequent pathology associated with ultraviolet light exposure and contact dermatitis (Depinay et al. 2006; Grimbaldeston et al. 2007). Mast cells have also been shown to control regulatory T cell–dependent peripheral tolerance of skin allografts (Lu et al. 2006).

3.2.3 Wound Healing

Mast cell numbers are high in the skin, which is constantly exposed to potential pathogens as well as injury. The wound-healing process is well defined in the skin, and can be broken down into several phases, including hemostasis and inflammation, proliferation, and scar formation/remodeling phases. Mast cells are actively involved in each of these stages. Immediately after injury, mast cells quickly degranulate (Figure 3.2), releasing a host of mediators that initiate an inflammatory response. These cells can also synthesize additional lipid mediators, cytokines, and growth factors that affect the inflammatory phase. Together, these preformed and newly synthesized mast cell–derived mediators are involved in the recruitment of circulating inflammatory and immune cells. Mast cells have been shown to be particularly important for aiding neutrophil infiltration into the wound (Egozi et al. 2003; Weller et al. 2006), and many mast cell mediators can activate immune cells to help prevent infection. During the proliferative phase of healing, keratinocytes proliferate and migrate to repair the breached epithelial barrier (reepithelialization) and endothelial cells actively form new blood vessels (angiogenesis). One study suggested that mast cells are involved in reepithelialization of large excisional wounds and that this is dependent on mast cell–derived histamine (Weller et al. 2006). In addition, mast cells release a plethora of pro-angiogenic molecules, including vascular endothelial growth factor (VEGF), platelet-derived growth factor (PDGF), and fibroblast growth factor-2 (FGF-2) that likely aid in wound angiogenesis. Mast cells can also stimulate fibroblasts, which are active during the proliferative phase and responsible for depositing and remodeling collagen during the scar formation/remodeling phase. Several studies have shown that mast cells are important for scar formation. Mast cells are capable of producing TGF-β1 (Gordon and Galli 1994), a potent pro-scarring molecule. Studies from our lab and others have shown that mast cells and mast cell activation are more prevalent in scar-forming wounds compared to those that heal without a scar (Schrementi and DiPietro 2005; Wulff et al. 2012; Mak et al. 2009). Mast cells have also been linked to both keloids and hypertrophic scars (Harunari et al. 2006; Kischer et al. 1978; Smith et al. 1987), and targeting mast cells with the degranulation inhibitor ketotifen has been shown to reduce scarring in a hypertrophic scar model (Gallant-Behm et al. 2008). More details about the interactions between mast cells and fibroblasts can be found in the section on fibrosis.

3.3 PATHOLOGICAL ROLE IN CHRONIC INFLAMMATORY DISEASES

As discussed previously, mast cells perform a variety of important physiological and protective functions (Weller et al. 2011); however, mast cells are also gaining attention for their potential contribution to nonallergic diseases associated with chronic inflammation, such as cancer, atherosclerosis, and fibrosis. A brief discussion highlighting the potential roles for mast cells in these diseases follows.

3.3.1 CANCER

The contribution of the microenvironment to the development, growth, and metastasis of tumors is well recognized. The prolonged presence of inflammatory cells and chronic inflammation are important for cancer development and progression, and the importance of inflammation has been demonstrated by the effective use of anti-inflammatory drugs as chemopreventive or chemotherapeutic agents (Coussens and Werb 2002; Grivennikov et al. 2010). Mast cells represent one of the inflammatory cell types present within the inflammatory tumor milieu (Ribatti and Crivellato 2009; Maltby et al. 2009). Ehrlich, in some of his initial studies of mast cells, noted that the tissue surrounding tumors was heavily populated with mast cells (Crivellato et al. 2003). Since then, a strong mast cell presence has been established in many different types of tumors, where they are most often localized to the tumor periphery (Figure 3.4a). Tumor cell–derived SCF, along with other mast cell chemoattractants released by tumor cells, is likely important for the enhanced mast cell number and activation in the area surrounding tumors (Huang et al. 2008). Increases in mast cell density have been linked to tumor invasiveness and poor prognosis in many different cancers (Duncan et al. 1998; Rojas et al. 2005; Molin et al. 2002; Ribatti et al. 2010; Ribatti et al. 2003; Takanami et al. 2000). The pro-inflammatory nature of mast cells and their prevalence in tumor stroma suggests that mast cells play an active role in this process. In fact, an important role for mast cells in tumorigenesis has been demonstrated using mast cell–deficient mice, which have been shown to be more resistant to the growth of several types of tumors (Coussens et al. 1999; Gounaris et al. 2007; Soucek et al. 2007; Wedemeyer and Galli 2005).

There are several ways that mast cells could promote tumorigenesis (Figure 3.4b). Given the large number of pro-inflammatory mediators produced by mast cells, they are likely important for initiating and maintaining a chronic inflammatory microenvironment, which is known to augment carcinogenesis (Coussens and Werb 2002; Grivennikov et al. 2010). Many of the pro-inflammatory mediators and growth factors produced by mast cells can affect tumor cells directly, leading to enhanced proliferation, invasion, or survival of tumor cells and eventually increased growth and/or spread of the tumor (Ribatti and Crivellato 2009; Theoharides and Conti 2004). As discussed before, mast cells are also capable of secreting a variety of immunosuppressive mediators, such as IL-10 and TGF-β, which could limit the detection and destruction of tumor cells by the immune system, ultimately leading to unchecked tumor growth and metastasis. Mast cells are also believed to regulate the invasion and spread of tumors by remodeling the tumor stroma (Maltby et al. 2009). Matrix metalloproteinase (MMP)-9 has been shown to be particularly important for tumor invasion (Coussens et al. 2000). Interestingly, mast cells can produce MMP-9, and mast cell–derived proteases (i.e., chymase) can activate latent pro-MMP-9 (Coussens et al. 1999; Kanbe et al. 1999). Perhaps the most studied mechanism for mast cell involvement in cancer is the augmentation of tumor angiogenesis, which is vital for tumor growth (Dvorak et al. 2011). Mast cells produce an array of pro-angiogenic growth factors and cytokines, including VEGF, FGF-2, PDGF, and IL-8 as

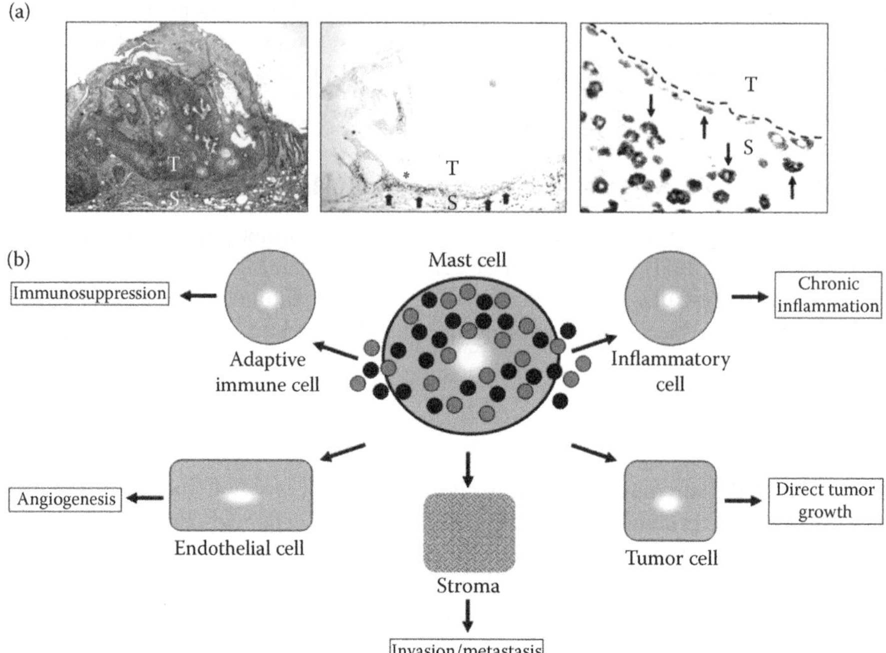

FIGURE 3.4 Tumor-associated mast cells. Mast cells are present in high numbers in tumors and are believed to be involved in the development, growth, and spread of several tumor types. (a) Mast cells are often located at the tumor (T)-stroma (S) interface. Serial sections of an ultraviolet light-induced murine skin tumor stained with hematoxylin and eosin (left) and toluidine blue (middle) are shown. Note the dark staining for mast cells (arrows) in the stroma adjacent to the tumor (middle). A higher magnification of the area indicated (*) in the toluidine-blue-stained section is also shown (right). A large number of granular mast cells (arrows) can be seen adjacent to the tumor (dotted line indicates tumor-stroma border). (b) Several possible mechanisms exist by which mast cells could promote carcinogenesis. Due to the large number of pro-inflammatory mediators produced by mast cells, they likely contribute to chronic inflammation, which is known to enhance tumor development and growth. Many mast cell–derived mediators, including several cytokines and growth factors, can directly stimulate tumor cells. Proteases released by mast cells can alter the tumor stroma, aiding in tumor cell invasion and metastasis. Mast cells are a rich source of growth factors and cytokines that promote tumor angiogenesis. Finally, mast cells can secrete immunosuppressive cytokines that could suppress the detection and destruction of the tumor by the immune system.

well as proteases that can aid in the sprouting of new vessels (Coussens et al. 1999). Furthermore, mast cell density correlates with angiogenesis in several types of human tumors (Takanami et al. 2000; Ribatti et al. 2003; Ribatti et al. 2010). The idea that mast cells contribute to tumor angiogenesis is supported by studies indicating a reduction in tumor angiogenesis in mast cell–deficient mice (Coussens et al. 1999; Soucek et al. 2007; Starky et al. 1988). Interestingly, different subpopulations of mast cells have been suggested to play different roles in carcinogenesis, with MC_T located within tumors stimulating angiogenesis and stromal MC_{TC} promoting extracellular matrix degradation and tumor invasion at the tumor-stroma interface (Rojas et al. 2005).

In contrast to the studies indicating a stimulatory role for mast cells in tumor growth, some studies have suggested that mast cells can limit tumor growth or are associated with a better prognosis (Dabiri et al. 2004; Hedström et al. 2007; Rajput et al. 2008; Sinnamon et al. 2008). Mast cells produce an assortment of mediators that can either promote or inhibit tumor growth, which may help explain these contradictory results (Ribatti and Crivellato 2009; Theoharides and Conti 2004). For example, one of the main mast cell mediators, histamine, was recently shown to restrict cancer growth (Yang et al. 2011). The conflicting data surrounding the role of mast cells in cancer suggests that these cells are more complex than originally thought and demonstrates the need for more studies to completely unravel the function of these cells within the tumor microenvironment.

3.3.2 Atherosclerosis

The development and progression of atherosclerosis has been strongly linked to chronic inflammation. Atherosclerotic lesions typically contain macrophages, T cells, and mast cells, which all interact to maintain a chronic inflammatory state (Kovanen 2009), and mast cells are thought to be important at both early and late stages of atherogenesis (Figure 3.5). Early on, mast cells participate in granule-mediated uptake of LDL by macrophages (Kovanen 1991). Heparin within released granules is able to bind low-density lipoprotein (LDL), which can then be modified by mast cell proteases. The complex containing modified LDL is then taken up by macrophages (Kovanen 1991). If LDL collects in the macrophages, cholesterol levels can increase, leading to the formation of foam cells and the development of a fatty streak. Mast cells are also involved in more advanced lesions. Large numbers of mast cells are present at the shoulder region of atheromas and are in close contact with foam cells (Jeziorska et al. 1997; Kovanen et al. 1995). Interestingly, the shoulder region where mast cells accumulate is the area of the plaque most susceptible to eventual erosion or rupture. There is also evidence of mast cell degranulation at these sites (Jeziorska et al. 1997; Kovanen et al. 1995). Tryptase and chymase released by degranulating mast cells have been shown to increase MMP-9 synthesis by macrophages and also activate latent MMPs. This contributes to destabilization of the plaque and increases the chances of plaque rupture (Johnson et al. 1998; Kaartinen et al. 1998).

The significance of mast cells in atherogenesis has been demonstrated *in vivo* using mouse models. Perivascular mast cells were shown to contribute to plaque progression and destabilization in apolipoprotein E (ApoE)-deficient mice. Treatment with either a mast cell stabilizing drug, which prevents degranulation, or a chymase inhibitor had therapeutic benefits in this model (Bot et al. 2007, Bot et al. 2011). In addition, Sun et al. (2007) showed that mast cells are directly involved in atherogenesis using low-density lipoprotein receptor-deficient (LDLR-/-) mice. When LDLR-/- mice were crossed to a mast cell–deficient strain, the mice had smaller lesions, less lipid deposition, and

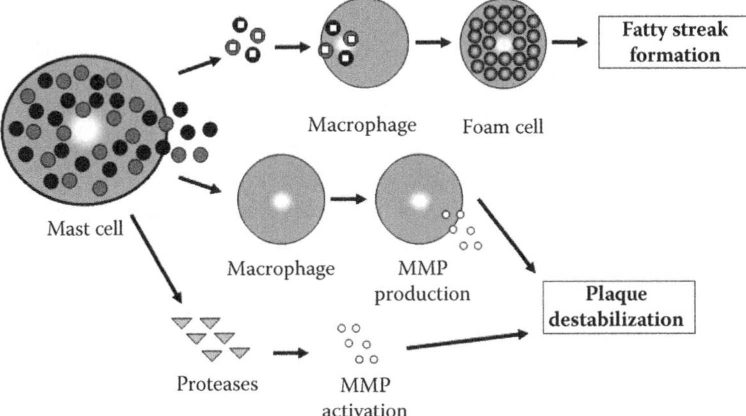

FIGURE 3.5 Mast cells in atherosclerosis. Inflammation drives atherogenesis, and mast cells can be influential at early and late stages. Mast cells can contribute at early phases of atherogenesis by participating in foam cell formation. LDL binds to proteoglycans present in released granules and is then modified by mast cell proteases. Macrophages then undergo granule-mediated uptake of LDL by phagocytosing granule-LDL complexes. Accumulation of LDL in the macrophages results in increased cholesterol levels and the formation of foam cells, driving fatty streak formation. Mast cells are also present and active at the shoulder region of more advanced plaques. Mast cell proteases released upon degranulation can increase MMP production by macrophages and activate latent MMPs, leading to plaque destabilization and increased risk of erosion or rupture.

reduced T cell and macrophage numbers compared to normal, mast cell–containing LDLR-/- mice. The authors also showed that mast cell–derived IL-6 and interferon-γ (IFN-γ) were important for atherogenesis by increasing the expression of matrix-degrading proteases (Sun et al. 2007). Overall, there is strong evidence that mast cells actively participate in atherosclerosis and that mast cells are a viable target for preventing the development and rupture of atherosclerotic plaques.

3.3.3 Fibrosis

Fibrosis is the replacement of normal tissue with scar tissue, which can impair proper organ function. For example, pulmonary fibrosis can interfere with respiration, glial scarring can inhibit axon regeneration, and cardiac fibrosis can reduce the pumping capacity of the heart. Mast cells are associated with various fibrotic conditions (Gruber 2003). They are prominent in Crohn's disease, where patients often develop intestinal fibrosis in the form of strictures (Andoh et al. 2006; Xu et al. 2004). In interstitial renal fibrosis, the number of tryptase-positive mast cells is associated with the degree of fibrosis (Kondo et al. 2001). There are high levels of SCF in scleroderma patients (Kihira et al. 1998), and changes in the number of MC_T as well as increased mast cell activation have been described in cutaneous scleroderma lesions (Irani et al. 1992).

Experimental evidence also suggests a role for mast cells in fibrosis. Mast cell numbers correlate with the degree of fibrosis and cirrhosis in a model of carbon tetrachloride-induced liver fibrosis (Jeong et al. 2002). In models of bleomycin-induced lung and fibrosis, mast cell–deficient mice showed reduced fibrosis (O'Brien-Ladner et al. 1993; Yamamoto et al. 1999). Several studies have examined mast cells in the tight-skin (tsk) mouse, which is used as a model of scleroderma. Walker and colleagues showed that tsk mice have high numbers of mast cells and degranulated mast cells (Walker et al. 1985). Furthermore, preventing mast cell degranulation reduced fibrosis (Walker et al. 1987), and mast cell–deficient tsk mice developed less fibrosis than their mast cell–containing counterparts (Everett et al. 1995).

Interactions between mast cells and fibroblasts are thought to be critically important for development of fibrosis (Figure 3.6). Activated mast cells release many pro-fibrotic mediators that are either preformed or produced *de novo*. Histamine is a stored mediator that has been shown to stimulate fibroblasts (Gailit et al. 2001; Kupietzky and Levi-Schaffer 1996). Mast cell proteases have also been shown to promote fibrotic responses in fibroblasts. Tryptase has chemotactic and mitogenic effects on fibroblasts and also stimulates collagen synthesis, contraction, and differentiation into myofibroblasts (Gailit et al. 2001; Gruber et al. 1997). Cleavage of procollagen type I by chymase can promote collagen fibril formation directly (Kofford et al. 1997). Mast cells can also produce TGF-β, PDGF, and many other pro-fibrotic factors upon stimulation (Gruber 2003). In addition to stimulating fibroblasts by releasing paracrine mediators, mast cells have also been shown to interact directly with fibroblasts. Mast cells and fibroblasts can form gap junctions, allowing direct intercellular communication. The formation of these heterocellular gap junctions is dependent on connexin-43 and has been shown to stimulate fibroblast proliferation, contraction, and differentiation into myofibroblasts (Pistorio and Ehrlich 2011). While these direct interactions have been demonstrated in experimental systems, their involvement in the development of fibrosis *in vivo* is not yet known.

The evidence linking mast cells to fibrosis suggests that they could be a viable target for anti-fibrotic therapies. Inhibiting mast cell degranulation or activation using mast cell stabilizing drugs such as cromolyn or ketotifen could represent one potential therapeutic avenue. Tyrosine kinase inhibitors, many of which inhibit c-kit signaling, may also be an option, considering SCF regulates mast cell recruitment (Huang et al. 2008). Finally, blocking the effects of mast cell mediators could be used as a therapeutic strategy. Recently, a mast cell chymase inhibitor was found to reduce fibrosis in tsk mice, possibly by inhibiting TGF-β1 activation (Shiota et al. 2005). Another chymase inhibitor was shown to reduce neutrophil infiltration and fibrosis in a model of silica-induced lung fibrosis (Takato et al. 2011). Additional studies will have to be performed to determine whether the

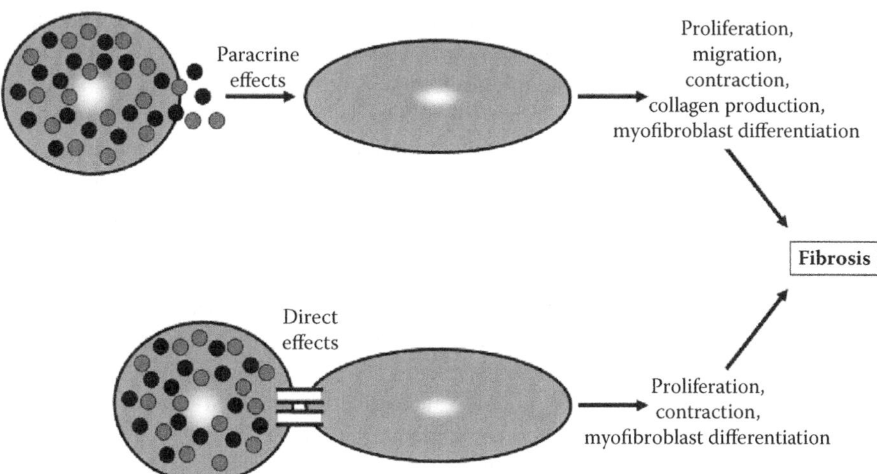

FIGURE 3.6 Mast cell–fibroblast interactions in fibrosis. Mast cells are frequently seen in fibrotic lesions and are believed to be important by stimulating fibroblasts. Mast cells can release many mediators that act in a paracrine manner to stimulate fibroblast proliferation, migration, contraction, collagen production, and differentiation into myofibroblasts. Mast cells and fibroblasts can also interact directly by forming gap junctions, inducing fibroblast proliferation, contraction, and myofibroblast differentiation. Both indirect and direct mast cell–fibroblast interactions and the subsequent changes in fibroblast behavior could be important for the development of fibrosis.

promising results observed in animal models will translate into clinical benefits for patients with fibrotic diseases.

3.4 CONCLUSIONS AND FUTURE DIRECTIONS

It is now becoming clear that mast cells are multifunctional cells that carry out a wide array of functions in physiological and pathological processes. Aside from being involved in allergic responses, where they can cause potentially deadly anaphylactic reactions, mast cells can protect the body from pathogens by regulating innate and adaptive immunity. In addition to stimulating nonspecific and specific immune reactions, a role for mast cells in limiting these responses or suppressing the immune system has now been shown. Mast cells are also active participants in chronic inflammation; as a result, mast cells have been implicated in various diseases resulting from chronic inflammation, such as cancer, atherosclerosis, and fibrosis.

From a basic science perspective, more work needs to be done to understand how both beneficial and detrimental functions of mast cells are regulated and to define the exact mechanisms by which mast cells influence chronic diseases. From a clinical point of view, the utility of targeting mast cells for therapeutic purposes must be determined. Specifically stimulating or inhibiting mast cells, depending on the circumstances, could be important clinically. While mast cell–deficient mouse models have offered a useful way to study the role of mast cells in various experimental disease models, it is still not clear whether these results can be translated to a clinical setting.

TAKE-HOME MESSAGES

- Mast cells are resident immune cells located at convenient anatomical sites for responding quickly to injury or infection. The classical role for mast cells is as an effector cell in hypersensitivity reactions; however, mast cells can help maintain homeostasis, regulate the immune system, and contribute to wound healing.

- Mast cells are involved in chronic inflammation, and there is increasing evidence supporting a role for these cells in promoting cancer, atherosclerosis, fibrosis, and other diseases that develop as a result of chronic inflammation.
- Mast cells are found in high numbers in tissue surrounding tumors and are believed to promote carcinogenesis by contributing to inflammation, tissue remodeling, and angiogenesis, or by stimulating tumor cells directly. Mast cells also have immunosuppressive properties, and could suppress the detection of tumors by the immune system. Although most of the literature suggests that mast cells stimulate the development and growth of tumors, some studies suggest that mast cells could inhibit tumor growth.
- There is a strong link between mast cells and atherosclerosis. Mast cells participate in chronic inflammation in this setting, contributing to foam cell and fatty streak formation. Mast cell proteases can increase the overall levels of active MMPs within plaques by inducing the macrophage MMP expression and activating pro-MMP molecules, which can lead to plaque destabilization and potential plaque rupture.
- Fibrotic tissue typically contains high numbers of mast cells. They likely enhance fibrosis as a result of both direct and indirect interactions with fibroblasts. Mast cells can produce many paracrine mediators that stimulate fibroblasts, and mast cells and fibroblasts can also communicate directly by forming gap junctions. Mast cell–derived mediators and direct mast cell–fibroblast contact can boost collagen production, contraction, migration, and proliferation of fibroblasts, as well as increase the differentiation of fibroblasts to myofibroblasts.
- Mast cells are beginning to be recognized as versatile cells capable of a variety of biological functions depending on the circumstances, demonstrating that they are much more than simply an effector cell in allergic responses. More work needs to be done to gain a better understanding of how these cells regulate both physiological processes as well as pathological processes that develop as a result of chronic inflammation.

REFERENCES

Abraham SN and St John AL. 2010. Mast cell-orchestrated immunity to pathogens. *Nat Rev Immunol* 10(6):440–52.

Andoh A, Deguchi Y, Inatomi O, Yagi Y, Bamba S, Tsujikawa T, et al. 2006. Immunohistochemical study of chymase-positive mast cells in inflammatory bowel disease. *Oncol Rep* 16(1):103–7.

Bot I, Bot M, van Heiningen SH, van Santbrink PJ, Lankhuizen IM, Hartman P, et al. 2011. Mast cell chymase inhibition reduces atherosclerotic plaque progression and improves plaque stability in ApoE-/- mice. *Cardiovasc Res* 89(1):244–52.

Bot I, de Jager SC, Zernecke A, Lindstedt KA, van Berkel TJ, Weber C, et al. 2007. Perivascular mast cells promote atherogenesis and induce plaque destabilization in apolipoprotein E-deficient mice. *Circulation* 115(19):2516–25.

Cindik ED, Maurer M, Hannan MK, Müller R, Hayes WC, Hovy L, et al. 2000. Phenotypical characterization of c-kit receptor deficient mouse femora using non-destructive high-resolution imaging techniques and biomechanical testing. *Technol Health Care* 8(5):267–75.

Claman HN, Choi KL, Sujansky W, and Vatter AE. 1986. Mast cell "disappearance" in chronic murine graft-vs-host disease (GVHD)-ultrastructural demonstration of "phantom mast cells." *J Immunol* 137(6):2009–13.

Coussens LM, Raymond WW, Bergers G, Laig-Webster M, Behrendtsen O, Werb Z, et al. 1999. Inflammatory mast cells up-regulate angiogenesis during squamous epithelial carcinogenesis. *Genes Dev* 13(11):1382–97.

Coussens LM, Tinkle CL, Hanahan D, and Werb Z. 2000. MMP-9 supplied by bone marrow-derived cells contributes to skin carcinogenesis. *Cell.* 103(3):481–90.

Coussens LM and Werb Z. 2002. Inflammation and cancer. *Nature.* 420(6917):860–7.

Crivellato E, Beltrami C, Mallardi F, and Ribatti D. 2003. Paul Ehrlich's doctoral thesis: A milestone in the study of mast cells. *Br J Haematol* 123(1):19–21.

Dabiri S, Huntsman D, Makretsov N, Cheang M, Gilks B, Bajdik C, et al. 2004. The presence of stromal mast cells identifies a subset of invasive breast cancers with a favorable prognosis. *Mod Pathol* 17(6):690–5.

Depinay N, Hacini F, Beghdadi W, Peronet R, and Mécheri S. 2006. Mast cell-dependent down-regulation of antigen-specific immune responses by mosquito bites. *J Immunol* 176(7):4141–6.

Di Nardo A, Vitiello A, and Gallo RL. 2003. Cutting edge: Mast cell antimicrobial activity is mediated by expression of cathelicidin antimicrobial peptide. *J Immunol* 170(5):2274–8.

Duncan LM, Richards LA, and Mihm MC Jr. 1998. Increased mast cell density in invasive melanoma. *J Cutan Pathol* 25(1):11–15.

Dvorak HF, Weaver VM, Tlsty TD, and Bergers G. 2011. Tumor microenvironment and progression. *J Surg Oncol* 103(6):468–74.

Echtenacher B, Männel DN, and Hültner L. 1996. Critical protective role of mast cells in a model of acute septic peritonitis. *Nature* 381(6577):75–7.

Egozi EI, Ferreira AM, Burns AL, Gamelli RL, and Dipietro LA. 2003. Mast cells modulate the inflammatory but not the proliferative response in healing wounds. *Wound Repair Regen* 11(1):46–54.

Everett ET, Pablos JL, Harley RA, LeRoy EC, and Norris JS. 1995. The role of mast cells in the development of skin fibrosis in tight-skin mutant mice. *Comp Biochem Physiol A Physiol* 110(2):159–65.

Gailit J, Marchese MJ, Kew RR, and Gruber BL. 2001. The differentiation and function of myofibroblasts is regulated by mast cell mediators. *J Invest Dermatol* 117(5):1113–9.

Gallant-Behm CL, Hildebrand KA, and Hart DA. 2008. The mast cell stabilizer ketotifen prevents development of excessive skin wound contraction and fibrosis in red Duroc pigs. *Wound Repair Regen* 16(2):226–33.

Galli SJ and Tsai M. 2008. Mast cells: Versatile regulators of inflammation, tissue remodeling, host defense and homeostasis. *J Dermatol Sci* 49(1):7–19.

Ghildyal N, McNeil HP, Gurish MF, Austen KF, and Stevens RL. 1992. Transcriptional regulation of the mucosal mast cell-specific protease gene, MMCP-2, by interleukin 10 and interleukin 3. *J Biol Chem* 267(12):8473–7.

Gordon JR and Galli SJ. 1994. Promotion of mouse fibroblast collagen gene expression by mast cells stimulated via the Fc epsilon RI. Role for mast cell-derived transforming growth factor beta and tumor necrosis factor alpha. *J Exp Med* 180(6):2027–37.

Gounaris E, Erdman SE, Restaino C, Gurish MF, Friend DS, Gounari F, et al. 2007. Mast cells are an essential hematopoietic component for polyp development. *Proc Natl Acad Sci USA* 104(50):19977–82.

Grimbaldeston MA, Nakae S, Kalesnikoff J, Tsai M, and Galli SJ. 2007. Mast cell–derived interleukin 10 limits skin pathology in contact dermatitis and chronic irradiation with ultraviolet B. *Nat Immunol* 8(10):1095–104.

Grivennikov SI, Greten FR, and Karin M. 2010. Immunity, inflammation, and cancer. *Cell* 19; 140(6):883–99.

Gruber BL. 2003. Mast cells in the pathogenesis of fibrosis. *Curr Rheumatol Rep* 5(2):147–53.

Gruber BL, Kew RR, Jelaska A, Marchese MJ, Garlick J, Ren S, et al. 1997. Human mast cells activate fibroblasts: Tryptase is a fibrogenic factor stimulating collagen messenger ribonucleic acid synthesis and fibroblast chemotaxis. *J Immunol* 158(5):2310–7.

Hart PH, Grimbaldeston MA, Swift GJ, Jaksic A, Noonan FP, and Finlay-Jones JJ. 1998. Dermal mast cells determine susceptibility to ultraviolet B-induced systemic suppression of contact hypersensitivity responses in mice. *J Exp Med* 187(12):2045–53.

Harunari N, Zhu KQ, Armendariz RT, Deubner H, Muangman P, Carrougher GJ, et al. 2006. Histology of the thick scar on the female, red Duroc pig: Final similarities to human hypertrophic scar. *Burns* 32(6):669–77.

Hedström G, Berglund M, Molin D, Fischer M, Nilsson G, Thunberg U, et al. 2007. Mast cell infiltration is a favourable prognostic factor in diffuse large B-cell lymphoma. *Br J Haematol* 138(1):68–71.

Huang B, Lei Z, Zhang GM, Li D, Song C, Li B, et al. 2008. SCF-mediated mast cell infiltration and activation exacerbate the inflammation and immunosuppression in tumor microenvironment. *Blood* 112(4):1269–79.

Irani AM, Gruber BL, Kaufman LD, Kahaleh MB, and Schwartz LB. 1992. Mast cell changes in scleroderma. Presence of MCT cells in the skin and evidence of mast cell activation. *Arthritis Rheum* 35(8):933–9.

Jeong WI, Lee CS, Park SJ, Chung JY, and Jeong KS. 2002. Kinetics of macrophages, myofibroblasts and mast cells in carbon tetrachloride-induced rat liver cirrhosis. *Anticancer Res* 22(2A):869–77.

Jeziorska M, McCollum C, and Woolley DE. 1997. Mast cell distribution, activation, and phenotype in atherosclerotic lesions of human carotid arteries. *J Pathol* 182(1):115–22.

Johnson JL, Jackson CL, Angelini GD, and George SJ. 1998. Activation of matrix-degrading metalloproteinases by mast cell proteases in atherosclerotic plaques. *Arterioscler Thromb Vasc Biol* 18(11):1707–15.

Kaartinen M, van der Wal AC, van der Loos CM, Piek JJ, Koch KT, Becker AE, et al. 1998. Mast cell infiltration in acute coronary syndromes: Implications for plaque rupture. *J Am Coll Cardiol* 32(3):606–12.

Kanakura Y, Thompson H, Nakano T, Yamamura T, Asai H, Kitamura Y, et al. 1988. Multiple bidirectional alterations of phenotype and changes in proliferative potential during the in vitro and in vivo passage of clonal mast cell populations derived from mouse peritoneal mast cells. *Blood* 72(3):877–85.

Kanbe N, Tanaka A, Kanbe M, Itakura A, Kurosawa M, and Matsuda H. 1999. Human mast cells produce matrix metalloproteinase 9. *Eur J Immunol* 29(8):2645–9.

Kandere-Grzybowska K, Letourneau R, Kempuraj D, Donelan J, Poplawski S, Boucher W, et al. 2003. IL-1 induces vesicular secretion of IL-6 without degranulation from human mast cells. *J Immunol* 171(9):4830–6.

Kihira C, Mizutani H, Asahi K, Hamanaka H, and Shimizu M. 1998. Increased cutaneous immunoreactive stem cell factor expression and serum stem cell factor level in systemic scleroderma. *J Dermatol Sci* 20(1):72–8.

Kirshenbaum AS and Metcalfe DD. 2006. Growth of human mast cells from bone marrow and peripheral blood-derived CD34+ pluripotent progenitor cells. *Methods Mol Biol* 315:105–12.

Kischer CW, Bunce H 3rd, and Shetlah MR. 1978. Mast cell analyses in hypertrophic scars, hypertrophic scars treated with pressure and mature scars. *J Invest Dermatol* 70(6):355–7.

Kofford MW, Schwartz LB, Schechter NM, Yager DR, Diegelmann RF, and Graham MF. 1997. Cleavage of type I procollagen by human mast cell chymase initiates collagen fibril formation and generates a unique carboxyl-terminal propeptide. *J Biol Chem* 272(11):7127–31.

Kondo S, Kagami S, Kido H, Strutz F, Müller GA, and Kuroda Y. 2001. Role of mast cell tryptase in renal interstitial fibrosis. *J Am Soc Nephrol* 12(8):1668–76.

Kovanen PT. 1991. Mast cell granule-mediated uptake of low density lipoproteins by macrophages: A novel carrier mechanism leading to the formation of foam cells. *Ann Med* 23(5):551–9.

Kovanen PT. 2009. Mast cells in atherogenesis: Actions and reactions. *Curr Atheroscler Rep* 11(3):214–9.

Kovanen PT, Kaartinen M, and Paavonen T. 1995. Infiltrates of activated mast cells at the site of coronary atheromatous erosion or rupture in myocardial infarction. *Circulation* 92(5):1084–8.

Kupietzky A and Levi-Schaffer F. 1996. The role of mast cell-derived histamine in the closure of an in vitro wound. *Inflamm Res* 45(4):176–80.

Lu LF, Lind EF, Gondek DC, Bennett KA, Gleeson MW, Pino-Lagos K, et al. 2006. Mast cells are essential intermediaries in regulatory T-cell tolerance. *Nature* 442(7106):997–1002.

Mak K, Manji A, Gallant-Behm C, Wiebe C, Hart DA, Larjava H, et al. 2009. Scarless healing of oral mucosa is characterized by faster resolution of inflammation and control of myofibroblast action compared to skin wounds in the red Duroc pig model. *J Dermatol Sci* 56(3):168–80.

Malaviya R, Ikeda T, Ross E, and Abraham SN. 1996. Mast cell modulation of neutrophil influx and bacterial clearance at sites of infection through TNF-alpha. *Nature* 381(6577):77–80.

Maltby S, Khazaie K, and McNagny KM. 2009. Mast cells in tumor growth: Angiogenesis, tissue remodelling and immune-modulation. *Biochim Biophys Acta* 1796(1):19–26.

Maurer M, Fischer E, Handjiski B, von Stebut E, Algermissen B, Bavandi A, et al. 1997. Activated skin mast cells are involved in murine hair follicle regression (catagen). *Lab Invest* 77(4):319–32.

Maurer M, Paus R, and Czarnetzki BM. 1995. Mast cells as modulators of hair follicle cycling. *Exp Dermatol* 4(4 Pt 2):266–71.

Metz M, Grimbaldeston MA, Nakae S, Piliponsky AM, Tsai M, and Galli SJ. 2007. Mast cells in the promotion and limitation of chronic inflammation. *Immunol Rev* 217:304–28.

Molin D, Edström A, Glimelius I, Glimelius B, Nilsson G, Sundström C, et al. 2002. Mast cell infiltration correlates with poor prognosis in Hodgkin's lymphoma. *Br J Haematol* 119(1):122–4.

Nagasaka A, Matsue H, Matsushima H, Aoki R, Nakamura Y, Kambe N, et al. 2008. Osteopontin is produced by mast cells and affects IgE-mediated degranulation and migration of mast cells. *Eur J Immunol* 38(2):489–99.

Nakano T, Sonoda T, Hayashi C, Yamatodani A, Kanayama Y, Yamamura T, et al. 1985. Fate of bone marrow-derived cultured mast cells after intracutaneous, intraperitoneal, and intravenous transfer into genetically mast cell-deficient W/Wv mice. Evidence that cultured mast cells can give rise to both connective tissue type and mucosal mast cells. *J Exp Med* 162(3):1025–43.

Nishida K, Yamasaki S, Ito Y, Kabu K, Hattori K, Tezuka T, et al. 2005. FcεRI-mediated mast cell degranulation requires calcium-independent microtubule-dependent translocation of granules to the plasma membrane. *J Cell Biol* 170(1):115–26.

O'Brien-Ladner AR, Wesselius LJ, and Stechschulte DJ. 1993. Bleomycin injury of the lung in a mast-cell-deficient model. *Agents Actions* 39(1-2):20–4.

Pistorio AL and Ehrlich HP. 2011. Modulatory effects of connexin-43 expression on gap junction intercellular communications with mast cells and fibroblasts. *J Cell Biochem* 112(5):1441–9.

Rajput AB, Turbin DA, Cheang MC, Voduc DK, Leung S, Gelmon KA, et al. 2008. Stromal mast cells in invasive breast cancer are a marker of favourable prognosis: A study of 4,444 cases. *Breast Cancer Res Treat* 107(2):249–57.

Rao KN and Brown MA. 2008. Mast cells: Multifaceted immune cells with diverse roles in health and disease. *Ann N Y Acad Sci* 1143:83–104.

Ribatti D and Crivellato E. 2009. The controversial role of mast cells in tumor growth. *Int Rev Cell Mol Biol* 275:89–131.

Ribatti D, Ennas MG, Vacca A, Ferreli F, Nico B, Orru S, et al. 2003. Tumor vascularity and tryptase-positive mast cells correlate with a poor prognosis in melanoma. *Eur J Clin Invest* 33(5):420–5.

Ribatti D, Guidolin D, Marzullo A, Nico B, Annese T, Benagiano V, et al. 2010. Mast cells and angiogenesis in gastric carcinoma. *Int J Exp Patho* 91(4):350–6.

Rojas IG, Spencer ML, Martínez A, Maurelia MA, Rudolph MI. 2005. Characterization of mast cell subpopulations in lip cancer. *J Oral Pathol Med* 34(5):268–73.

Schrementi ME and DiPietro LA. 2005. Site-specific alterations in the early inflammatory process of wounds. *Wound Repair Regen* 13:A43.

Shiota N, Kakizoe E, Shimoura K, Tanaka T, and Okunishi H. 2005. Effect of mast cell chymase inhibitor on the development of scleroderma in tight-skin mice. *Br J Pharmacol* 145(4):424–31.

Silberstein R, Melnick M, Greenberg G, and Minkin C. 1991. Bone remodeling in W/Wv mast cell deficient mice. *Bone*. 12(4):227–36.

Sinnamon MJ, Carter KJ, Sims LP, Lafleur B, Fingleton B, and Matrisian LM. 2008. A protective role of mast cells in intestinal tumorigenesis. *Carcinogenesis* 29(4):880–6.

Skokos D, Le Panse S, Villa I, Rousselle JC, Peronet R, David B, et al. 2001. Mast cell-dependent B and T lymphocyte activation is mediated by the secretion of immunologically active exosomes. *J Immunol* 166(2):868–76.

Smith CJ, Smith JC, and Finn MC. 1987. The possible role of mast cells (allergy) in the production of keloid and hypertrophic scarring. *J Burn Care Rehabil* 8(2):126–31.

Soucek L, Lawlor ER, Soto D, Shchors K, Swigart LB, and Evan GI. 2007. Mast cells are required for angiogenesis and macroscopic expansion of Myc-induced pancreatic islet tumors. *Nat Med* 13(10):1211–8.

Starkey JR, Crowle PK, and Taubenberger S. 1988. Mast-cell-deficient *W/W^v* mice exhibit a decreased rate of tumor angiogenesis. *Int J Cancer* 42:48–52.

Sun J, Sukhova GK, Wolters PJ, Yang M, Kitamoto S, Libby P, et al. 2007. Mast cells promote atherosclerosis by releasing proinflammatory cytokines. *Nat Med* 13(6):719–24.

Takanami I, Takeuchi K, and Naruke M. 2000. Mast cell density is associated with angiogenesis and poor prognosis in pulmonary adenocarcinoma. *Cancer* 88(12):2686–92.

Takato H, Yasui M, Ichikawa Y, Waseda Y, Inuzuka K, Nishizawa Y, et al. 2011. The specific chymase inhibitor TY-51469 suppresses the accumulation of neutrophils in the lung and reduces silica-induced pulmonary fibrosis in mice. *Exp Lung Res* 37(2):101–8.

Theoharides TC, Alysandratos KD, Angelidou A, Delivanis DA, Sismanopoulos N, Zhang B, et al. 2011. Mast cells and inflammation. *Biochim Biophys Acta* [Epub ahead of print] doi:10.1016/j.bbadis.2010.12.014

Theoharides TC and Conti P. 2004. Mast cells: The Jekyll and Hyde of tumor growth. *Trends Immunol* 25(5):235–41.

Toru H, Eguchi M, Matsumoto R, Yanagida M, Yata J, and Nakahata T. 1998. Interleukin-4 promotes the development of tryptase and chymase double-positive human mast cells accompanied by cell maturation. *Blood* 91(1):187–95.

Tsai M, Grimbaldeston M, and Galli SJ. 2011. Mast cells and immunoregulation/immunomodulation. *Adv Exp Med Biol* 716:186–211.

Van Nassauw L, Adriaensen D, and Timmermans JP. 2007. The bidirectional communication between neurons and mast cells within the gastrointestinal tract. *Auton Neurosci* 133(1):91–103.

Walker M, Harley R, Maize J, DeLustro F, and LeRoy EC. 1985. Mast cells and their degranulation in the Tsk mouse model of scleroderma. *Proc Soc Exp Biol Med* 180(2):323–8.

Walker MA, Harley RA, and LeRoy EC. 1987. Inhibition of fibrosis in TSK mice by blocking mast cell degranulation. *J Rheumatol* 14(2):299–301.

Wedemeyer J and Galli SJ. 2005. Decreased susceptibility of mast cell-deficient Kit(W)/Kit(W-v) mice to the development of 1,2-dimethylhydrazine-induced intestinal tumors. *Lab Invest* 85(3):388–96.

Weller CL, Collington SJ, Williams T, and Lamb JR. 2011. Mast cells in health and disease. *Clin Sci (Lond)* 120(11):473–84.

Weller K, Foitzik K, Paus R, Syska W, and Maurer M. 2006. Mast cells are required for normal healing of skin wounds in mice. *FASEB J* 20(13):2366–8.

Wulff BC, Parent AE, Meleski MA, DiPietro LA, Schrementi ME, and Wilgus TA. 2012. Mast cells contribute to scar formation during fetal wound healing. *J Invest Dermatol* 132(2):458–65.

Xu X, Rivkind A, Pikarsky A, Pappo O, Bischoff SC, and Levi-Schaffer F. 2004. Mast cells and eosinophils have a potential profibrogenic role in Crohn disease. *Scand J Gastroenterol* 39(5):440–7.

Yamamoto T, Takahashi Y, Takagawa S, Katayama I, and Nishioka K. 1999. Animal model of sclerotic skin. II. Bleomycin induced scleroderma in genetically mast cell deficient WBB6F1-W/W(V) mice. *J Rheumatol* 26(12):2628–34.

Yang XD, Ai W, Asfaha S, Bhagat G, Friedman RA, Jin G, et al. 2011. Histamine deficiency promotes inflammation-associated carcinogenesis through reduced myeloid maturation and accumulation of CD11b+Ly6G+ immature myeloid cells. *Nat Med* 17(1):87–95.

4 Hypoxia and Hypoxia-Inducible Factor in Inflammation

Kiichi Hirota
Department of Anesthesia, Kyoto University Hospital
Kyoto, Japan

CONTENTS

4.1 INFLAMMATION

Inflammation is part of the complex biological response to harmful stimuli, such as pathogens or damaged cells. Acute inflammation is a short-term process, usually appearing within a few minutes or hours and ceasing upon the removal of the injurious stimulus. It is classically characterized by four cardinal signs: dolor (pain), calor (heat), rubor (redness), and tumor (swelling). Inflammation is also a protective process of the organism to remove the injurious stimuli and to initiate the healing process. Inflammation is not always accompanied with infection, even in cases where inflammation is caused by infection. Although infection is caused by a microorganism, inflammation is one of the responses of the organism to the pathogen. It is considered as a mechanism of innate immunity, as compared to adaptive immunity, which is specific for each pathogen.

Wounds and infections would never subside without the inflammation process. However, chronic inflammation can also lead to a host of diseases, such as allergy, atherosclerosis, rheumatoid arthritis, neurodegenerative disease, and diabetes mellitus.

Inflammation can be classified as either acute or chronic (Dehne and Brüne 2009). Acute inflammation is the initial response of the body to harmful stimuli and is achieved by the increased movement of plasma and white blood cells from the bloodstream into the injured tissues. A cascade of biochemical events propagates and facilitates the inflammatory response, involving the local vascular system, the immune system, and various kinds of cells within the injured tissue. Prolonged inflammation, known as chronic inflammation, leads to a progressive shift in the types of cells present at the site of inflammation and is characterized by simultaneous destruction and healing of the tissue from the inflammatory process.

4.2 INTERACTION BETWEEN HYPOXIC ENVIRONMENT AND INFLAMMATION

4.2.1 Hypoxia-Induced Inflammation

The concept that hypoxia can induce inflammation has gained general acceptance from a line of studies of the hypoxia signaling pathway (Dehne and Brüne 2009; Eltzschig and Carmeliet 2011; Semenza 2007; Weir et al. 2005). The development of inflammation in response to hypoxia is clinically relevant. Ischemia in organ grafts increases the risk of inflammation and graft failure or rejection (Kruger et al. 2009). In patients undergoing kidney transplantation, the renal expression of toll-like receptor (TLR) 4, which constitutes a receptor system for bacterial lipopolysaccharide (LPS), was shown to correlate with the degree of ischemic injury. Increases in pulmonary cytokine levels and TLR expression were shown to correlate with greater ischemic injury of transplanted lungs and loss of graft function. The relationship is also observed in acute respiratory distress syndrome (ARDS) and chronic obstructive pulmonary disease (COPD) models (Figure 4.1). In the setting of obesity, an imbalance between the supply of and demand for oxygen in enlarged adipocytes causes tissue hypoxia and an increase in inflammatory adipokines in fat (Suganami and Ogawa 2010). The resultant infiltration by macrophages and chronic low-grade systemic inflammation promote insulin resistance. Taken together, these clinical studies indicate that hypoxia promotes inflammation.

4.2.2 Inflammation-Induced Hypoxia

Acute focus of tissue inflammation, which is generated in response to infection, injury, noxious agents, or stimuli, present a unique microenvironment (Eltzschig and Carmeliet 2011). Hypoxia (low oxygen) or anoxia (complete lack of oxygen), hypoglycemia (low blood glucose), acidosis (high H^+ concentration), and abundant free reactive oxygen species (ROS) are characteristic features of inflamed tissues. In addition, activated myeloid cells such as neutrophils and macrophages gather into inflammatory tissue. Contributors to tissue hypoxia during inflammation include an increase in the metabolic demands of immune cells and a reduction in metabolic substrates caused by thrombosis, trauma, compression, or atelectasis. Moreover, multiplication of intracellular pathogens can deprive infected cells of oxygen. Thus, in the case of inflamed tissue, dysregulation of oxygen metabolism can influence the environment of the tissue, particularly by regulating oxygen-dependent gene expression (Figure 4.2).

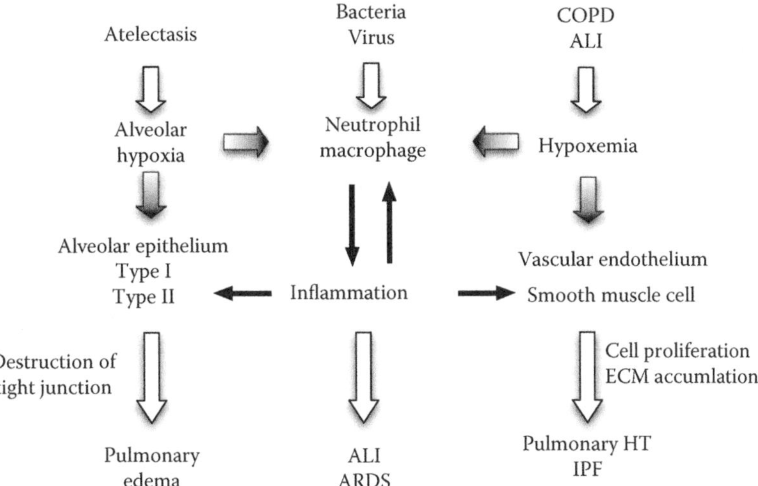

FIGURE 4.1 The interplay of inflammation and hypoxia in lung diseases. In acute and chronic inflammatory conditions of the lung, hypoxia and infection with subsequent inflammation are coincidental occurrences. The presence of microbes and hypoxia in the inflamed lung leads to the expression of genes involved in innate immunity and inflammation. Thus the interplay between inflammation and dysregulation of oxygen metabolism contributes to the establishment of lung diseases such as pulmonary edema, chronic obstructive pulmonary disease (COPD), acute lung injury (ALI), acute respiratory distress syndrome (ARDS), pulmonary hypertension, and idiopathic pulmonary fibrosis (IPF).

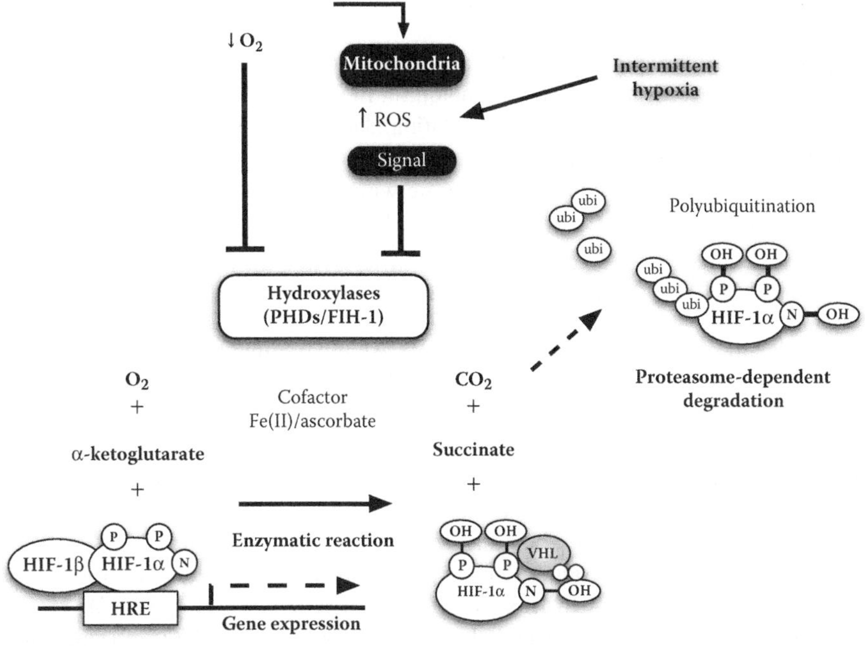

FIGURE 4.2 Under normoxic conditions, hydroxylation of hypoxia-inducible factor 1α (HIF-1α) by prolyl hydroxylases (PHDs) occurs in an O_2-dependent manner at amino acid residues 402 and 564. This results in polyubiquitination of the HIF1α protein by von Hippel–Lindau tumor suppressor protein (VHL) and ultimately in the degradation of HIF1α by proteasomes. The asparaginyl hydroxylase factor inhibiting HIF-1 (FIH-1; also known as HIF1AN) functions in conjunction with prolyl hydroxylation, although in this case to hydroxylate an asparagine residue in the carboxy-terminal domain of HIF1α. As all of these post-translational events depend on intracellular oxygen, they are inhibited by intracellular oxygen deprivation.

4.3 CANONICAL SIGNALING PATHWAY TO HYPOXIA-INDUCIBLE FACTOR ACTIVATION

4.3.1 Induction of HIF Activation by Continuous Hypoxia

Hypoxia-inducible factor 1 (HIF-1) was identified and purified as a nuclear factor that was induced in hypoxic cells and bound to the *cis*-acting hypoxia response element (HRE) located in the 3'-flanking region of the human EPO gene, which encodes erythropoietin (Semenza and Wang 1992). HIF-1 is a heterodimeric transcription factor composed of an HIF-1α subunit and an HIF-1α subunit (Wang et al. 1995). Both HIF-1 subunits are members of the basic helix-loop-helix (HLH)-containing PER-ARNT-SIM (PAS)-domain family of transcription factors. The HLH and PAS domains mediate heterodimer formation between the HIF-1α subunit and HIF-1α subunits, which is necessary for DNA binding by the basic domains (Hirota and Semenza 2005).

In humans (and other mammals), the HIF1A, EPAS1, and HIF3A genes have been shown to encode HIF-1α and the structurally related proteins HIF-2α and HIF-3α, respectively. HIF-1α and HIF-2α show the greatest structural and functional similarity, as each of these proteins is hypoxia induced, dimerizes with HIF-1β, and mediates HRE-dependent transcriptional activity, although they regulate distinct groups of target genes *in vivo*. In contrast, HIF-3α (also known as IPAS) appears to function as an inhibitor that is involved in the negative regulation of transcriptional responses to hypoxia.

Whereas HIF-1β is constitutively expressed, HIF-1α expression increases exponentially as O_2 concentration declines (Hirota and Semenza 2005). In order to respond rapidly to hypoxia, cells continuously synthesize, ubiquitinate, and degrade HIF-1α protein under non-hypoxic conditions. Under hypoxic conditions, the degradation of HIF-1α is inhibited, resulting in accumulation of the protein, dimerization with HIF-1β, binding to HREs within target genes, and activation of transcription via recruitment of the coactivators p300 and CBP (Figure 4.2).

Hydroxylation of two prolyl residues (Pro402 and Pro564 in human HIF-1α) mediates interactions with the von Hippel–Lindau (VHL) E3 ubiquitin ligase complex that targets HIF-1α (as well as HIF-2α and HIF-3α) for proteasomal degradation (Jaakkola et al. 2001; Maxwell et al. 1999). These hydroxylated residues are present within a conserved motif that is recognized by HIF-1α prolyl hydroxylases, which utilize O_2 as a substrate with a Km that is slightly above atmospheric concentration, such that enzymatic activity is modulated by changes in O_2 concentration under physiological conditions (Schofield and Ratcliffe 2004). A family of three human HIF-1α prolyl hydroxylases, designated prolyl hydroxylase domain-containing proteins (PHDs) or HIF-1α prolyl hydroxylases (HPHs) 1, 2, and 3, was identified and shown to be encoded by the EGLN2, EGLN1, and EGLN3 genes, respectively (Epstein et al. 2001). An asparaginyl residue in the transactivation domain of HIF-1α (Asn803 in human HIF-1α) is hydroxylated by factor inhibiting HIF-1 (FIH-1; Lando et al. 2002; Mahon et al. 2001). Hydroxylation of Asn803 blocks interaction of the HIF-1α transactivation domain with the transcriptional coactivators CBP and p300.

The prolyl and asparaginyl hydroxylation reactions require O_2, Fe (II), and α-ketoglutarate (also known as 2-oxoglutarate), and generate succinate and CO_2 as by-products (Hirota and Semenza 2005; Schofield and Ratcliffe 2005). The prolyl hydroxylases and FIH-1 possess a double-stranded α-helix core and Fe (II)-binding residues that are present in other members of the dioxygenase family such as the procollagen prolyl 4-hydroxylases (McNeill et al. 2002) (Figure 4.2).

The expression of more than 50 genes is known to be activated at the transcriptional level by HIF-1 as determined by the most stringent criteria, including the induction of gene expression in response to hypoxia, the presence of a functionally essential HIF-1 binding site in the gene, and an effect of HIF-1 gain of function or loss of function on expression of the gene (Semenza 2007). However, a recent study of global gene expression using DNA microarrays indicates that more than 2% of all human genes are directly or indirectly regulated by HIF-1 in arterial endothelial cells (Manalo et al. 2005).

4.3.2 Induction of HIF by Intermittent Hypoxia

Systemic hypoxia can be either continuous or intermittent. Chronic intermittent hypoxia (CIH) is a common and life-threatening condition that occurs in many different diseases, including sleep-disordered breathing, manifested as recurrent apneas (Shahar et al. 2001). IH increases HIF-1α as well as HIF-1–mediated transcriptional activation in a stimulus-dependent manner in PC12 cells. Under intermittent hypoxia, Ca^{2+}-dependent activation of calcium-calmodulin protein kinase (CaMK) stimulates HIF-1 transcriptional activity by phosphorylation of p300 (Semenza and Prabhakar 2007). HIF-1α protein levels also are induced by intermittent hypoxia. The molecular mechanisms underlying the increase in HIF-1α levels have not been determined but may involve increased translation of HIF-1α mRNA resulting from activation of a signal-transduction pathway involving the mammalian target of rapamycin (mTOR). IH also increases tyrosine hydroxylase (TH) enzyme activity (the rate-limiting enzyme in catecholamine synthesis) in PC12 cells, and this effect is mediated by increased serine phosphorylation involving activation of protein kinase A as well as calcium/calmodulin-dependent protein kinase (Prabhakar et al. 2007). These studies suggest that IH stimulates transcriptional as well as post-translational mechanisms. Interestingly, treatment with antioxidants or ROS scavengers suppresses the IH-induced HIF-1 activation, indicating a critical involvement of NADPH oxidase-derived ROS in this process.

4.4 OXYGEN TENSION-INDEPENDENT ACTIVATION OF HIF

Physiological stimuli other than hypoxia can also induce HIF-1 activation and the transcription of hypoxia-inducible genes under non-hypoxic conditions. Signaling via the HER2/neu or IGF-1 receptor tyrosine kinase induces HIF-1 expression by an oxygen-independent mechanism that increases the rate of HIF-1α protein synthesis (Laughner et al. 2001; Treins et al. 2002). IGF-1-induced HIF-1α synthesis is dependent upon both the PI3K and MAP kinase (MAPK) pathways. In addition, mutation and deletion of some tumor suppressor genes such as VHL, PTEN, and p53 or abberant activation of oncogenes such as Ras also result in constitutive HIF activation under normoxic conditions.

4.4.1 Pro-Inflammatory Cytokines and Chemokines and HIF

In addition to growth factors, cytokines and chemokines are another group of messengers released during inflammation. They are involved in orchestration of the microbicidal activities of phagocytes, contribute to the recruitment of leukocytes, enhance hematopoiesis, and induce fever. TNF-α is rapidly released after infection and is, together with IL-1β, highly important for immune cell activation and, thus, the successful pathogen defense. It is reported that both TNF-α and IL-1β activate HIF-1 (Hellwig-Burgel et al. 1999). TNF-α activates HIF-1 by multiple pathways including ROS and nitric oxide (NO) production, and PI3K and/or NF-κB activation. IL-1β also stimulates HIF-1α translation. In addition, small molecule mediators of the inflammatory microenvironment, such as adenosine and LPS, may also work as positive factors for HIF activation. HIF-1α protein is upregulated after stimulation of the adenosine receptor in a PI3K-dependent manner (Dehne and Brüne 2009). PI3K-dependent HIF-1 activation was not further investigated in immune cells, but growth factors, oncogenes, or inactive tumor suppressors activate multiple phosphorylation events that may either result in translation of HIF-1α or transactivation of HIF-1. In this context, the PI3K/Akt pathway is central for cap-dependent translation, and an activation of Akt is evident in several tumors. Though the steady state of HIF-1α protein expression is mainly determined by hydroxylation-mediated degradation, increased protein translation may shift the balance by overwhelming the degradation system.

A line of studies indicates that NO, exogenously added or endogenously produced, accumulates HIF-1α protein and causes transactivation of HIF-1 under normoxia. NO attenuates HIF-1α ubiquitination in an *in vitro* assay and decreases PHD activity, implying that hypoxia and NO use overlapping signaling pathways to stabilize HIF-1α. However, increased PI3K-dependent HIF-1α protein expression in response to NO has also been proposed (Kasuno et al. 2004). In contrast, when HIF-1α expression is analyzed under hypoxic conditions, NO reduces accumulation of HIF-1α protein (Hagen et al. 2003). This paradox can be explained by the observation that NO competes with O_2 for binding to mitochondrial cytochrome oxidase, which consumes most of the oxygen within a cell (Hagen et al. 2003).

4.5 CROSS TALK BETWEEN NF-κB AND HIF

Members of the nuclear factor κB (NF-κB) family of transcription factors regulate inflammation and orchestrate immune responses and tissue homeostasis. Members of this family interact with members of the PHD–HIF pathway in ways that link inflammation to hypoxia (Cummins et al. 2006; Dehne and Brüne 2009; Rius et al. 2008). Studies of a mouse model of inflammatory bowel disease indicate that PHDs have a regulatory role in the antiapoptotic effects of NF-κB in intestinal inflammation (Colgan and Taylor 2010). The hypoxia of intestinal ischemia reperfusion activates NF-κB in intestinal epithelial cells, which in turn increases the production of TNF-α, but simultaneously attenuates intestinal epithelial apoptosis. Additional interactions between hypoxia and inflammation are seen in the IκB kinase complex, a regulatory component of NF-κB and in the regulation of HIF-1α transcription by NF-κB before and during inflammation. Hypoxia amplifies the NF-κB pathway by increasing the expression and signaling of TLRs, which enhance the production of antimicrobial factors and stimulate phagocytosis, leukocyte recruitment, and adaptive immunity.

NF-κB activity is controlled by inhibitors of NF-κB (IκB) kinases (IKKs), mainly IKKβ, which carry out the phosphorylation-dependent degradation of IκB inhibitors in response to infectious or inflammatory stimuli. HIF was shown to mediate NF-κB activation in neutrophils under anoxic conditions and to promote the expression of NF-κB-regulated cytokines in macrophages stimulated by LPS in a TLR4-dependent manner (Walmsley et al. 2005). Interestingly, hypoxia itself can stimulate NF-κB activation by inhibiting prolyl hydroxylases that negatively modulate IKKβ catalytic activity (Cummins et al. 2006; Cummins and Taylor 2005).

NF-κB was found to contribute to increased HIF1α mRNA transcription under hypoxic conditions. The activation of *HIF1A* transcription by bacteria or LPS under normoxic as well as hypoxic conditions has been recently verified in a study using mice deficient in IKKβ (Rius et al. 2008). Macrophages infected with gram-positive or gram-negative bacteria, and mice subjected to hypoxia, reveal a marked defect in HIF1α expression following deletion of the gene encoding IKKβ (Rius et al. 2008). These results confirm that transcriptional activation of *HIF1A* by IKKβ-responsive NF-κB is a crucial precursor to post-transcriptional stabilization and accumulation of HIF-1α protein.

4.6 IMMUNOCOMPETENT CELLS AND HIF

4.6.1 Macrophages and Dendritic Cells

Evidence is mounting showing that HIF is a key regulator of the intrinsic immune and inflammatory responses in various cell types, including tissue epithelial cells and other specialized leukocytes (Figure 4.3).

Effector cells of the innate immune system must maintain their viability and physiologic functions in a hypoxic microenvironment. Monocytes circulating in the bloodstream differentiate into macrophages. During this process, cells acquire the ability to exert effects at hypoxic sites of inflammation.

Macrophage differentiation of THP-1 cells or monocytes from peripheral blood induces increased expression of both HIF-1α and HIF-1β as well as increased HIF-1 transcriptional activity leading to

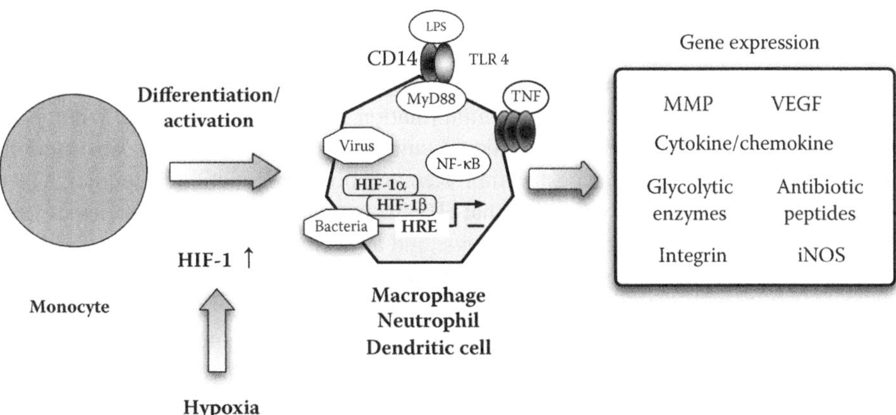

FIGURE 4.3 Inflammatory lesions is not only hypoxic but also is abundant in cytokines; chemokines that induce myeroid cells acting as major players for innate immunity. Mononuclear phagocytes are recruited in large numbers as primary monocytes from the circulation to diseased tissues, where they accumulate within ischemic/hypoxic sites terminally differentiating into inflammatory macrophages or myeloid dendritic cells (DCs). Thus, mononuclear phagocytes including neutrophil responses that ensue at pathological sites begin in the setting of reduced pO_2. In the last five years, extensive work from several groups has been carried out to characterize hypoxia-mediated changes in mononuclear phagocyte gene expression and functional properties under different pathologic situations, demonstrating that oxygen availability is a critical regulator of their functional behavior.

increased expression of HIF-1 target genes (Oda et al. 2006). The increased HIF-1 activity in differentiated THP-1 cells resulted from the combined effect of increased HIF-1α mRNA levels and increased HIF-1α protein synthesis. Differentiation-induced HIF-1α protein and mRNA and HIF-1-dependent gene expression was blocked by treating cells with an inhibitor of the protein kinase C or MAPK signaling pathway. THP-1 cell differentiation was also associated with increased phosphorylation of the translational regulatory proteins p70 S6 kinase, S6 ribosomal protein, eukaryotic initiation factor 4E, and 4E binding protein 1, thus providing a possible mechanism for the modulation of HIF-1α protein synthesis.

Dendritic cells (DC) are professional antigen presenting cells that represent an important link between innate and adaptive immunity. External alert signals such as toll-like receptor (TLR) agonists induce maturation of DCs leading to a T cell–mediated adaptive immune response. It is reported that exogenous as well as endogenous inflammatory stimuli for TLR4 and TLR2 induce the expression of HIF-1α in human monocyte-derived DCs under normoxic conditions (Jantsch et al. 2008). On the functional level, inhibition of HIF-1α using small molecule inhibitors of HIF such as YC-1 and digoxin lead to no consistent effect on MoDC maturation, or cytokine secretion, despite having the common effect of blocking HIF-1α stabilization or activity through different mechanisms. In addition, we could show that TLR stimulation resulted in an increase of HIF-1α controlled VEGF secretion. Hypoxia alone does not induce maturation of DCs, but is able to augment maturation after TLR ligation.

4.6.2 Neutrophils

Neutrophils are also key mediators of the innate immune response. Because these cells rely on glycolysis to generate adenosine triphosphate (ATP), they appear to be adapted uniquely to function at sites of inflammation. Cramer et al. created deletions of the HIF-1α transcription factor via crosses into a background of Cre expression driven by the lysosome M promoter, which allows specific deletion of this factor in myeloid cell lineage (2003). This murine model has made it possible to dissect the role of HIF-1α in myeloid cell migration and activation *in vivo*. Interestingly, these animals have a normal phenotype and viability, and normal total white cell and neutrophil counts. The HIF-1α knockout mice display a dramatic reduction in cellular ATP pools and a profound impairment in myeloid cell aggregation, motility, invasiveness, and reduced bacterial killing (Cramer et al. 2003). This is manifest *in vivo* as a loss of inflammation in paucivascular sites, with ablation of phorbol 12-myristate 13-acetate (PMA)-mediated cutaneous inflammation and also inhibition of rheumatoid serum-induced cartilage destruction. Remarkably, the reduction in cellular ATP is also seen under normoxic conditions, suggesting that HIF-1α is required for the maintenance of energy in myeloid cells even in oxygenated environments, and as such this is postulated to be the mechanism through which HIF-1α regulates the pro-inflammatory myeloid response (Cramer and Johnson 2003; Cramer et al. 2003).

The role of HIF in regulating neutrophil apoptosis at inflamed sites *in vivo* remains to be determined. Apoptosis itself has been implicated in the resolution of inflammation, and excessive neutrophil activation and prolonged survival have been described in several disease settings, including the ARDS, nonresolving pneumonias, ischemic-reperfusion injury, and rheumatoid arthritis (Serhan and Savill 2005). Considering the exaggerated oxygen gradient that exists at many inflamed sites, a role for HIF in the regulation of neutrophil apoptosis in these settings seems highly probable.

Inflammatory regions are characterized by hypoxia and the dramatic recruitment of myeloid cells. The recruitment of myeloid cells to sites of inflammation is coordinated by the β2 integrin family of adhesion receptors. The β2 integrins are heterodimeric glycoproteins that exist in four forms. Each form is composed of a unique α-subunit, encoded by the CD11a, CD11b, CD11c, or CD11d gene, noncovalently associated with a common β-subunit encoded by the CD18 gene. Interestingly, β2 integrin expression is under regulation by HIF-1 (Kong et al. 2004). Analysis of these mutant mice demonstrated that HIF-1α is critically important for successful inflammatory responses mediated by myeloid cells. The disruption of HIF-1α did not influence myeloid cell differentiation or development. As described, HIF-1α deletion did result in significant metabolic defects manifest as profound impairment of myeloid cell motility, bacterial phagocytosis, and aggregation (Cramer and Johnson 2003; Cramer et al. 2003). Interestingly, these functional responses are also dependent on β2 integrin expression.

4.6.3 T Lymphocytes

T cell differentiation into distinct functional effector and inhibitory subsets is regulated, in part, by the cytokine environment present at the time of antigen recognition. It is demonstrated that HIF-1 regulates the balance between regulatory T cell (Treg) and T_H17 differentiation. HIF-1 enhances T_H17 development through direct transcriptional activation of RORγt and via tertiary complex formation with RORγt and p300 recruitment to the IL-17 promoter (Dang et al. 2011). Concurrently, HIF-1 attenuates Treg development by binding Foxp3 and targeting it for proteasomal degradation. Interestingly, this regulation occurs under both normoxic and hypoxic conditions. Mice with HIF-1α-deficient T cells are resistant to induction of T_H17-dependent experimental autoimmune encephalitis associated with diminished T_H17 and increased Treg cells. These findings highlight the importance of metabolic cues in T cell fate determination and suggest that metabolic modulation could ameliorate certain T cell–based immune pathologies (Dang et al. 2011; Shi et al. 2011; Woodman 2011).

Peripheral T lymphocytes undergo activation by antigenic stimulation and function in hypoxic areas of inflammation. It is demonstrated that CD3-positive human T cells accumulating in inflammatory tissue express HIF-1α, indicating a role of hypoxia-mediated signals in regulation of T cell function (Makino et al. 2003). Surprisingly, accumulation of HIF-1α in human T cells required not only hypoxia but also TCR/CD3-mediated activation (Makino et al. 2003). Moreover, hypoxia repressed activation-induced cell death (AICD) by TCR/CD3 stimulation, resulting in an increased survival of the cells.

4.6.4 B Lymphocytes

In addition to T cells, HIF-1 is also involved in regulation of B cells. It is reported that an HIF-1α gene deficiency caused abnormal B cell development and autoimmunity (Kojima et al. 2002). The key role of HIF-1α–enabled glycolysis in bone marrow B cells was demonstrated by glucose deprivation during *in vitro* bone marrow cell culture and by using a glycolysis inhibitor in the bone marrow cell culture. These findings indicate that glucose dependency differs at different B cell develpmental stages and that HIF-1α plays an important role in B cell development (Kojima et al. 2002; Kojima et al. 2010).

4.6.5 Mast Cells

Mast cells are specialized granulocytic cells that are resident in the skin and the mucosa of the respiratory and gastrointestinal tracts. Their roles in allergy are well studied, and they are increasingly recognized to function in both innate and adaptive immune responses. Activation of HIF in human mast cells leads to release of pro-inflammatory cytokines such as IL-8 and TNF-α (Lee et al. 2008). HIF activation in mast cells of the bronchial epithelium stimulates VEGF expression, leading to increased vascular permeability, protein extravasation into the alveolar space, and airway edema. Moreover, HIF activation stimulates histidine decarboxylase expression by human mast cells, catalyzing the formation of histamine, a potent inflammatory mediator (Lee et al. 2008).

4.7 IMPACT OF ANTI-INFLAMMATORY DRUGS ON HIF ACTIVITY

The synthetic glucocorticoid (GC) dexamethasone is a widely used drug to suppress the development of local heat, redness, swelling, and tenderness—the symptoms of inflammation. Besides these beneficial effects, dexamethasone causes unwanted side effects such as impaired wound healing and immunosuppression. Indeed, dexamethasone attenuates HIF-1 activity in a GR-dependent manner (Wagner et al. 2008). It became evident that dexamethasone led to an unusual distribution of hypoxically induced HIF-1α within the cell by comparing cytosolic protein extracts with nuclear protein extracts. Instead of accumulating exclusively in the nucleus, HIF-1α amounts increased in the cytosol after dexamethasone treatment, explaining the reduced nuclear HIF-1α protein. The evidence shows the importance of HIF-1 during wound healing. Although dexamethasone treatment of patients has the unwanted systemic side effect, it might be a promising approach to improve HIF-1 function in wounds of such patients.

Nonsteroidal anti-inflammatory drugs (NSAIDs) block prostaglandin synthesis and impair healing of gastrointestinal ulcers and growth of colonic tumors, in part, by inhibiting angiogenesis. The mechanisms of this inhibition are not completely explained. Both nonselective (indomethacin and ibuprofen) and COX-2-selective NSAIDs inhibit hypoxia-induced *in vitro* angiogenesis in gastric microvascular endothelial cells (Palayoor et al. 2003). The NSAIDs increased expression of the von Hippel–Lindau (VHL) tumor suppressor, which targets proteins for ubiquitination leading to reduced accumulation of HIF-1α and, as a result, reduced expression of VEGF and its specific receptor Flt-1.

4.8 INFLAMMATORY DISORDERS AND HIF

4.8.1 Infection

Increased levels of HIF are observed in macrophages and neutrophils stimulated by various bacterial species. Acute infection with viruses is also generally found to induce HIF protein stabilization in target cells. Moreover, *Toxoplasma gondii* rapidly induces HIF expression by infected fibroblast,

TABLE 4.1
Inflammatory Mediators Activating HIF

Mediators and Pathogens	Mechanism	References
Hypoxia	Inhibition of PHD	(Li et al. 2007)
Intermittent hypoxia	Inhibition of PHD by ROS	(Semenza and Prabhakar 2007)
	Induction of HIF-1α transcription	(Prabhakar et al. 2007)
ROS	Oxidative modulation of PHD	(Gerald et al. 2004)
	Oxidation of Fe^{2+}	(Gerald et al. 2004)
NO	Increase of HIF-1α translation	(Kasuno et al. 2004)
	Inhibition of PHD	(Brune and Zhou 2007)
	S-nitrosylation of HIF-1α	(Li et al. 2007)
Cytokines (TNF-α, IL-1β)	Increase of HIF-1α translation	(Hellwig-Burgel et al. 1999)
Chemokines (MIF, SDF-1)	Increase of HIF-1α translation	(Pan et al. 2006)
	Stabilization of HIF-1α	(Oda et al. 2008)
Thrombin	Increase of HIF-1α translation	(Gorlach et al. 2001)
PGE_2	Increase of HIF-1α translation	(Fukuda et al. 2003)
LPS	Increase of HIF-1α translation	(Nishi et al. 2008)
	Induction of HIF-1α transcription	(Nishi et al. 2008)
Monocyte differentiation	Increase of HIF-1α translation	(Oda et al. 2006)
	Induction of HIF-1α transcription	(Zinkernagel et al. 2007)
Bacteria	Inhibition of PHD	(Zinkernagel et al. 2007)
	Increase of HIF-1α translation	(Zinkernagel et al. 2007)
Virus	Increase of PHD degradation	(Zinkernagel et al. 2007)
	Increase of HIF-1α translation	(Zinkernagel et al. 2007)
Acidosis	Increase of HIF-1α stabilization	(Ohh et al. 2000)

Bacteria: *S. pyrogenes, S. agalactiae, P. aeruginosa, S. typhimurium, C. pneumoniae*
Viruses: EBV, human papilloma virus, hepatitis C virus, hepatitis B virus, KSHV, respiratory syncytial virus

which leads to upregulation of genes encoding glycolytic enzymes, glucose transporters, and VEGF under hypoxic conditions.

A line of infectious pathogens and reagents affect HIF protein level and transcriptional activity (Table 4.1).

4.8.2 Sepsis

Sepsis, one of the leading causes of death in intensive care units, reflects a detrimental host response to infection in which bacteria or LPS act as potent activators of immune cells, including monocytes and macrophages. So far, no single agent or treatment strategy has shown sufficient promise for use in routine clinical practice to block the aberrant cytokine activation patterns of sepsis.

It is shown that LPS raises the level of the transcriptional regulator HIF-1α in macrophages, increasing HIF-1α and decreasing prolyl hydroxylase mRNA production in a TLR4-dependent fashion (Peyssonnaux et al. 2007). Murine conditional gene targeting of HIF-1α in the myeloid lineage model demonstrates that HIF-1α is a critical determinant of the sepsis phenotype (Peyssonnaux et al. 2007). HIF-1α promotes the production of inflammatory cytokines, including TNF-α, IL-1, IL-4, IL-6, and IL-12, that reach harmful levels in the host during early sepsis. HIF-1α deletion in macrophages is protective against LPS-induced mortality and blocks the development of clinical markers, including hypotension and hypothermia. Inhibition of HIF-1α activity may thus represent a novel therapeutic target for LPS-induced sepsis.

4.8.3 Wound Healing

Because of localized vascular damage and increased tissue oxygen demand, wound healing occurs in a relatively hypoxic microenvironment (Sen and Roy 2008). These features are particularly relevant to wound healing and fibrosis in chronic inflammatory conditions. In the last decade, its activity in the context of wound healing has been the object of increasing investigation (Eming et al. 2007). On the molecular level, HIF-1 transcriptional target products have been shown to regulate the process of endothelial cell survival, migration, and proliferation (VEGF, ANGPT-1, ANGPT-2, ANGPT-4, FGF-2, PlGF, PDGF-B, RGC-32), vascular smooth muscle cell migration and proliferation (FGF-2, EGF, PDGF, thrombospondin), and mobilization of circulating angiogenic cells to the periphery (SFD-1/CXCR4). Studies on the effect of HIF-1 on the expression and activity of extracellular cell matrix modifying enzymes, such as MMPs and prolidase, have been conducted in the context of tumor angiogenesis and metastasis, and have resulted in controversial findings. A growing body of evidence suggests that HIF-1 also affects reepithelialization of the wound bed, through increasing keratinocyte migration, but decreasing their proliferation. Diminished HIF-1 levels and activity have been documented in conditions of impaired wound healing, such as wound healing in aged and in diabetic mice (Albina et al. 2001; Wagner et al. 2008).

However, there has not been a systematic exploration of the relationship between HIF activity and hypoxia in the burn wound (Jeschke et al. 2011; Schwacha et al. 2008). The time course of the appearance of hypoxia and the increased activity of HIF and appearance of HIF's downstream transcription products has not been described. Hypoxia was found in the healing margin of burn wounds beginning at 48 hours after burn and peaking at day three after burn. On sequential sections of the same tissue block, positive staining of HIF-1α, SDF-1, and vascular endothelial growth factor all occurred at the leading margin of the healing area and peaked at day three, as did hypoxia. Immunohistochemical analysis was used to explore the characteristics of the hypoxic region of the wound.

4.9 CONCLUSIONS AND FUTURE DIRECTIONS

HIF and the hypoxic response are deeply involved with the regulatory pathways of innate immune defense. The key implication of these findings is that the nature and magnitude of host bactericidal and inflammatory activities are deeply dependent on factors in the local tissue or niche microenvironment such as oxygen metabolism. Through HIF control of immune cell energetics and gene expression pathways, antimicrobial activities can be focused and amplified where they are needed most, namely, foci of tissue infection, which are harsh and threatening microenvironments where oxygen and nutrients are limiting and cytotoxic molecules abound. A detailed understanding of the relationships between HIF, pathways of innate immune signal transduction such as TLR–NF-κB signaling, and the deployment of various immune effector molecules will provide a clearer and more physiological understanding of infectious and inflammatory disease pathogenesis. Because of the short half-life and well-understood mechanism for post-translational regulation of HIF levels, HIF is an attractive pharmacological target to fine tune immune cell functions for the treatment of human disease. The recent discovery of a new family of oxygen sensors—including prolyl hydroxylase domain-containing proteins 1–3 (PHD1–3)—has yielded exciting novel insights into how cells sense oxygen and keep oxygen supply and consumption in balance (Fraisl et al. 2009; Smith and Talbot 2010). Advances in understanding of the role of these oxygen sensors in hypoxia tolerance, ischemic

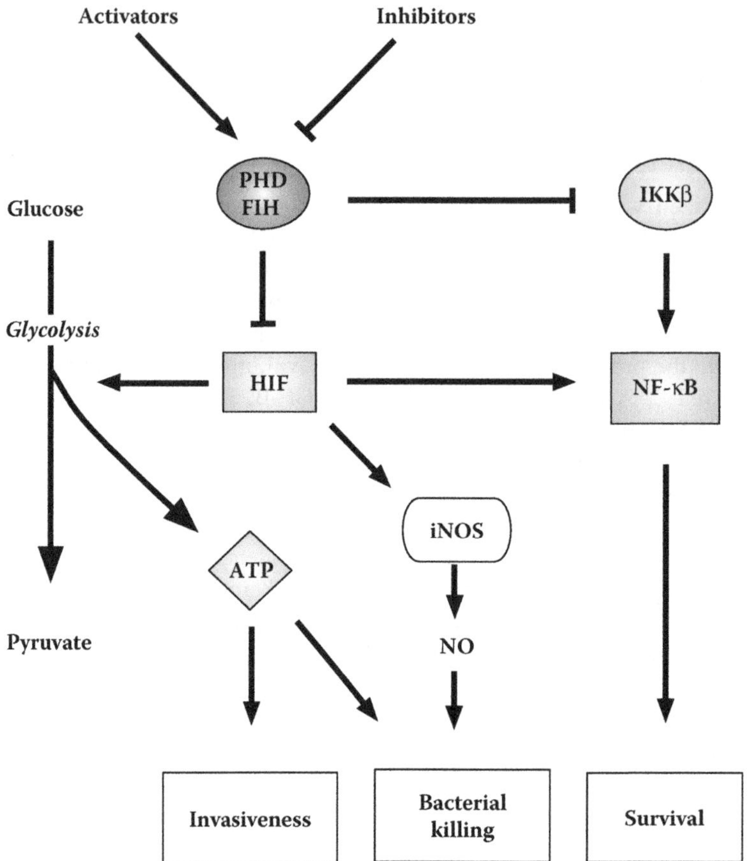

FIGURE 4.4 Inhibition of the PHD-HIF pathway in inflammation. Several studies suggest that PHD inhibition offers beneficial effects in various diseases. Therapeutic induction of hypoxia/ischemia tolerance, hibernation, or hypoxic/ischemic preconditioning could create considerable opportunities to treat numerous ischemic disorders, such as myocardial infarction and stroke. Many other conditions are associated with some sort or degree of acute or chronic ischemia, ranging from inflammation and neurodegeneration to organ transplantation and surgical interventions. Blockade of prolyl hydroxylase domain-containing proteins (PHDs) with an inhibitor in macrophages activates the HIF pathway, which subsequently activates inducible nitric oxide (iNOS) and produces bactericidal nitric oxide (NO). In addition, HIF-mediated stimulation of anaerobic glucose utilization (glycolysis; glucose to pyruvate conversion) ensures sufficient production of ATP to support macrophage functions (such as invasiveness and bacterial killing) in an oxygen-depleted environment. The PHD/HIF pathway also contributes to enhanced neutrophil survival. Inhibition of PHDs stabilizes HIF and induces IκB kinase β (IKKβ) function; both will contribute to activation of nuclear factor κB (NF-κB), thereby promoting neutrophil survival.

preconditioning, and inflammation are creating new opportunities for pharmacological interventions for inflammatory diseases (Figure 4.4).

TAKE-HOME MESSAGES

- Systemic and local hypoxia can induce inflammation without bacterial or viral infection.
- In turn, systemic or local inflammation often becomes hypoxic due to thrombosis and edema.
- The oxygen-sensing system consisting of hypoxia-inducible factor (HIF) and two classes of hydroxylase plays a critical role in the interdependence between hypoxia and inflammation.

- Pro-inflammatory mediators activate HIF even under non-hypoxic conditions, and HIF induces the transcription of a set of genes expressing for inflammation an process.
- HIF modulate function of immunocompetent cells such as macrophages, dendritic cells, neutrophils, and lymphocytes.
- The HIF-PHD/FIH-1 system can be one of the most critical targets for the regulation of inflammation.

REFERENCES

Albina, J.E., Mastrofrancesco, B., Vessella, J.A., et al. (2001). HIF-1 expression in healing wounds: HIF-1alpha induction in primary inflammatory cells by TNF-alpha. *Am J Physiol Cell Physiol* 281, C1971–1977.

Brune, B., and Zhou, J. (2007). Hypoxia-inducible factor-1alpha under the control of nitric oxide. *Methods Enzymol* 435, 463–478.

Colgan, S.P., and Taylor, C.T. (2010). Hypoxia: An alarm signal during intestinal inflammation. *Nat Rev Gastroenterol Hepatol* 7, 281–287.

Cramer, T., and Johnson, R.S. (2003). A novel role for the hypoxia inducible transcription factor HIF-1alpha: Critical regulation of inflammatory cell function. *Cell Cycle* 2, 192–193.

Cramer, T., Yamanishi, Y., Clausen, B.E., et al. (2003). HIF-1alpha is essential for myeloid cell-mediated inflammation. *Cell* 112, 645–657.

Cummins, E.P., Berra, E., Comerford, K.M., et al. (2006). Prolyl hydroxylase-1 negatively regulates IkappaB kinase-beta, giving insight into hypoxia-induced NFkappaB activity. *Proc Natl Acad Sci USA* 103, 18154–18159.

Cummins, E.P., and Taylor, C.T. (2005). Hypoxia-responsive transcription factors. *Pflugers Arch* 450, 363–371.

Dang, E.V., Barbi, J., Yang, H.Y., et al. (2011). Control of T(H)17/T(reg) balance by hypoxia-inducible factor 1. *Cell* 146, 772–784.

Dehne, N., and Brüne, B. (2009). HIF-1 in the inflammatory microenvironment. *Exp Cell Res* 315, 1791–1797.

Eltzschig, H.K., and Carmeliet, P. (2011). Hypoxia and inflammation. *N Engl J Med* 364, 656–665.

Eming, S.A., Krieg, T., and Davidson, J.M. (2007). Inflammation in wound repair: Molecular and cellular mechanisms. *Journal of Investigative Dermatology* 127, 514–525.

Epstein, A., Gleadle, J., McNeill, L., et al. (2001). *C. elegans* EGL-9 and mammalian homologs define a family of dioxygenases that regulate HIF by prolyl hydroxylation. *Cell* 107, 43–54.

Fraisl, P., Aragones, J., and Carmeliet, P. (2009). Inhibition of oxygen sensors as a therapeutic strategy for ischaemic and inflammatory disease. *Nat Rev Drug Discov* 8, 139–152.

Fukuda, R., Kelly, B., and Semenza, G.L. (2003). Vascular endothelial growth factor gene expression in colon cancer cells exposed to prostaglandin E2 is mediated by hypoxia-inducible factor 1. *Cancer Res* 63, 2330–2334.

Gerald, D., Berra, E., Frapart, Y.M., et al. (2004). JunD reduces tumor angiogenesis by protecting cells from oxidative stress. *Cell* 118, 781–794.

Gorlach, A., Diebold, I., Schini-Kerth, V.B., et al. (2001). Thrombin activates the hypoxia-inducible factor-1 signaling pathway in vascular smooth muscle cells: Role of the p22(phox)-containing NADPH oxidase. *Circ Res* 89, 47–54.

Hagen, T., Taylor, C.T., Lam, F., and Moncada, S. (2003). Redistribution of intracellular oxygen in hypoxia by nitric oxide: Effect on HIF1alpha. *Science* 302, 1975–1978.

Hellwig-Burgel, T., Rutkowski, K., Metzen, E., Fandrey, J., and Jelkmann, W. (1999). Interleukin-1beta and tumor necrosis factor-alpha stimulate DNA binding of hypoxia-inducible factor-1. *Blood* 94, 1561–1567.

Hirota, K., and Semenza, G.L. (2005). Regulation of hypoxia-inducible factor 1 by prolyl and asparaginyl hydroxylases. *Biochem Biophys Res Commun* 338, 610–616.

Jaakkola, P., Mole, D.R., Tian, Y.M., et al. (2001). Targeting of HIF-alpha to the von Hippel-Lindau ubiquitylation complex by O2-regulated prolyl hydroxylation. *Science* 292, 468–472.

Jantsch, J., Chakravortty, D., Turza, N., et al. (2008). Hypoxia and hypoxia-inducible factor-1 alpha modulate lipopolysaccharide-induced dendritic cell activation and function. *J Immunol* 180, 4697–4705.

Jeschke, M.G., Gauglitz, G.G., Kulp, G.A., et al. (2011). Long-term persistance of the pathophysiologic response to severe burn injury. *PLoS ONE* 6, e21245.

Kasuno, K., Takabuchi, S., Fukuda, K., et al. (2004). Nitric oxide induces hypoxia-inducible factor 1 activation that is dependent on MAP kinase and phosphatidylinositol 3-kinase signaling. *J Biol Chem* 279, 2550–2558.

Kojima, H., Gu, H., Nomura, S., et al. (2002). Abnormal B lymphocyte development and autoimmunity in hypoxia-inducible factor 1alpha -deficient chimeric mice. *Proc Natl Acad Sci USA* 99, 2170–2174.

Kojima, H., Kobayashi, A., Sakurai, D., et al. (2010). Differentiation stage-specific requirement in hypoxia-inducible factor-1alpha-regulated glycolytic pathway during murine B cell development in bone marrow. *J Immunol* 184, 154–163.

Kong, T., Eltzschig, H.K., Karhausen, J., Colgan, S.P., and Shelley, C.S. (2004). Leukocyte adhesion during hypoxia is mediated by HIF-1-dependent induction of beta2 integrin gene expression. *Proc Natl Acad Sci USA* 101, 10440–10445.

Kruger, B., Krick, S., Dhillon, N., et al. (2009). Donor Toll-like receptor 4 contributes to ischemia and reperfusion injury following human kidney transplantation. *Proc Natl Acad Sci USA* 106, 3390–3395.

Lando, D., Peet, D.J., Gorman, J.J., et al. (2002). FIH-1 is an asparaginyl hydroxylase enzyme that regulates the transcriptional activity of hypoxia-inducible factor. *Genes Dev* 16, 1466–1471.

Laughner, E., Taghavi, P., Chiles, K., Mahon, P.C., and Semenza, G.L. (2001). HER2 (neu) signaling increases the rate of hypoxia-inducible factor 1alpha (HIF-1alpha) synthesis: Novel mechanism for HIF-1-mediated vascular endothelial growth factor expression. *Mol Cell Biol* 21, 3995–4004.

Lee, K.S., Kim, S.R., Park, S.J., et al. (2008). Mast cells can mediate vascular permeability through regulation of the PI3K-HIF-1alpha-VEGF axis. *Am J Respir Crit Care Med* 178, 787–797.

Li, F., Sonveaux, P., Rabbani, Z.N., et al. (2007). Regulation of HIF-1alpha stability through S-nitrosylation. *Mol Cell* 26, 63–74.

Mahon, P.C., Hirota, K., and Semenza, G.L. (2001). FIH-1: A novel protein that interacts with HIF-1a and VHL to mediate repression of HIF-1 transcriptional activity. *Genes Dev* 15, 2675–2686.

Makino, Y., Nakamura, H., Ikeda, E., et al. (2003). Hypoxia-inducible factor regulates survival of antigen receptor-driven T cells. *J Immunol* 171, 6534–6540.

Manalo, D.J., Rowan, A., Lavoie, T., et al. (2005). Transcriptional regulation of vascular endothelial cell responses to hypoxia by HIF-1. *Blood* 105, 659–669.

Maxwell, P.H., Wiesener, M.S., Chang, G.W., et al. (1999). The tumour suppressor protein VHL targets hypoxia-inducible factors for oxygen-dependent proteolysis. *Nature* 399, 271–275.

McNeill, L.A., Hewitson, K.S., Gleadle, J.M., et al. (2002). The use of dioxygen by HIF prolyl hydroxylase (PHD1). *Bioorg Med Chem Lett* 12, 1547–1550.

Nishi, K., Oda, T., Takabuchi, S., et al. (2008). LPS induces hypoxia-inducible factor 1 activation in macrophage-differentiated cells in a reactive oxygen species-dependent manner. *Antioxidants & Redox Signaling* 10, 983–996.

Oda, S., Oda, T., Nishi, K., et al. (2008). Macrophage migration inhibitory factor activates hypoxia-inducible factor in a p53-dependent manner. *PLoS ONE* 3, e2215.

Oda, T., Hirota, K., Nishi, K., et al. (2006). Activation of hypoxia-inducible factor 1 during macrophage differentiation. *Am J Physiol Cell Physiol* 291, C104–C113.

Ohh, M., Park, C.W., Ivan, M., et al. (2000). Ubiquitination of hypoxia-inducible factor requires direct binding to the beta-domain of the von Hippel-Lindau protein. *Nat Cell Biol* 2, 423–427.

Palayoor, S.T., Tofilon, P.J., and Coleman, C.N. (2003). Ibuprofen-mediated reduction of hypoxia-inducible factors HIF-1alpha and HIF-2alpha in prostate cancer cells. *Clin Cancer Res* 9, 3150–3157.

Pan, J., Mestas, J., Burdick, M.D., et al. (2006). Stromal derived factor-1 (SDF-1/CXCL12) and CXCR4 in renal cell carcinoma metastasis. *Mol Cancer* 5, 56.

Peyssonnaux, C., Cejudo-Martin, P., Doedens, A., et al. (2007). Cutting edge: Essential role of hypoxia inducible factor-1alpha in development of lipopolysaccharide-induced sepsis. *J Immunol* 178, 7516–7519.

Prabhakar, N.R., Kumar, G.K., Nanduri, J., and Semenza, G.L. (2007). ROS signaling in systemic and cellular responses to chronic intermittent hypoxia. *Antioxidants & Redox Signaling* 9, 1397–1403.

Rius, J., Guma, M., Schachtrup, C., et al. (2008). NF-kappaB links innate immunity to the hypoxic response through transcriptional regulation of HIF-1alpha. *Nature* 453, 807–811.

Schofield, C.J., and Ratcliffe, P.J. (2004). Oxygen sensing by HIF hydroxylases. *Nat Rev Mol Cell Biol* 5, 343–354.

Schofield, C.J., and Ratcliffe, P.J. (2005). Signalling hypoxia by HIF hydroxylases. *Biochem Biophys Res Commun* 338, 617–626.

Schwacha, M.G., Nickel, E., and Daniel, T. (2008). Burn injury-induced alterations in wound inflammation and healing are associated with suppressed hypoxia inducible factor-1alpha expression. *Mol Med* 14, 628–633.

Semenza, G.L. (2007). Life with oxygen. *Science* 318, 62–64.

Semenza, G.L., and Prabhakar, N.R. (2007). HIF-1-dependent respiratory, cardiovascular, and redox responses to chronic intermittent hypoxia. *Antioxidants & Redox Signaling* 9, 1391–1396.

Semenza, G.L., and Wang, G.L. (1992). A nuclear factor induced by hypoxia via de novo protein synthesis binds to the human erythropoietin gene enhancer at a site required for transcriptional activation. *Mol Cell Biol* 12, 5447–5454.

Sen, C.K., and Roy, S. (2008). Redox signals in wound healing. *Biochim Biophys Acta* 1780, 1348–1361.

Serhan, C.N., and Savill, J. (2005). Resolution of inflammation: The beginning programs the end. *Nat Immunol* 6, 1191–1197.

Shahar, E., Whitney, C.W., Redline, S., et al. (2001). Sleep-disordered breathing and cardiovascular disease: Cross-sectional results of the Sleep Heart Health Study. *Am J Physiol Cell Physiol* 163, 19–25.

Shi, L.Z., Wang, R., Huang, G., et al. (2011). HIF1alpha-dependent glycolytic pathway orchestrates a metabolic checkpoint for the differentiation of TH17 and Treg cells. *J Exp Med* 208, 1367–1376.

Smith, T.G., and Talbot, N.P. (2010). Prolyl hydroxylases and therapeutics. *Antioxidants & Redox Signaling* 12, 431–433.

Suganami, T., and Ogawa, Y. (2010). Adipose tissue macrophages: Their role in adipose tissue remodeling. *J Leukoc Biol* 88, 33–39.

Treins, C., Giorgetti-Peraldi, S., Murdaca, J., Semenza, G.L., and Van Obberghen, E. (2002). Insulin stimulates hypoxia-inducible factor 1 through a phosphatidylinositol 3-kinase/target of rapamycin-dependent signaling pathway. *J Biol Chem* 277, 27975–27981.

Wagner, A.E., Huck, G., Stiehl, D.P., Jelkmann, W., and Hellwig-Burgel, T. (2008). Dexamethasone impairs hypoxia-inducible factor-1 function. *Biochem Biophys Res Commun* 372, 336–340.

Walmsley, S.R., Print, C., Farahi, N., et al. (2005). Hypoxia-induced neutrophil survival is mediated by HIF-1alpha-dependent NF-kappaB activity. *J Exp Med* 201, 105–115.

Wang, G., Jiang, B., Rue, E., and Semenza, G. (1995). Hypoxia-inducible factor 1 is a basic-helix-loop-helix-PAS heterodimer regulated by cellular O_2 tension. *Proc Natl Acad Sci USA* 92, 5510–5514.

Weir, E.K., Lopez-Barneo, J., Buckler, K.J., and111 Archer, S.L. (2005). Acute oxygen-sensing mechanisms. *N Engl J Med* 353, 2042–2055.

Woodman, I. (2011). T cells: A metabolic sHIFt to turn 17. *Nature Reviews Immunology* 11, 503.

Zinkernagel, A.S., Johnson, R.S., and Nizet, V. (2007). Hypoxia inducible factor (HIF) function in innate immunity and infection. *J Mol Med (Berl)* 85, 1339–1346.

5 Bioactive Phospholipid Mediators of Inflammation

Sainath R. Kotha, Jordan D. Secor, Smitha Malireddy, Gowrishankar Gnanasekaran, and Narasimham L. Parinandi
The Ohio State University College of Medicine
Columbus, Ohio

CONTENTS

5.1 SALIENT FEATURES OF INFLAMMATION

The *Oxford English Dictionary* describes inflammation in general as the action of inflaming or setting on fire or catching fire or the condition of being in flames, conflagration. However, the dictionary defines the pathological state of inflammation as a morbid process affecting some organ or part of the body, characterized by excessive heat, swelling, pain, and redness. Inflammation in a tissue of a living body arises in response to local injury, leading to the buildup of fluid and blood cells. Inflammation is a necessary and highly useful/essential process that animals have evolved to defend against pathogenic microbes. Stress causes tissue injury, which in turn leads to inflammation through a variety of signal mediators. In general, any injury to the vascularized tissue leads to a local inflammatory response. Inflammation is broadly divided into two major types: (1) acute inflammation (immediate and nonspecific response), and (2) chronic inflammation (delayed and highly specific response; Figure 5.1) [1]. Immediately following the tissue injury caused by stimuli (physical, chemical, and biological), vasodilation, vascular leak, edema, and recruitment of leukocytes (polymorphonuclear leukocytes; PMNs) result in the injured tissue area. Vascular leak is responsible for protein exudation and edema.

The vascular endothelial cell (EC) lining plays a pivotal role in the manifestation of vascular leak. Histamine, bradykinin, leukotrienes, pro-inflammatory cytokines (interleukin-1, IL-1) and tumor-necrosis factor (TNF) cause endothelial tight junction alterations through cytoskeletal rearrangements and thus lead to vascular hyperpermeability in the injured tissue site. In the early stages of inflammation, the microvasculature is the most affected, contributing to the vascular hyperpermeability, macromolecular exudation, and edema in the interstitial regions. Tissue injuries of

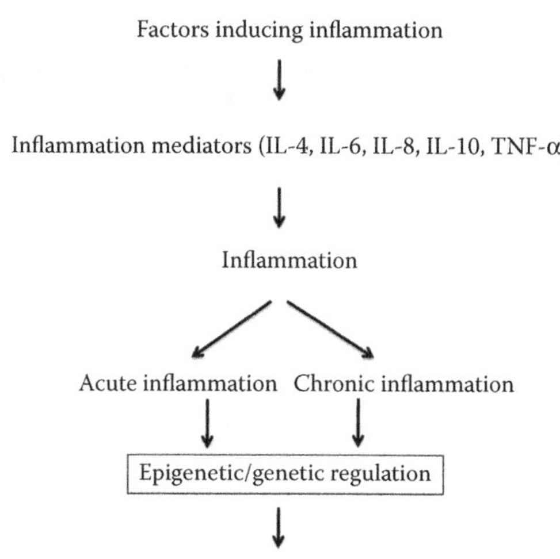

FIGURE 5.1 Inflammation inducers causing acute and chronic inflammation through mediators. Inducers of inflammation (microbes, microbial toxins, allergens, metabolic disturbances, stress, toxicants) cause inflammation, and mediators of inflammation modulate/regulate the process through complex cascades of signaling at the cellular and genetic levels.

severe nature cause EC damage or death contributing to prolonged vascular leak. During the acute phase of inflammation, leukocytes adhere to the endothelium, causing endothelial damage and inducing vascular leak through oxidant and other signaling mechanisms. Leukocytes in circulation escape the blood vessel by rolling, migration, adhesion, activation, and chemotaxis and thus participate in tissue injury. During vascular leak, using the cell adhesion molecules, leukocytes accumulate along the endothelial surface of the vasculature (adhesion). The leukocyte and endothelial adhesion is mediated by specific cell adhesion molecules called selectins. Cytokines play a crucial role in up regulating these cell adhesion molecules. Leukocytes including the PMNs, lymphocytes, and monocytes undergo diapedesis (transmigration) during the initial stages of inflammation. Leukocytes thus are attracted to the region of injury through chemotaxis, wherein molecules such as the pro-inflammatory cytokines (IL-8), complement proteins, and eicosanoids play important roles. Following chemotaxis and activation, leukocytes are engaged in phagocytosis in the injured tissue. Phagocytosis carried out by leukocytes is not only critical for the microbicidal activity but also highly essential for the clearance of the damaged tissue as a part of tissue repair process through engulfment and phagolysosome formation. Generation of reactive oxygen species (oxidative burst), oxidative stress, and formation of arachidonic acid metabolites (eicosanoids) also result at this stage of inflammation [2]. Overall, the activated leukocytes contribute to inflammation at the site of tissue injury through a multitude of mechanisms initiated and regulated by several mediators of inflammation.

5.1.1 Disorders and Diseases and Inflammation

Inflammation appears to be unavoidable and imminent in any state of disorder or disease. Although injury and inflammation are not clearly demarcated, both are intertwined whether inflammation is the cause or effect of tissue injury in a condition of disorder/disease, and therefore it is highly crucial to understand the mechanism of inflammation in a particularly focused disorder or disease for proper and effective therapeutic intervention. Beginning with microbial and parasitic infections, conditions including allergen exposures, cancer, cardiovascular disorders, neuronal injuries and diseases, asthma, wound healing, obesity, and many more diseases to name a few are associated with inflammation [3–8].

5.2 MEDIATORS OF INFLAMMATION

Inflammatory cells respond to mediators in a highly complex manner and operate the cascades of inflammation [9]. These lipid mediators are emerging as either pro-inflammatory or anti-inflammatory molecules throughout the inflammatory stages, leading toward either exacerbation of the injury or repair or remodeling or healing of the injured tissue, depending on the nature of the mediator, its mechanism(s) of action, type of the cell/tissue at the injured site, and nature of the disease/disorder. The generation and turnover of these mediators in the cells or extracellular environments in the body during tissue injury are apparently critical. For convenience, here, the mediators of inflammation are broadly divided into two groups: (1) non-lipid mediators, and (2) lipid mediators (Figure 5.2).

5.2.1 Non-Lipid Mediators of Inflammation

The most important non-lipid mediators of inflammation are histamine, kinins, cytokines, and chemokines [9]. Histamine, being a vasodilator and vascular smooth muscle constrictor, causes severe vascular permeability that sets the stage for subsequent inflammatory events. Through the metabolic decarboxylation of the amino acid, histidine, in the mast cells and basophils, histamine is generated and stored in secretory granules, and upon sensitization, histamine is released along with the binding of inflammation inducers (allergens) to immunoglobulin E (IgE) [9]. Several different cells such as the vascular smooth muscle cells, neurons, exocrine cells, endocrine cells, immunocytes, and blood cells also respond to histamine. Kinins fall under the category of peptide hormones [9]. Under the conditions of inflammation, kinins are synthesized in the body fluids and tissues from kininogens by enzyme-catalyzed proteolysis. The mechanisms of inflammatory actions of kinins include contraction of smooth muscle, vasodilation, and enhanced circulation in the capillaries. Cytokines are the most extensively studied mediators as the most important players in inflammation. Several cytokines, including the interleukins (IL), tumor necrosis factor-α (TNF-α), and granulocyte-macrophage colony stimulating factor (GM-CSF), are established to regulate the

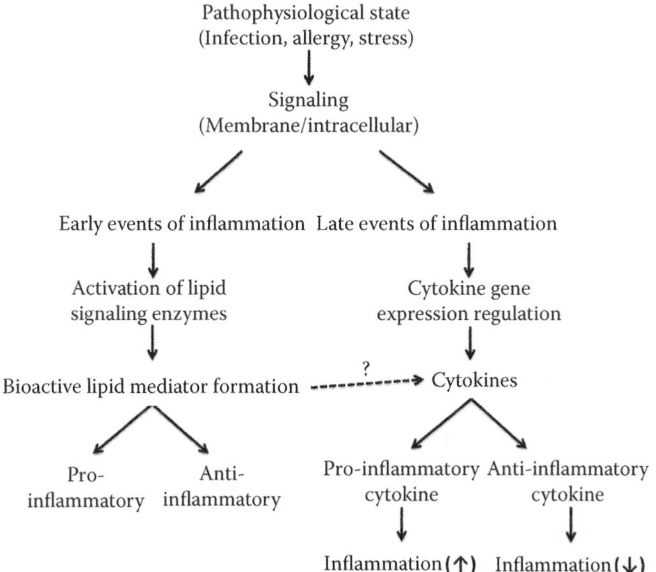

FIGURE 5.2 Role of non-lipid and lipid mediators in inflammation. During inflammation, lipid signaling enzymes generate bioactive lipids that act as mediators of inflammation. Non-lipid mediators such as cytokines also participate in the inflammation process. Some of these mediators may act either as pro-inflammatory or anti-inflammatory regulators.

actions of diverse immune cells such as the macrophages, T-cells, B-cells, and eosinophils during inflammation through a wide variety of molecular mechanisms [9]. Cytokines including IL-8, macrophage inflammatory protein-1α (MIP-1α), macrophage chemoattractant proteins, and RANTES are known to possess chemoattractant properties and hence are called *chemokines*. These chemokines act as chemoattractants for macrophages and other leukocytes and participate in the inflammatory processes.

5.2.2 Lipid Mediators of Inflammation

Membranes of living cells including the plasma membrane constitute lipids of diverse types. The most crucial types of those cellular bilayer membrane lipids are phospholipids followed by other minor lipidic constituents such as glycolipids, sphingolipids, gangliosides, and cholesterol. The composition of the membrane lipids varies from one type of cell to another. Within the same type of cell, depending on the metabolic state of the cell (including age), the lipid composition varies. Hence, lipids of the membrane should not be merely considered as the structural entities of the biological cell but are emerging to play critical roles in cellular physiological, biochemical, and molecular functions. Inflammation is not an exception to this as cell membrane lipids have been shown as important players therein. Platelet-activating factor (PAF) has been known for a long time as a potent phospholipid autocoid and innate immunity modulator [10]. The roles of cyclooxygenase (COX)- and lipoxygenase (LOX)-generated arachidonic acid (AA) metabolites (eicosanoids such as prostaglandins and leukotrienes) in inflammation have been thoroughly investigated and established [2,11]. The release of AA from the *SN-2* position of the membrane phospholipids upon the action of phospholipase A_2 through hydrolysis is the most critical regulatory step in the conversion of free AA into eicosanoids by either COX or LOX [12]. Thus, the eicosanoids have been established as potent bioactive lipid mediators in modulation/regulation of inflammation [9]. However, there are a few other critical bioactive lipid mediators of membrane phospholipid origin that are being identified also as potent mediators or modulators of inflammation. They include phosphatidic acid (PA), lysophosphatidic acid (LPA), anadamides, and sphingosine-1-phosphate (S1P) [13,14]. This chapter highlights the inflammation-mediating/modulatory actions of these specific newly emerging bioactive phospholipid mediators of inflammation, PA and LPA.

5.3 PHOSPHATIDIC ACID (PA) AS A BIOACTIVE PHOSPHOLIPID MEDIATOR OF INFLAMMATION

PA is an important component of the membranes including the mammalian cells. Although it is a metabolic intermediate in the biosynthesis of different phospholipids, PA acts a potent bioactive lipid signal mediator. Phospholipase D (PLD) is a membrane phospholipid-hydrolyzing enzyme that cleaves the head group of the phospholipid (phosphatidylcholine) and generates PA (Figure 5.3) [15]. Two isoforms of PLD, PLD1 and PLD2, have been identified in mammalian cells that take part in cellular signaling events mediated by the PLD-generated PA. Enhanced activity (upregulation) of PLD has been linked with the mechanism of action of GM-CSF through PA generation in the fMLP-stimulated neutrophils, emphasizing the priming effect of GM-CSF [16]. PLD activation has been identified as the initial step in the integrin-induced human neutrophil phagocytosis [17]. Activation of PLD has been shown to play an important role in the ATP-induced death of virulent *Mycobacterium tuberculosis* in the infected human macrophages [18]. In macrophages, PA has been observed to regulate the systemic inflammatory responses through modulation of the Akt-mammalian target of rapamycin-p70 S6 kinase 1 signaling cascade [19]. Neutrophil priming and subsequent PLD activation have been noticed in cardiopulmonary bypass and inhibition of PLD activation has been observed to inhibit enzyme activity with the attenuation of cardiopulmonary bypass-induced inflammation [20]. This study underscores the importance of PLD and

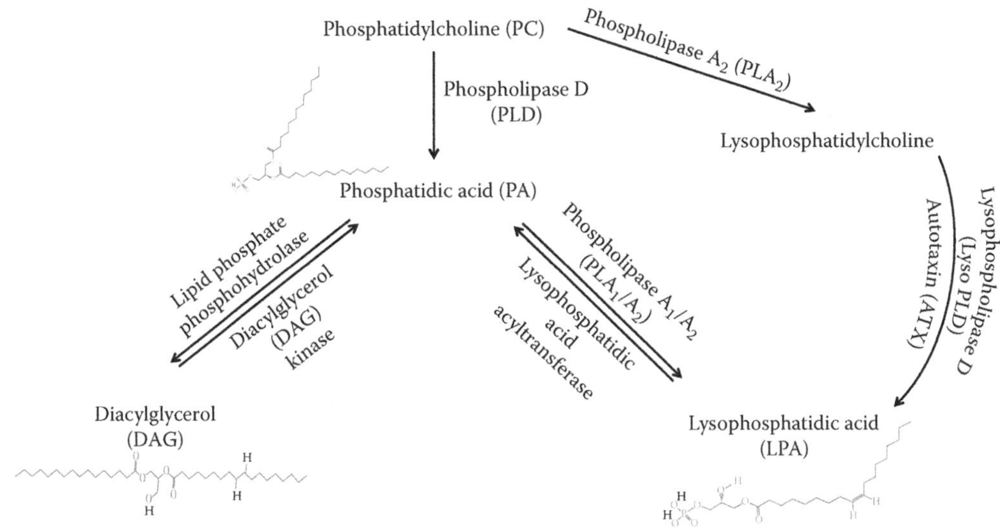

FIGURE 5.3 Generation of bioactive phospholipid mediators, phosphatidic acid (PA), and lysophosphatidic acid (LPA) in cells during inflammation. Membrane phospholipid-hydrolyzing enzymes such as phospholipase D (PLD) generates PA from membrane phospholipids (phosphatidylcholine), which can act as a phospholipid mediator of inflammation. PA can also undergo hydrolytic cleavage by phospholipase A_1/A_2 to form LPA, which is also emerging as a potent bioactive phospholipid mediator of inflammation.

PLD-generated bioactive lipid, PA, in the cardiopulmonary bypass-induced inflammation through neutrophil activity.

The roles of both PLD1 and PLD2 isoforms have been established as the coordinate regulators of phagocytosis in macrophages [21]. Stimulation of PLD1 by phagocyte adhesion and regulation of early stages of phagocyte adhesion by PLD1 have been established [22], emphasizing the importance of PLD1 in phagocyte adhesion. In human monocytes, PLD1 has been noticed to play an important role in the TNF-α-mediated signaling pathway leading toward inflammation [23]. PLD1 has emerged as a novel target for the treatment of inflammatory conditions/diseases including sepsis, respiratory ailments, and autoimmune diseases in which TNF-α is a critical player [23]. PLD-generated microvesicles loaded with phospholipids from macrophages have been identified to mediate inflammatory cascades in macrophages, and thus the phospholipids in microvesicles are suggested to mediate inflammation in a variety of diseases [24]. PLD1 has been shown to regulate the secretion of the pro-coagulant von Willebrand factor (VWF) from endothelial cells (ECs) [25]. In this study, PA generated by PLD1 has been observed as a critical player in the histamine-induced VWF secretion by the ECs. Furthermore, this study also suggests that PLD1 may play an important role in the endothelial-regulated vascular permeability that is important in acute inflammation. An enzyme that catalyzes the metabolism of PA, PA phosphohydrolase has been reported to regulate the inflammatory signaling [27]. The vital role of PLD2-generated PA and participation of GTPase have been shown to be important in cell migration, enhanced leukocyte chemotaxis, and inflammation, thus emphasizing the crucial role of PLD signaling in inflammation and cancer [27]. Overall, these studies so far have revealed that PLD and the PLD-generated bioactive phospholipid (PA) appear to play important roles in mediation and regulation of inflammation.

5.4 LYSOPHOSPHATIDIC ACID (LPA) AS A BIOACTIVE PHOSPHOLIPID MEDIATOR OF INFLAMMATION

LPA is emerging as a potent bioactive phospholipid mediator in different physiological and pathophysiological states including inflammation. In many cells, LPA acts as a chemoattractant and mitogen. LPA is formed in the mammalian cells/tissues through different mechanisms. One of the mechanisms involves PLD, which forms PA from membrane phospholipids, which subsequently undergoes hydrolysis by either phospholipase A1 (PLA1) or phospholipase A2 (PLA2) into LPA (Figure 5.3). A second mechanism of LPA generation involves, especially in tumor development and inflammation, the autotoxin (ectonucleotide pyrophosphatase/phosphodiesterase-2, ENPP2) that is a secretory lysophospholipase D [28]. Autotaxin hydrolyzes lysophosphatidylcholine into LPA. A third mechanism of generation of LPA involves the action of lysophospholipase D, which converts lysophospholipids in cells into LPA. LPA exerts its biological actions involving specific G-protein-coupled receptors (GPCRs) such as LPA-1, LPA-2, LPA-3, LPA-4, LPA-5, and P2Y5 [29]. It has been suggested that LPA plays an important role in airway inflammation by modulating the expression of either pro-inflammatory or anti-inflammatory genes [29]. LPA has been shown to induce potent stimulation of polarization, motility, and metabolic burst of human neutrophils, suggesting that LPA formation induced by thrombin in platelets may regulate inflammation [30]. Activation of the transcription factor, nuclear factor-κB (NF-κB), by LPA has been reported in fibroblasts utilizing many pathways [31]. Since cytokines act as targets for NF-κB, LPA regulation of NF-κB could be highly important in inflammation. Vascular EC (human aortic EC) and leukocyte (monocyte and neutrophil) interaction has been shown to be regulated by LPA through its receptor(s) involvement, emphasizing the role of LPA and its receptors in inflammation [32].

LPA has been observed to improve healing of acute cutaneous wounds in mice through enhanced reepithelialization but not by modulating inflammation, suggesting that LPA acts as a mitogen in wound healing and repair [33]. LPA has been described as a platelet-derived bioactive lipid growth factor that plays a crucial role in the activation and differentiation of leukocytes, and this has implications in several inflammatory disorders such as granulomatous disorder, chronic inflammation (arthritis), and atherosclerosis [34]. LPA at nanomolar doses has been shown to activate human monocytes involving the LPA receptors, and hence a role for LPA in the mediation of inflammation has been suggested [35]. Pro-inflammatory gene expression has been noticed to be modulated by the nuclear LPA-1 receptor in microvascular ECs, hepatoma cells, and liver tissue emphasizing the role of the cell nucleus as a pivotal site for LPA-mediated signaling toward modulation of expression of pro-inflammatory genes [36]. This study clearly highlights the nuclear site of action of LPA in the regulation of inflammation.

It has been revealed that LPA induces chemotaxis, oxidant generation, CD11b upregulation, calcium mobilization, and actin cytoskeletal reorganization in the human neutrophils through the involvement of G proteins [37]. This study suggests that LPA appears to have a role in diseases involving eosinophilic inflammation, such as atopic diseases, by acting as a chemotaxic agent and activator of pro-inflammatory processes. LPA has been reported to stimulate inflammatory cascade in the airway epithelial cells through modulation of cytokine production [38]. It has been observed that LPA attenuates endotoxemia (lipopolysaccharide, LPS)-induced organ injury through the GPCRs and peroxisome proliferator-activated receptor-γ (PPAR-γ) [39]. This study underscores the utilization of LPA as a therapeutic agent in the treatment of shock. The presence of LPA has been reported in the human bronchoalveolar lavage (BAL) fluid following allergen exposure as determined by liquid chromatography-mass spectrometry [40]. From this study, the authors have concluded that LPA does not appear to act as a predominant eosinophil chemoattractant but may enhance epithelial barrier integrity during allergic airway inflammation.

In the human ECs, LPA has been shown to regulate inflammation-associated genes by involving the LPA-1 and LPA-3 receptors, suggesting those LPA receptors as targets for pharmacological intervention in the treatment of severe inflammation [41]. Inhibition of endotoxin (LPS)-induced pro-inflammatory conditions by exhibiting anti-inflammatory actions has been shown in *in vitro* and *in vivo* mouse models [42]. This investigation has also revealed the involvement of signaling pathways operated by extracellular-regulated kinase ½ (ERK ½), serine/threonine phosphatases, and PI3 kinase. As autotaxin (lysophospholipase D) is also responsible for the generation of LPA in the extracellular milieu, it has been suggested that the pharmacological inhibitors of autotaxin may be beneficial in the treatment of cardiovascular diseases and tumor metastasis [43]. LPA has also been established as a potent bioactive phospholipid mediator of airway epithelial signaling in the regulation of inflammatory mediators in the airway [44]. The involvement of LPA and LPA-2 receptors in the progression of airway inflammation induced by the *Schistosoma mansoni* soluble egg antigen in the mouse asthma model has been reported [45]. From a study with the bleomycin-induced mouse lung fibrosis model, it has been revealed that a novel oral LPA-1 receptor antagonist, AM966, offers inhibition of lung fibrosis [46]. This study further promises the use of AM966 as an effective, selective, and potent LPA-1 receptor antagonist in not only the treatment of lung damage and fibrosis but also could be beneficial in treating inflammatory conditions. In a translational study, liquid chromatography-mass spectrometry analysis of saliva and gingival crevicular fluid of human subjects has revealed elevated levels of different molecular species of LPA (with different fatty acid composition) during the state of periodontal diseases [47]. This study has clearly indicated the involvement of LPA in periodontal diseases. Overall, these studies so far have unequivocally revealed the role of LPA and LPA receptors in the pathogenesis of several disease states.

5.5 CONCLUSIONS AND FUTURE DIRECTIONS

Inflammation is obligatory and painful in almost every disease and disorder state. However, it is driven by complex mechanistic operations being intertwined and regulated tightly at the physiological, metabolic, biochemical, and molecular levels in the cells and tissues at different microcosms, finally manifesting at the organismal level. The bioactive phospholipid mediators such as PA and LPA are just emerging as the most critical membrane-derived lipid mediators of inflammation. The challenges ahead are myriad. The first challenge is to carefully identify the temporal and spatial identification and determination of PA and LPA and their origin and fate in the cells and tissues at the injured and inflammatory sites. The second challenge involves the use of the right/appropriate animal model for a chosen disease/disorder that appropriately addresses the human disease/disorder in question. Translational studies with human subjects in this regard are mandatory. PA and LPA should also be analyzed in the body fluids and tissue biopsies of human subjects during inflammation, injury, and treatment as biomarkers of inflammation and injury. The cellular and molecular targets for PA and LPA have to be clearly identified in cellular and animal models and human subjects, which will lead to identifying specific drug targets for combatting inflammation under a particular state of disorder/disease. Finally, from both pathological and pharmacological perspectives, the intertwined operations between non-lipid mediators (e.g., cytokines and chemotaxins) and lipid mediators (PA, LPA, and eicosanoids) have to be established at different system levels that overlap with each other (e.g., cellular, nuclear, molecular, tissue, and organismal).

TAKE-HOME MESSAGES

- Inflammation is associated almost with every disorder/disease.
- Although inflammation is essential for repair/healing of tissue injury, the outcome is painful and detrimental.
- Mediators of inflammation are broadly divided into two categories: (1) non-lipid mediators, and (2) lipid mediators.
- Among lipid mediators of inflammation, the membrane phospholipid-derived species, phosphatidic acid (PA) and lysophosphatidic acid (LPA), are emerging as critical players in inflammation in several diseases/disorders.
- PA and LPA exert their biological actions (pro- or anti-inflammatory) through complex signaling cascades at the cellular and genomic levels through their specific receptors.
- Understanding the targets for PA and LPA during inflammation at the target tissue will offer drug targets for therapeutic interventions.

ACKNOWLEDGMENT

This work was supported by funding from the National Institutes of Health (HL 093463).

REFERENCES

1. Ryan GB, Majno G: Acute inflammation: A review. *Am J Pathol* 1977, 86:183–276.
2. Basu S: Bioactive eicosanoids: Role of prostaglandin $F_{2\alpha}$ and F_2 isoprostanes in inflammation and oxidative stress related pathology. *Mol Cells* 2010, 30:383–391.
3. Aliberti J, Bafica A: Anti-inflammatory pathways as a host evasion mechanism for pathogens. *Prostaglandins Leukot Essent Fatty Acids* 2005, 73:283–288.
4. Karp CL, Flick LM, Yang R, Uddin J, Petasis NA: Cystic fibrosis and lipoxins. *Prostaglandins Leukot Essent Fatty Acids* 2005, 73:263–270.
5. Gronert K: Lipoxins in the eye and their role in wound healing. *Prostaglandins Leukot Essent Fatty Acids* 2005, 73:221–229.
6. Greene ER, Huang S, Serhan CN, Panigrahy D: Regulation of inflammation in cancer by eicosanoids. *Prostaglandins Other Lipid Mediat* 2011, 96:27–36.
7. Prescott D, McKay DM: Aspirin-triggered lipoxin enhances macrophage phagocytosis of bacteria while inhibiting inflammatory cytokine production. *Am J Physiol Gastrointest Liver Physiol* 2011, 301:G487–G497.
8. Gonzalez-Periz A, Claria J: Resolution of adipose tissue inflammation. *Scientific World Journal* 2010, 10:832–856.
9. White M: Mediators of inflammation and the inflammatory process. *J Allergy Clin Immunol* 1999, 103:S378–S381.
10. McIntyre TM, Prescott SM, Stafforini DM: The emerging roles of PAF acetylhydrolase. *J Lipid Res* 2009, 50 (Supplement):S255–S259.
11. Carlo T, Levy BD: Molecular circuits of resolution in airway inflammation. *Scientific World Journal* 2010, 10:1386–1399.
12. Liu NK, Xu XM: Phospholipase A2 and its molecular mechanism after spinal cord injury. *Mol Neurobiol* 2010, 41:197–205.
13. Chalfant CE, Spiegel S: Sphingosine 1-phosphate and ceramide 1-phosphate: Expanding roles in signaling. *J Cell Sci* 2005, 118:4605–4612.
14. Zhao J, He D, Su Y, Berdyshev EV, Chun J, Natarajan V, Zhao Y: Lysophosphatidic acid receptor 1 modulates lipopolysaccharide-induced inflammation in alveolar epithelial cells and murine lungs. *Am J Physiol Lung Cell Mol Physiol* 2011, 301:L547–L556.
15. Patel RB, Kotha SR, Sherwani SI, Sliman SM, Gurney TO, Loar B, et al.: Pulmonary fibrosis inducer, bleomycin, causes redox-sensitive activation of phospholipase D and cytotoxicity through formation of bioactive lipid signal mediator, phosphatidic acid, in lung microvascular endothelial cells. *Int J Toxicol* 2011, 30:69–90.
16. Bourgoin S, Plante E, Gaudry M, Naccache PH, Borgeat P, Poubelle PE: Involvement of phospholipase D in the mechanism of action of granulocyte-macrophage colony-stimulating factor (GM-CSF): Priming of human neutrophils in vitro with GM-CSF is associated with accumulation of phosphatidic acid and diradylglycerol. *J Exp Med* 1990, 172:767–777.
17. Serrander L, Fallman M, Stendahl O: Activation of phospholipase D is an early event in integrin-mediated signaling leading to phagocytosis in human neutrophils. *Inflammation* 1996, 20:439–450.
18. Kusner DJ, Adams J: ATP-induced killing of virulent *Mycobacterium tuberculosis* within human macrophages requires phospholipase D. *J Immunol* 2000, 164:379–388.
19. Lim HK, Choi YA, Park W, Lee T, Ryu SH, Kim SY, et al.: Phosphatidic acid regulates systemic inflammatory responses by modulating the Akt-mammalian target of rapmycin-p70 S6 kinase 1 pathway. *J Biol Chem* 2003, 278:45117–45127.
20. Wu M, Lu YB, Chen RK: Effects of phospholipase D on cardiopulmonary bypass-induced neutrophil priming. *Chin J Traumatol* 2004, 7:70–75.
21. Iyer SS, Barton JA, Bourgoin S, Kusner DJ: Phospholipase D1 and D2 coordinately regulate macrophage phagocytosis. *J Immunol* 2004, 173:2615–2623.
22. Iyer SS, Agrawal RS, Thompson CR, Thompson S, Barton JA, Kusner DJ: Phospholipase D1 regulates phagocyte adhesion. *J Immunol* 2006, 176:3686–3696.
23. Sethu S, Pushparaj PN, Melendez AJ: Phospholipase D1 mediates TNFalpha-induced inflammation in a murine TNFalpha-induced peritonitis. *PLoS One* 2010, 5:e10506.

24. Thomas LM, Salter RD: Activation of macrophages by P2X7-induced microvesicles from myeloid cells is mediated by phospholipids and is partially dependent on TLR4. *J Immunol* 2010, 185:3740–3749.
25. Disse J, Vitale N, Bader MF, Gerke V: Phospholipase D1 is specifically required for regulated secretion of von Willebrand factor from endothelial cells. *Blood* 2009, 113:973–980.
26. Grkovich A, Dennis EA: Phosphatidic acid phosphohydrolase in the regulation of inflammatory signaling. *Adv Enzyme Regul* 2009, 49:114–120.
27. Gomez-Cambronero J: New concepts in phospholipase D signaling in inflammation and cancer. *Scientific World Journal* 2010, 10:1356–1369.
28. Hausmann J, Kamtekar S, Christodoulou E, Day JE, Wu T, Fulkerson Z, et al.: Structural basis of substrate discrimination and integrin binding by autotaxin. *Nat Struct Mol Biol* 2011, 18:198–204.
29. Matsuzaki S, Ishizuka T, Hisada T, Aoki H, Komachi M, Ichimonji I, et al.: Lysophosphatidic acid inhibits CC chemokine ligand 5/RANTES production by blocking IRF-1-mediated gene transcription in human bronchial epithelial cells. *J Immunol* 2010, 18:4863–4872.
30. Chettibi S, Lawrence AJ, Stevenson RD, Young JD: Effect of lysophosphatidic acid on motility, polarization, and metabolic burst of human neutrophils. *FEMS Immunol Med Microbiol* 1994, 8:271–281.
31. Shahrestanifar M, Fan X, Manning DR: Lysophosphatidic acid activates NF-kappaB in fibroblasts. A requirement for multiple inputs. *J Biol Chem* 1999, 274:3828–3833.
32. Rizza C, Leitinger N, Yue J, Fischer DJ, Wang DA, Shih PT, et al.: Lysophosphatidic acid as a regulator of endothelial/leukocyte interaction. *Lab Invest* 1999, 79:1227–1235.
33. Demoyer JS, Skalak TC, Durieux ME: Lysophosphatidic acid enhances healing of acute cutaneous wounds in the mouse. *Wound Repair Regen* 2000, 8:530–537.
34. Lee H, Liao JJ, Graeler M, Huang MC, Goetzl EJ: Lysophospholipid regulation of mononuclear phagocytes. *Biochim Biophys Acta* 2002, 1582:175–177.
35. Fueller M, Wang DA, Tigyi G, Siess W: Activation of human monocytic cells by lysophosphatidic acid and sphingosine-1-phosphate. *Cell Signal* 2003, 15:367–375.
36. Gobeil F Jr, Bernier SG, Vazquez-Tello A, Brault S, Beauchamp MH, Quiniou C, et al.: Modulation of pro-inflammatory gene expression by nuclear lysophosphatidic acid receptor type-1. *J Biol Chem* 2003, 278:38875–38883.
37. Idzko M, Laut M, Panther E, Sorichter S, Durk T, Fluhr JW, et al.: Lysophosphatidic acid induces chemotaxis, oxygen radical production, CD11b up-regulation, Ca2+ mobilization, and actin reorganization in human eosinophils via pertussis toxin-sensitive G proteins. *J Immunol* 2004, 172:4480–4485.
38. Barekzi E, Roman J, Hise K, Georas S, Steinke JW: Lysophosphatidic acid stimulates inflammatory cascade in airway epithelial cells. *Prostaglandins Leukot Essent Fatty Acids* 2006, 74:357–363.
39. Murch O, Collin M, Thiemermann C: Lysophosphatidic acid regulates the organ injury caused by endotoxemia–A role for G protein-coupled receptors and peroxisome proliferator-activated receptor-gamma. *Shock* 2007, 27:48–54.
40. Georas SN, Berdyshev E, Hubbard W, Gorshkova IA, Usatyuk PV, Saatian B, et al.: Lysophosphatidic acid is detectable in human bronchoalveolar lavage fluids at baseline and increased after segmental allergen challenge. *Clin Exp Allergy* 2007, 37:311–322.
41. Lin CI, Chen CN, Lin PW, Chang KJ, Hsieh FJ, Lee H: Lysophosphatidic acid regulates inflammation-related genes in human endothelial cells through LPA1 and LPA3. *Biochem Biophys Res Commun* 2007, 363:1001–1008.
42. Fan H, Zingarelli B, Harris V, Tempel GE, Halushka PV, Cook JA: Lysophosphatidic acid inhibits bacterial endotoxin-induced pro-inflammatory response: Potential anti-inflammatory signaling pathways. *Mol Med* 2008, 14:422–428.
43. Federico L, Pamuklar Z, Smyth SS, Morri AJ: Therapeutic potential of autotaxin/lysophospholipase D inhibitors. *Curr Drug Targets* 2008, 9:698–708.
44. Zhao Y, Natarajan V: Lysophosphatidic acid signaling in airway epithelium: Role in airway inflammation and remodeling. *Cell Signal* 2009, 21:367–377.
45. Zhao Y, Tong J, He D, Pendyala S, Evgeny B, Chun J, et al.: Role of lysophosphatidic acid receptor LPA2 in the development of allergic airway inflammation in a murine model of asthma. *Respir Res* 2009, 10:114.
46. Swaney JS, Chapman C, Correa LD, Stebbins KJ, Bundey RA, Prodanovich PC, et al.: A novel, orally active LPA(1) receptor antagonist inhibits lung fibrosis in the mouse bleomycin model. *Br J Pharmacol* 2010, 160:1699–1713.
47. Bathena SP, Huang J, Nunn ME, Miyamoto T, Parrish LC, Lang MS, et al.: Quantitative determination of lysophosphatidic acids (LPAs) in human saliva and gingival crevicular fluid (GCF) by LC-MS/MS. *J Pharm Biomed Anal* 2011, 56:402–407.

6 Hematopoietic Stem Cells in Atherosclerotic Development and Resolution

Reeva Aggarwal, Vincent J. Pompili, and Hiranmoy Das
The Wexner Medical Center at The Ohio State University
Columbus, Ohio

CONTENTS

6.1 INTRODUCTION

Cardiovascular pathologies are the leading cause of morbidity and mortality worldwide. It is a multifactorial disease and its major underlying cause is the inflammation of large blood vessels (atherosclerosis), leading to thrombus formation, plaque rupture, and blockage of blood and nutrient supply to the adjacent cardiac tissues. The development of atherosclerosis has a strong inflammatory component, and involves a number of risk factors such as elevated serum levels of modified low-density lipoprotein (ox-LDL), lower levels of high-density lipoprotein (HDL), smoking, sedentary lifestyle, genetics, age, a fatty acid–rich diet, and cholesterol.

Atherosclerosis is initiated by migration and recruitment of the immune cells *via* secretion of various cytokines and chemokines, whose expression is upregulated in injured endothelium, activated monocytes, and proliferating vascular smooth muscle cells (vSMCs), resulting in reduced lumen size and blood flow obstruction (Libby and Aikawa 2002; Figure 6.1). Although serum level of HDL is a strong predictor of atherosclerosis development, newer studies have shown that the

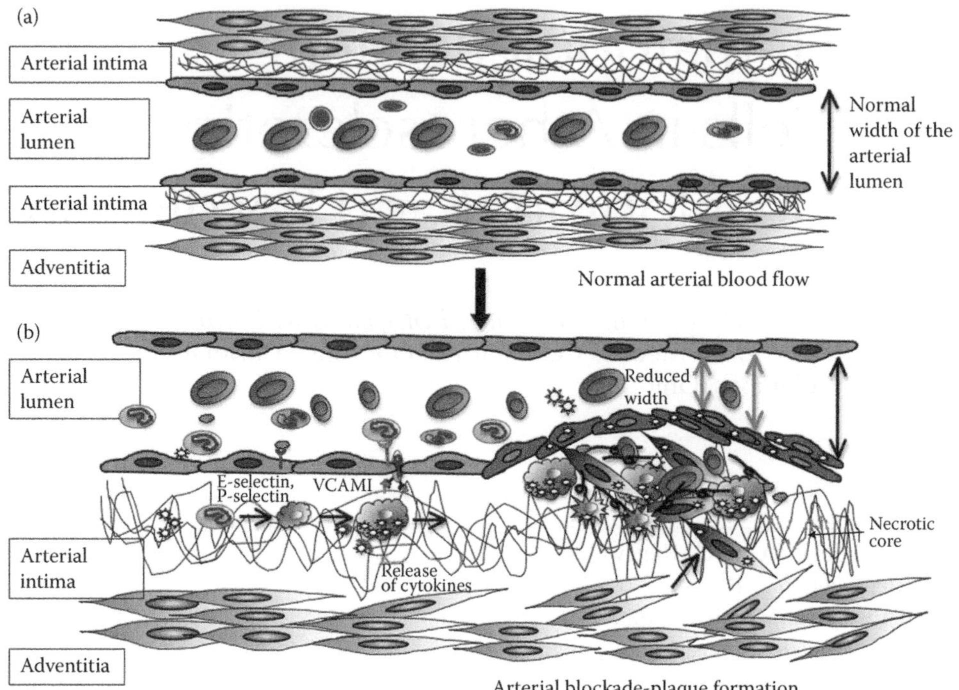

FIGURE 6.1 (see color insert) Schematics of normal artery and development of atherosclerosis. (a) Schematics of normal arterial sectional view, where blood flows without any obstruction. The lumen of the artery is lined by the endothelium followed by the arterial intima and the lining of vascular smooth muscle cells followed by adventitia. Normally a small number of circulating immune cells and progenitor cells and a large number of red blood cells along with white blood cells containing blood circulate within the lumen. (b) Atherosclerosis is an inflammatory disorder that develops within the arteries due to the injury caused to the endothelium in the presence of a high-cholesterol diet and other risk factors. Modified lipids such as oxidized low-density lipoprotein (ox-LDL) interact with the endothelium. The injured endothelium upregulates chemotactic proteins and adhesion molecules such as VCAM1, ICAM1, and selectins. These molecules attract the circulating leukocytes and platelets, which migrate into the arterial intima. In the arterial intima, ox-LDL is phagocytosed by macrophages that express scavenger receptor (SR) and convert into foam cells. The vSMCs also become pro-inflammatory and migrate to the intima, secrete extracellular matrix, and form lipid core that starts to obstruct the arterial blood flow. Pro-inflammatory endothelial cells, platelets, foam cells, and other immune cells (B, T, DC, and mast cells) also get trapped in the lipid-necrotic core forming a plaque, which if it ruptures, will block the blood supply, causing ischemia in adjacent tissues.

number of circulating progenitor cells are also good predictor of developing cardiac pathologies (Chong et al. 2004).

Studies have shown that endogenous hematopoietic stem cells give rise to immune cells, endothelial progenitor cells (EPCs), and vSMC progenitors. Although the role of immune cells is mostly recognized as inflammatory in nature, the role of EPCs and vSMCs in atherosclerosis is largely unknown. Findings suggest that levels of progenitor cells in the cardiovascular patients are thought to be novel predictors of the cardiovascular risk factors. Recent findings show that existing treatments for coronary heart disease (CHD) alleviate cardiac pathologies not just by lowering the LDL levels or interfering with immune interaction, but by recruiting endogenous stem cells to the lesion site. Stem cells were shown to cause re-endothelialization, stabilization of atherosclerotic plaque, and reversal of ischemic cardiac tissue and restoration of cardiac functions. However, pre-clinical and clinical application is greatly impeded by the available number of stem cells. Our lab has established an *ex vivo* expansion technology of human cord blood-derived hematopoietic stem cells

(HSCs) and shown therapeutic potential of expanded stem cells in myocardial infarction and hind limb ischemia models. This chapter focuses on the development of atherosclerosis and the role of HSCs and vascular progenitor cells in atherosclerosis and highlights the therapeutic potential of progenitor cells in the resolution of atherosclerotic disease.

6.2 HSC REGULATION OF SELF-RENEWAL

The fine balance between self-renewal and differentiation is critical for maintaining homeostasis of the HSC pool and blood cells in the vascular system. The life span of mature blood cells is limited and continuous replacement of blood cells is required in the healthy condition as well as in inflammation and disease states. Numerous intrinsic and extrinsic signals, guided by the microenvironment, regulate self-renewal activities, maintain quiescence of HSCs, and inhibit their differentiation. Factors such as Wnt, Notch, hedgehog, Homeobox proteins (HoxB4), cell cycle regulator proteins, stromal derived factor (SDF-1), c-kit, stem cell factor (SCF), thrombopoietin (TPO), fetal liver kinase (Fl), and interleukins regulate self-renewal and proliferation of HSCs (Reya 2003).

Self-renewal activities are regulated by HSC niches such as the osteoblastic niche and endosteal niche. HSCs are anchored to the surrounding cells (such as osteoblasts or vascular cells) via adhesion molecules, such as cadherin and integrins, especially N-cadherin, VCAM-1, α1β2 integrin, angiopoietin-1 (Ang-1), and osteopontin, which are expressed on the osteoblastic cells and bind through their cognate receptors expressed on HSCs to maintain quiescence and survival of HSCs (Arai et al. 2004). Osteoblasts were shown to regulate HSCs through various signaling pathways, such as bone morphogenetic pathways (BMP), and Notch signaling by parathyroid hormone receptors, CD150 and CXCR4 (Sugiyama et al. 2006; Calvi et al. 2003). Recent evidence from osteoblast ablation experiments has shown that in the absence of osteoblasts, HSCs could still be maintained in the spleen or liver. Thus, osteoblast-HSC interaction might be more crucial for maintaining committed HSCs compared to their uncommitted counterparts (Adams et al. 2006). It is believed that not only osteoblasts or endothelial cells but also other cell types, such as osteoclasts, regulate homeostatic and stress-induced (inflammation or bleeding) migration and localization of HSCs (Kollet et al. 2006). However, it is not yet clear why HSCs are located near the bone marrow sinusoids and other extramedullary tissues.

6.3 DIFFERENTIATION OF HSCS

HSCs differentiate into mature blood lineages to maintain normal homeostasis of the hematopoietic system. Various pathological conditions also trigger differentiation of HSCs, such as hematological stress conditions, bleeding, inflammatory and hematopoietic regeneration in response to infection, wound healing, cytotoxic agents, or pathological diseases such as atherosclerosis (Libby and Aikawa 2002). A continuous supply of blood cells is needed to sustain normal functions of the organism. On the other hand, HSCs' quiescence (i.e., HSCs exist in G0/G1 phase) is also equally important to maintain the stem cell pool and to provide radioprotection, as HSCs in proliferative phase (in S and M phase) are more susceptible to apoptosis, when exposed to cytotoxic agents.

6.3.1 Lymphoid Differentiation

HSCs could be differentiated into various lineages such as B, T, and NK cells. NK cells and a subset of T cells belong to the innate immune system, and B and a majority of T cells belong to the adaptive immune system. B cell development and effector function are regulated both by intrinsic and extrinsic signals including transcription factors, chemokines, cytokines, and so forth. B cell maturation also depends on the signaling from T cells. T cells mature in the thymus and their development depends on the transcription factors such as Pu.1, Pax5, GATA-3, members of the Runx,

E2A/HEB, and members of the Ikaros family. T cells could develop into either cytotoxic, helper, or memory T cells. T cells function through T cell receptors (TCR) and activate other immune cells. Although NK cells represent the innate immune system, they share some of the properties with the T cells. In particular, NK cells share a common killing mechanism with CD4+ and CD8+ cytotoxic T lymphocytes (CTL), that is, using interferon-γ (IFN- γ), perforin, and granzymes, respectively. NK cell development is governed by transcription factors such as Ets-1 and myeloid Elf-1–like factor (MEF; Rothenberg 2011; Choi et al. 2011).

6.3.2 Myeloid Differentiation

HSC precursors give rise to myeloid lineages such as monocyte-macrophage, granulocytes, and erythroid-megakarocyte lineages. Common myeloid lineages also give rise to dendritic cells (DCs) and mast cells. Monocyte-macrophage lineages provide immune defense against foreign pathogens and represent an adaptive immune system. Macrophages are activated due to tissue injury or poor oxidation and thus unregulate the production of inflammatory cytokines attracting more monocyte from the circulation. Specifically, the role of monocyte-macrophage lineages is predominantly seen in initiation of cardiovascular diseases (Libby and Aikawa 2002). DCs are important regulators of the adaptive immune system and are specialized antigen-presenting cells. Although macrophages could behave as antigen presenting cells, the important difference is that DCs can specifically migrate to the regions where T cells are present to interact with them. Several transcription factors are involved in differentiation of HSCs to DCs. Ikaros dominant-negative homozygous mice lack all cells of lymphoid origin, including T, B, and NK cells, and also show a defect in the development of DCs (Wu et al. 1997). Mast cells represent the innate immune system and are known to have a principal role in inflammatory diseases such as atherosclerosis and myocardial infarction (Laine et al. 1999). Transcriptional factors such as GATA-2 were shown to be required for survival and terminal differentiation of mast cells; however, they were not critical for erythroid and macrophages (Tsai and Orkin 1997). Myeloid progenitor cells also give rise to red blood cells (RBC, erythroid lineages) and megakaryocytes (mature to form platelets). Erythroid cells progress from nucleated to mature enucleated RBCs and are important in tissue oxygenation.

Cross-talk among all the immune cells protects the body against infections and is involved in reparative mechanisms in wound healing and inflammatory responses in diseases such as atherosclerosis. Thus it is important to understand the transcriptional regulation during differentiation and the maturation of immune cells so as to develop effective immune-based therapies for disease prevention and treatment.

6.4 DEVELOPMENT OF ATHEROSCLEROSIS

Atherosclerosis is a multifactorial inflammatory disease caused by elevated levels of LDL cholesterol leading to blockage of arteries. Atherosclerosis is the most common pathological process that leads to cardiovascular disease (CVD). Formation of atherosclerotic plaques consists of necrotic cores, calcified regions accumulated with modified lipids, inflamed smooth muscle cells (SMCs), endothelial cells (ECs), leukocytes, and foam cells (Libby and Aikawa 2002). Earlier, atherosclerosis was considered as a passive disease consisting of accumulation of LDL accompanied by injury of endothelial cells, dysfunction of vascular smooth muscle cells, leading to activation of platelets, immune cells, and formation of atherosclerotic plaques that obstruct the blood flow through the blood vessels, resulting in distal tissue ischemia and myocardial infarction or stroke. Recently, however, involvement of progenitor cells, immune cells, and inflammatory mechanisms in the progression of atherosclerosis was identified (Hansson 2005). The following section focuses on the role of inflammatory cells in atherosclerosis and their mechanisms, as well as the resolution, prevention of atherosclerosis, and regeneration of damaged tissue resulted from ischemia.

6.4.1 Role of Stem Cells and Progenitor Cells in Atherosclerosis

Atherosclerosis progression and lesions involve a heterogeneous population of cells including progenitor cells of endothelial lineages (EPC), inflammatory cells and their progenitors, vascular progenitors, and vascular smooth muscle cells (vSMCs). Each stage of lesion progression involves cross talk between these cells. The earliest stages of atherosclerosis are triggered by endothelial dysfunction and poor endothelial regeneration, followed by endothelial apoptosis and infiltration of inflammatory cells and vSMCs, finally leading to ischemia (Libby and Aikawa 2002). Progenitor cells present in peripheral blood (circulating cells), adventitia, or media of the atherosclerotic arteries were reported to be involved either in progression or prevention of atherosclerosis (Hu et al. 2004). Human studies have indicated reparative roles for circulating progenitor cells; however, involvement in progression of lesion formation is not clear (Simper et al. 2002). This section will focus on some of the potential findings that link the presence of circulating progenitor cells with vascular pathologies such as atherosclerosis (Table 6.1).

6.4.1.1 Vascular Stem Cells

Vascular stem cells are undifferentiated, multipotent stem cells that can differentiate into functional mature cells such as endothelial cells or vSMCs.

6.4.1.1.1 Endothelial Progenitor Cells

It has been shown that EPCs and HSCs originate from common hemangioblast cells and HSCs could differentiate to endothelial lineage and smooth muscle cells (Asahara et al. 1997; Das, George et al. 2009). Endogenous circulating EPC (cEPCs) derived from bone marrow were shown to express both hematopoietic markers (AC133, CD34) as well as an endothelial cell marker (VEGF) and give rise to mature endothelial cells and line the blood vessels. Cardiovascular risk factors were shown to negatively influence EPC numbers and functioning (Vasa et al. 2001). Endothelial dysfunction is believed to be the first step in the manifestation of atherosclerotic disease and leads to a cascade

TABLE 6.1
Origin and Role of the Endogenous Stem and Progenitor Cells in Atherosclerosis

Types of Progenitor and Stem Cells	Origin	Differentiation Potential	Role in Cardiovascular Pathology
Endogenous Hematopoietic Stem Cells (HSCs)	May originate from common hemangioblast	Differentiate into committed HSCs, EPCs, and or vSMCs in lesion sites	Home to site of atherosclerotic lesions and participate in inflammatory cascade in atherosclerotic development.
Committed HSCs	Originate from uncommitted multipotent HSCs	Can give rise to either myeloid, lymphoid, or erythroid lineages	Monocyte-macrophage lineage releases pro-inflammatory cytokines that recruit additional inflammatory cells from circulation. T cells release anti-inflammatory cytokines. Exogenous HSCs regenerate cells in ischemic region via angiogenesis and improve organ function.
Endothelial Progenitor Cells (EPCs)	May originate from bone marrow precursor cells or from common hemangioblast	May give rise to endothelial cells	Repair, replace, and regenerate the injured endothelium.
Vascular Smooth Muscle (vSMC) Progenitor Cells	May originate from bone marrow or HSCs	May give rise to smooth muscle cells	Differentiate to pro-inflammatory vSMCs and contribute to atherosclerotic lesions, or may stabilize plaque and avoid plaque rupture.

of events that involves infiltration of inflammatory cells, proliferation of vSMCs, and formation of thrombotic plaques. There are conflicting views about the role of EPCs, especially in plaque angiogenesis. Some reports suggest that plaque angiogenesis is pathologically increased by EPC while others report protective effects in atherosclerosis by contributing toward endothelial repair (Schober et al. 2005). This is supported by the fact that in atherosclerotic disease both endogenous HSCs and circulating EPC levels decrease and vulnerability for development of atherosclerotic disease increases (Liu et al. 2009). Thus it becomes very important to repair and restore the functioning of endothelial cells through various therapeutic strategies (discussed in Section 6.5).

6.4.1.1.2 Vascular Smooth Muscle Progenitor Cells

The appearance of SMCs in the intima is one of the early events in the pathogenesis of atherosclerosis. Vascular smooth muscle progenitor cells differentiate into smooth muscle cells and are believed to expand the bulk of atherosclerotic lesions, reducing the lumen size. Conversely, it was reported that although vSMC progenitors migrate to lesion sites, they stabilize the plaque and prevent plaque rupture (Zoll et al. 2008). It is believed that the phenotype of SMCs within atherosclerotic lesions differs from that of the medial cells, which are contractile and secretory SMCs. This difference is considered to be essential to the migration and proliferation of SMCs in the pathogenesis of atherosclerosis. These SMCs display a pro-inflammatory phenotype and express genes sharing a similarity with proliferating stem cells as well as a number of hematopoietic lineage markers (Miyamoto et al. 1997). But it is not clear whether EPC and vSMC progenitors have a common vascular stem cell origin or not. The vSMCs' progenitors were shown to originate in the media of the atherosclerotic arteries either from the bone marrow or circulating progenitor cells or from fusion of progenitor cells with the SMCs present in the lesions (Saiura et al. 2001; Sata et al. 2002). Thus, it could be possible that vascular stem cells and progenitors may be recruited for the purpose of repair of injured endothelium and at the same time may also lead to pathological remodeling of blood vessels.

6.4.1.2 HSC Progenitors and Differentiated Immune Cells

Evidence shows that endogenous bone marrow–derived HSCs home to the sites of atherosclerotic lesions and may contribute to the inflammatory cascades (Sata et al. 2002). It is believed that endogenous HSCs participate in the healing process of the injured organs but may also contribute to the pathological remodeling. Thus, migration to the sites of endothelial injury and atherogenic vessels may be one of the key mechanisms that initiate the progression of the atherosclerosis. Another mechanism is that once the bone marrow–derived endogenous HSC home to the sites of the atherosclerosis, they were, surprisingly, shown to differentiate into SMC and EC in vascular lesions. Thus, it is possible that bone marrow cells could be recruited to atherogenic sites for repair and instead differentiate toward putative vascular cells due to microenvironmental effects and ultimately contribute to lesion progression. Mechanistically, the role of receptors and adenosine triphosphate binding cassette (ABC) transporters in the plasma membrane, expressed by HSPCs and its differentiated cells such as monocytes and macrophages, is also believed to play a role in atherosclerosis development.

A number of clinical studies have shown a rise in leukocyte number in the lesion of atherosclerotic patients and coronary heart disease patients (Coller 2005). Even in animal studies it was shown that an atherogenic diet such as a high-fat diet or lower levels of high-density lipoprotein (HDL) resulted in an increase in leukocytes and monocytes, causing increased monocyte entry into atherosclerotic lesions. However, the observation of a higher count of immune cells was not linked to HDL levels until recently. The mechanisms that make HDL an effective anti-atherosclerotic agent is due to its ability to cause cholesterol efflux from macrophage foam cells in atherosclerotic lesions. This involves the use of ABC transporters (ABCA1 and ABCG1), whose expression is regulated by nuclear receptors (liver X receptor, LXR), activated by oxidized cholesterols. ABC transporters are expressed on HSCs and hematopoietic multipotential progenitor cells. This was supported by the fact that Abcal$^{-/-}$ Abcg1$^{-/-}$ mice had elevated infiltration of macrophage-laden foam cells and very high levels of leukocytes, monocytes, and other immune cells into the lesion (Yvan-Charvet et al. 2007). Thus the absence of cholesterol efflux mechanisms and the absence of HDL cause an increased HSCs proliferation and progression of atherosclerosis (Yvan-Charvet et al. 2010). Not only ABC transporters, the role of Toll-like receptors (TLRs) is also critical to the inflammatory biology of atherosclerosis. TLRs are important for the recognition of microbial moieties by innate immune cells and are also expressed on HSPCs. Activation of these receptors by lipid moieties (lipopolysaccharides, oxidized LDL) can facilitate rapid differentiation of progenitors into macrophages and other immune cells. Receptors such as TLR2 and TLR4 were specifically found to be expressed in the atherosclerotic lesions especially by macrophages and endothelial cells (Xu et al. 2001). Since TLRs were linked to proliferation of progenitor cells, it could therefore increase the population of inflammatory cells in arterial plaques. However, an emerging body of evidence suggests that mesenchymal stem cells suppress lymphocyte proliferation at the site of inflammation (Meirelles Lda et al. 2009).

6.4.1.2.1 *Role of Lymphoid and Myeloid Lineages in the Development and Progression of Atherosclerosis*

Circulating immune cells belonging to the innate and adaptive immune system were implicated in the progression of atherosclerotic diseases. Immune cells belong to monocyte-macrophage lineages and are the critical cells that migrate into arterial tissue in response to locally produced chemokines by injured endothelium. Once circulating free monocyte and T cells (lymphoid lineages) come in contact with the injured endothelium, they adhere to the endothelium, which expresses vascular cell adhesion molecules (VCAM1) and intercellular adhesion molecules (ICAM1; Ross 1999). Monocytes then differentiate into macrophages that express scavenger receptors engulfing ox-LDL and transforming into foam cells. Simultaneously, T cells, mast cells, and neutrophils also get activated and produce pro-inflammatory cytokines, further amplifying the inflammatory response.

Macrophages express an array of receptors such as TLRs or pattern-recognition receptors (PRRs) that phagocytose ox-LDL particles and become foam cells. Although the aforementioned cells are critical to atherosclerotic progression, other immune cells, such as mast cells, dendritic cells, and B cells or plasma cells, are also found in the lesions (Libby and Aikawa 2002). Ox-LDL during the late stage of monocyte differentiation gives rise to phenotypically mature DCs that secrete IL-12 but not IL-10 (anti-atherogenic), supporting T cell stimulation. However, their exact role in plaque formation and lesion progression is still not clear. Apoptosis of foam cells is also known to play a role in attenuation of lesions and necrosis of adjacent inflammatory cells. At the same time, this influx of cells causes severe pro-inflammatory responses, further promoting increased proliferation of inflammatory SMCs and leukocytes within the plaques.

6.4.2 Role of Chemokines, Cytokines, and Interleukins in Development and Progression of Atherosclerosis

Cytokines or chemokines are chemotactic molecules that represent a vast family of structurally related small molecular weight proteins. The chemokines are divided into four families based on the configuration of cysteines (CC, CXC, CX3C, XC). These molecules play important roles in cell mobilization, proliferation, and apoptosis, thus affecting the progression of atherosclerotic disease. Chemokines induce their biological functions via binding to receptors (G-coupled transmembrane proteins) present on their target cells. The highly expressed chemokine in atherosclerotic lesions, monocyte chemotactic protein (MCP-1/CCL2), belongs to the CC subfamily, is secreted by injured EC and dysfunctional SMC in the atherosclerosis, and is responsible for further recruitment of the monocyte to the lesions. The MCP-1 receptor CCR2 is expressed by monocytes, T cells, and vSMC, thus causing their infiltration into intimal lesions. Recruitment of leukocytes to the injured vessel requires an upregulation of VCAM1, ICAM1, and selectins for increased leukocyte rolling, adhesion, and migration to the vessel wall. Furthermore, chemokines that belong to the family of CXC, such as CXCL12/CXCR4, CXCL1/CXCR2 and CXCL7, have also been implicated in mobilization of progenitors from bone marrow to the inflamed vascular tissues. Mobilization occurs in response to the release of stromal-derived factor (SDF-1/CXCL12), CXCL1, and CXCL7, and matrix metalloproteinase 9 (MMP9) from either attached platelets within the atherosclerotic lesions, injured endothelium, or ischemic arterial tissues. This interaction between ligand and receptor helps in endothelial recovery and repair (Massberg et al. 2006).

A recently identified chemokine belonging to the family of CX3C called fractalkine/CXC3CL1, expressed by atherosclerotic arteries only, was reported to be a potent chemoattractant for monocytes and T cells (Greaves et al. 2001). Furthermore, IL-1, IL-2, IL-4, IL-6, and TNF-α were shown to exert pro-inflammatory effects; however, IL-10 exerted anti-inflammatory effects on the progression of atherosclerosis. IL-10 is produced by activated monocytes and lymphocytes and exerts its anti-inflammatory effects through inhibition of leukocyte-EC interaction, inhibition of pro-inflammatory cytokine and adhesion molecules, and affects differentiation of immune cells, specifically T cells. Thus, results from the cytokine biology reflect an important pattern of cross talk among the various immune cells similar to that observed in other auto-immune/inflammatory disorders (von der Thusen et al. 2003); however, their exact role in lesion formation in the early phase of atherogenesis is still unresolved.

6.5 THERAPIES FOR ATHEROSCLEROSIS

Several treatment options have been in practice to limit the progression of atherosclerosis. However, prevention is still thought to be the best cure for this disease: following a healthy and active lifestyle and consuming foods that elevate good cholesterol (HDL) and lower bad cholesterol (LDL) levels. The endogenous reparative mechanisms are greatly affected by risk factors such as age, familial inheritance, smoking, or hypertension. As discussed earlier, the role of the secretory chemokines and ligands has been found to be significant for the recruitment of the endogenous stem cells/progenitor cells to the sites of lesions. Thus, therapies that interfere with the recruitment and mobilization of the progenitor cells are being investigated (Figure 6.2). Many factors, such as levels of VEGF, SDF, nitric oxide (NO), erythropoietin, G-CSF, statins, thiazolidinediones, and exercise, have also been shown to exert mobilizing action on EPCs from bone marrow to atherosclerotic lesions, thus inhibiting lesion progression (Wang et al. 2006; Aicher et al. 2003). Newer insights into progenitor cells' therapeutic effects suggest that circulating endogenous progenitor vSMC cells may promote pathological arterial remodeling; however, local inhibition of binding of vSMCs could attenuate the progression of coronary disease. Thus, therapies that involve inhibition of recruitment of endogenous SMCs and simultaneously promote recruitment of EPC to the injured blood vessels could be beneficial for stabilizing the plaque and reversing atherosclerotic progression.

Contrarily, infusion of progenitor vSMCs halted the development of atherosclerotic lesions due to increased formation of collagen content, impaired macrophage infiltration, and caused plaque stabilization (Zoll et al. 2008). Furthermore, involvement of stem cells in cardiac diseases may provide opportunities for their genetic modification with genes that enhance neovascularization (such as VEGF/PDGF), anti-inflammatory factors (IL-10), and/or endothelial cell survival. However, multiple parameters would have to be investigated before this therapeutic approach becomes a standard of care for cardiac patients. Parameters such a gene dosage, gene transduction vectors, stem cell infusion strategies, disease severity, and age of patients are some criteria that need further validation.

Lipid-lowering drugs such as statins have been used for patients—these drugs block the key rate-limiting enzyme in the biosynthesis of the cholesterol, thus lowering the level of blood

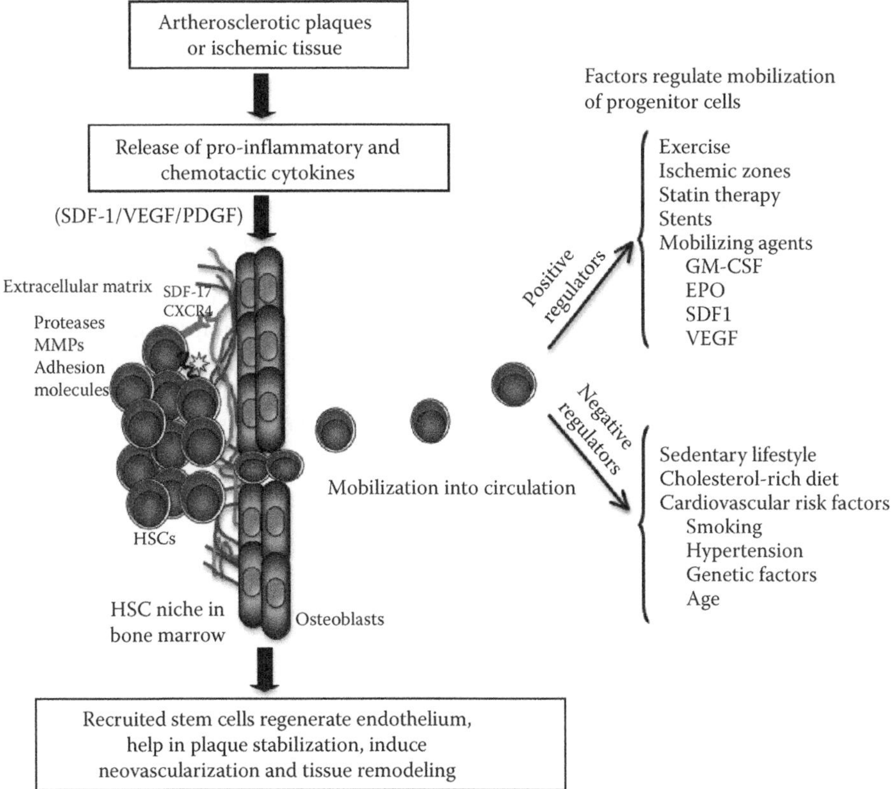

FIGURE 6.2 (see color insert) Mobilization and recruitment of progenitor cells. Injured vessel and ischemic zones release chemotactic factors (SDF-1, VEGF, PDGF, etc.) into the circulation. Release of mobilizing factors causes egress of progenitor cells (HSC, EPC, or vSMC) from bone marrow and recruits proteases such as MMP9, which mediates cleavage of c kit+ HSCs from the bone marrow. The positive regulators of mobilization of progenitor cells include ischemic zones, exercise, mobilizing agents, etc. Factors such as smoking, hypertension, lifestyle choices, and certain cardiovascular conditions cause decreased levels of circulating progenitor cells. The progenitor cells replace and regenerate injured endothelial cells and pro-inflammatory smooth muscle cells and may lead to intimal hyperplasia. In ischemic zones, progenitor cells help regenerate damaged tissues by inducing neovascularization and tissue remodeling.

LDL-cholesterol (Bonetti et al. 2003). Statins have been found to be associated with the effects related to progenitor cells. Besides having lipid-lowering effects, statins exert anti-inflammatory effects that may limit the progression of atherosclerosis and hence reduce ischemia (Pasterkamp and van Lammeren 2010). Also, statins were shown to affect the inflammatory cells that destabilize the plaque in atherosclerotic diseases. Mechanistically, statins affect both progenitor and mature endothelial cells by activation of survival/proliferation signals via upregulation of PI3K/Akt pathways, induce the expression of eNOS, and activate anti-inflammatory transcription factors such as KLF2, which inhibits the expression of adhesion molecules responsible for monocyte adhesion (Lin et al. 2010).

Particularly, statins reduced the expression of IL6 and MCP1 in EC and vSMC, thereby reducing monocyte chemotaxis. Blocking either the secretion or functionality of the pro-inflammatory cytokines may be able to reduce inflammatory cell recruitment in atherosclerotic lesions. However, specificity and off-target effects are still a major challenge. In relation to the chemokine therapies, MCP-1 neutralizing antibodies were shown to be effective in progression of atherosclerosis. However, it seemed to be not so beneficial for myocardial infarction. The MCP-1 blocking effects were found to be associated with decreased and delayed replacement of injured cardiomyocytes and downregulation of pro- and anti-inflammatory cytokines. Thus, chemokine inhibition seems to have different effects in the different stages of the cardiovascular pathologies (Dewald et al. 2005).

One of the therapeutic approaches for the treatment of atherosclerotic plaque uses surgical interventions that employ the use of stents to remove blockage from the arteries in cardiovascular patients. Various types of stents have been used to avoid restenosis, such as bare metal stents, polymer-coated stents, and drug-eluting stents. Drug-eluting stents or polymer-coated stents generally use various drugs such as sirolimus/paclitaxel/batimastat, which inhibit migration of hyperproliferative SMCs and leukocytes or block proliferation of the surrounding inflammatory cells. Development of instant restenosis is the major complication that leads to the failure of the stent-based therapies. Restenosis leads to repeated surgical procedures, which exposes the patients to recurrent stroke and risks. Some of the clinical trials using drug-eluting stents failed to show any improvements (Investigators 2004). However, recently, CD34 antibody–coated stents were used to capture circulating EPCs, and it was observed that the stent was completely re-endothelialized within 48 hours of the surgical placement of the stent (Tanabe et al. 2004). The role of progenitor cells in the treatment of atherosclerosis still needs further evaluation (Table 6.2).

TABLE 6.2
Therapeutic Regimens for Atherosclerosis/Ischemia and Their Mechanisms of Action

Therapeutic Target	Mechanisms of Action/ Biological Function	Type of Therapy	Biomarker/Optimum Dose Indicator
VCAM1, ICAM1	Adhesion molecules	Antagonist	Level of soluble VCAM1 or ICAM1
CCR2, CX3CR1	Leukocyte recruitment	Antagonist	None
CCR7	Leukocyte recruitment	Agonist	Augmentation of EPC in circulation and recovery of damaged endothelium
MMP9 and Cathepsins	Extracellular matrix turnover and plaque rupture	Inhibitor	MMP activity Cathepsin activity
CETP	Lipoprotein metabolism	Inhibitor	Serum levels of LDL cholesterol and HDL cholesterol
LXR	Lipoprotein metabolism	Agonist	ABCA1 expression
PCSK9	Lipoprotein metabolism	Antisense oligonucleotide, siRNA	Serum levels of LDL cholesterol
Endothelial lipase	Lipoprotein metabolism	Inhibitor	Serum levels of HDL cholesterol
PPAR α and γ	Lipoprotein metabolism	Agonist (Fibrates)	Serum levels of LDL cholesterol and HDL cholesterol, ABCA1 expression
Apolipoprotein B	Lipoprotein metabolism	Inhibitor (Niacin)	Serum levels of HDL cholesterol
Infusion of *ex vivo* expanded stem cells	Regeneration of injured endothelium; regeneration of infarcted myocardium; anti-inflammatory effects	Cellular regeneration and restoration of normal functioning	Improvement in cardiovascular functions
Ox-LDL or Apolipoprotein B	Progression of atherosclerotic lesions or lipoprotein metabolism	Vaccination	Serum levels of LDL

Abbreviations: ABCA1, ATP-binding cassette transporter; CCR, Chemokine (C-C motif) receptor; CETP, Cholesteryl ester transfer protein; CX3CR1, Chemokine (CXXXC) motif receptor/fractalkine; EPC, Endothelial progenitor cells; HDL, High density lipoprotein; ICAM1, Inter-cellular adhesion molecule 1; LDL, Low density lipoprotein; LXR, Liver X receptor; MMP, Matrix metalloproteinase; PCSK9, Proprotein convertase subtilisin/kexin type 9; Ox-LDL, Oxidized low density lipoprotein; PPAR (α and γ), Peroxisome proliferator-activated receptors (alpha and gamma); siRNA, Small interfering RNA; VCAM1, Vascular cell adhesion molecule1.

6.6 DEVELOPMENT OF ISCHEMIA AND THERAPY

During the course of atherosclerosis, the vascular tissues or cardiac tissue progress to a disease state due to the formation of ischemia and necrotic core, thus impairing myocyte functions. Usually, in this diseased state, adverse left ventricular (LV) remodeling occurs, causing an increase in the LV size and decrease in ejection fraction (EF). Thus, drugs such as beta-blockers and angiotensin converting enzyme (ACE) inhibitors were used to promote LV function and remodeling. Statins, known as a lipid-lowering class of drugs, were shown to affect endothelial nitric oxide synthase (eNOS), inhibit LV remodeling and dysfunction, and reduce cardiomyocyte hypertrophy in murine models of MI and clinical studies (Landmesser et al. 2004). Furthermore, complete inhibition of pathways mediated via innate immunity receptors such as TLRs, inflammatory pathways, or MMPs were also shown to decrease LV dysfunction and improve cardiac functioning (Shishido et al. 2003). However, revival of the cardiac infarct zone and improvement of fibrotic scar in the myocardium remains incurable. This is due to the limited reparative capacity of the cardiac tissue. Evidence has shown that cardiomyocytes may be able to reenter the cell cycle and may cause limited regeneration. However, these intrinsic repair mechanisms are overwhelmed by substantial damage to the myocardium from the artherosclerotic plaque ruptures (Beltrami et al. 2001). Thus, therapies involving generation of cardiac cells are currently under investigation. These regenerative therapies employ stem cells (e.g., HSCs) that could migrate to the site of infarction and differentiate into various cells and promote neovascularization. However, the limiting factor is the availability of a sufficient number of stem cells for therapeutic application in preclinical and clinical settings.

Our lab evaluated the therapeutic potential of *ex vivo* expanded human umbilical cord blood-derived HSCs (CD 133/34) in myocardial infarction as well as hind limb ischemia models. The stem cells migrated to the ischemic site in both models and were able to restore neovascularization, resulting in improvement of heart and limb functions without any tumorigenic transformation (Das, Abdulhameed et al. 2009; Das, George et al. 2009). Thus, our studies suggest that regeneration of ischemic tissues is plausible due to either paracrine or autocrine effects of stem cells in the ischemic zones. These cells could act by causing vasculogenesis, myocardium regeneration, remodeling, and activation of resident cardiac cells or improved angiogenesis (Figure 6.3). Also, it has been reported that progenitor cells are able to degrade the extracellular matrix by secretion of proteolytic enzymes, beneficial in homing of progenitor cells to assist in neovascularization in ischemic zones (Urbich et al. 2005).

6.7 CONCLUSIONS AND PERSPECTIVES

Although atherosclerosis is a disease of the vascular system and was thought to originate in the walls of large arteries, emerging evidence suggests a systemic pattern of disease progression. Recruitment of injured endothelial cells, SMCs in the vessel wall, circulating immune cells, and endogenous bone marrow progenitor cells seems to be crucial for the atherosclerotic etiology. The progression of atherosclerosis leads to severe forms of cardiovascular disease, such as myocardial ischemia and even congestive heart failure leading to sudden death. In the wake of these diseases, the importance of stem cells has taken center stage due to their unique capability of regeneration and differentiation into multiple lineages. However, stem cell biology is still in a nascent stage that needs further investigation to become a therapy of choice. It is generally accepted in the field of cardiovascular disease that endogenous EPCs have the ability to repair the injured endothelium; however, vSMC progenitors become pro-inflammatory in nature and are implicated in atherosclerotic progression. Pertinent to this, however, the role of endogenous HSCs is still debatable in atherosclerosis. Moreover, identification of homing mechanisms of progenitor cells to the site of injured endothelium, atherosclerotic lesions, or infracted ischemic cardiac zones would provide clues to therapeutic interventions. Do HSCs proliferate and migrate to lesion sites

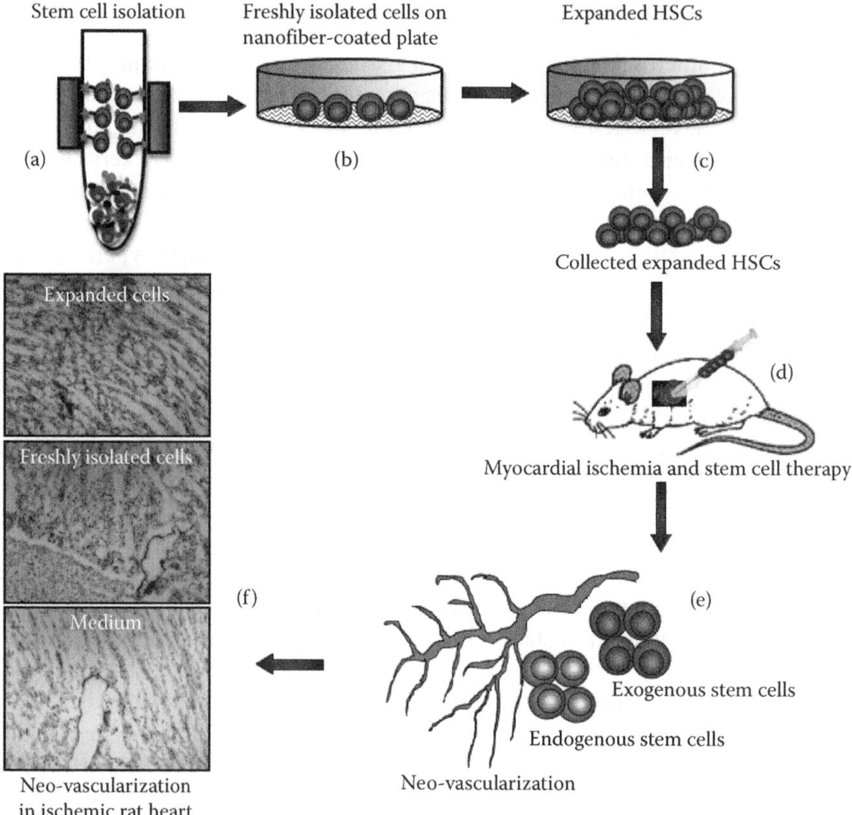

FIGURE 6.3 (see color insert) Stem cell therapy–mediated neovascularization in myocardial ischemia. (a) Schematics of autoMACS-mediated isolation of CD133+ cells. (b) Seeding of CD133+ isolated stem cells on the nanofiber (polyether sulfone, PES)-coated plates. (c) Expansion of stem cells over 10 days of culture in serum-free media supplemented with cytokines and growth factors. (d) Injection of either freshly isolated stem cells or nanofiber-expanded stem cells in immunocompromised rat model of myocardial ischemia via intra cardio-ventricular route. Media was used as control. (e) Besides the paracrine effect, exogenous and endogenous host stem cells take part in the neovascularization process. (f) After 4 weeks of therapy, the rat was sacrificed and cardiac tissues were stained for detection of blood vessels using alkaline phosphatase staining. In ischemic condition like myocardial ischemia, exogenous HSCs as well as circulating endogenous HSCs give rise to appropriate cell type, increase angiogenesis, and reduce fibrosis, resulting in recovery of cardiac function.

to repair the injured endothelium? Do they differentiate into immune cells at the site of lesions to cause repair but end up causing plaque formation due to the micro-environmental effects? Is contribution to inflammation in lesions largely due to activation and recruitment of circulating immune cells or in part due to large involvement of endogenous progenitor cells? Exogenous transplanted stem cells have always been shown to present anti-inflammatory effects, regenerate functional myocardial tissues, and restore neovascularization. Also, there is lot of speculation about the role of genetically modified stem cells in cardiovascular diseases, which may be used as an alternative treatment option for cardiac patients with genetic deficiencies. The mode of genetic engineering, safety of stem cell therapies in terms of dosage and infusion, and stage of a patient's disease are some of the essential parameters that need careful experimentation. It seems very likely that future treatment for ischemia could use a combination of current drugs and stem cells for better therapeutic effects.

TAKE-HOME MESSAGES

- Atherosclerosis is an inflammatory disease, which could be slowed down, contained, or alleviated by the use of stem cell therapies.
- Serum levels of circulating stem cells may be considered as a surrogate biological marker for developing cardiovascular disease, such as atherosclerosis, even in asymptomatic patients.
- Hematopoietic stem cells (HSCs) were shown to differentiate into immune cells, endothelial progenitor cells (EPCs), and vascular smooth muscle cells (vSMCs).
- Not only arteries or cardiac tissue seems to be affected by the risk factors of atherosclerosis but involvement of endogenous bone marrow also seems to be crucial in atherosclerosis.
- Current lipid-lowering drugs were also shown to have strong mobilizing effects on endogenous stem cell recruitment, especially endogenous EPCs.
- Vascular progenitors were shown to have differential effects on the atherosclerotic plaque stability and stem cells and may act as a two-edged sword.
- Improvement in cardiac function after stem cell therapy may not be reflective of an active ventricular remodeling, or full regeneration of endothelium/myocardium, and thus therapies need to be devised for cardiac improvements.
- Exogenous stem cells could induce their effects via paracrine effects by secreting either angiogenic factors or anti-inflammatory factors, or may self-renew and differentiate into cells of host tissue for regeneration.
- Exogenous stem cells could be genetically modified and used as a strong tool for anti-inflammatory effects in atherosclerosis, stabilizing plaque rupture, enhancing angiogenic effects in myocardial ischemia.
- Limited availability of stem cells for therapeutic purposes calls urgently need for the development of a universal protocol that could *ex vivo* expand stem cells while maintaining their stem cell characters.

ACKNOWLEDGMENTS

This work was supported in part by National Institutes of Health grants K01 AR054114 (NIAMS), SBIR R44 HL092706-01 (NHLBI), R21 CA143787 (NCI), and The Ohio State University start-up fund for stem cell research. The funders had no role in study design, data collection and analysis, decision to publish, or preparation of the manuscript.

REFERENCES

Adams, G. B., K. T. Chabner, I. R. Alley, D. P. Olson, Z. M. Szczepiorkowski, M. C. Poznansky, et al. 2006. Stem cell engraftment at the endosteal niche is specified by the calcium-sensing receptor. *Nature* 439 (7076):599–603.

Aicher, A., C. Heeschen, C. Mildner-Rihm, C. Urbich, C. Ihling, K. Technau-Ihling, et al. 2003. Essential role of endothelial nitric oxide synthase for mobilization of stem and progenitor cells. *Nat Med* 9 (11):1370–6.

Arai, F., A. Hirao, M. Ohmura, H. Sato, S. Matsuoka, K. Takubo, et al. 2004. Tie2/angiopoietin-1 signaling regulates hematopoietic stem cell quiescence in the bone marrow niche. *Cell* 118 (2):149–61.

Asahara, T., T. Murohara, A. Sullivan, M. Silver, R. van der Zee, T. Li, et al. 1997. Isolation of putative progenitor endothelial cells for angiogenesis. *Science* 275 (5302):964–7.

Beltrami, A. P., K. Urbanek, J. Kajstura, S. M. Yan, N. Finato, R. Bussani, et al. 2001. Evidence that human cardiac myocytes divide after myocardial infarction. *N Engl J Med* 344 (23):1750–7.

Bonetti, P. O., L. O. Lerman, C. Napoli, and A. Lerman. 2003. Statin effects beyond lipid lowering—are they clinically relevant? *Eur Heart J* 24 (3):225–48.

Calvi, L. M., G. B. Adams, K. W. Weibrecht, J. M. Weber, D. P. Olson, M. C. Knight, et al. 2003. Osteoblastic cells regulate the haematopoietic stem cell niche. *Nature* 425 (6960):841–6.

Choi, H. J., Y. Geng, H. Cho, S. Li, P. K. Giri, K. Felio, et al. 2011. Differential requirements for the Ets transcription factor Elf-1 in the development of NKT cells and NK cells. *Blood* 117 (6):1880–7.

Chong, A. Y., A. D. Blann, J. Patel, B. Freestone, E. Hughes, and G. Y. Lip. 2004. Endothelial dysfunction and damage in congestive heart failure: Relation of flow-mediated dilation to circulating endothelial cells, plasma indexes of endothelial damage, and brain natriuretic peptide. *Circulation* 110 (13):1794–8.

Coller, B. S. 2005. Leukocytosis and ischemic vascular disease morbidity and mortality: Is it time to intervene? *Arterioscler Thromb Vasc Biol* 25 (4):658–70.

Das, H., N. Abdulhameed, M. Joseph, R. Sakthivel, H. Q. Mao, and V. J. Pompili. 2009. Ex vivo nanofiber expansion and genetic modification of human cord blood-derived progenitor/stem cells enhances vasculogenesis. *Cell Transplant* 18 (3):305–18.

Das, H., J. C. George, M. Joseph, M. Das, N. Abdulhameed, A. Blitz, et al. 2009. Stem cell therapy with overexpressed VEGF and PDGF genes improves cardiac function in a rat infarct model. *PLoS One* 4 (10):e7325.

Dewald, O., P. Zymek, K. Winkelmann, A. Koerting, G. Ren, T. Abou-Khamis, et al. 2005. CCL2/monocyte chemoattractant protein-1 regulates inflammatory responses critical to healing myocardial infarcts. *Circ Res* 96 (8):881–9.

Greaves, D. R., T. Hakkinen, A. D. Lucas, K. Liddiard, E. Jones, C. M. Quinn, et al. 2001. Linked chromosome 16q13 chemokines, macrophage-derived chemokine, fractalkine, and thymus- and activation-regulated chemokine, are expressed in human atherosclerotic lesions. *Arterioscler Thromb Vasc Biol* 21 (6):923–9.

Hansson, G. K. 2005. Inflammation, atherosclerosis, and coronary artery disease. *N Engl J Med* 352 (16):1685–95.

Hu, Y., Z. Zhang, E. Torsney, A. R. Afzal, F. Davison, B. Metzler, et al. 2004. Abundant progenitor cells in the adventitia contribute to atherosclerosis of vein grafts in ApoE-deficient mice. *J Clin Invest* 113 (9):1258–65.

Investigators, SSYLVIA Study. 2004. Stenting of Symptomatic Atherosclerotic Lesions in the Vertebral or Intracranial Arteries (SSLVIA): Study results. *Stroke* 35:1388–1392.

Kollet, O., A. Dar, S. Shivtiel, A. Kalinkovich, K. Lapid, Y. Sztainberg, et al. 2006. Osteoclasts degrade endosteal components and promote mobilization of hematopoietic progenitor cells. *Nat Med* 12 (6):657–64.

Laine, P., M. Kaartinen, A. Penttila, P. Panula, T. Paavonen, and P. T. Kovanen. 1999. Association between myocardial infarction and the mast cells in the adventitia of the infarct-related coronary artery. *Circulation* 99 (3):361–9.

Landmesser, U., N. Engberding, F. H. Bahlmann, A. Schaefer, A. Wiencke, A. Heineke, et al. 2004. Statin-induced improvement of endothelial progenitor cell mobilization, myocardial neovascularization, left ventricular function, and survival after experimental myocardial infarction requires endothelial nitric oxide synthase. *Circulation* 110 (14):1933–9.

Libby, P., and M. Aikawa. 2002. Stabilization of atherosclerotic plaques: New mechanisms and clinical targets. *Nat Med* 8 (11):1257–62.

Lin, Z., V. Natesan, H. Shi, A. Hamik, D. Kawanami, C. Hao, et al. 2010. A novel role of CCN3 in regulating endothelial inflammation. *J Cell Commun Signal* 4 (3):141–53.

Liu, P., B. Zhou, D. Gu, L. Zhang, and Z. Han. 2009. Endothelial progenitor cell therapy in atherosclerosis: A double-edged sword? *Ageing Res Rev* 8 (2):83–93.

Massberg, S., I. Konrad, K. Schurzinger, M. Lorenz, S. Schneider, D. Zohlnhoefer, et al. 2006. Platelets secrete stromal cell-derived factor 1alpha and recruit bone marrow-derived progenitor cells to arterial thrombi in vivo. *J Exp Med* 203 (5):1221–33.

Meirelles Lda, S., A. M. Fontes, D. T. Covas, and A. I. Caplan. 2009. Mechanisms involved in the therapeutic properties of mesenchymal stem cells. *Cytokine Growth Factor Rev* 20 (5-6):419–27.

Miyamoto, T., Y. Sasaguri, T. Sasaguri, S. Azakami, H. Yasukawa, S. Kato, et al. 1997. Expression of stem cell factor in human aortic endothelial and smooth muscle cells. *Atherosclerosis* 129 (2):207–13.

Pasterkamp, G., and G. W. van Lammeren. 2010. Pleiotropic effects of statins in atherosclerotic disease. *Expert Rev Cardiovasc Ther* 8 (9):1235–7.

Reya, T. 2003. Regulation of hematopoietic stem cell self-renewal. *Recent Prog Horm Res* 58:283–95.

Ross, R. 1999. Atherosclerosis—an inflammatory disease. *N Engl J Med* 340 (2):115–26.

Rothenberg, E. V. 2011. T cell lineage commitment: identity and renunciation. *J Immunol* 186 (12):6649–55.

Saiura, A., M. Sata, Y. Hirata, R. Nagai, and M. Makuuchi. 2001. Circulating smooth muscle progenitor cells contribute to atherosclerosis. *Nat Med* 7 (4):382–3.

Sata, M., A. Saiura, A. Kunisato, A. Tojo, S. Okada, T. Tokuhisa, et al. 2002. Hematopoietic stem cells differentiate into vascular cells that participate in the pathogenesis of atherosclerosis. *Nat Med* 8 (4):403–9.

Schober, A., R. Hoffmann, N. Opree, S. Knarren, E. Iofina, G. Hutschenreuter, et al. 2005. Peripheral CD34+ cells and the risk of in-stent restenosis in patients with coronary heart disease. *Am J Cardiol* 96 (8):1116–22.

Shishido, T., N. Nozaki, S. Yamaguchi, Y. Shibata, J. Nitobe, T. Miyamoto, et al. 2003. Toll-like receptor-2 modulates ventricular remodeling after myocardial infarction. *Circulation* 108 (23):2905–10.

Simper, D., P. G. Stalboerger, C. J. Panetta, S. Wang, and N. M. Caplice. 2002. Smooth muscle progenitor cells in human blood. *Circulation* 106 (10):1199–204.

Sugiyama, T., H. Kohara, M. Noda, and T. Nagasawa. 2006. Maintenance of the hematopoietic stem cell pool by CXCL12-CXCR4 chemokine signaling in bone marrow stromal cell niches. *Immunity* 25 (6):977–88.

Tanabe, K., E. Regar, C. H. Lee, A. Hoye, W. J. van der Giessen, and P. W. Serruys. 2004. Local drug delivery using coated stents: New developments and future perspectives. *Curr Pharm Des* 10 (4):357–67.

Tsai, F. Y., and S. H. Orkin. 1997. Transcription factor GATA-2 is required for proliferation/survival of early hematopoietic cells and mast cell formation, but not for erythroid and myeloid terminal differentiation. *Blood* 89 (10):3636–43.

Urbich, C., C. Heeschen, A. Aicher, K. Sasaki, T. Bruhl, M. R. Farhadi, et al. 2005. Cathepsin L is required for endothelial progenitor cell-induced neovascularization. *Nat Med* 11 (2):206–13.

Vasa, M., S. Fichtlscherer, A. Aicher, K. Adler, C. Urbich, H. Martin, et al. 2001. Number and migratory activity of circulating endothelial progenitor cells inversely correlate with risk factors for coronary artery disease. *Circ Res* 89 (1):E1–7.

von der Thusen, J. H., J. Kuiper, T. J. van Berkel, and E. A. Biessen. 2003. Interleukins in atherosclerosis: Molecular pathways and therapeutic potential. *Pharmacol Rev* 55 (1):133–66.

Wang, C. H., S. Verma, I. C. Hsieh, Y. J. Chen, L. T. Kuo, N. I. Yang, et al. 2006. Enalapril increases ischemia-induced endothelial progenitor cell mobilization through manipulation of the CD26 system. *J Mol Cell Cardiol* 41 (1):34–43.

Wu, L., A. Nichogiannopoulou, K. Shortman, and K. Georgopoulos. 1997. Cell-autonomous defects in dendritic cell populations of Ikaros mutant mice point to a developmental relationship with the lymphoid lineage. *Immunity* 7 (4):483–92.

Xu, X. H., P. K. Shah, E. Faure, O. Equils, L. Thomas, M. C. Fishbein, et al. 2001. Toll-like receptor-4 is expressed by macrophages in murine and human lipid-rich atherosclerotic plaques and upregulated by oxidized LDL. *Circulation* 104 (25):3103–8.

Yvan-Charvet, L., T. Pagler, E. L. Gautier, S. Avagyan, R. L. Siry, S. Han, et al. 2010. ATP-binding cassette transporters and HDL suppress hematopoietic stem cell proliferation. *Science* 328 (5986):1689–93.

Yvan-Charvet, L., M. Ranalletta, N. Wang, S. Han, N. Terasaka, R. Li, et al. 2007. Combined deficiency of ABCA1 and ABCG1 promotes foam cell accumulation and accelerates atherosclerosis in mice. *J Clin Invest* 117 (12):3900–8.

Zoll, J., V. Fontaine, P. Gourdy, V. Barateau, J. Vilar, A. Leroyer, et al. 2008. Role of human smooth muscle cell progenitors in atherosclerotic plaque development and composition. *Cardiovasc Res* 77 (3):47180.

7 Inflammation as a Confounding Factor in Regenerative Medicine

Myron Allukian
University of Pennsylvania School of Medicine
Philadelphia, Pennsylvania

Kenneth W. Liechty
The University of Mississippi Medical Center
Jackson, Mississippi

CONTENTS

7.1 INFLAMMATION

Inflammation is an essential component to the host defense against invading microbes. Infection initiates an amplified cascade of signals that leads to the recruitment of inflammatory cells. Neutrophils and macrophages phagocytose these infectious organisms, release cytokines and chemokines, and activate lymphocytes, which results in adaptive immunity. Since tissue injury frequently involves exposure to the outside environment and potential infection, it is intuitive that inflammation developed as a critical component of wound repair.

Until the early 20th century, most wounds were traumatic and contaminated with microbes. Our ancestors likely benefited from the protection of an exuberant inflammatory response that neutralized invading microbes, but resulted in scar formation, which was an acceptable alternative to overwhelming sepsis and death. The advent of antibiotics and aseptic surgical technique has resulted in decreased wound contamination, with the majority of wounds being devoid of microbes, and as a result may have made the survival benefit of a prolific inflammatory response obsolete.

While few debate the necessary and protective role of inflammation to effectively heal contaminated wounds, inflammation does not occur without negative sequelae. Inflammation produces reactive oxygen species (ROS), proteases, and growth factors that can amplify the inflammatory response and cause further destruction of the injured tissue and cell death. Devitalized tissue is debrided by macrophages and neutrophils, which leaves the wound bed devoid of matrix and results in tissue fibrosis and scar formation. Scar tissue is aesthetically, structurally, and functionally inferior to native tissue. Depending on the location of the injured tissue, scar formation may result in disfigurement, contracture, bowel obstruction, or even heart failure.

Wound repair in the adult follows an orderly and well-defined overlapping sequence of events: inflammation, proliferation, and remodeling. In the adult, wound healing is reparative and results in scar formation. Unlike reparative wound healing, regenerative healing is characterized by the reconstitution of cellular architecture and restoration of original tissue function. In contrast to adult reparative healing, the fetus can heal wounds of the dermis [1], tendon [2], and heart [3] in a regenerative and scarless manner. Scarless wound healing is associated with a significantly decreased inflammatory response [4, 5]. Following myocardial infarction in adult sheep there is a brisk inflammatory response with numerous CD45+ cells demonstrated within the infarct at 2 weeks (Figure 7.1A), and is associated with a progressive decline in ventricular function. In contrast, the numbers of CD45+ inflammatory cells are dramatically decreased in the fetal infarct at 1 week (Figure 7.1B), and was associated with restoration of cardiac function [3]. At 1 month following infarction, adult infarcts demonstrated persistence of this inflammatory cell infiltrate (Figure 7.1C), whereas the inflammatory cell infiltrate in the fetal infarct had completely resolved (Figure 7.1D).

The role of inflammation in wound healing is complex. Early landmark studies concluded that inflammation was a necessary component of proper wound repair [6, 7]. However, recent studies have shown that selective modulation of the inflammatory response improves postnatal wound

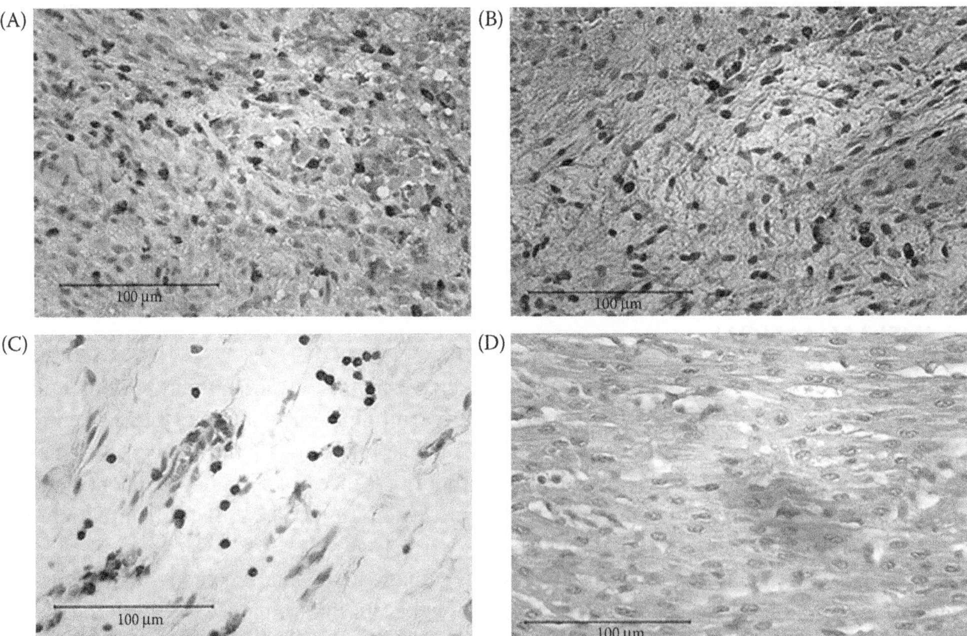

FIGURE 7.1 (see color insert) CD45 immunohistochemistry following myocardial infarction demonstrates markedly less cellular inflammatory response in fetal versus adult hearts. One week following myocardial infarction, the fetal heart (B) shows minimal numbers of inflammatory cells while the adult heart (A) demonstrates a significantly increased inflammatory infiltrate. One month following infarction, the inflammatory cell infiltrate in the fetal myocardium (D) has decreased, while it persists within the adult myocardium (C).

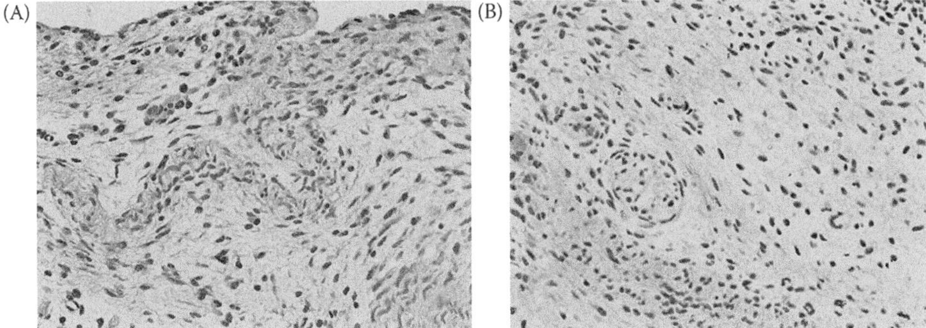

FIGURE 7.2 (see color insert) Immunohistochemical staining for CD45, the common leukocyte antigen, in 8mm (A) or 2mm (B) dermal wounds in fetal sheep 7 days after injury. Note: Increased wound size was associated with a dramatic increase in the number of inflammatory cells in the wound and subsequent scar formation.

healing outcomes [8–10]. In addition, other studies have shown that increasing the inflammatory response in fetal wounds with irradiated bacteria, chemical irritants, or larger wound size results in conversion to the adult wound phenotype of scar formation [11–14]. Figure 7.2 demonstrates immunohistochemical staining for CD45, the common leukocyte antigen, in 8mm (Panel A) or 2mm (Panel B) dermal wounds in fetal sheep 7 days after injury. Increased wound size was associated with a dramatic increase in the number of inflammatory cells in the wound and subsequent scar formation. Therefore a balance must exist between the constructive and destructive roles inflammation plays in wound healing. The field of regenerative medicine aims to tip the balance in favor of replacing damaged tissue with equally functional tissue while accelerating the healing process in adults.

This chapter will review the key elements of the inflammatory response and the potential confounding roles they play in the wound healing response. Specifically, the potential interactions among inflammatory cells, cytokines, and growth factors and the extracellular matrix (ECM) will be examined, as well as the roles they play in determining the resultant wound phenotype. Understanding the pleiotropic effects and roles these factors play, as well as the redundancy of the inflammatory signaling pathways, will help to illuminate the confounding relationship between inflammation and regenerative wound healing.

7.2 CELLS

7.2.1 Platelets

Platelets are the first cells recruited to the site of injury (Table 7.1). Exposed collagen induces platelet activation and aggregation. Activated platelets adhere to the exposed connective tissue forming a hemostatic plug while releasing α-granules, which contain pro-inflammatory mediators that stimulate the inflammatory cascade (Table 7.2) [15]. Several platelet-derived factors and cytokines, including transforming growth factor-β (TGF-β) and platelet-derived growth factor (PDGF), may increase inflammation by recruiting activated leukocytes to the wound [16, 17].

The paucity of an acute inflammatory cell infiltrate in scarless fetal wound healing may partially be attributed to differences in fetal and adult platelet function. Although fetal and adult platelets appear similar in morphology and quantity, fetal platelet aggregation is significantly decreased compared to the adult after exposure to collagen and adenosine diphosphate (ADP). Decreased platelet degranulation and release of PDGF and TGF-β may partially explain the scarless wound healing observed in fetal dermal wounds, as exogenous application of PDGF and TGF-β to fetal wounds recapitulates the inflammatory response observed in adults [18, 19]. Interestingly, the age-dependent transition from impaired to normal platelet aggregation in the fetus occurs temporally with the transition of wound healing from the regenerative to the reparative phenotype [20].

TABLE 7.1
Cells Involved in Acute Inflammatory Response

Cell Type	Recruited From	Role
Mast Cell	Tissue resident Diapedesis from circulation	Vascular permeability
Platelet	Circulation	Hemostasis Vascular permeability
Neutrophil	Diapedesis from circulation	Phagocytosis (antimicrobial) Propagate inflammatory response Tissue repair
Macrophage	Tissue resident (very few) Diapedesis from circulation	Endothelial activation Phagocytosis (neutrophil and tissue debridement) Propagate inflammatory response Suppress inflammation Fibrolysis Fibroplasia

Platelets were once thought to play an essential role in initiating the proliferative response because many studies demonstrated improved wound healing following the application of exogenous platelet-derived growth factors (most notably PDGF and TGF-β) [21–26]. However, hemostatic dermal wounds in thrombocytopenic mice do not display differences in the proliferative aspects of wound repair, including wound closure, angiogenesis, and collagen synthesis [27] (Table 7.3). Depletion of platelets in hemostatic wounds may not produce a deficiency of growth factors required for wound repair because they are also secreted by other cells, which may limit the targeting of platelets as a potential therapeutic agent to decrease inflammation and subsequent scar formation.

TABLE 7.2
Mediators of Acute Inflammation

Mediator	Source	Action
Histamine and Serotonin	Mast cell, platelet	Increased vascular permeability
C3a	Plasma	Increased vascular permeability
C5a	Macrophage	Increased vascular permeability Leukocyte chemotaxis
Reactive Oxygen Species	Leukocytes, other cells	Increased vascular permeability Microbicidal Tissue destruction
IL1 and TNFα	Leukocytes, other cells	Chemotaxis Endothelial activation Propagate inflammatory response
Chemokines	Leukocytes	Leukocyte activation Chemotaxis
NO	Macrophage, endothelium	Increased vascular permeability Vasodilation Chemotaxis Cytotoxic

TABLE 7.3
Summary of Experiments Depleting Inflammatory Components and Effects on Wound Repair

Target	Method	Inflammatory Cell Population			Wound Phenotype	Reference
		PMN	MΦ	T-Cell		
Platelet depletion	AS	=	↑	↑	Normal	[27]
Neutrophil depletion	AS	↓	=	na	Normal, faster healing	[6] [34]
Macrophage depletion	AS, steroids	=	↓	na	Delayed closure	[7]
Mast cell depletion	KO	↓	=	=	Normal	[28]
PU.1	KO	↓	↓	na	Reduced scar	[34]
TNF-Rp55	KO	↓	↓	=	Faster closure	[49]
IL-1ra	KO	↓	na	na	Delayed closure	[50]
IL-6	KO	↓	↓	=	Delayed closure	[51]
IL-10	KO	↓	↑	na	Accelerated closure	[70]
MCP-1	KO	na	=	na	Delayed closure	[71]
MIP-1a	KO	=	=	=	Normal	[71]
MIF	KO	=	=	=	Normal	[72]
IFN-g	KO	↓	↓	↓	Accelerated closure	[73]
Smad3	KO	↓	↓	na	Accelerated closure	[74]

AS, antisera; MΦ, macrophage; MCP, monocyte chemoattractant protein; MIP, macrophage inflammatory protein; na, not assessed; PMN, polymorphonuclear leukocyte; TNF, tumor necrosis factor

7.2.2 Mast Cells

Mast cells reside within connective tissue with frequency in areas commonly exposed to the external environment. Following injury, mast cells degranulate and release cytokines and other inflammatory mediators that increase vascular permeability and leukocyte infiltration (Table 7.1). Although the resident population is depleted of mast cells immediately following injury, their numbers return to normal around 48 hours post injury as tissue repair progresses [28].

Mast cells produce many leukocyte chemoattractants, and are considered key regulators in the inflammatory phase of healing. Pro-inflammatory mediators (TNF-α, MIP-2, IL-8) released from activated mast cells have been implicated in neutrophil recruitment to the wound. Studies in wild-type and mast cell–deficient mice demonstrate no difference in reepithelialization, collagen deposition, or angiogenesis during wound healing. However, decreased numbers of neutrophils were recruited to the mast cell–deficient wound 24 hours following injury [28]. Therapeutic strategies to stabilize the mast cell membrane have the potential to decrease neutrophil recruitment and promote an earlier resolution of the inflammatory response.

7.2.3 Neutrophils

The first nucleated cell recruited to the wound bed is the neutrophilic granulocyte or neutrophil. An early component of the innate immune response, the neutrophil migrates to the wound bed from the circulating blood within a few hours of injury (Table 7.1). While augmenting the inflammatory response by producing a variety of growth factors, including IL-8 [29] and VEGF [30], neutrophils debride tissue and destroy invading microbes in phagolysosomes and "extruded nets" of DNA and histones [31]. Although several of the antimicrobial substances generated (ROS, cationic peptides, eicosanoids, and proteases [32]) by neutrophils aid in the destruction of invading organisms, they may also damage healthy cells.

Several studies have demonstrated improved wound healing associated with reduced neutrophil recruitment. Minimal neutrophil infiltrate has been observed not only in scarless fetal wounds, but also in adult wounds of the oral mucosa, which heal rapidly and with decreased scar formation [33]. Neutropenic animal wound healing models have also demonstrated no significant differences in collagen deposition, wound disruption strength, or macrophage density, but have exhibited significantly faster reepithelialization than controls (Table 7.3) [6, 34]. The reduction of neutrophil populations in sterile wounds may minimize the collateral damage of healthy tissue in the wound bed and result in accelerated wound healing.

7.2.4 Macrophages

Macrophages are present at all stages of the wound healing process. While resident macrophages exist at low density within non-wounded tissues [35], the majority of macrophages found in wounds are differentiated monocytes recruited from the blood to the wound bed by the inflammatory response (Table 7.1). Once in the wound bed, the macrophage phenotype is dictated by its surrounding microenvironment. Experimental models depleting macrophages at various stages of wound healing have given insight into the various roles macrophages play.

Macrophage numbers increase during the inflammatory phase, peak during the proliferative phase, and decline during wound maturation [36]. Recruitment of macrophages to the wound begins with emigration of monocytes from the blood. The release of pro-inflammatory cytokines (CCL-2, IL-6, IL-8) from endothelial cells induces monocyte chemotaxis, which peaks about 42 hours after injury. Other factors that recruit monocytes to the wound bed are MIP-1a (CCL3), CCL5, TGF-β, PDGF, and VEGF, TGF-α, fibronectin, elastin, C5a, C3a, nerve growth factor, and ECM components [37].

Upon entering the wound, several cytokines (IL4, IL-10, IFN-γ, and IL-13), bacterial products (LPS), and ECM components cause the monocyte to differentiate into a macrophage. The functional phenotype of the macrophage is dictated by the wound microenvironment. The classical macrophage phenotype (M1) is induced by inflammatory cytokines (IFN-γ), bacterial products (LPS), and the presence of neutrophils, which also release inflammatory cytokines (IL-1a, IL-1b, IL-6, and TNF-α) [38]. M1 macrophages exert pro-inflammatory functions such as antigen presentation, phagocytosis, and production of inflammatory mediators [39]. Classically activated macrophages exhibit other antimicrobial properties and release several inflammatory mediators (TNF-α, NO, and IL-6) that can cause serious collateral damage and exacerbate the inflammatory response [40]. It is estimated that 85% of the wound macrophage population exhibits the M1 phenotype one day after wounding [41].

Over time, changes in the wound microenvironment convert the most prevalent macrophage phenotype from classical (M1) to alternatively activated macrophages (M2) by day five. Expression of IL-4 and IL-13 by Th2 lymphocytes has been shown to induce the M2 phenotype that displays decreased reactivity to inflammatory mediators. The M2 macrophage is thought to suppress inflammation through expression of IL-10 and IL-1ra. M2 macrophages are also a prominent source of TGFβ, which promotes wound repair by contraction, angiogenesis, and ECM deposition [42]. Alternatively activated macrophages play an important role in wound healing and angiogenesis, but they have also been implicated in several pathologic conditions including fibrosis, atherosclerosis, and cancer [42, 43].

Early studies using the depletion of macrophages with antisera and steroids demonstrated a significant delay in wound healing and concluded that macrophages were essential to wound repair [7]. However, recent studies have cast doubt on the necessity of macrophages to orchestrate all stages of tissue repair in sterile wounds. Depletion of resident macrophages demonstrated neither delayed wound healing nor attenuated inflammatory response [44]. PU.1-knockout neonatal mice, which are deficient in macrophages and neutrophils (notably also B cells, mast cells, and eosinophils), heal with a scarless fetal phenotype in the same time course as their wild-type scar-forming siblings [10]. Additionally, the lack of myeloid lineage cells in these mutant mice suggests that inflammatory cells may not be essential for efficient tissue repair; however, the neonatal microenvironment may also contribute to the superior wound healing phenotype. Interestingly, blocking the initial burst of macrophages in the adult murine dermal wound results in decreased scar formation, but mid-stage depletion resulted in hemorrhage and regression of mature granulation tissue. In addition, late-stage macrophage depletion did not significantly impact tissue maturation [45], highlighting the importance of the temporal relationships required during the wound healing response.

Increased neutrophil and macrophage recruitment does not always correlate with increased scar formation. Wounds in elderly humans demonstrate neutrophil infiltration with a significantly increased macrophage population compared to younger controls. Despite increased numbers of inflammatory cells, the elderly have been shown to have decreased scar formation with greater regeneration of the dermal elements [46]. Optimizing the timing and extent of macrophage recruitment to dermal wounds may play a role in reducing scar formation. Modulating the dominant macrophage phenotype within wounds may also increase the capacity of injured adult tissue to regenerate or heal with decreased scarring, as macrophage phenotype has been shown to differentially regulate proliferation and differentiation of progenitor cells involved in regeneration of colonic epithelium and skeletal muscle following injury [47, 48].

7.2.5 Mediators of Inflammation (Cytokines, Chemokines, and Growth Factors)

Inflammatory mediators are produced by multiple cell types in response to injury (Table 7.2). Expression of these small proteins is tightly regulated and their biologic function is generally pleiotropic, acting on multiple different cell types. A large degree of functional redundancy exists, with multiple mediators performing similar roles in the modulation of the inflammatory response. An individual cytokine or growth factor may have both positive and negative regulatory functions.

TNF and IL-1 are two major cytokines that mediate inflammation. They are secreted in response to many inflammatory stimuli including wounding and infection. TNF primes neutrophils and induces the release of proteolytic enzymes from mesenchymal cells, causing additional tissue injury and loss. Along with IL-6, TNF and IL-1 induce systemic acute phase responses to injury.

Mice deficient for the TNF receptor, TNF-Rp55, demonstrate decreased leukocyte recruitment in response to dermal wounding, which was also associated with accelerated wound closure [49]. In addition, knockout of the IL-1 inhibitor IL-1ra results in an increased neutrophil infiltrate and impaired dermal wound healing [50]. These studies lend support to the premise that decreased inflammation improves wound healing while increased inflammation can be detrimental to wound repair.

The recruitment of inflammatory cells is also regulated by a large number of other growth factors, cytokines, and chemokines, such as interleukin-6 (IL-6), interleukin-8 (IL-8), and monocyte chemotactic protein-1 (MCP-1/CCL2), which are produced by fibroblasts, endothelial cells, and macrophages in response to injury. These pro-inflammatory cytokines recruit inflammatory cells, including neutrophils, monocytes, and macrophages into the wound where they become activated and produce more inflammatory cytokines, giving rise to additional inflammation and cytokine production.

Elimination of genes considered to be pro-inflammatory, however, does not always result in decreased inflammatory cell infiltrate or improved wound healing. Dermal wounds in IL-6 knockout mice demonstrate a significantly reduced inflammatory infiltrate, but this was also associated with delayed reepithelialization and decreased granulation tissue [51]. In addition to its pro-inflammatory properties, IL-6 is also known to exert a mitogenic effect on keratinocytes [52], and without the mitogenic effect of IL-6, wound healing is impaired. In contrast, increasing IL-6 levels in fetal wounds converts the scarless fetal wound healing response to that of scar formation [4]. A balance must be reached when modulating IL-6 levels in wound repair, as it performs different roles on the various cells within the wound bed (Table 7.3). These studies highlight the complex and confounding relationship between the inflammatory response and wound healing phenotype.

Gene knockout studies have also exposed redundancy in inflammatory pathways. MCP-1 is regarded as one of the major chemoattractants of macrophages in response to wounding. However, recruitment of macrophages to the wound in MCP-1 knockout mice was not decreased, implicating activation of compensatory recruitment pathways (Table 7.3). Interestingly, the wound repair in MCP-1 knockouts was impaired, which suggests MCP-1 may also be involved in activating macrophages. Understanding the redundancies in the inflammatory pathways will be crucial to developing targeted therapies to improve wound healing.

The transforming growth factor-β family has been implicated in many diverse processes of wound healing including cell recruitment (leukocytes, fibroblasts, and keratinocytes), cell function (apoptosis and differentiation), and ECM production [53]. TGF-β directly induces collagen production by MSCs and fibroblasts, inhibits collagenases and MMPs, and thus promotes fibrosis [36]. TGF-β has three isotypes—β1, β2, and β3—which all stimulate inflammatory cell infiltrate. A large amount of evidence suggests decreasing TGF-β1 activity would reduce scar formation in adult wounds. Low levels of TGF-β1 found in fetal [54] and adult wounds depleted of TGF-β1 display decreased scarring [55], while the exogenous application of PDGF and TGF-β to fetal wounds recapitulates the inflammatory response observed in adults [18, 19]. In addition, blockade of TGF-β transmembrane signal in Smad3 knockout mice is associated with accelerated reepithelialization, decreased monocyte recruitment, and inhibition of fibrosis [8]. Manipulation of the functional TGF-β concentration and isoform distribution within the wound environment has the potential to modulate the fibrotic response; however, redundant pathways in cytokine signaling also need to be considered.

7.3 EXTRACELLULAR MATRIX

The extracellular matrix (ECM) is a scaffold consisting of collagen, proteoglycans, and glycosaminoglycans. The structure, orientation, and concentration of specific ECM molecules alter the wound microenvironment during the wound healing response. By regulating inflammation, cell adhesion, migration, proliferation, and the availability of growth factors, the ECM not only provides a structural lattice for wound healing but also may determine whether the wound phenotype is regenerative or reparative.

Hyaluronic acid (HA) is a glycosaminoglycan that plays a critical role during each stage of wound healing [56]. During the inflammatory phase, HA likely acts as a promoter of early inflammation, with a direct correlation between increasing amounts of HA and the pro-inflammatory cytokines TNF-α and IL-1β in cultured uterine fibroblasts [57]. Due to its size and biochemical structure, HA is a potent scavenger of oxygen free radicals and as such may also act as a moderator of inflammation throughout the remainder of the wound healing period [56]. This effect becomes more potent the higher the molecular weight of the hyaluronic acid [58–60]. In addition, low- and intermediate-weight HA fragments increase the expression of multiple pro-inflammatory chemokines vital for initiating inflammation, including macrophage inflammatory protein-1α (MIP-1α) and monocyte chemotactic protein (MCP-1).

It has long been known that high levels of HA are present in the developing fetus [61]. More recently it has been shown that the majority of this is HMW-HA [62]. Similar to adults, fetal levels of HA increase in response to wounding. However, HMW-HA has been shown to increase to a greater extent than LMW-HA in response to injury [63]. These facts suggest that HMW-HA may play a key role in fetal regenerative healing. This has been further suggested by *in vitro* experiments that successfully created scarless repair in an HMW-HA rich environment in tissue that would normally have scarred [64].

Matricellular proteins are another group of ECM molecule that are upregulated during wound healing and modulate cell-matrix interactions crucial to efficient tissue repair such as cell migration. This group of proteins includes galectins, osteopontin, SPARC, tenascins, thrombospondins,

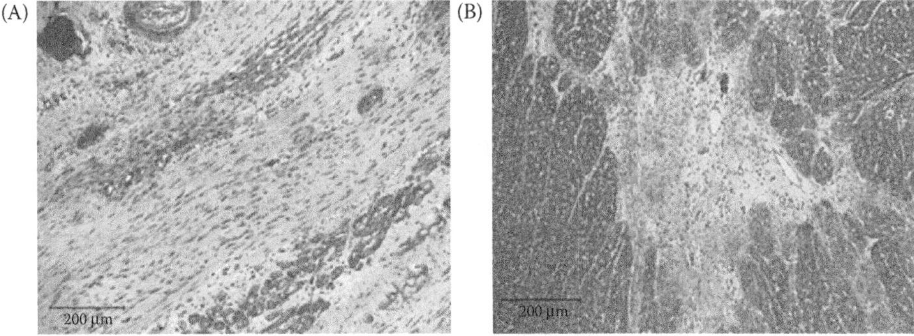

FIGURE 7.3 (see color insert) Trichrome staining of adult (A) and fetal (B) myocardium demonstrates increased collagen deposition and scar formation in the adult compared to the fetus 1 month following myocardial infarction.

and vitronectins. Although these proteins associate with the ECM, they do not perform a structural role within these tissues [65]. The matricellular protein osteopontin has been shown to have both pro- and anti-inflammatory properties during different phases of wound repair. Mice unable to express osteopontin have diminished macrophage recruitment to excisional dermal wounds and kidney injuries. However, in the later stages of wound healing, osteopontin-deficient mice have extensive fibrosis and calcification, as they are unable to suppress inflammation [66].

During the progression of wound healing, the ECM composition, matricellular protein concentration, and cell populations vary greatly. This variation is thought to explain the differences in osteopontin activity [67]. Matricellular proteins from the same family can also have opposing effects on inflammation. Galectin-3 null mice exhibit decreased infiltration and activation of inflammatory cells, which suggests a pro-inflammatory role. However, exogenous galectin-1 prevents mast cell degranulation and NF-kB activation, thereby limiting the inflammatory response [68].

Matricellular proteins provide attractive therapeutic targets. Their expression is tightly regulated in response to wounding in the adult; they are specific, can be applied topically, and degrade naturally. However, the variability of protein function with changes in ECM and cell population significantly complicates therapeutic strategies at this point [65].

Collagen is the most abundant structural protein in the ECM. Persistent fibroblast stimulation results in the abnormal accumulation, organization, and orientation characteristic of scar formation or fibrosis. Following myocardial infarction, the adult heart demonstrates a prolific inflammatory cell infiltrate (Figure 7.1A), which is associated with scar formation and a progressive decline in function [69]. This decline in function is associated with increased abnormal collagen deposition as demonstrated in Figure 7.3A. In contrast, the fetal regenerative response to infarction with restoration of cardiac function is associated with resolution of the inflammatory cell response and minimal abnormal collagen deposition (Figure 7.3B). The development of therapeutic strategies to modulate inflammation and decrease persistent fibroblast stimulation may hold tremendous promise in preventing fibrosis following injury.

7.4 CONCLUSIONS AND FUTURE DIRECTIONS

The wound healing phenotype is determined by the complex interactions between the composition, cellularity, and organization of the wound. While certain components of the inflammatory response appear essential in adult wound healing, there is ample evidence that inflammation also impedes it. Currently, proper wound healing requires a balance between the destructive and constructive properties of the inflammatory response. While the ultimate goal of regenerating injured tissue in the adult is a lofty goal, modulation of the inflammatory response may tip the scales in favor of a more functional healing phenotype.

While the results of many of these studies are promising, one must remember several caveats. First, many of the experiments in which a single gene was knocked out were performed in mice. The effect of eliminating this gene during human wound healing remains unknown. Second, significant redundancy exists in the function of cells, cytokines, and other biologically active molecules. Therefore elimination of a single component of inflammation may not become clinically relevant as other pathways compensate. Finally, the function of an inflammatory mediator on wound healing must be kept in context. During the wound healing response, there is a large flux in the composition of cells, cytokines, and the ECM present in the wound bed. The inflammatory mediator that was associated with negative sequelae at one phase of wound healing may be beneficial to another.

Much of the work in regenerative medicine has been done in lower organisms such as the newt or zebra fish, or during the fetal period when the organism is developing. One potentially confounding variable is the increased numbers of tissue specific stem cells that may be present during this developmental period, which may aid in local tissue regeneration, whereas these cells are much lower in the adult. This is of particular importance when dealing with relatively acellular tissues such as the adult tendon.

Increased numbers of stem cells in the fetus may also play a role in modulating the inflammatory response to injury. Regeneration of local tissue from the resident stem cell populations may halt or diminish the systemic inflammatory signaling and decrease the amplification of the inflammatory response. Conversely, decreased inflammation in the fetus may help to create an environment conducive to stem cell proliferation, migration, and tissue regeneration.

Inflammation remains a significant and confounding factor in the field of regenerative medicine. The term itself is very broad and encompasses multiple processes that can alter the balance between fibrosis and regeneration following injury. There appears to be a threshold of inflammation that, once exceeded, results in fibrosis or scar formation. The development of a better understanding of the interaction of the multiple components involved in the inflammatory response to injury is needed to develop potential therapeutic targets to promote regeneration. However, this will likely require multimodal therapy targeting cytokines, inflammatory cells, resident stem cell populations, and the systemic inflammatory response.

TAKE-HOME MESSAGES

- The wound healing phenotype is determined by the complex interactions between the composition, cellularity, and organization of the wound.
- Adult mammalian wounds heal with exuberant inflammatory reaction and scar formation.
- Scar tissue is structurally, aesthetically, and functionally inferior to unwounded tissue.
- Scarless dermal wound healing in the fetus is associated with minimal inflammation and restoration of tissue function.
- Modulation of the inflammatory response may reduce scar formation and its associated morbidity.
- Some components of inflammation appear necessary for proper wound repair in adults.
- The wound microenvironment is dynamic; rapid changes in protein expression, matrix composition, and cell population can cause anti-inflammatory components to act as pro-inflammatory mediators.
- Redundant functions of cells and inflammatory mediators create obstacles to therapeutic modulation of specific aspects of the inflammatory response.

REFERENCES

1. Rowlatt U, Intrauterine wound healing in a 20 week human fetus. *Virchows Arch A Pathol Anat Histol*, 1979. 381(3): p. 353–61.

2. Beredjiklian PK, Favata M, Cartmell JS, Flanagan CL, Crombleholme TM, et al., Regenerative versus reparative healing in tendon: A study of biomechanical and histological properties in fetal sheep. *Ann Biomed Eng*, 2003. 31(10): p. 1143–52.
3. Herdrich BJ, Danzer E, Davey MG, Allukian M, Englefield V, et al., Regenerative healing following foetal myocardial infarction. *Eur J Cardiothorac Surg,* 38(6): p. 691–8.
4. Liechty KW, Adzick NS, Crombleholme TM, Diminished interleukin 6 (IL-6) production during scarless human fetal wound repair. *Cytokine,* 2000. 12(6): p. 671–6.
5. Mast BA, Krummel TM, Acute inflammation if fetal wound healing, in *Fetal Wound Healing*, eds. N.S. Adzick, Longaker MT, Editor. 1992, Elsevier: New York. p. 227–40.
6. Simpson DM, Ross R, The neutrophilic leukocyte in wound repair a study with antineutrophil serum. *J Clin Invest,* 1972. 51(8): p. 2009–23.
7. Leibovich SJ, Ross R, The role of the macrophage in wound repair. A study with hydrocortisone and antimacrophage serum. *Am J Pathol,* 1975. 78(1): p. 71–100.
8. Roberts AB, Russo A, Felici A, Flanders KC, Smad3: A key player in pathogenetic mechanisms dependent on TGF-beta. *Ann N Y Acad Sci,* 2003. 995: p. 1–10.
9. Peranteau WH, Zhang L, Muvarak N, Badillo AT, Radu A, et al., IL-10 overexpression decreases inflammatory mediators and promotes regenerative healing in an adult model of scar formation. *J Invest Dermatol,* 2008. 128(7): p. 1852–60.
10. Martin P, D'Souza D, Martin J, Grose R, Cooper L, et al., Wound healing in the PU.1 null mouse—tissue repair is not dependent on inflammatory cells. *Curr Biol,* 2003. 13(13): p. 1122–8.
11. Liechty KW, Kim HB, Adzick NS, Crombleholme TM, Fetal wound repair results in scar formation in interleukin-10-deficient mice in a syngeneic murine model of scarless fetal wound repair. *J Pediatr Surg,* 2000. 35(6): p. 866–72; discussion 872–3.
12. Frantz FW, Bettinger DA, Haynes JH, Johnson DE, Harvey KM, et al., Biology of fetal repair: The presence of bacteria in fetal wounds induces an adult-like healing response. *J Pediatr Surg,* 1993. 28(3): p. 428–33; discussion 433–4.
13. Kumta S, Ritz M, Hurley JV, Crowe D, Romeo R, et al., Acute inflammation in foetal and adult sheep: The response to subcutaneous injection of turpentine and carrageenan. *Br J Plast Surg,* 1994. 47(5): p. 360–8.
14. Herdrich BJ, Danzer E, Davey MG, Bermudez DM, Radu A, et al., Fetal tendon wound size modulates wound gene expression and subsequent wound phenotype. *Wound Repair Regen.* 18(5): p. 543–9.
15. Mast B, The skin, in *Wound Healing; Biochemical and Clinical Aspects,* eds. IK Cohen, RF Diegelmann, and WJ Lindblad, Chap 22. 1992, Philadelphia: Saunders. 344–55.
16. Klinger MH, Platelets and inflammation. *Anat Embryol (Berl),* 1997. 196(1): p. 1–11.
17. Deuel TF, Senior RM, Huang JS, Griffin GL, Chemotaxis of monocytes and neutrophils to platelet-derived growth factor. *J Clin Invest,* 1982. 69(4): p. 1046–9.
18. Krummel TM, Michna BA, Thomas BL, Sporn MB, Nelson JM, et al., Transforming growth factor beta (TGF-beta) induces fibrosis in a fetal wound model. *J Pediatr Surg,* 1988. 23(7): p. 647–52.
19. Haynes JH, Johnson DE, Mast BA, Diegelmann RF, Salzberg DA, et al., Platelet-derived growth factor induces fetal wound fibrosis. *J Pediatr Surg,* 1994. 29(11): p. 1405–8.
20. Olutoye OO, Alaish SM, Carr ME, Jr., Paik M, Yager DR, et al., Aggregatory characteristics and expression of the collagen adhesion receptor in fetal porcine platelets. *J Pediatr Surg,* 1995. 30(12): p. 1649–53.
21. Knighton DR, Ciresi KF, Fiegel VD, Austin LL, Butler EL, Classification and treatment of chronic non-healing wounds. Successful treatment with autologous platelet-derived wound healing factors (PDWHF). *Ann Surg,* 1986. 204(3): p. 322–30.
22. Mustoe TA, Pierce GF, Thomason A, Gramates P, Sporn MB, et al., Accelerated healing of incisional wounds in rats induced by transforming growth factor-beta. *Science,* 1987. 237(4820): p. 1333–6.
23. Pierce GF, Mustoe TA, Senior RM, Reed J, Griffin GL, et al., In vivo incisional wound healing augmented by platelet-derived growth factor and recombinant c-sis gene homodimeric proteins. *J Exp Med,* 1988. 167(3): p. 974–87.
24. Steed DL, Clinical evaluation of recombinant human platelet-derived growth factor for the treatment of lower extremity diabetic ulcers. Diabetic Ulcer Study Group. *J Vasc Surg,* 1995. 21(1): p. 71–8; discussion 79–81.
25. Heldin CH, Westermark B, Wasteson A, Platelet-derived growth factor. Isolation by a large-scale procedure and analysis of subunit composition. *Biochem J,* 1981. 193(3): p. 907–13.
26. Assoian RK, Komoriya A, Meyers CA, Miller DM, Sporn MB, Transforming growth factor-beta in human platelets. Identification of a major storage site, purification, and characterization. *J Biol Chem,* 1983. 258(11): p. 7155–60.

27. Szpaderska AM, Egozi EI, Gamelli RL, DiPietro LA, The effect of thrombocytopenia on dermal wound healing. *J Invest Dermatol, 2003.* 120: p. 1130–7.
28. Egozi EI, Ferreira AM, Burns AL, Gamelli RL, Dipietro LA, Mast cells modulate the inflammatory but not the proliferative response in healing wounds. *Wound Repair Regen,* 2003. 11(1): p. 46–54.
29. Rennekampff HO, Hansbrough JF, Kiessig V, Dore C, Sticherling M, et al., Bioactive interleukin-8 is expressed in wounds and enhances wound healing. *J Surg Res,* 2000. 93(1): p. 41–54.
30. McCourt M, Wang JH, Sookhai S, Redmond HP, Proinflammatory mediators stimulate neutrophil-directed angiogenesis. *Arch Surg,* 1999. 134(12): p. 1325–31; discussion 1331–2.
31. Brinkmann V, Reichard U, Goosmann C, Fauler B, Uhlemann Y, et al., Neutrophil extracellular traps kill bacteria. *Science,* 2004. 303(5663): p. 1532–5.
32. Weiss SJ, Tissue destruction by neutrophils. *N Engl J Med,* 1989. 320(6): p. 365–76.
33. Szpaderska AM, Zuckerman JD, DiPietro LA, Differential injury responses in oral mucosal and cutaneous wounds. *J Dent Res,* 2003. 82(8): p. 621–6.
34. Dovi JV, He LK, DiPietro LA, Accelerated wound closure in neutrophil-depleted mice. *J Leukoc Biol,* 2003. 73(4): p. 448–55.
35. DiPietro LA, Wound healing: The role of the macrophage and other immune cells. *Shock,* 1995. 4(4): p. 233–40.
36. Martin P, Leibovich SJ, Inflammatory cells during wound repair: the good, the bad and the ugly. *Trends Cell Biol,* 2005. 15(11): p. 599–607.
37. Delavary BM, van der Veer WM, van Egmond M, Niessen FB, Beelen RH, Macrophages in skin injury and repair. *Immunobiology.* 216(7): p. 753–62.
38. Mosser DM, Edwards JP, Exploring the full spectrum of macrophage activation. *Nat Rev Immunol,* 2008. 8(12): p. 958–69.
39. Mantovani A, Sozzani S, Locati M, Allavena P, Sica A, Macrophage polarization: Tumor-associated macrophages as a paradigm for polarized M2 mononuclear phagocytes. *Trends Immunol,* 2002. 23(11): p. 549–55.
40. Mosser DM, The many faces of macrophage activation. *J Leukoc Biol,* 2003. 73(2): p. 209–12.
41. Daley JM, Brancato SK, Thomay AA, Reichner JS, Albina JE, The phenotype of murine wound macrophages, in *J Leukoc Biol*, p. 59–67.
42. Gordon S, Alternative activation of macrophages. *Nat Rev Immunol,* 2003. 3(1): p. 23–35.
43. Gordon S, Martinez FO, Alternative activation of macrophages: Mechanism and functions. *Immunity,* 2010. 32(5): p. 593–604.
44. MacDonald KP, Palmer JS, Cronau S, Seppanen E, Olver S, et al., An antibody against the colony-stimulating factor 1 receptor depletes the resident subset of monocytes and tissue- and tumor-associated macrophages but does not inhibit inflammation. *Blood,* 116(19): p. 3955–63.
45. Lucas T, Waisman A, Ranjan R, Roes J, Krieg T, et al., Differential roles of macrophages in diverse phases of skin repair. *J Immunol,* 184(7): p. 3964–77.
46. Ashcroft GS, Horan MA, Ferguson MW, Aging alters the inflammatory and endothelial cell adhesion molecule profiles during human cutaneous wound healing. *Lab Invest,* 1998. 78(1): p. 47–58.
47. Pull SL, Doherty JM, Mills JC, Gordon JI, Stappenbeck TS, Activated macrophages are an adaptive element of the colonic epithelial progenitor niche necessary for regenerative responses to injury. *Proc Natl Acad Sci USA,* 2005. 102(1): p. 99–104.
48. Arnold L, Henry A, Poron F, Baba-Amer Y, van Rooijen N, et al., Inflammatory monocytes recruited after skeletal muscle injury switch into antiinflammatory macrophages to support myogenesis. *J Exp Med,* 2007. 204(5): p. 1057–69.
49. Mori R, Kondo T, Ohshima T, Ishida Y, Mukaida N, Accelerated wound healing in tumor necrosis factor receptor p55-deficient mice with reduced leukocyte infiltration. *Faseb J,* 2002. 16(9): p. 963–74.
50. Ishida Y, Kondo T, Kimura A, Matsushima K, Mukaida N, Absence of IL-1 receptor antagonist impaired wound healing along with aberrant NF-kappaB activation and a reciprocal suppression of TGF-beta signal pathway. *J Immunol,* 2006. 176(9): p. 5598–606.
51. Gallucci RM, Simeonova PP, Matheson JM, Kommineni C, Guriel JL, et al., Impaired cutaneous wound healing in interleukin-6-deficient and immunosuppressed mice. *Faseb J,* 2000. 14(15): p. 2525–31.
52. Sato M, Sawamura D, Ina S, Yaguchi T, Hanada K, et al., In vivo introduction of the interleukin 6 gene into human keratinocytes: Induction of epidermal proliferation by the fully spliced form of interleukin 6, but not by the alternatively spliced form. *Arch Dermatol Res,* 1999. 291(7-8): p. 400–4.
53. Schiller M, Javelaud D, Mauviel A, TGF-beta-induced SMAD signaling and gene regulation: Consequences for extracellular matrix remodeling and wound healing. *J Dermatol Sci,* 2004. 35(2): p. 83–92.

54. Hsu M, Peled ZM, Chin GS, Liu W, Longaker MT, Ontogeny of expression of transforming growth factor-beta 1 (TGF-beta 1), TGF-beta 3, and TGF-beta receptors I and II in fetal rat fibroblasts and skin. *Plast Reconstr Surg,* 2001. 107(7): p. 1787–94; discussion 1795–6.
55. Shah M, Foreman DM, Ferguson MW, Control of scarring in adult wounds by neutralising antibody to transforming growth factor beta. *Lancet,* 1992. 339(8787): p. 213–4.
56. Chen WY, Abatangelo G, Functions of hyaluronan in wound repair. *Wound Repair Regen,* 1999. 7(2): p. 79–89.
57. Brecht M, Mayer U, Schlosser E, Prehm P, Increased hyaluronate synthesis is required for fibroblast detachment and mitosis. *Biochem J,* 1986. 239(2): p. 445–50.
58. Presti D, Scott JE, Hyaluronan-mediated protective effect against cell damage caused by enzymatically produced hydroxyl (OH.) radicals is dependent on hyaluronan molecular mass. *Cell Biochem Funct,* 1994. 12(4): p. 281–8.
59. Kvam BJ, Fragonas E, Degrassi A, Kvam C, Matulova M, et al., Oxygen-derived free radical (ODFR) action on hyaluronan (HA), on two HA ester derivatives, and on the metabolism of articular chondrocytes. *Exp Cell Res,* 1995. 218(1): p. 79–86.
60. Fukuda K, Takayama M, Ueno M, Oh M, Asada S, et al., Hyaluronic acid inhibits interleukin-1-induced superoxide anion in bovine chondrocytes. *Inflamm Res,* 1997. 46(3): p. 114–7.
61. Longaker MT, Chiu ES, Adzick NS, Stern M, Harrison MR, et al., Studies in fetal wound healing. V. A prolonged presence of hyaluronic acid characterizes fetal wound fluid. *Ann Surg,* 1991. 213(4): p. 292–6.
62. Girish KS, Kemparaju K, The magic glue hyaluronan and its eraser hyaluronidase: A biological overview. *Life Sci,* 2007. 80(21): p. 1921–43.
63. Tammi R, Pasonen-Seppanen S, Kolehmainen E, Tammi M, Hyaluronan synthase induction and hyaluronan accumulation in mouse epidermis following skin injury. *J Invest Dermatol,* 2005. 124(5): p. 898–905.
64. Donello JE, Loeb JE, Hope TJ, Woodchuck hepatitis virus contains a tripartite posttranscriptional regulatory element. *J Virol,* 1998. 72(6): p. 5085–92.
65. Bornstein P, Sage EH, Matricellular proteins: Extracellular modulators of cell function. *Curr Opin Cell Biol,* 2002. 14(5): p. 608–16.
66. Denhardt DT, Giachelli CM, Rittling SR, Role of osteopontin in cellular signaling and toxicant injury. *Annu Rev Pharmacol Toxicol,* 2001. 41: p. 723–49.
67. Midwood KS, Williams LV, Schwarzbauer JE, Tissue repair and the dynamics of the extracellular matrix. *Int J Biochem Cell Biol,* 2004. 36(6): p. 1031–7.
68. Rabinovich GA, Rubinstein N, Toscano MA, Role of galectins in inflammatory and immunomodulatory processes. *Biochim Biophys Acta,* 2002. 1572(2-3): p. 274–84.
69. Jiang B, Liao R, The paradoxical role of inflammation in cardiac repair and regeneration. *J Cardiovasc Transl Res,* 3(4): p. 410–6.
70. Eming SA, Werner S, Bugnon P, Wickenhauser C, Siewe L, et al., Accelerated wound closure in mice deficient for interleukin-10. *Am J Pathol,* 2007. 170: p 188–202.
71. Low QE, Drugea IA, Duffner LA, Quinn DG, Cook DN, et al., Wound healing in MIP-1alpha(-/-) and MCP-1(-/-) mice. *Am J Pathol,* 2001. 159: p. 457–463.
72. Ashcroft GS, Mills SJ, Lei K, Gibbons L, Jeong MJ, et al., Estrogen modulates cutaneous wound healing by downregulating macrophage migration inhibitory factor. *J Clin Invest,* 2003. 111: p. 1309–18.
73. Ishida Y, Kondo T, Takayasu T, Iwakura Y, Mukaida N, The essential involvement of cross-talk between IFN-gamma and TGF-beta in the skin wound-healing process. *J Immunol,* 2004. 172: p. 1848–55.
74. Ashcroft GS, Yang X, Glick AB, Weinstein M, Letterio JL, et al., Mice lacking Smad3 show accelerated wound healing and an impaired local inflammatory response. *Nat Cell Biol,* 1999. 1: p. 260–66.

8 NOX in the CNS
Inflammation and Beyond

Annadora J. Bruce-Keller
Louisiana State University
Baton Rouge

CONTENTS

8.1 NOX HISTORY AND STRUCTURE

Nicotinamide adenine dinucleotide phosphate (NAPDH) oxidase (NOX) is a membrane-associated, multi-subunit enzyme system that catalyzes the single-electron reduction of molecular oxygen to superoxide. NOX is distinguished by its dedication to the *specific and deliberate* production of reactive oxygen species (ROS), as opposed to other pro-oxidant systems such as mitochondrial electron transport, xanthine oxidases, cyclooxygenases, and monoamine oxidases, which produce free radicals as secondary by-products. NOX was sequentially purified and described in a series of publications in the 1970s and 1980s in which NOX was identified as the main component of the lymphocytic oxidative burst [1]. These data built upon earlier reports of a "respiratory burst" [2] describing how neutrophils dramatically increased oxygen uptake when phagocytosing bacteria, and indeed, subsequent studies revealed the necessity for oxygen in the killing of engulfed microbes [3]. However, NOX activation does not reflect true cellular respiration, and the phrase "extra respiration of phagocytosis" is actually incorrect, as NOX-based oxygen consumption is preserved when mitochondria are poisoned.

The first identified and most widely studied NOX complex is expressed in phagocytic lymphocytes such as neutrophils, eosinophils, and macrophages [1, 4]. This canonical NOX consists of individual membrane (gp91phox and p22phox) and cytosolic (p47phox, p67phox, and p40phox) components. The membrane components gp91phox and p22phox form a stable heterodimer referred to as flavocytochrome b_{558}. Likewise, in resting cells, p47phox, p40phox, and p67phox exist in a stoichiometric, cytosolic complex that is stabilized by intramolecular, autoinhibitory SH3 domain interactions within p47phox [5]. Upon activation, p47phox becomes heavily phosphorylated, and the entire cytosolic complex migrates to the membrane where the subunits assemble to form the active oxidase. Phosphorylation of p47phox occurs primarily in the polybasic domain and C terminus, and triggers conformational changes that unmask the SH3 domains, which then associate with C-terminal, proline-rich regions of p22phox to stabilize the active complex at the membrane [5]. The fully assembled complex also contains the small GTPase Rac, which resides in the cytoplasm prior to cell stimulation. The protein gp91phox is the catalytic core of the enzyme responsible for the electron transfer from NADPH to molecular oxygen. Glucose-6-phosphate dehydrogenase is a major source of the substrate NADPH within the cytosol, and electrons from NADPH are transferred across the membranes down an electrochemical gradient via intermediate flavin adenine dinucleotide (FAD) and heme prosthetic groups, and then bound to oxygen in the extracellular space or in the lumen of intracellular organelles. Superoxide anion is generally thought to be the primary product of the electron transfer, but other downstream ROS, in particular, hydrogen peroxide (H_2O_2), are also thought to be generated (reviewed in [6]).

Recent expansion of genome databases has led to identification of several homologues of the catalytic subunit gp91phox. These proteins constitute the NOX family of oxidases, and have been identified in fungi, plants, fruit flies, nematodes, sea urchins, and in multiple organs in higher animals [7]. The human genome contains 5 NOX members: NOX1 through NOX5, with gp91phox as NOX2. These electron-transporting molecules all have a structure similar to that described for gp91phox, and conserved structural properties of all NOX family members include a C-terminus NADPH-binding site with proximal FAD-binding regions, six conserved transmembrane domains, and four highly conserved heme-binding histidines in the third and fifth transmembrane domains. In addition to the 5 NOX family members, longer homologs of gp91phox with "DUAL functions" have been identified as DUOX 1 and 2 [8], so-named because they possess an extracellular peroxidase-like domain in addition to their C-terminal NOX-like portions.

The potential role of NOX proteins in central nervous system (CNS) physiology and pathophysiology has received considerable attention over the past several years (reviewed in [9]). The nervous system accounts for more than 20% of the oxygen consumed by the body, and as a result produces large quantities of ROS. Additionally, the nervous system is particularly sensitive to oxidative stress because of enrichment of polyunsaturated fatty acids in many of the membranes. In total brain mRNA, NOX2 is easily detected, while the presence of NOX1 and NOX4 transcripts have also been reported [9]. NOX2 and NOX4 expression in the brain have been detected by immunohistochemistry and in situ hybridization, with evidence pointing to expression in microglia, astrocytes, and neurons [7, 9, 10]. Further evidence of the expression of NOX isoforms in specific cell types of the CNS comes from studies on primary cultures, in which it appears that NOX1, NOX2, and NOX4 are present in neurons, astrocytes, and microglia, whereas little is known of the potential localization and function of NOX3 and NOX5 in the CNS [9]. Likewise, little is known about the localization/function of DUOX1 and 2 in the CNS. Indeed, the relative amount of different NOX enzymes and their functional activity in different brain cells has not been systematically or comparatively studied. However, the widespread expression of these NOX subunits has led to the recognition that deliberate ROS production by NOX plays an important role in biologic signaling events in addition to established roles in host defense. Interestingly, the detrimental actions of NOX appear to be most strongly associated with age-related chronic diseases, including Alzheimer's disease, Parkinson's disease, atherosclerosis, hypertension, and different kinds of stroke. Based on these observations, an elegant concept termed "antagonistic pleiotropy" has been proposed to explain the dual role of NOX in the brain, describing a scenario in which the physiologic production of ROS garners an advantage in early life, but the sustained or aberrant activation of NOX results in harmful effects later in life [11].

8.2 CELL TYPE SPECIFIC EXPRESSION AND FUNCTION IN BRAIN

8.2.1 Microglia

NOX proteins are found in all phagocytes, including microglia and macrophages, in addition to neutrophils (PMN), eosinophils, and monocytes. Published data show that NOX2 is very highly expressed in microglia at both the protein and mRNA levels, while NOX1 and NOX4 mRNA can also be detected [9]. The importance of phagocytic NOX to host immunity is clearly demonstrated by the genetic disorder chronic granulomatous disease (CGD), which is caused by any of several genetic defects in essential NOX components, resulting in an inactive oxidase [12]. Patients with CGD experience severe, recurrent bacterial and fungal infections and often develop granulomas formed by the fusion of monocytes and macrophages.

In brain, microglia assume sentinel positions and are the resident macrophages of the CNS. Several physiologic processes involved in the activation of microglia are regulated by NOX-dependent ROS production, including proliferation, cytokine release, and phagocytosis, and indeed, most reports of NOX-induced brain injury relate to microglial NOX (reviewed in [13]). Activation of NOX is a characteristic feature of microglial activation both *in vitro* and *in vivo*, and experimental evidence suggests that ROS generated by activated microglia could directly contribute to brain injury by inducing lipid peroxidation, DNA fragmentation, and protein oxidation in surrounding cells—a phenomenon called "bystander lysis" [14]. In particular, when activation of NOX is combined with generation of nitric oxide, the resulting peroxynitrite production may cause extensive neuronal death, and such excessive free radical generation by microglia might contribute to neuronal damage after stroke or in neurodegenerative diseases (reviewed in [13]).

In addition to direct, free radical-mediated injury, NOX has been shown to drive intracellular inflammatory signaling and the promulgation of the inflammatory cascade in the brain, contributing to local concentrations of neurotoxic inflammatory mediators. For instance, NOX activity has been specifically implicated in activation of NFκB and synthesis of TNFα [15, 16] which may be important mediators of inflammatory neuronal damage. Likewise, NOX-driven oxidative burst activity has also been shown to be critical for microglial release of glutamate [17, 18], suggesting that the excitotoxic neuropathology associated with many neurodegenerative diseases might originate in part with NOX-induced activation of microglial glutamate release. NOX-based ROS production in activated microglia is generally associated with NOX2, but roles for NOX1 and NOX4 cannot yet be ruled out.

Interestingly, while NOX activity has been repeatedly associated with pro-inflammatory actions, compelling support for an anti-inflammatory role of NOX can be found in several recent publications. Indeed, the phrase "chronic granulomatous disease" was coined to refer to the hyperinflammation observed in the absence of NOX2, as chronic granulomas in CGD patients are thought to be sterile complications [19]. Furthermore, patients with CGD suffer from a variety of inflammatory conditions, including inflammatory bowel disease, systemic lupus erythematosus, chorioretinitis, and obstructive inflammatory lesions of the esophagus, gastrointestinal tract, and urinary tract [20, 21], which in some instances may be the first clinical manifestations of CGD. This is also observed in CGD mice, in which a sterile hyperinflammation can be caused by injection of sterilized *Aspergillus fumigatus* extracts [22]. Hyperinflammation in NOX-deficient mice has also been observed in models of *Helicobacter* gastritis, influenza, arthritis, demyelinating disease, and even sunburn [23, 24, 25]. Finally, genetic evaluations suggest that mutations in p47phox might underlie the inflammatory phenotype observed in mice and rats genetically prone to arthritis [26]. Thus, there is significant evidence for an important role of NOX in limiting aberrant or excessive inflammatory responses. The underlying molecular mechanisms await further studies, but illustrate again the dynamic and complex role of ROS in intra- and inter-cellular signaling (see Section 8.3).

8.2.2 Astrocytes

Astrocytes are large and ramified glial cells that not only provide scaffolding and nutrient support for neurons and regulate their activity but also participate in brain inflammation (reviewed in [27]). Although NOX expression was first described in phagocytes, it is now well established that several NOX isoforms are expressed in astrocytes, and published data show that NOX1 and NOX2 are expressed in astrocytes at both the protein and mRNA levels, while NOX4 mRNA has also been detected [9]. NOX activity in astrocytes is activated by PKC and changes in intracellular calcium, produces detectable ROS, and is upregulated in models of astrocyte activation [28]. In spite of these data, the role that NOX plays in astrocyte physiology is not well understood. There is evidence that NOX plays a role in astrocyte survival [29], although there is some controversy in this regard [30]. Most available data support a scenario in which the production of ROS by astroglial NOX contributes to neurotoxic inflammatory processes in the brain. For example, in a manner similar to that observed in microglia, NOX activity in astrocytes regulates NFκB activation and the resultant expression of pro-inflammatory genes, including TNF, COX-2, iNOS, and secreted forms of PLA2 [31, 32]. Likewise, other studies show that amyloid peptides and cytokines can trigger a neurotoxic phenotype in glial cells through NOX activation [33]. In addition to direct toxicity, astrocytic NOX appears to facilitate the progression of immune cell infiltration into the brain. For example, NOX has been implicated in the upregulation of astroglial adhesion molecules and chemokines in models of HIV encephalitis [34, 35].

8.2.3 Oligodendrocytes

Oligodendrocytes function primarily to physically support axons and to produce the insulating myelin sheath. Neither NOX expression nor activity has been documented in oligodendrocytes, although it is well known that these cells are quite sensitive to oxidative damage [36], and thus may be damaged by excessive NOX activity in neighboring cells. Indeed, NOX activity has been implicated in demyelinating diseases like multiple sclerosis (MS) [37]. However, evidence also supports a role for NOX in limiting demyelination and brain inflammation in MS [38], reflecting again the dichotomous role of NOX in inflammation.

8.2.4 Neurons

While neuronal NOX expression was originally thought improbable in light of the absence of dedicated immune function in neurons and the high neuronal susceptibility to oxidative damage, it is now well known that neurons express NOX2 as well as NOX4 and NOX1 [9, 10, 39]. This widespread pattern of NOX expression has led to the recognition that NOX participates in neuronal signaling [39] in addition to host defense. While the exact roles of NOX in neuronal function are under investigation, three general functions of neuronal NOX have been proposed: neuronal differentiation and development, modulation of neuronal activity, and oxidative neuronal injury.

8.2.4.1 Neuronal Differentiation

Evidence that ROS might be involved in neuronal differentiation and development is found in studies showing NOX involvement in growth factor signaling and neuronal differentiation of PC12 cells [40, 41]. Furthermore, data indicate that NOX might regulate selected aspects of neurite outgrowth [40, 42].

8.2.4.2 Synaptic Physiology

There is substantial evidence that ROS in general are important signaling molecules involved in synaptic plasticity, and that NOX specifically is a key regulator of both plasticity and memory formation (reviewed in [43]). Indeed, all elements of the NOX are found in hippocampal neurons [39], and NOX2 appears involved in NMDA receptor signaling [44]. Furthermore, published reports have documented cognitive dysfunction in human CGD patients [45], as well as in mice deficient in either gp91phox or p47phox [46]. The degree of learning and memory impairment in these mice is mild, however, suggesting that NOX plays only a modulatory role or that various NOX isoforms might have redundant function. Interestingly, some experimental data appear to indicate a delicate balance of ROS required for signaling, with either too little or too much ROS resulting in impairments in long-term potentiation (LTP) and memory [47]. Furthermore, data show that antioxidants can impair LTP in young mice but preserve LTP and memory in aged mice [48, 49], suggesting an age-related shift in the role of ROS in memory.

8.2.4.3 Neuronal Death

Most studies on the role of NOX in neurons have focused on oxidative cell injury (reviewed in [9]). However, in many studies, the role of NOX is inferred from protective action of NOX inhibitors, and NOX inhibitors are notoriously nonspecific (see Section 8.4). Conversely, genetic deletion of NOX2 was recently shown to attenuate neurovascular dysfunction and cognitive decline in transgenic mice overexpressing the Swedish mutation of the human amyloid precursor protein [50]. Given that neurons also express other NOX isoforms, and indeed NOX4 has been shown to be upregulated in mouse models of stroke [51] and Alzheimer's disease [52], the role of alternate NOX isoforms in neuronal death cannot be excluded.

8.2.5 Cerebrovascular Cells

While cerebrovascular cells are not specifically "brain resident," the involvement of NOX enzymes in vascular physiology and pathophysiology has historically attracted enormous attention, with literally thousands of papers published. Many excellent reviews are available and detail roles for vascular oxidases in vascular tone and in pathological conditions related to endothelial dysfunction, arterial remodeling, and vascular reactivity/inflammation (reviewed in [53]).

8.3 PHYSIOLOGIC MECHANISMS OF NOX SIGNALING

8.3.1 Oxidative Burst: Direct Antimicrobial Activity in Immune Cells

The infection rate in CGD patients illustrates the important role of NOX in host defense, and ROS engage bacteria in the isolated niches of cellular phagosomes. Although superoxide is the species most directly produced by NOX enzymes, it is not clear whether superoxide itself is directly involved in killing of microorganisms as superoxide is a rather "weak" ROS. However, formation of the highly reactive protonated form of superoxide is enhanced in nonpolar environments close to cell membranes or at low pH, and thus superoxide could participate in antimicrobial activity under these conditions. Once generated, superoxide dismutates into H_2O_2, either spontaneously or facilitated by superoxide dismutase. Therefore, bacterial killing by H_2O_2 derived from superoxide may be a mechanism of NOX anti-microbial activity. Indeed, the toxicity of H_2O_2 is well known, as it reacts quickly with a wide range of biologically important compounds and produces derivatives that are usually far more reactive than the parent compounds. A number of peroxidases exist in the body, which differ in terms of structure, synthesis, and localization, but they have one thing in common: they dramatically increase the rate of H_2O_2-dependent reactions. For example, the combined effect of H_2O_2 and myeloperoxidase has been extensively studied, and it is clear that this system is powerful in killing bacteria and in neutralizing bacterial pathogenicity (reviewed in [54]). However, while myeloperoxidase-deficient individuals are at an increased risk of infection, most are healthy, unlike the more severe immunosuppression found in CGD patients. Thus, while myeloperoxidase likely amplifies NOX2-dependent killing mechanisms, it appears not to be the only mechanism. Indeed, nitric oxide and other reactive nitrogen species (RNS) have an important role in antimicrobial functions of monocytes and macrophages. For example, data show that RNS are sufficient to control *Leishmania donovani* infection in murine visceral macrophages even in the absence of ROS [55], suggesting that while ROS and RNS likely act together in the early stage of infection to regulate antimicrobial activity, RNS alone may be both necessary and sufficient for eventual control of visceral infection.

8.3.2 Regulation of pH and Ion Concentration

While most data indicate that NOX-based ROS mediate direct killing of microbes (see Section 8.3.1), additional data support a role for NOX-dependent changes of phagosomal pH and ion concentrations in host defense. NOX activity, particularly mediated via NOX2, leads to a rise in phagosomal pH [56], and such alkalinization facilitates optimal neutral protease function [57] and antigen processing [58]. Additionally, as an electron transporter, activation of NOX enzymes leads to a charge build-up that requires compensation, potentially occurring through H^+ channels [59] or K^+ fluxes [60]. Such K^+ fluxes could contribute to bacterial killing through changes in phagosomal osmolarity and activation of cationic proteases [60]. Thus, the contribution of NOX2 to microbial killing lies in both direct ROS effects and indirect effects through modulation of pH and ion homeostasis. Indeed, studies on the relative contribution of direct ROS killing of bacteria versus K^+ flux-dependent killing suggest that at low NOX2 activity K^+ flux is important, while at high NOX2 activity direct ROS-dependent killing is predominant [61].

There is also some evidence that the electrogenic effects of NOX could participate in redox signaling in non-phagocytes. The transfer of electrons across biological membranes is an electrogenic process, which generates currents as charges are separated and moved across the lipid bilayer. The amplitude of the electron current carried by the phagocyte NOX has been calculated as equivalent to an "electronic current" of 16 pA [62], which if not compensated, is expected to depolarize the plasma membrane, and indeed, experiments have shown that NOX activation depolarizes cells to voltages exceeding 0 mV [59]. With specific reference to the CNS, NOX activity involves electron transport across biologic membranes and, hence, causes local depolarization. For each electron transported across the membrane, one H+ ion is left in the cytoplasm. Thus, to avoid H+ accumulation and cytosolic acidification, H+ extrusion occurs throughout proton channels. These mechanisms have been studied on NOX2 in phagocytes. However, they might also be important in neurons, because neuronal activity is dependent on the plasma membrane potential. Expression of voltage-gated proton channels has been reported in neurons from several species (reviewed in [63]). Interestingly, it has been observed that expression of proton channels and NOX2 often occurs in parallel [64, 63]. However, further analysis must be done to better understand the involvement of NOX enzymes in the regulation of membrane potential and H+ fluxes in neurons.

8.3.3 Redox Modulation of Protein Function

While ROS were historically considered ballistic signals lacking specific or dedicated targets, recent data clearly support the theory that ROS can function as discrete second messengers (reviewed in [6]). An important mechanism whereby this function is carried is by the targeted oxidation of specific protein residues, resulting in structural and functional changes in that impact function. NOX produces superoxide and hydrogen peroxide, and these two species both have highly favorable chemical profiles as signaling molecules.

As superoxide has only limited mobility across biologic membranes, it is confined within organelles such as mitochondria, endosomes, and, classically, phagosomes. While this confinement limits the activity of superoxide as a signaling mediator, superoxide is rather versatile chemically as it can act both as an oxidant and as a reductant. The protonated form of superoxide (HO_2•) acts primarily as an oxidant, but only a small fraction of superoxide is protonated at physiologic pH. Alternatively, the charged anionic form superoxide is highly attracted to iron-sulfur (Fe-S) centers within proteins. As these Fe-S centers are stable under multiple oxidation states, they are able to participate in a variety of intracellular redox reactions. One highly studied such reaction is their involvement in the mitochondrial electron-transport chain, and NADH dehydrogenase is one of the biggest multi–Fe-S proteins known. Furthermore, Fe-S proteins are also involved in non–electron-transfer functions, such as substrate binding and catalysis, in which Fe-S clusters can lead to polarization of surrounding groups to function as active sites of enzymes [65]. For example, destabilization of the Fe-S cluster in aconitase by superoxide can inhibit aconitase activity, thereby limiting mitochondrial respiration [66]. Although the exact contribution of superoxide/Fe-S interactions to CNS physiology awaits investigation, multiple proteins contain Fe-S centers and may thus be amenable to modulation by superoxide.

A better understood chemical alteration utilized by ROS to impact signaling without causing cytotoxicity involves reversible thiol oxidation of cysteine (Cys) residues [67], generally mediated by hydrogen peroxide. Hydrogen peroxide is a two-electron oxidant that acts as an electrophile and can react with protein thiol moieties to produce different sulfur oxidation states including disulfides and sulfenic (–SOH) moieties, both of which can alter protein structure and function [68]. However, further oxidation of sulfenic acid by ROS can lead to formation of sulfinic ($-SO_2H$), or sulfonic ($-SO_3H$) acid products that appear to be irreversible modifications [69]. The best-documented targets of this type of Cys modification are protein tyrosine phosphatases (PTPs), whose enzymatic activity is abolished by oxidation of a Cys residue in their active site [68]. All PTPs are characterized by an active-site motif that consists of Cys and arginine (Arg) separated by five residues (I/V-C-XX-G-X-X-R-S/T), where X is any amino acid [70]. The proximity of the basic Arg residue creates a microenvironment that facilitates thiolate formation, and in the presence of H_2O_2, the active-site Cys residue is oxidized to a sulfenic acid intermediate followed by rapid intraprotein conversion to a cysteine sulfenyl-amide, which results in an active-site conformational change that inhibits substrate binding [68]. Indeed, data show that NOX activation results in specific decreases in PTP activity and increased levels of intracellular protein tyrosine phosphorylation [71]. While this pathway has not been well studied in the brain, there is evidence that NOX-based regulation of PTP activity extends beyond phagocytes. For example, insulin-sensitive adipocytes express high levels of NOX4, which appears to enhance insulin signal transduction via the oxidative inhibition of PTP1B [72].

Finally, evidence suggests that protein kinases themselves may be potential direct targets of ROS, as ~80% of known PTKs contain a conserved C-terminal MXXCW motif (where X is any residue), and oxidation of the highly conserved cysteine in this motif can trigger the catalytic activity of the enzyme [73]. Data have shown that ROS produced by UV radiation facilitates the dimerization and activation of the receptor tyrosine kinase Ret though oxidation of C-terminal Cys residues [74]. Given that similar Cys residues are conserved in other protein tyrosine kinases, such as Src, Abl, and Lck, one might speculate that these kinases also are subject to oxidation-induced activation, which could be important in CNS signaling [75, 76].

8.4 CONCLUSIONS AND UNANSWERED QUESTIONS

It is now very clear that NOX, as an integral component of glial and neuronal physiology, has the ability to modulate CNS function and responses to injury. However, while NOX has been repeatedly implicated in CNS disease, neither the extent nor the physiologic mechanisms of NOX involvement in CNS dysfunction are fully established. Indeed, the continuing identification of signaling pathways that are directly influenced by NOX clearly illustrates that ROS production participates in a variety of biological processes, including developmental and differentiation processes, innate immunity, and intracellular signaling responses [9]. Furthermore, the recognition that multiple NOX isoforms are expressed in both neurons and glia combined with observations that NOX activity in neurons plays an important role in synaptic plasticity and memory formation [43, 44, 39] precludes simplistic conclusions and one-dimensional declarations of the role of NOX in brain injury.

Thus, significant effort exists, in both basic and clinical research enterprises, to critically evaluate and understand the mechanisms by which NOX can affect brain inflammation and cognitive function. Genetic deletions or mutations in NOX enzymes that occur naturally in humans and can be generated in animals have helped tremendously in the understanding of NOX biology, as summarized in this review, but many questions remain. A very important issue that cannot be resolved with existing knockout models is the cell type specific role of NOX in the CNS. In this regard, the generation of conditional knockout or ever-expressing systems would be exceptionally helpful to elucidate the role of neuronal NOX from that expressed in glia or cerebrovascular cells. In addition, there is a highly urgent need to develop novel pharmaceutical agents that directly and specifically modulate NOX activity, either through binding to specific protein subunits or by binding in a cell type specific manner. Although existing compounds such as diphenyleneiodonium, apocynin, atorvastatin, and 4-(2-aminoethyl) benzenesulfonyl fluoride have been used to inhibit NOX, these compounds are rather nonspecific. Novel NOX inhibitors are being sought in both synthetic and botanical systems, and novel inhibitors may soon proceed to clinical trials [77]. However, before NOX inhibitors can successfully serve as effective drugs for the treatment of CNS disorders, the complex and pleiotrophic nature of NOX in both brain inflammation and in learning and memory needs to be much better understood.

TAKE-HOME MESSAGES

- NOX proteins are highly expressed in microglia, astroglia, and neurons.
- NOX participates in glial inflammation and innate immune reactions in the brain.
- NOX has poorly understood roles in the resolution of inflammation and the prevention of autoimmunity.
- NOX participates in synaptic plasticity and has pleiotropic roles in learning and memory.
- Reactive oxygen species produced by NOX are directly antimicrobial.
- NOX orchestrates cell behavior through discrete, local changes in intracellular pH and ion concentration.
- NOX modulates intracellular signal transduction pathways through redox modulation of cysteine moieties on key enzymes.

REFERENCES

1. Babior, B.M., The respiratory burst oxidase and the molecular basis of chronic granulomatous disease. *Am. J. Hematol.,* 1991. 37: p. 263–6.
2. Baldridge, C.W. and R.W. Gerard, The extra respiration of phagocytosis. *Am. J. Physiol.,* 1933. 103: p. 235–36.
3. Selvaraj, R.J. and A.J. Sbarra, The role of the phagocyte in host-parasite interactions. VII. Di- and triphosphopyridine nucleotide kinetics during phagocytosis. *Biochim. Biophys. Acta.,* 1967. 141: p. 243–9.
4. Babior, B.M., NADPH oxidase. *Curr. Opin. Immunol.,* 2004. 16: p. 42–7.
5. Ago, T., H. Nunoi, T. Ito, and H. Sumimoto, Mechanism for phosphorylation-induced activation of the phagocyte NADPH oxidase protein p47phox-triple replacement of serines 303, 304, and 328 with aspartates disrupts the SH3 domain-mediated intramolecular interaction in p47phox, thereby activating the oxidase. *J. Biol. Chem.,* 1999. 274: p. 33644–53.

6. Brown, D.I. and K.K. Griendling, Nox proteins in signal transduction. *Free Radic. Biol. Med.,* 2009. 47: p. 1239–53.
7. Lambeth, J.D., NOX enzymes and the biology of reactive oxygen. *Nat. Rev. Immunol.,* 2004. 4: p. 181–189.
8. De Deken, X., D. Wang, M.C. Many, S. Costagliola, F. Libert, G. Vassart, et al., Cloning of two human thyroid cDNAs encoding new members of the NADPH oxidase family. *J. Biol. Chem.,* 2000. 275: p. 23227–33.
9. Sorce, S. and K.H. Krause, NOX enzymes in the central nervous system: From signaling to disease. *Antioxid. Redox Signal.,* 2009. 11: p. 2481–504.
10. Bruce-Keller, A.J., S. Gupta, T.E. Parrino, A.G. Knight, P.J. Ebenezer, A.M. Weidner, et al., NOX activity is increased in mild cognitive impairment. *Antioxid. Redox Signal.,* 2010. 12: p. 1371–82.
11. Lambeth, J.D., Nox enzymes, ROS, and chronic disease: An example of antagonistic pleiotropy. *Free Radic. Biol. Med.,* 2007. 43: p. 332–47.
12. Jurkowska, M., E. Bernatowska, and J. Bal, Genetic and biochemical background of chronic granulomatous disease. *Arch. Immunol. Ther. Exp. (Warsz).* 2004. 52: p. 113–20.
13. Brown, G.C. and J.J. Neher, Inflammatory neurodegeneration and mechanisms of microglial killing of neurons. *Mol. Neurobiol.,* 2010. 41: p. 242–7.
14. McGeer, P.L. and E.G. McGeer, The role of the immune system in neurodegenerative disorders. *Mov. Disord.,* 1997. 12: p. 855–8.
15. Turchan-Cholewo, J., V.M. Dimayuga, S. Gupta, R.M. Gorospe, J.N. Keller, and A.J. Bruce-Keller, NADPH oxidase drives cytokine and neurotoxin release from microglia and macrophages in response to HIV-Tat. *Antioxid. Redox. Signal.,* 2009. 11: p. 193–204.
16. Kaul, N. and H.J. Forman, Activation of NF kappa B by the respiratory burst of macrophages. *Free Radic. Biol. Med.,* 1996. 21: p. 401–5.
17. Gupta, S., A.G. Knight, S. Gupta, P.E. Knapp, K.F. Hauser, J.N. Keller, et al., HIV-Tat elicits microglial glutamate release: Role of NAPDH oxidase and the cystine-glutamate antiporter. *Neurosci. Lett.,* 2010. 485: p. 233–6.
18. Barger, S.W., M.E. Goodwin, M.M. Porter, and M.L. Beggs, Glutamate release from activated microglia requires the oxidative burst and lipid peroxidation. *J. Neurochem.,* 2007. 101: p. 1205–13.
19. Cale, C.M., A.M. Jones, and D. Goldblatt, Follow up of patients with chronic granulomatous disease diagnosed since 1990. *Clin. Exp. Immunol.,* 2000. 120: p. 351–355.
20. Foster, C.B., T. Lehrnbecher, F. Mol, S.M. Steinberg, D.J. Venzon, T.J. Walsh, D. et al., Host defense molecule polymorphisms influence the risk for immune-mediated complications in chronic granulomatous disease. *J. Clin. Invest.,* 1998. 102: p. 2146–55.
21. Winkelstein, J.A., M.C. Marino, R.B.J. Johnston, J. Boyle, J. Curnutte, J.I. Gallin, et al., Chronic granulomatous disease. Report on a national registry of 368 patients. *Medicine,* 2000. 79: p. 155–169.
22. Morgenstern, D.E., M.A. Gifford, L.L. Li, C.M. Doerschuk, and M.C. Dinauer, Absence of respiratory burst in X-linked chronic granulomatous disease mice leads to abnormalities in both host defense and inflammatory response to *Aspergillus fumigatus. J. Exp. Med.,* 1997. 185: p. 207–18.
23. Blanchard, T.G., F. Yu, C.L. Hsieh, and R.W. Redline, Severe inflammation and reduced bacteria load in murine *Helicobacter* infection caused by lack of phagocyte oxidase activity. *J. Infect. Dis.,* 2003. 187: p. 1609–15.
24. Hultqvist, M., P. Olofsson, J. Holmberg, B.T. Backstrom, J. Tordsson, and R. Holmdahl, Enhanced autoimmunity, arthritis, encephalomyelitis in mice with a reduced oxidative burst due to a mutation in the Ncf1 gene. *Proc. Natl. Acad. Sci. U.S.A.,* 2004. 101: p. 12646–51.
25. Komatsu, J., H. Koyama, N. Maeda, and Y. Aratani, Earlier onset of neutrophil-mediated inflammation in the ultraviolet-exposed skin of mice deficient in myeloperoxidase and NADPH oxidase. *Inflamm. Res.,* 2006. 55: p. 200–6.
26. Hultqvist, M. and R. Holmdahl, Ncf1 (p47phox) polymorphism determines oxidative burst and the severity of arthritis in rats and mice. *Cell Immunol.,* 2005. 233: p. 97–101.
27. Sofroniew, M.V. and H.V. Vinters, Astrocytes: Biology and pathology. *Acta Neuropathol.,* 2010. 119: p. 7–35.
28. Abramov, A.Y., J. Jacobson, F. Wientjes, J. Hothersall, L. Canevari, and M.R. Duchen, Expression and modulation of an NADPH oxidase in mammalian astrocytes. *J. Neurosci.,* 2005. 25: p. 9176–84.
29. Liu, Q., J.H. Kang, and R.L. Zheng, NADPH oxidase produces reactive oxygen species and maintains survival of rat astrocytes. *Cell Biochem. Funct.* 2005. 23: p. 93–100.
30. Choi, J.W., C.Y. Shin, B.K. Yoo, M.S. Choi, W.J. Lee, B.H. Han, et al., Glucose deprivation increases hydrogen peroxide level in immunostimulated rat primary astrocytes. *J. Neurosci. Res.,* 2004. 75: p. 722–731.

31. Pawate, S., Q. Shen, F. Fan, and N.R. Bhat, Redox regulation of glial inflammatory response to lipopolysaccharide and interferongamma. *J. Neurosci. Res.*, 2004. 77: p. 540–51.
32. Jensen, M.D., W. Sheng, A. Simonyi, G.S. Johnson, A.Y. Sun, and G.Y. Sun, Involvement of oxidative pathways in cytokine-induced secretory phospholipase A2-IIA in astrocytes. *Neurochem. Int.*, 2009. 55: p. 362–8.
33. Qin, L., Y. Liu, C. Cooper, B. Liu, B. Wilson, and J.S. Hong, Microglia enhance beta-amyloid peptide-induced toxicity in cortical and mesencephalic neurons by producing reactive oxygen species. *J. Neurochem.*, 2002. 83: p. 973–83.
34. Song, H.Y., S.M. Ju, W.Y. Seo, A.R. Goh, J.K. Lee, Y.S. Bae, et al., Nox2-based NADPH oxidase mediates HIV-1 Tat-induced up-regulation of VCAM-1/ICAM-1 and subsequent monocyte adhesion in human astrocytes. *Free Radic. Biol. Med.*, 2010. 50: p. 576–84.
35. Williams, R., H. Yao, F. Peng, Y. Yang, C. Bethel-Brown, and S. Buch, Cooperative induction of CXCL10 involves NADPH oxidase: Implications for HIV dementia. *Glia,* 2010. 58: p. 611–21.
36. Smith, K.J., R. Kapoor, and P.A. Felts, Demyelination: the role of reactive oxygen and nitrogen species. *Brain Pathol.,* 1999. 9: p. 69–92.
37. Vladimirova, O., F.M. Lu, L. Shawver, and B. Kalman, The activation of protein kinase C induces higher production of reactive oxygen species by mononuclear cells in patients with multiple sclerosis than in controls. *Inflamm. Res.,*1999. 48: p. 412–6.
38. Mossberg, N., C. Movitz, K. Hellstrand, T. Bergström, S. Nilsson, and O. Andersen, Oxygen radical production in leukocytes and disease severity in multiple sclerosis. *J. Neuroimmunol.,* 2009. 213: p. 131–4.
39. Tejada-Simon, M.V., F. Serrano, L.E. Villasana, B.I. Kanterewicz, G.Y. Wu, M.T. Quinn, et al., Synaptic localization of a functional NADPH oxidase in the mouse hippocampus. *Mol. Cell Neurosci.,* 2005. 29: p. 97–106.
40. Ibi, M., M. Katsuyama, C. Fan, K. Iwata, T. Nishinaka, T. Yokoyama, et al., NOX1/NADPH oxidase negatively regulates nerve growth factor-induced neurite outgrowth. *Free Radic. Biol. Med.,* 2006. 40: p. 1785–95.
41. Suzukawa, K., K. Miura, J. Mitsushita, J. Resau, K. Hirose, R. Crystal, et al., Nerve growth factor-induced neuronal differentiation requires generation of Rac1-regulated reactive oxygen species. *J. Biol. Chem.,* 2000. 275: p. 13175–8.
42. Munnamalai, V. and D.M. Suter, Reactive oxygen species regulate F-actin dynamics in neuronal growth cones and neurite outgrowth. *J. Neurochem.,* 2009. 108: p. 644–61.
43. Kishida, K.T. and E. Klann, Sources and targets of reactive oxygen species in synaptic plasticity and memory. *Antioxid. Redox Signal.,* 2007. 9: p. 233–44.
44. Kishida, K.T., M. Pao, S.M. Holland, and E. Klann, NADPH oxidase is required for NMDA receptor-dependent activation of ERK in hippocampal area CA1. *J Neurochem.,* 2005. 94: p. 299–306.
45. Pao, M., E.A. Wiggs, M.M. Anastacio, J. Hyun, E.S. DeCarlo, J.T. Miller, et al., Cognitive function in patients with chronic granulomatous disease: A preliminary report. *Psychosomatics,* 2004. 45: p. 230–4.
46. Kishida, K.T., C.A. Hoeffer, D. Hu, M. Pao, S.M. Holland, and E. Klann, Synaptic plasticity deficits and mild memory impairments in mouse models of chronic granulomatous disease. *Mol. Cell Biol.,* 2006. 26: p. 5908–20.
47. Knapp, L.T. and E. Klann, Role of reactive oxygen species in hippocampal long-term potentiation: Contributory or inhibitory? *J. Neurosci. Res.,* 2002. 70: p. 1–7.
48. Hu, D., F. Serrano, T.D. Oury, and E. Klann, Aging-dependent alterations in synaptic plasticity and memory in mice that overexpress extracellular superoxide dismutase. *J. Neurosci.,* 2006. 26: p. 3933–41.
49. Liu, R., I.Y. Liu, X. Bi, R.F. Thompson, S.R. Doctrow, B. Malfroy, et al., Reversal of age-related learning deficits and brain oxidative stress in mice with superoxide dismutase/catalase mimetics. *Proc. Natl. Acad. Sci. U.S.A.,* 2003. 100: p. 8526–31.
50. Park, L., P. Zhou, R. Pitstick, C. Capone, J. Anrather, E.H. Norris, et al., Nox2-derived radicals contribute to neurovascular and behavioral dysfunction in mice overexpressing the amyloid precursor protein. *Proc. Natl. Acad. Sci. U.S.A.,* 2008. 105: p. 1347–52.
51. Vallet, P., Y. Charnay, K. Steger, E. Ogier-Denis, E. Kovari, F. Herrmann, et al., Neuronal expression of the NADPH oxidase NOX4, its regulation in mouse experimental brain ischemia. *Neuroscience* 2005. 132: p. 233–8.
52. Bruce-Keller, A.J., S. Gupta, A.G. Knight, T.L. Beckett, J.M. McMullen, P.R. Davis, et al., Cognitive impairment in humanized APP×PS1 mice is linked to Aβ(1-42) and NOX activation. *Neurobiol. Dis,* 2011. Epub ahead of print.
53. Touyz, R.M., A.M. Briones, M. Sedeek, D. Burger, and A.C. Montezano, NOX isoforms and reactive oxygen species in vascular health. *Mol. Interv.,* 2011. 11: p. 27–35.

54. van der Veen, B.S., M.P. de Winther, and P. Heeringa, Myeloperoxidase: Molecular mechanisms of action and their relevance to human health and disease. *Antioxid. Redox Signal.,* 2009. 11: p. 2899–937.
55. Murray, H.W. and C.F. Nathan, Macrophage microbicidal mechanisms in vivo: Reactive nitrogen versus oxygen intermediates in the killing of intracellular visceral *Leishmania donovani. J. Exp. Med.,* 1999. 189: p. 741–6.
56. Segal, A.W., M. Geisow, R. Garcia, A. Harper, and R. Miller, The respiratory burst of phagocytic cells is associated with a rise in vacuolar pH. *Nature,* 1981. 290: p. 406–9.
57. Segal, A.W., How neutrophils kill microbes. *Annu. Rev. Immunol.,* 2005. 23: p. 197–223.
58. Savina, A., C. Jancic, S. Hugues, P. Guermonprez, P. Vargas, I. Moura, et al., NOX2 controls phagosomal pH to regulate antigen processing during crosspresentation by dendritic cells. *Cell Biochem. Funct.,* 2006. 126: p. 205–18.
59. Henderson, L.M., J.B. Chappell, and O.T. Jones, The superoxide-generating NADPH oxidase of human neutrophils is electrogenic and associated with an H+ channel. *Biochem. J.,* 1987. 246: p. 325–9.
60. Reeves, E.P., H. Lu, H.L. Jacobs, C.G. Messina, S. Bolsover, G. Gabella, et al., Killing activity of neutrophils is mediated through activation of proteases by K+ flux. *Nature,* 2002. 416: p. 291–7.
61. Rada, B.K., M. Geiszt, K. Kaldi, C. Timar, and E. Ligeti, Dual role of phagocytic NADPH oxidase in bacterial killing. *Blood.,* 2004. 104: p. 2947–53.
62. Demaurex, N. and G.L. Petheö, Electron and proton transport by NADPH oxidases. *Philos. Trans. R. Soc. Lond. B. Biol. Sci.,* 2005. 1464: p. 2315–25.
63. DeCoursey, T.E., Voltage-gated proton channels. *Cell. Mol. Life. Sci.,* 2008. 65: p. 2554–73.
64. Henderson, L.M. and R.W. Meech, Evidence that the product of the human X-linked CGD gene, gp91-phox, is a voltage-gated H(+) pathway. *J. Gen. Physiol.,* 1999. 114: p. 771–86.
65. Beinert, H., Iron-sulfur proteins: Ancient structures, still full of surprises. *J. Biol. Inorg. Chem.,* 2000. 5: p. 2–15.
66. Gardner, P.R., I. Raineri, L.B. Epstein, and C.W. White, Superoxide radical and iron modulate aconitase activity in mammalian cells. *J. Biol. Chem.,* 1995. 270: p. 13399–405.
67. Parasassi, T., R. Brunelli, G. Costa, M. De Spirito, E. Krasnowska, T. Lundeberg, et al., Thiol redox transitions in cell signaling: A lesson from N-acetylcysteine. *ScientificWorldJournal,* 2010. 10: p. 1192–202.
68. Denu, J.M. and K.G. Tanner, Specific and reversible inactivation of protein tyrosine phosphatases by hydrogen peroxide: Evidence for a sulfenic acid intermediate and implications for redox regulation. *Biochemistry,* 1998. 37: p. 5633–42.
69. Salmeen, A., J.N. Andersen, M.P. Myers, T.C. Meng, J.A. Hinks, N.K. Tonks, et al., Redox regulation of protein tyrosine phosphatase 1B involves a sulphenyl-amide intermediate. *Nature,* 2003. 423: p. 769–73.
70. Barford, D., Z. Jia, and N.K. Tonks, Protein tyrosine phosphatases take off. *Nat. Struct. Biol.,* 1995. 2: p. 1043–53.
71. Zor, U., E. Ferber, P. Gergely, K. Szucs, V. Dombradi, and R. Goldman, Reactive oxygen species mediate phorbol ester-regulated tyrosine phosphorylation and phospholipase A_2 activation: Potentiation by vanadate. *Biochem. J.,* 1993. 295: p. 879–88.
72. Mahadev, K., H. Motoshima, X. Wu, J.M. Ruddy, R.S. Arnold, G. Cheng, et al., The NAD(P)H oxidase homolog Nox4 modulates insulin-stimulated generation of H2O2 and plays an integral role in insulin signal transduction. *Mol. Cell. Biol.,* 2004. 24: p. 1844–54.
73. Nakashima, I., K. Takeda, Y. Kawamoto, Y. Okuno, M. Kato, and H. Suzuki, Redox control of catalytic activities of membrane-associated protein tyrosine kinases. *Arch. Biochem. Biophys.,* 2005. 434: p. 3–10.
74. Kato, M., T. Iwashita, A.A. Akhand, W. Liu, K. Takeda, K. Takeuchi, et al., Molecular mechanism of activation and superactivation of Ret tyrosine kinases by ultraviolet light irradiation. *Antioxid. Redox Signal.,* 2000. 2: p. 841–9.
75. Kalia, L.V. and M.W. Salter, Interactions between Src family protein tyrosine kinases and PSD-95. *Neuropharmacology,* 2003. 45: p. 720–8.
76. Omri, B., I.P. Crisant, M.C. Marty, F. Alliot, R. Fagard, T. Molina, et al., The Lck tyrosine kinase is expressed in brain neurons. *J. Neurochem.,* 1996. 67: p. 1360–4.
77. Choi, D.K., S. Koppula, M. Choi, and K. Suk, Recent developments in the inhibitors of neuroinflammation and neurodegeneration: Inflammatory oxidative enzymes as a drug target. *Expert Opin. Ther. Pat.,* 2010. 20: p. 1531–46.

9 Resolution of Inflammation

Amitava Das and Sashwati Roy
The Ohio State University
Columbus, Ohio

CONTENTS

9.1 INTRODUCTION

Inflammation is the body's primary response to infection or injury and aims not only to eliminate the causative agent, but also to restore tissue structure and function [1]. Dysregulated inflammatory response leads to non-resolving chronic wounds that represent a major and increasing socioeconomic threat affecting more than 6.5 million people in the United States, costing in excess of $25 billion annually [2, 3]. The onset of inflammation is marked by the release of an array of mediators including cytokines, chemokines, lipid mediators, and bioactive amines that are secreted by resident tissue cells—primarily macrophages, dendritic cells, and mast cells [4]. Inflammation is essentially a beneficial response that normally resolves with the restoration of normal tissue homeostasis. However, when inflammation persists (chronic inflammation), it can cause tissue damage and loss of function [1]. The problem with inflammation is not how often it starts, but how often it fails to resolve. Non-resolving inflammation contributes significantly to the pathogenesis of a wide array of disorders, including atherosclerosis, obesity, cancer, chronic obstructive pulmonary disease, asthma, inflammatory bowel disease, neurodegenerative disease, multiple sclerosis, and rheumatoid arthritis [5]. Chronic wounds fail to progress through the normal phases of healing and enter a state of persistent non-resolving inflammation [6].

9.2 CASCADES OF WOUND HEALING

Wound healing is a well-orchestrated reparative event that occurs in response to injury and its microenvironment [7]. Based on the time of closure, wounds can be classified into two categories: acute wounds, which advance through the process of healing in a stepwise manner and close within days, and chronic wounds that are derailed from the physiological healing cascade and remain open for more than four weeks. The process of wound healing is divided into four functional phases, namely, hemostasis, inflammation, proliferation, and remodeling. All of these phases take place in an overlapping series of programmed events with the aim of barrier function of the skin.

Hemostasis is the physiological response following an injury in which the blood components are exposed to the subendothelial layers of the vessel wall. The process of hemostasis is intended to prevent blood loss and begins with the formation of a fibrin plug that also lays the foundation for subsequent inflammation and healing processes [8]. The fibrin plug and the surrounding wound tissues release pro-inflammatory cytokines and growth factors such as transforming growth factor (TGF-β), platelet-derived growth factor (PDGF), fibroblast growth factor (FGF), and epidermal growth factor (EGF) to the injury site. Fibrin also provides the structural support for the cellular constituents of inflammation.

An acute phase inflammatory response, meant to prepare the wound site for subsequent wound closure, is elicited following tissue injury. Inflammation involves a sequence of responses of vascularized tissues of the body to injury, and is characterized by four cardinal signs: rubor (redness), tumor (swelling), calor (heat), and dolor (pain). During normal healing, the inflammatory response is characterized by spatially and temporally changing patterns of specific leukocyte subsets. Polymorphonuclear neutrophils (PMN) are the first leukocytes to pull in at the inflammatory site, primarily to scavenge invading pathogens and set the inflammatory process on track. Upon completion of their tasks, the PMNs undergo apoptotic cell death provoking dead cell clearance (efferocytosis) by the infiltrating macrophages, which reach the inflammatory site as the level of PMNs falls. Impairment in macrophage function at the wound site derails the resolution of inflammation [9]. Besides carrying out efferocytosis, macrophages release an array of growth, angiogenic, and inflammatory factors that mark the transition of the inflammatory phase to the proliferative phase of wound healing [10].

The proliferative phase usually overlaps with the inflammatory phase and is characterized by the formation of new blood vessels, the influx of fibroblasts, and laying down of the extracellular matrix (reepithelialization). While fibroblasts are limited to cellular replication and migration in the initial stages of the proliferative phase, collagen synthesis takes place in the later half, followed by cross linking of collagen, which is responsible for vascular integrity and mechanical strength of new capillary beds. It is in the later stage that fibroblasts start attaching to the fibronectin and collagen in the extracellular matrix (ECM).

Remodeling represents the concluding phase of wound healing and is well coordinated by the balance between the synthesis and breakdown of the extracellular matrix components. This stage often continues even after months of wound closure and bears an influence on the scar outcomes of the healed wound.

9.3 RESOLUTION OF INFLAMMATION

Inflammation response following an injury is essential for the repair process. This response, however, is only beneficial if it is transient and resolves in a timely manner. Complications with wound healing may arise due to dysregulated inflammation [9]. Resolution of inflammation is regulated by a number of key factors including cessation of further leukocyte recruitment, apoptosis of the leukocytes at the injury site, followed by removal of apoptotic cells by professional phagocytes. Clearance of apoptotic cells by macrophages (also known as efferocytosis) from the site of injury is a prerequisite for the resolution of wound inflammation [9]. Recent studies have provided convincing evidence that uptake of apoptotic cells by macrophages serves as one of the signals that drives the switching of macrophages from a pro- to an anti-inflammatory state [11, 12]. Any impairment in macrophage function will derail the resolution of inflammation, which will adversely affect the wound healing [9]. Lipid mediators, such as the lipoxins, resolvins, protectins, and maresins, have emerged as a novel class of potent pro-resolutionary factors that counter-regulate excessive acute inflammation [13]. Recent studies have indicated an essential role of miRNAs in the immune responses [14].

9.3.1 Macrophage Phenotype

Macrophages are plastic, dynamic, and heterogeneous cells assigned to two groups: classically activated or type I macrophages (M1), which are pro-inflammatory effectors (the prototypical activating stimuli are IFNγ and LPS), and alternatively activated or type II macrophages (M2) [15, 16]. M2 macrophages are further subdivided into M2a (after exposure to IL-4 or IL-13), M2b (immune complexes in combination with IL-1β or LPS), and M2c (IL-10, TGF-β or glucocorticoids). While M1 macrophages possess potent microbicidal properties and support IL-12-mediated Th1 responses, M2 supports Th2-related effector functions [17]. A better understanding of the mechanisms and cues that are implicated in change in macrophage phenotype is essential. Such understanding will help in designing and implementing strategies that can restore macrophage dysfunction and thus promote resolution of inflammation in chronic wounds. In the inflammatory condition, it is unclear whether the type II reparative macrophages that predominate during the healing phase originate from a subset of entirely new attracted monocytes or from a switch in the activation state of the already existing pro-inflammatory macrophages. A study by Porcheray et al. clearly demonstrated that pro-inflammatory macrophages may change its phenotype to pro-resolutionary state [18]. Diabetic wound macrophages display dysfunctional inflammatory responses [9]. The unrelenting inflammatory state of diabetic wound macrophages may be due to impairment in the apoptotic cell clearance activity of these macrophages [9]. Phospholipase C β2 (PLCβ2), the enzyme that catalyzes the formation of inositol 1,4,5-trisphosphate and diacylglycerol from phosphatidylinositol 4,5-bisphosphate, has been shown to play a novel and critical role in switching of macrophages from an inflammatory (M1) phenotype to an angiogenic (M2-like) phenotype, suggesting that regulation of this pathway might provide an additional target for the regulation of inflammation, wound healing, and fibroproliferative processes [19].

The protein known as SHIP (SH2-containing inositol phosphatase) is indicated to be associated with the phenotype of macrophages. Because of a 10-fold higher level of arginase-I in the SHIP−/− macrophage L-arginine, the iNOS substrate is redirected from NO to ornithine production, resulting in 5–10-fold less NO in SHIP−/− macrophages than their wild-type counterparts. This suggests that the chronically elevated PI-3,4,5-P3 levels in SHIP−/− mice may switch NO-producing M1 macrophages to reparative M2 macrophages, which produce ornithine to promote host cell growth and collagen formation [20].

Macrophages with dissimilar phenotypes (M1 and M2) are reported to be present in the cervical region and are most likely associated with the postpartum repair of tissue. However, elevated Csfr1 mRNA expression and expression of other markers of alternatively activated macrophages during labor or shortly postpartum suggest a role of M2 macrophages in postpartum tissue repair [21]. In CNS injury, M1 macrophages are found to be neurotoxic while M2 macrophages are responsible for endorsing a regenerative growth in response in adult sensory axons. These observations suggest that CNS repair could be augmented by preferential polarization of the differentiation of resident microglia and infiltrating blood monocytes toward alternatively activated M2 macrophage phenotype [15]. In the relapsing experimental autoimmune encephalomyelitis (EAE) model of multiple sclerosis, expression profiles of M1/M2 macrophages in brain showed that M1/M2 equilibrium in CNS favors mild EAE, but if the equilibrium shifts more toward M1, relapsing EAE is favored [22]. Recently, characterization of distinct circulating monocyte populations that migrate into wounds was performed. Intriguingly, the phenotype of macrophages isolated from murine wounds partially reflected those of their precursor monocytes, changed with time, and did not conform to current macrophage classifications, that is, M1 or M2 [23].

9.3.2 Dead Cell Clearance

Unique downstream consequences and distinctive morphologic features caused by the phagocytosis of apoptotic cells persuaded deCathelineau and Henson [24] to coin the term *efferocytosis* (from the Latin *effero* meaning to take to the grave or to bury). Efferocytosis consists of at least four distinct steps (Figure 9.1): (1) secreted "find-me" signals help attract phagocytes at the site rich in apoptotic cells; (2) recognition ("eat-me") of apoptotic cells by recruited phagocytes through a number of receptors and molecules; (3) engulfment of apoptotic cells by a distinctive uptake process; and (4) processing of engulfed cells within phagocytes [25].

"Find-me" signals are released by apoptotic cells that attract phagocytes to their location [26]. The supernatants of apoptotic cells possess chemoattractant activity for monocytes and primary human macrophages [26]. Caspase-3-mediated activation of Ca^{2+}-independent phospholipase A2 (iPLA2) in apoptotic cells results in secretion of signals like lysophosphatidylcholine (LPC) [26].

Apoptotic cells present themselves for removal by exposing "eat-me" signals on their surface, which helps the macrophages to recognize and distinguish them from viable cells [26]. Flipping of phosphatidylserine (PS) on the cell surface is a characteristic of apoptosis [27, 28]. Oxidation of PS by NADPH oxidase followed by externalization has been established in a wide range of apoptotic

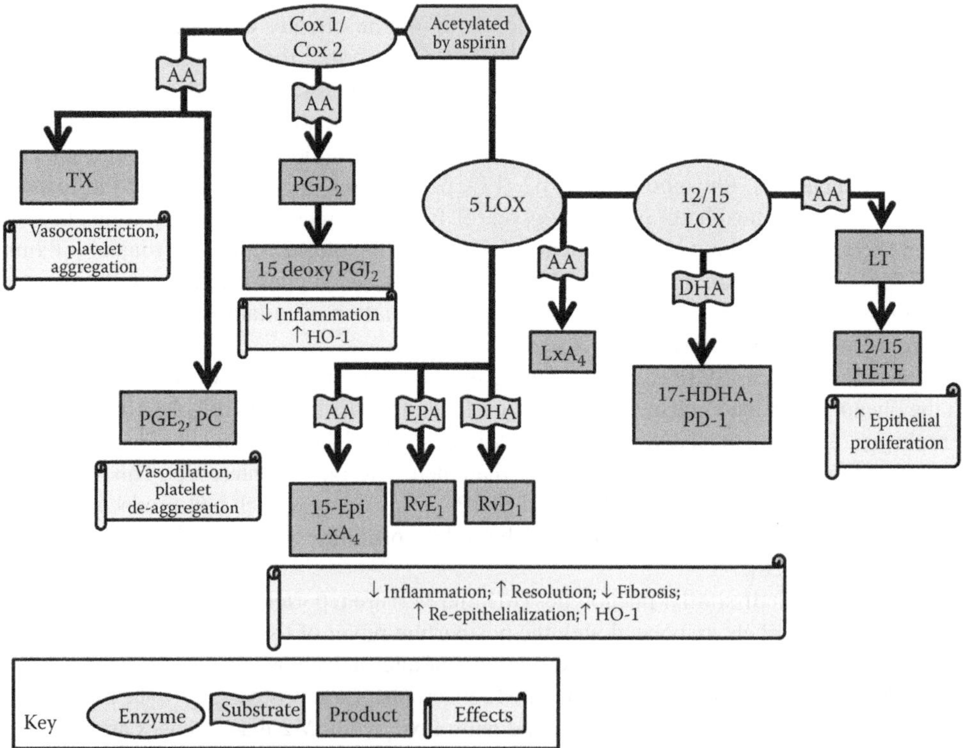

FIGURE 9.1 Resolution of wound inflammation. Polymorphonuclear neutrophils (PMNs) are the first leukocytes to arrive at the site of injury during the early inflammatory phase and are eliminated from the site via apoptotic cell death. Next, an influx of monocytes/macrophages to the site occurs, which effectively performs the clearance of apoptotic PMNs. A number of molecules have been implicated in the recognition and engulfment of apoptotic cells by macrophages. Such engulfment of apoptotic cells results in attenuation of pro-inflammatory and increase in anti-inflammatory molecule production, thus facilitating resolution of inflammation. Macrophages are assigned to two groups: type I macrophages (M1, pro-inflammatory) and type II macrophages (M2, anti-inflammatory). The uptake of apoptotic cells by macrophages has been proposed as one of the cues that drives the switching of macrophages toward an anti-inflammatory state. MFGE8, milk fat globulin E8; PS, phosphatidyl serine.

models [29]. Flipping of oxidized PS along with its non-oxidized counterpart serves as an "eat me" signal [26].

Most receptors present on macrophages do not directly bind to phospholipids, but attach through soluble bridging proteins [30]. These soluble factors opsonize apoptotic cells and establish bridges between a specific component of the apoptotic cells and surface of the phagocyte and includes annexin I (Anx I or lipocortin), TSP, milk-fat-globule-EGF-factor 8 (MFG-E8), Del-1, β2-glycoprotein I, protein S, and growth arrest specific gene 6 (Gas6) [30, 31]. MerTK has been proposed to aid phagocytosis of apoptotic cells and downregulate activation in macrophages [32]. Complements may also be involved in the uptake of apoptotic cells through direct binding of bridging factors in some physiological conditions involving opsonization and engagement of the complement receptors.

Knowledge about the final step of apoptotic cell clearance—engulfment and processing of engulfed cells—is limited. Information related to the regulation of apoptotic body processing, and its difference from the processing of classically opsonized or microbial cells that employ a route of degradation from phagosomes to lysosomes, needs to be augmented.

9.3.3 Lipid Mediators

Stress, injury, or inflammatory stimuli result in the rapid release of polyunsaturated fatty acids (PUFAs). Several studies have demonstrated that the ω-3 PUFAs eicosapentaenoic (EPA; i.e., ω-3, C20:5) and docosahexaenoic acid (DHA; i.e., ω-3, C22:6) are transformed, in a manner similar to arachidonic acid metabolism, by COX-2 and LOX enzymes (Figure 9.2) to generate novel classes of

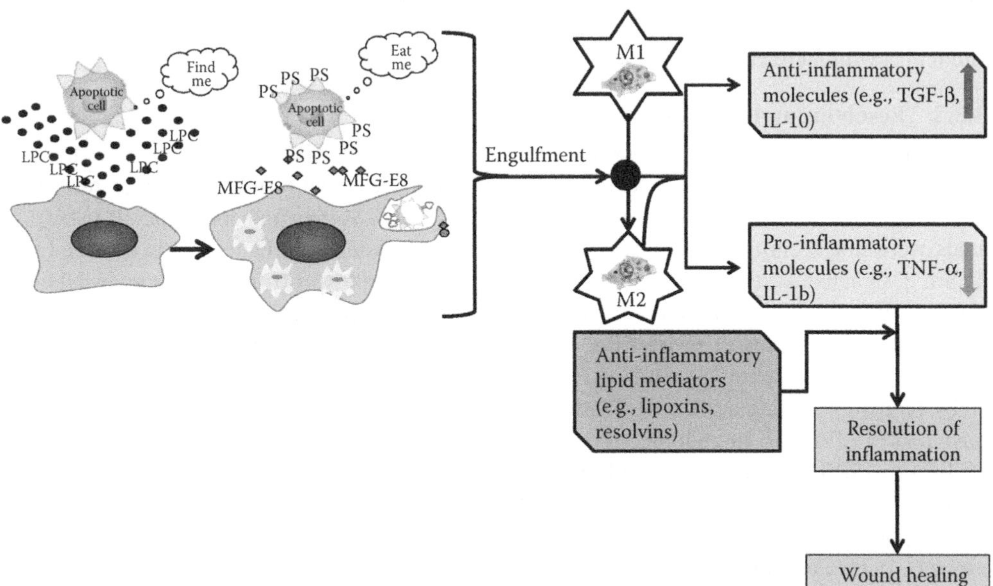

FIGURE 9.2 (see color insert) Lipid biosynthesis pathways that regulate inflammation and resolution. The pro- as well and anti-inflammatory lipid mediators regulate both initiation as well as the resolution phases of wound inflammation. These lipids are synthesized from polyunsaturated fatty acids, such as arachidonic acid (AA), eicosapentaenoic acid (EPA), and docosahexaenoic acid (DHA) by the enzymes cyclooxygenase (Cox) and/or lipoxygenase (LO). AA is converted to pro-inflammatory thromboxanes (TX), prostacyclin (PC), and prostaglandins (PG) via the cyclooxygenase (COX) pathway. In the LO pathway, hydroperoxyeicosatetraenoic acids (HPETEs) are produced that can be further enzymatically reduced to the hydroxylated form (HETE). LO enzymes are also involved in the production of anti-inflammatory lipoxins (LX), resolvins (RvEs and RvDs), and protectins (PD) from AA, EPA, and DHA. Lipoxins and resolvins are generated from AA and EPA/DHA by the enzymatic actions of both LO and acetylated COX-2.

endogenous lipid autacoids with anti-inflammatory and protective activities [10]. Lipidomic studies suggest that DHA is present in relatively copious amounts in murine skin (in full thickness; 300–3000 ng/g skin tissue). Gel-controlled release of DHA and three other essential fatty acids to wounds significantly promoted wound healing, unveiling a therapeutic potential for DHA derivatives in wound healing [33]. In the following section, we provide a brief review of the major lipid autacoids with known anti-inflammatory activity.

9.3.3.1 Cyclopentenone Prostaglandins

Cyclopentenone prostaglandin (15dPGJ2) is biosynthesized by the *in vivo* and *in vitro* dehydration of PGD2 by COX-2. This prostaglandin is reported to inhibit TNF-α-induced expression of vascular cell adhesion molecule 1 (VCAM1) and intercellular adhesion molecule 1 (ICAM1) by human endothelial cells. In addition, 15dPGJ2 also inhibits monocyte binding to human aortic endothelial cells. Though 15dPGJ2 does not influence neutrophil adhesion, it blocks adhesion-dependent oxidative bursts in neutrophils [34].

9.3.3.2 Lipoxins

First isolated in 1984, lipoxins are formed through transcellular biosynthesis during cell-cell interactions. Lipoxins A4 (LXA4) and B4 (LXB4) are generated by the enzyme platelet 12-lipoxygenase on neutrophil leukotrienes A4 (LTA4) [35]. Aspirin treatment results in acetylation of cyclooxygenase-2 [36], leading to the biosynthesis of carbon 15 epimers of LXs (i.e., 15-epi-LXs). It has recently been reported that 15-epi-LXA4 mediates local anti-inflammatory actions of low-dose aspirin in healthy subjects [37]. This class of lipid autocoids is reported to inhibit PMN chemotaxis [38], PMN adhesion to and transmigration through endothelial cells, as well as PMN-regulated increases in vascular permeability [39]. They also draw monocytes and stimulate monocyte adherence to vascular endothelium [40] without releasing reactive oxygen species [41].

9.3.3.3 Resolvins and Protectins

In addition to arachidonic acid, docosahexaenoic acid (DHA) and EPA serve as substrates to potent bioactive mediators that possess anti-inflammatory properties. The term *resolvins* or *resolution-phase interaction products* was coined by professor Charles N. Serhan and colleagues as these compounds were first identified in resolving inflammatory exudates [42]. Compounds obtained from EPA are designated as resolvins of the E series, while those derived from the precursor DHA are denoted as either resolvins of the D series and protectins (neuroprotectins) [43]. Resolvins originate from endogenous sources of DHA/EPA via 15-lipoxygenase (15-LO) and 5-lipoxygenase (5-LO) interactions. E-series member resolvin E1 blocks human neutrophil transendothelial migration, thereby reducing inflammation [42]. Plasma resolvin E1 levels are elevated in individuals taking aspirin and/or EPA [44]. In endothelial cells, aspirin treatment acetylates COX-2, which then converts EPA to 15R HEPE and 18R HEPE, both known to potently inhibit transendothelial migration of PMN [42]. Bioactive members from DHA-containing conjugated triene structures or docosatrienes that possess neuroprotective and immunoregulatory actions are collectively termed as *neuroprotectins*. In studies addressing resolvin formation in brain tissue in response to aspirin treatment, it was shown that new docosatrienes initially termed *neuroprotectins* are produced. In recognition of the fact that the protective actions of these docosanoids are not restricted to neural tissue, it has been suggested that the more generic term *protectins* be used instead [45]. Since this potent chemical mediator has a broad array of activities for nonneuronal local biosynthesis and actions, the name protectinD1 (PD1) was created [46].

9.3.3.4 Maresins

Maresins (14S-HDHA) are a class of anti-inflammatory lipid mediators primarily produced by macrophages through the 14-lipoxygenase pathway [47]. Addition of either DHA or 14S-hydroperoxydocosa-4Z,7Z,10Z,12E,16Z,19Z-hexaenoic acid (14S-HpDHA) to human or murine macrophages converts these substrates to dihydroxy-containing products that have potent anti-inflammatory and pro-resolving activity, which is in the range of the actions of RvE1 and PD1 [47].

9.4 ROLE OF MICRORNAS IN INFLAMMATION

Wound healing is highly dependent on injury-inducible protein-coding genes that act as modulators of an intrinsic tissue repair program in order to restore structural and functional integrity of the injured tissue. Separating a protein coding gene from its corresponding protein involves two fundamental steps: the DNA harboring the gene of interest should transcribe to mRNA, followed by the latter translated to protein. Work carried out in the past decade suggests that both of these vital steps are subject to regulation by microRNAs (miRNAs; 19–22 nucleotides long), which are non-coding RNAs found in all eukaryotic cells. miRNAs carry out post-transcriptional gene silencing through mRNA stabilization as well as translational repression. According to the miRbase database, the human genome encodes 1048 miRNAs, and the count is rapidly growing. These miRNAs may control 30%–50% of all protein-coding genes and virtually all biological processes—wound healing is no exception [48]. A thorough understanding of microRNA (miRNA or miR) biology is critically important to develop a comprehensive understanding of the molecular mechanisms that regulate wound healing [49–53]. Disruption of miRNA biogenesis has a profound impact on the overall immune system. Emerging studies indicate that miRNAs, especially miR-21, miR-146a/b, and miR-155, play an important role in regulating several phases that orchestrate the inflammatory process [54]. Specific miRNAs have been shown to be regulated by resolvins, which eventually target genes involved in resolution of inflammation and establish a novel resolution circuit involving RvD1 receptor-dependent regulation of specific miRNAs [55]. The brain-specific microRNA-124 can control inflammation by turning off activated microglial cells and macrophages [56].

9.5 LINKING MICRORNAS AND LIPID MEDIATORS IN RESOLUTION OF INFLAMMATION

Recent studies conducted in the laboratory of Dr. Charles Serhan have investigated the miRNAs and resolvin-dependent miRNAs that are regulated in a self-limited acute inflammatory response. The pro-resolving miRNA candidates studied included miR-21, miR-146b, miR-208a, and miR-219, which are regulated through GPCR by resolvin D1 (RvD1). RvD1 dampened neutrophil infiltration into the peritoneum 25%–50% induced by zymosan and abridged the resolution interval by 4 h. In peritonitis model, at 12 h time point, RvD1 upregulated miR-21, miR-146b, and miR-219 and downregulated miR-208a in vivo. Low concentrations (10 nM) of RvD1 significantly regulated these miRNAs in same manner in human macrophages, overexpressing recombinant RvD1 receptors ALX/FPR2 or GPR32. RvD1-miRNA networks target cytokines and proteins involved in the immune system. The miRNA–NF-κB axis was suggested as a key component in the RvD1-GPCR downstream signaling pathways [55].

TAKE-HOME MESSAGES

- Chronic wounds fail to progress through the normal phases of healing and enter a state of persistent non-resolving inflammation.
- Resolution of inflammation is an active coordinated process that initiates immediately after the start of the inflammatory response.
- Engulfment of the apoptotic cells by macrophages, that is, efferocytosis, is a prerequisite for resolution of inflammation, resulting in the restoration of normal tissue physiology and function.

- Impaired clearance of apoptotic cells may result in chronic inflammation. Macrophages from diabetic wounds are severely impaired in the apoptotic cell engulfment activity and display characteristics of inflammatory macrophages.
- Macrophages are plastic, dynamic, and heterogeneous cells assigned to two groups: type I macrophages (M1; pro-inflammatory) and type II macrophages (M2; anti-inflammatory).
- While M1 macrophages initiate and amplify the process of inflammation, M2 macrophages are more reparative in function.
- Macrophages derived from mouse wounds changed with time, and did not conform to current macrophage classifications, that is, M1 or M2.
- Metabolism of ω-3 fatty acids generates a novel class of anti-inflammatory lipid mediators known as resolvins and protectins.
- miRNAs, small non-coding RNAs involved in post-transcriptional gene silencing, play a vital role in regulation of inflammatory response.

REFERENCES

1. Rodriguez-Vita J, Lawrence T: The resolution of inflammation and cancer, *Cytokine Growth Factor Rev* 2010, 21:61–65.
2. Singer AJ, Clark RA: Cutaneous wound healing, *N Engl J Med* 1999, 341:738–746.
3. Crovetti G, Martinelli G, Issi M, Barone M, Guizzardi M, Campanati B, et al.: Platelet gel for healing cutaneous chronic wounds, *Transfus Apher Sci* 2004, 30:145–151.
4. Lawrence T, Fong C: The resolution of inflammation: Anti-inflammatory roles for NF-kappaB, *Int J Biochem Cell Biol* 2010, 42:519–523.
5. Nathan C, Ding A: Nonresolving inflammation, *Cell* 2010, 140:871–882.
6. Menke NB, Ward KR, Witten TM, Bonchev DG, Diegelmann RF: Impaired wound healing, *Clin Dermatol* 2007, 25:19–25.
7. Martin P: Wound healing—aiming for perfect skin regeneration, *Science* 1997, 276:75–81.
8. Broughton G, II, Janis JE, Attinger CE: Wound healing: An overview, *Plast Reconstr Surg* 2006, 117:1e-S–32e-S.
9. Khanna S, Biswas S, Shang Y, Collard E, Azad A, Kauh C, et al.: Macrophage dysfunction impairs resolution of inflammation in the wounds of diabetic mice, *PLoS One* 2010, 5:e9539.
10. Gronert K: Lipid autacoids in inflammation and injury responses: A matter of privilege, *Mol Interv* 2008, 8:28–35.
11. Voll RE, Herrmann M, Roth EA, Stach C, Kalden JR, Girkontaite I: Immunosuppressive effects of apoptotic cells, *Nature* 1997, 390:350–351.
12. Fadok VA, Bratton DL, Konowal A, Freed PW, Westcott JY, Henson PM: Macrophages that have ingested apoptotic cells in vitro inhibit proinflammatory cytokine production through autocrine/paracrine mechanisms involving TGF-beta, PGE2, and PAF, *J Clin Invest* 1998, 101:890–898.
13. Spite M, Serhan CN: Novel lipid mediators promote resolution of acute inflammation: Impact of aspirin and statins, *Circ Res* 2010, 107:1170–1184.
14. Asirvatham AJ, Gregorie CJ, Hu Z, Magner WJ, Tomasi TB: MicroRNA targets in immune genes and the Dicer/Argonaute and ARE machinery components, *Mol Immunol* 2008, 45:1995–2006.
15. Kigerl KA, Gensel JC, Ankeny DP, Alexander JK, Donnelly DJ, Popovich PG: Identification of two distinct macrophage subsets with divergent effects causing either neurotoxicity or regeneration in the injured mouse spinal cord, *J Neurosci* 2009, 29:13435–13444.
16. Martinez FO, Helming L, Gordon S: Alternative activation of macrophages: An immunologic functional perspective, *Annu Rev Immunol* 2009, 27:451–483.
17. Laskin DL: Macrophages and inflammatory mediators in chemical toxicity: A battle of forces, *Chem Res Toxicol* 2009, 22:1376–1385.
18. Porcheray F, Viaud S, Rimaniol AC, Leone C, Samah B, Dereuddre-Bosquet N, et al.: Macrophage activation switching: An asset for the resolution of inflammation, *Clin Exp Immunol* 2005, 142:481–489.
19. Grinberg S, Hasko G, Wu D, Leibovich SJ: Suppression of PLCbeta2 by endotoxin plays a role in the adenosine A(2A) receptor-mediated switch of macrophages from an inflammatory to an angiogenic phenotype, *Am J Pathol* 2009, 175:2439–2453.

20. Rauh MJ, Sly LM, Kalesnikoff J, Hughes MR, Cao LP, Lam V, Krystal G: The role of SHIP1 in macrophage programming and activation, *Biochem Soc Trans* 2004, 32:785–788.
21. Timmons BC, Fairhurst AM, Mahendroo MS: Temporal changes in myeloid cells in the cervix during pregnancy and parturition, *J Immunol* 2009, 182:2700–2707.
22. Mikita J, Dubourdieu-Cassagno N, Deloire MS, Vekris A, Biran M, Raffard G, et al.: Altered M1/M2 activation patterns of monocytes in severe relapsing experimental rat model of multiple sclerosis. Amelioration of clinical status by M2 activated monocyte administration, *Mult Scler* 2011, 17:2–15.
23. Brancato SK, Albina JE: Wound macrophages as key regulators of repair: Origin, phenotype, and function, *Am J Pathol* 2011, 178:19–25.
24. deCathelineau AM, Henson PM: The final step in programmed cell death: Phagocytes carry apoptotic cells to the grave, *Essays Biochem* 2003, 39:105–117.
25. Erwig LP, Henson PM: Clearance of apoptotic cells by phagocytes, *Cell Death Differ* 2008, 15:243–250.
26. Grimsley C, Ravichandran KS: Cues for apoptotic cell engulfment: Eat-me, don't eat-me and come-get-me signals, *Trends Cell Biol* 2003, 13:648–656.
27. Bratton DL, Fadok VA, Richter DA, Kailey JM, Guthrie LA, Henson PM: Appearance of phosphatidylserine on apoptotic cells requires calcium-mediated nonspecific flip-flop and is enhanced by loss of the aminophospholipid translocase, *J Biol Chem* 1997, 272:26159–26165.
28. Fadok VA, deCathelineau A, Daleke DL, Henson PM, Bratton DL: Loss of phospholipid asymmetry and surface exposure of phosphatidylserine is required for phagocytosis of apoptotic cells by macrophages and fibroblasts, *J Biol Chem* 2001, 276:1071–1077.
29. Kagan VE, Borisenko GG, Serinkan BF, Tyurina YY, Tyurin VA, Jiang J, et al.: Appetizing rancidity of apoptotic cells for macrophages: Oxidation, externalization, and recognition of phosphatidylserine, *Am J Physiol Lung Cell Mol Physiol* 2003, 285:L1–17.
30. Zullig S, Hengartner MO: Cell biology. Tickling macrophages, a serious business, *Science* 2004, 304:1123–1124.
31. Wu Y, Tibrewal N, Birge RB: Phosphatidylserine recognition by phagocytes: A view to a kill, *Trends Cell Biol* 2006, 16:189–197.
32. Scott RS, McMahon EJ, Pop SM, Reap EA, Caricchio R, Cohen PL, et al.: Phagocytosis and clearance of apoptotic cells is mediated by MER, *Nature* 2001, 411:207–211.
33. Tian H, Lu Y, Shah SP, Hong S: Novel 14S,21-dihydroxy-docosahexaenoic acid rescues wound healing and associated angiogenesis impaired by acute ethanol intoxication/exposure, *J Cell Biochem* 2010, 111:266–273.
34. Lawrence T, Willoughby DA, Gilroy DW: Anti-inflammatory lipid mediators and insights into the resolution of inflammation, *Nat Rev Immunol* 2002, 2:787–795.
35. Kilfeather S: 5-lipoxygenase inhibitors for the treatment of COPD, *Chest* 2002, 121:197S–200S.
36. Serhan CN: Resolution phase of inflammation: Novel endogenous anti-inflammatory and proresolving lipid mediators and pathways, *Annu Rev Immunol* 2007, 25:101–137.
37. Morris T, Stables M, Hobbs A, de Souza P, Colville-Nash P, Warner T, et al.: Effects of low-dose aspirin on acute inflammatory responses in humans, *J Immunol* 2009, 183:2089–2096.
38. Levy BD, Clish CB, Schmidt B, Gronert K, Serhan CN: Lipid mediator class switching during acute inflammation: Signals in resolution, *Nat Immunol* 2001, 2:612–619.
39. Serhan CN, Takano T, Clish CB, Gronert K, Petasis N: Aspirin-triggered 15-epi-lipoxin A4 and novel lipoxin B4 stable analogs inhibit neutrophil-mediated changes in vascular permeability, *Adv Exp Med Biol* 1999, 469:287–293.
40. Maddox JF, Serhan CN: Lipoxin A4 and B4 are potent stimuli for human monocyte migration and adhesion: Selective inactivation by dehydrogenation and reduction, *J Exp Med* 1996, 183:137–146.
41. Jozsef L, Zouki C, Petasis NA, Serhan CN, Filep JG: Lipoxin A4 and aspirin-triggered 15-epi-lipoxin A4 inhibit peroxynitrite formation, NF-kappa B and AP-1 activation, and IL-8 gene expression in human leukocytes, *Proc Natl Acad Sci USA* 2002, 99:13266–13271.
42. Serhan CN, Clish CB, Brannon J, Colgan SP, Chiang N, Gronert K: Novel functional sets of lipid-derived mediators with antiinflammatory actions generated from omega-3 fatty acids via cyclooxygenase 2-nonsteroidal antiinflammatory drugs and transcellular processing, *J Exp Med* 2000, 192:1197–1204.
43. Serhan CN, Hong S, Gronert K, Colgan SP, Devchand PR, Mirick G, et al.: Resolvins: A family of bioactive products of omega-3 fatty acid transformation circuits initiated by aspirin treatment that counter proinflammation signals, *J Exp Med* 2002, 196:1025–1037.
44. Arita M, Yoshida M, Hong S, Tjonahen E, Glickman JN, Petasis NA, et al.: Resolvin E1, an endogenous lipid mediator derived from omega-3 eicosapentaenoic acid, protects against 2,4,6-trinitrobenzene sulfonic acid-induced colitis, *Proc Natl Acad Sci USA* 2005, 102:7671–7676.

45. Serhan CN, Arita M, Hong S, Gotlinger K: Resolvins, docosatrienes, and neuroprotectins, novel omega-3-derived mediators, and their endogenous aspirin-triggered epimers, *Lipids* 2004, 39:1125–1132.
46. Serhan CN, Gotlinger K, Hong S, Lu Y, Siegelman J, Baer T, et al.: Anti-inflammatory actions of neuroprotectin D1/protectin D1 and its natural stereoisomers: Assignments of dihydroxy-containing docosatrienes, *J Immunol* 2006, 176:1848–1859.
47. Serhan CN, Yang R, Martinod K, Kasuga K, Pillai PS, Porter TF, et al.: Maresins: Novel macrophage mediators with potent antiinflammatory and proresolving actions, *J Exp Med* 2009, 206:15–23.
48. Banerjee J, Chan YC, Sen CK: MicroRNAs in skin and wound healing, *Physiol Genomics* 2011, 43:543–556.
49. Biswas S, Roy S, Banerjee J, Hussain SR, Khanna S, Meenakshisundaram G, et al.: Hypoxia inducible microRNA 210 attenuates keratinocyte proliferation and impairs closure in a murine model of ischemic wounds, *Proc Natl Acad Sci U S A* 2010, 107:6976–6981.
50. Sen CK, Roy S: miRNA: Licensed to kill the messenger, *DNA Cell Biol* 2007, 26:193–194.
51. Sen CK, Roy S: Redox signals in wound healing, *Biochim Biophys Acta* 2008, 1780:1348–1361.
52. Shilo S, Roy S, Khanna S, Sen CK: MicroRNA in cutaneous wound healing: A new paradigm, *DNA Cell Biol* 2007, 26:227–237.
53. Shilo S, Roy S, Khanna S, Sen CK: Evidence for the involvement of miRNA in redox regulated angiogenic response of human microvascular endothelial cells, *Arterioscler Thromb Vasc Biol* 2008, 28:471–477.
54. Roy S, Sen CK: MiRNA in innate immune responses: Novel players in wound inflammation, *Physiol Genomics* 2011, 43:557–565.
55. Recchiuti A, Krishnamoorthy S, Fredman G, Chiang N, Serhan CN: MicroRNAs in resolution of acute inflammation: Identification of novel resolvin D1-miRNA circuits, *FASEB J* 2011, 25:544–560.
56. Ponomarev ED, Veremeyko T, Barteneva N, Krichevsky AM, Weiner HL: MicroRNA-124 promotes microglia quiescence and suppresses EAE by deactivating macrophages via the C/EBP-alpha-PU.1 pathway, *Nat Med* 2011, 17:64–70.

10 H_2S in Inflammation

Hyun-Ock Pae
Wonkwang University School of Medicine
Iksan, Republic of Korea

Hun-Taeg Chung
University of Ulsan
Ulsan, Republic of Korea

CONTENTS

10.1 INTRODUCTION

Recently, the diverse physiologic actions of nitric oxide (NO), carbon monoxide (CO), and hydrogen sulfide (H_2S) and their roles in different diseases have attracted a great deal of interest. Although initially viewed as toxic substances, NO, CO, and H_2S are now recognized as important signaling molecules acting in a variety of functional capacities in autocrine, paracrine, or juxtacrine fashions. All three gases have been proposed to play a role in inflammation [1–3]. However, the mechanisms whereby these gases exert pro-inflammatory and anti-inflammatory effects are not simple and are still largely unknown. Unlike NO and CO, the part played by H_2S in inflammation has yet to be thoroughly investigated, and this yielded rather controversial data with respect to the pro-inflammatory or anti-inflammatory properties of H_2S [1]. This chapter reviews the current literature on the therapeutic potential of H_2S, with a special focus on its regulation of inflammation. We begin this chapter by pointing out basic concepts regarding synthesis and functions of H_2S.

In mammalian tissues, H_2S is produced from the sulfur-containing amino acid *L*-cysteine by cystathionine-β-synthase (CBS) and cystathionine-γ-lyase (CSE), both using pyridoxal 5'-phosphate as a cofactor [3]. Cysteine aminotransferase and cysteine lyase can also convert *L*-cysteine to H_2S [1]. CBS and CSE, however, appear to be responsible for the majority of the endogenous production of H_2S in mammalian tissues. It should be noted that both CBS and CSE are also responsible for metabolism of *L*-methionine into *L*-cysteine, which is, in turn, used for synthesis of H_2S [4]. CBS and CSE are widely distributed in mammalian tissues; however, CBS is highly expressed in the hippocampus and cerebellum in mammalian brain [1]. In contrast, CSE is most likely the predominant H_2S-producing enzyme in the vasculature [1]. Relatively large amounts of CSE protein are determined in both vascular smooth muscle cells (VSMCs) and endothelial cells [2]. In some tissues, such as the liver and kidney, both CBS and CSE contribute to H_2S production. CBS exists as separate and distinct isoforms, whereas CSE occurs as a single transcript with no splice variant in mouse brain, liver, kidney, lung, and heart [1]. The activity of CBS is regulated presumably at transcriptional level, and glucocorticoids and insulin stimulate and inhibit CBS gene expression, respectively [1, 2]. Although little is known about the regulation of CSE activity, NO donor and lipopolysaccharide (LPS) have been shown to stimulate CSE mRNA expression in VSMCs and macrophages, respectively [3]. The physiological significance of NO in the regulation of H_2S synthesis is supported by the observation that circulating H_2S levels as well as CSE gene expression and enzymatic activity in the cardiovascular system are reduced in rats chronically treated with NO synthase (NOS) inhibitor [5]. Circulatory plasma levels of H_2S have been reported to be 10–50 μM in rats and 10–100 μM in humans [6]. Local tissue levels of H_2S may be higher than its circulatory levels. H_2S is involved in the regulation of vascular tone, myocardial contractility, neurotransmission, and insulin secretion [7]. Decreased levels of H_2S were observed in various animal models of arterial and pulmonary hypertension, Alzheimer's disease, gastric mucosal injury, and liver cirrhosis [7]. CBS deficiency is the leading cause of homocystinuria in humans [8]. A growing body of evidence suggests that vascular disorders found in one-third of the patients with premature arterial disease or cerebrovascular disease are the result of mild hyperhomocysteinemia, some of which may result from heterozygous CBS deficiency [8]. There is an experimental study showing that in CBS/apolipoprotein E double-knockout mice, hyperhomocysteinemia promotes differentiation of inflammatory monocyte subsets and their accumulation in atherosclerotic lesions [9]. CSE is also linked to homocystinuria. Mice genetically deficient in CSE expression exhibited age-dependent increased blood pressure, decreased endogenous H_2S level, severe hyperhomocysteinemia, and impaired endothelium-dependent vasorelaxation [10]. Excessive production of H_2S may contribute to the pathogenesis of mental retardation in patients with Down syndrome [8]. H_2S was oxidized in mitochondria to thiosulfate, sulfite, and sulfate, and then excreted in the urine. Sulfate (SO_4^{2-}) is the major end product of H_2S. H_2S binds to methemoglobin to form sulfhemoglobin. H_2S stimulates ATP-sensitive potassium channels in VSMCs, neurons, cardiomyocytes, and pancreatic beta cells [6]. In addition, H_2S may react with reactive oxygen and/or nitrogen species, limiting their toxic effects [11].

10.2 DYNAMIC INTERPLAY OF H_2S WITH NO AND CO IN INFLAMMATORY CONDITIONS

NO, CO, and H_2S can be produced naturally within mammalian cells. NO is synthesized by constitutive NOS (cNOS) or inducible NOS (iNOS) [12]. CO is synthesized by constitutive heme oxygenase-2 (HO-2) or inducible heme oxygenase-1 (HO-1) [12]. Unlike cNOS and HO-2, iNOS and HO-1 can be markedly expressed during inflammation and synthesize large amounts of NO and CO, respectively [13]. Interestingly, the production of H_2S is also increased during inflammation [13]. Thus, the cells involved, at least, in inflammation are exposed to these three gases at the same time, implying that three gases could interact one with the others in a number of ways including affecting each other's biosynthesis and biological responses. A well-known example is a possible interaction between NO and CO during inflammation. The NO/iNOS pathway has been reported to induce HO-1 expression in different types of cells [13, 14]. The induction of HO-1 expression by NO is independent of cyclic guanosine monophosphate (cGMP) production, since exogenous cGMP has no effect, and involves the activation of transcription factors, including nuclear factor-erythroid 2-related factor-2 (Nrf2) [14]. CO has been reported to modulate the NO/iNOS system [13]. CO can directly bind to hemoprotein of iNOS and inactivate iNOS, decreasing NO production [12]. Importantly, the CO/HO-1 system can also inhibit NO production through suppression of iNOS expression [3], indicating the possible presence of functional interaction between the NO/iNOS and

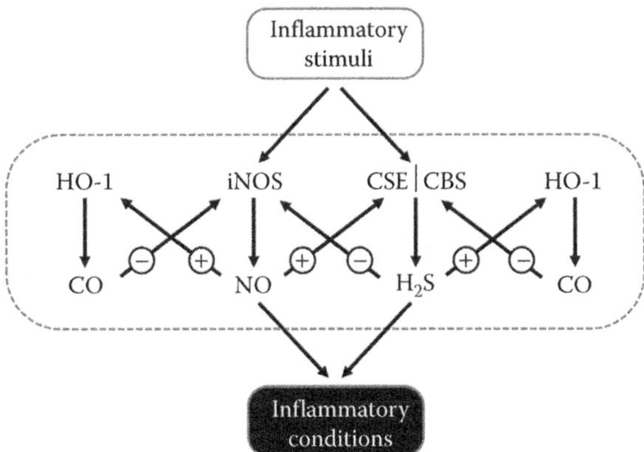

FIGURE 10.1 The possible interplay of H2S/CSE/CBS with NO/iNOS *via* CO/HO-1 in inflammation. Inflammatory stimuli induce iNOS expression, resulting in increased NO production that stimulates inflammatory pathways for killing and eradicating pathogens. When the levels of NO production are so high that NO could exert toxic effects on the host, HO-1 expression is induced in response to the high doses of NO, resulting in increased CO production that inhibits both NO/iNOS and inflammatory pathways. Pathogens also induce CSE/CBS expression, resulting in increased H_2S production that could stimulate inflammatory pathways. H_2S that can be synthesized from CSE and CBS enzymes could inhibit NO/iNOS pathway, probably *via* stimulation of CO/HO-1 pathway at least in macrophages. CO can inhibit H_2S production by binding to hemoprotein of CBS. The arrows represent the targeting pathway and the circled signs, "plus" and "minus," indicate "positive" and "negative" effects, respectively.

CO/HO-1 systems in regulating inflammation (Figure 10.1). There is an excellent study supporting the negative feedback regulation of the pro-inflammatory NO/iNOS system by the anti-inflammatory CO/HO-1 system. Ashino and colleagues [15] performed time-course experiments to characterize the expression pattern of HO-1 and iNOS in macrophages stimulated with LPS, and found an apparent time lag between iNOS and HO-1 expression for 6 h; iNOS expression reached a peak level at 12 h after LPS treatment and followed HO-1 expression peak level at 6 h later. Using iNOS-deficient macrophages, they showed that LPS failed to induce HO-1 expression, suggesting that NO/iNOS is implicated in HO-1 expression. In other experimental sets, they investigated iNOS and HO-1 expression by LPS using Nrf2-deficient macrophages. Whereas LPS failed to induce HO-1 expression, LPS strongly induced iNOS expression in Nrf2-deficient macrophages, supporting the notion that the CO/HO-1 system inhibits iNOS expression.

NO can also regulate the endogenous production of H_2S in VSMCs by increasing CSE and CBS expression [16]. In the same cells, H_2S can enhance IL-1β-induced NO production by increasing iNOS expression [17]. These findings suggest that as with NO and CO, there may be the possible interaction between NO and H_2S; at least in the vascular system, NO enhances the H_2S/CSE pathway and H_2S potentiates the NO/iNOS pathway. However, the nature of this interaction may be different from that of the interplay between NO and H_2S in inflammatory states—perhaps because the NO/iNOS and H_2S/CSE pathways are fully activated in inflammatory conditions, but not in normal physiological conditions. There are two notable studies demonstrating that NO and H_2S can interact with each other in inflammatory conditions. Firstly, Anuar and colleagues determined whether exogenous NO could affect H_2S production in an animal model of systemic inflammation [18]. In their study, nitroflurbiprofen was used as the NO donor, because the administration of nitroflurbiprofen to animals results in esterbond cleavage of the parent molecule, flurbiprofen, and leads to the slow release of NO. The administration of nitroflurbiprofen resulted in a dose-dependent inhibition of the LPS-mediated increase in liver and kidney H_2S production and CSE mRNA expression, whereas flurbiprofen had no effect. These results show that NO downregulates the biosynthesis of H_2S in endotoxic shock models. Secondly, Oh and colleagues determined whether H_2S could affect the NO/iNOS pathway in cultured RAW264.7 macrophages stimulated with LPS [19]. They showed that exogenous H_2S inhibited LPS-induced activation of NF-κB and, in turn, blocked the increase in iNOS expression and NO production. To test the role of endogenous H_2S in iNOS-mediated NO production, they blocked CSE activity by using a specific inhibitor, and the results suggested that H_2S produced from CSE can inhibit the further activation of the NO/iNOS pathway. Interestingly, a possible mechanism responsible for these effects of H_2S involves its up-regulation of the HO-1/CO pathway, perhaps indicating that the H_2S/CSE system may also interact with the HO-1/CO system. Collectively, two interesting studies provide the evidence supporting the existence of interaction between NO and H_2S in inflammation; the NO/iNOS and H_2S/CSE pathways inhibit each other in pathological conditions.

10.3 JANUS FACE OF H_2S: ITS PRO-INFLAMMATORY AND ANTI-INFLAMMATORY EFFECTS

Conflicting data are available on the effects of H_2S on the regulation of inflammation; interestingly, H_2S has been shown to exert both pro-inflammatory and anti-inflammatory effects [11]. In animal models of systemic inflammation, inhibition of endogenous synthesis of H_2S by propargylglycine (PAG), an irreversible inhibitor of CSE, attenuated inflammatory reactions, as evidenced by decreased production of the pro-inflammatory cytokines and chemokines and reduced leukocyte activation and trafficking, whereas administration of NaHS, an H_2S donor, augmented these inflammatory reactions [20–25]. Despite these data suggesting the pro-inflammatory effects of H_2S, its role during inflammatory diseases is still a matter of debate. Kang and colleagues applied PAG and NaHS to investigate their effects on the severity of liver injury induced by ischemia-reperfusion [26]. Administration of NaHS significantly attenuated the severity of liver injury and inhibited the production of lipid peroxidation, serum inflammatory factors, and cell apoptosis, whereas PAG aggravated them. Zanardo and colleagues observed that NaHS inhibited aspirin-induced leukocyte adherence in mesenteric venules, whereas inhibition of endogenous H_2S synthesis elicited leukocyte adherence [27]. Carrageenan-induced paw edema was suppressed by NaHS and enhanced by an inhibitor of H_2S synthesis [27]. In mice with acute lung injury induced by burn and smoke inhalation, H_2S inhibited IL-1β levels, increased the concentration of the anti-inflammatory cytokine IL-10, and attenuated protein oxidation in lung tissue [28]. Li and colleagues examined the possible role of H_2S in the pathogenesis of oleic acid-induced acute lung injury and its regulatory effects on the inflammatory response [29]. As compared to controls, significantly increased IL-6, IL-8, and IL-10 levels together with decreased H_2S levels were observed in the plasma and lung tissue of oleic acid-treated rats. Administration of NaHS into oleic acid-treated rats decreased IL-6 and IL-8 levels but increased IL-10 levels in the plasma and lung tissues [27]. In RAW264.7 macrophages stimulated with LPS, H_2S inhibited NF-kB activation and, in turn, blocked the increase in iNOS expression and NO production [19]. Similarly, H_2S inhibited LPS-induced iNOS expression and TNF-α production in microglia [30]. Most interestingly, GYY4137, a slow-releasing H_2S donor, inhibited LPS-mediated systemic inflammation, while NaHS, a fast-releasing H_2S donor, enhanced the inflammation [31]. Other slow-releasing H_2S donors, such as *S*-diclofenac and *S*-mesalamine, also exhibited anti-inflammatory activity *in vivo* [32, 33]. In the context of these contradictory findings, H_2S may exert both pro-inflammatory and anti-inflammatory effects depending on a variety of factors, including the concentration of the gas achieved at the inflamed site or released from donors administered H_2S. It is most likely that H_2S at physiological concentrations may be anti-inflammatory, whereas at higher concentrations produced endogenously in certain circumstances, this gas may be pro-inflammatory. The concept of H_2S being anti-inflammatory at physiological concentrations but pro-inflammatory at nonphysiological concentrations bears similarity to the situation with NO; NO can exert both pro-inflammatory and anti-inflammatory effects depending on its concentrations [14].

10.4 PERSPECTIVES

10.4.1 Clinical Science

Recently, there have been many studies and reviews published on the involvement of H_2S in inflammatory events. No definitive conclusions, however, have been made as to whether H_2S exerts pro-inflammatory or anti-inflammatory properties. In addition to the question of dosing and timing, the preferred route of H_2S administration for its clinical application remains to be settled. Nevertheless, H_2S-releasing nonsteroidal anti-inflammatory drugs (NSAIDs) have been developed since 2007, and some of them have been shown to exhibit enhanced anti-inflammatory activity as compared to the parent drugs [31–33]. Additionally, H_2S-releasing NSAIDs have been reported to cause less gastrointestinal and cardiovascular injury than the parent NSAIDs in preclinical models [34]. H_2S-releasing NSAIDs represent examples of new anti-inflammatory drugs created through the exploitation of the anti-inflammatory effects of an H_2S. While the development of an H_2S-releasing anti-inflammatory drug is in its infancy, the preclinical data available thus far provide cause for optimism.

10.4.2 Basic Science

It is becoming clearer that NO, CO, and H_2S are synthesized naturally in the body, often by the same cells within the same organs, and that all three gases exert essentially similar biological effects even if *via* different mechanisms. Each of these molecules acts, *via* well-established molecular targets, to effect physiological and/or pathophysiological functions within the body. All three gases exhibit remarkably similar effects in inflammation. It is most likely that inflammation is regulated not by the activity of a single gas working in isolation but by the concerted activity of all three of these gases working together (Figure 10.1). If this is the case, it would no longer be good enough simply to study the role of one gas in the regulation of inflammation. Conversely, it would be necessary to investigate a possible interaction among three gases in the regulation of inflammation. A study had tried to determine a possible interplay of H_2S/CSE with NO/iNOS and CO/HO-1 in RAW264.7 macrophages stimulated with LPS [19]. This study showed that H_2S downregulated iNOS expression and NO production by upregulating HO-1 expression and CO production. Additional research is needed to further characterize the different nature of the interaction of H_2S with other gases in different conditions. It is hoped that active research on the possible interplay of gaseous molecules will help address several unanswered questions in inflammatory diseases.

TAKE-HOME MESSAGES

- H_2S, together with NO and CO, makes up a family of biologically active gases with a range of physiological and pathophysiological effects.
- H_2S, NO, and CO are synthesized naturally in the body, often by the same cells within the same organs, and all three gases exhibit remarkably similar effects in inflammation.
- H_2S, NO, and CO act, *via* well-established molecular targets, to effect physiological and/or pathophysiological functions within the body.
- H_2S is synthesized from *L*-cysteine mainly by CBS and CSE in mammalian tissues.
- Mutation or deficiency of CBS and CSE results in increased homocysteine and cystathionine, leading to inflammatory atherosclerosis.
- H_2S exhibits potent vasodilator activity both *in vitro* and *in vivo* most probably by stimulating ATP-sensitive potassium channels.
- H_2S, like NO, has been shown to exert both pro-inflammatory and anti-inflammatory effects.

- Abnormal metabolism and functions of the H_2S pathway have been shown to be linked to inflammatory diseases.
- H_2S, NO, and CO have been shown to interact one with the other in a number of ways including affecting each other's biosynthesis and biological responses.

H_2S-releasing nonsteroidal anti-inflammatory drugs that exploit the anti-inflammatory and protective effects of H_2S have been recently developed, and some of them have been shown to exhibit enhanced anti-inflammatory activity as compared to the parent drugs.

ACKNOWLEDGMENTS

This work was supported by the Korea Research Foundation Grant funded by the Korean Government (MOEHRD) (BRL 2010-0001199).

REFERENCES

1. Kajimura, M., Fukuda, R., Bateman, R.M., Yamamoto, T., and M. Suematsu. 2010. Interactions of multiple gas-transducing systems: Hallmarks and uncertainties of CO, NO, and H_2S gas biology. *Antioxidants & Redox Signaling* 13:157–92.
2. Baumgart, K., Radermacher, P., and F. Wagner. 2009. Applying gases for microcirculatory and cellular oxygenation in sepsis: Effects of nitric oxide, carbon monoxide, and hydrogen sulfide. *Current Opinion in Anaesthesiology* 22:168–76.
3. Pae, H.O., Lee, Y.C., Jo, E.K., and H.T. Chung. 2009. Subtle interplay of endogenous bioactive gases (NO, CO and H_2S) in inflammation. *Archives of Pharmacal Research* 32:1155–62.
4. Mani, S., Yang, G., and R. Wang. 2011. A critical life-supporting role for cystathionine g-lyase in the absence of dietary cysteine supply. *Free Radical Biology & Medicine* 50:1280–7.
5. Zhong, G., Chen, F., Cheng, Y., Tang, C., and J. Du. 2003. The role of hydrogen sulfide generation in the pathogenesis of hypertension in rats induced by inhibition of nitric oxide synthase. *Journal of Hypertension* 21:1879–85.
6. Wang, R. 2002. Two's company, three's a crowd: Can H_2S be the third endogenous gaseous transmitter? *FASEB Journal* 16:1792–8.
7. Łowicka, E., and Bełtowski, J. 2007. Hydrogen sulfide (H_2S) – the third gas of interest for pharmacologists. *Pharmacological Reports* 59:4–24.
8. Kashiba, M., Kajimura, M., Goda, N., and M. Suematsu. 2002. From O_2 to H_2S: A landscape view of gas biology. *The Keio Journal of Medicine* 51:1–10.
9. Zhang, D., Jiang, X., Fang, P., Yan, Y., Song, J., Gupta, S., et al. Hyperhomocysteinemia promotes inflammatory monocyte generation and accelerates atherosclerosis in transgenic cystathionine b-synthase-deficient mice. *Circulation* 120:1893–902.
10. Yang, G., Wu, L., Jiang, B., Yang, W., Qi, J., Cao, K., et al. 2008. H_2S as a physiologic vasorelaxant: Hypertension in mice with deletion of cystathionine γ-lyase. *Science* 322:587–90.
11. Li, L., Hsu, A., and P.K. Moore. 2009. Actions and interactions of nitric oxide, carbon monoxide and hydrogen sulphide in the cardiovascular system and in inflammation—a tale of three gases! *Pharmacology & Therapeutics* 123:386–400.
12. Pae, H.O., Son, Y., Kim, N.H., Jeong, H.J., Chang, K.C., and H.T. Chung. 2010. Role of heme oxygenase in preserving vascular bioactive NO. *Nitric Oxide* 23:251–7.
13. Pae, H.O., and H.T. Chung. 2009. Heme oxygenase-1: Its therapeutic roles in inflammatory diseases. *Immune Network* 9:12–9.
14. Pae, H.O., Lee, Y.C., and H.T. Chung. 2008. Heme oxygenase-1 and carbon monoxide: Emerging therapeutic targets in inflammation and allergy. *Recent Patents on Inflammation & Allergy Drug Discovery* 2:159–65.
15. Ashino, T., Yamanaka, R., Yamamoto, M., Shimokawa, H., Sekikawa, K., Iwakura, Y., et al. 2008. Negative feedback regulation of lipopolysaccharide-induced inducible nitric oxide synthase gene expression by heme oxygenase-1 induction in macrophages. *Molecular Immunology* 45:2106–15.
16. Zhao, W., Zhang, J., Lu, Y., and R. Wang. 2001. The vasorelaxant effect of H_2S as a novel endogenous gaseous K_{ATP} channel opener. *EMBO Journal* 20:6008–16.

17. Jeong, S.O., Pae, H.O., Oh, G.S., Jeong, G.S., Lee, B.S., Lee, S., et al. 2006. Hydrogen sulfide potentiates interleukin-1β-induced nitric oxide production *via* enhancement of extracellular signal-regulated kinase activation in rat vascular smooth muscle cells. *Biochemical and Biophysical Research Communications* 345:938–44.
18. Anuar, F., Whiteman, M., Siau, J.L., Kwong, S.E., Bhatia, M., and P.K. Moore. 2006. Nitric oxide-releasing flurbiprofen reduces formation of proinflammatory hydrogen sulfide in lipopolysaccharide-treated rat. *British Journal of Pharmacology* 147:966–74.
19. Oh, G.S., Pae, H.O., Lee, B.S., Kim, B.N., Kim, J.M., Kim, H.R., et al. 2006. Hydrogen sulfide inhibits nitric oxide production and nuclear factor-κB *via* heme oxygenase-1 expression in RAW264.7 macrophages stimulated with lipopolysaccharide. *Free Radical Biology & Medicine* 41:106–19.
20. Collin, M., Anuar, F.B., Murch, O., Bhatia, M., Moore, P.K., and C. Thiemermann. 2005. Inhibition of endogenous hydrogen sulfide formation reduces the organ injury caused by endotoxemia. *British Journal of Pharmacology* 146:498–505.
21. Li, L., Bhatia, M., Zhu, Y.Z., Zhu, Y.C., Ramnath, R.D., Wang, Z.J., et al. 2005. Hydrogen sulfide is a novel mediator of lipopolysaccharide-induced inflammation in the mouse. *FASEB Journal* 19:1196–8.
22. Zhang, H., Zhi, L., Moore, P.K., and M. Bhatia. 2006. Role of hydrogen sulfide in cecal ligation and puncture-induced sepsis in the mouse. *American Journal of Physiology-Lung Cellular and Molecular Physiology* 290:L1193–L1201.
23. Zhang, H., Zhi, L., Moochhala, S.M., Moore, P.K., and M. Bhatia. 2007. Endogenous hydrogen sulfide regulates leukocyte trafficking in cecal ligation and puncture-induced sepsis. *Journal of Leukocyte Biology* 82:894–905.
24. Zhang, H., Zhi, L., Moochhala, S., Moore, P.K., and M. Bhatia. 2007. Hydrogen sulfide acts as an inflammatory mediator in cecal ligation and puncture-induced sepsis in mice by upregulating the production of cytokines and chemokines *via* NF-κB. *American Journal of Physiology-Lung Cellular and Molecular Physiology* 292:L960–L971.
25. Zhang, H., Moochhala, S.M., and M. Bhatia. Endogenous hydrogen sulfide regulates inflammatory response by activating the ERK pathway in polymicrobial sepsis. *Journal of Immunology* 181:4320–31.
26. Kang, K., Zhao, M., Jiang, H., Tan, G., Pan, S., and X. Sun. 2009. Role of hydrogen sulfide in hepatic ischemia-reperfusion-induced injury in rats. *Liver Transplantation* 15:1306–14.
27. Zanardo, R.C., Brancaleone, V., Distrutti, E., Fiorucci, S., Cirino, G., and J.L. Wallace. 2006. Hydrogen sulfide is an endogenous modulator of leukocyte-mediated inflammation. *FASEB Journal* 20:2118–20.
28. Esechie, A., Kiss, L., Olah, G., Horváth, E.M., Hawkins, H., Szabo, C., et al. 2008. Protective effect of hydrogen sulfide in a murine model of acute lung injury induced by combined burn and smoke inhalation. *Clinical Science (London)* 115:91–7.
29. Li, T., Zhao, B., Wang, C., Wang, H., Liu, Z., Li, W., et al. 2008. Regulatory effects of hydrogen sulfide on IL-6, IL-8 and IL-10 levels in the plasma and pulmonary tissue of rats with acute lung injury. *Experimental Biology and Medicine (Maywood)* 233(9):1081–7.
30. Hu, L.F., Wong, P.T., Moore, P.K., and J.S. Bian. 2007. Hydrogen sulfide attenuates lipopolysaccharide-induced inflammation by inhibition of p38 mitogen-activated protein kinase in microglia. *Journal of Neurochemistry* 100:1121–8.
31. Whiteman, M., Li, L., Rose, P., Tan, C.H., Parkinson, D.B., and P.K. Moore. 2010. The effect of hydrogen sulfide donors on lipopolysaccharide-induced formation of inflammatory mediators in macrophages. *Antioxidants & Redox Signaling* 12:1147–54.
32. Sidhapuriwala, J., Li, L., Sparatore, A., Bhatia, M., and P.K. Moore. 2007. Effect of *S*-diclofenac, a novel hydrogen sulfide releasing derivative, on carrageenan-induced hindpaw oedema formation in the rat. *European Journal of Pharmacology* 569:149–54.
33. Fiorucci, S., Orlandi, S., Mencarelli, A., Caliendo, G., Santagada, V., Distrutti, E., et al. 2007. Enhanced activity of a hydrogen sulphide-releasing derivative of mesalamine (ATB-429) in a mouse model of colitis. *British Journal of Pharmacology* 150:996–1002.
34. Fiorucci, S., and L. Santucci. 2011. Hydrogen sulfide-based therapies: Focus on H_2S releasing NSAIDs. *Inflammation & Allergy Drug Targets* 10:133–40.

Section II

Pathologies Associated with Inflammation

Section II

11 Is There a Connection between Inflammation and Oxidative Stress?

Chandrakala Aluganti Narasimhulu, Xueting Jiang, Zhaohui Yang, Krithika Selvarajan, and Sampath Parthasarathy
University of Central Florida
Orlando, Florida

CONTENTS

11.1 INTRODUCTION

Oxidative stress and inflammation play major roles in both acute and chronic diseases (Table 11.1). The former attracted attention due to its relevance to nutrition and the expectation that common antioxidants present in food would prove to be antidotes to popular maladies. The latter presented itself as a major player due to its association with infection and immune functions. In addition, the association of "inflammatory cells," such as neutrophils, eosinophils, basophils, lymphocytes, and macrophages, with the human pathophysiology has elevated the status of inflammation as a more physiological disease marker. In other words, oxidative stress is expected to be attenuated by the use of antioxidants affording benefits to humans while the inflammatory stress is considered purely as a key to identifying biomarkers with little or no expectation that anti-inflammatory agents would prevent chronic diseases.

11.2 ATHEROSCLEROSIS

One of the major examples of a disease wherein both oxidative stress and inflammation have been implicated is atherosclerosis. Atherosclerosis is a chronic, progressive disease characterized by the formation of cholesterol-rich plaques in the intima of medium- and large-sized arteries [1–3]. Current views suggest that the disease is initiated by dyslipidemia with subsequent recruitment of monocytes to the intima. Pathological observations that macrophages are abundantly distributed from early stage fatty streak lesion to the late stage fibrous plaques indicate that these cells play a critical role in the development of the disease [4–7]. Although these cells were seen as scavengers

TABLE 11.1
List of Inflammatory and Oxidative Diseases

Oxidative Stress Implicated Diseases	Inflammatory Diseases
• Alzheimer's and other neurodegenerative diseases	• Acne vulgaris
• Anemia or iron overload	• Alzheimer's
• Arthritis	• Arthritis
• Atherosclerosis	• Asthma
• Autoimmune diseases	• Atherosclerosis
• Cancer	• Autoimmune diseases
• Cataractogenesis	• Cancer
• Diabetes	• Celiac disease
• Endometriosis	• Crohn's disease
• Hypertension	• Dermatitis
• Ischemia-reperfusion injury	• Diabetes
• Infection and inflammation	• Diverticulitis
• Macular degeneration	• Endometriosis
• Multiple sclerosis	• Hay fever
• Muscular dystrophy	• Hepatitis
• Parkinson's disease	• Irritable bowel syndrome
• Progeria and accelerated aging	• Ischemia reperfusion injury
	• Lupus erythematosus
	• Multiple sclerosis
	• Nephritis
	• Parkinson's disease
	• Sarcoidosis
	• Ulcerative colitis
	• Vasculitis

of modified lipoproteins [8–12], particularly of oxidized lipoproteins [13–17], somewhere in the schema of events the concept that oxidative stress might function independently or in conjunction with inflammatory stress to affect disease progression evolved (Figure 11.1). There are many unanswered questions. There has been very little effort to understand the connection between dyslipidemia and oxidative stress. Why should increased cholesterol lead to increased oxidative stress? Perhaps cholesterol-enriched cell membranes and lipoproteins are seen as pathogens and monocyte/macrophages eliminate them by oxidative assimilation. The inflammation and immune response could be the result of the damage resulting from oxidation to cell membranes and lipoproteins. Would attenuating the immune response prevent the disease? Is the enhanced immune and inflammatory response not the result of direct oxidation but the result of molecules that are presented as antigens as the result of damage due to oxidation?

Following activation, the recruited monocytes differentiate into macrophages, which take up modified low-density lipoprotein (LDL) through scavenger receptors, thereby promoting cholesterol loading and foam cell formation [18–20]. The later stages of the disease are also characterized by calcium deposits, dying cells, accumulation of bound and free iron and other trace metals, and the presence of extracellular cholesterol crystals [21–25]. While atherosclerotic plaque in the advanced state could cause partial or total stenosis [26], the composition and stability of the plaque determines the clinical outcome of the disease. Lipid-laden macrophages produce multiple pro-inflammatory mediators and reactive oxygen species (ROS) contributing to plaque vulnerability and rupture [27, 28]. They also produce procoagulants that promote local inflammation, platelet adherence, and

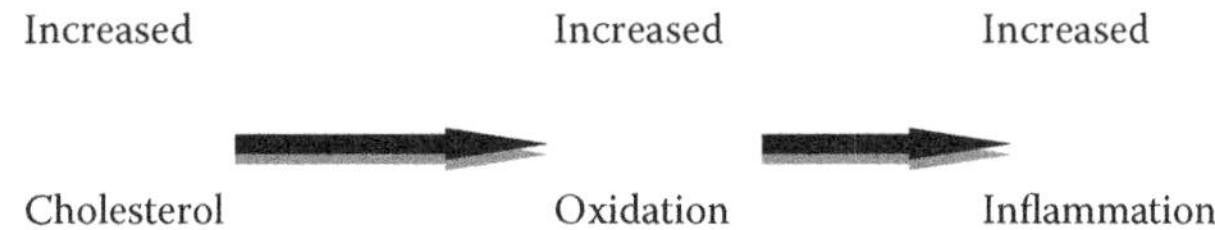

FIGURE 11.1 Oxidation, inflammation and atherosclerosis.

thrombosis. It is thus currently believed that oxidative stress might not only initiate plaque formation but also might be involved in plaque progression by way of promoting smooth muscle cell proliferation and in causing plaque rupture by way of promoting biochemical and physical disruption of plaque. The specific role of inflammation on these stages of atherosclerosis is still not yet clearly understood. It is just assumed that inflammation is harmful and that increase in inflammatory mediators during disease progression represents worsening of the disease. Numerous *in vitro* and animal studies have documented that deleting or overexpressing specific molecules associated with inflammation affected atherosclerosis, mostly in a negative manner. Only very recently have macrophage and lymphocyte subtypes been identified in the artery, and they seem to vary between early and late stages of atherosclerosis. Surprisingly more pro-inflammatory macrophages are found in early lesions while anti-inflammatory macrophages seem to abound in late stages [29]. Their connection to oxidative stress has not been explored.

11.3 OXIDATIVE STRESS AND ATHEROSCLEROSIS

For reasons that are not obvious, it was always felt that oxidative stress is deleterious and oxidative metabolites are pathological. As a result, the etiology of almost every major human pathological condition has been linked to oxidative stress (Table 11.1). Oxidative stress has long been associated with cardiovascular disease (CVD) in general and atherosclerosis in particular. The biochemical processes that contribute to the formation of early atherosclerotic lesions, the fatty streak lesions, are still under debate. The LDL oxidation hypothesis was put forward in the 1980s to explain the formation of fatty streak lesions [30–34]. Countless reviews and over 5,000 articles have appeared on the topic to date providing evidence for the involvement of oxidative processes in animal and human atherosclerotic disease. The oxidation of high-density lipoprotein (HDL) also has been noted to affect its biological properties. For example, oxidized HDL has been noted to be a poor promoter of reverse cholesterol transport [35]. While there is a general consensus for the involvement of an oxidative process, there is still vagueness as to its origin, target, function, and even effects. Numerous oxidants have been implicated. These include free and bound redox metals [36, 37], specific oxidases that generate reactive oxygen species (ROS) such as the NADPH oxidase [38, 39], enzymes that act specifically on lipids, for example, lipoxidase [40–42], heme and heme-related proteins [43, 44], peroxidases, specifically leukocyte-derived myeloperoxidase [45–48], and nitric oxide–derived oxidants such as peroxynitrite [49, 50]. Superoxide radicals, hydrogen peroxide, other oxygen radicals, hypochlorous acid, and many other oxidant species have been implicated. Accordingly, the targets of oxidation also are nonspecific that include small molecules (including antioxidants), lipids, amino acids associated with proteins, and even DNA bases. In addition, many of these and their products (e.g., lipid peroxide–derived aldehydes) are highly reactive and "modify" other molecules.

In contrast to the source of inflammatory molecules, the origin of the oxidative stress is still not understood. Endothelium, the medial smooth muscle cells, monocyte-macrophages, and even acellular components have been suggested to be involved. While the vascular biologists see the endothelium and smooth muscle cell–derived oxidants as important, macrophage-derived myeloperoxidase has received undue attention [51]. This could be attributed to the cell's ability to oxidize as well as to scavenge the oxidatively damaged particles. These cells also develop and express a variety of scavenger receptors that recognize various forms of oxidized lipids and lipoproteins [52–54]. In addition, they are suggested to respond to the oxidative stress by secreting mediators of inflammation and immune function [55–57]. However, studies have not gone beyond immunostaining for the presence of myeloperoxidase (MPO) or identifying products that are assumed to be MPO specific [58]. Studies knocking out or overexpressing MPO in mice have neither confirmed nor disproved the hypothesis as proponents of the hypothesis seem to suggest that somehow human MPO is different from mouse MPO [59, 60]. It is hard to imagine how the enzyme activity could be different in terms of function among species.

11.4 INFLAMMATION AND ATHEROSCLEROSIS

As mentioned earlier, many diseases that are associated with increased oxidative stress are also associated with increased inflammation. Numerous inflammatory molecules have been found to be abundantly expressed in human and animal atherosclerotic tissue (Table 11.2) and many genetically modified animal models have validated the importance of inflammation. C-reactive protein (CRP), a marker for chronic inflammatory state, has been found to be an independent predictor of CVD [61–63]. Similarly, serum amyloid A (SAA), an inflammatory marker, has not only been found at elevated levels in patients but was also found to be an important inducer of many different cytokines and adhesion molecules. Recently, other inflammatory molecules and their metabolic modulators such as platelet activating factor acetyl hydrolase (PAF-AcH), secretory phospholipase A_2 ($sPLA_2$) lipoprotein associated phospholipase A_2 (Lp-PLA_2), and paraoxonase (PON) have been identified as biomarkers of atherosclerosis-specific inflammation, providing information about plaque inflammation and stability [64, 65].

The progression of the disease is accelerated by a variety of risk factors including age [66–68], gender [69], diabetes mellitus [70–78], poor diet and environment [44], dyslipidemia [79–81], hypertension [81–85], obesity [86–87], smoking [88–93, 43], and a sedentary lifestyle (Figure 11.2). Increased oxidative stress and increased inflammatory stress have been associated with these risk factors.

The beneficial cultural, social, and nutritional trends around the world have been attributed to antioxidative aspects. For example, aspects of the Mediterranean diet, yoga, vegetarianism, curry powder, red wine and chocolate consumption, physical activity, selenium or lack of it in the soil, and so forth have been attributed to their abilities to affect oxidative stress and thereby influence the disease process. Likewise, there are constant attempts to link proven therapies and lifestyle modalities to their potential antioxidant effects. Exercise, statins [94, 95], many antihypertensive drugs, drugs that contain a phenolic hydroxyl group or thiol function and so forth [96–111] have been suggested to have antioxidant effects (Table 11.3).

TABLE 11.2
Inflammatory Molecules

- Adhesion molecules (ICAM and P-selectin)
- C-reactive protein (hs-CRP)
- Chemokines of CXC & CC subgroups
- Cytokines and their receptors (IL-1 , 2, 6, 8, 12, 18, TNF-α , IFN-y, GDF15)
- Isoprostanes
- Myeloperoxidase
- Oxidized low-density lipoproteins (MM-LDL, FO-LDL, Ox-Mod-LDL)
- Matrix metalloproteinases (MMPs) and MMPs tissue inhibitors
- PAF
- Plasma malondialdehyde
- SAA
- sPLA2

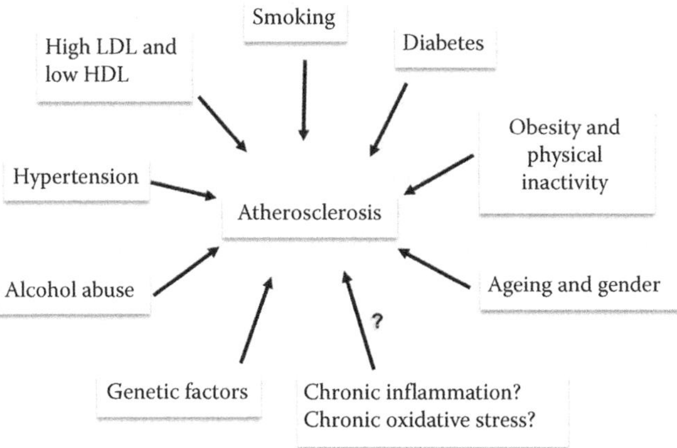

FIGURE 11.2 Risk factors of atherosclerosis.

TABLE 11.3
Antioxidants That Inhibit the Oxidation of LDL

Natural Antioxidants	Synthetic Antioxidants
• Vitamin E and other tocopherol derivatives	• Probucol and related compounds
• Ascorbic acid	• BHT, BHA, etc.
• Estradiol and related phytoestrogens	• Spin traps
• β-Carotene	• Nitric oxide donors
• Glutathione, Cysteine	• RU-486 and amino steroids
• Lipoic acid	• Tamoxifen and raloxifene
• Arginine	• Pyrrolidine dithiocarbamate (PDTC)
• DHEA	• Amino guanidine
• Curcumin	• Captopril
• Caffeic acid, Ferulic acid	• Calcium channel blockers
• Ferritin	• EDTA and other metal chelators
• MUFA-Oleic acid	• Dipyridamole
• Plasmalogens, Sphingomyelin	• DPPD
• Folic acid	• Fibric acid derivatives
• Red wine, resveratrol, and other polyphenols	• Statins
• Garlic extract	• Angiotensin receptor blockers
• Quercetin	• Lipoxygenase inhibitors
• Boldine	• NDGA
• Licorice	• SOD mimics
• Sesamol	*and many others*
• Green tea components	
• Nitric oxide	
• Ginko extracts	
• Blackberry	
• Melatonin	
• Anthocyanins	
• Pychnogenol	
• HDL	
• Paraoxonase	
• Broccoli	
and many others	

11.5 WOULD ANTI-INFLAMMATORIES AND ANTIOXIDANTS PREVENT ATHEROSCLEROSIS?

Is there any scientific evidence to suggest that antioxidants prevent chronic diseases? The verdict is unclear. Countless review articles have appeared regarding the "antioxidant trials" that have failed to document the benefits of long-term consumption of antioxidant vitamins to affect cardiovascular diseases [112]. Regardless of the reasons, the conclusions are undeniable. One question remains unanswered. Did the antioxidant supplements attenuate inflammation? In other words, did the suppression of oxidative stress have an impact on the inflammatory stress or did the latter prevail and not affect the course of the disease despite a reduction in oxidative stress?

A clear-cut answer can't be forthcoming as the inflammatory process is as complex as the oxidative process. First, we need to explore whether there are reasons to expect that attenuation of oxidative stress would have an effect on inflammation. Some of the common markers of inflammation are shown in Table 11.2. These include the so-called high-sensitive CRP, serum amyloid A, many interleukins, TNF-α, MCP-1, adhesion molecules, enzymes such as PAF-acetyl hydrolase and phospholipases, myeloperoidase, and many others. In addition, there are numerous lipid inflammatory mediators, such as the prostaglandins. The anti-inflammatory molecules are vague and are definitely not macrophage derived. For example, apoA1 (and HDL) and paraoxonases are suggested to be anti-inflammatory.

Whether oxidative stress directly influences the synthesis of the genes and proteins related to inflammation remains to be established. Currently, oxidative stress is believed to influence specific translational events mediated by NFkB [113–126], AP-1 [127–129], ARE [130], and others (Table 11.4). It is understood that if the genes of interest possess these response elements, they would be affected or influenced by oxidative stress. Some of the inflammatory genes that might be affected by oxidative stress are listed in Table 11.4.

Inflammation could have both beneficial and negative effects (Figure 11.3). Many anti-inflammatory agents have been shown to affect atherosclerosis and other cardiovascular end points (Table 11.5). It is a common fallacy to assume that anything that is elevated under a pathological condition represents a condition that needs to be attenuated and these agents affected inflammation under the experimental conditions and thereby prevented atherosclerosis. The body's adaptive response is often ignored. Besides, some of the inflammatory molecules could be atheroprotective. For example, IL-10 exerts its atheroprotective effect on plaque progression,

TABLE 11.4
Inflammatory Molecules and Their Mediators

	Regulatory Factors			
Inflammatory Genes	NFkB in the Promoter Region of	AP1-Binding Site in the Promoter Region of	AP2, AP3 Binding Site in the Promoter Region of	ARE Through Nrf2
Pro-inflammatory cytokines	MCP-1, M-CSF, G-CSF, GM-CSF, TF, CYP1A1, CYP1B1, IL 1, IL 6, TNF-α			VEGF
Matrix metalloproteinases	MMP 1, MMP 3, MMP 9			
Chemokines	IL 8, MIP 1 α, MCP-1, Eotaxin	IL-8, MCP-1		
Inflammatory enzymes	COX-2, 5LOX, iNOS, cytosolic PLA2		COX2, cSPLA2	
Adhesion molecules	ICAM-1, VCAM-1, ELAM-1, E-selectin	ELAM-1, ICAM1 VCAM1		

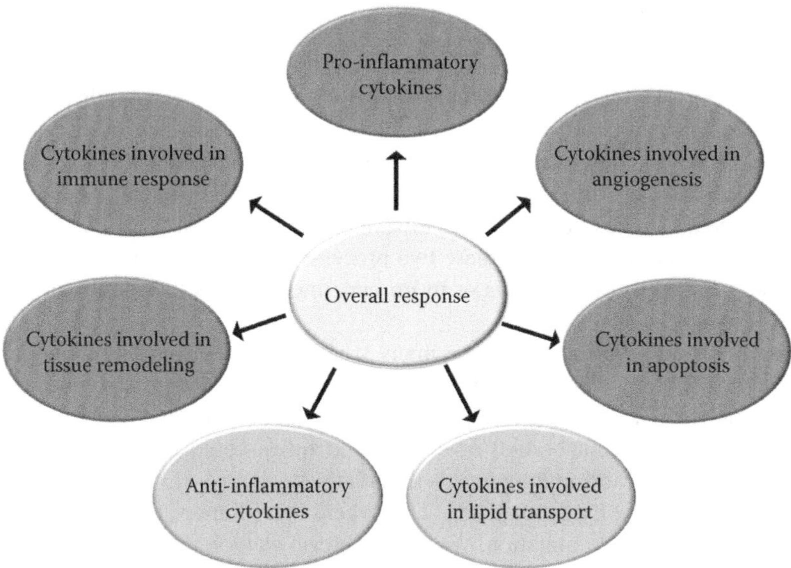

FIGURE 11.3 Overall response of inflammation.

rupture, or thrombosis throughout the different stages of atherosclerosis by influencing the local inflammatory process within the atherosclerotic lesion. As an anti-inflammatory, cytokine IL-10's atheroprotective effects are exerted mainly by inhibiting various cell processes including the production of inflammatory mediators, matrix metalloproteinases (MMPs) and tissue factor (TF) production, and apoptosis.

TABLE 11.5
Anti-inflammatory Drugs and Atherosclerosis

Category	Anti-inflammatory Drugs	Atherosclerosis (ATH)
Sialyc acid derivative	Aspirin	Prevents ATH in type 2 diabetic patients
		Inhibits progression of ATH
		Inhibits ATH in rabbits
		Attenuates the initiation but not the progression
		Continuous administration of aspirin attenuates ATH
		Prevents ATH in monkey
Propionic acid derivative	Ibuprofen	Reduced atherosclerosis
		Protection of LDL from oxidation
Acetic acid derivative	Indomethacin	Reduced ATH
		Reduced ATH in apo E null mice
		Suppression of oxidative stress
		Reduced ATH in LDL r(-/-) mice
Glucocorticoid	Dexamethasone	Suppresses ATH in rabbits
		Reduced ATH in rabbits
		Reduced PMN chemotaxis
		Inhibits accumulation of macrophages
Glucocorticoid	Prednisolone	Did not show inhibition in ATH of RA patients
	Phenylbutazone,	Reduces ATH in patients
	Flufenamic acid	Reduces ATH in patients
	Oxyphenylbutazone	Reduces ATH in patients

11.6 CONCLUSIONS

In summary, the pathways of oxidative stress that contribute to chronic diseases are poorly understood. Inflammation is seen as an independent risk factor for diseases such as atherosclerosis. While there is enough evidence to link these two, the hypothesis that oxidative stress drives inflammation needs to be tested. However, whether attenuating either of these independently is feasible and would have an effect on disease processes remains to be established. Clinicians and scientists tend to shy away from antioxidant studies due to the negative outcome of clinical trials. A similar fate might befall anti-inflammatory agent trials if these two processes are closely linked. Besides, neither of these is currently linked to dyslipidemia or to immune mechanisms that might play an early and late role in the disease.

During the 1990s and 2000s, there were many studies that described the potential beneficial roles of oxidants [131, 132]. These included the induction of antioxidant enzymes and mediators of vascular functions [133]. In addition, we recently pointed out that some of the oxidation-derived breakdown products (e.g., azelaic acid) of peroxidized lipids could be beneficial in being anti-inflammatory as well in retarding the progression of atherosclerosis [134]. The generation of such compounds is also inhibited by antioxidants. Thus, there is a lot more to be learned about oxidative stress as well as about inflammation. We must exercise caution in abandoning antioxidants, if oxidative stress and inflammation are closely related and if they play a key role in chronic diseases. On the other hand, inflammatory cells and processes are vital and expansive to the body's defense mechanisms such that we need to understand specific mechanisms and connections to dyslipidemia before we start using generic antioxidants and anti-inflammatory molecules to control major diseases.

TAKE-HOME MESSAGES

- Oxidative stress has been associated with many chronic diseases.
- Many chronic diseases also have been known to be associated with elevated inflammatory markers.
- Antioxidants that inhibit oxidative stress also attenuate inflammatory stress.
- Many genes for inflammatory proteins appear to be influenced by antioxidants.
- However, whether the attenuation of either oxidative stress or inflammation would have an effect on the disease process needs to be addressed.
- Considering the findings that antioxidants, in general, did not affect the progression of cardiovascular disease in humans is of concern as atherosclerotic cardiovascular disease is now considered an inflammatory disease.
- The connection between oxidative stress, inflammation, and dyslipidemia as seen in atherosclerotic diseases is unexplored.

ACKNOWLEDGMENT

This study was supported by National Institutes of Health grants HL69038 and DK056353.

REFERENCES

1. Glass CK, Witztum JL. Atherosclerosis. The road ahead. *Cell* 104:503–516, 2001.
2. Libby P. Inflammation in atherosclerosis *Nature* 420:868–874, 2002.
3. Hansson GK. Inflammation, atherosclerosis, and coronary artery disease *N Engl J Med* 352:1685–1695, 2005.
4. Steinberg D, Witztum JL. Lipoproteins and atherogenesis; current concepts. *J Am Med Assoc* 264:3047–3052, 1990
5. Ross R. The pathogenesis of atherosclerosis; a perspective for the 1990s. *Nature* 362:801–809, 1993.

6. Goldstein JL, Ho YK, Basu SK, Brown MS. Binding site on macrophages that mediates uptake and degradation of acetylated low density lipoprotein, producing massive cholesterol deposition. *Proc Natl Acad Sci USA*.76:333–337, 1979.
7. Brown MS, Basu SK, Falck JR, Ho YK, Goldstein JL. The scavenger cell pathway for lipoprotein degradation: Specificity of the binding site that mediates the uptake of negatively-charged LDL by macrophages. *J Supramol Struct*. 13:67–81, 1980.
8. Haberland ME, Fogelman AM, Edwards PA. Specificity of receptor-mediated recognition of malondialdehyde modified low density lipoproteins *Proc Natl Acad Sci* 79:1712–1716, 1982.
9. Haberland ME, Fless GM, Scanu AM, Fogelman AM. Malondialdehyde modification of lipoprotein (a) produces avid uptake by human monocyte-macrophages. *J Biol Chem*. 267:4143–4151, 1992.
10. Fogelman AM, Shechter I, Seager J, Hokom M, Child JS, Edwards PA. Malondialdehyde alteration of low density lipoproteins leads to cholesteryl ester accumulation in human monocyte-macrophages. *Proc Natl Acad Sci USA* 77:2214–2218, 1980.
11. Shechter I, Fogelman AM, Haberland ME, Seager J, Hokom M, Edwards PA. The metabolism of native and malondialdehyde-altered low density lipoproteins by human monocyte-macrophages. *J Lipid Res* 22:63–71,1981.
12. Hoff HF, Cole TB. Macrophages uptake by low density lipoprotein modified by 4-hydroxy nonenal an ultrastructure study. *Lab Invest* 64:254–264, 1991.
13. Ferretti G, Bacchetti T, Negre-Salvayre A, Salvayre R, Dousset N, Curatola G. Structural modifications of HDL and functional consequences. *Atherosclerosis* 184:1–7, 2006.
14. Kervinen K, Horkko S, Beltz WF, Kesaniemi A. Modification of VLDL apoprotein B by acetaldehyde alters apoprotein B metabolism. *Alcohol* 12:189–194, 1995.
15. Nagano Y, Arai H, Kita T. High density lipoprotein loses its effect to stimulate efflux of cholesterol from foam cells after oxidative modification. *Proc Natl Acad Sci USA* 88:6457–6461, 1991.
16. Ghiselli G, Giorgini L, Gelati M, Musanti R. Oxidatively modified HDLs are potent inhibitors of cholesterol biosynthesis in human skin fibroblasts. *Arterioscler Thromb* 12:929–935, 1992.
17. Van Lenten BJ, Wagner AC, Nayak DP, Hama S, Navab M, Fogelman AM. High-density lipoprotein loses its anti-inflammatory properties during acute influenza infection. *Circulation* 103:2283–2288, 2001.
18. Fowler SH, Shio H, Haley NJ. Characteristics of lipid-laden aortic cells from cholesterol-fed rabbits: IV. Investigation of macrophage-like properties of aortic cell populations. *Lab Invest* 41:372–378, 1979.
19. Schaffner T, Taylor K, Bartucci EJ, Fischen-Dzoga JH, Beeson S, Glagov S, Wissler RW. Arterial foam cells with distinctive immunomorphologic and histochemical features of macrophages. *Am J Pathol* 100:57–80, 1980.
20. Garrity RG. The role of the monocyte in atherogenesis. *Am J Pathol* 103:181–190, 1981.
21. Ross R. Atherosclerosis—an inflammatory disease. *N Engl J Medicine* 340:115–126, 1999.
22. Stary H, Chandler A, Glagov S, Guyton J, Insull W Jr, Rosenfeld M, Schaffer SA, Schwartz CJ, Wagner WD, Wissler RW. A definition of initial, fatty streak, and intermediate lesions of atherosclerosis. A report from the Committee on Vascular Lesions of the Council on Arteriosclerosis, American Heart Association. *Arterioscler Thromb* 14:840–856, 1994.
23. Eggen DA, Strong JP, McGill HC. Coronary calcification: Relationship to clinically significant coronary lesions and race, sex and topographic distribution. *Circulation* 32:948–955, 1965.
24. Virmani R, Ladich ER, Burke AP, Kolodgie FD. Histopathology of carotid atherosclerotic disease. *Neurosurgery* 59:S219–S227, 2006.
25. Virmani R, Burke AP, Farb A, Kolodgie FD. Pathology of the vulnerable plaque. *J Am Coll Cardiol* 47:C13–C18, 2006.
26. Ni M, Wang Y, Zhang M, Zhang PF, Ding SF, Liu CX, et al. Atherosclerotic plaque disruption induced by stress and lipopolysaccharide in apolipoprotein E knockout mice. *AJP - Heart* 296:H1598–H1606, 2009.
27. Spagnoli LG, Mauriello A, Sangiorgi G, et al. Extracranial thrombotically active carotid plaque as a risk factor for ischemic stroke. *JAMA* 292:1845–1852, 2004.
28. Libby P. Current concepts of the pathogenesis of the acute coronary syndromes. *Circulation* 104:365–372, 2001.
29. Khallou-Laschet J, Varthaman A, Fornasa G, Compain C, Gaston AT, Clement M, et al. Macrophage plasticity in experimental atherosclerosis. *PLoS One* 5(1):e8852, 2010.
30. Steinberg D, Parthasarathy S, Crew TE, Khoo JC, Witztum JL. Beyond cholesterol: Modification of low-density lipoprotein that increase its atherogenecity. *N Engl J Med* 320:915–924, 1989.

31. Parthasarathy S. *Modified Lipoproteins in the Pathogenesis of Atherosclerosis.* Austin, Texas: R. G. Landes Co. 1994.
32. Chisolm GM, Steinberg D. The oxidative modification hypothesis of atherogenesis: An overview. *Free Radic Biol Med* 28:1815–1826, 2000.
33. Berliner J. Introduction. Lipid oxidation products and atherosclerosis. *Vascul Pharmacol* 38:187–191, 2002.
34. Matsuura E, Kobayashi K, Tabuchi M, Lopez LR. Oxidative modification of low-density lipoprotein and immune regulation of atherosclerosis. *Prog Lipid Res* 45:466–486, 2006.
35. Bergt C, Pennathur S, Fu X, Byun J, O'Brien K, McDonanld TO, et al. The myeloperoxidase product hypochlorous acid oxidizes HDL in the human artery wall and impairs ABCA1-dependent cholesterol transport. *Proc Natl Acad Sci USA* 101:13032–13037, 2004.
36. Lamb DJ, Leake DS. Iron released from transferrin at acidic pH can catalyse the oxidation of low density lipoprotein. *FEBS Lett* 352:15–18, 1994.
37. Lamb DJ, Hider RC, Leake DS. Hydroxypyridinones and desferrioxamine inhibit macrophage mediated LDL oxidation by iron but not by copper. *Biochem Soc Trans* 21:234S, 1993.
38. Natarajan R, Gerrity RG, Gu JL, Lanting L, Thomas L, Nadler JL. Role of 12-lipoxygenase and oxidant stress in hyperglycaemia-induced acceleration of atherosclerosis in a diabetic pig model. *Diabetologia* 45:125–133, 2002.
39. Natarajan R, Gu JL, Rossi J, Gonzales N, Lanting L, Xu L, Nadler J. Elevated glucose and angiotensin II increase 12-lipoxygenase activity and expression in porcine aortic smooth muscle cells. *Proc Natl Acad Sci USA* 90:4947–4951, 1993.
40. Shen J, Herderick E, Cornhill JF, Zsigmond E, Kim HS, Kuhn H, et al. Macrophage-mediated 15-lipoxygenase expression protects against atherosclerosis development. *J Clin Investig* 98:2201–2208, 1996.
41. Harats D, Shaish A, George J, Mulkins M, Kurihara H, Levkovitz H, et al. Overexpression of 15-lipoxygenase in vascular endothelium accelerates early atherosclerosis in LDL receptor-deficient mice. *Arterioscler Thromb Vasc Biol* 20:2100–2105, 2000.
42. Huo Y, Zhao L, Hyman MC, Shashkin P, Harry BL, Burcin T, et al. Critical role of macrophage 12/15-lipoxygenase for atherosclerosis in apolipoprotein E-deficient mice. *Circulation* 110:2024–2031, 2004.
43. Frei B, Forte TM, Ames BN, Cross CE. Gas phase oxidants of cigarette smoke induce lipid peroxidation and changes in lipoprotein properties in human blood plasma. Protective effects of ascorbic acid. *Biochem J* 277:133–138, 1991.
44. Bhatnagar A. Cardiovascular pathophysiology of environmental pollutants. *Am J Physiol Heart Circ Physiol* 286:H479–H485, 2004.
45. Brennan ML, Anderson MM, Shih DM, Qu XD, Wang X, Mehta AC, et al. Increased atherosclerosis in myeloperoxidase-deficient mice. *J Clin Invest* 107:419–430, 2001.
46. McMillen TS, Heinecke JW, LeBoeuf RC. Expression of human myeloperoxidase by macrophages promotes atherosclerosis in mice. *Circulation* 111:2798–2804, 2005.
47. Tribble DL, Gong EL, Leeuwenburgh C, Heinecke JW, Carlson EL, Verstuyft JG, et al. Fatty streak formation in fat-fed mice expressing human copper-zinc superoxide dismutase. *Arterioscler Thromb Vasc Biol* 17:1734–1740, 1997.
48. Santanam N, Parthasarathy S. Paradoxical actions of antioxidants in the oxidation of low density lipoprotein by peroxidases. *J Clin Invest* 95:2594–2600, 1995.
49. Gieseg S, Duggan S, Gebicki JM. Peroxidation of proteins before lipids in U937 cells exposed to peroxyl radicals. *Biochem J* 350:215–218, 2000.
50. Patel R, Diczfalusy U, Dzeletovic S, Wilson M, Darley-Usmar V. Formation of oxysterols during oxidation of low density lipoprotein by peroxynitrite, myoglobin, and copper. *J Lipid Res* 37:2361–2371, 1996.
51. Ross R: The pathogenesis of atherosclerosis: A perspective for the 1990s. *Nature* 362:801–809, 1993.
52. Nick Platt N, Gordon S. Is the class A macrophage scavenger receptor (SR-A) multifunctional?—the mouse's tale. *J Clin Invest* 108:649–654, 2001.
53. Rahaman SO, Lennon DJ, Febbraio M, Podrez EA, Hazen SL, Silverstein RL. A CD36-dependent signaling cascade is necessary for macrophage foam cell formation. *Cell Metab* 4:211–221, 2006.
54. Fraser I, Hughes D, Gordon S. Divalent cation-independent macrophage adhesion inhibited by monoclonal antibody to murine scavenger receptor. *Nature* 364:343–346, 1993.
55. Aderem A, Underhill DM. Mechanisms of phagocytosis in macrophages. *Annu Rev Immunol* 17:593–623, 1999.

56. Erwig LP, Henson PM. Clearance of apoptotic cells by phagocytes. *Cell Death Differ* 15:243–50, 2008.
57. Gordon S. The macrophage: Past, present and future. *Eur J Immunol* 37:S9–S17, 2007.
58. Nicholls SJ, Hazen SL. Myeloperoxidase and cardiovascular disease. *Arterioscler Thromb Vasc Biol* 25:1102–11, 2005.
59. Brennan ML, Anderson MM, Shih DM, Qu XD, Wang X, Mehta AC, et al. Increased atherosclerosis in myeloperoxidase-deficient mice. *J Clin Invest* 107:419–430, 2001.
60. McMillen TS, Heinecke JW, LeBoeuf RC. Expression of human myeloperoxidase by macrophages promotes atherosclerosis in mice. *Circulation* 111:2798–2804, 2005.
61. Braunwald E, Bristow MR. Congestive heart failure: Fifty years of progress. *Circulation* 102:IV14–IV23, 2000.
62. Clerico A, Emdin M. Diagnostic accuracy and prognostic relevance of the measurement of the cardiac natriuretic peptides: A review. *Clin Chem* 50:33–50, 2004.
63. Doust JA, Glasziou PP, Pietrzak E, Dobson AJ. A systematic review of the diagnostic accuracy of natriuretic peptides for heart failure. *Arch Intern Med* 164:1978–1984, 2004.
64. Biasucci LM, Biasillo G, Stefanelli A. Inflammatory markers, cholesterol and statins: Pathophysiological role and clinical importance. *Clin Chem Lab Med* 48(12):1685–1691, 2010.
65. Braun LT, Davidson MH. Lp-PLA2: A new target for statin therapy. *Curr Atheroscler Rep* 12(1):29–33, 2010.
66. Stary HC. Evolution and progression of atherosclerotic lesions in coronary arteries of children and young adults. *Arteriosclerosis* 9:19–132, 1989.
67. Homma S, Troxclair DA, Zieske AW, Malcom GT, Strong JP. The Pathobiological Determinants of Atherosclerosis in Youth (PDAY) Research Group. Histological topographical comparisons of atherosclerosis progression in juveniles and young adults. *Atherosclerosis* 2007.
68. McMahan CA, Gidding SS, Malcom GT, Schreiner PJ, Strong JP, Tracy RE, et al. Pathobiological Determinants of Atherosclerosis in Youth (PDAY) Research Group. Comparison of coronary heart disease risk factors in autopsied young adults from the PDAY Study with living young adults from the CARDIA study. *Cardiovasc Pathol* 16:151–158, 2007.
69. Isles CG, Hole DJ, Hawthorne VM, Lever AF. Relation between coronary risk and coronary mortality in women of the Renfrew and Paisley survey: Comparison with men. *Lancet* 339:702–706, 1992.
70. Bemeur C, Ste-Marie L, Montgomery J. Increased oxidative stress during hyperglycemic cerebral ischemia. *Neurochem Int* 50:890–904, 2007.
71. Ceriello A. Oxidative stress and diabetes-associated complications. *Endocr Pract* 12:S60–S62, 2006.
72. Rolo AP, Palmeira CM. Diabetes and mitochondrial function: Role of hyperglycemia and oxidative stress. *Toxicol Appl Pharmacol* 212:167–178 2006.
73. Bonnefont-Rousselot D. Glucose and reactive oxygen species. *Curr Opin Clin Nutr Metab Care* 5:561–568, 2002.
74. Hunt JV, Dean RT, Wolff SP. Hydroxyl radical production and autoxidative glycosylation. Glucose autoxidation as the cause of protein damage in the experimental glycation model of diabetes mellitus and ageing. *Biochem J* 256:205–212, 1988.
75. Masaki H, Okano Y, Sakurai H. Generation of active oxygen species from advanced glycation endproducts (AGE) under ultraviolet light A (UVA) irradiation. *Biochem Biophys Res Commun* 235: 306–310, 1997.
76. Schmidt AM, Hori O, Brett J, Yan SD, Wautier JL, Stern D. Cellular receptors for advanced glycation end products. Implications for induction of oxidant stress and cellular dysfunction in the pathogenesis of vascular lesions. *Arterioscler Thromb* 14:1521–1528, 1994.
77. Al-Abed Y, Liebich H, Voelter W, Bucala R. Hydroxyalkenal formation induced by advanced glycosylation of low density lipoprotein. *J Biol Chem* 271:2892–2896, 1996.
78. Bucala R, Makita Z, Koschinsky T, Cerami A, Vlassara H. Lipid advanced glycosylation: Pathway for lipid oxidation in vivo. *Proc Natl Acad Sci USA* 90:6434–6438, 1993.
79. Libby P. The vascular biology of atherosclerosis. In: Bonow RO, Mann DL, Zipes DP, Libby P, eds. *Braunwald's Heart Disease: A Textbook of Cardiovascular Medicine*, 9th ed. Philadelphia, PA: Saunders Elsevier; chap 43, 2011.
80. Savransky V, Jun J, Li J, Nanayakkara A, Fonti S, Moser AB, Steele KE, et al. Dyslipidemia and atherosclerosis induced by chronic intermittent hypoxia are attenuated by deficiency of stearoyl coenzyme A desaturase *Circulation Research* 103:1173–1180, 2008.
81. Mehta PK, Griendling KK. Angiotensin II cell signaling: Physiological and pathological effects in the cardiovascular system. *Am J Physiol Cell Physiol* 292:C82–97, 2007.
82. Ushio-Fukai M, Alexander RW. Reactive oxygen species as mediators of angiogenesis signaling: Role of NAD(P)H oxidase. *Mol Cell Biochem* 264:85–97, 2004.

83. Ushio-Fukai, M. Redox signaling in angiogenesis: Role of NADPH oxidase. *Cardiovasc Res* 71:226–235, 2006.
84. Touyz RM. Reactive oxygen species as mediators of calcium signaling by angiotensin II: Implications in vascular physiology and pathophysiology. *Antioxid Redox Signal* 7:1302–1314, 2005.
85. Beckman JS, Beckman TW, Chen J, Marshall PA, Freeman BA. Apparent hydroxyl radical production by peroxynitrite: Implications for endothelial injury from nitric oxide and superoxide. *Proc Natl Acad Sci USA* 87:1620–1624, 1990.
86. Bouloumie A, Marumo T, Lafontan M, Busse R. Leptin induces oxidative stress in human endothelial cells. *FASEB J* 13:1231–1238, 1999.
87. Penumetcha M, Parthasarathy S. (unpublished observations)
88. Chalmers A. Smoking and oxidative stress. *Am J Clin Nutr* 69:572, 1999.
89. Cross CE, Van der Vliet A, Eiserich JP. Cigarette smokers and oxidant stress: A continuing mystery. *Am J Clin Nutr* 67:184–185, 1998.
90. Burke A, Fitzgerald GA. Oxidative stress and smoking-induced vascular injury. *Prog Cardiovasc Dis* 46:79–90, 2003.
91. Yokode M, Ueyama K, Arai NH, Ueda Y, Kita T. Modification of high- and low-density lipoproteins by cigarette smoke oxidants. *Ann NY Acad Sci* 15:245–251, 1996.
92. Yamaguchi Y, Matsuno S, Kagota S, Haginaka J, Kunitomo M. Oxidants in cigarette smoke extract modify low-density lipoprotein in the plasma and facilitate atherogenesis in the aorta of Watanabe heritable hyperlipidemic rabbits. *Atherosclerosis* 156:109–117, 2001.
93. Bloomer RJ. Decreased blood antioxidant capacity and increased lipid peroxidation in young cigarette smokers compared to nonsmokers: Impact of dietary intake. *Nutr J* 6:39, 2007.
94. Rikitake Y, Kawashima S, Takeshita S, Yamashita T, Azumi H, Yasuhara M, et al. Anti-oxidative properties of fluvastatin, an HMG-CoA reductase inhibitor, contribute to prevention of atherosclerosis in cholesterol-fed rabbits. *Atherosclerosis* 154:87–96, 2001.
95. Stoll LL, McCormick ML, Denning GM, Weintraub NL. Antioxidant effects of statins. *Drugs Today (Barc)* 40:975–990, 2004.
96. Cominacini L, Fratta Pasini A, Garbin U, Pastorino AM, Davoli A, Nava C, et al. Antioxidant activity of different dihydropyridines. *Biochem Biophys Res Commun* 302:679–684, 2003.
97. Inouye M, Mio T, Sumino K. Nilvadipine protects low-density lipoprotein cholesterol from in vivo oxidation in hypertensive patients with risk factors for atherosclerosis. *Eur J Clin Pharmacol* 5:35–41, 2000.
98. Kritz H, Oguogho A, Aghajanian AA, Sinzinger H. Semotiadil, a new calcium antagonist, is a very potent inhibitor of LDL-oxidation, prostaglandins. *Leukot Essent Fatty Acids* 61:183–188, 1999.
99. Napoli C, Salomone S, Godfraind T, Palinski W, Capuzzi DM, Palumbo G, et al. 1,4-Dihydropyridine calcium channel blockers inhibit plasma and LDL oxidation and formation of oxidation-specific epitopes in the arterial wall and prolong survival in stroke-prone spontaneously hypertensive rats. *Stroke* 30:1907–1915, 1999.
100. Lupo E, Locher R, Weisser B, Vetter W. In vitro antioxidant activity of calcium antagonists against LDL oxidation compared with alpha-tocopherol. *Biochem Biophys Res Commun* 203:1803–1808, 1994.
101. Khan BV, Navalkar S, Khan QA, Rahman ST, Parthasarathy S. Irbesartan, an angiotensin type 1 receptor inhibitor, regulates the vascular oxidative state in patients with coronary artery disease. *J Am Coll Cardiol* 38:1662–1667, 2001.
102. Godfrey EG, Stewart J, Dargie HJ, Reid JL, Dominiczak M, Hamilton CA, et al. Effects of ACE inhibitors on oxidation of human low density lipoprotein. *Br J Clin Pharmacol* 37:63–66, 1994.
103. Ziedén B, Wuttge DM, Karlberg BE, Olsson AG. Effects of in vitro addition of captopril on copper-induced low density lipoprotein oxidation. *Br J Clin Pharmacol* 39:201–203, 1995.
104. Hayek T, Attias J, Smith J, Breslow JL, Keidar S. Antiatherosclerotic and antioxidative effects of captopril in apolipoprotein E-deficient mice. *J Cardiovasc Pharmacol* 31:540–544, 1998.
105. Van Antwerpen P, Boudjeltia KZ, Babar S, Legssyer I, Moreau P, Moguilevsky N, et al. Thiol-containing molecules interact with the yeloperoxidase/H_2O_2/chloride system to inhibit LDL oxidation. *Biochem Biophys Res Commun* 337:82–88, 2005.
106. Van Antwerpen P, Legssyer I, Zouaoui Boudjeltia K, Babar S, Moreau P, Moguilevsky N, et al. Captopril inhibits the oxidative modification of apolipoprotein B-100 caused by myeloperoxidase in a comparative in vitro assay of angiotensin converting enzyme inhibitors. *Eur J Pharmacol* 537:31–36, 2006.
107. Miura T, Muraoka S, Ogiso T. Antioxidant activity of adrenergic agents derived from catechol. *Biochem Pharmacol* 55:2001–2016, 1998.

108. Hogg N, Struck A, Goss SP, Santanam N, Joseph J, Parthasarathy S, Kalyanaraman B. Inhibition of macrophage-dependent low density lipoprotein oxidation by nitric-oxide donors. *J Lipid Res* 36:1756–1762, 1995.
109. Hogg N, Kalyanaraman B, Joseph J, Struck A, Parthasarathy S. Inhibition of low-density lipoprotein oxidation by nitric oxide, potential role in atherogenesis. *FEBS Lett* 334:170–174, 1993.
110. Hermann M, Kapiotis S, Hofbauer R, Exner M, Seelos C, Held I, et al. Salicylate inhibits LDL oxidation initiated by superoxide/nitric oxide radicals. *FEBS Lett* 445:212–214, 1999.
111. Whiteman M, Kaur H, Halliwell B. Protection against peroxynitrite dependent tyrosine nitration and alpha 1-antiproteinase inactivation by some anti-inflammatory drugs and by the antibiotic tetracycline. *Ann Rheum Dis* 55:383–387, 1996.
112. Williams KJ, Fisher EA. Oxidation, lipoproteins, and atherosclerosis: Which is wrong, the antioxidants or the theory? *Curr Opin Clin Nutr Metab Care* 8:139–146, 2005.
113. Dwarakanath RS, Sahar S, Reddy MA, Castanotto D, Rossi JJ, Natarajan R. Regulation of monocyte chemoattractant protein-1 by the oxidized lipid, 13-hydroperoxyoctadecadienoic acid, in vascular smooth muscle cells via nuclear factor–kappa B (NF-κB). *J Mol Cell Cardio* 36: 585–595, 2004.
114. Baeuerle PA, Henkee T. Function and activation of NF-kappa B in the immune system. *Annu Rev Immunol* 12:141–179, 1994.
115. Brennan FM, Maini RN, Feldmann M. Cytokine expression in chronic inflammatory disease. *Br Med Bull* 51:368–384, 1995.
116. Ward PA. Role of complement, chemokines and regulatory cytokines in acute lung injury. *Ann N Y Acad Sci* 796:104–112, 1996.
117. Akira S, Kishimoto A. NF-IL6 and NF-kB in cytokine gene regulation. *Adv Immunol* 65:1–46, 1997.
118. Bond, M., Chase AJ, Baker AH, Newby AC. Inhibition of transcription factor NFkB#61547; B reduces matrix metalloproteinase-1, -3 and -9 production by vascular smooth muscle cells. *Cardiovasc Res* 50:556–565, 2001.
119. Leonardo MJ, Baltimore D. NF-KB: A pleiotropic mediator of inducible and tissue-specific gene control *Cell* 58:227–229, 1998.
120. Baeuerle PA, Baltimore PA. NF-kB: Ten years after. *Cell* 87:13–20, 1996.
121. Appleby SB, Ristimäki A, Neilson K, Narko K, Hla T. Structure of the human cyclo-oxygenase-2 gene. *Biochem J* 302:723–727, 1994.
122. Chiang-Wen Lee, Chih-Chung Lin, I-Ta Lee, Hui-Chun Lee, Chuen-Mao Yang. Activation and induction of cytosolic phospholipase A2 by TNF-mediated through Nox2, MAPKs, NF-κB, and p300 in human tracheal smooth muscle cells. *J Cell Physiol* 226:2103–2114, 2011.
123. Xie QW, Kashiwabara Y, Nathan C. Role of transcription factor NFkB/ Rel in induction of nitric oxide synthase. *J Biol Chem* 269:4705–4708, 1994.
124. Osborn L, Hession C, Tizzard R, Vassallo C, Luhowskyj S, Chi-Rosso G, et al. Direct expression cloning of vascular cell adhesion molecule 1, a cytokine induced endothelial protein that binds to lymphocytes *Cell* 59:1203–1211, 1989.
125. Neish AS, William AJ, Palmer HJ, Whitley MZ, Collins T. Functional analysis of the human vascular cell adhesion molecule 1 promoter. *J Exp Med* 176:1583–1593, 1992.
126. Pradyumna ET, Chen X-L, Sundell CL, Laursen JB, Hammes CP, Alexander RW, et al. *Circulation.* 100:1223–1229, 1999.
127. Mukaida N, Mahe Y, Matsushima K. Cooperative interaction of nuclear factor-kB- and cis-regulatory enhancer binding protein-like factor binding elements in activating the interleukin-8 gene by pro-inflammatory cytokines. *J Biol Chem* 265:21128–21133, 1990.
128. Roebuck KA, Rahman A, Lakshminarayanan V, Janakidevi K, Malik AB. H2O2 and tumour necrosis factor-alpha activate intercellular adhesion molecule 1 (ICAM-1) gene transcription through distinct cis-regulatory elements within the ICAM-1 promoter. *J Biol Chem* 270:18966–18974, 1995.
129. Ahmad M, Theofanidis P, Medford RM. Role of activating protein-1 in the regulation of the vascular cell adhesion molecule-1 gene expression by tumor necrosis factor-alpha. *J Biol Chem* 273:4616–4621, 1998.
130. Afonyushkin T, Oskolkova OV, Philippova M, Resink TJ, Erne P, Binder BR, et al. Oxidized phospholipids regulate expression of ATF4 and VEGF in endothelial cells via NRF2-dependent mechanism: Novel point of convergence between electrophilic and unfolded protein stress pathways. *Arteriosclerosis, Thrombosis, and Vascular Biology* 30:1007–1013, 2010.
131. Parthasarathy S, Santanam N, Ramachandran S, Meilhac O. Oxidants and antioxidants in atherogenesis: An appraisal. 40:2143–2157, 1999.

132. Ramasamy S, Parthasarathy S, Harrison DG. Regulation of endothelial nitric oxide synthase gene expression by oxidized linoleic acid. *J Lipid Res* 39:268–276, 1998.
133. Parthasarathy S, Merchant NK, Penumetcha M, Santanm N. Oxidation and cardiovascular disease-potential role of oxidants in inducing antioxidant defense enzymes. *J Nucl Cardiol* 8:379–389, 2001.
134. Litvinov D, Selvarajan K, Garelnabi M, Brophy L, Parthasarathy S. Anti-atherosclerotic actions of azelaic acid, an end product of linoleic acid peroxidation, in mice. *Atherosclerosis* 209:449–454, 2010.

12 Chronic Inflammation and Cancer

A Matter of Lifestyle

Subash C. Gupta, Ji Hye Kim, Sahdeo Prasad, and Bharat B. Aggarwal
The University of Texas MD Anderson Cancer Center
Houston, Texas

CONTENTS

12.1 WHAT IS INFLAMMATION?

The term *inflammation* is derived from the Latin word *inflammare* meaning "to set on fire." Inflammation is usually indicated by adding the suffix *-itis* to the organ where inflammation is induced (Table 12.1), as in cervicitis (inflammation of the cervix) or bronchitis (inflammation of the bronchi of the lung). Some conditions, such as inflammatory bowel disease and prostate inflammatory atrophy, do not follow this convention. A variety of inducers of inflammation have been reported, some of which are listed in Table 12.1. Inflammation is a part of the host defense system that counteracts insults incurred by internal or external stimuli. The clinical and fundamental signs of inflammation include redness, swelling, heat, pain, and loss of function. Roman physician Cornelius Celsus (ca. 30 BCE–38 CE) was the first to describe the first four signs of inflammation, while loss of function was added by German physician Rudolf Virchow in 1870 [1].

Inflammation can be classified as acute or chronic. Acute inflammation is a means by which plasma and leukocytes leave the blood and enter the tissue around an injured or infected site. Acute inflammation is an immediate response of the body and is required to ward off harmful pathogens. When inflammation persists for a longer time, the type of cells at the site of damage change, leading

TABLE 12.1
Inflammation as a Risk Factor for Cancer

Inducer	Inflammation	Cancer type	Reference
Human papillomavirus	Cervicitis	Cervical	137
Bacteria, GBS	Cholecystitis	Gall bladder	138
Allergy	Dermatitis	Skin	139
Tobacco	Bronchitis	Lung	140
GA, alcohol, tobacco	Esophagitis	Esophageal	141
Epstein-Barr virus	Mononucleosis	Burkitt's lymphoma, Hodgkin's disease	142
Gut pathogens	IBD	Colorectal	143
Infections, STD	PIA	Prostate	144
Infections, radiation	Nephritis	Renal	145
Infections	Encephalitis	Brain	146
Chemo-radiation therapy	Mucositis	Oral	147
Infections	Thyroiditis	Thyroid	148
Infections, radiation	Myositis	Sarcoma	149
Infection, radiation	Otitis	Nasopharyngeal	150
Helicobacter pylori	Gastritis	Gastric	151
Obstruction	Appendicitis	Appendiceal	152
Infection	Laryngitis	Laryngeal	153
Hepatitis B and C virus	Hepatitis	Hepatocellular carcinoma	154
Infection	Orchitis	Testicular	155
Tobacco	Pancreatitis	Pancreatic	156
Viral or bacterial infection	Tonsillitis	Tonsillar	157
Infection	Lymphadenitis	Lymphatic leukemia	158

GA, gastric acid; GBS, gall bladder stones; IBD, inflammatory bowel disease; PIA, prostate inflammatory atrophy; STD, sexually transmitted disease.

to chronic inflammation. Chronic inflammation is a delayed response and has been associated with numerous human chronic diseases, including cardiovascular, pulmonary, autoimmune, and degenerative diseases, cancer, diabetes, and Alzheimer's disease [2]. Research over the past half century has indicated that most of these inflammatory conditions are caused by perturbations in lifestyle factors and thus might be potentially preventable.

In this chapter, we discuss the association of chronic inflammation with cancer. The various lifestyle factors that contribute to inflammation and cancer are reviewed, as are opportunities for preventing inflammation and cancer by modifying these factors.

12.2 REGULATION OF INFLAMMATION AT THE MOLECULAR LEVEL

At the molecular level, inflammation is regulated by numerous molecules and factors, including cytokines (IL-1, IL-2, IL-6, IL-12, TNF-α, TNF-β), chemokines (monocyte chemoattractant protein 1, IL-8), pro-inflammatory transcription factors (NF-κB, STAT3), pro-inflammatory enzymes (COX-2, 5-LOX, 12-LOX, MMPs), prostate-specific antigen (PSA), C-reactive protein, adhesion molecules (intercellular adhesion molecule [ICAM-1], vascular cell adhesion molecule [VCAM-1], endothelial-leukocyte adhesion molecule [ELAM-1]), vascular endothelial growth factor (VEGF), and TWIST (Figure 12.1) [2]. Among all these mediators, NF-κB is the central regulator of inflammation [2, 3]. It has been shown to activate more than 500 genes, most of which are implicated in inflammation [4, 5].

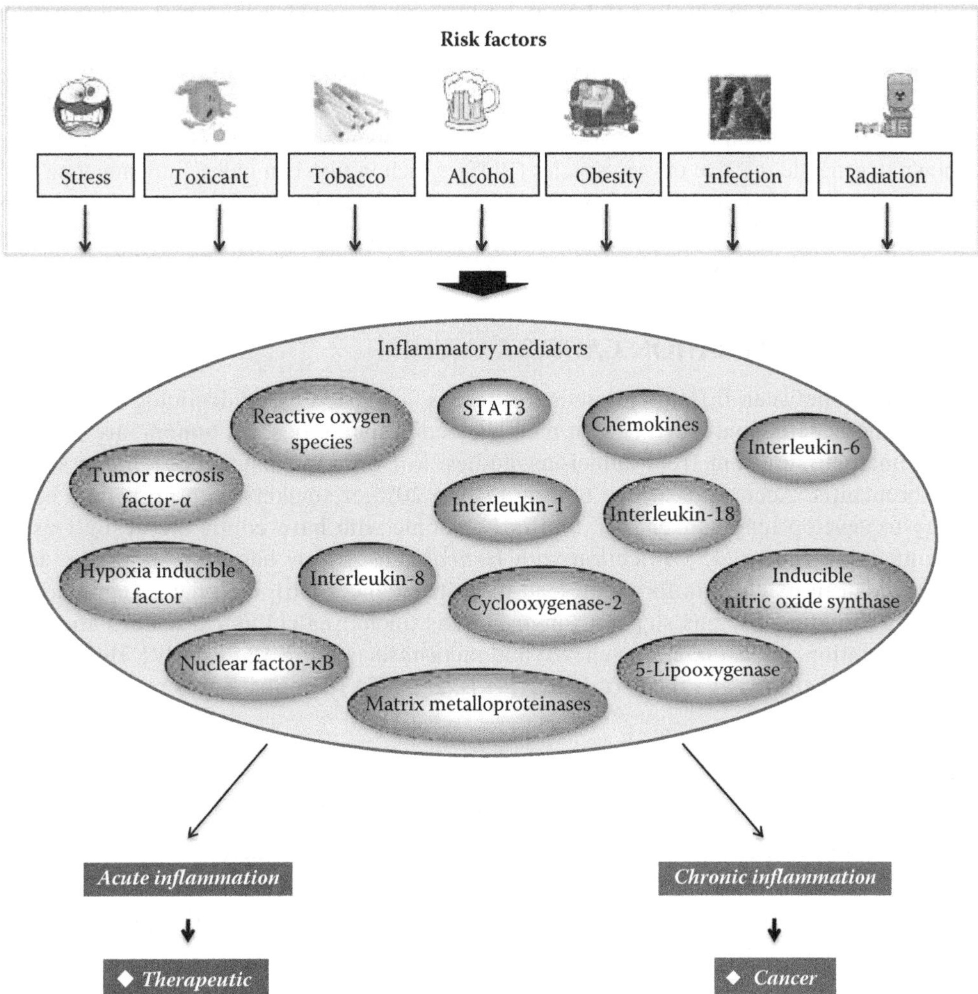

FIGURE 12.1 Risk factors and molecular mediators of inflammation. The molecular mediators of inflammation include cytokines, chemokines, pro-inflammatory transcription factors, and numerous other molecules.

12.3 LIFESTYLE FACTORS THAT ACTIVATE INFLAMMATION

Over the past several years, attempts have been made to identify the causes of inflammation. It is now known that lifestyle factors such as stress, toxicants, tobacco, alcohol, obesity, infectious agents, and radiation can activate inflammatory pathways. For example, diets rich in omega-6 essential fatty acids and low-density lipoproteins are known to induce inflammation. Foods high in low-density lipoproteins are known to induce inflammation in the arteries. The well-known omega-6 essential fatty acids are linoleic acid and arachidonic acid, both of which are consumed all over the world as vegetable oils. While omega-6 essential fatty acids have the potential to induce inflammation, omega-3 fatty acids calm down the inflammation. According to one report, the diet of our ancestors included an omega-6 to omega-3 ratio of 1:1, while the ratio of our current diet is somewhere between 10:1 and 25:1. Other well-known dietary inducers of inflammation are dairy protein (casein) and wheat protein (gluten). Environmental toxicants such as synthetic fibers, latex, glues, adhesives, plastics, air fresheners, and cleaning products have also been shown to induce inflammation. Changes in estrogen, progesterone, and testosterone levels have been shown to play a role in inducing inflammation. Similarly, excessive alcohol consumption can lead to pancreatitis, hepatitis, and colitis and is one of the major risk factors for liver cancer, esophageal cancer, and pancreatic cancer [6].

How lifestyle factors contribute to chronic inflammation has been investigated over the years. These factors are known to modulate numerous pro-inflammatory molecules, including chemokines, cytokines, pro-inflammatory transcription factors, enzymes, and others [7]. Lifestyle factors are known to generate reactive oxygen species (ROS), which in turn can induce inflammation [8, 9, 159]. Cyclooxygenase-2, inflammatory cytokines, chemokines, and pro-inflammatory transcription factors are all known to be regulated by ROS [10]. Furthermore, mitochondrial ROS has been demonstrated to play a major role in inducing chronic inflammation [11, 12].

12.4 HOW INFLAMMATION CAUSES CANCER

The association between inflammation and cancer is supported by epidemiological, pharmacological, and genetic studies [13]. That most cancers, especially solid tumors, are preceded by inflammation is evident from numerous studies. For instance, smokers are more susceptible to bronchitis. According to one report, 15% to 20% of smokers with bronchitis have a tendency to develop lung cancer [14]. Similarly, people who have colitis are at high risk of developing colon cancer [15]. Infection with *Helicobacter pylori* has been shown to induce gastritis, which in its chronic form can lead to gastric cancer [16]. Chronic inflammation is now known to induce various steps of tumorigenesis, including cellular transformation, survival, proliferation, invasion, angiogenesis, and metastasis (Figure 12.2) [9, 17]. How inflammatory molecules, especially TNF-α, ILs, NF-κB, and STAT3 contribute to tumorigenesis is discussed in this section.

12.4.1 Role of Inflammatory Molecules in Cellular Transformation

Cellular transformation is a process whereby normal cells acquire properties of malignant cells. The underlying causes of malignant transformation are gain-of-function mutations in oncogenes and loss-of-function mutations in tumor suppressor genes. The mutation(s) lead to perturbations of

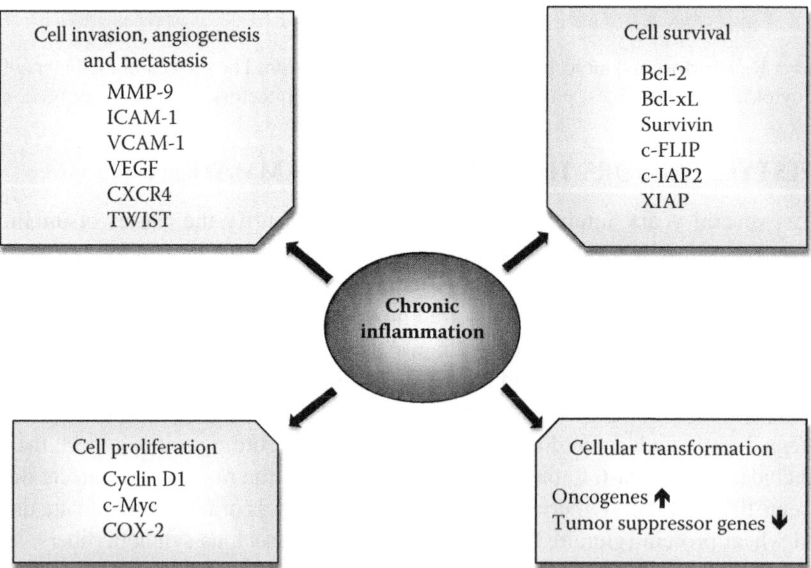

FIGURE 12.2 The targets of inflammatory molecules during tumorigenesis. Inflammatory molecules can target the survival, proliferation, invasion, angiogenesis, and metastasis steps of tumorigenesis.

a number of signaling molecules, including p53, Raf, Rb, PP2A, telomerase, Ral-GEFs, PI3K, Ras, Rac, Myc, STAT3, NF-κB, and hypoxia-inducible factor (HIF)-1α [18].

Mounting evidence over the past several years has indicated an association between inflammatory molecules and malignant transformation. For instance, TNF-α has been shown to play a role in transformation of mouse fibroblasts to malignant cells that was partially suppressed by antioxidants [19]. TNF-α also enhanced cellular transformation initiated by 3-methylcholanthrene in fibroblasts [20]. Matrix metalloproteinase-3 is a stromal enzyme that is upregulated in many breast tumors and has been shown to induce ROS, DNA damage, genomic instability, and transformation of mouse mammary epithelial cells to malignant cells [21].

The transformation of cells by various oncogenes, protein tyrosine kinases, and viruses accompanies activation of NF-κB and STAT3. For example, oncogenic Ras, which is constitutively active in several tumor types, is regulated by NF-κB [22, 23]. Similarly, transformation of cells by Src protein kinase is mediated through activation of STAT3 [24]. The transformation of T cells by human T-cell lymphotropic virus I (HTLV-1) was also mediated through activation of STAT3 [25]. Hepatitis C virus core protein has been shown to transform cells through activation of STAT3 [26].

12.4.2 Role of Inflammatory Molecules in Tumor Cell Survival

Under normal physiological conditions, the human body maintains homeostasis by eliminating unwanted, damaged, aged, and misplaced cells. Homeostasis is carried out via a genetically programmed process referred to as apoptosis (programmed cell death) [27, 28]. One of the chief characteristics of cancer cells is their inherent capacity to evade apoptosis and survive. Cancer cells gain these characteristics by constitutively expressing pro-inflammatory transcription factor NF-κB. NF-κB in turn regulates numerous cell survival proteins, including cFLIP [29], Bcl-xL [30], Bcl-2 [31], XIAP [32], c-IAP1 [33], cIAP-2 [33], and survivin [34]. NF-κB has been linked to anti-apoptotic function in tumors such as T-cell lymphoma, melanoma, pancreatic cancer, bladder cancer, and breast cancer, and in tumor-related cell types such as B cells, T cells, granulocytes, macrophages, neuronal cells, smooth muscle cells, and osteoclasts [2].

Like NF-κB, STAT3 can suppress apoptosis by upregulating cell survival proteins such as Bcl-xL [35], Bcl-2 [36], survivin [37], Mcl-1 [38], and cIAP2 [39]. Most tumor cells that exhibit constitutive activation of STAT3 also express these cell survival gene products [40]. These findings imply a close association between inflammatory molecules and tumor survival. Agents that suppress NF-κB and STAT3 activation can suppress expression of all these cell survival gene products and thus promote cancer cell death [41].

12.4.3 Role of Inflammatory Molecules in Tumor Cell Proliferation

Uncontrolled proliferation is one of the chief characteristics of tumor cells. In normal cells, proliferation is regulated by a robust balance between growth signals and antigrowth signals. Cancer cells, however, acquire the ability to generate their own growth signals and become insensitive to antigrowth signals [42, 43].

Research over the years has indicated that several genes that mediate cell proliferation are regulated by NF-κB [3]. For instance, cyclin D1, a cell cycle-regulatory protein, is known to be regulated by NF-κB [44]. Prostaglandin E_2 (PGE_2) has been shown to induce proliferation of some tumor cells. The synthesis of COX-2, which controls PGE_2 production, is also regulated by NF-κB activation [45]. Growth factors such as epidermal growth factor (EGF) and platelet-derived growth factor (PDGF) are known to induce proliferation of tumor cells through activation of NF-κB [46, 47].

STAT3 activation has also been shown to promote proliferation of certain tumor cells. The ability of STAT3 to promote proliferation depends upon its ability to induce expression of cyclin D1 [48]. Other growth-promoting genes known to be regulated by STAT3 include c-*Myc* [49] and *pim-1*

[50]. Paradoxically, some reports also suggest that STAT3 can activate expression of the cell cycle inhibitor p21 (waf1) [51]. This implies that STAT3 can also block cell cycle progression and prevent abnormal cell proliferation.

The cytokine TNF-α has been shown to be a growth factor for glioblastoma [44, 52] and cutaneous T-cell lymphoma [53]. TNF-α has also been shown to induce amphiregulin, which in turn can induce proliferation of cervical cancer cells [54]. TNF-α induces expression of EGF receptor and transforming growth factor alpha (TGF-α) and mediates proliferation of pancreatic cancer cells [55]. Similarly, IL-1β is known to be a growth factor for acute myelogenous leukemia [56], while IL-6 is a growth factor for multiple myeloma [57] and head and neck squamous cell carcinoma [58].

12.4.4 Role of Inflammatory Molecules in Tumor Cell Invasion, Angiogenesis, and Metastasis

Tumor cell invasion, angiogenesis, and metastasis are interrelated processes that represent the final, most devastating stage of malignancy. The process involves cell growth, adhesion, and migration; proteolytic degradation of tissue barriers; and formation of new blood vessels [59]. Several proteolytic enzymes such as the MMPs [60] and ICAM participate in degradation of these barriers [61]. Other molecules involved in this process are serine proteases such as urokinase-type plasminogen activator (u-PA), u-PA receptor, VEGF, VEGF receptors (VEGFR), EGF, PDGF, fibroblast growth factors, ephrins, angiopoietins, endothelins, integrins, cadherins, and transcription factors (NF-κB, STAT3) [62, 63].

The involvement of NF-κB in tumor angiogenesis and metastasis is supported by numerous lines of evidence. Using an adenovirus that overexpresses the inhibitory subunit I kappaB alpha (IκBα), Bond et al. found that NF-κB activation was absolutely required for upregulation of MMP-9 [160]. Urokinase-type plasminogen activator, the critical protease in tumor invasion and metastasis, is transcriptionally activated by phorbol myristate acetate, IL-1, and TNF-α. The activation also requires induction of NF-κB activity and decay of IκBα [64]. In pancreatic tumor cells, u-PA is overexpressed and requires constitutive RelA activity [65].

STAT3 has been reported to play major roles in tumor cell invasion, angiogenesis, and metastasis through numerous mechanisms [66, 67]. It upregulates transcription of MMP-2 through direct interaction with the MMP-2 promoter. The role of STAT3 in tumor metastasis is further supported by observations that blockage of activated STAT3 in highly metastatic cells significantly suppressed the invasiveness of the tumor cells and prevented metastasis in nude mice. Furthermore, overexpression of activated STAT3 correlated with the invasion and metastasis of cutaneous squamous cell carcinoma [68]. STAT3 also controls expression of the *MUC1* gene, which can mediate tumor invasion [69]. Constitutive STAT3 has been suggested to upregulate VEGF expression and thus tumor angiogenesis [70]. Most tumor cells that exhibit constitutive STAT3 also express VEGF [71]. The metastasis of human melanoma to brain has been linked to STAT3 activation [72]. STAT3 has been shown to regulate TWIST, another mediator of tumor metastasis [73]. Ironically, STAT3 is also known to upregulate tissue inhibitors of metalloproteinase 1, a cytokine known to block metalloproteinases and decrease invasiveness of certain cancer types [74].

The metastatic potential of chemokines has been attributed to their ability to induce expression of MMPs [75]. The role of chemokines in tumor invasion is supported by observations that the silencing of endogenous chemokine receptors abolishes the adhesive and invasive nature of salivary gland mucoepidermoid carcinoma cells [76]. One study found a close association between expression of IL-8 by human melanoma and ovarian cancer cells and their metastatic potential [77].

TNF-α has been shown to confer an invasive phenotype on mammary epithelial cells [78] and can induce angiogenesis in malignant glioma cells [79]. TNF-α also stimulates epithelial tumor cell motility, a critical function in embryonic development, tissue repair, and tumor invasion [80], and has been reported to mediate macrophage-induced angiogenesis [81]. The role of TNF-α in mediating invasiveness of some carcinomas is supported by *in vivo* studies [82].

12.5 CONTROLLING INFLAMMATION AND CANCER

It is clear from this discussion that NF-κB, STAT3, and other inflammatory molecules contribute to tumor development. Thus agents with potential to suppress these inflammatory pathways might offer promise for prevention and treatment of cancer. A wide variety of drugs based upon these targets has been designed and developed (Table 12.2). Most of these drugs modulate a single specific target, but because cancer is caused by dysregulation of multiple genes, these drugs are less likely to be effective than multi-targeted agents. Most of these drugs produce numerous side effects and cannot be consumed for long duration. Therefore, there is an urgent need for development of agents that are multi-targeted, cost-effective, and immediately available. In this regard, agents derived from natural sources, often called nutraceuticals, seem to possess enormous potential [83, 84]. Nutraceuticals are derived from spices, vegetables, fruits, pulses, nuts, and cereals (Figure 12.3).

TABLE 12.2
Partial List of FDA-Approved Anti-Cancer Drugs That Work through Modulation of Inflammatory Pathways

Target	Drug	Year	Cancer Type
NF-κB[a]	Rituximab	1997	Non-Hodgkin's lymphoma, CLL
	Bortezomib	2003	Mantle cell lymphoma, MM
	Lenalidomide	2005	MM, Myelodysplastic syndrome
	Sunitinib malate	2006	Renal, Gastrointestinal
STAT3[a]	Rituximab	1997	Non-Hodgkin's lymphoma, CLL
Interleukin-2	Aldesleukin	1998	Melanoma, Renal
	Denileukin diftitox	1999	Cutaneous T-cell lymphoma
EGFR	Gefitinib	2003	Lung
	Cetuximab	2004	Colorectal, Head and neck
	Erlotinib	2004	Prostate, Lung
	Panitumumab	2006	Colorectal
	Lapatinib ditosylate	2007	Breast
	Vandetanib	2011	Thyroid
VEGFR	Bevacizumab	2004	Colorectal, Renal, Lung, Glioblastoma
	Sorafenib tosylate	2005	Renal, Liver
	Pazopanib	2009	Renal
	Vandetanib	2011	Thyroid
PDGFR	Sorafenib tosylate	2005	Renal, Liver
	Sunitinib malate	2006	Renal, Gastrointestinal
	Dasatinib	2006	CML, ALL
	Nilotinib	2007	CML

[a] Not an actual target, but these drugs have been shown to modulate NF-κB and STAT3 pathways.
ALL, acute lymphoblastic leukemia; CLL, chronic lymphocytic leukemia; CML, chronic myelogenous leukemia; EGFR, epidermal growth factor receptor; FDA, United States Food and Drug Administration; MM, multiple myeloma; NF-κB, nuclear factor kappa-light-chain-enhancer of activated B cells; PDGFR, platelet derived growth factor receptor; STAT3, signal transducer and activator of transcription 3; VEGFR, vascular endothelial growth factor receptor.

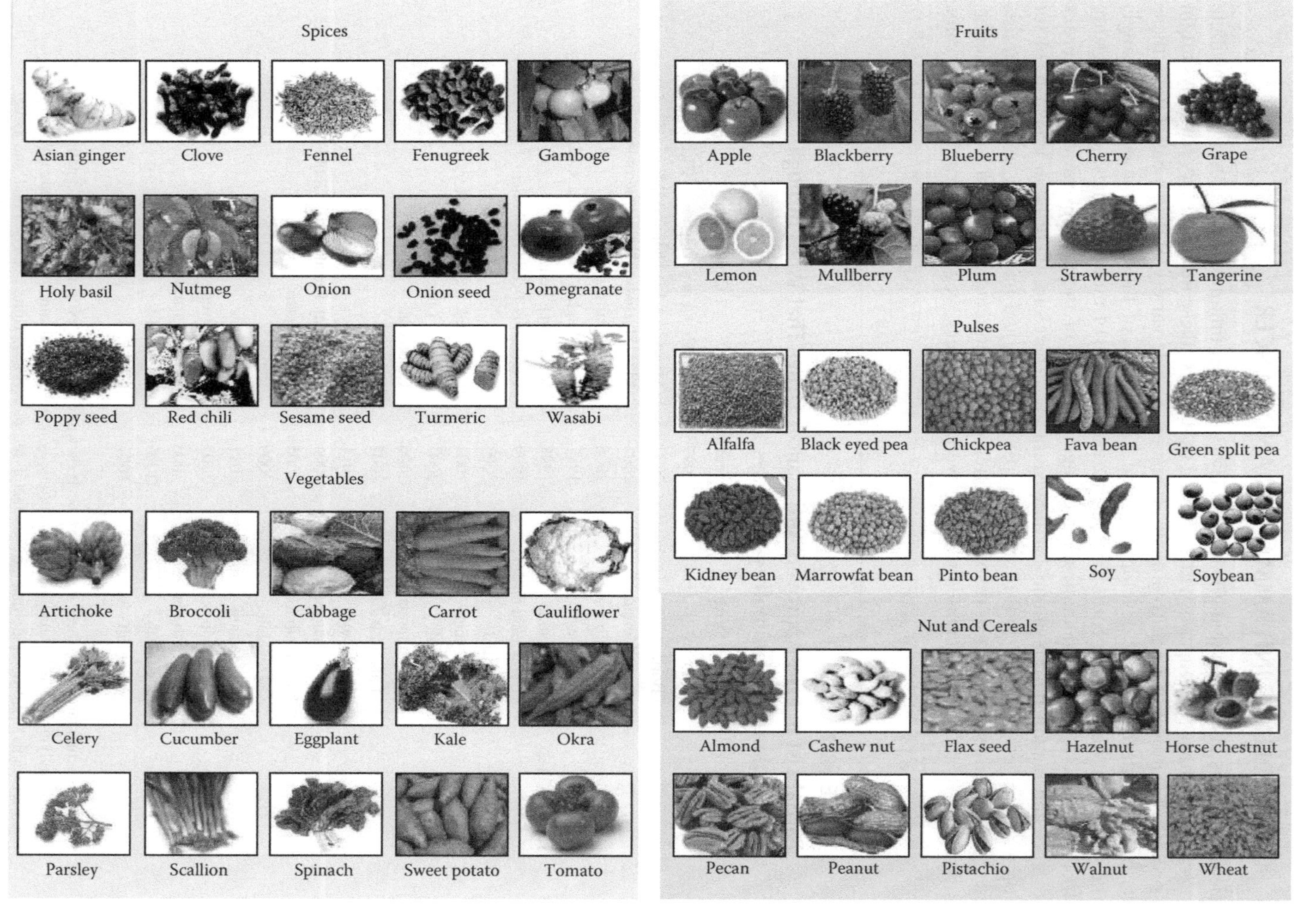

FIGURE 12.3 (see color insert) Common sources of plant-derived nutraceuticals. These sources include spices, vegetables, fruits, pulses, nuts, and cereals.

TABLE 12.3
Partial List of Nutraceuticals with Potential Activity against Inflammatory Molecules

Molecule	Nutraceuticals
NF-κB	γ-Tocotrienol, ACA, anethole, berberine, betulinic acid, butein, CAPE, capsaicin, crotepoxide, curcumin, embelin, emodin, fisetin, flavopiridol, gambogic acid, gossypin, isodeoxyelephantopin, morin, nimbolide, noscapine, oleandrin, pentamethoxyflavone, piceatannol, picroliv, pinitol, plumbagin, resveratrol, sanguinarine, sesamin, silymarin, thiocolchicoside, thymoquinone, ursolic acid, withanolides, xanthohumol, zerumbone
STAT3	γ-Tocotrienol, atiprimod, betulinic acid, boswellic acid, butein, capsaicin, curcumin, gambogic acid, guggulsterone, ursolic acid
IL-6	Diallyl sulfide, phytic acid, piperine, nimbolide
IL-8	Allicin, phytic acid
IL-1β	Allicin, apigenin, eugenol, gingerol, humulene, kaempferol, phytic acid, piperine
TNF	Ajoene, allicin, apigenin, curcumin, eugenol, gingerol, humulene, kaempferol, nimbolide, piperine, zingerone

ACA, 1'-acetoxychavicol acetate; CAPE, caffeic acid phenethyl ester; IL, interleukin; NF-κB, nuclear factor kappa-light-chain-enhancer of activated B cells; STAT3, signal transducer and activator of transcription 3; TNF, tumor necrosis factor.

Numerous lines of evidence from preclinical and clinical studies have shown that nutraceuticals can target NF-κB and STAT3 and numerous other inflammatory pathways (Table 12.3). Nutraceuticals are chemically diverse and can target one or more steps of tumorigenesis (Table 12.4). According to one estimate, more than 63% of anticancer drugs introduced over the past 25 years are natural products or can be traced back to a natural product source [85]. Moreover, some dietary agents have shown potential to inactivate inflammatory molecules by direct binding. Curcumin is one of the most widely studied dietary agents that can bind to a number of inflammatory molecules [86]. In one study, curcumin docked at the receptor-binding sites of TNF-α [87].

TABLE 12.4
Partial List of Nutraceuticals with Potential Activity against Different Stages of Tumorigenesis

Stage	Nutraceuticals
Cell survival	β-Escin, γ-tocotrienol, ACA, anacardic acid, anethole, bharangin, berberine, betulinic acid, butein, capsaicin, celastrol, coronarin, curcumin, deguelin, EGCG, embelin, emodin, evodiamine, fisetin, flavopiridol, gambogic acid, garcinol, genistein, indirubin, indole-3-carbinol, isodeoxyelephantopin, noscapine, oleandrin, plumbagin, resveratrol, sanguinarine, silymarin, sulforaphane, thymoquinone, withanolides, xanthohumol, zerumbone
Cell proliferation	β-Escin, ACA, anacardic acid, bharangin, berberine, betulinic acid, boswellic acid, butein, capsaicin, celasterol, coronarin, curcumin, deguelin, diosgenin, emodin, evodiamine, fisetin, flavopiridol, gambogic acid, genistein, gossypol, guggulsterone, isodeoxyelephantopin, morin, nimbolide, noscapine, piceatannol, pinitol, quercetin, silibinin, sulforaphane, thymoquinone, tubocapsanolide A, ursolic acid, zerumbone
Cell invasion, angiogenesis, metastasis	β-Carotene, γ-tocotrienol, bharangin, [6]-gingerol, 3,3' -diindolylmethane, AITC, allicin, alliin, apigenin, berberine, butein, caffeic acid, capsaicin, carnosol, catechin gallate, celastrol, crocetin, curcumin, diallyl sulfide, EGCG, evodiamine, fisetin, flavopiridol, gambogic acid, ganoderic acid, genistein, indole-3-carbinol, kaempferol, luteolin, lycopene, myricetin, nimbolide, perillyl alcohol, piperine, quercetin, resveratrol, rosmarinic acid, S-allylcysteine, sanguinarine, silibinin, sulforaphane, taxol, ursolic acid, vanillin, zerumbone

ACA, 1'-acetoxychavicol acetate; AITC, allyl isothiocyanate; EGCG, epigallocatechin gallate

Curcumin exhibited direct interaction with TNF-α by both noncovalent and covalent interactions [87]. Curcumin has also been shown to inhibit COX-1, COX-2, and MMP activities by direct binding [88–90].

During the past two decades, we and other research groups have shown that nutraceuticals can exert anticancer activity by suppressing one or more steps of the NF-κB signaling pathway. The most common targets of nutraceuticals in the NF-κB signaling pathway include I-kappaB kinase (IKK) activation, IκBα phosphorylation and degradation, p65 nuclear translocation, p65 phosphorylation, p65 acetylation, and p65 DNA binding. The most widely studied nutraceuticals having demonstrated potential to inhibit NF-κB activation include curcumin [91], guggulsterone [92], resveratrol [93–95], capsaicin [96], sanguinarine [97], emodin [98], caffeic acid phenethyl ester (CAPE) [99], and epigallocatechin gallate (EGCG) [100, 101].

Some nutraceuticals act by inhibiting IKK activation. We were first to demonstrate that curcumin has the potential to inhibit IKK in a human myeloid leukemic cell line [91]. Among other nutraceuticals with demonstrated ability to target IKK are guggulsterone [92] and EGCG [101]. Nutraceuticals with the potential to prevent phosphorylation and degradation of IκBα, the most important steps in NF-κB activation, include curcumin [57, 92, 102], guggulsterone [92], capsaicin [96, 103], resveratrol [95], sanguinarine [97], emodin [98], and EGCG [101]. Curcumin has also been reported to suppress the TNF-α-induced nuclear translocation of NF-κB in a human myeloid leukemia cell line [91]. Other nutraceuticals having potential to prevent nuclear translocation of NF-κB include resveratrol [95], capsaicin [96, 103, 104] and EGCG [101].

Some nutraceuticals have an ability to inhibit p65-DNA binding. Curcumin inhibited the DNA-binding ability of NF-κB in a human myeloid leukemic cell line [91]. Caffeic acid phenethyl ester suppressed NF-κB activation by suppressing the binding of the p50-p65 complex directly to DNA [99]. Emodin oxidized the redox-sensitive site on NF-κB and prevented NF-κB binding to the target DNA in HeLa cells [105]. Plumbagin inhibited the DNA-binding activity of NF-κB in breast cancer cells [106, 107].

More recently, EGCG was found to abrogate p65 acetylation *in vitro* and *in vivo* by diverse stimuli [108]. Gallic acid obtained from natural products such as gallnuts, sumac, oak bark, and green tea was reported to possess anti-histone acetyltransferase activity, thus showing potential to downregulate NF-κB activation [108]. Anacardic acid derived from traditional medicinal plants can inhibit NF-κB activation by inhibiting p65 acetylation [109].

We have identified a number of nutraceuticals from natural sources that target one or more steps in the NF-κB activation pathway to sensitize and induce apoptosis in a variety of cancer cells. The most common among these are 1'-acetoxychavicol acetate [110], evodiamine [111], noscapine [112], indirubin [113], isodeoxyelephantopin [114], anacardic acid [109], coronarin [115], thymoquinone [116], γ-tocotrienol [117], β-escin [118], and withanolides [119].

Similarly, a number of nutraceuticals have shown potential to inhibit survival of tumor cells through inhibition of the STAT3 pathway. Muto et al. showed that emodin can induce apoptosis in human myeloid cells through downregulation of STAT3 [120]. Capsaicin has been reported to induce apoptosis in multiple myeloid cells through downregulation of STAT3 [121]. Adult T-cell leukemia is an aggressive malignancy of peripheral T cells infected with HTLV-1. Deguelin induced apoptosis in HTLV-1-transformed T cells via inhibition of STAT3 phosphorylation through the ubiquitin/proteasome pathway [122]. In an orthotopic murine model of ovarian cancer, curcumin inhibited tumor growth that correlated with inhibition in the STAT3 activation pathway [123]. Caffeic acid suppression of STAT3-mediated HIF-1 and VEGF expression correlated with inhibition of vascularization and angiogenesis in mice bearing Caki-I human renal carcinoma cells [124]. Quercetin inhibition of hypoxia-induced VEGF expression in NCI-H157 cells correlated with suppression of STAT3 tyrosine phosphorylation, suggesting that inhibition of STAT3 function may play a role in inhibition of angiogenesis [125].

Nutraceuticals have been found to target numerous other inflammatory molecules, including IL-6, IL-8, IL-1β, and TNF-α. Some of these nutraceuticals are listed in Table 12.4.

The potential of plant-derived nutraceuticals in suppressing inflammatory pathways and cancer is evident from clinical studies as well. For example, green tea has become a popular beverage because of the potential health benefits of one of its components, the polyphenolic compound EGCG [126]. The tea polyphenols have been shown to decrease serum levels of PSA, hepatocyte growth factor (HGF), and VEGF in patients with prostate cancer [127, 128]. In a Chinese study, the risk of prostate adenocarcinoma decreased with increasing frequency, duration, and quantity of green tea consumption [129]. In another study, drinking black tea decreased levels of inflammatory biomarkers in patients with colon cancer [130]. The consumption of tea polyphenols has been shown to decrease the risk of gastric cancer [131], breast cancer [132], and lung cancer [133]. Another natural product, pomegranate, has been used for centuries for medicinal purposes. The fruit is known for its isoflavonoid compounds, such as quercetin, kaempferol, and luteolin [134]. In a phase II clinical trial evaluating the effects of pomegranate juice consumption in men with rising PSA after surgery or radiotherapy for prostate cancer, mean PSA doubling time significantly increased after treatment with pomegranate juice, from a mean of 15 months at baseline to 54 months after treatment. A decrease in cell proliferation and an increase in apoptosis were observed in the patients who consumed pomegranate [135]. Selenium supplementation has been found to reduce the incidence of prostate, colorectal, and lung cancers [136].

The use of nutraceuticals seems promising in reducing the risk of cancer. Besides their potential efficacy, their safety and immediate availability warrant the use of nutraceuticals for prevention and treatment of cancer.

12.6 CONCLUSION AND FUTURE DIRECTIONS

Chronic inflammation is a critical mediator of cancer and affects every facet of tumor development. Thus agents that can suppress pro-inflammatory pathways might have potential in the prevention and treatment of cancer. The mono-targeted drugs are unlikely to cure cancer because of the multigenic nature of the disease. Moreover, mono-targeted drugs produce numerous side effects and therefore cannot be taken over long time periods. On the other hand, some lifestyle factors play a major role in the development of inflammation and cancer. In contrast to drugs, therefore, robust attention to these lifestyle factors might provide the best solution for the prevention and treatment of cancer. A number of dietary elements, or nutraceuticals, have already shown potential as anti-inflammatory agents in cell culture and animal studies. Well-controlled clinical trials are required to realize the full potential of these molecules. Future studies focused on deciphering the clinical efficacy of these fascinating molecules will, we hope, lead to development of novel cancer therapeutics.

TAKE-HOME MESSAGES

- Acute inflammation is a part of the host defense system that counteracts the insults incurred by internal or external stimuli.
- Chronic inflammation increases the risk of chronic diseases, including cancer.
- Pro-inflammatory transcription factors and their gene products provide a molecular link between inflammation and cancer.
- Lifestyle factors such as stress, toxicants, tobacco, alcohol, obesity, infectious agents, and radiation are known to activate inflammatory pathways.
- Inflammatory pathways have the potential to affect every step of tumor development.
- Targeting inflammatory pathways provides opportunities for cancer prevention and treatment.
- Mono-targeted drugs are less likely than multi-targeted agents to be effective against inflammatory pathways and cancer.

- Agents derived from natural sources (nutraceuticals) have enormous potential against inflammatory pathways and cancer.
- Nutraceuticals have potential activity against every step of tumor development.
- Chronic inflammation and cancer can be prevented by changing lifestyle.

ACKNOWLEDGMENTS

We thank Kathryn Hale of the University of Texas MD Anderson Cancer Center Department of Scientific Publications for carefully editing the manuscript and providing valuable comments. Dr. Aggarwal is the Ransom Horne, Jr., Professor of Cancer Research. This work was supported by a core cancer center grant from the National Institutes of Health (CA16672), a program project grant from the National Institutes of Health (NIH CA124787-01A2), and a grant from the Center for Targeted Therapy at MD Anderson Cancer Center.

REFERENCES

1. Heidland A, Klassen A, Rutkowski P, Bahner U: The contribution of Rudolf Virchow to the concept of inflammation: What is still of importance?, *J Nephrol* 2006, 19 Suppl 10:S102–109.
2. Aggarwal BB: Nuclear factor-kappaB: The enemy within, *Cancer Cell* 2004, 6:203–208.
3. Ahn KS, Aggarwal BB: Transcription factor NF-kappaB: A sensor for smoke and stress signals, *Ann N Y Acad Sci* 2005, 1056:218–233.
4. Gupta SC, Kim JH, Prasad S, Aggarwal BB: Regulation of survival, proliferation, invasion, angiogenesis, and metastasis of tumor cells through modulation of inflammatory pathways by nutraceuticals, *Cancer Metastasis Rev* 2010, 29:405–434.
5. Gupta SC, Sundaram C, Reuter S, Aggarwal BB: Inhibiting NF-kappaB activation by small molecules as a therapeutic strategy, *Biochim Biophys Acta* 2010, 1799:775–787.
6. Yadav D, Whitcomb DC: The role of alcohol and smoking in pancreatitis, *Nat Rev Gastroenterol Hepatol* 2010, 7:131–145.
7. Aggarwal BB, Shishodia S, Sandur SK, Pandey MK, Sethi G: Inflammation and cancer: How hot is the link?, *Biochem Pharmacol* 2006, 72:1605–1621.
8. Dickinson BC, Chang CJ: Chemistry and biology of reactive oxygen species in signaling or stress responses, *Nat Chem Biol* 2011, 7:504–511.
9. Mantovani A: Cancer: Inflammation by remote control, *Nature* 2005, 435:752–753.
10. Hussain SP, Harris CC: Inflammation and cancer: An ancient link with novel potentials, *Int J Cancer* 2007, 121:2373–2380.
11. Warburg O: On respiratory impairment in cancer cells, *Science* 1956, 124:269–270.
12. Aggarwal BB, Sung B: The relationship between inflammation and cancer is analogous to that between fuel and fire, *Oncology (Williston Park)* 2011, 25:414–418.
13. Mantovani A, Allavena P, Sica A, Balkwill F: Cancer-related inflammation, *Nature* 2008, 454:436–444.
14. Wingo PA, Ries LA, Giovino GA, Miller DS, Rosenberg HM, Shopland DR, Thun MJ, et al.: Annual report to the nation on the status of cancer, 1973–1996, with a special section on lung cancer and tobacco smoking, *J Natl Cancer Inst* 1999, 91:675–690.
15. Itzkowitz SH, Yio X: Inflammation and cancer IV. Colorectal cancer in inflammatory bowel disease: The role of inflammation, *Am J Physiol Gastrointest Liver Physiol* 2004, 287:G7–17.
16. Peter S, Beglinger C: Helicobacter pylori and gastric cancer: The causal relationship, *Digestion* 2007, 75:25–35.
17. Coussens LM, Werb Z: Inflammation and cancer, *Nature* 2002, 420:860–867.
18. Ralph SJ, Rodriguez-Enriquez S, Neuzil J, Saavedra E, Moreno-Sanchez R: The causes of cancer revisited: "Mitochondrial malignancy" and ROS-induced oncogenic transformation—why mitochondria are targets for cancer therapy, *Mol Aspects Med* 2010, 31:145–170.
19. Yan B, Peng Y, Li CY: Molecular analysis of genetic instability caused by chronic inflammation, *Methods Mol Biol* 2009, 512:15–28.
20. Komori A, Yatsunami J, Suganuma M, Okabe S, Abe S, Sakai A, et al.: Tumor necrosis factor acts as a tumor promoter in BALB/3T3 cell transformation, *Cancer Res* 1993, 53:1982–1985.

21. Radisky DC, Levy DD, Littlepage LE, Liu H, Nelson CM, Fata JE, et al.: Rac1b and reactive oxygen species mediate MMP-3-induced EMT and genomic instability, *Nature* 2005, 436:123–127.
22. Mayo MW, Norris JL, Baldwin AS: Ras regulation of NF-kappa B and apoptosis, *Methods Enzymol* 2001, 333:73–87.
23. Balmain A, Pragnell IB: Mouse skin carcinomas induced in vivo by chemical carcinogens have a transforming Harvey-ras oncogene, *Nature* 1983, 303:72–74.
24. Yu CL, Meyer DJ, Campbell GS, Larner AC, Carter-Su C, Schwartz J, et al.: Enhanced DNA-binding activity of a Stat3-related protein in cells transformed by the Src oncoprotein, *Science* 1995, 269:81–83.
25. Migone TS, Lin JX, Cereseto A, Mulloy JC, O'Shea JJ, Franchini G, et al.: Constitutively activated Jak-STAT pathway in T cells transformed with HTLV-I, *Science* 1995, 269:79–81.
26. Yoshida T, Hanada T, Tokuhisa T, Kosai K, Sata M, Kohara M, et al.: Activation of STAT3 by the hepatitis C virus core protein leads to cellular transformation, *J Exp Med* 2002, 196:641–653.
27. Green DR: Apoptotic pathways: Paper wraps stone blunts scissors, *Cell* 2000, 102:1–4.
28. Steller H: Mechanisms and genes of cellular suicide, *Science* 1995, 267:1445–1449.
29. Kreuz S, Siegmund D, Scheurich P, Wajant H: NF-kappaB inducers upregulate cFLIP, a cycloheximide-sensitive inhibitor of death receptor signaling, *Mol Cell Biol* 2001, 21:3964–3973.
30. Zong WX, Edelstein LC, Chen C, Bash J, Gelinas C: The prosurvival Bcl-2 homolog Bfl-1/A1 is a direct transcriptional target of NF-kappaB that blocks TNFalpha-induced apoptosis, *Genes Dev* 1999, 13:382–387.
31. Tamatani M, Che YH, Matsuzaki H, Ogawa S, Okado H, Miyake S, et al.: Tumor necrosis factor induces Bcl-2 and Bcl-x expression through NFkappaB activation in primary hippocampal neurons, *J Biol Chem* 1999, 274:8531–8538.
32. Stehlik C, de Martin R, Kumabashiri I, Schmid JA, Binder BR, Lipp J: Nuclear factor (NF)-kappaB-regulated X-chromosome-linked iap gene expression protects endothelial cells from tumor necrosis factor alpha-induced apoptosis, *J Exp Med* 1998, 188:211–216.
33. Wang CY, Mayo MW, Korneluk RG, Goeddel DV, Baldwin AS, Jr.: NF-kappaB antiapoptosis: Induction of TRAF1 and TRAF2 and c-IAP1 and c-IAP2 to suppress caspase-8 activation, *Science* 1998, 281:1680–1683.
34. Zhu L, Fukuda S, Cordis G, Das DK, Maulik N: Anti-apoptotic protein survivin plays a significant role in tubular morphogenesis of human coronary arteriolar endothelial cells by hypoxic preconditioning, *FEBS Lett* 2001, 508:369–374.
35. Catlett-Falcone R, Landowski TH, Oshiro MM, Turkson J, Levitzki A, Savino R, et al.: Constitutive activation of Stat3 signaling confers resistance to apoptosis in human U266 myeloma cells, *Immunity* 1999, 10:105–115.
36. Zushi S, Shinomura Y, Kiyohara T, Miyazaki Y, Kondo S, Sugimachi M, et al.: STAT3 mediates the survival signal in oncogenic ras-transfected intestinal epithelial cells, *Int J Cancer* 1998, 78:326–330.
37. Mahboubi K, Li F, Plescia J, Kirkiles-Smith NC, Mesri M, Du Y, Carroll JM, et al.: Interleukin-11 up-regulates survivin expression in endothelial cells through a signal transducer and activator of transcription-3 pathway, *Lab Invest* 2001, 81:327–334.
38. Liu H, Ma Y, Cole SM, Zander C, Chen KH, Karras J, Pope RM: Serine phosphorylation of STAT3 is essential for Mcl-1 expression and macrophage survival, *Blood* 2003, 102:344–352.
39. Bhattacharya S, Schindler C: Regulation of Stat3 nuclear export, *J Clin Invest* 2003, 111:553–559.
40. Aoki Y, Feldman GM, Tosato G: Inhibition of STAT3 signaling induces apoptosis and decreases survivin expression in primary effusion lymphoma, *Blood* 2003, 101:1535–1542.
41. Konnikova L, Kotecki M, Kruger MM, Cochran BH: Knockdown of STAT3 expression by RNAi induces apoptosis in astrocytoma cells, *BMC Cancer* 2003, 3:23.
42. Hanahan D, Weinberg RA: Hallmarks of cancer: The next generation, *Cell* 2011, 144:646–674.
43. Hanahan D, Weinberg RA: The hallmarks of cancer, *Cell* 2000, 100:57–70.
44. Mukhopadhyay A, Banerjee S, Stafford LJ, Xia C, Liu M, Aggarwal BB: Curcumin-induced suppression of cell proliferation correlates with down-regulation of cyclin D1 expression and CDK4-mediated retinoblastoma protein phosphorylation, *Oncogene* 2002, 21:8852–8861.
45. Yamamoto K, Arakawa T, Ueda N, Yamamoto S: Transcriptional roles of nuclear factor kappa B and nuclear factor-interleukin-6 in the tumor necrosis factor alpha-dependent induction of cyclooxygenase-2 in MC3T3-E1 cells, *J Biol Chem* 1995, 270:31315–31320.
46. Habib AA, Chatterjee S, Park SK, Ratan RR, Lefebvre S, Vartanian T: The epidermal growth factor receptor engages receptor interacting protein and nuclear factor-kappa B (NF-kappa B)-inducing kinase to activate NF-kappa B. Identification of a novel receptor-tyrosine kinase signalosome, *J Biol Chem* 2001, 276:8865–8874.

47. Romashkova JA, Makarov SS: NF-kappaB is a target of AKT in anti-apoptotic PDGF signalling, *Nature* 1999, 401:86–90.
48. Masuda M, Suzui M, Yasumatu R, Nakashima T, Kuratomi Y, Azuma K, et al.: Constitutive activation of signal transducers and activators of transcription 3 correlates with cyclin D1 overexpression and may provide a novel prognostic marker in head and neck squamous cell carcinoma, *Cancer Res* 2002, 62:3351–3355.
49. Kiuchi N, Nakajima K, Ichiba M, Fukada T, Narimatsu M, Mizuno K, et al.: STAT3 is required for the gp130-mediated full activation of the c-myc gene, *J Exp Med* 1999, 189:63–73.
50. Shirogane T, Fukada T, Muller JM, Shima DT, Hibi M, Hirano T: Synergistic roles for Pim-1 and c-Myc in STAT3-mediated cell cycle progression and antiapoptosis, *Immunity* 1999, 11:709–719.
51. Bellido T, O'Brien CA, Roberson PK, Manolagas SC: Transcriptional activation of the p21(WAF1,CIP1,SDI1) gene by interleukin-6 type cytokines. A prerequisite for their pro-differentiating and anti-apoptotic effects on human osteoblastic cells, *J Biol Chem* 1998, 273:21137–21144.
52. Aggarwal BB, Schwarz L, Hogan ME, Rando RF: Triple helix-forming oligodeoxyribonucleotides targeted to the human tumor necrosis factor (TNF) gene inhibit TNF production and block the TNF-dependent growth of human glioblastoma tumor cells, *Cancer Res* 1996, 56:5156–5164.
53. Giri DK, Aggarwal BB: Constitutive activation of NF-kappaB causes resistance to apoptosis in human cutaneous T cell lymphoma HuT-78 cells. Autocrine role of tumor necrosis factor and reactive oxygen intermediates, *J Biol Chem* 1998, 273:14008–14014.
54. Woodworth CD, McMullin E, Iglesias M, Plowman GD: Interleukin 1 alpha and tumor necrosis factor alpha stimulate autocrine amphiregulin expression and proliferation of human papillomavirus-immortalized and carcinoma-derived cervical epithelial cells, *Proc Natl Acad Sci USA* 1995, 92:2840–2844.
55. Schmiegel W, Roeder C, Schmielau J, Rodeck U, Kalthoff H: Tumor necrosis factor alpha induces the expression of transforming growth factor alpha and the epidermal growth factor receptor in human pancreatic cancer cells, *Proc Natl Acad Sci USA* 1993, 90:863–867.
56. Estrov Z, Thall PF, Talpaz M, Estey EH, Kantarjian HM, Andreeff M, et al.: Caspase 2 and caspase 3 protein levels as predictors of survival in acute myelogenous leukemia, *Blood* 1998, 92:3090–3097.
57. Bharti AC, Donato N, Singh S, Aggarwal BB: Curcumin (diferuloylmethane) down-regulates the constitutive activation of nuclear factor-kappa B and IkappaBalpha kinase in human multiple myeloma cells, leading to suppression of proliferation and induction of apoptosis, *Blood* 2003, 101:1053–1062.
58. Kato T, Duffey DC, Ondrey FG, Dong G, Chen Z, Cook JA, et al.: Cisplatin and radiation sensitivity in human head and neck squamous carcinomas are independently modulated by glutathione and transcription factor NF-kappaB, *Head Neck* 2000, 22:748–759.
59. Fan TP, Yeh JC, Leung KW, Yue PY, Wong RN: Angiogenesis: From plants to blood vessels, *Trends Pharmacol Sci* 2006, 27:297–309.
60. Sternlicht MD, Werb Z: How matrix metalloproteinases regulate cell behavior, *Annu Rev Cell Dev Biol* 2001, 17:463–516.
61. Aimes RT, Quigley JP: Matrix metalloproteinase-2 is an interstitial collagenase. Inhibitor-free enzyme catalyzes the cleavage of collagen fibrils and soluble native type I collagen generating the specific 3/4- and 1/4-length fragments, *J Biol Chem* 1995, 270:5872–5876.
62. Nerlov C, Rorth P, Blasi F, Johnsen M: Essential AP-1 and PEA3 binding elements in the human urokinase enhancer display cell type-specific activity, *Oncogene* 1991, 6:1583–1592.
63. Aggarwal BB, Vijayalekshmi RV, Sung B: Targeting inflammatory pathways for prevention and therapy of cancer: short-term friend, long-term foe, *Clin Cancer Res* 2009, 15:425–430.
64. Novak U, Cocks BG, Hamilton JA: A labile repressor acts through the NFkB-like binding sites of the human urokinase gene, *Nucleic Acids Res* 1991, 19:3389–3393.
65. Wang W, Abbruzzese JL, Evans DB, Chiao PJ: Overexpression of urokinase-type plasminogen activator in pancreatic adenocarcinoma is regulated by constitutively activated RelA, *Oncogene* 1999, 18:4554–4563.
66. Zhao S, Venkatasubbarao K, Lazor JW, Sperry J, Jin C, Cao L, et al.: Inhibition of STAT3 Tyr705 phosphorylation by Smad4 suppresses transforming growth factor beta-mediated invasion and metastasis in pancreatic cancer cells, *Cancer Res* 2008, 68:4221–4228.
67. Xiong H, Zhang ZG, Tian XQ, Sun DF, Liang QC, Zhang YJ, et al.: Inhibition of JAK1, 2/STAT3 signaling induces apoptosis, cell cycle arrest, and reduces tumor cell invasion in colorectal cancer cells, *Neoplasia* 2008, 10:287–297.
68. Suiqing C, Min Z, Lirong C: Overexpression of phosphorylated-STAT3 correlated with the invasion and metastasis of cutaneous squamous cell carcinoma, *J Dermatol* 2005, 32:354–360.

69. Gaemers IC, Vos HL, Volders HH, van der Valk SW, Hilkens J: A stat-responsive element in the promoter of the episialin/MUC1 gene is involved in its overexpression in carcinoma cells, *J Biol Chem* 2001, 276:6191–6199.
70. Niu G, Wright KL, Huang M, Song L, Haura E, Turkson J, et al.: Constitutive Stat3 activity up-regulates VEGF expression and tumor angiogenesis, *Oncogene* 2002, 21:2000–2008.
71. Wei D, Le X, Zheng L, Wang L, Frey JA, Gao AC, et al.: Stat3 activation regulates the expression of vascular endothelial growth factor and human pancreatic cancer angiogenesis and metastasis, *Oncogene* 2003, 22:319–329.
72. Xie TX, Huang FJ, Aldape KD, Kang SH, Liu M, Gershenwald JE, et al.: Activation of stat3 in human melanoma promotes brain metastasis, *Cancer Res* 2006, 66:3188–3196.
73. Cheng GZ, Zhang WZ, Sun M, Wang Q, Coppola D, Mansour M, et al.: Twist is transcriptionally induced by activation of STAT3 and mediates STAT3 oncogenic function, *J Biol Chem* 2008, 283:14665–14673.
74. Dien J, Amin HM, Chiu N, Wong W, Frantz C, Chiu B, et al.: Signal transducers and activators of transcription-3 up-regulates tissue inhibitor of metalloproteinase-1 expression and decreases invasiveness of breast cancer, *Am J Pathol* 2006, 169:633–642.
75. Lu H, Ouyang W, Huang C: Inflammation, a key event in cancer development, *Mol Cancer Res* 2006, 4:221–233.
76. Wen DS, Zhu XL, Guan SM, Wu YM, Yu LL, Wu JZ: Silencing of CXCR4 inhibits the proliferation, adhesion, chemotaxis and invasion of salivary gland mucoepidermoid carcinoma Mc3 cells in vitro, *Oral Oncol* 2008, 44:545–554.
77. Xu L, Fidler IJ: Acidic pH-induced elevation in interleukin 8 expression by human ovarian carcinoma cells, *Cancer Res* 2000, 60:4610–4616.
78. Montesano R, Soulie P, Eble JA, Carrozzino F: Tumour necrosis factor alpha confers an invasive, transformed phenotype on mammary epithelial cells, *J Cell Sci* 2005, 118:3487–3500.
79. Nabors LB, Suswam E, Huang Y, Yang X, Johnson MJ, King PH: Tumor necrosis factor alpha induces angiogenic factor up-regulation in malignant glioma cells: a role for RNA stabilization and HuR, *Cancer Res* 2003, 63:4181–4187.
80. Rosen EM, Goldberg ID, Liu D, Setter E, Donovan MA, Bhargava M, et al.: Tumor necrosis factor stimulates epithelial tumor cell motility, *Cancer Res* 1991, 51:5315–5321.
81. Leibovich SJ, Polverini PJ, Shepard HM, Wiseman DM, Shively V, Nuseir N: Macrophage-induced angiogenesis is mediated by tumour necrosis factor-alpha, *Nature* 1987, 329:630–632.
82. Yoshida S, Ono M, Shono T, Izumi H, Ishibashi T, Suzuki H, et al.: Involvement of interleukin-8, vascular endothelial growth factor, and basic fibroblast growth factor in tumor necrosis factor alpha-dependent angiogenesis, *Mol Cell Biol* 1997, 17:4015–4023.
83. Jensen GL: Inflammation as the key interface of the medical and nutrition universes: A provocative examination of the future of clinical nutrition and medicine, *J Parenter Enteral Nutr* 2006, 30:453–463.
84. Jensen GL, Roubenoff R: Introduction: Nutrition and inflammation: Research Makes The Connection—Intersociety Research Workshop, Chicago, February 8–9, 2008, *J Parenter Enteral Nutr* 2008, 32:625.
85. Newman DJ, Cragg GM: Natural products as sources of new drugs over the last 25 years, *J Nat Prod* 2007, 70:461–477.
86. Gupta SC, Prasad S, Kim JH, Patchva S, Webb LJ, Priyadarsini IK, et al.: Multitargeting by curcumin as revealed by molecular interaction studies, *Natural Product Reports* 2011, 28:1937–1955.
87. Wua ST, Suna JC, Leeb KJ, Sunc YM: Docking Prediction for Tumor Necrosis Factor-α and Five Herbal Inhibitors, *Internat J Engin Sci Tech* 2010, 2:4263–4277.
88. Selvam C, Jachak SM, Thilagavathi R, Chakraborti AK: Design, synthesis, biological evaluation and molecular docking of curcumin analogues as antioxidant, cyclooxygenase inhibitory and anti-inflammatory agents, *Bioorg Med Chem Lett* 2005, 15:1793–1797.
89. Padhye S, Banerjee S, Chavan D, Pandye S, Swamy KV, Ali S, et al.: Fluorocurcumins as cyclooxygenase-2 inhibitor: Molecular docking, pharmacokinetics and tissue distribution in mice, *Pharm Res* 2009, 26:2438–2445.
90. Girija CR, Karunakar P, Poojari CS, Begum NS, Syed AA: Molecular docking studies of curcumin derivatives with multiple protein targets for procarcinogen activating enzyme inhibition, *J Proteomics Bioinform* 2010, 3:200–203.
91. Singh S, Aggarwal BB: Activation of transcription factor NF-kappa B is suppressed by curcumin (diferuloylmethane) [corrected], *J Biol Chem* 1995, 270:24995–25000.

92. Shishodia S, Aggarwal BB: Nuclear factor-kappaB: A friend or a foe in cancer?, *Biochem Pharmacol* 2004, 68:1071–1080.
93. Mouria M, Gukovskaya AS, Jung Y, Buechler P, Hines OJ, Reber HA, et al.: Food-derived polyphenols inhibit pancreatic cancer growth through mitochondrial cytochrome C release and apoptosis, *Int J Cancer* 2002, 98:761–769.
94. Banerjee S, Bueso-Ramos C, Aggarwal BB: Suppression of 7,12-dimethylbenz(a)anthracene-induced mammary carcinogenesis in rats by resveratrol: Role of nuclear factor-kappaB, cyclooxygenase 2, and matrix metalloprotease 9, *Cancer Res* 2002, 62:4945–4954.
95. Bhardwaj A, Sethi G, Vadhan-Raj S, Bueso-Ramos C, Takada Y, Gaur U, et al.: Resveratrol inhibits proliferation, induces apoptosis, and overcomes chemoresistance through down-regulation of STAT3 and nuclear factor-kappaB-regulated antiapoptotic and cell survival gene products in human multiple myeloma cells, *Blood* 2007, 109:2293–2302.
96. Han SS, Keum YS, Seo HJ, Chun KS, Lee SS, Surh YJ: Capsaicin suppresses phorbol ester-induced activation of NF-kappaB/Rel and AP-1 transcription factors in mouse epidermis, *Cancer Lett* 2001, 164:119–126.
97. Chaturvedi MM, Kumar A, Darnay BG, Chainy GB, Agarwal S, Aggarwal BB: Sanguinarine (pseudochelerythrine) is a potent inhibitor of NF-kappaB activation, IkappaBalpha phosphorylation, and degradation, *J Biol Chem* 1997, 272:30129–30134.
98. Kumar A, Dhawan S, Aggarwal BB: Emodin (3-methyl-1,6,8-trihydroxyanthraquinone) inhibits TNF-induced NF-kappaB activation, IkappaB degradation, and expression of cell surface adhesion proteins in human vascular endothelial cells, *Oncogene* 1998, 17:913–918.
99. Natarajan K, Singh S, Burke TR, Jr., Grunberger D, Aggarwal BB: Caffeic acid phenethyl ester is a potent and specific inhibitor of activation of nuclear transcription factor NF-kappa B, *Proc Natl Acad Sci USA* 1996, 93:9090–9095.
100. Nomura M, Ma W, Chen N, Bode AM, Dong Z: Inhibition of 12-O-tetradecanoylphorbol-13-acetate-induced NF-kappaB activation by tea polyphenols, (-)-epigallocatechin gallate and theaflavins, *Carcinogenesis* 2000, 21:1885–1890.
101. Afaq F, Adhami VM, Ahmad N, Mukhtar H: Inhibition of ultraviolet B-mediated activation of nuclear factor kappaB in normal human epidermal keratinocytes by green tea Constituent (-)-epigallocatechin-3-gallate, *Oncogene* 2003, 22:1035–1044.
102. Philip S, Kundu GC: Osteopontin induces nuclear factor kappa B-mediated promatrix metalloproteinase-2 activation through I kappa B alpha /IKK signaling pathways, and curcumin (diferulolylmethane) down-regulates these pathways, *J Biol Chem* 2003, 278:14487–14497.
103. Park KK, Chun KS, Yook JI, Surh YJ: Lack of tumor promoting activity of capsaicin, a principal pungent ingredient of red pepper, in mouse skin carcinogenesis, *Anticancer Res* 1998, 18:4201–4205.
104. Aggarwal BB, Shishodia S: Suppression of the nuclear factor-kappaB activation pathway by spice-derived phytochemicals: Reasoning for seasoning, *Ann N Y Acad Sci* 2004, 1030:434–441.
105. Jing Y, Yang J, Wang Y, Li H, Chen Y, Hu Q, et al.: Alteration of subcellular redox equilibrium and the consequent oxidative modification of nuclear factor kappaB are critical for anticancer cytotoxicity by emodin, a reactive oxygen species-producing agent, *Free Radic Biol Med* 2006, 40:2183–2197.
106. Ahmad A, Banerjee S, Wang Z, Kong D, Sarkar FH: Plumbagin-induced apoptosis of human breast cancer cells is mediated by inactivation of NF-kappaB and Bcl-2, *J Cell Biochem* 2008, 105:1461–1471.
107. Sandur SK, Ichikawa H, Sethi G, Ahn KS, Aggarwal BB: Plumbagin (5-hydroxy-2-methyl-1,4-naphthoquinone) suppresses NF-kappaB activation and NF-kappaB-regulated gene products through modulation of p65 and IkappaBalpha kinase activation, leading to potentiation of apoptosis induced by cytokine and chemotherapeutic agents, *J Biol Chem* 2006, 281:17023–17033.
108. Choi KC, Lee YH, Jung MG, Kwon SH, Kim MJ, Jun WJ, et al.: Gallic acid suppresses lipopolysaccharide-induced nuclear factor-kappaB signaling by preventing RelA acetylation in A549 lung cancer cells, *Mol Cancer Res* 2009, 7:2011–2021.
109. Sung B, Pandey MK, Ahn KS, Yi T, Chaturvedi MM, Liu M, et al.: Anacardic acid (6-nonadecyl salicylic acid), an inhibitor of histone acetyltransferase, suppresses expression of nuclear factor-kappaB-regulated gene products involved in cell survival, proliferation, invasion, and inflammation through inhibition of the inhibitory subunit of nuclear factor-kappaBalpha kinase, leading to potentiation of apoptosis, *Blood* 2008, 111:4880–4891.
110. Ichikawa H, Takada Y, Murakami A, Aggarwal BB: Identification of a novel blocker of I kappa B alpha kinase that enhances cellular apoptosis and inhibits cellular invasion through suppression of NF-kappa B-regulated gene products, *J Immunol* 2005, 174:7383–7392.

111. Takada Y, Kobayashi Y, Aggarwal BB: Evodiamine abolishes constitutive and inducible NF-kappaB activation by inhibiting IkappaBalpha kinase activation, thereby suppressing NF-kappaB-regulated anti-apoptotic and metastatic gene expression, up-regulating apoptosis, and inhibiting invasion, *J Biol Chem* 2005, 280:17203–17212.
112. Sung B, Ahn KS, Aggarwal BB: Noscapine, a benzylisoquinoline alkaloid, sensitizes leukemic cells to chemotherapeutic agents and cytokines by modulating the NF-kappaB signaling pathway, *Cancer Res* 2010, 70:3259–3268.
113. Sethi G, Ahn KS, Sandur SK, Lin X, Chaturvedi MM, Aggarwal BB: Indirubin enhances tumor necrosis factor-induced apoptosis through modulation of nuclear factor-kappa B signaling pathway, *J Biol Chem* 2006, 281:23425–23435.
114. Ichikawa H, Nair MS, Takada Y, Sheeja DB, Kumar MA, Oommen OV, et al.: Isodeoxyelephantopin, a novel sesquiterpene lactone, potentiates apoptosis, inhibits invasion, and abolishes osteoclastogenesis through suppression of nuclear factor-kappaB (nf-kappaB) activation and nf-kappaB-regulated gene expression, *Clin Cancer Res* 2006, 12:5910–5918.
115. Kunnumakkara AB, Ichikawa H, Anand P, Mohankumar CJ, Hema PS, Nair MS, et al.: Coronarin D, a labdane diterpene, inhibits both constitutive and inducible nuclear factor-kappa B pathway activation, leading to potentiation of apoptosis, inhibition of invasion, and suppression of osteoclastogenesis, *Mol Cancer Ther* 2008, 7:3306–3317.
116. Sethi G, Ahn KS, Aggarwal BB: Targeting nuclear factor-kappa B activation pathway by thymoquinone: Role in suppression of antiapoptotic gene products and enhancement of apoptosis, *Mol Cancer Res* 2008, 6:1059–1070.
117. Ahn KS, Sethi G, Krishnan K, Aggarwal BB: Gamma-tocotrienol inhibits nuclear factor-kappaB signaling pathway through inhibition of receptor-interacting protein and TAK1 leading to suppression of antiapoptotic gene products and potentiation of apoptosis, *J Biol Chem* 2007, 282:809–820.
118. Harikumar KB, Sung B, Pandey MK, Guha S, Krishnan S, Aggarwal BB: Escin, a pentacyclic triterpene, chemosensitizes human tumor cells through inhibition of nuclear factor-kappaB signaling pathway, *Mol Pharmacol* 2010, 77:818–827.
119. Ichikawa H, Takada Y, Shishodia S, Jayaprakasam B, Nair MG, Aggarwal BB: Withanolides potentiate apoptosis, inhibit invasion, and abolish osteoclastogenesis through suppression of nuclear factor-kappaB (NF-kappaB) activation and NF-kappaB-regulated gene expression, *Mol Cancer Ther* 2006, 5:1434–1445.
120. Muto A, Hori M, Sasaki Y, Saitoh A, Yasuda I, Maekawa T, et al.: Emodin has a cytotoxic activity against human multiple myeloma as a Janus-activated kinase 2 inhibitor, *Mol Cancer Ther* 2007, 6:987–994.
121. Bhutani M, Pathak AK, Nair AS, Kunnumakkara AB, Guha S, Sethi G, et al.: Capsaicin is a novel blocker of constitutive and interleukin-6-inducible STAT3 activation, *Clin Cancer Res* 2007, 13:3024–3032.
122. Ito S, Oyake T, Murai K, Ishida Y: Deguelin suppresses cell proliferation via the inhibition of survivin expression and STAT3 phosphorylation in HTLV-1-transformed T cells, *Leuk Res* 2010, 34:352–357.
123. Lin YG, Kunnumakkara AB, Nair A, Merritt WM, Han LY, Armaiz-Pena GN, et al.: Curcumin inhibits tumor growth and angiogenesis in ovarian carcinoma by targeting the nuclear factor-kappaB pathway, *Clin Cancer Res* 2007, 13:3423–3430.
124. Jung JE, Kim HS, Lee CS, Park DH, Kim YN, Lee MJ, et al.: Caffeic acid and its synthetic derivative CADPE suppress tumor angiogenesis by blocking STAT3-mediated VEGF expression in human renal carcinoma cells, *Carcinogenesis* 2007, 28:1780–1787.
125. Anso E, Zuazo A, Irigoyen M, Urdaci MC, Rouzaut A, Martinez-Irujo JJ: Flavonoids inhibit hypoxia-induced vascular endothelial growth factor expression by a HIF-1 independent mechanism, *Biochem Pharmacol* 2010, 79:1600–1609.
126. Khan N, Afaq F, Mukhtar H: Cancer chemoprevention through dietary antioxidants: progress and promise, *Antioxid Redox Signal* 2008, 10:475–510.
127. Bettuzzi S, Brausi M, Rizzi F, Castagnetti G, Peracchia G, Corti A: Chemoprevention of human prostate cancer by oral administration of green tea catechins in volunteers with high-grade prostate intraepithelial neoplasia: A preliminary report from a one-year proof-of-principle study, *Cancer Res* 2006, 66:1234–1240.
128. Chow HH, Cai Y, Hakim IA, Crowell JA, Shahi F, Brooks CA, et al.: Pharmacokinetics and safety of green tea polyphenols after multiple-dose administration of epigallocatechin gallate and polyphenon E in healthy individuals, *Clin Cancer Res* 2003, 9:3312–3319.
129. Jian L, Xie LP, Lee AH, Binns CW: Protective effect of green tea against prostate cancer: A case-control study in southeast China, *Int J Cancer* 2004, 108:130–135.

130. Sun CL, Yuan JM, Koh WP, Yu MC: Green tea, black tea and colorectal cancer risk: A meta-analysis of epidemiologic studies, *Carcinogenesis* 2006, 27:1301–1309.
131. Mu LN, Lu QY, Yu SZ, Jiang QW, Cao W, You NC, et al.: Green tea drinking and multigenetic index on the risk of stomach cancer in a Chinese population, *Int J Cancer* 2005, 116:972–983.
132. Seely D, Mills EJ, Wu P, Verma S, Guyatt GH: The effects of green tea consumption on incidence of breast cancer and recurrence of breast cancer: A systematic review and meta-analysis, *Integr Cancer Ther* 2005, 4:144–155.
133. Arts IC: A review of the epidemiological evidence on tea, flavonoids, and lung cancer, *J Nutr* 2008, 138:1561S–1566S.
134. Gil MI, Tomas-Barberan FA, Hess-Pierce B, Holcroft DM, Kader AA: Antioxidant activity of pomegranate juice and its relationship with phenolic composition and processing, *J Agric Food Chem* 2000, 48:4581–4589.
135. Pantuck AJ, Leppert JT, Zomorodian N, Aronson W, Hong J, Barnard RJ, et al.: Phase II study of pomegranate juice for men with rising prostate-specific antigen following surgery or radiation for prostate cancer, *Clin Cancer Res* 2006, 12:4018–4026.
136. Duffield-Lillico AJ, Reid ME, Turnbull BW, Combs GF, Jr., Slate EH, Fischbach LA, et al.: Baseline characteristics and the effect of selenium supplementation on cancer incidence in a randomized clinical trial: a summary report of the Nutritional Prevention of Cancer Trial, *Cancer Epidemiol Biomarkers Prev* 2002, 11:630–639.
137. Castle PE, Hillier SL, Rabe LK, Hildesheim A, Herrero R, Bratti MC, et al.: An association of cervical inflammation with high-grade cervical neoplasia in women infected with oncogenic human papillomavirus (HPV), *Cancer Epidemiol Biomarkers Prev* 2001, 10:1021–1027.
138. Kanoh K, Shimura T, Tsutsumi S, Suzuki H, Kashiwabara K, Nakajima T, et al.: Significance of contracted cholecystitis lesions as high risk for gallbladder carcinogenesis, *Cancer Lett* 2001, 169:7–14.
139. Cruickshank CN: Occupational dermatitis and industrial skin cancer, *Proc R Soc Med* 1952, 45:611–612.
140. Martey CA, Pollock SJ, Turner CK, O'Reilly KM, Baglole CJ, Phipps RP, et al.: Cigarette smoke induces cyclooxygenase-2 and microsomal prostaglandin E2 synthase in human lung fibroblasts: Implications for lung inflammation and cancer, *Am J Physiol Lung Cell Mol Physiol* 2004, 287:L981–991.
141. Murphy SJ, Anderson LA, Johnston BT, Fitzpatrick DA, Watson PR, Monaghan P, et al.: Have patients with esophagitis got an increased risk of adenocarcinoma? Results from a population-based study, *World J Gastroenterol* 2005, 11:7290–7295.
142. Okano M, Gross TG: From Burkitt's lymphoma to chronic active Epstein-Barr virus (EBV) infection: An expanding spectrum of EBV-associated diseases, *Pediatr Hematol Oncol* 2001, 18:427–442.
143. Vagefi PA, Longo WE: Colorectal cancer in patients with inflammatory bowel disease, *Clin Colorectal Cancer* 2005, 4:313–319.
144. Nelson WG, De Marzo AM, DeWeese TL, Isaacs WB: The role of inflammation in the pathogenesis of prostate cancer, *J Urol* 2004, 172:S6–11; discussion S11–12.
145. Reich HN, Gladman DD, Urowitz MB, Bargman JM, Hladunewich MA, Lou W, et al.: Persistent proteinuria and dyslipidemia increase the risk of progressive chronic kidney disease in lupus erythematosus, *Kidney Int* 2011, 79:914–920.
146. Shuper A, Michovitz S, Amir J, Kornreich L, Boikov O, Yaniv Y, et al.: Idiopathic granulomatous encephalitis mimicking malignant brain tumor, *Pediatr Neurol* 2006, 35:280–283.
147. Murphy BA, Beaumont JL, Isitt J, Garden AS, Gwede CK, Trotti AM, et al.: Mucositis-related morbidity and resource utilization in head and neck cancer patients receiving radiation therapy with or without chemotherapy, *J Pain Symptom Manage* 2009, 38:522–532.
148. Wortsman J, Dietrich J, Apesos J, Folse JR: Hashimoto's thyroiditis simulating cancer of the thyroid, *Arch Surg* 1981, 116:386–388.
149. Galloway HR, Dahlstrom JE, Bennett GM: Focal myositis, *Australas Radiol* 2001, 45:347–349.
150. Jin YT, Tsai ST, Li C, Chang KC, Yan JJ, Chao WY, et al.: Prevalence of human papillomavirus in middle ear carcinoma associated with chronic otitis media, *Am J Pathol* 1997, 150:1327–1333.
151. Peek RM, Jr., Crabtree JE: Helicobacter infection and gastric neoplasia, *J Pathol* 2006, 208:233–248.
152. Ruggiero RP, Costantino G: Adenocarcinoma of the appendix in older patients, *J Am Geriatr Soc* 1982, 30:245–247.
153. Manohar MB, Saleem M, McArthur P, Tulbah A: Laryngeal inflammation mimicking laryngeal carcinoma, *J Laryngol Otol* 1997, 111:568–570.
154. Di Bisceglie AM: Hepatitis C and hepatocellular carcinoma, *Hepatology* 1997, 26:34S–38S.
155. Kulkarni S, Coup A, Kershaw JB, Buchholz NP: Metastatic appendiceal adenocarcinoma presenting late as epididymo-orchitis: A case report and review of literature, *BMC Urol* 2004, 4:1.

156. Garcea G, Dennison AR, Steward WP, Berry DP: Role of inflammation in pancreatic carcinogenesis and the implications for future therapy, *Pancreatology* 2005, 5:514–529.
157. Lukes P, Pacova H, Kucera T, Vesely D, Martinek J, Astl J: Expression of endothelial and inducible nitric oxide synthase and caspase-3 in tonsillar cancer, chronic tonsillitis and healthy tonsils, *Folia Biol (Praha)* 2008, 54:141–145.
158. Bairey O, Huminer D, Sandbank J, Pitlik S: Bacterial lymphadenitis complicating chronic lymphatic leukemia, *Isr J Med Sci* 1994, 30:835–837.
159. Gupta SC, Hevia D, Patchva S, Park B, Koh W, Aggarwal BB: Upsides and downsides of reactive oxygen species for cancer: The roles of reactive oxygen species in tumorigenesis, prevention, and therapy, *Antioxid Redox Signal* 2012, DOI:10.1089/ars.2011.4414.
160. Bond M, Fabunmi RP, Baker AH, Newby AC: Synergistic upregulation of metalloproteinase-9 by growth factors and inflammatory cytokines: An absolute requirement for transcription factor NF-kappa B. *FEBS Lett* 1998, 435:29–34.

13 Chronic Wounds and Inflammation

Jaideep Banerjee and Chandan K. Sen
Wexner Medical Center at The Ohio State University
Columbus, Ohio

CONTENTS

13.1 INTRODUCTION

Wounds can be clinically classified into two categories based on the duration to closure. Acute wounds are those that progress through the process of healing in a stepwise manner and achieve closure within days. A chronic wound is defined as a wound or any interruption in the continuity of the body's surface that requires a prolonged time to heal (more than four weeks), does not heal, or recurs [1]. Chronic wounds arise in a great variety of situations and may be associated with a number of pathological processes.

Chronic wounds are common and constitute a significant health problem. It has been estimated that 1% of the population of industrialized countries will experience a leg ulcer at some time [2]. In the United States alone, chronic wounds affect 6.5 million patients [3, 4]. The immense economic and social impact of wounds in our society calls for enhancing our understanding of the biological mechanisms underlying cutaneous wound complications [5]. Chronic wound care is expensive, and therefore treatment options that are both clinically effective and cost effective are vital.

Tissue damage inevitably invokes an inflammatory reaction. This is imperative to our survival, as it fights infection and provides signals that direct the repair process; however, inflammatory cells and their secreted mediators also have many negative side effects that contribute to clinical problems ranging from scarring to cancer formation. We lack a clear understanding of the positive versus negative aspects of wound-associated inflammation. Chronic wounds fail to progress through the normal phases of healing and therefore enter a state of prolonged pathologic inflammation [6]. This chapter aims to explain the role of chronic inflammation in delayed wound healing.

13.2 MOST COMMON TYPES OF CHRONIC WOUNDS

13.2.1 Diabetic Foot Ulcer

One particular type of chronic wound often associated with ischemia is the foot ulcer associated with diabetes. In diabetes mellitus, the development of foot ulcers is usually the result of peripheral neuropathy and/or peripheral vascular disease.

13.2.2 Venous Ulcer

Venous ulcers (also known as varicose or stasis ulcers) are caused by venous reflux or obstruction resulting in high venous pressure.

13.2.3 Arterial Ulcer

Arterial ulcers are the result of impaired perfusion to the feet or legs and are viewed as one clinical sign of general arteriosclerosis.

13.2.4 Pressure Ulcers

Pressure ulcers (also known as pressure sores, decubitus ulcers, and bed sores) may present as persistently hyperemic, broken, or necrotic skin, most often extending to the underlying tissue, including muscles and bone. They are caused by unrelieved pressure or friction and can be found predominantly below the waist and at bony prominences (sacrum, heels, and hips).

13.3 PHASES OF HEALING

The process of wound healing is well regulated and for the ease of understanding is divided into specific functional phases, namely, hemostasis, inflammation, proliferation, and remodeling (Figure 13.1). All these phases of healing take place in an overlapping series of programmed events to promptly reestablish barrier function of the skin.

13.3.1 Hemostasis

The first step following injury is the formation of the fibrin plug. This process also lays the foundation for subsequent inflammation and healing processes [7]. The fibrin plug and the surrounding wound tissue formed during this stage releases pro-inflammatory cytokines and growth factors such as transforming growth factor (TGF-β), platelet derived growth factor (PDGF), fibroblast growth factor (FGF), and epidermal growth factor (EGF). Fibrin also provides the structural support for the cellular constituents of inflammation.

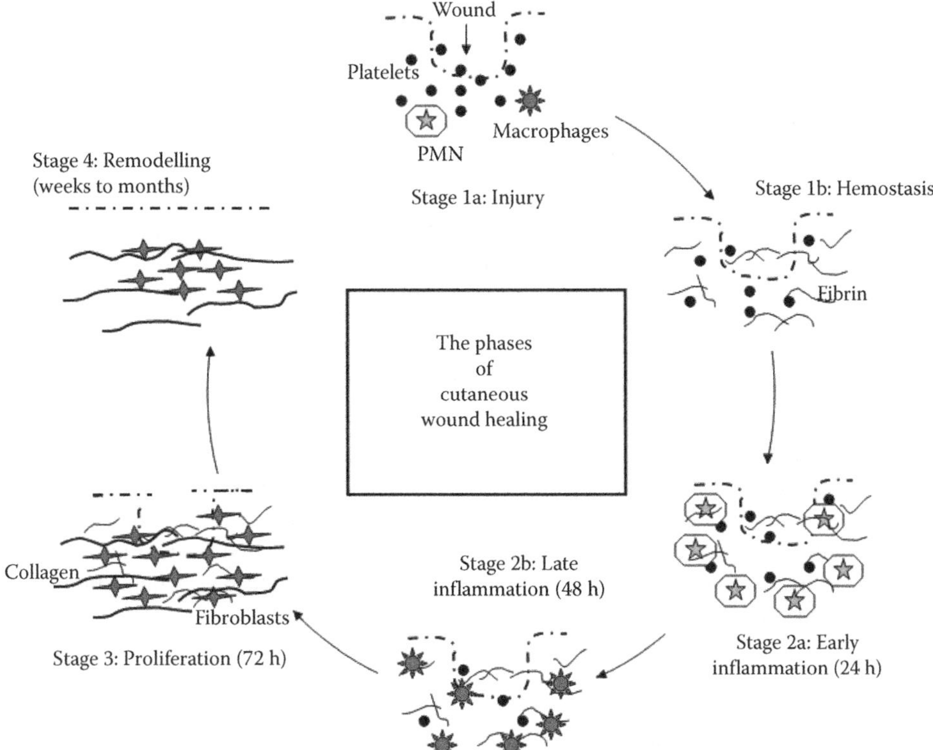

FIGURE 13.1 Stages of wound healing. At the site of injury, lots of platelets accumulate. In the hemostasis stage, fibrin clots are formed at the wound site. During the early inflammatory phase, PMN cells predominate, while in the late inflammatory phase, there is an abundance of macrophages. This is followed by the proliferative phase where collagen is formed and fibroblasts migrate to the wound site. The final phase is the remodeling whereby the new skin layer is re-formed. (Modified from S. R. Beanes, C. Dang, C. Soo, and K. Ting *Expert Reviews in Molecular Medicine*, 5, 21, 2003.)

13.3.2 Inflammation

The inflammatory phase is characterized by the release of cytokines, chemokines, and growth factors [8]. The inflammatory phase begins with the influx of leukocytes to the site of injury. The first leukocytes to arrive at the site of injury are polymorphonuclear leukocytes (PMN). Macrophages represent the second wave of leukocyte effectors at the injured tissue. Thorough release of an array of growth, angiogenic, and inflammatory factor macrophages mark the transition of the inflammatory phase to the proliferative phase of wound healing [9].

13.3.3 Proliferation

The proliferative phase is generally characterized by epithelial proliferation and migration over the provisional matrix within the wound (reepithelialization). The primary signal for the migration of wound fibroblasts is the numerous chemotactic signals and growth factors released at the wound site. Once in the wound, fibroblasts proliferate profusely and produce matrix proteins, which further support the cell migration and are essential for the repair process. In the early proliferative phase, fibroblasts are limited to cellular replication and migration, while collagen synthesis occurs in the later stages followed by cross linking of collagen, which is responsible for vascular integrity and mechanical strength of new capillary beds. At this stage, fibroblasts start attaching to the fibronectin and collagen in the extracellular matrix (ECM). Wound contraction is a crucial step in the reparative process that helps to draw the wound edges together and promote the rapid closure of the wound [10].

13.3.4 Remodeling

This phase is characterized by a balance between the synthesis and breakdown of the extracellular matrix components. Collagen deposition reaches a peak by the third week after wounding. Collagen acts as a framework on which new tissues are laid. As this framework becomes more organized and hydroxylated, it increases the tensile strength of the wound tissue. Remodeling continues even after months of wound closure and influences the scar outcomes of the healed wound.

13.4 WOUND INFLAMMATORY RESPONSE

The inflammatory response in wounds is tightly regulated by signals that initiate, maintain, or resolve inflammation [11]. An imbalance between these signals may cause chronic inflammation, derailing the healing cascade. The primary aim of the inflammatory response is to fight wound infection. However, unlike adult cutaneous wounds, fetal wounds heal in a scarless manner with an attenuated inflammatory response [12]. While the signals that initiate and maintain wound inflammation have been extensively studied [11, 13, 14], the signals that resolve wound inflammation remain poorly understood [15, 16].

13.4.1 The Inflammatory Cells

The inflammatory response to wounding involves active recruitment of circulating immune cells, such as neutrophils and macrophages. These cells are attracted to sites of tissue damage by a combination of factors including serum, foreign epitopes of invading microorganisms, and growth factor and cytokine signals emanating from both immune and nonimmune cells resident within the wounded tissue (e.g., mast cells, T cells, Langerhans cells, keratinocytes, and fibroblasts). *Platelets* are the first cells visiting the site of injury as a result of direct spill from injured vessels to initiate the coagulation process. Platelets aggregate at the ends of damaged blood vessels, convert fibrinogen to fibrin, and prevent loss of blood from damaged vessels. Next to arrive are the *neutrophils*. Peak recruitment takes hours, and a lower level of recruitment may continue for several days. The primary role of neutrophils is to cleanse microbes invading the open wound. They are also a source of pro-inflammatory cytokines, including interleukins (IL-1α and β) and tumor necrosis factor alpha (TNF-α), which provide some of the earliest signals activating local fibroblasts and keratinocytes. During phagocytosis, neutrophils generate copious amounts of reactive oxygen species (ROS) by way of respiratory burst. ROS can not only kill pathogens, it is also a major player in redox signaling. Excessive ROS, however, is deleterious for the regenerating host tissue as is observed in chronic wound situations.

Neutrophils also help recruit *macrophages* to the wound site [17]. Macrophages are the predominant cell types in a healing wound, peaking around 3–5 days following injury. The primary function of wound macrophages is to operate as phagocytes cleansing the wound of all matrix and cell debris including fibrin and apoptotic neutrophils. Macrophages also produce a range of cytokines and growth and angiogenic factors that play key roles in the regulation of fibroblast proliferation and angiogenesis [11, 17, 18]. Initially, macrophages are thought to mainly take the form of classically activated, pro-inflammatory M1 macrophages that amplify the inflammatory response. As the repair process progresses, alternatively activated M2 macrophages predominate, which have anti-inflammatory characteristics and contribute to the resolution of the wound-induced inflammatory response. (The role of macrophages in the resolution of inflammation has been reviewed in Chapter 9, this book.) Another type of cell that accumulates at the site of injury during the inflammatory cells is the *mast cell* [19]. Although these cells are best known for their role in allergic reactions, they degranulate and release a variety of prestored mediators from their granules after injury [20]. Mast cells are critical regulators of neutrophil infiltration into the wound; wounds from mast cell–deficient (KitW/KitW-v) mice contain fewer wound-site neutrophils as compared to wild-type mice [19, 21]. However, the exact role of mast cells in modifying wound-induced inflammatory response remains to be further investigated.

13.4.2 Cytokines

Cytokines can be pro-inflammatory as well as anti-inflammatory. Some of the pro-inflammatory cytokines are IL-1α, IL-1β, IL-6, TNF-α, and IL-18, which are prominently upregulated during the repair process [14]. The major anti-inflammatory cytokines are IL-4, IL-10, IL-11, and IL-13. IL-6 knockout animals take up to three times longer to heal than those of wild-type controls [22], suggesting a key role of this cytokine in driving the wound repair process. Persistent expression of the inflammatory cytokines IL-1α and TNF-α was observed in an excisional wound healing model in diabetic (db/db) mice [15, 23]. Depending on the concentration, length of exposure, and presence of other cytokines, TNF-α can be beneficial or deleterious [24]. *In vitro*, cytotoxic and growth inhibitory effects of TNFα have been demonstrated in endothelial cells and fibroblasts [25, 26]. Subcutaneous injection of TNF-α increases collagen deposition and enhances wound disruption strength (WDS) in adriamycin-treated animals [27]. Lowering of the functionally available levels of the pro-inflammatory cytokine TNF-α using anti-TNF-α therapy directed at managing activated macrophages restores diabetic wound healing in ob/ob mice [28]. Studies using TNF-α null mice demonstrated that lack of TNF-α potentiates Smad-mediated fibrogenic reaction in the healing dermis, potentially leading to fibrosis, abnormal contraction, and eventually organ dysfunction [29]. IL-10, which is an anti-inflammatory cytokine [30], attenuates the expression of pro-inflammatory cytokines in fetal wounds, resulting in minimized matrix deposition and scar-free healing [31]. Increased levels of the pro-inflammatory cytokines TNF-α and IL-6 and a decreased level of IL-10 were reported in diabetic wound tissue compared to non-diabetic healing wound tissue (Figure 13.2).

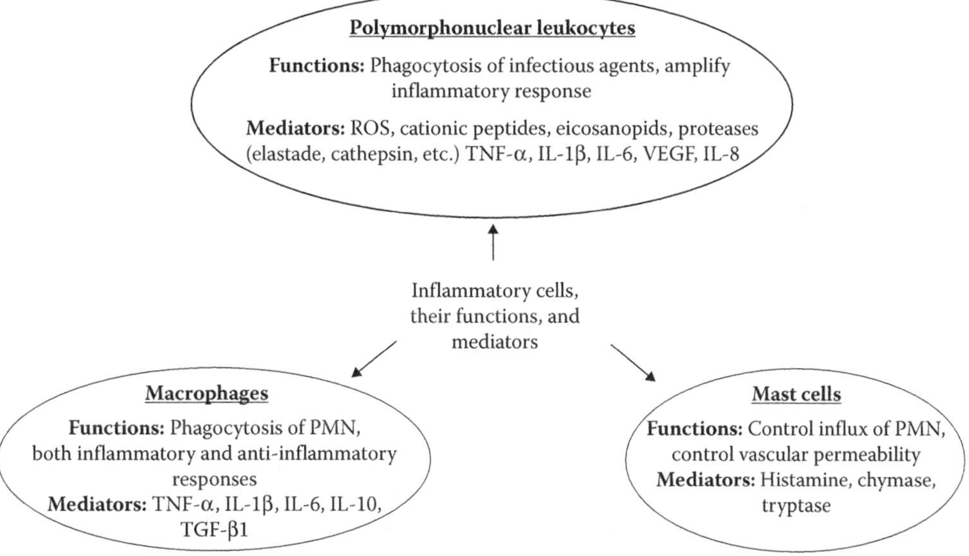

FIGURE 13.2 Inflammatory cells, their functions, and mediators.

13.4.3 Lipid Mediators

Lipid mediators such as eicosanoids are derived from oxygenation of arachidonic acid (i.e., ω-6 C20:4), and are released from membrane phospholipids by phospholipases A2 in response to inflammatory stimuli [9]. Eicosanoids consist of a family of biologically active metabolites, including prostaglandins, prostacyclin, thromboxanes, leukotrienes, and lipoxins. Free arachidonic acid may be metabolized through the cyclooxygenase (COX) pathway, involving COX-1 and COX-2, along with terminal synthases, to generate prostaglandins, prostacycilns, or thromboxanes. Induction of COX-2 represents one of the earliest responses following cutaneous injury. Consequent deployment of pro- and anti-inflammatory prostaglandin signaling mechanisms drives progression of the healing response [13]. COX-2 is the primary enzyme responsible for increased production of the pro-inflammatory mediator PGE2 in wounded skin [32]. Elevated COX-2 protein expression and prostaglandin production in chronic venous leg ulcers may contribute to the failure of these wounds to properly resolve inflammation and close in a timely manner [32].

Alternatively, arachidonic acid may be oxidized by the lipoxygenase pathway—5-lipoxygenase, 12/15-lipoxygenase, LTA4 hydrolase, and LTC4 synthase—to produce leukotrienes and lipoxins [33]. These eicosanoids initiate, amplify, and perpetuate inflammation in both acute as well as chronic wounds [34].

Some of the lipid autacoids having anti-inflammatory properties include ω-3 polyunsaturated fatty acids (PUFAs)—eicosapentaenoic (EPA; i.e., ω-3, C20:5) and docosahexaenoic acid (DHA; i.e., ω-3, C22:6); lipoxins A4 and B4 (LXA4 and LXB4), which are generated by the action of platelet 12-lipoxygenase on neutrophil leukotrienes A4 (LTA4) [35]; cyclopentenone prostaglandins (15dPGJ2), which are formed by the *in vivo* and *in vitro* dehydration of PGD2 by COX-2; resolvins and protectins, which are formed from DHA and EPA [40]; and maresins (14S-HDHA), which are primarily generated by macrophages and are novel metabolites of the 14-lipoxygenase pathway [44].

13.4.4 Mechanisms of Inflammatory Resolution

Successful repair after tissue injury requires resolution of the inflammatory response. While knowledge about mechanisms and molecules inducing and perpetuating the inflammatory response is well established, mechanisms that limit and downregulate this activity have been less appreciated. Such mechanisms might include downregulation of chemokine expression by anti-inflammatory cytokines such as IL-10 [45] or TGF-β1 [46, 47]; or upregulation of anti-inflammatory molecules like IL-1 receptor antagonist or soluble TNF receptor; resolution of the inflammatory response mediated by the cell surface receptor for hyaluronan CD44 [48]; apoptosis [49]; and receptor unresponsiveness or downregulation by high concentrations of ligands (Figure 13.3). The mechanisms of inflammatory resolution have been addressed in detail in Chapter 9, this book.

13.5 EXCESS INFLAMMATION IS ASSOCIATED WITH IMPAIRED WOUND HEALING

The prolonged inflammatory phase in nonhealing wounds is attributed to both local stimuli and underlying systemic defects [50]. An example of a systemic defect affecting inflammatory reaction and wound healing response is pyoderma gangrenosum; effective treatment is only achieved by immune suppression. Factors intrinsic to underlying metabolic disease, such as diabetic hyperglycemia or increased hydrostatic pressure associated with venous disease, can enhance and perpetuate the inflammatory response. Tissue hypoxia, bacterial components, foreign bodies, and fragments of necrotic tissue are powerful local stimuli that are capable of sustaining a continued influx of neutrophils and macrophages [3, 51]. Chronic inflammation is often a result of unbalanced proteolytic activity, which overwhelms local tissue protective mechanisms in nonhealing wounds [52–55]. Cells in the wound site, such as activated keratinocytes at the wound edge,

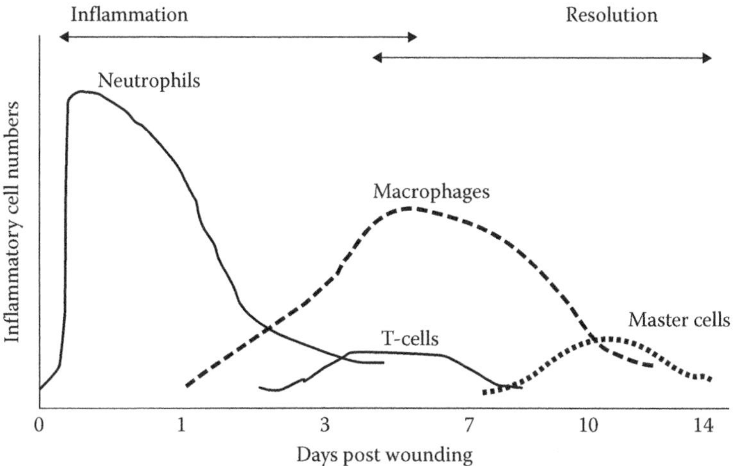

Inflammation	**Resolution**
• Chemokines (MCP, MIP, RANTES)	• Cytokines (IL-10, TGF-β, IL-1 ra)
• Growth factors (TNF-α, IL-1, IL-6, PDGF, TGF-β)	• Apoptosis mediators (CD44, caspases)
• Eicosanoids (PGs, LTs)	• Transcription factors (Nrf-2, NF-κB)
• ECM fragments (fibronectin, elastin, collagen)	• ECM fragments (fibronectin, elastin, collagen)
• ROS	• Proteolysis of chemokines
• Nitric oxide	
• Bacterial derived chemoattractants	

FIGURE 13.3 Time kinetics of the arrival of inflammatory cells. First the neutrophils arrive at the wound site, later followed by the macrophages. There is also a simultaneous influx of T cells, and then mast cells in the final stage when resolution sets in.

fibroblasts, and endothelial cells, invading neutrophils, and macrophages are considered to be the major source of numerous proteases. The expression and activity of various MMP classes, including collagenases (MMP-1, MMP-8), gelatinases (MMP-2, MMP-9), and stromelysins (MMP-3, MMP-10, and MMP-11), as well as the membrane type MMP (MT1-MMP), have been shown to be highly upregulated in chronic venous stasis ulcers [54, 56]. Pro-inflammatory cytokines are potent inducers of MMP expression and downregulate the expression of tissue inhibitor of metalloproteinases in chronic wounds. In addition, elevated levels of serine proteinases, particularly of neutrophil origin such as cathepsin G, urokinase-type plasminogen activator, and neutrophil elastase [55, 57, 58], have been found at the chronic wound site. As a consequence of the highly proteolytic microenvironment, mediators crucial for repair, such as a1-proteinase inhibitor and a2-macroglobulin, as well as components of the provisional wound matrix, such as fibronectin and vitronectin, become targets of wound proteases and are degraded and inactivated within the chronic wound environment [57, 58]. Growth factors pivotal for repair, such as platelet-derived growth factor or vascular endothelial growth factor, are also targets of wound proteases, and they are inactivated by proteolytic cleavage [59–62].

The chronic wound is a highly prooxidant microenvironment [63–66]. The disturbed oxidant/antioxidant balance within the chronic wound microenvironment is a major factor that amplifies the persistent inflammatory state of nonhealing wounds. Potential sources of ROS (superoxide anion, hydroxyl radicals, singlet oxygen, hydrogen peroxide) in a wound environment include leukocytes, especially neutrophils [67], endothelial cells, and fibroblasts, in particular senescent fibroblasts that are prominent in chronic wounds [64, 68]. In addition to direct damage of cell membranes and structural proteins of the extracellular matrix, ROS can selectively affect signaling pathways, leading to the activation of transcription factors that control the expression of pro-inflammatory cytokines (IL-1, -6, and TNF-α), chemokines, and proteolytic enzymes including MMPs and serine proteinases [65].

13.6 MICRORNAS IN CHRONIC INFLAMMATION

MicroRNAs (miRNAs) are an abundant class of short (~22 nucleotides), nonprotein-coding RNAs that regulate the expression of protein-coding genes at the posttranscriptional level [69]. Two of the most common chronic inflammatory disorders of the skin are psoriasis and atopic eczema, both of which are characterized by infiltration of inflammatory cells into the epidermis and altered keratinocyte differentiation. MiRNAs have recently been implicated in the pathogenesis of psoriasis and atopic eczema. In particular, miR-203 is specifically overexpressed in psoriasis, while miR-146a and miR-125b are involved in the regulation of innate immune responses. Tumor necrosis factor (TNF)-α pathway is also deregulated in psoriasis and atopic eczema [70]. One of the targets of miR-203 for posttranscriptional suppression is the suppressor of cytokine signalling-3 (SOCS-3). This protein is a negative regulator of the STAT3 pathway, which is activated by inflammatory cytokines (e.g., interleukin-6, interferon-c) and has important functions in the regulation of both innate and adaptive immunity, and also in cell growth, survival, and differentiation. The increased expression of miR-203 leads to decreased SOCS-3 levels in psoriatic skin, which may consequently result in sustained activation of the STAT3 signaling pathway [71]. Therefore, miR-203 may contribute to increased/prolonged skin inflammation in response to T cell-derived cytokines, due to a defective negative feedback mechanism of cytokine signaling in keratinocytes. MiR-146a targets TRAF6 and IRAK, which are proteins involved in regulating the TNF-α signaling pathway [71]. MiR-125b also targets TNF-α directly for posttranscriptional repression [72]. Therefore, downregulation of miR-125b may contribute to elevated TNF-α production during skin inflammation.

Since chronic inflammation is a hallmark of chronic nonhealing wounds, it is very likely that miRNAs regulating inflammation pathways will play a role in the pathology of chronic wounds. Transgenic mice overexpressing SOCS-3 in keratinocytes show severe impairment in acute wound healing [73]. Keratinocyte-specific overexpression of SOCS-3 leads to atrophied wound-margin epithelia and an increased inflammatory response by an increase in chemokine (MIP-2) and inflammatory enzyme (COX-2 and iNOS) expression [73]. In addition, wound tissue of transgenic mice showed a prolonged persistence of neutrophils and macrophages, all of which are characteristics of chronic wounds. Similarly, excessive amounts of pro-inflammatory cytokines such as TNF-α are associated with inflammatory diseases, including chronic wounds. A more detailed commentary on the involvement of miRNAs in chronic inflammation has been addressed in Chapter 9, this book.

13.7 BIOFILM AND CHRONIC WOUND INFLAMMATION

Along with elevated pro-inflammatory cytokines, high protease levels (matrix metalloprotease and elastase), and excessive neutrophils, chronic wounds are also characterized by the presence of biofilm, which is considered to play a significant role in chronic inflammation.

Biofilms are surface adherent, environmentally resistant groups of bacteria (often multiple genotypes) held together by extracellular polymeric substances like polysaccharides, proteins, and DNA. [74].

Chronic cutaneous wounds have specific biochemical and cellular similarities in spite of their dissimilar etiologies, which may be explained by the presence of biofilms [75]. A muted initial immune response in a chronic wound allows bacteria to establish a biofilm community in the wound, which is difficult to eradicate and perpetuates inflammation, promoting chronicity. Host immunity is ineffective against biofilm infections because white blood cells, antibodies, and complement are often unsuccessful in resolving the biofilm challenge. An increase in genetic diversity within a biofilm is commonly associated with an increase in its ability to withstand environmental stress. A wound biofilm that demonstrates a single predominant pathogenic species is generally less robust and easier to suppress than a genetically diverse biofilm [76]. In contrast, diverse biofilm as seen in chronic wounds must possess a "functional equivalence" that allows bacteria to attach, organize, and maintain a hyper-inflammatory wound environment [77]. Nonpathogenic bacterial species may act symbiotically as part of a unified biofilm community to promote a chronic wound biofilm infection. Pathogens like *Staphylococcus aureus* and *Pseudomonas aeruginosa* may be sufficient to initiate and sustain a wound biofilm, but to be functionally effective, a diverse population with several cooperating species is required to achieve the same degree of infection. One species may attach to host epitopes while another may self-secrete or organize host components into a protective matrix; a third species may be perpetually releasing lipopolysaccharides, inducing a perpetual inflammatory state, while a fourth species may be co-aggregating the community.

Studies have demonstrated that bacteria can upregulate pro-inflammatory cytokines in order to manipulate the host's innate immune response. For example, Shigella expresses a plasma virulence gene that increases expression of pro-inflammatory cytokines [78]. Quorum-sensing molecules from *Pseudomonas aeruginosa* can act directly on host cells to induce expression of pro-inflammatory cytokines [79]. *Staphylococcus aureus* (and other pathogens) express "modulins" and superantigens (enterotoxins, toxic shock syndrome toxin-1) that can induce massive and sustainable release of pro-inflammatory cytokines [80]. Additionally, the type III secretion systems of pathogens inject protein effectors into host immune cells, dampening phagocytosis and reducing the lethal actions of neutrophils, thereby increasing the release of pro-inflammatory cytokines into the wound [81].

A major component of gram-negative bacteria cell walls is lipopolysaccharide, which is a potent inducer of continued chemotaxis of neutrophils to the site of the wound biofilm. In a healthy immune response, neutrophils begin apoptosis after engulfing a pathogen. They express proteinaceous surface molecules, such as phosphatidylserine, which macrophages interpret as a request for phagocytosis [82–84]. Macrophages recognize, engulf, and then degrade these functionally terminal neutrophils, preventing necrotic disintegration of the neutrophil *in situ*, a mechanism that is required to control inflammation [83–88]. Lipopolysaccharide can, however, interfere with the neutrophils' membrane-associated phosphatidylserine. A localized accumulation of neutrophils at the infected site is a healthy host response, but this accumulation must be accompanied by the orderly elimination of neutrophils by macrophages in order to control the inflammatory response [88]. When bacterial products like lipopolysaccharide interfere with the phosphatidylserine, the macrophages may not recognize the neutrophils. As a result, they are inefficiently cleared from the site of infection where they degrade, resulting in localized release of proteolytic enzymes (elastase, metalloproteases, and inflammatory mediators) into the wound bed [89–91]. The excess protease activity in the chronic wounds favors biofilm formation and thus sets up a vicious cycle. For example, in the case of *Staphylococcus epidermidis*, host proteases proteolytically process accumulation-associated protein, which activates the bacteria's adhesion function and initiates the biofilm formation [53, 92]. Elastase then degrades the neutrophil's CXCR1 receptors, fragments of which stimulate TLR2 receptors to produce additional pro-inflammatory cytokines that feed the inflammatory cycle and recruit additional neutrophils. This perpetual cycle produces and sustains elevated levels of inflammation [93], which discourages healing. Proteases and elastases induce apoptosis in other host cell types, including local tissue, that should be involved in the wound repair process [94].

13.8 CONCLUSION AND FUTURE DIRECTIONS

To achieve successful tissue repair, there needs to be a fine-tuned balance between the numerous pro- and anti-inflammatory mediators involved in wound healing. Dysregulation of the critical parameters of these interactions will result in pathologic and chronic inflammatory disease states that are associated with nonhealing chronic wounds (Figure 13.4). Unraveling pro- and anti-inflammatory

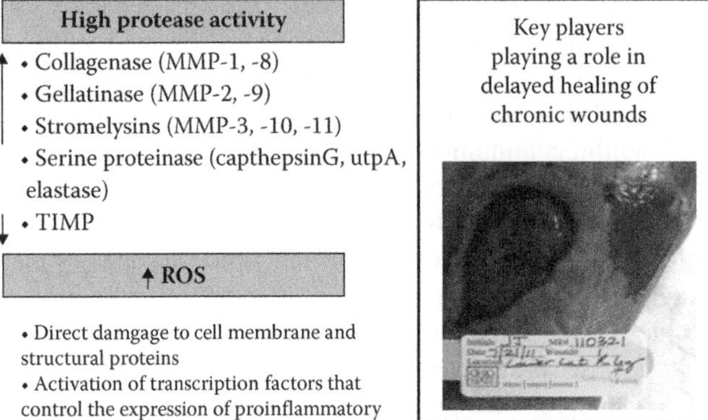

FIGURE 13.4 The key players involved in the delayed wound healing response of chronic wounds.

pathways in tissue repair might be an important avenue to develop protective strategies, which shield the regenerative tissue from damage caused by the chronically inflamed microenvironment of the nonhealing wound. The excessive and unbalanced inflammation characterizing the chronic wound suggests a promising target for future therapeutic interventions. One of the commonly practiced solutions for chronic inflammation includes wound debridement, which removes dead, damaged, or infected tissue to improve the healing potential of the remaining healthy tissue [16]. Other methods aimed at altering the inflammatory cascade involve: (1) use of exogenous cytokines and growth factors to shift the degradative disequilibrium found in a chronic wound toward a more synthetic mode, (2) the use of anti-inflammatory drugs such as nonsteroidal anti-inflammatory drugs (NSAIDs) [96], and (3) nutritional interventions like using PUFA supplementation. Thus, a better understanding of the mediators and mechanisms that are central to the initiation and resolution of wound inflammation will help design improved strategies to manage persistent non-resolving inflammation commonly associated with chronic wounds.

TAKE-HOME MESSAGES

- A chronic wound is defined as a wound or any interruption in the continuity of the body's surface that requires a prolonged time to heal (more than four weeks), does not heal, or recurs.
- Chronic wounds fail to progress through the normal phases of healing and therefore enter a state of prolonged pathologic inflammation.
- Pro-inflammatory cytokines are potent inducers of MMP expression and downregulate the expression of tissue inhibitor of metalloproteinases in chronic wounds.

- The disturbed oxidant/antioxidant balance within the chronic wound microenvironment is a major factor that amplifies the persistent inflammatory state of nonhealing wounds.
- ROS can selectively affect signaling pathways, leading to the activation of transcription factors that control the expression of pro-inflammatory cytokines (IL-1, -6, and TNF-α), chemokines, and proteolytic enzymes including MMPs and serine proteinases.
- Since chronic inflammation is a hallmark of chronic nonhealing wounds, it is very likely that miRNAs regulating inflammation pathways will play a role in the pathology of chronic wounds.
- Along with elevated pro-inflammatory cytokines, high protease levels (matrix metalloprotease and elastase), and excessive neutrophils, chronic wounds are also characterized by the presence of biofilm, which is considered to play a significant role in chronic inflammation.

REFERENCES

1. Wysocki, A.B. Wound fluids and the pathogenesis of chronic wounds. *J Wound Ostomy Continence Nurs* 23, 283–290 (1996).
2. Baker, S.R., Stacey, M.C., Jopp-McKay, A.G., Hoskin, S.E. & Thompson, P.J. Epidemiology of chronic venous ulcers. *Br J Surg* 78, 864–867 (1991).
3. Singer, A.J. & Clark, R.A. Cutaneous wound healing. *N Engl J Med* 341, 738–746 (1999).
4. Crovetti, G., et al. Platelet gel for healing cutaneous chronic wounds. *Transfus Apher Sci* 30, 145–151 (2004).
5. Sen, C.K., Gordillo, G.M., Roy, S., Kirsner, R., Lambert, L., Hunt, T.K., et al. Human skin wounds: A major and snowballing threat to public health and the economy. *Wound Repair Regen* 17, 763–771 (2009).
6. Menke, N.B., Ward, K.R., Witten, T.M., Bonchev, D.G. & Diegelmann, R.F. Impaired wound healing. *Clin Dermatol* 25, 19–25 (2007).
7. Broughton, G., 2nd, Janis, J.E. & Attinger, C.E. Wound healing: An overview. *Plast Reconstr Surg* 117, 1e-S–32e-S (2006).
8. Schreml, S., Szeimies, R.M., Prantl, L., Landthaler, M. & Babilas, P. Wound healing in the 21st century. *J Am Acad Dermatol* (2010).
9. Gronert, K. Lipid autacoids in inflammation and injury responses: A matter of privilege. *Mol Interv* 8, 28–35 (2008).
10. Velnar, T., Bailey, T. & Smrkolj, V. The wound healing process: An overview of the cellular and molecular mechanisms. *J Int Med Res* 37, 1528–1542 (2009).
11. Eming, S.A., Krieg, T. & Davidson, J.M. Inflammation in wound repair: Molecular and cellular mechanisms. *J Invest Dermatol* 127, 514–525 (2007).
12. Colwell, A.S., Longaker, M.T. & Lorenz, H.P. Fetal wound healing. *Front Biosci* 8, s1240–1248 (2003).
13. Oberyszyn, T.M. Inflammation and wound healing. *Front Biosci* 12, 2993–2999 (2007).
14. Werner, S. & Grose, R. Regulation of wound healing by growth factors and cytokines. *Physiol Rev* 83, 835–870 (2003).
15. Khanna, S., Biswas, S., Shang, Y., Collard, E., Azad, A., Kauh, C., et al. Macrophage dysfunction impairs resolution of inflammation in the wounds of diabetic mice. *PLoS One* 5, e9539 (2010).
16. Roy, S. Resolution of Inflammation in Wound Healing: Significance of Dead Cell Clearance, in *Wound Healing Society Year Book*, Vol. 1 (ed. Sen, C.K.) 253–258 (Mary Ann Liebert, Inc. publishers, New Rochelle, NY, 2010).
17. Martin, P. & Leibovich, S.J. Inflammatory cells during wound repair: The good, the bad and the ugly. *Trends Cell Biol* 15, 599–607 (2005).
18. Rappolee, D.A., Mark, D., Banda, M.J. & Werb, Z. Wound macrophages express TGF-alpha and other growth factors in vivo: Analysis by mRNA phenotyping. *Science* 241, 708–712 (1988).
19. Egozi, E.I., Ferreira, A.M., Burns, A.L., Gamelli, R.L. & Dipietro, L.A. Mast cells modulate the inflammatory but not the proliferative response in healing wounds. *Wound Repair Regen* 11, 46–54 (2003).
20. Puxeddu, I., Piliponsky, A.M., Bachelet, I. & Levi-Schaffer, F. Mast cells in allergy and beyond. *Int J Biochem Cell Biol* 35, 1601–1607 (2003).
21. Weller, K., Foitzik, K., Paus, R., Syska, W. & Maurer, M. Mast cells are required for normal healing of skin wounds in mice. *FASEB J* 20, 2366–2368 (2006).
22. Gallucci, R.M., Simeonova, P.P., Matheson, J.M., Kommineni, C., Guriel, J.L., Sugawara, T., et al. Impaired cutaneous wound healing in interleukin-6-deficient and immunosuppressed mice. *FASEB J* 14, 2525–2531 (2000).

23. Wetzler, C., Kampfer, H., Stallmeyer, B., Pfeilschifter, J. & Frank, S. Large and sustained induction of chemokines during impaired wound healing in the genetically diabetic mouse: Prolonged persistence of neutrophils and macrophages during the late phase of repair. *J Invest Dermatol* 115, 245–253 (2000).
24. Sander, A.L., Henrich, D., Muth, C.M., Marz, I., Barker, J.H., & Frank, J.M. In vivo effect of hyperbaric oxygen on wound angiogenesis and epithelialization. *Wound Repair Regen* 17, 179–184 (2009).
25. Frater-Schroder, M., Risau, W., Hallmann, R., Gautschi, P. & Bohlen, P. Tumor necrosis factor type alpha, a potent inhibitor of endothelial cell growth in vitro, is angiogenic in vivo. *Proc Natl Acad Sci USA* 84, 5277–5281 (1987).
26. Mauviel, A., Daireaux, M., Rédini, F., Galera, P., Loyau, G., & Pujoi, J.-P. Tumor necrosis factor inhibits collagen and fibronectin synthesis in human dermal fibroblasts. *FEBS Lett* 236, 47–52 (1988).
27. Mooney, D.P., O'Reilly, M. & Gamelli, R.L. Tumor necrosis factor and wound healing. *Ann Surg* 211, 124–129 (1990).
28. Goren, I., Müller, E., Schiefelbein, D., Christen, U., Pfeilschifter, J., Mühl, H., et al. Systemic anti-TNFalpha treatment restores diabetes-impaired skin repair in ob/ob mice by inactivation of macrophages. *J Invest Dermatol* 127, 2259–2267 (2007).
29. Shinozaki, M., Okada, Y., Kitano, A., Ikeda, K. & Saika, S. Impaired cutaneous wound healing with excess granulation tissue formation in TNFalpha-null mice. *Arch Dermatol Res* 301, 531–537 (2009).
30. Moore, K.W., de Waal Malefyt, R., Coffman, R.L. & O'Garra, A. Interleukin-10 and the interleukin-10 receptor. *Annu Rev Immunol* 19, 683–765 (2001).
31. Liechty, K.W., Kim, H.B., Adzick, N.S. & Crombleholme, T.M. Fetal wound repair results in scar formation in interleukin-10-deficient mice in a syngeneic murine model of scarless fetal wound repair. *J Pediat Surg* 35, 866–872; discussion 872–863 (2000).
32. Abd-El-Aleem, S.A., Ferguson, M.W., Appleton, I., Bhowmick, A., McCollum, C.N., & Ireland, G.W. Expression of cyclooxygenase isoforms in normal human skin and chronic venous ulcers. *J Pathol* 195, 616–623 (2001).
33. Haeggstrom, J.Z., Rinaldo-Matthis, A., Wheelock, C.E. & Wetterholm, A. Advances in eicosanoid research, novel therapeutic implications. *Biochem Biophys Res Commun* 396, 135–139 (2010).
34. Broughton, G., 2nd, Janis, J.E. & Attinger, C.E. The basic science of wound healing. *Plast Reconstr Surg* 117, 12S–34S (2006).
35. Kilfeather, S. 5-lipoxygenase inhibitors for the treatment of COPD. *Chest* 121, 197S–200S (2002).
36. Levy, B.D., Clish, C.B., Schmidt, B., Gronert, K. & Serhan, C.N. Lipid mediator class switching during acute inflammation: signals in resolution. *Nat Immunol* 2, 612–619 (2001).
37. Maddox, J.F. & Serhan, C.N. Lipoxin A4 and B4 are potent stimuli for human monocyte migration and adhesion: Selective inactivation by dehydrogenation and reduction. *J Exp Med* 183, 137–146 (1996).
38. Jozsef, L., Zouki, C., Petasis, N.A., Serhan, C.N. & Filep, J.G. Lipoxin A4 and aspirin-triggered 15-epi-lipoxin A4 inhibit peroxynitrite formation, NF-kappa B and AP-1 activation, and IL-8 gene expression in human leukocytes. *Proc Natl Acad Sci USA* 99, 13266–13271 (2002).
39. Lawrence, T., Willoughby, D.A. & Gilroy, D.W. Anti-inflammatory lipid mediators and insights into the resolution of inflammation. *Nat Rev Immunol* 2, 787–795 (2002).
40. Serhan, C.N., Clish, C.B., Brannon, J., Colgan, S.P., Chiang, N., & Gronert, K. Novel functional sets of lipid-derived mediators with antiinflammatory actions generated from omega-3 fatty acids via cyclooxygenase 2-nonsteroidal antiinflammatory drugs and transcellular processing. *J Exp Med* 192, 1197–1204 (2000).
41. Serhan, C.N., Hong, S., Gronert, K., Colgan, S.P., Devchand, P.R., Mirick, G., et al. Resolvins: A family of bioactive products of omega-3 fatty acid transformation circuits initiated by aspirin treatment that counter proinflammation signals. *J Exp Med* 196, 1025–1037 (2002).
42. Arita, M., Yoshida, M., Hong, S., Tjonahen, E., Glickman, J.N., Petasis, N.A., et al. Resolvin E1, an endogenous lipid mediator derived from omega-3 eicosapentaenoic acid, protects against 2,4,6-trinitrobenzene sulfonic acid-induced colitis. *Proc Natl Acad Sci USA* 102, 7671–7676 (2005).
43. Serhan, C.N., Arita, M., Hong, S. & Gotlinger, K. Resolvins, docosatrienes, and neuroprotectins, novel omega-3-derived mediators, and their endogenous aspirin-triggered epimers. *Lipids* 39, 1125–1132 (2004).
44. Serhan, C.N., Yang, R., Martinod, K., Kasuga, K., Pillai, P.S., Porter, T.F., et al. Maresins: Novel macrophage mediators with potent antiinflammatory and proresolving actions. *J Exp Med* 206, 15–23 (2009).
45. Sato, K., Nagayama, H., Tadokoro, K., Juji, T. & Takahashi, T.A. Extracellular signal-regulated kinase, stress-activated protein kinase/c-Jun N-terminal kinase, and p38mapk are involved in IL-10-mediated selective repression of TNF-alpha-induced activation and maturation of human peripheral blood monocyte-derived dendritic cells. *J Immunol* 162, 3865–3872 (1999).

46. Ashcroft, G.S. Bidirectional regulation of macrophage function by TGF-beta. *Microbes Infect* 1, 1275–1282 (1999).
47. Ashcroft, G.S., Yang, X., Glick, A.B., Weinstein, M., Letterio, J.J., Mizel, D.E., et al. Mice lacking Smad3 show accelerated wound healing and an impaired local inflammatory response. *Nat Cell Biol* 1, 260–266 (1999).
48. Teder, P., Vandivier, R.W., Jiang, D., Liang, J., Cohn, L., Puré, E., et al. Resolution of lung inflammation by CD44. *Science* 296, 155–158 (2002).
49. Greenhalgh, D.G. The role of apoptosis in wound healing. *Int J Biochem Cell Biol* 30, 1019–1030 (1998).
50. Loots, M.A., Lamme, E.N., Zeegelaar, J., Mekkes, J.R., Bos, J.D., & Middelkoop, E. Differences in cellular infiltrate and extracellular matrix of chronic diabetic and venous ulcers versus acute wounds. *J Invest Dermatol* 111, 850–857 (1998).
51. Eming, S.A., Smola, H. & Krieg, T. Treatment of chronic wounds: State of the art and future concepts. *Cells Tissues Organs* 172, 105–117 (2002).
52. Palolahti, M., Lauharanta, J., Stephens, R.W., Kuusela, P. & Vaheri, A. Proteolytic activity in leg ulcer exudate. *Exp Dermatol* 2, 29–37 (1993).
53. Harris, I.R., Yee, K.C., Walters, C.E., Cunliffe, W.J., Kearney, J.N., Wood, E.J., et al. Cytokine and protease levels in healing and non-healing chronic venous leg ulcers. *Exp Dermatol* 4, 342–349 (1995).
54. Saarialho-Kere, U.K. Patterns of matrix metalloproteinase and TIMP expression in chronic ulcers. *Arch Dermatol Res* 290 Suppl, S47–54 (1998).
55. Barrick, B., Campbell, E.J. & Owen, C.A. Leukocyte proteinases in wound healing: Roles in physiologic and pathologic processes. *Wound Repair Regen* 7, 410–422 (1999).
56. Norgauer, J., Hildenbrand, T., Idzko, M., Panther, E., Bandemir, E., Hartmann, M., et al. Elevated expression of extracellular matrix metalloproteinase inducer (CD147) and membrane-type matrix metalloproteinases in venous leg ulcers. *Br J Dermatol* 147, 1180–1186 (2002).
57. Grinnell, F., Ho, C.H. & Wysocki, A. Degradation of fibronectin and vitronectin in chronic wound fluid: Analysis by cell blotting, immunoblotting, and cell adhesion assays. *J Invest Dermatol* 98, 410–416 (1992).
58. Grinnell, F. & Zhu, M. Fibronectin degradation in chronic wounds depends on the relative levels of elastase, alpha1-proteinase inhibitor, and alpha2-macroglobulin. *J Invest Dermatol* 106, 335–341 (1996).
59. Wlaschek, M., Peus, D., Achterberg, V., Meyer-Ingold, W. & Scharffetter-Kochanek, K. Protease inhibitors protect growth factor activity in chronic wounds. *Br J Dermatol* 137, 646 (1997).
60. Lauer, G., Sollberg, S., Cole, M., Flamme, I., Stürzebecher, J., Mann, K., et al. Expression and proteolysis of vascular endothelial growth factor is increased in chronic wounds. *J Invest Dermatol* 115, 12–18 (2000).
61. Lauer, G., Sollberg, S., Cole, M., Krieg, T. & Eming, S.A. Generation of a novel proteolysis resistant vascular endothelial growth factor165 variant by a site-directed mutation at the plasmin sensitive cleavage site. *FEBS Lett* 531, 309–313 (2002).
62. Roth, D., Piekarek, M., Paulsson, M., Christ, H., Bloch, W., Krieg, T., et al. Plasmin modulates vascular endothelial growth factor-A-mediated angiogenesis during wound repair. *Am J Pathol* 168, 670–684 (2006).
63. James, T.J., Hughes, M.A., Cherry, G.W. & Taylor, R.P. Evidence of oxidative stress in chronic venous ulcers. *Wound Repair Regen* 11, 172–176 (2003).
64. Mendez, M.V., Stanley, A., Park, H.Y., Shon, K., Phillips, T., & Menzoian, J.O. Fibroblasts cultured from venous ulcers display cellular characteristics of senescence. *J Vasc Surg* 28, 876–883 (1998).
65. Wenk, J., Foitzik, A., Achterberg, V., Sabiwalsky, A., Dissemond, J., Meewes, C., et al. Selective pick-up of increased iron by deferoxamine-coupled cellulose abrogates the iron-driven induction of matrix-degrading metalloproteinase 1 and lipid peroxidation in human dermal fibroblasts in vitro: A new dressing concept. *J Invest Dermatol* 116, 833–839 (2001).
66. Wlaschek, M. & Scharffetter-Kochanek, K. Oxidative stress in chronic venous leg ulcers. *Wound Repair Regen* 13, 452–461 (2005).
67. Weiss, S.J. Tissue destruction by neutrophils. *N Engl J Med* 320, 365–376 (1989).
68. Campisi, J. Replicative senescence: An old lives' tale? *Cell* 84, 497–500 (1996).
69. Banerjee, J., Chan, Y.C. & Sen, C.K. MicroRNAs in skin and wound healing. *Physiol Genomics* 43, 543–556 (2011).
70. Roy, S. & Sen, C.K. MiRNA in innate immune responses: Novel players in wound inflammation. *Physiol Genomics* 43, 557–565 (2011).
71. Sonkoly, E., Stahle, M. & Pivarcsi, A. MicroRNAs: novel regulators in skin inflammation. *Clin Exp Dermatol* 33, 312–315 (2008).
72. Tili, E., Michaille, J.J., Cimino, A., Costinean, S., Dumitru, C.D., Adair, B., et al. Modulation of miR-155 and miR-125b levels following lipopolysaccharide/TNF-alpha stimulation and their possible roles in regulating the response to endotoxin shock. *J Immunol* 179, 5082–5089 (2007).

73. Linke, A., Goren, I., Bosl, M.R., Pfeilschifter, J. & Frank, S. Epithelial overexpression of SOCS-3 in transgenic mice exacerbates wound inflammation in the presence of elevated TGF-beta1. *J Invest Dermatol* 130, 866–875 (2010).
74. Stoodley, P., Sauer, K., Davies, D.G. & Costerton, J.W. Biofilms as complex differentiated communities. *Annu Rev Microbiol* 56, 187–209 (2002).
75. James, G.A., Swogger, E., Wolcott, R., Pulcini, E., Secor, P., Sestrich, J., et al. Biofilms in chronic wounds. *Wound Repair Regen* 16, 37–44 (2008).
76. Ehrlich, G.D., Hu, F.Z., Shen, K., Stoodley, P. & Post, J.C. Bacterial plurality as a general mechanism driving persistence in chronic infections. *Clin Orthop Relat Res* 437, 20–24 (2005).
77. Wolcott, R.D., Rhoads, D.D. & Dowd, S.E. Biofilms and chronic wound inflammation. *J Wound Care* 17, 333–341 (2008).
78. D'Hauteville, H., Khan, S., Maskell, D.J., Kussak, A., Weintraub, A., Mathison, J., et al. Two msbB genes encoding maximal acylation of lipid A are required for invasive Shigella flexneri to mediate inflammatory rupture and destruction of the intestinal epithelium. *J Immunol* 168, 5240–5251 (2002).
79. Jahoor, A., Patel, R., Bryan, A., Do, C., Krier, J., Watters, C., et al. Peroxisome proliferator-activated receptors mediate host cell proinflammatory responses to Pseudomonas aeruginosa autoinducer. *J Bacteriol* 190, 4408–4415 (2008).
80. Llewelyn, M. & Cohen, J. Superantigens: Microbial agents that corrupt immunity. *Lancet Infect Dis* 2, 156–162 (2002).
81. Cornelis, G.R. The type III secretion injectisome, a complex nanomachine for intracellular "toxin" delivery. *Biol Chem* 391, 745–751 (2010).
82. Fadok, V.A., Bratton, D.L., Guthrie, L. & Henson, P.M. Differential effects of apoptotic versus lysed cells on macrophage production of cytokines: Role of proteases. *J Immunol* 166, 6847–6854 (2001).
83. Fadok, V.A., de Cathelineau, A., Daleke, D.L., Henson, P.M. & Bratton, D.L. Loss of phospholipid asymmetry and surface exposure of phosphatidylserine is required for phagocytosis of apoptotic cells by macrophages and fibroblasts. *J Biol Chem* 276, 1071–1077 (2001).
84. Fadok, V.A., Voelker, D.R., Campbell, P.A., Cohen, J.J., Bratton, D.L., & Henson, P.M. Exposure of phosphatidylserine on the surface of apoptotic lymphocytes triggers specific recognition and removal by macrophages. *J Immunol* 148, 2207–2216 (1992).
85. Vandivier, R.W., Fadok, V.A., Hoffmann, P.R., Bratton, D.L., Penvari, C., Brown, K.K., et al. Elastase-mediated phosphatidylserine receptor cleavage impairs apoptotic cell clearance in cystic fibrosis and bronchiectasis. *J Clin Invest* 109, 661–670 (2002).
86. Baran, J., Guzik, K., Hryniewicz, W., Ernst, M., Flad, H.D., & Pryjma, J. Apoptosis of monocytes and prolonged survival of granulocytes as a result of phagocytosis of bacteria. *Infect Immun* 64, 4242–4248 (1996).
87. Guzik, K., Bzowska, M., Smagur, J., Krupa, O., Sieprawska, M., Travis, J., et al. A new insight into phagocytosis of apoptotic cells: proteolytic enzymes divert the recognition and clearance of polymorphonuclear leukocytes by macrophages. *Cell Death Differ* 14, 171–182 (2007).
88. Coxon, A., Tang, T. & Mayadas, T.N. Cytokine-activated endothelial cells delay neutrophil apoptosis in vitro and in vivo. A role for granulocyte/macrophage colony-stimulating factor. *J Exp Med* 190, 923–934 (1999).
89. Baumann, R., Casaulta, C., Simon, D., Conus, S., Yousefi, S., & Simon, H. Macrophage migration inhibitory factor delays apoptosis in neutrophils by inhibiting the mitochondria-dependent death pathway. *FASEB J* 17, 2221–2230 (2003).
90. Gamberale, R., Giordano, M., Trevani, A.S., Andonegui, G. & Geffner, J.R. Modulation of human neutrophil apoptosis by immune complexes. *J Immunol* 161, 3666–3674 (1998).
91. Anwar, S. & Whyte, M.K. Neutrophil apoptosis in infectious disease. *Exp Lung Res* 33, 519–528 (2007).
92. Rohde, H., Burdelski, C., Bartscht, K., Hussain, M., Buck, F., Horstkotte, M.A., et al. Induction of Staphylococcus epidermidis biofilm formation via proteolytic processing of the accumulation-associated protein by staphylococcal and host proteases. *Mol Microbiol* 55, 1883–1895 (2005).
93. Hartl, D., Latzin, P., Hordijk, P., Marcos, V., Rudolph, C., Woischnik, M., et al. Cleavage of CXCR1 on neutrophils disables bacterial killing in cystic fibrosis lung disease. *Nat Med* 13, 1423–1430 (2007).
94. Yang, J.J., Kettritz, R., Falk, R.J., Jennette, J.C. & Gaido, M.L. Apoptosis of endothelial cells induced by the neutrophil serine proteases proteinase 3 and elastase. *Am J Pathol* 149, 1617–1626 (1996).
95. Sosroseno, W. & Herminajeng, E. The role of macrophages in the induction of murine immune response to Actinobacillus actinomycetemcomitans. *J Med Microbiol* 51, 581–588 (2002).
96. Khanapure, S.P., Garvey, D.S., Janero, D.R. & Letts, L.G. Eicosanoids in inflammation: Biosynthesis, pharmacology, and therapeutic frontiers. *Curr Top Med Chem* 7, 311–340 (2007).

14 Multiphasic Roles for TGF-Beta in Scarring

Implications for Therapeutic Intervention

Praveen R. Arany and George X. Huang
Harvard University
Cambridge, Massachusetts

Woo Seob Kim
Harvard University
Cambridge, Massachusetts

Chung-Ang University
Seoul, Republic of Korea

CONTENTS

14.1 INTRODUCTION

Wound healing is the process by which damaged tissues are repaired through the ordered replacement of damaged cells and simultaneous remodeling of the extracellular matrix (ECM), and is critical for survival [1]. An ideal end point for healing is regeneration that entails complete restoration of original tissue form and function. But the process of scarring interferes with regenerative or normal healing sequelae, preventing normal restoration of anatomical form and, when extensive, can impair physiological function. Specifically, scarring of internal organs—pulmonary fibrosis, liver cirrhosis, renal glomerulosclerosis, and cardiac fibrosis—often cause organ failure and mortality [1]. Furthermore, cutaneous scars are most commonly associated with cosmetic disfigurement while organ fibroses can be accompanied by pruritis, dysesthesia, and pain [2].

Normal wound healing occurs in three phases: inflammation, proliferation, and maturation. Following injury, hemostasis is achieved by the formation of a fibrin clot and the release of potent cytokines (TGF-β, EGF, IGF, PDGF, among others) during degranulation of platelets. These factors drive the inflammatory phase by facilitating recruitment of neutrophils, mast cells, and macrophages to the injury site that phagocytose debris. This inflammatory phase can last from a few hours up to five to seven days. This is followed by the proliferative phase (which lasts three to six weeks) that includes epidermal closure and dermal remodeling by fibroblasts that reestablishes tissue integrity through synthesizing new ECM (composed primarily of procollagen, elastin, and proteoglycans). Endothelial cells are also recruited to the wound site and participate in neoangiogenesis while myofibroblasts initiate wound contraction. Finally, during the maturation phase (which can last several months), procollagen at the wound site is remodeled to Type I collagen, and excess ECM components are degraded by collagenases, proteoglycanases, and metalloproteinases [2, 3]. The highly complex mechanisms regulating wound healing are still being elucidated. Similarly, the precise etiology of abnormal scar formation is not well understood, although it is postulated that exaggerated cytokine activation leads to excessive accumulation of ECM and the ultimate formation of the permanent fibroses as discussed in the following sections.

The two most common cutaneous scarring sequelae are hypertrophic scars and keloids. The incidence of hypertrophic scarring is estimated to occur 39%–68% following surgery and 33%–91% following burns [4]. The incidence of keloidal scarring ranges from 4.5% to 20% in black, Hispanic, and East Asian populations [5] (Figure 14.1). Both scar types are characterized initially by rich vasculature and erythema, high mesenchymal cell density, excessive deposition of collagen in the dermis and subcutaneous tissues, and increased thickness of the epidermal layer [5, 6]. They can be distinguished by their sources of primary injury, growth pattern, clinical course, and collagen organization. Hypertrophic scars (1) develop in response to major skin wounds such as trauma, surgery, or second- and third-degree burns; (2) remain confined to the boundaries of the initial wound, and scar growth occurs by pushing the margins of the scar outward rather than invasion of the surrounding tissue; (3) grow in a claw-like fashion and regress after several months, improving naturally over two to five years following initial injury; and (4) exhibit nodular structures in the dermis consisting of fibroblasts, small blood vessels, and collagen bundles parallel to the epithelial surface [2, 5–7]. In contrast, keloids (1) develop in response to minor skin wounds such as acne, insect bites, or piercing; (2) grow beyond the boundaries of the original wound by invasion of surrounding tissue; (3) grow indefinitely and do not resolve naturally; and (4) exhibit haphazard deposition of thick, hyaline-like collagen fibers in the dermis that are randomly oriented with respect to the epithelial surface. Notably, keloids have a higher recurrence rate following surgical excision

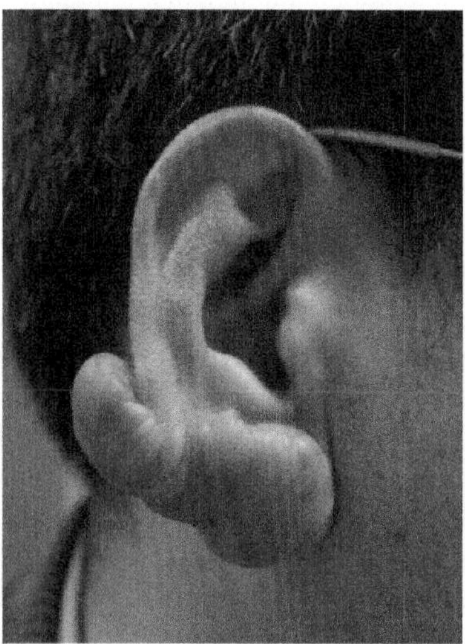
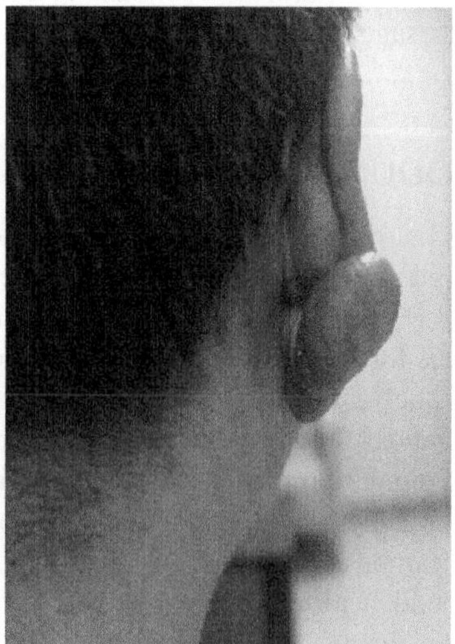

FIGURE 14.1 Common presentation of keloid in ear lobes.

TABLE 14.1
Characteristic Features of Scar Tissue

	Characteristics	Hypertrophic Scars	Keloids
Differences	Clinical margins	Stays within original wound	Extends well beyond original wound
	Clinical course	Soon after surgery/injury	Months after injury
	Incidence	Common	Rare (Ethnic susceptibility)
	Anatomical sites	Skin creases, joints	Ear lobes, sternum, shoulders
	Causal factors	Substantial injury	Minor and major injury
	Surgical management	Relatively easy, improves after treatment	Difficult, often gets worse following surgery
	Recurrences	Rare	Common
	Collagen organization	Wavy, thickened, and orient parallel to epidermal surface	Dense, nodular, and organized haphazardly
	Collagen content	About 3 times more than adjacent tissues	About 20 times more than adjacent tissues
	Cellularity	Hyper-cellular	Hypo-cellular
	Vascularity	Small vessels evident	Few vessels
Similarities	Appearance	Raised, itchy, rarely painful	
	Inciting agent	Injury, surgery	
	Collagen content	Higher than adjacent tissue	

of 45% to 100%. These key similarities and differences between hypertrophic scars and keloids are summarized in Table 14.1. It is interesting to note that the anatomically restricted hypertrophic scars (within the wound margins) are readily amenable to therapy while, in contrast, excessively enlarging keloids (extending beyond the wound margin) recur commonly, suggesting an underlying progressive spectrum of unregulated biological responses in the latter.

The number of available treatment modalities for hypertrophic scars and keloids is increasing, with many treatments targeting different stages of scar formation [8]. The clinical strategy for prevention of cutaneous scars involves generous debridement and irrigation of the wound site, rapid primary closure of the wound under minimal tension, and limiting exposure to foreign bodies (such as silk-based sutures) in order to curtail inflammation [3, 7]. Common strategies to mitigate the development of hypertrophic scars and keloids are silicone gel sheeting and compression therapy. It is thought that silicone sheeting improves hydration of the wound site and decreases delivery of pro-inflammatory cytokines through capillaries, thereby reducing collagen deposition. The efficacy of this silicone sheeting is significantly limited by patient compliance, because the sheets must be worn for 12 hours per day, often for several months. Compression therapy, during which the patient wears pre-sized garments delivering at least 15 mmHg of constant pressure for 8–24 hours per day for ~6 months, is thought to accelerate scar maturation but the underlying mechanism is unknown. A limitation of compression therapy is that the compressive garments can cause further trauma to the wound site; patient discomfort also reduces compliance.

Existing hypertrophic scars can be successfully treated by complete surgical excision followed by primary closure or skin grafting. Due to their high recurrence rate, excision is combined with adjuvant therapy to prevent recurrence in keloids. The most common adjuvant used in combinatorial therapy is the injection of synthetic corticosteroids such as triamcinolone, dexamethasone, and hydrocortisone acetate following surgical resection (Figure 14.2). It is believed that these corticosteroids downregulate the production of pro-inflammatory cytokines, ultimately inhibiting proliferation of fibroblasts and decreasing the deposition of fibrotic ECM [3]. While corticosteroid injections following surgical resection have been shown to prevent recurrence of keloids by 80%, limitations of this therapy are that painful injections must

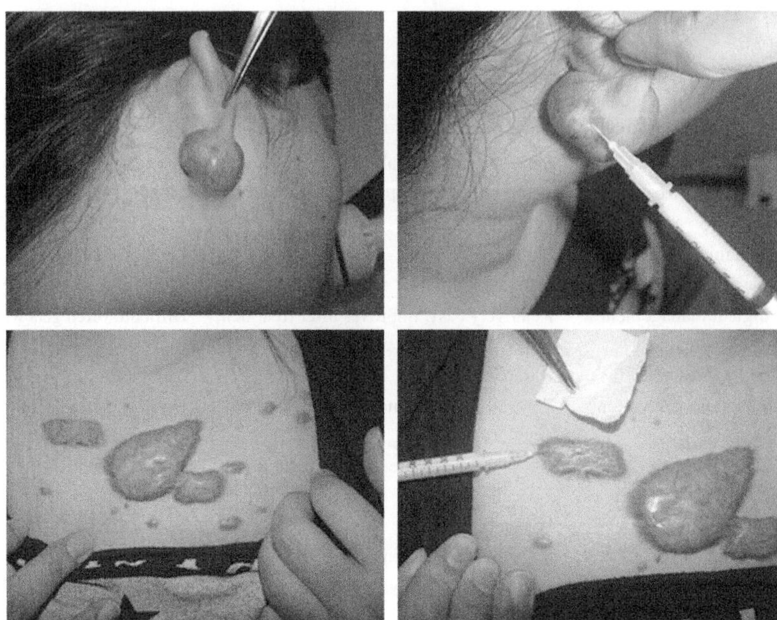

FIGURE 14.2 Scar management with intralesionsal steroid injections.

be administered every four to six weeks for up to one year, and side effects of the corticosteroids include menstrual dysfunction, suppression of adrenal cortical function, and glaucoma. An alternative adjuvant is radiation therapy (usually a dose of 15–20 Gy over a series of five to six treatments), which is thought to decrease collagen deposition through inducing apoptosis of fibroblasts; however, this remains controversial due to the theoretical risk of producing malignant tumors. Ultimately, current treatments for hypertrophic scars and keloids are inconsistent and unsatisfactory, most likely because these empirically directed treatments do not target the precise scar etiopathogenesis, which remains elusive. In order to improve existing treatments and develop new modalities for treating cutaneous scars, greater attention should be devoted to understanding the pathophysiology of scar formation. Besides the basic molecular mechanisms of scar development, reevaluating how existing combinatorial therapies interact with the kinetics of biological mechanisms responsible for fibrosis could be an excellent avenue to unravel effective scar management strategies.

14.2 ETIOPATHOGENESIS OF SCAR GENERATION

In a recent article, an attempt to rationalize interventional strategies classified the causal factors contributing to the complex etiopathogenesis of scar generation into three main categories: scar triggers, scar modulators, and scar effectors [9]. These factors contribute to specific phases of scar development in a sequential, yet overlapping, manner. The *scar triggers* contribute to the initiation and establishment of the increased fibrosis. The *scar modulators* have a critical role in either progressively increasing the fibroses process or accelerating its pathological accumulation (preventing turnover or remodeling). The *scar effectors* are the active biological entities that promote the tissue fibroses; these may be directly induced by the scar triggers or by subsequent scar modulators. While these effectors are placed last in the sequence of scar etiopathogenetic factors, they are probably active in all phases and are critical as biological mediators inducing and maintaining the fibrotic milieu characteristic of scar tissues.

14.2.1 Scar Triggers

The possible inciting causative agents leading to scar generation has been broadly described as an autoimmune process due to genetic susceptibility, sebum reaction hypothesis, and mechanical disequilibrium. Genetic susceptibility of individuals is based on the premise of an aberrant, exaggerated fibrotic response to the normal healing sequelae. The predilection of keloids in dark-skinned individuals and, often, familial tendencies has been suggested as a plausible genetic bias to scar formation in these individuals. Further, particular human leukocyte antigen subtypes B14, 21, BW35, DR5, and DQW3, along with blood group A, have been shown to be associated with individuals with higher incidence of keloids [10–13].

Keloid formation has been suggested to be an autoimmune connective tissue disease based on the presence of circulating non-complement fixing anti-fibroblast antibodies that bind fibroblasts and stimulate proliferation as well as induce collagen production [14, 15]. Interestingly, keloid explants placed in nude mice initially grow and vascularize but eventually regress, implicating a key role for the inflammatory response in its etiopathogenesis [16, 17]. One of the peculiar anatomical predilections of keloids is its presence in anatomical sites of high pilo-sebaceous glands. The disruption of sebaceous glands at the edges of the wound margins and along margins of an enlarging keloid releases the sebum to systemic circulation, inciting a potent inflammatory reaction leading to keloid propagation [18–20]. The presence of sebaceous glands in humans is cited as a possible reason of its unique incidence of keloids among all mammals. Further, a sebum vaccine has been shown to reduce recurrence of keloids in patients.

Finally, the role of tissue forces as scar-inciting agents, perhaps more relevant to hypertrophic scars than keloids, is the tensional hypothesis based on both *in vivo* observations and *in vitro* data. The increased scar incidence of wounds across normal skin tensional lines, the reduced incidence rate in elderly patients, and the linearly oriented collagen fibers noted in scars suggest a key role for tissue forces in hypertrophic scar development [21, 22]. *In vitro* experiments have demonstrated the ability of mechanical tension in driving fibroblast proliferation and collagen synthesis [23, 24]. Perhaps the most insightful evidence for the role of physical tissue forces is not only from these experimental models but also from the abundant clinical evidence of effective routine scar management strategies using silicone gels or sheets, surgical revisions, and suturing techniques that relieve tissue tension as described previously.

14.2.2 Scar Modulators

Some factors contributing to the scarring processes are low oxygen tension and angiogenesis, which are often correlated. The role of vascular changes as a predisposing factor to a fibroproliferative disease has been proposed in many specific scenarios such as idiopathic lung fibroses, cutaneous scleroderma, and diabetic retinopathy, among others. The key role of proangiogenic and antiangiogenic factors regulates the effective angiogenic milieu that may be disrupted, leading to hypoxia and subsequent fibroses. Interestingly, scar tissue has a paucity of vascular elements, and the mechanisms resulting in the decrease or loss of vasculature are being carefully elucidated in many studies investigating specific organ fibroses. Some potential targets involve the role of the CXC chemokine family in lung fibroses and circulating endothelial progenitors cells [25–27]. The role of vascular abnormalities leading to changes in the local oxygen environment leading to scarring is perhaps best illustrated by work in ophthalmology in the area of ocular fibroses such as age related macular degeneration and diabetic retinopathy. Many end-point fibrotic pathologies are characterized by inhibition of normal retinal vascular development associated with microvascular degeneration. The resultant ischemia and hypoxia lead to a compensatory excessive, but abnormal, vasculature leading to fibrous scar tissue formation [28]. A recent study demonstrated decreased oxygen tension in horse limbs post-wounding that are very susceptible to exuberant granulation tissue, akin to keloidal tissue, using a novel *in vivo* near-infrared spectroscopic technique [29].

14.2.3 Scar Effectors

The effectors can be broadly divided into cellular components that principally produce the ECM and the biomolecules involved in providing them with instructional cues. The cell lineages involved in increased matrix synthesis leading to scar formation are predominantly mesenchymal, and fibroblasts in the dermis, specifically. The roles of epithelium, endothelium, and inflammatory cells are also important via their direct regulation and interactions with the dermal compartment as well as indirect modulation of the scar microenvironment via secreted factors. These secreted factors elucidated from both *in vitro* and *in vivo* experiments have been characterized as proteins predominantly classified as growth factors, cytokines, chemokines, and enzymes, including proteinases. Growth factors are proteins that can modulate proliferation, inhibition, growth, or function of cells. A subset of these growth factors can also transmit instructional cues to cells secreting them (autocrine, feedback loop) or surrounding (paracrine) and distant cells (endocrine). These factors are called cytokines. The predominant groups of cytokines are interleukins and interferons. The former is secreted by cells in both physiological and pathological conditions, while the latter are secreted specifically in response to a foreign body, including microorganisms or tumor cells. Interleukins were also previously classified as lymphokines (secreted by lymphocytes) and chemokines (acting as chemoattractants) but this usage is discontinued due to significant redundancy in source and functions of interleukins. While these proteins are key regulatory nodes in mediating the increased fibroses, their expression and secretion are tightly regulated by intracellular mechanisms. Recent research has focused on understanding these regulatory mechanisms and has unraveled key roles for a new class of regulatory molecules termed short-hairpin (shRNA) and micro-RNAs (miRNA). These sh and miRNA are 22 nucleotides or less and have complementary sequences to multiple mRNA and hence bind and sequester the message, preventing downstream expression of several factors. These biomolecules, along with their potential biological functions in mediating fibroses, are summarized in Table 14.2.

14.3 TGF-β1 AS A PLAUSIBLE LINK

The central role of one particular growth factor, transforming growth factor-β (TGF-β), in mediating all three phases of scar pathophysiology is clearly evident. TGF-β is a multifaceted growth

TABLE 14.2
List of Scar Effectors Currently Being Evaluated for Therapy

Group of Biomolecules	Candidates	Functions
Regulatory molecules	miR-29b, miR-29c, miR-200a, miR-141, and miR-199a	Upstream modulators of multiple effectors (e.g., TGF-β)
Growth factors	TGF-β1, PDGF, CTGF, HGF, and VEGF	Major modulators of matrix synthesis
Cytokines and antagonists	IL-1b, IL-5, IL-6, IL-13, IL-21, IL-4R, IL-13Ral, GM-CSF, TNF-a, oncostatin M, WISP-1, IFN-g, IFN-a/b, IL-12, and IL-10	Paracrine, autocrine, and endocrine inducers of cell signaling
Chemokines and antagonists	CXCL1, CXCL2, CXCL12, CCL2, CCL3, CCL6, CCL17, CCL18, CXCL10, CXCL11, CCR2, CCR3, CCR5, CCR7, CXCR2, and CXCR4	Homing and trafficking of potent cell populations, including fibrocytes and other mesenchymal stem cells
Immune modulators	TLR3, TLR4, and TLR9	Modulate local inflammatory milieu
Angiogenic modulators	Adenosine deaminase, ANG II, ACE, aldosterone, and ET-1	Modulate local vasculature
Matrix modulators, including enzymes and proteases	MMPs, TIMPs, Proxyl Hydroxylase, α1β1 and αvβ6 integrins, integrin linked kinase, ICAM-1, and VCAM-1	Modulate matrix remodeling and disruptions result in excessive ECM accumulations

factor that plays key roles in many biological processes, ranging from development to physiological homeostasis to malignant transformations [30–32]. Since its first description as a secreted factor from transformed cells, and hence its name, TGF-βs has been isolated from almost all known cell types [33, 34]. TGF-β is a superfamily of more than 30 members, including BMPs, activins, inhibins, GDFs, Mullerian inhibiting substance, and Nodal, among others.

The broad roles of TGF-β in mediating the inflammatory microenvironment as well as angiogenesis and ECM syntheses, all key aspects of scar development, have been well established. TGF-β can contribute to either initiation or resolution of an immune response, based on the cellular subtypes present and the overall milieu. The ability of TGF-β to selectively induce FoxP3 positive CD4+ CD25+ Regulatory T cells has brought forth its key role as an immunomodulatory factor [35–37]. TGF-β has potent effects on hematopoetic cells such as eosinophils, mast cells, neutrophils (PMNs), lymphocytes, and monocyte-macrophages that play key roles in the inflammatory phase of scar development. Further, TGF-β also has distinct roles in the normal epithelial cell biology, inhibiting their normal proliferative response and promoting their migration. In contrast, TGF-β promotes the proliferation and synthetic functions of mesenchymal cells.

The process of formation of new vessels is termed *angiogenesis* and involves endothelial cell activation, migration, proliferation, lumenization, branching, and maturation of intercellular junctions and the surrounding basement membrane [38]. While all blood vessels begin as simple endothelial lined tubes, they undergo maturation and are surrounded by pericytes and smooth muscle cells. TGF-β can modulate all three cell responses including effects on their individual differentiation and function as well as their interactions. Recent data have also indicated that TGF-β is capable of mediating endothelial-mesenchymal transdifferentiation, adding another layer of complexity to its regulation in angiogenesis [39]. The key role of TGF-β family members in angiogenesis is evident from the embryonic lethal knockout mice phenotypes for several TGF-β family signaling components with angiogenesis defects, while an increasing number of human pathologies with vascular dysfunction implicate roles for TGF-β such as Marfan and Loey-Dietz syndromes, preeclampsia, and hereditary hemorrhagic telangiectasia, among others.

There are five known isoforms of the prototypical TGF-βs, namely TGF-β1, TGF-β 2, and TGF-β3 in mammals, and TGF-β4 and TGF-β5 in amphibians. The three mammalian TGF-β isoforms have many similarities in receptor binding and biological functions but show distinct tissue distributions. All three isoforms have been extensively studied, and their knockouts have low life expectancy due to severe immune disorders, supporting their key roles in mediating normal development and physiology [40–43]. The interplay of cells, matrix, and soluble factors in wound healing best highlights the complex, pleiotropic role of TGF-β on each of these heterogeneous components. TGF-βs are known potent inducers of extracellular matrix expression as demonstrated by their key roles in fibrotic disease as well as normal wound healing [44]. Exogenous supplementation of TGF-β1 was shown to promote wound healing. Surprisingly, the TGF-β1 knockout also demonstrated faster epithelial closure of dermal wounds [45, 46].

The absence of smad3, the cytoplasmic signal transducer in TGF-β signaling, recapitulated this accelerated epidermal healing phenotype [47, 48]. In contrast to the accelerated healing observed in dermal wounds, matrix-unsupported ear wounds in the smad3 knockout mice demonstrated a poor healing phenotype with a progressive enlargement of the ear-punch wound [49]. This phenomenon was dissected carefully to evaluate contributions from the epithelium, inflammatory-immune milieu, and the matrix components. It was observed that the primary causative factor contributing to the exacerbated healing phenotype is the abnormal matrix organization in the smad3 knockout mice. The aberrant matrix resulted in altered mechanical properties of the wound edges and tissue forces, resulting in subsequent matrix remodeling and exacerbation of the ear wound. Interestingly, when the TGF-β1 knockout was crossed with SCID mice in an attempt to abrogate the excessive inflammation seen in the former mice, the rapid epidermal healing phenotype was also reversed to a delayed, poor healing response [50]. These studies further highlight the distinct roles of TGF-βs in mediating various components in the healing scenario.

While most actions of TGF-β3 have been similar to TGF-β1, two characteristic features separate these two isoforms. First, TGF-β3 is about four times more biologically potent than TGF-β1 at equimolar concentrations, while both β1 and 3 are about 50 times more potent than β2 in specific biological contexts [51]. These differences have been attributed to the individual affinity of the isoforms to TGF-β Receptors I and II [52]. Second, in contrast to the abundance of TGF-β1 in scar and keloidal tissue, TGF-β3 has been shown to promote a scarless healing phenotype [53–55]. The use of the two isoform-specific neutralizing antibodies has further confirmed the intriguing individual roles of TGF-β1 and 3 in mediating biological responses [56].

In vivo overexpression of systemic levels of active TGF-β1 using a transgenic approach produced widespread organ fibrosis. Surprisingly, this did not result in a cutaneous scarring (following epidermal incision) phenotype, even though it led to a significant increase in matrix deposition in dermal healing (polyvinyl alcohol sponge implantation model) [57]. This was attributed to the increased TGF-β3 and TGF-βRII expression evident at these healing sites. The ability of TGF-β1 to induce β3 in an autocrine manner further highlights the critical role of the isoform regulation and context dependency of this growth factor–signaling pathway.

14.4 TARGETING TGF-β1 IN SCAR MANAGEMENT

This chapter indicates the multifaceted roles of TGF-β in mediating scar development and progression. Interventional strategies aimed at modulating TGF-β include antisense (decoy) oligonucleotides, small molecule inhibitors, soluble antibodies, and small molecule inhibitors, among others [58, 59]. The current elaborate scar management strategies, including conventional and alternative management modalities, have been previously outlined. Steroids play a central role in these protocols. The basic rationale for using steroids is to modulate the inflammatory response, thus, eliminating a potential causal scar-promoting factor. Interestingly, steroids have also been shown to have direct effects on decreasing collagen synthesis as well as altering cytokine secretion by local cells especially for bFGF and TGF-β [60].

This chapter highlights the unique temporal-spatial roles of TGF-β in the phases of normal and perturbed wound healing leading to scar generation. Hence, the use of steroids based on its specific roles of TGF-β in various stages of scarring may be a suitable rationale in scar management. Further, additional or concurrent inhibition of TGF-β may be employed in a combinatorial approach to increase efficacy of anti-scar treatments. These phase-specific roles include the role of TGF-β as a trigger modulating acute inflammation, or modulator of angiogenesis and chronic inflammation, or its most widely implicated role in increasing ECM, leading to fibrosis as outlined in Figure 14.3. While the use of steroids after routine surgery is usually contraindicated due to its effects on normal wound healing, careful use and timing of steroids postsurgery, as well as careful evaluation of clini-

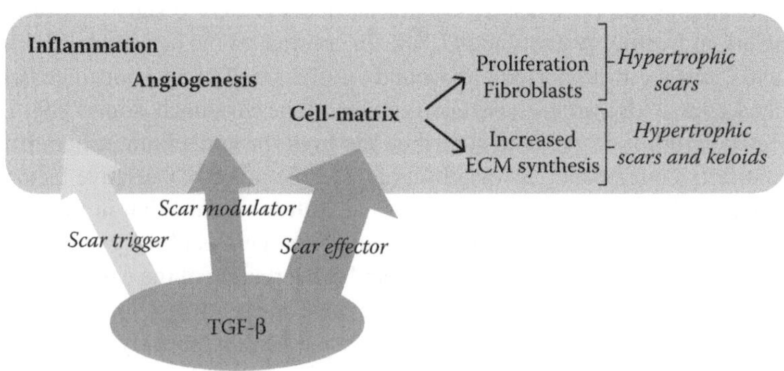

FIGURE 14.3 Phase specific role for TGF-β in scar etiopathogenesis.

cal presentations, and if feasible, histological or molecular assessment of specific scar phase might aid in most effective management strategies.

14.5 CONCLUSIONS AND FUTURE DIRECTIONS

Scarring is a complex pathophysiological process involving sequential and multiple factors that play distinct and often overlapping roles during progression. Many current management strategies are deemed inadequate, with often unsatisfying results for both the patient and the surgeon. The careful elucidation of scar development into modular components as scar triggers, modulators, and effectors may aid rationalization of targeted singular and combinatorial approaches for clinical scar management. The central role of TGF-β in these processes is highlighted in this chapter and suggests its key targeting in anti-scar therapy using conventional and newer approaches might be warranted. The use of certain conventional therapies, specifically the use of steroids, may also be potentially tailored to increase the efficacy and potency of anti-scar treatment if their timing is carefully revised, with emphasis on clinical presentations of the distinct scar phases.

Future research could be focused on both unraveling the basic mechanisms leading to scar formation and progression as well as clinical research on timing and presentation of scars. The latter research can provide significant benefit if combined with effective diagnostic and imaging modalities to unravel the distinct phases of scar progression so appropriate interventional strategies can be adopted.

TAKE-HOME MESSAGES

- The key factors contributing to scar development can be broadly divided into scar triggers, modulators, or effectors, each playing key roles in formation and progression of scars.
- TGF-β has a key role in various phases of scar development and can act as all three—a scar trigger, scar modulator, or scar effector.
- Interventional strategies targeting TGF-β may be better rationalized by targeting its specific spatio-temporal roles in scar pathogenesis.
- Steroids are routinely used in scar management predominantly for their direct effect on modulating inflammatory sequelae. Their use can be further tailored for their effects on the scar-promoting cytokine milieu, especially with respect to TGF-βs.

REFERENCES

1. Wynn TA: Common and unique mechanisms regulate fibrosis in various fibroproliferative diseases, *J Clin Invest* 2007, 117:524–529.
2. Urioste SS, Arndt KA, Dover JS: Keloids and hypertrophic scars: Review and treatment strategies, *Semin Cutan Med Surg* 1999, 18:159–171.
3. Slemp AE, Kirschner RE: Keloids and scars: A review of keloids and scars, their pathogenesis, risk factors, and management, *Curr Opin Pediatr* 2006, 18:396–402.
4. Niessen FB, Spauwen PH, Schalkwijk J, Kon M: On the nature of hypertrophic scars and keloids: A review, *Plast Reconstr Surg* 1999, 104:1435–1458.
5. Atiyeh BS, Costagliola M, Hayek SN: Keloid or hypertrophic scar: The controversy: Review of the literature, *Ann Plast Surg* 2005, 54:676–680.
6. Tuan TL, Nichter LS: The molecular basis of keloid and hypertrophic scar formation, *Mol Med Today* 1998, 4:19–24.
7. Ogawa R: The most current algorithms for the treatment and prevention of hypertrophic scars and keloids, *Plast Reconstr Surg* 2010, 125:557–568.
8. Mustoe TA, Cooter RD, Gold MH, Hobbs FD, Ramelet AA, Shakespeare PG, et al.: International clinical recommendations on scar management, *Plast Reconstr Surg* 2002, 110:560–571.
9. Arany P.R., T.L. Tuan: Targeted approaches to scar management based on our current understanding of triggers, modulators and effectors of abberant healing, *Advances in Wound Care* 2011, 2:5.

10. Cohen IK, McCoy BJ, Mohanakumar T, Diegelmann RF: Immunoglobulin, complement, and histocompatibility antigen studies in keloid patients, *Plast Reconstr Surg* 1979, 63:689–695.
11. Ramakrishnan KM, Thomas KP, Sundararajan CR: Study of 1,000 patients with keloids in South India, *Plast Reconstr Surg* 1974, 53:276–280.
12. Castagnoli C, Peruccio D, Stella M, Magliacani G, Mazzola G, Amoroso A, et al.: The HLA-DR beta 16 allogenotype constitutes a risk factor for hypertrophic scarring, *Hum Immunol* 1990, 29:229–232.
13. Laurentaci G, Dioguardi D: HLA antigens in keloids and hypertrophic scars, *Arch Dermatol* 1977, 113:1726.
14. Kazeem AA: The immunological aspects of keloid tumor formation, *J Surg Oncol* 1988, 38:16–18.
15. Bloch EF, Hall MG, Jr., Denson MJ, Slay-Solomon V: General immune reactivity in keloid patients, *Plast Reconstr Surg* 1984, 73:448–451.
16. Shetlar MR, Shetlar CL, Hendricks L, Kischer CW: The use of athymic nude mice for the study of human keloids, *Proc Soc Exp Biol Med* 1985, 179:549–552.
17. Estrem SA, Domayer M, Bardach J, Cram AE: Implantation of human keloid into athymic mice, *Laryngoscope* 1987, 97:1214–1218.
18. Yagi KI, Dafalla AA, Osman AA: Does an immune reaction to sebum in wounds cause keloid scars? Beneficial effect of desensitisation, *Br J Plast Surg* 1979, 32:223–225.
19. Fong EP, Bay BH: Keloids—the sebum hypothesis revisited, *Med Hypotheses* 2002, 58:264–269.
20. Fasika OM: Keloids: A study of the immune reaction to sebum, *East Afr Med J* 1992, 69:114–116.
21. Peacock EE, Jr., Madden JW, Trier WC: Biologic basis for the treatment of keloids and hypertrophic scars, *South Med J* 1970, 63:755–760.
22. Courtiss EH, Longarcre JJ, Destefano GA, Brizic L, Holmstrand K: The placement of elective skin incisions, *Plast Reconstr Surg* 1963, 31:31–44.
23. Sussman MD: Effect of increased tissue traction upon tensile strength of cutaneous incisions in rats, *Proc Soc Exp Biol Med* 1966, 123:38–41.
24. Curtis AS, Seehar GM: The control of cell division by tension or diffusion, *Nature* 1978, 274:52–53.
25. Strieter RM, Gomperts BN, Keane MP: The role of CXC chemokines in pulmonary fibrosis, *J Clin Invest* 2007, 117:549–556.
26. Varga J, Abraham D: Systemic sclerosis: a prototypic multisystem fibrotic disorder, *J Clin Invest* 2007, 117:557–567.
27. Del Papa N, Quirici N, Soligo D, Scavullo C, Cortiana M, Borsotti C, et al.: Bone marrow endothelial progenitors are defective in systemic sclerosis, *Arthritis Rheum* 2006, 54:2605–2615.
28. Sapieha P, Joyal JS, Rivera JC, Kermorvant-Duchemin E, Sennlaub F, Hardy P, et al.: Retinopathy of prematurity: Understanding ischemic retinal vasculopathies at an extreme of life, *J Clin Invest* 2010, 120:3022–3032.
29. Celeste CJ, Deschene K, Riley CB, Theoret CL: Regional differences in wound oxygenation during normal healing in an equine model of cutaneous fibroproliferative disorder, *Wound Repair Regen* 2011, 19:89–97.
30. Sporn MB, Roberts AB, Shull JH, Smith JM, Ward JM, Sodek J: Polypeptide transforming growth factors isolated from bovine sources and used for wound healing in vivo, *Science* 1983, 219:1329–1331.
31. Massague J: TGFbeta in Cancer, *Cell* 2008, 134:215–230.
32. Blobe GC, Schiemann WP, Lodish HF: Role of transforming growth factor beta in human disease, *N Engl J Med* 2000, 342:1350–1358.
33. Moses HL, Branum EL, Proper JA, Robinson RA: Transforming growth factor production by chemically transformed cells, *Cancer Res* 1981, 41:2842–2848.
34. Roberts AB, Anzano MA, Lamb LC, Smith JM, Sporn MB: New class of transforming growth factors potentiated by epidermal growth factor: Isolation from non-neoplastic tissues, *Proc Natl Acad Sci USA* 1981, 78:5339–5343.
35. Letterio JJ, Roberts AB: Regulation of immune responses by TGF-beta, *Annu Rev Immunol* 1998, 16:137–161.
36. Li MO, Flavell RA: TGF-beta: A master of all T cell trades, *Cell* 2008, 134:392–404.
37. Wahl SM: Transforming growth factor beta (TGF-beta) in inflammation: A cause and a cure, *J Clin Immunol* 1992, 12:61–74.
38. Goumans MJ, Liu Z, ten Dijke P: TGF-beta signaling in vascular biology and dysfunction, *Cell Res* 2009, 19:116–127.
39. Goumans MJ, van Zonneveld AJ, ten Dijke P: Transforming growth factor beta-induced endothelial-to-mesenchymal transition: A switch to cardiac fibrosis? *Trends Cardiovasc Med* 2008, 18:293–298.

40. Kulkarni AB, Huh CG, Becker D, Geiser A, Lyght M, Flanders KC, et al.: Transforming growth factor beta 1 null mutation in mice causes excessive inflammatory response and early death, *Proc Natl Acad Sci USA* 1993, 90:770–774.
41. Shull MM, Ormsby I, Kier AB, Pawlowski S, Diebold RJ, Yin M, et al.: Targeted disruption of the mouse transforming growth factor-beta 1 gene results in multifocal inflammatory disease, *Nature* 1992, 359:693–699.
42. Kaartinen V, Voncken JW, Shuler C, Warburton D, Bu D, Heisterkamp N, Groffen J: Abnormal lung development and cleft palate in mice lacking TGF-beta 3 indicates defects of epithelial-mesenchymal interaction, *Nat Genet* 1995, 11:415–421.
43. Proetzel G, Pawlowski SA, Wiles MV, Yin M, Boivin GP, Howles PN, et al.: Transforming growth factor-beta 3 is required for secondary palate fusion, *Nat Genet* 1995, 11:409–414.
44. Roberts AB, Sporn MB, Assoian RK, Smith JM, Roche NS, Wakefield LM, et al.: Transforming growth factor type beta: Rapid induction of fibrosis and angiogenesis in vivo and stimulation of collagen formation in vitro, *Proc Natl Acad Sci USA* 1986, 83:4167–4171.
45. Koch RM, Roche NS, Parks WT, Ashcroft GS, Letterio JJ, Roberts AB: Incisional wound healing in transforming growth factor-beta1 null mice, *Wound Repair Regen* 2000, 8:179–191.
46. Beck LS, Deguzman L, Lee WP, Xu Y, McFatridge LA, Amento EP: TGF-beta 1 accelerates wound healing: Reversal of steroid-impaired healing in rats and rabbits, *Growth Factors* 1991, 5:295–304.
47. Falanga V, Schrayer D, Cha J, Butmarc J, Carson P, Roberts AB, et al.: Full-thickness wounding of the mouse tail as a model for delayed wound healing: Accelerated wound closure in Smad3 knock-out mice, *Wound Repair Regen* 2004, 12:320–326.
48. Ashcroft GS, Yang X, Glick AB, Weinstein M, Letterio JL, Mizel DE, et al.: Mice lacking Smad3 show accelerated wound healing and an impaired local inflammatory response, *Nat Cell Biol* 1999, 1:260–266.
49. Arany PR, Flanders KC, Kobayashi T, Kuo CK, Stuelten C, Desai KV, et al.: Smad3 deficiency alters key structural elements of the extracellular matrix and mechanotransduction of wound closure, *Proc Natl Acad Sci USA* 2006, 103:9250–9255.
50. Crowe MJ, Doetschman T, Greenhalgh DG: Delayed wound healing in immunodeficient TGF-beta 1 knockout mice, *J Invest Dermatol* 2000, 115:3–11.
51. Cheifetz S, Hernandez H, Laiho M, ten Dijke P, Iwata KK, Massague J: Distinct transforming growth factor-beta (TGF-beta) receptor subsets as determinants of cellular responsiveness to three TGF-beta isoforms, *J Biol Chem* 1990, 265:20533–20538.
52. Lyons RM, Miller DA, Graycar JL, Moses HL, Derynck R: Differential binding of transforming growth factor-beta 1, -beta 2, and -beta 3 by fibroblasts and epithelial cells measured by affinity cross-linking of cell surface receptors, *Mol Endocrinol* 1991, 5:1887–1896.
53. Bush J, Duncan JA, Bond JS, Durani P, So K, Mason T, et al.: Scar-improving efficacy of avotermin administered into the wound margins of skin incisions as evaluated by a randomized, double-blind, placebo-controlled, phase II clinical trial, *Plast Reconstr Surg* 2010, 126:1604–1615.
54. Laverty HG, Occleston NL, Johnson M, Little J, Jones R, Fairlamb D, et al.: Effects of avotermin (transforming growth factor beta3) in a clinically relevant pig model of long, full-thickness incisional wounds, *J Cutan Med Surg* 2010, 14:223–232.
55. Ferguson MW, Duncan J, Bond J, Bush J, Durani P, So K, et al.: Prophylactic administration of avotermin for improvement of skin scarring: Three double-blind, placebo-controlled, phase I/II studies, *Lancet* 2009, 373:1264–1274.
56. Shah M, Foreman DM, Ferguson MW: Neutralisation of TGF-beta 1 and TGF-beta 2 or exogenous addition of TGF-beta 3 to cutaneous rat wounds reduces scarring, *J Cell Sci* 1995, 108 (Pt 3):985–1002
57. Shah M, Revis D, Herrick S, Baillie R, Thorgeirson S, Ferguson M, et al.: Role of elevated plasma transforming growth factor-beta1 levels in wound healing, *Am J Pathol* 1999, 154:1115–1124.
58. Cutroneo KR: TGF-beta-induced fibrosis and SMAD signaling: Oligo decoys as natural therapeutics for inhibition of tissue fibrosis and scarring, *Wound Repair Regen* 2007, 15 Suppl 1:S54–60.
59. Bonafoux D, Lee WC: Strategies for TGF-beta modulation: A review of recent patents, *Expert Opin Ther Pat* 2009, 19:1759–1769.
60. Carroll LA, Hanasono MM, Mikulec AA, Kita M, Koch RJ: Triamcinolone stimulates bFGF production and inhibits TGF-beta1 production by human dermal fibroblasts, *Dermatol Surg* 2002, 28:704–709.

15 Natural Vitamin E Tocotrienol against Neuroinflammation and Oxidative Stress

Cameron Rink and Savita Khanna
Wexner Medical Center at The Ohio State University
Columbus, Ohio

CONTENTS

15.1 CENTRAL NERVOUS SYSTEM IMMUNOLOGY

The seminal discovery by Sir Peter Medawar in 1948 that the brain does not reject foreign tissue grafts set the stage for defining the unique immunological and inflammatory processes of the central nervous system (CNS) [1]. Further characterization would identify that the CNS is deficient of classical antigen-presenting cells (i.e., dendritic cells), lacks constitutive major histocompatibility complex (MHC) I and II expression on parenchymal cells, and does not have lymphatic circulation [2]. Taken together, these findings led to the concept of "immune privilege" in the CNS, a term used to describe the perceived independence of CNS immunology from that of the periphery. Current literature, however, suggests that CNS immunology is not independent from the periphery, only refined in immunological and inflammatory response. Indeed, given the highly specialized, non-regenerative characteristics of neural cells in the CNS, unique immunological and inflammatory processes have been identified.

The CNS consists of highly differentiated cell types that include neural cells (i.e., Purkinje neurons, granule cells, motoneurons), glial cells (i.e., astrocytes, oligodendrocytes, ependymal cells, resident microglia), and endothelial cells that comprise part of the blood brain barrier (BBB). Unlike peripheral tissues, the CNS lacks resident natural killer cells, T and B lymphocytes, and has limited leukocyte extravasation across the BBB [3]. As such, major histocompatibility complex (MHC) class I and II proteins and antigen presentation is primarily a function of "activated" resident microglia [4]. Along with endothelial cells that form tight junctions, pericytes and astrocytic end-feet compose the BBB and serve to tightly regulate proteins, small molecules, and circulating cells from crossing the CNS under physiological conditions [5]. By limiting the paracellular flux of hydrophilic molecules, the BBB maintains homeostatic differences between blood and CSF constituents, and through selective transport acts as a highly selective barrier to maintain the ionic concentrations of the CNS within a narrow physiological range [3]. Less hindered interaction between CNS and peripheral blood is limited to circumventricular organs (CVO), in which fenestrations of the BBB enable regulation of metabolic and endocrine function [6]. There is increasing evidence that cross talk between CVO and peripheral immunity exists, such that pro-inflammatory cytokines produced in the periphery can feed back to the brain and vice versa [7–10]. The strict definition of "immune privilege" has softened in recognition of immune-surveillance and inflammatory responses in the CNS that are associated with circulating leukocytes [2, 11]. The growing consensus is that while CNS immunity is adapted to support the post-mitotic neural cell population, it is both immune competent and interacts with the peripheral immune system.

15.1.1 Defining Neuroinflammation

Inflammation in peripheral tissue serves two general purposes: (1) homeostasis by removal of dead tissue, and promoting rapid death of injured cells; and (2) tissue defense by production of cytotoxins, as well as cytokine and chemokine synthesis for immune cell recruitment and coordination of the immune response [11]. In the CNS both functions are required, however—the inflammatory response of the CNS is attuned to protect terminally differentiated post-mitotic neural cells. As CNS function is dependent on activity-driven development of precise neural circuitry, inflammation-mediated remodeling of neural circuitry is not necessarily adaptive or beneficial to the healing response [11]. Immune processes of the CNS, therefore, do not typically result in abundant pro-inflammatory T-cell and mononuclear phagocyte recruitment at the injury site as is the case in peripheral tissues such as skin. For example, intradermal delivery of heat-killed mycobacterium Bacillus Calmette-Guérin (10^5 organisms in 1μl) results in rapid recruitment of MHC II+ mononuclear phagocytes and activated T-cells within the first seven days [12]. The same amount of mycobacterium delivered to brain parenchyma, however, does not initiate T cell activation, and there is limited mononuclear phagocyte recruitment [12].

That is not to say, however, that the CNS is passive in its inflammatory response to pathogens or injury. The innate immune response of astrocytes and resident microglia, which comprise >50% of the total CNS cell population, contributes to inflammatory processes. In response to CNS insult, astrocytes become "reactive," undergoing cellular hypertrophy and producing pro-inflammatory cytokines including tumor necrosis factor α (TNF-α) and interleukin-6 (IL-6) [13]. Furthermore, reactive astrocytes form a scar border around injured CNS tissue, protecting non-affected CNS tissue from being exposed to dead cell clearance and necrotic factors. While the barrier function of the astrogliotic scar is beneficial in the short term, such scar formation prohibits axonal regrowth and neuroregeneration at the injured site [14].

Resident microglia also contribute to the innate neuroinflammatory response of the CNS. Unlike neurons and astrocytes, which are neuroectodermal in origin, resident microglia of the CNS are of mesenchymal origin [15]. Their role in tissue repair after injury has been compared to that of resident macrophages in peripheral organs [4]. Microglia activation occurs at the onset of CNS injury, often preceding reactions of other cell types, and is known to be induced by a number of cytokines and small molecules associated with the extracellular milieu at the injury site [16]. Once activated, microglia proliferate, change their morphology into an amoeboid shape, and phagocytose cells that are pathologically injured. As professional phagocytes, activated microglia act as first responders to destroy invading microorganisms, remove cellular debris, and restore homeostasis by secreting neurotrophic factors [4, 16]. In the relative absence of classical antigen-presenting cells (i.e., dendritic cells, B-cells), activated microglia also play a key role in the adaptive immune response following injury in the CNS [17]. Indeed, microglia that are activated by pro-inflammatory cytokines have been shown to express MHC class II and co-stimulatory molecules capable of activating CD4+ T cells [17].

The term *neuroinflammation* is relatively new, not appearing in the scientific literature before 1995 [18], and broadly describes the aforementioned specialized inflammatory processes of the CNS in response to or implicit in injury (acute) and disease (chronic). Regarding disease, a neuroinflammatory response is associated with a number of chronic neurodegenerative disorders, including Alzheimer's disease (AD), amyotrophic lateral sclerosis (ALS), multiple sclerosis (MS), and Parkinson's disease (PD). As such, the definition of neuroinflammation has been narrowed by many to denote the role of inflammatory processes in the pathophysiology of chronic neurodegenerative disease [18, 19].

15.1.2 PUFA Oxidation and Neuroinflammation

Common to both acute and chronic neuroinflammation is underlying oxidative stress, a hallmark of neuroinflammatory processes caused by an imbalance of reactive oxygen species (ROS) and the CNS's ability to neutralize them. Neural tissue is prone to oxidative stress and ROS-induced injury for a number of reasons. First, the human brain is one of the most metabolically active organs in the body. While representing ~2% of total body mass, the brain consumes ~20% of total oxygen used by the body [20]. Second, the lipid-rich brain tissue is highly concentrated in polyunsaturated fatty acids (PUFAs) that are vulnerable to lipid peroxidation. Finally, as compared to other organ systems, the neural tissue of the CNS has lower antioxidant capacity [20]. The consequence of acute or chronic injury in neural tissue enriched with PUFAs is the accumulation of lipid peroxidation species, which themselves are pro-inflammatory in nature and feed forward to cause neurodegeneration. In ALS, for example, chronic activation of astrocytes and microglia is associated with COX-2–catalyzed oxidation of PUFA arachidonic acid, subsequent production of reactive oxygen species as a by-product of COX-2 peroxidase activity, and neurodegeneration [21, 22]. This chapter reviews neuroinflammation associated with oxidation of arachidonic acid. In that light, the therapeutic role of natural vitamin E tocotrienol as a modulator of neuroinflammation will be addressed.

15.2 TOCOTRIENOL VITAMIN E

Herbert M. Evans is credited with the discovery of vitamin E in 1922, when he found that the chlorophyll-rich fraction of oil from lettuce leaves was essential for reproduction in rats kept on a rancid lard diet [23]. Knowing that this fertility factor was lipid soluble, he experimentally validated that his supplement was unique from the other two known lipid-soluble vitamins of the time—vitamins A and D. In 1936, Evans published the chemical formula of vitamin E in the *Journal of Biological Chemistry* [24] and appropriately named it *tocopherol* from the Greek words *tocos* meaning childbirth and *pheros* meaning to support.

Today, the natural vitamin E family is known to be composed of eight distinct isomers divided evenly into two family groups—tocopherols and tocotrienols [25, 26]. Common to all eight family members is a chromanol head and a 16-carbon long hydrocarbon tail (Figure 15.1). Tocopherols possess a saturated phytyl tail with three chiral carbons, while tocotrienols have a farnesyl tail with three unsaturated bonds at carbon positions 3', 7', and 11'. The chromanol head of tocopherols and tocotrienols bears their well-known antioxidant function [26]. The position and degree of methylation on the chromanol head is denoted by their designation as α, β, γ, or δ [25].

Compared to tocopherols, tocotrienols are less ubiquitous in the plant kingdom [26]. Tocotrienols are the major form of vitamin E in seeds of most monocots and a limited number of dicots. Analyses of tocotrienol enrichment in plants revealed that the presence of this natural vitamin E is localized

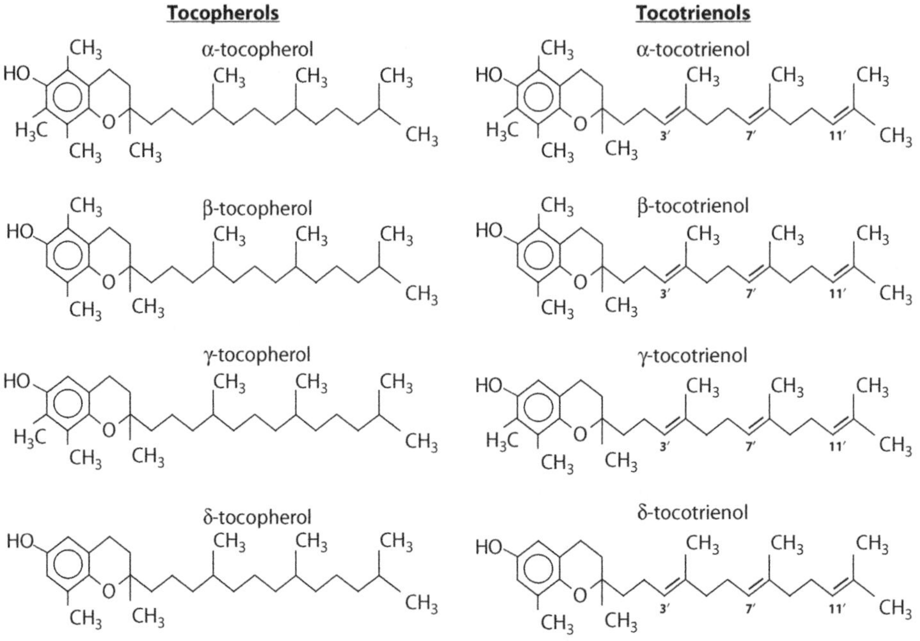

FIGURE 15.1 Chemical structure of natural vitamin E family members. Tocopherols and tocotrienols are designated α, β, γ, and δ on the basis of position and degree of methyl groups found on the chromanol head. Tocopherols possess a saturated phytyl tail while tocotrienols have three unsaturations on carbons 3', 7', and 11'.

in non-photosynthetic tissue [27]. Natural dietary sources of tocotrienols include cereal grains such as oat, rye, and barley [28]. Palm oil, a common cooking oil in Eastern diets, is one of the most abundant natural sources of tocotrienols, with crude palm oil (also referred to as the "tocotrienol-rich fraction") containing up to 800 mg/kg of α- and γ-tocotrienol isotypes [26]. The distribution of vitamin E in palm oil is 30% tocopherols and 70% tocotrienols. In contrast, other commonly used dietary vegetable oils—including corn, olive, peanut, sesame, soybean, and sunflower—contain tocopherols exclusively [26].

While all eight vitamin E family members are metabolized by the human body, one isoform, α-tocopherol, is preferentially transported to tissues. Consequently, as the most bioavailable form of vitamin E, and with a well-characterized and selective transport mechanism in mammals, the vast majority of vitamin E research to date has focused on α-tocopherol [26, 28]. Comparatively, less than 2% of the peer-reviewed literature on vitamin E addresses tocotrienol family members [28]. The disparity in vitamin E research on the basis of tissue distribution has led to a public and scientific misconception that vitamin E and α-tocopherol are synonymous. Indeed, the Recommended Dietary Allowance for vitamin E as established by the Institute of Medicine is based solely on α-tocopherol [29]. Furthermore, title claims of numerous preclinical and clinical trials testing "vitamin E" in various states of health and disease largely address α-tocopherol alone [30].

A growing body of literature, however, suggests that the lesser-characterized tocotrienol vitamin E family members possess unique biological activity not shared by their tocopherol counterparts. For example, α-tocotrienol is a potent neuroprotective agent at nanomolar concentration [25]. On a concentration basis, this represents the most potent biological function of all natural forms of vitamin E [28]. While oral supplementation of α-tocotrienol does reach brain tissue in sufficient quantity to confer neuroprotection, mechanisms of delivery remain to be elucidated. The well-characterized protein carrier for α-tocopherol, known as the tocopherol transfer protein (TTP), has an 8.5-fold higher affinity for α-tocopherol over α-tocotrienol [31]. True to the Greek name bestowed by Evans, TTP knockout mice are unable to bear offspring as a result of α-tocopherol deficiency [32]. Interestingly, oral supplementation of α-tocotrienol to TTP knockout mice restores fertility, suggesting alternative mechanisms of transport [33]. Indeed, oral α-tocotrienol supplementation has been reported to reach a number of vital organs, including heart, lung, liver, spinal cord, and brain [33].

15.3 ARACHIDONIC ACID CASCADE IN NEUROINFLAMMATORY DISEASE

Phospholipid membranes are highly enriched with n-3 and n-6 PUFAs. Taken together, arachidonic acid (AA; 20:4n-6 PUFA) and docosahexaenoic acid (DHA; 22:6n-3) account for one-fifth of all fatty acids in the mammalian brain [34]. Both are nutritionally essential to early development and postnatal brain physiology, including membrane fluidity, signal transduction, and gene transcription [35–39]. A number of chronic neuroinflammatory conditions in the human brain are associated with disturbed PUFA metabolism of AA, including AD, PD, and ALS [40, 41]. In such pathological states, AA that is liberated from membrane phospholipids undergoes oxidative metabolism, forming pro-inflammatory metabolites. Uncontrolled enzymatic or nonenzymatic oxidative metabolism of AA is referred to as the AA cascade and amplifies the overall production of free radicals and oxidative damage to lipids, proteins, and nucleic acids (Figure 15.2). Natural vitamin E α-tocotrienol has been

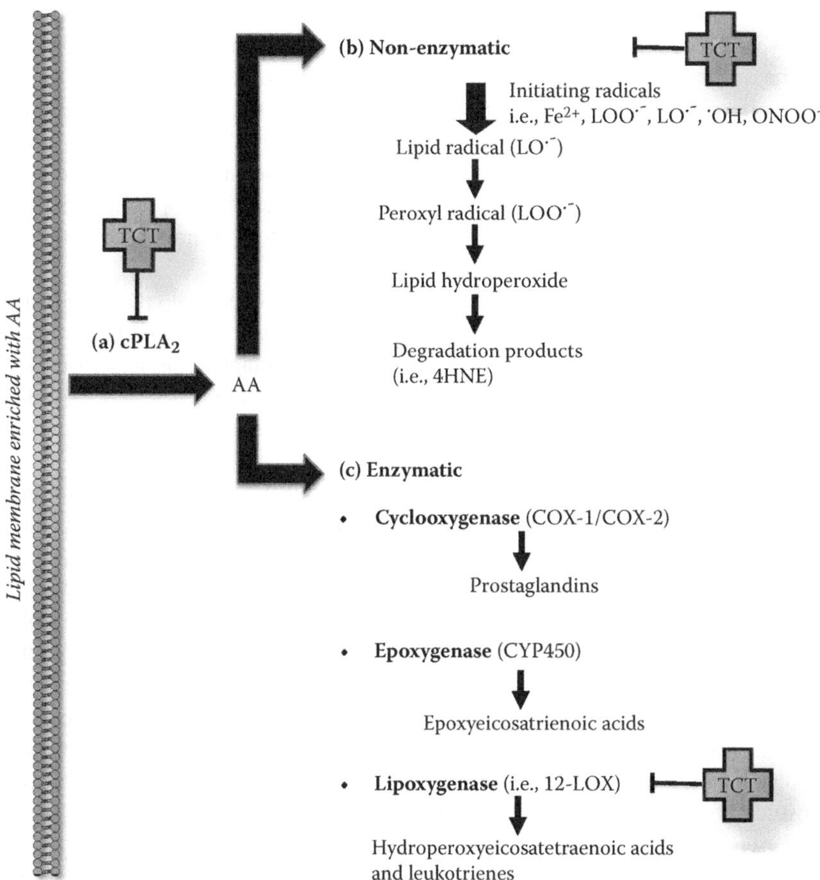

FIGURE 15.2 The arachidonic acid cascade. Arachidonic acid (AA) is liberated from phospholipids by cytosolic phospholipase A_2 ($cPLA_2$) under pathological conditions associated with neuroinflammation (a). Free AA undergoes nonenzymatic (b) or enzymatic (c) oxidative metabolism. Natural vitamin E tocotrienol (TCT) inhibits $cPLA_2$-mediated release and oxidative metabolism of AA at multiple levels of the cascade.

reported to disrupt the AA cascade on the basis of both antioxidant-dependent and independent mechanisms.

15.3.1 Phospholipase A_2

The phospholipase A_2 (PLA_2) family is characterized by a common function—the enzymatic hydrolysis of the sn-2 ester bond of glycerophospholipids producing a free fatty acid (i.e., AA) and lysophospholipid (i.e., lysophosphatidylcholine, LPC). There are five classes of PLA_2: (1) secreted small molecular weight $sPLA_2$, (2) larger cytosolic calcium-dependent $cPLA_2$, (3) calcium-independent $iPLA_2$, (4) platelet-activating factor acetylhydrolases (PAFA), and (5) lysosomal PLA_2 isozymes. Currently, only $sPLA_2$ and $cPLA_2$ have defined roles in neuroinflammatory AA metabolism [42].

PLA_2 activity is induced by a rise in intracellular calcium [43–46]. Indeed, calcium dyshomeostasis is a hallmark of acute and chronic neuroinflammation. In AD, increased intracellular calcium elicits characteristic lesions, accumulation of amyloid-β, hyperphosphorylation of TAU protein, and neural cell death [47]. In stroke, unregulated extracellular glutamate release overstimulates NMDA receptors, leading to excessive influx of intracellular calcium and $cPLA_2$ activation [48]. The $sPLA_2$s are characterized by the requirement of histidine in their active site, calcium for catalysis, and the presence of six conserved disulfide bonds [49]. Under pathological conditions of ischemic stroke, $sPLA_2$ mRNA and protein expression is significantly upregulated [50, 51] and activity is induced by inflammatory cytokine tumor necrosis factor-alpha (TNF-α) [52]. The $cPLA_2$s are the only PLA_2 that demonstrate a preference for AA in the sn-2 position of phospholipids [53]. Localized predominantly in gray matter, they lack the disulfide bonding network of $sPLA_2$s and function through the action of a serine/aspartic acid dyad [54]. Under pathological conditions, $cPLA_2$ subunit mRNA and protein expression is elevated [46]. Intracellular Ca^{2+} accumulation mediates $cPLA_2$ subunit translocation to the membrane phospholipid bilayer [55], and activity is induced by phosphorylation of serine residue 505 by mitogen-activated protein kinase [56].

Once released by PLA_2, free AA has three potential fates: reincorporation into phospholipids, diffusion outside the cell, and metabolism. In a pathological setting associated with neuroinflammation, the accumulation and oxidation of free AA produces pro-inflammatory prostaglandins, leukotrienes, thromboxanes, isoprostanes, and other nonenzymatic lipid peroxidation species [57]. The AA cascade amplifies the overall production of free radicals, both reactive oxygen and nitrogen species, and subsequently oxidative damage to lipids, proteins, and nucleic acids.

It has recently been shown that α-tocotrienol attenuates $cPLA_2$ activity under conditions of glutamate-mediated toxicity in neural cells [58]. Glutamate is the most abundant neurotransmitter of the CNS. Atypical glutamate clearance at the synaptic cleft as found in ALS, AD, and stroke patients contributes to a pro-inflammatory phenotype and neurodegeneration. Glutamate activates $cPLA_2$ in neurons in a calcium-dependent manner leading to the hydrolysis of AA from phospholipids [58]. Both phosphorylation and translocation of $cPLA_2$ are inhibited with nanomolar concentrations of α-tocotrienol, a level previously demonstrated to be readily achievable by dietary supplementation [33, 59]. Targeting PLA_2 to prevent neuroinflammatory processes associated with AD is being actively investigated [60]. Of note, mixed vitamin E (tocopherols and tocotrienols) but not tocopherols alone were found to reduce the risk of AD in clinical investigation [61]. Taken together, further study of tocotrienol-mediated attenuation of PLA_2 activity for prevention of chronic inflammatory disease (i.e., AD) is warranted.

15.3.2 Nonenzymatic Oxidative Metabolism of Arachidonic Acid

Free radical species react with double bonds of PUFAs producing alkyl radicals, which in turn react with molecular oxygen to form a peroxyl radical ($ROO^{\cdot}$). Importantly, peroxyl radicals can abstract hydrogen from adjacent PUFAs to produce a lipid hydroperoxide (ROOH) and a second alkyl radical, thereby feeding forward a chain reaction of lipid oxidation [62]. Lipid peroxides degrade and give rise to α,β-unsaturated aldehydes that include 4-hydroxynonenal (4HNE). Pathophysiological links have been made between 4HNE and a number of chronic neuroinflammatory diseases, including PD, ALS, and AD. In AD, for example, 4HNE forms aldehyde-protein adducts and hastens Aβ aggregation and toxicity, which in turn induces greater oxidative stress and lipid peroxidation [63].

Vitamin E is best known as a lipophilic antioxidant and the first line of defense in protecting lipid membranes from peroxidation [64]. Compared to tocopherols, the antioxidant capacity of tocotrienols is believed to be significantly greater [65, 66]. Because of the unsaturated nature of their phytyl tail, tocotrienols possess greater flexibility in the side chain that increases curvature stress on phospholipid membranes [67]. It is believed that the unsaturated side chain of tocotrienol also enables more efficient penetration into tissues with saturated fatty layers such as brain [68].

15.3.3 Enzymatic Oxidative Metabolism of Arachidonic Acid

15.3.3.1 Cyclooxygenase

Following PLA_2 liberation, the primary effectors of enzymatic AA oxidation are cyclooxygenases (COX), lipoxygenases, and epoxygenases. The role of cyclooxygenase in neuroinflammation and neurodegenerative disease remains clouded. Cyclooxygenase is expressed in two forms: COX-1 is widely distributed in a number of cell types and believed to mediate physiological responses, while COX-2 is widely recognized as an inducible form rapidly expressed in response to cytokines in a pro-inflammatory state. As such, selective COX-2 inhibition has been the focus of many trials to attenuate inflammation without affecting the physiological function of COX-1. In the CNS, however, there is evidence to support constitutive expression of COX-2 in hippocampal and cortical glutamatergic neurons for homeostatic function that includes neurovascular coupling and synaptic plasticity [69, 70]. It is not surprising then that outcomes of preclinical and clinical trials to test COX-2 inhibitors in neuroinflammatory disease have been mixed. In models of cerebral ischemia, COX-2 inhibition has been reported to attenuate acute neuronal injury [71, 72]. Conversely, in models of primary inflammation (i.e., lipopolysaccharide injection) COX-2 knockdown is reportedly associated with increased glial activation and inflammation [73]. While the specific role of COX in acute and chronic neuroinflammatory pathology remains to be elucidated, changes in expression patterns of COX and selective inhibition ameliorating disease have been reported [74]. For example, COX-1, but not COX-2, inhibitors have been shown to reduce the risk of AD. Conversely, in preclinical models of PD and ALS, COX-2 inhibitors are neuroprotective. To date, literature on the effects of tocotrienols on cyclooxygenase expression and activity are scant. Tocotrienols have been reported to lower peripheral macrophage expression of COX-2 following LPS stimulation [75]. Given the unique expression pattern and functional significance of COX in the CNS, however, it remains to be seen whether tocotrienols have any effect on COX-mediated neuroinflammation.

15.3.3.2 Epoxygenase

Cytochrome P450 (CYP) epoxygenases metabolize AA into epoxyeicosatrienoic acids (EETs). Compared to COX and LOX enzyme systems, CYP activity in CNS health and disease is generally viewed in the light of anti-inflammatory effects and neuroprotection [76, 77]. EETs have been shown to protect neurons and astrocytes from ischemic brain injury and inhibit cytokine-induced endothelial cell adhesion molecule expression. The biological significance of CYP and EETs in the CNS, however, has been examined less than COX and LOX enzyme systems—owing in part to low expression levels of CYP in CNS tissue as compared to peripheral organs [78]. Likewise, the biological significance of tocotrienol vitamin E on CYP function in the CNS remains to be examined in detail.

15.3.3.3 Lipoxygenase

The lipoxygenase (LOX) enzyme family is composed of four members (5-, 8-, 12-, and 15-LOX) that are differentiated on the basis of the carbon position in which they oxidize arachidonic acid [20]. While each LOX isoform carries out the same general reaction, hydrogen abstraction and insertion of oxygen in AA, each has a unique gene structure, amino acid sequence, and tissue distribution profile. Only three forms of LOX are found in brain tissue: 5-, 12-, and 15-LOX [57]. Of these, 12-LOX is the most abundant in the brain [79], with highest expression in cortical neurons, astrocytes, and oligodendrocytes [80]. Once AA is cleaved from the sn-2 position of glycerophospholipids, LOX metabolism generates pro-inflammatory hydroperoxyeicosatetraenoic acids (HPETEs) and leukotrienes [20]. The HPETE products are quickly reduced to corresponding hydroxy derivatives (i.e., 12-HETE) by cellular peroxidase activity.

The significance of 12-LOX-mediated neuroinflammation and neurodegeneration is best documented in response to acute ischemic stroke. Following stroke, 12-LOX enzyme activity and metabolites are increased in the stroke-affected hemisphere [81–83], and 12-LOX metabolism of AA is known to damage mitochondria in the brain [84]. 12-LOX–deficient mice are resistant to ischemic stroke injury [85], and at nanomolar concentrations α-tocotrienol vitamin E is a known 12-LOX inhibitor in neural cells subjected to glutamate-mediated neurotoxicity [86]. *In silico* studies suggest the presence of an α-tocotrienol binding site on 12-LOX that prohibits access to AA [86]. Furthermore, in small- and large-animal preclinical models of acute ischemic stroke, prophylactic oral supplementation of α-tocotrienol attenuates stroke-induced lesion volume [30, 81] and reduces the evolution of edema in the first 48 hours after stroke [30].

12-LOX catalytic function is regulated by tyrosine kinases. In response to high levels of extracellular glutamate, as is the case in acute brain injury [20], 12-LOX undergoes tyrosine phosphorylation by c-Src kinase [85]. Indeed, rapid activation of c-Src plays a critical step in neurodegeneration [87, 88] such that c-Src knockdown or blockade of activity in mice protects neocortex from ischemic stroke-induced injury [89]. Importantly, extracellular glutamate-induced c-Src activity is completely blocked by nanomolar levels of α-tocotrienol but not α-tocopherol [88].

In chronic neuroinflammatory disorders, the significance of LOX metabolism is less clear. Recently, 5-LOX inhibition has been shown to limit cuprizone-induced demyelination and neuroinflammation in a mouse model of MS [90]. Furthermore, 12/15-LOX protein and activity are reported to be elevated in AD human brain as compared to age-matched control brain [91]. Whether tocotrienols attenuate LOX-mediated injury in chronic neuroinflammatory disorders such as MS or AD, however, remains to be investigated.

15.4 CONCLUSION AND FUTURE DIRECTIONS

The highly specialized cellular environment of the CNS, enriched with PUFAs and terminally differentiated neural cells, is susceptible to neuroinflammation and neurodegeneration via uncontrolled oxidative metabolism of AA. Thus far, research efforts and drug development have focused on specific perpetrators of oxidative injury in the AA cascade that appear to be disease dependent. Given multiple pathways of oxidative AA metabolism, however, targeting a single protein may result in shunting of substrate to another. This conclusion is supported in part by the limited efficacy of pharmacological immunotherapies that while relatively specific in their action have largely failed in clinical trials. Therapeutic intervention that targets multiple levels of the AA cascade are therefore more likely to succeed in clinical applications. A growing body of literature suggests that the lesser-characterized vitamin E tocotrienol may be such an intervention. As a naturally occurring vitamin with a long history of safe dietary consumption, the risks of toxicity and other side effects for tocotrienol are low. To that end, the FDA has recently approved GRAS (Generally Recognized As Safe) status for tocotrienols to be included as a supplement in functional foods. Successful outcomes of preclinical and clinical pilot studies now set the stage for larger-scale clinical testing of tocotrienol vitamin E against neuroinflammatory disease.

TAKE-HOME MESSAGES

- Immunological responses of the CNS are unique from, but not exclusive of, the periphery.
- Inflammation of the CNS occurs under both homeostatic (controlled) and pathological (uncontrolled) conditions.
- *Neuroinflammation* is a term used to denote pathological (uncontrolled) immune response, often in the context of neurodegenerative disease.
- Oxidative stress and uncontrolled metabolism of arachidonic acid are hallmarks of neuroinflammation.
- The uncontrolled metabolism of arachidonic acid is implicated in both acute and chronic neuroinflammatory diseases, including ischemic stroke, Alzheimer's, Parkinson's, amyotrophic lateral sclerosis, and multiple sclerosis.
- Tocotrienols represent one-half of the natural vitamin E family, and possess unique biological function as compared to the better-characterized tocopherols.
- Orally supplemented tocotrienols are delivered to vital organs, including brain and spinal cord, in sufficient quantity to exert biological effect.
- Tocotrienols inhibit oxidative metabolism at multiple levels of the arachidonic acid cascade by both antioxidant-dependent and independent mechanisms.
- Natural vitamin E α-tocotrienol is a 12-lipoxygenase inhibitor and protects brain tissue from stroke-induced neuroinflammation and cell death.
- As a natural dietary supplement with a history of safe consumption, tocotrienol vitamin E represents a low-risk/high-reward candidate for clinical testing against neuroinflammatory disease.

REFERENCES

1. Medawar PB. Immunity to homologous grafted skin; the fate of skin homografts transplanted to the brain, to subcutaneous tissue, and to the anterior chamber of the eye. *Br J Exp Pathol.* 1948;29:58–69.
2. Engelhardt B, Ransohoff RM. The ins and outs of t-lymphocyte trafficking to the CNS: Anatomical sites and molecular mechanisms. *Trends Immunol.* 2005;26:485–495.
3. Antony J, Power C. Neuro-inflammation: An emerging therapeutic target in neurological disease. In: Kilpatrick T, Ransohoff R, Wesselingh S, eds. *Inflammatory diseases of the central nervous system.* New York: Cambridge University Press; 2010:245–260.
4. Kreutzberg GW. Microglia: A sensor for pathological events in the CNS. *Trends Neurosci.* 1996;19: 312–318.
5. Ballabh P, Braun A, Nedergaard M. The blood-brain barrier: An overview: Structure, regulation, and clinical implications. *Neurobiol Dis.* 2004;16:1–13.
6. Ganong WF. Circumventricular organs: Definition and role in the regulation of endocrine and autonomic function. *Clin Exp Pharmacol Physiol.* 2000;27:422–427.
7. Agostinho P, Cunha RA, Oliveira C. Neuroinflammation, oxidative stress and the pathogenesis of Alzheimer's disease. *Curr Pharm Des.* 2010;16:2766–2778.
8. Banks WA, Kastin AJ, Broadwell RD. Passage of cytokines across the blood-brain barrier. *Neuroimmunomodulation.* 1995;2:241–248.
9. Paton JF, Waki H, Abdala AP, Dickinson J, Kasparov S. Vascular-brain signaling in hypertension: Role of angiotensin ii and nitric oxide. *Curr Hypertens Rep.* 2007;9:242–247.
10. Shi P, Raizada MK, Sumners C. Brain cytokines as neuromodulators in cardiovascular control. *Clin Exp Pharmacol Physiol.* 2010;37:e52–e57.

11. Carson MJ, Doose JM, Melchior B, Schmid CD, Ploix CC. CNS immune privilege: Hiding in plain sight. *Immunol Rev*. 2006;213:48–65.
12. Matyszak MK, Perry VH. A comparison of leucocyte responses to heat-killed bacillus Calmette-Guerin in different CNS compartments. *Neuropathol Appl Neurobiol*. 1996;22:44–53.
13. Araque A, Carmignoto G, Haydon PG. Dynamic signaling between astrocytes and neurons. *Annu Rev Physiol*. 2001;63:795–813.
14. Sofroniew MV. Reactive astrocytes in neural repair and protection. *Neuroscientist*. 2005;11:400–407.
15. Carson MJ, Thrash JC, Walter B. The cellular response in neuroinflammation: The role of leukocytes, microglia and astrocytes in neuronal death and survival. *Clin Neurosci Res*. 2006;6:237–245.
16. Nakamura Y. Regulating factors for microglial activation. *Biol Pharm Bull*. 2002;25:945–953.
17. Olson JK, Miller SD. Microglia initiate central nervous system innate and adaptive immune responses through multiple TLRs. *J Immunol*. 2004;173:3916–3924.
18. Streit WJ, Mrak RE, Griffin WS. Microglia and neuroinflammation: A pathological perspective. *J Neuroinflammation*. 2004;1:14.
19. O'Callaghan JP, Sriram K, Miller DB. Defining "neuroinflammation." *Ann N Y Acad Sci*. 2008;1139: 318–330.
20. Rink C, Khanna S. Significance of brain tissue oxygenation and the arachidonic acid cascade in stroke. *Antioxid Redox Signal*. 2011;14:1889–1903.
21. McGeer PL. COX-2 and ALS. *Amyotroph Lateral Scler Other Motor Neuron Disord*. 2001;2:121–122.
22. McGeer PL, McGeer EG. Inflammatory processes in amyotrophic lateral sclerosis. *Muscle Nerve*. 2002;26:459–470.
23. Kiple KF, Ornelas KC. *The Cambridge world history of food*. Cambridge, UK; New York: Cambridge University Press; 2000.
24. Evans HM, Emerson OH, Emerson GA. The isolation from wheat germ oil of an alcohol, alpha-tocopherol, having the properties of vitamin E. *Journal of Biological Chemistry*. 1936;113:319–332.
25. Sen CK, Khanna S, Rink C, Roy S. Tocotrienols: The emerging face of natural vitamin E. *Vitam Horm*. 2007;76:203–261.
26. Sen CK, Rink C, Khanna S. Palm oil-derived natural vitamin E alpha-tocotrienol in brain health and disease. *J Am Coll Nutr*. 2010;29:314S–323S.
27. Horvath G, Wessjohann L, Bigirimana J, Jansen M, Guisez Y, Caubergs R, et al. Differential distribution of tocopherols and tocotrienols in photosynthetic and non-photosynthetic tissues. *Phytochemistry*. 2006;67:1185–1195.
28. Sen CK, Khanna S, Roy S. Tocotrienols in health and disease: The other half of the natural vitamin E family. *Mol Aspects Med*. 2007;28:692–728.
29. Institute of Medicine (U.S.). Panel on Dietary Antioxidants and Related Compounds. *Dietary reference intakes for vitamin C, vitamin E, selenium, and carotenoids: A report of the panel on dietary antioxidants and related compounds, subcommittees on upper reference levels of nutrients and of interpretation and use of dietary reference intakes, and the standing committee on the scientific evaluation of dietary reference intakes, Food and Nutrition Board, Institute of Medicine*. Washington, D.C.: National Academy Press; 2000.
30. Rink C, Christoforidis G, Khanna S, Peterson L, Patel Y, Abduljalil A, et al. Tocotrienol vitamin E protects against preclinical canine ischemic stroke by inducing arteriogenesis. *J Cereb Blood Flow Metab*. 2011.
31. Hosomi A, Arita M, Sato Y, Kiyose C, Ueda T, Igarashi O, et al. Affinity for alpha-tocopherol transfer protein as a determinant of the biological activities of vitamin E analogs. *FEBS Lett*. 1997;409: 105–108.
32. Khanna S, Patel V, Rink C, Roy S, Sen CK. Delivery of orally supplemented alpha-tocotrienol to vital organs of rats and tocopherol-transport protein deficient mice. *Free Radic Biol Med*. 2005;39: 1310–1319.
33. Khanna S, Patel V, Rink C, Roy S, Sen CK. Delivery of orally supplemented alpha-tocotrienol to vital organs of rats and tocopherol-transport protein deficient mice. *Free Radic Biol Med*. 2005;39: 1310–1319.
34. Contreras MA, Greiner RS, Chang MC, Myers CS, Salem N, Jr., Rapoport SI. Nutritional deprivation of alpha-linolenic acid decreases but does not abolish turnover and availability of unacylated docosahexaenoic acid and docosahexaenoyl-coa in rat brain. *J Neurochem*. 2000;75:2392–2400.
35. Carlson SE, Werkman SH, Peeples JM, Cooke RJ, Tolley EA. Arachidonic acid status correlates with first year growth in preterm infants. *Proc Natl Acad Sci USA*. 1993;90:1073–1077.
36. Eichberg J. *Phospholipids in nervous tissues*. New York: Wiley; 1985.

37. Jones CR, Arai T, Rapoport SI. Evidence for the involvement of docosahexaenoic acid in cholinergic stimulated signal transduction at the synapse. *Neurochem Res*. 1997;22:663–670.
38. Neuringer M, Connor WE, Lin DS, Barstad L, Luck S. Biochemical and functional effects of prenatal and postnatal omega 3 fatty acid deficiency on retina and brain in rhesus monkeys. *Proc Natl Acad Sci USA*. 1986;83:4021–4025.
39. Rapoport SI, Chang MC, Spector AA. Delivery and turnover of plasma-derived essential PUFAs in mammalian brain. *J Lipid Res*. 2001;42:678–685.
40. Farooqui AA, Horrocks LA, Farooqui T. Modulation of inflammation in brain: A matter of fat. *J Neurochem*. 2007;101:577–599.
41. Rapoport SI. Arachidonic acid and the brain. *J Nutr*. 2008;138:2515–2520.
42. Adibhatla RM, Hatcher JF. Phospholipase a(2), reactive oxygen species, and lipid peroxidation in CNS pathologies. *BMB Rep*. 2008;41:560–567.
43. Clemens JA, Stephenson DT, Smalstig EB, Roberts EF, Johnstone EM, Sharp JD, et al. Reactive glia express cytosolic phospholipase a2 after transient global forebrain ischemia in the rat. *Stroke*. 1996;27:527–535.
44. Lauritzen I, Heurteaux C, Lazdunski M. Expression of group ii phospholipase a2 in rat brain after severe forebrain ischemia and in endotoxic shock. *Brain Res*. 1994;651:353–356.
45. Saluja I, Song D, O'Regan MH, Phillis JW. Role of phospholipase a2 in the release of free fatty acids during ischemia-reperfusion in the rat cerebral cortex. *Neurosci Lett*. 1997;233:97–100.
46. Stephenson D, Rash K, Smalstig B, Roberts E, Johnstone E, Sharp J, et al. Cytosolic phospholipase a2 is induced in reactive glia following different forms of neurodegeneration. *Glia*. 1999;27:110–128.
47. LaFerla FM. Calcium dyshomeostasis and intracellular signalling in Alzheimer's disease. *Nat Rev Neurosci*. 2002;3:862–872.
48. Braughler JM, Hall ED. Central nervous system trauma and stroke. I. Biochemical considerations for oxygen radical formation and lipid peroxidation. *Free Radic Biol Med*. 1989;6:289–301.
49. Burke JE, Dennis EA. Phospholipase a2 biochemistry. *Cardiovasc Drugs Ther*. 2009;23:49–59.
50. Adibhatla RM, Hatcher JF, Larsen EC, Chen X, Sun D, Tsao FH. Cdp-choline significantly restores phosphatidylcholine levels by differentially affecting phospholipase a2 and ctp: Phosphocholine cytidylyltransferase after stroke. *J Biol Chem*. 2006;281:6718–6725.
51. Lin TN, Wang Q, Simonyi A, Chen JJ, Cheung WM, He YY, et al. Induction of secretory phospholipase a2 in reactive astrocytes in response to transient focal cerebral ischemia in the rat brain. *J Neurochem*. 2004;90:637–645.
52. Anthonsen MW, Solhaug A, Johansen B. Functional coupling between secretory and cytosolic phospholipase a2 modulates tumor necrosis factor-alpha- and interleukin-1beta-induced NF-kappa b activation. *J Biol Chem*. 2001;276:30527–30536.
53. Burke JE, Dennis EA. Phospholipase a2 structure/function, mechanism, and signaling. *J Lipid Res*. 2009;50 Suppl:S237–242.
54. Stephenson DT, Manetta JV, White DL, Chiou XG, Cox L, Gitter B, et al. Calcium-sensitive cytosolic phospholipase a2 (cpla2) is expressed in human brain astrocytes. *Brain Res*. 1994;637:97–105.
55. Yoshihara Y, Watanabe Y. Translocation of phospholipase a2 from cytosol to membranes in rat brain induced by calcium ions. *Biochem Biophys Res Commun*. 1990;170:484–490.
56. Lin LL, Wartmann M, Lin AY, Knopf JL, Seth A, Davis RJ. Cpla2 is phosphorylated and activated by map kinase. *Cell*. 1993;72:269–278.
57. Phillis JW, Horrocks LA, Farooqui AA. Cyclooxygenases, lipoxygenases, and epoxygenases in CNS: Their role and involvement in neurological disorders. *Brain Res Rev*. 2006;52:201–243.
58. Khanna S, Parinandi NL, Kotha SR, Roy S, Rink C, Bibus D, et al. Nanomolar vitamin E alpha-tocotrienol inhibits glutamate-induced activation of phospholipase a2 and causes neuroprotection. *J Neurochem*.112:1249–1260.
59. Patel V, Khanna S, Roy S, Ezziddin O, Sen CK. Natural vitamin E alpha-tocotrienol: Retention in vital organs in response to long-term oral supplementation and withdrawal. *Free Radic Res*. 2006;40: 763–771.
60. Sanchez-Mejia RO, Mucke L. Phospholipase a2 and arachidonic acid in Alzheimer's disease. *Biochim Biophys Acta*. 2010;1801:784–790.
61. Mangialasche F, Kivipelto M, Mecocci P, Rizzuto D, Palmer K, Winblad B, et al. High plasma levels of vitamin E forms and reduced Alzheimer's disease risk in advanced age. *J Alzheimers Dis*. 2010;20:1029–1037,
62. Beal MF, Howell N, Bodis-Wollner I. *Mitochondria and free radicals in neurodegenerative diseases*. New York: Wiley-Liss; 1997.

63. Siegel SJ, Bieschke J, Powers ET, Kelly JW. The oxidative stress metabolite 4-hydroxynonenal promotes Alzheimer protofibril formation. *Biochemistry*. 2007;46:1503–1510.
64. Kamal-Eldin A, Appelqvist LA. The chemistry and antioxidant properties of tocopherols and tocotrienols. *Lipids*. 1996;31:671–701.
65. Serbinova EA, Packer L. Antioxidant properties of alpha-tocopherol and alpha-tocotrienol. *Methods Enzymol*. 1994;234:354–366.
66. Kamat JP, Devasagayam TP. Tocotrienols from palm oil as potent inhibitors of lipid peroxidation and protein oxidation in rat brain mitochondria. *Neurosci Lett*. 1995;195:179–182.
67. Sen CK, Khanna S, Rink C, Roy S. Tocotrienols: The emerging face of natural vitamin E. *Vitam Horm*. 2007;76:203–261.
68. Suzuki YJ, Tsuchiya M, Wassall SR, Choo YM, Govil G, Kagan VE, et al. Structural and dynamic membrane properties of alpha-tocopherol and alpha-tocotrienol: Implication to the molecular mechanism of their antioxidant potency. *Biochemistry*. 1993;32:10692–10699.
69. Niwa K, Araki E, Morham SG, Ross ME, Iadecola C. Cyclooxygenase-2 contributes to functional hyperemia in whisker-barrel cortex. *J Neurosci*. 2000;20:763–770.
70. Yang H, Chen C. Cyclooxygenase-2 in synaptic signaling. *Curr Pharm Des*. 2008;14:1443–1451.
71. Iadecola C, Niwa K, Nogawa S, Zhao X, Nagayama M, Araki E, et al. Reduced susceptibility to ischemic brain injury and n-methyl-d-aspartate-mediated neurotoxicity in cyclooxygenase-2-deficient mice. *Proc Natl Acad Sci USA*. 2001;98:1294–1299.
72. Nogawa S, Zhang F, Ross ME, Iadecola C. Cyclo-oxygenase-2 gene expression in neurons contributes to ischemic brain damage. *J Neurosci*. 1997;17:2746–2755.
73. Aid S, Langenbach R, Bosetti F. Neuroinflammatory response to lipopolysaccharide is exacerbated in mice genetically deficient in cyclooxygenase-2. *J Neuroinflammation*. 2008;5:17.
74. Choi SH, Aid S, Bosetti F. The distinct roles of cyclooxygenase-1 and -2 in neuroinflammation: Implications for translational research. *Trends Pharmacol Sci*. 2009;30:174–181.
75. Yam ML, Abdul Hafid SR, Cheng HM, Nesaretnam K. Tocotrienols suppress proinflammatory markers and cyclooxygenase-2 expression in raw264.7 macrophages. *Lipids*. 2009;44:787–797.
76. Campbell WB. New role for epoxyeicosatrienoic acids as anti-inflammatory mediators. *Trends Pharmacol Sci*. 2000;21:125–127.
77. Node K, Huo Y, Ruan X, Yang B, Spiecker M, Ley K, et al. Anti-inflammatory properties of cytochrome p450 epoxygenase-derived eicosanoids. *Science*. 1999;285:1276–1279.
78. Strobel HW, Thompson CM, Antonovic L. Cytochromes p450 in brain: Function and significance. *Curr Drug Metab*. 2001;2:199–214.
79. Hambrecht GS, Adesuyi SA, Holt S, Ellis EF. Brain 12-hete formation in different species, brain regions, and in brain microvessels. *Neurochem Res*. 1987;12:1029–1033.
80. Bendani MK, Palluy O, Cook-Moreau J, Beneytout JL, Rigaud M, Vallat JM. Localization of 12-lipoxygenase mRNA in cultured oligodendrocytes and astrocytes by in situ reverse transcriptase and polymerase chain reaction. *Neurosci Lett*. 1995;189:159–162.
81. Khanna S, Roy S, Slivka A, Craft TK, Chaki S, Rink C, et al. Neuroprotective properties of the natural vitamin e alpha-tocotrienol. *Stroke*. 2005;36:2258–2264.
82. Moskowitz MA, Kiwak KJ, Hekimian K, Levine L. Synthesis of compounds with properties of leukotrienes c4 and d4 in gerbil brains after ischemia and reperfusion. *Science*. 1984;224:886–889.
83. van Leyen K, Kim HY, Lee SR, Jin G, Arai K, Lo EH. Baicalein and 12/15-lipoxygenase in the ischemic brain. *Stroke*. 2006;37:3014–3018.
84. Schewe T, Halangk W, Hiebsch C, Rapoport SM. A lipoxygenase in rabbit reticulocytes which attacks phospholipids and intact mitochondria. *FEBS Lett*. 1975;60:149–152.
85. Khanna S, Roy S, Slivka A, Craft TK, Chaki S, Rink C, et al. Neuroprotective properties of the natural vitamin E alpha-tocotrienol. *Stroke*. 2005;36:2258–2264.
86. Khanna S, Roy S, Ryu H, Bahadduri P, Swaan PW, Ratan RR, et al. Molecular basis of vitamin E action: Tocotrienol modulates 12-lipoxygenase, a key mediator of glutamate-induced neurodegeneration. *J Biol Chem*. 2003;278:43508–43515.
87. Khanna S, Venojarvi M, Roy S, Sen CK. Glutamate-induced c-src activation in neuronal cells. *Methods Enzymol*. 2002;352:191–198.
88. Sen CK, Khanna S, Roy S, Packer L. Molecular basis of vitamin E action. Tocotrienol potently inhibits glutamate-induced pp60(c-src) kinase activation and death of ht4 neuronal cells. *J Biol Chem*. 2000;275: 13049–13055.
89. Paul R, Zhang ZG, Eliceiri BP, Jiang Q, Boccia AD, Zhang RL, et al. Src deficiency or blockade of src activity in mice provides cerebral protection following stroke. *Nat Med*. 2001;7:222–227.

90. Yoshikawa K, Palumbo S, Toscano CD, Bosetti F. Inhibition of 5-lipoxygenase activity in mice during cuprizone-induced demyelination attenuates neuroinflammation, motor dysfunction and axonal damage. *Prostaglandins Leukot Essent Fatty Acids*. 2011;85:43–52.
91. Pratico D, Zhukareva V, Yao Y, Uryu K, Funk CD, Lawson JA, et al. 12/15-lipoxygenase is increased in Alzheimer's disease: Possible involvement in brain oxidative stress. *Am J Pathol*. 2004;164: 1655–1662.

16 Inflammatory Cascades in Autoimmune Disease

Amita Aggarwal and Arpita Myles
Sanjay Gandhi Postgraduate Institute of Medical Sciences
Lucknow, India

CONTENTS

16.1 INTRODUCTION

Autoimmune diseases are characterized by persistent inflammation mediated by perturbation in the immune system. They are usually accompanied by the presence of autoantibodies or autoreactive T cells. However, in some diseases the defect may lie in the innate immune system like auto-inflammatory diseases. The classical example of autoimmune disease is systemic lupus erythematosus (SLE), where an array of autoantibodies is associated with a multisystem involvement. In contrast, in rheumatoid arthritis (RA) the immune attack is more restricted and the synovium is the major target. Autoreactive B and T cells alone are not sufficient to cause autoimmune disease; they need to have access to their target site, initiate an inflammatory response through recruitment of other immune cells, soluble mediators like cytokines and chemokines, to cause uncontrolled inflammation and tissue damage resulting in disease. In this chapter we take two prototype diseases, one that is mainly initiated by the presence of autoantibodies and another mediated by T cells, and discuss the inflammatory cascades involved.

16.2 RHEUMATOID ARTHRITIS

Rheumatoid arthritis is a chronic autoimmune inflammatory disorder predominantly affecting the joints. Though autoantibodies are present in a majority of patients, T cells are supposed to play the major role in disease pathogenesis. Earlier, RA was thought to be a Th1-mediated disease, but now the role of Th17 cells in its pathogenesis is thought to be more crucial. Apart from T cells, the RA synovium is also rich in monocytes and neutrophil infiltrates, which contribute further to disease pathology. All these immune cells initiate a strong inflammatory reaction in the synovium, which ultimately leads to cartilage destruction and bone erosion. Hence, it is important to understand the inflammatory cascade that leads to the recruitment of immune cells resulting in tissue damage in order to design molecules that can block the disease at this step.

16.2.1 Adhesion

During adhesion the circulating leukocytes tether to endothelium transiently and "roll" when this bond is broken by the shearing force of flowing blood. In response to inflammatory stimuli, the endothelium becomes activated and starts expressing cell surface molecules that overcome the shearing force and cause firm adhesion of leukocytes. These surface molecules are the key players in adhesion.

Selectins and their ligands are important initiators of adhesion (Ley and Tedder 1995). The three main members of the highly conserved selectin family are E-, P-, and L-selectins, expressed by endothelial cells, platelets and endothelial cells, and leukocytes, respectively. P-selectins are transported to the cell surface from granules inside platelets and endothelial cells during initial phase of inflammation and are crucial for leukocyte rolling (Nolte et al. 1994). E-selectin is expressed in response to pro-inflammatory cytokines like TNF-α, while L-selectin is constitutively expressed. Selectins generally have a low affinity for their ligands, hence, leukocytes are able to "roll" along the endothelium instead of being firmly attached to it (Kansas 1996). As mentioned earlier, an important aspect of selectin binding is that it takes place in a dynamic environment and is affected by shear forces of the blood. If this shear force is below a certain level, then leukocytes remain bound to the endothelium. Once this threshold is crossed, the selectin-ligand binding breaks and the leukocyte is free to roll until new bonds are formed (Thomas 2006).

Integrins are another family of molecules involved in leukocyte rolling, adhesion, and more importantly in transmigration from periphery into extravascular tissue. They also mediate rolling but are less efficient than selectins and are not constitutively expressed (Springer 1995). Integrins are formed of two structural subunits α and β that link non-covalently to form heterodimers. Although 18α and 8β subunits have been identified thus far, the actual number of heterodimers is much more as one α subunit can interact with more than one β subunit (Luo et al. 2007). Integrin subfamilies are grouped according to their subunits. In relation to autoimmunity, the α4 integrin has been most studied. It aids migration of T cells into the target organ after interaction with VCAM-1.

Integrins bind to cellular adhesion molecules (CAMs), like intercellular cell adhesion molecules (ICAM), vascular cell adhesion molecules (VCAM), and the less commonly discussed junctional adhesion molecules (JAMs). Leukocytes engage VCAM-1 (CD106) by their α-4 integrin receptor (Elices et al. 1990). PECAM-1 (platelet endothelial cell adhesion molecule) or CD31 is crucial for leukocytes to migrate across the endothelium (Muller et al. 1993). It has been shown that CD31 engagement causes upregulation of integrins, especially $\alpha_6\beta_1$ that mediates leukocyte transmigration (Dangerfield et al. 2002). Apart from these, certain integrin binding sites may also be exposed during inflammation (Smith 2008), thereby resulting in a positive feedback recruitment of more leukocytes.

The ICAMs are type I transmembrane glycoproteins belonging to the immunoglobulin superfamily and comprise five members (Lawson and Wolf 2009). ICAM1 is expressed by most cells upon pro-inflammatory stimulus. Interaction of ICAM1 with its cognate ligand leads to an increase in intracellular calcium ions, which ultimately causes expansion of endothelial junctions, thereby facilitating transmigration. This interaction results in a motion termed "crawling of leukocytes over the endothelium" and is known to be crucial for migration (Phillipson et al. 2006). While ICAM1 forms homodimers, ICAM2 is a monomer and is constitutively expressed on platelets and endothelial cells. Its expression is not affected by inflammatory cytokines. ICAM3 is expressed predominantly by leukocytes and ICAM4 by erythrocytes. ICAM4 is important in erythrophagocytosis seen in some autoimmune diseases like systemic onset juvenile arthritis or systemic lupus erythematosus. ICAM5 is widely distributed on cells of the central nervous system.

VCAMs are expressed only upon stimulation of the endothelium. VCAM-1 mediates the adhesion of lymphocytes, monocytes, eosinophils, and basophils to vascular endothelium. It also functions in leukocyte-endothelial cell signal transduction, and it may play a role in the development of atherosclerosis and rheumatoid arthritis.

JAMs are crucial for endothelial transmigration of leukocytes. Like ICAMs they are also type I transmembrane proteins, but JAMs are unusual in that their homologues are found in primitive invertebrates (Bradfield et al. 2007). JAMs are generally clustered at endothelial cell borders. To date, five members of the JAM family have been identified: JAM-A, JAM-B, JAM-C, JAM-4, and JAM-like (JAML). JAM-C is upregulated in patients with osteoarthritis and RA and is important for leukocyte recruitment and adhesion to the arthritic synovium (Rabquer et al. 2008). Thus, JAMs may be an important therapeutic target for RA.

16.2.2 Migration

All steps prior to transendothelial migration (TEM) are reversible, that is, leukocytes can reenter circulation after initiating rolling and adhesion. However, once diapedesis occurs, the leukocyte cannot go back into circulation (Muller 2009). Ligation of integrins increases intracellular Ca^{2+} ions, as well as p38 mitogen-activated protein kinase (MAPK) and RAS homologue (RHO) GTPase. Together, these cause endothelial-cell contractions, facilitating transmigration. Leukocytes can cross the endothelial barrier in two ways—they can either squeeze through gap junctions in the endothelium itself (paracellular migration), or they can migrate transcellularly. PECAM1, JAM-A, and ICAM2 are involved in paracellular migration. This pathway is initiated with ICAM-1 ligand binding and is thought to be the normal mode of transmigration. However, if crawling is impaired, then the transcellular pathway is followed (Phillipson et al. 2006). This pathway is also implicated in causing inflammation (Engelhardt and Wolburg 2004). Transcellular migration involves formation of intracellular channels by linking together of caveoli.

The final step in migration is penetration of basement membrane and pericyte sheath. Integrins are again involved in this process along with other factors like matrix metalloproteases (MMP-3). The basement membrane has certain permissive sites where the expression of laminin and collagen is very low (Wang et al. 2006). It is hypothesized that these sites are also more sensitive to chemoattractants and, thereby, to the development of a chemoattractant gradient.

16.2.3 Leukocyte Adhesion and Migration and RA

The interaction of leukocytes with synovial endothelium is a pivotal point in the development of joint inflammation. RA synovial fluid and serum have high levels of soluble E- and P- selectin, ICAM-1, VCAM-1, and PECAM-1 (Palmer et al. 2006). These molecules aid the adhesion and transmigration of pathogenic leukocytes into the RA joint. Blockade of $\alpha 4\beta 1$/VCAM-1 and $\alpha 6\beta 2$/ICAM-1 interactions in animal models has been shown to suppress immune cell migration to the joint and ameliorate disease (Issekutz et al. 1996). Similarly, antibody-mediated blockade of PECAM-1 signaling in mice reduced arthritis (Ishikaw et al. 2002). However, clinical trials in human subjects have not shown promising results. Use of anti–ICAM-1 monoclonal antibodies in an open-label study in RA patients had only limited benefits (Kavanaugh et al. 1996). Yet another study that used anti-sense oligonucleotides to block ICAM-1 reported no efficacy in RA patients (Maksymowych et al. 2002). Thus, while molecules involved in leukocyte adhesion and migration are overexpressed in RA and are potential therapeutic targets, inhibition of single molecules is not a successful treatment strategy. This could be due to functional redundancy among various members of the leukocyte recruitment pathway.

16.2.4 Role of Chemokines and Their Receptors in Adhesion and Migration

Chemokines (chemotactic cytokines) and chemokine receptors play an important role in leukocyte extravasation by regulating the expression on integrins and selectins. Receptors on tethered leukocytes sense the chemokines bound to the endothelial cell membrane, leading to firm adhesion (Langer and Chavakis 2009), which precedes endothelial transmigration. Chemokines are rapidly synthesized when endothelial cells are activated by pro-inflammatory cytokines (Ley et al. 2007). Several important chemokines like CXCL4, CXCL7, and CCL5, and chemokine receptors like CXCR4 and CXCR5 are released during degranulation of platelets and mast cells and transported to endothelium (von Hundelshausen et al. 2001). Receptor-ligand binding can trigger downstream signaling by two pathways (Imhof and Aurrand-Lions 2004). In the first pathway, a GTPase RAP is converted from its inert GDP-bound form to active GTP-RAP complex. This induces adhesion and migration mediated by $\alpha4\beta1$ integrin (McLeod et al. 2004).

The second signaling pathway involves PI3K, leading to subsequent activation of atypical protein kinase C-ζ. The role of PKC-ζ is to increase integrin avidity and mobility, causing the formation of integrin clusters (Giagulli et al. 2004). In the next step, PYK2 (protein tyrosine kinase 2β) is activated and causes polarization of integrin clusters to facilitate migration (Avraham et al. 2000).

The CXC chemokines mainly act on neutrophils and lymphocytes, whereas the CC chemokines mainly act on monocytes and lymphocytes without affecting neutrophils (Iwamoto et al. 2008). Table 16.1 lists some chemokines and their functions.

16.2.5 Local inflammation

Once the T cells have gained access to their target site, they either release pro-inflammatory cytokines themselves or instigate a cascade of events leading to inflammation. However, the presence of autoreactive T cells alone is not sufficient to cause autoimmune diseases (Mackay et al. 2008). Development of the latter usually depends upon the response of the target tissue to initial damage and inflammation (Hill et al. 2007). This section discusses how the target tissue itself can contribute to either remission or exacerbation of an autoimmune disease.

Similar to immune cells and endothelium, the target tissue itself can produce chemokines that can worsen the inflammation. These locally produced chemokines have been shown to be important for pathogenesis of several autoimmune diseases. In RA, locally produced chemokines are well known to exacerbate inflammation and tissue damage. The synovium and chondrocytes both secrete numerous cytokines (Patel et al. 2001; Kanbe et al. 2002). Apart from this, the infiltrating cells express high levels of the chemokine receptors CXCR3, CCR5, CCR3, CCR2, and CXCR2 (Katschke et al. 2001). CCR-2, the receptor for CCL-2 (also called monocyte chemoattractant protein-1/MCP-1) is highly expressed by RA synovial macrophages and its levels correlate significantly with levels of IL-1β, IL-6, and IL-8/CXCL8 in culture supernatants of synovium from RA patients. CXCR6 is expressed more in RA synovium than in peripheral blood. Treatment of mice with collagen-induced arthritis (CIA) with anti-CXCR6 antibody reduced infiltration of inflammatory cells, bone destruction, and overall arthritis score (Nanki et al. 2005).

Apart from recruiting leukocytes to the synovial compartment, how else does the chemokine system influence the pathogenesis of RA? Chemokines stimulate chondrocytes and fibroblast-like synoviocytes (FLS) to produce cartilage degrading factors like MMPs and pro-inflammatory cytokines (Iwamoto et al. 2008). CCL2, CXCL12, CXCL10, CXCL9, CCL13, and several other

TABLE 16.1
Cellular Targets of Different Chemokines

Chemokine	T-cell	Neutrophils	Monocytes	DCs	Eosinophils	Basophils	Mast cells	B cells	NK
CCL1			+	+				+	+
CCL2 (MCP-1)	+		+	+		+			+
CCL3		+							+
CCL4	+		+						
CCL5	+				+	+			+
CCL7			+						
CCL8	+		+		+	+	+		+
CCL9				+					
CCL10									+
CCL11					+				
CCL12	+		+		+			+	
CCL13	+		+		+	+	+		
CCL14			+						
CCL15	+	+	+					+	
CCL16	+		+					+	
CCL17	+								
CCL18	+							+	
CCL19	+			+				+	
CCL20	+	+						+	
CCL21	+							+	
CCL22			+	+					+
CCL23	+	+	+						
CCL24	+	+			+				
CCL25			+	+					
CXCL1		+							
CXCL2		+							
CXCL3			+						
CXCL4		+	+						
CXCL5		+							
CXCL6		+							
CXCL8		+	+				+		
CXCL9	+								
CXCL10	+		+	+					+
CX3CL1	+		+						

Note: + denotes cells that secrete the chemokine

chemokines induce cell proliferation, thereby causing synovial hyperplasia. Fractalkine, which is a chemoattractant for monocytes and T cells and inducer of adhesion and trans-endothelial migration, is also overexpressed in RA synovium, and its soluble form is present at higher levels in the synovial fluid than the blood (Murphy et al. 2008). Lastly, certain chemokines, known as the ELR chemokines because they possess the ELR (Glu–Leu–Arg) motif near the N-terminus before the first cysteine, are known to be angiogenic (Strieter et al. 1995).

Cytokines produced by target tissue also contribute to inflammation by promoting survival and proliferation of recruited leukocytes and also regulating the expression of adhesion molecules. This is of importance in RA, because not only are cytokines known to be highly expressed in the synovium, but blockade of cytokine signaling using biologicals has now been successfully incorporated in the therapeutic regimen of patients with RA. To further aid the development of new

TABLE 16.2
Cytokines in Autoimmune Diseases

Cytokine	Effects	Therapeutic Antagonist
TNF-α	Induction of other proinflammatory cytokines and chemokines, angiogenesis, activation of endothelial cells, chondrocytes and osteoclasts	Infliximab, etanercept, adalimumab
IL-6	Maturation and activation of B and T cells, macrophages, osteoclasts, chondrocytes, and endothelial cells	Tocilizumab
IL-18	IL-32 induction	
IFN-γ	Promotes NK cell activity and phagocytosis by macrophages, promotes antigen presentation, TH1 development, and leukocyte adhesion and migration	
IL-23	Survival and proliferation of Th17 cells	
IL-15	Survival of autoreactive T cells, secretion of TNF-α and Th17	HuMax
IL-22	Stimulates inflammatory responses, often coexpressed with IL-17	
IL-1 family	Activate synovial monocytes, chondrocytes, cause MMP-3 release by fibroblasts	Anakinra, AMG 108
IL-33	Activation of mast cells, neutrophils migration to the synovium	
IL-17	Upregulation of NF-κB, HLA class I, and several proinflammatory cytokines, osteoclastogenesis, monocyte migration to the synovium	In phase II trial
IL-12	Survival and proliferation of Th1 cells, induction of MMPx	

treatment strategies, it is essential to know which are the prominent cytokines involved in disease pathogenesis and the mechanism of their action. Most of these cytokines are produced by activated macrophages and are summarized in Table 16.2. In addition, B cells produce Blys and APRIL, which support the maintenance of autoreactive B-cell population. Belimumab and atacicept are the antagonists currently undergoing phase II clinical trials.

The rheumatoid synovium has been shown to be more than just a mere victim of a runaway immune system. The RA synovial fibroblasts (SFs) contribute to disease exacerbation by upregulating expression of adhesion molecules, which helps them to invade and degrade cartilage and bone by producing MMPs (Pap et al. 2000). A recent study reports that apart from perpetuating joint damage, RASFs can migrate, through the bloodstream, from affected to non-affected joints and transfer their pathogenic properties to the target cell (Lefevre et al. 2009). This is a prime example of how the target tissue can itself contribute to further worsening of disease.

T cell invasion of the synovium plays a crucial role in the loss of the articular cartilage and adjacent bone. Elegant studies have demonstrated that interplay between FLS, T cells, adhesion molecules, and cytokine contribution to RA pathogenesis (Muller-Ladner and Neumann 2009). The percentage of adherent cells increases upon stimulation of FLS with IFNγ, TNFα, or IL-1β. These cytokines also upregulate expression of adhesion molecule ICAM-1, and the adherent T cell subset displayed a higher expression of the ICAM-1 counter-receptor LFA-1 (Matsuoka et al. 1991; Krzesicki et al. 1991). Co-culture experiments have shown that T cells and FLS can activate each other, but again, this is dependent on the expression of LFA-1/ICAM-1 interactions (Nakatsuka et al. 1997). FLS stimulated for 24 hrs with T cells has increased production of MMPs, IL-6, and IL-8-factors important in joint inflammation and joint destruction (Yamamura et al. 2001). Another study documented production of ICAM-1, IL-8, IL-6, and IL-15 by FLS co-cultured for 96 hours with T cells (Miranda-Carus et al. 2004).

16.2.6 Tissue Damage

The various mechanisms underlying tissue damage in RA are not clear. However, some commonly implicated signaling pathways are those mediated by matrix metalloproteinases, RANKL-OPG interaction, and the WNT pathway.

16.2.6.1 Matrix Metalloproteinases

The primary mechanism of tissue damage in RA appears to be because of MMPs. Members of the MMP family digest a nonidentical but overlapping group of connective tissue components. MMPs cleave not only ECM components, but also each other, thereby initiating a chain of events in which pro-MMPs are successively converted to their active forms. MMPs produced by RA SF include MMP-1, 3, 9, 10, and 13 (Abeles and Pillinger, 2006). MMP1 and MMP3 are important for RA pathogenesis, and MMP3 is also instrumental in aiding transmigration of inflammatory cells into the synovium (Tolboom et al. 2002). MMP1, also known as collagenase 1, is one of the primary enzymes responsible for the degradation of type II collagen (Goldring 2000). MMP3 (stromelysin 1) is active against cartilage matrix components, such as proteoglycan and fibronectin, and can activate pro-MMPs (Nagase 1997).

The secretion of MMPs is increased in response to pro-inflammatory cytokines like TNF and IL-1 (Brinckerhoff, 1991). Evidence proves that overproduction of MMPs facilitates joint destruction in RA cartilage explants (Martel-Pelletier et al. 1993). Increased MMP3 levels correlate with radiographic damage in RA (Posthumus et al. 2000).

16.2.6.2 Wnt Pathway

"Wnt" comprises a family of secreted glycoprotein ligands that bind to a group of cell surface receptors called "frizzled" or "Fz." Among its several other functions, the wnt-fz pathway is involved in osteogenesis. The binding of Wnt proteins to plasma membrane receptors on mesenchymal cells induces the differentiation of these cells into the osteoblast lineage and thereby supports bone formation. It is thought that following joint injury or stress, the wnt-fz pathway may be responsible for activating RA FLS, thereby contributing to RA pathogenesis (Sen 2005).

Wnt signaling leads to stabilization of β-catenin, which then translocates to the nucleus, induces the gene needed for osteoblast development, and thus plays a critical role in bone remodeling. Mice that do not express β-catenin on differentiated osteoblasts develop severe osteopenia associated with increased numbers of activated osteoclasts resulting from an increased RANKL:OPG ratio; interestingly, the functional dysregulation of osteoblasts seen in these mice resembles the osteoblast phenotype found at sites of focal bone loss in inflammatory arthritis, suggesting that inhibition of Wnt signaling may be a mechanism whereby bone formation is compromised in inflammatory states (Glass et al. 2005). Sclerostin and Dickkopf-1 (Dkk-1), both expressed by osteoblasts and osteocytes, prevent Wnt signaling by interacting with Wnt co-receptors. In an elegant paper by Diarra et al. (2007), it was shown that blocking DKK1 in a CIA model resulted in the prevention of bone erosion. Later, in patients with RA, high levels of DKK were seen in serum, which normalized after treatment with anti-TNF treatment, and the levels correlated with disease activity (Diarra et al. 2007).

Members of the wnt-fz family, like Wnt1, Wnt5a, and Fz5, are overexpressed in RA synovial biopsy (Sen et al. 2000). This increased expression is also associated with high levels of IL-6, IL-8, IL-15, fibronectin, and MMP3, all three of which are known to contribute to inflammation in RA (Sen et al. 2001; Sen et al. 2002).

16.2.6.3 RANK Pathway

Receptor activator of NF-κB (RANK), a member of the TNF receptor family, causes bone degradation and osteoclast maturation upon binding RANK ligand (RANKL). In RA synovium, RANKL is highly expressed at sites of bone erosion, and RASFs facilitate RANK/RANKL interaction by actively producing RANKL (Lee et al. 2006). RANKL is expressed by all cell types in the RA synovial tissue—T cells, neutrophils, chondrocytes, endothelial cells, and activated macrophages. In addition to a cell-bound RANKL form, activated T lymphocytes produce a secreted RANKL form, which can also be generated by proteolytic cleavage of the membrane-bound form (Hofbauer et al. 2001). RANK is expressed on osteoclast precursor cells, mature osteoclasts, and dendritic cells. Osteoprotegerin (OPG) is a decoy receptor produced by osteoclasts and endogenous antagonist, which binds to RANKL and prevents bone resorption (Simonet et al. 1997). OPG also decreases the production of cytokines, such as IL-6 and IL-11, in response to DC stimulation by RANKL. RANKL is expressed by osteoblasts and bone marrow stromal cells, whereas its receptor RANK is expressed in preosteoclasts and other cells of this lineage (Vega et al. 2007).

Besides bone remodelling, RANK-RANKL interaction induces proliferation of antigen-presenting dendritic cells and T cells (Neumann et al. 2005). Activated T cells express more of RANKL and thereby induce chronic bone remodeling (Kotake et al. 2001). This is yet another mechanism by which T cells contribute to tissue damage in RA. IL-17 produced by Th17 cells is a potent inducer of RANK and osteoclastogenesis.

Several therapeutic agents currently in use act by inhibiting the RANK- RANKL pathway. Methotrexate, sulfasalazine, and infliximab inhibit the expression of RANKL in RASFs in a dose-dependent manner and increase the synthesis of osteoprotegerin (OPG), a RANKL antagonist, in RASF supernatants (Lee et al. 2004). Even the efficacy of cyclosporine A in RA patients has been explained by a drug-induced inhibition of T cell activation, resulting in decreased RANKL production (Hofbauer et al. 2001). The role for the RANK signaling pathway was further confirmed by reports that RA synovial fluid has lower OPG levels than OA (Takayanagi et al. 2000). These studies indicate that the RANK-RANKL-OPG pathway is a good therapeutic target for arthritis.

Thus in RA, Th1 and Th17 cells infiltrate the joint and produce cytokines and chemokines, which attract more immune cells like neutrophils and monocytes, and also stimulate these cells to secrete IL-1, IL-6, and TNF-α. T cells initiate and maintain activation of synovial fibroblasts, which are important players in RA pathology. The inflammatory milieu further supports the increased expression of MMPs and increased WNT and RANK signaling, leading to osteoclast differentiation and bone loss. Thus, targeting the migration of T cells into the RA synovium can be an effective therapeutic tool.

16.3 SYSTEMIC LUPUS ERYTHEMATOSUS (SLE)

SLE affects young females and presents with involvement of skin, joints, kidney, brain, and blood cells. The hallmark of this disease is the presence of antinuclear antibodies directed against various components of the nucleus, for example, DNA, RNA associated proteins. The precise mechanism of generation of autoantibodies is not clear but is presumed to be due to defective clearance of apoptotic bodies by the macrophages, thus increasing the load of nuclear autoantigens including nucleosomes. The presence of autoantigens leads to a break in tolerance and generation of pathogenic autoantibodies in a genetically susceptible host (Gualtierotti et al. 2010).

16.3.1 How Do These Autoantibodies Cause Inflammation and Tissue Damage?

Autoantibodies can cause tissue or cell injury in many ways. Antibodies can bind to cell surface expressed antigens or antigens adsorbed on the cell membrane from circulation and affect cell function or survival. Antibodies can bind to circulating antigens and form immune complexes (ICs). Immune complexes get trapped in the capillaries and initiate complement activation, release of chemo-attractants, egress of inflammatory cells, and consequent inflammation resulting in tissue injury (Toong et al. 2011).

Antibodies can bind to cell bound antigen and then antibody coated cells are taken up by macrophages in the spleen and destroyed. In SLE, antibodies directed against RBC and platelet antigens coat these blood cells, and as the blood is filtered through the spleen the antibody-coated blood cells are taken up by Fc mediated mechanisms and destroyed, leading to cytopenias.

Antibodies can also bind to other cells and modulate their function. Anti-DNA antibodies interact with actinin on podocytes, thus initiating dysfunction of podocytes (Migliorini et al. 2002).

Anti-DNA antibodies can directly bind to cells like renal tubular cells and cause production of pro-inflammatory cytokines like TNF and IL-6, thus playing a role in tubule-interstitial inflammation (Yuang et al. 2005). Circulating DNA or nucleosomes can bind to basement membrane of the glomerulus, and then autoantibodies directed against DNA can bind to the antigens in situ rather than in circulation, leading to formation of ICs.

Most often antibodies bind to circulating antigens to form immune complexes (ICs). Some ICs are pathogenic, while others are not. Pathogenic ICs are intermediate in size and usually have cationic charge. In SLE, free DNA and nucleosomes are increased in circulation due to increased apoptosis as well as decreased clearance (Bartoloni et al. 2011). Anti-nucleosomal antibodies or anti-dsDNA antibodies bind the circulating DNA and form ICs. Further, due to congenital or acquired complement defects there is decreased solubilization of ICs. Thus, intermediate-sized circulating immune complexes are trapped in the capillaries of organs with high blood flow like kidney, brain, and lung. Kidney biopsies from SLE patients show IC deposition both in subendothelial and mesangial areas (Toong et al. 2011).

16.3.2 Immune Complex Mediated Inflammation

Immune complexes bind complement and cause complement activation. In SLE predominantly classical pathway is activated although some role of MBL and alternative pathway is also described (Ceribelli et al. 2009; Sekine et al. 2011). Classical pathway activation leads to release of anaphylotoxins like C4a, C3a, and C5a. These anaphylotoxins bind to mast cells and release histamine and other soluble mediators. These soluble mediators help in recruitment of inflammatory cells by increasing capillary permeability. C3a, C5a, and C5b67 also induce monocytes and neutrophils to adhere to vascular endothelial cells, extravasate through the endothelial lining of the capillary, and migrate toward the site of complement activation in the tissues. The process of adhesion and migration is similar to that seen in rheumatoid arthritis.

In the MRL/lpr mice model of lupus, IC deposition is followed by production of chemokines like MCP-1 and RANTES in the glomeruli, primarily by mesangial cells (Perez et al. 2001). This precedes even the inflammatory cell infiltrate. These chemokines attract mononuclear cells like lymphocytes and monocytes to the site of IC deposition. Further, the circulating ICs cause release of cytokines through activation of the classical pathway of complement activation (Rönnelid et al. 2008). Products of complement activation including membrane attack complex can activate endothelial cells, leading to expression of adhesion molecules, thus favoring migration of cells and release of cytokines.

16.3.3 Activation of the Innate Immune System

Anti-dsDNA containing ICs binds to Fc receptors and gets internalized, and then the DNA contained in ICs can bind to TLR9 in the endosomal compartment leading to a cascade of events culminating in inflammation (Leadbetter et al. 2002). Similarly, ICs containing RNA can bind to TLR7. Ronnblom and Alm (2003) have demonstrated the ability of both types of complexes to stimulate IFN-α production from PBMCs in vitro, and this activity was abolished on treatment of ICs with nucleases, suggesting that the nucleic acids drive the IFN-α production probably through TLRs. Similarly, RNA and DNA containing ICs bind to plasmacytoid dendritic cells via FCRII receptor and cause production of IFN-α, a major mediator in SLE (Means et al. 2005). Inhibitors of both TLR7 and TLR9 signaling inhibit IFN production from human pDCs in response to stimulation by ICs isolated from lupus patients, suggesting that TLRs are involved in IFN production by pDCs (Barrat et al. 2005). Treatment of lupus-prone mice with TLR7 and 9 inhibitor leads to amelioration of disease (Barrat et al. 2007). In SLE the level of IFN-alpha is increased in serum as well as gene expression profile of PBMC, suggesting an IFN signature (Baechler et al. 2003). Further, patients treated with IFN for hepatitis infection can develop lupus-like symptoms. Tubular TLR-9 activation by DNA contained in ICs has a pathogenetic role in tubule-interstitial inflammation and damage in experimental and human lupus nephritis (Benigni et al. 2007).

Macrophage stimulation leads to local cytokine and chemokine secretion, especially IL-1 and IL-6, that contributes to local inflammation and leukocyte accumulation. The process of leukocyte accumulation is similar to that seen in RA. In contrast, maturation of dendritic cells leads to emigration from the site of injury to the regional lymph nodes to convey antigens and pro-inflammatory signals to T cells, which is a prerequisite for the adaptive immune response.

Activation of TLRs without appropriate control can lead to substantial inflammation resulting in tissue damage and autoimmunity. ICs can form a positive feedback loop with increased IFN production, induction of IFN responsive genes, inflammation, activation of B cells, and increased antibody production, again leading to IC formation.

16.3.4 Activation of B Cells

DNA–anti-DNA IgG ICs stimulate autoantibody production in mice by a process involving TLR9 (Leadbetter et al. 2002). Immune complexes bind to B cell receptors on B cells and get internalized, leading to internalization of DNA bound to anti-DNA antibodies. The dual stimulation, one via B cell receptor and another via TLR9, leads to production of more pathogenic antibodies by B cells independent of need for T cell help.

16.3.5 Role of Local Factors

We have seen from the previous discussion the role of antibodies in the pathogenesis of SLE; however, the wide array of manifestations seen in SLE is unlikely due to only antibody-mediated damage. Further, the heterogeneity seen among different patients probably suggests a role of local factors like cytokines, chemokines in tissue injury. Despite the presence of IC deposition in the kidney, there is no pathological lesion seen in class I nephritis. In addition, some patients have marked inflammation but no fibrosis and others have more fibrosis but less inflammation. About one-third of patients with SLE have anti-phospholipid antibodies, but only a minority develop thrombosis, usually related to an additional local insult like trauma (Schonfield et al. 2008).Another factor that may account for heterogeneity may be genetics. In NZM mice some genes predispose to anti-DNA mediated glomerulonephritis and some contribute to chronic glomerulosclerosis (Waters et al. 2004).

16.3.6 Cytokines

Several studies have shown increased serum levels of pro-inflammatory cytokines like IL-6 and TNF-alpha in SLE (Aringer and Smolen 2004; Kyttaris et al. 2005). Patients with SLE have increased concentrations of IL-6 and MCP-1. These cytokines are associated with increased inflammation, BMI, and adverse lipid profiles. IL-6 is associated with burden of atherosclerosis in SLE (Asanuma et al. 2006).

Low producers of TNF have less risk of lupus nephritis as well, as patients treated with anti-TNF therapies can develop ANA and lupus-like symptoms, suggesting that TNF has more of an immune-regulatory than pro-inflammatory role in SLE (Williams et al. 2009; Jacob and Stohl 2011). Serum IL-10 levels are elevated in SLE and have the best correlation with disease activity. In a recent study SLE patients had lower levels of TGF-β1 and IL-1β as compared to healthy controls. Higher disease activity correlated with low TGF-β1 levels (Becker-Merok et al. 2010). The low TGF-β may lead to fewer regulatory T cells and thus continued inflammation. Use of tolerogenic peptide in SLE lowered TNF and IFN-γ but increased TGF-β and Foxp3 expression, thus leading to more T regulatory cells, again suggesting that TGF-β is important for the generation of T regulatory cells (Sthoeger et al. 2009).

IL-4, IL-10, and TGF-β have been linked to fibrosis in kidney, and the same is true of lupus nephritis (Kanai et al. 1994). Though TGF-β levels are low in serum, in the local milieu of kidney TGF-β is increased, resulting in increased urinary excretion. In animal lupus models progression from inflammatory lesion to fibrosing lesion is characterized by expression of TGF-β.

Interferon-α acts to amplify the immune inflammation, whereas TNF mediates organ inflammation in kidney (Jacob and Stohl 2011). Reduced IL-2 production is a central finding in SLE due to intrinsic T cell defect; this results in fewer T regulatory cells (Gomez-Martin et al. 2009).

The kallikrein system has also been linked to renal disease and KLK1 and KLK3 gene polymorphisms have been found in SLE. The resultant reduced kallikrein production could cause more severe nephritis (Liu et al. 2009). Kallikrein regulates fibrosis, inflammation, and blood pressure and thus has a renoprotective effect. Indeed, delivery of kallikrein gene reverses renal fibrosis and inflammation (Bledsoe et al. 2006).

16.3.7 Role of T Cells in SLE

Even though in SLE the inflammation is mainly mediated by antibodies, IC deposition, and complement activation, T cells also play a role. Though multiple T cell abnormalities are known in SLE, recently Th17 cells that differentiate in the presence of TGF-β and IL-6 are being implicated (Shin et al. 2011). Patients with SLE have high plasma levels of IL-17 and IL-23 and increased frequency of Th17 cells (Xing et al. 2011; Chen et al. 2010). IL-17 causes infiltration of neutrophils, thus perpetuating inflammation and damage. Increased production of total IgG, anti-dsDNA IgG, and IL-6 by peripheral blood mononuclear cells of patients with lupus nephritis was observed when they were cultured with IL-17, thus linking T cells with humoral immunity.

To summarize, in SLE the inflammation is mediated by autoantibodies and ICs stimulating the innate immune system through TLRs or causing complement activation leading to release of anaphylotoxins, endothelial activation, and release of cytokines and chemokines. Further local factors like TGF and kallikrein lead to tissue fibrosis and chronic damage in lupus nephritis (Figure 16.1). Strategies to block key cytokines and TLRs may prove useful in SLE.

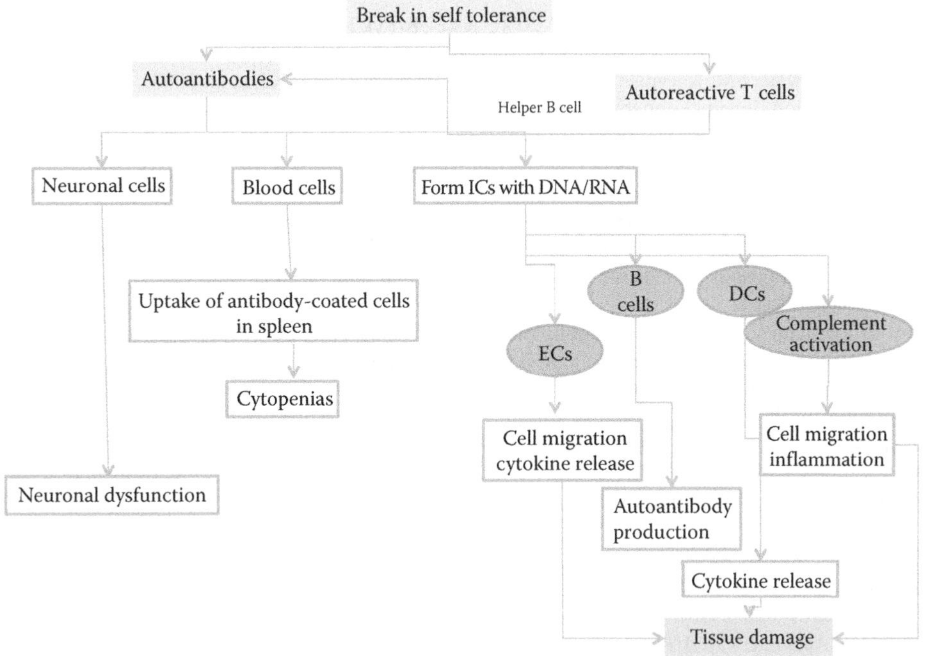

FIGURE 16.1 Cascade of events in SLE leading to cell or tissue damage.

TAKE-HOME MESSAGES

- Autoimmunity is an immune response directed against an antigen within the body of the host.
- Autoimmune diseases are associated with T lymphocyte activation and/or auto-antibodies produced by the dysregulated immune system.
- Autoreactive B and T cells alone are not sufficient to cause autoimmune disease; they need to have access to their target site, and initiate an inflammatory response through recruitment of other immune cells, soluble mediators like cytokines and chemokines, to cause uncontrolled inflammation and tissue damage resulting in multi-organ chronic complex diseases.
- To design molecules that can inhibit the inflammatory cascade of a specific disease, it is important to elucidate the critical roles of adhesion molecules and chemokines associated with the recruitment of immune cells at the local site.
- The alternative approach to disease inhibition could be to manipulate the pathologic T and/or B lymphocytes.
- At the local site of pathology, underlying mechanisms of tissue damage also play a critical role in the disease process. Some commonly implicated signaling pathways are those mediated by matrix metalloproteinases, RANKL-OPG interaction. NGF/NGF-R and the WNT pathway are currently under investigation as possible targets for drug development.

REFERENCES

Abeles AM, Pillinger MH. 2006. The role of the synovial fibroblast in rheumatoid arthritis: Cartilage destruction and the regulation of matrix metalloproteinases. *Bull NYU Hosp Jt Dis* 64:20–4.

Aringer M, Smolen JS. 2004. Tumour necrosis factor and other proinflammatory cytokines in systemic lupus erythematosus: A rationale for therapeutic intervention. *Lupus* 13:344–7.

Asanuma Y, Chung CP, Oeser A, Shintani A, Stanley E, Raggi P, et al. 2006, Increased concentration of proatherogenic inflammatory cytokines in systemic lupus erythematosus: Relationship to cardiovascular risk factors. *J Rheumatol* 33:539–45.

Avraham H, Park SY, Schinkmann K, et al. 2000. RAFTK/Pyk2-mediated cellular signalling. *Cell Signal* 12:123–33.

Baechler EC, Batliwalla FM, Karypis G, Gaffney PM, Ortmann WA, Espe KJ, et al. 2003. Interferon-inducible gene expression signature in peripheral blood cells of patients with severe lupus. *Proc Natl Acad Sci USA* 100:2610–5.

Barrat FJ, Meeker T, Chan JH, Guiducci C, Coffman RL. 2007. Treatment of lupus-prone mice with a dual inhibitor of TLR7 and TLR9 leads to reduction of autoantibody production and amelioration of disease symptoms. *Eur J Immunol* 37:3582–6.

Barrat FJ, Meeker T, Gregorio J, Chan JH, Uematsu S, Akira S, et al. 2005. Nucleic acids of mammalian origin can act as endogenous ligands for Toll-like receptors and may promote systemic lupus erythematosus. *J Exp Med* 202:1131–9.

Bartoloni E, Ludovini V, Alunno A, Pistola L, Bistoni O, Crinò L, et al. 2011. Increased levels of circulating DNA in patients with systemic autoimmune diseases: A possible marker of disease activity in Sjogren's syndrome. *Lupus* May 25. [Epub ahead of print.]

Becker-Merok A, Eilertsen GØ, Nossent JC. 2010. Levels of transforming growth factor-beta are low in systemic lupus erythematosus patients with active disease. *J Rheumatol* 37:2039–45.

Benigni A, Caroli C, Longaretti L, Gagliardini E, Zoja C, Galbusera M, et al. 2007. Involvement of renal tubular Toll-like receptor 9 in the development of tubulointerstitial injury in systemic lupus. *Arthritis Rheum* 56:1569–78.

Bledsoe G, et al. 2006. Reversal of renal fibrosis, inflammation, and glomerular hypertrophy by kallikrein gene delivery. *Hum. Gene Ther* 17:545–55.

Bradfield PF, Nourshargh S, Aurrand-Lions M, et al. 2007. JAM family and related proteins in leukocyte migration (Vestweber series). *Arterioscler Thromb Vasc Biol* 27:2104–12.

Brinckerhoff CE. 1991. Joint destruction in arthritis: Metalloproteinase in the spotlight. *Arthritis Rheum* 34:1073–5.

Ceribelli A, Andreoli L, Cavazzana I, Franceschini F, Radice A, Rimoldi L, et al. 2009. Complement cascade in systemic lupus erythematosus: Analyses of the three activation pathways. *Ann N Y Acad Sci* 1173:427–34.

Chen XQ, Yu YC, Deng HH, Sun JZ, Dai Z, Wu YW, et al. 2010. Plasma IL-17A is increased in new-onset SLE patients and associated with disease activity. *J Clin Immunol* 30:221–5.

Dangerfield J, Larbi KY, Huang MT. 2002. PECAM-1 (CD31) homophilic interaction up-regulates α 6 β 1 on transmigrated neutrophils in vivo and plays a functional role in the ability of α 6 integrins to mediate leukocyte migration through the perivascular basement membrane. *J Exp Med* 196:1201–21.

Diarra D, Stolina M, Polzer K, Zwerina J, Ominsky MS, Dwyer D, et al. 2007. Dickkopf-1 is a master regulator of joint remodeling. *Nat Med* 13:156–63.

Elices MJ, Osborn L, Takada Y, et al. 1990. VCAM-1 on activated endothelium interacts with the leukocyte integrin VLA-4 at a site distinct from the VLA-4/fibronectin binding site. *Cell* 60:577–84.

Engelhardt B, Wolburg H. 2004. Mini-review: Transendothelial migration of leukocytes: Through the front door or around the side of the house? *Eur J Immunol* 34:2955–63.

Giagulli C, Scarpini E, Ottoboni L, et al. 2004. RhoA and ζ-PKC control distinct modalities of LFA-1 activation by chemokines. Critical role of LFA-1 affinity triggering in lymphocyte in vivo homing. *Immunity* 20:25–35.

Glass DA 2nd, Bialek P, Ahn JD et al. 2005. Canonical Wnt signaling in differentiated osteoblasts controls osteoclast differentiation. *Dev Cell* 8:751–64.

Goldring MB. 2000. The role of the chondrocyte in osteoarthritis. *Arthritis Rheum* 43:1916–26.

Gomez-Martin D, Diaz-Zamudio M, Crispin JC, Alcocer-Varela J. 2009. Interleukin 2 and systemic lupus erythematosus: Beyond the transcriptional regulatory net abnormalities. *Autoimmun Rev* 9:34–9.

Gualtierotti R, Biggioggero M, Penatti AE, Meroni PL. 2010. Updating on the pathogenesis of systemic lupus erythematosus. *Autoimmun Rev* 10:3–7.

Hill NJ, Hultcrantz M, Sarvetnick N, et al. 2007. The target tissue in autoimmunity—an influential niche. *Eur J Immunol* 3:589–97.

Hofbauer LC, Heufelder AE, Erben RG. 2001. Osteoprotegerin, RANK, and RANK ligand: The good, the bad, and the ugly in rheumatoid arthritis. *J Rheumatol* 28:685–7.

Imhof BA, Aurrand-Lions M. 2004. Adhesion mechanisms regulating the migration of monocytes. *Nat Rev Immunol* 4:432–44.

Ishikaw J, Okada Y, Bird IN. 2002. Use of anti-platelet-endothelial cell adhesion molecule-1 antibody in the control of disease progression in established collagen-induced arthritis in DBA/1J mice. *Jpn J Pharmacol* 88:332–40.

Issekutz AC, Ayer L, Miyasaka M. 1996. Treatment of established adjuvant arthritis in rats with monoclonal antibody to CD18 and very late activation antigen-4 integrins suppresses neutrophils and T-lymphocyte migration to the joints and improves clinical disease. *Immunology* 88:569–76.

Iwamoto T, Okamoto H, Toyama Y, et al. 2008. Molecular aspects of rheumatoid arthritis: Chemokines in the joints of patients. *FEBS J.* 275:4448–55.

Jacob N, Stohl W. 2011 Cytokine disturbances in systemic lupus erythematosus. *Arthritis Res Ther* 13:228.

Kanai H, Mitsuhashi H, Ono K, Yano S, Naruse T. 1994. Increased excretion of urinary transforming growth factor beta in patients with focal glomerular sclerosis. *Nephron* 66:391–5.

Kanbe K, Takagishi K, Chen Q. 2002. Stimulation of matrix metalloprotease 3 release from human chondrocytes by the interaction of stromal cell-derived factor 1and CXC chemokine receptor 4. *Arthritis Rheum* 46:130–7.

Kansas GS. 1996. Selectins and their ligands: Current concepts and controversies. *Blood* 88:3259–87.

Katschke KJ Jr, Rottman JB, Ruth JH, et al. 2001. Differential expression of chemokine receptors on peripheral blood, synovial fluid, and synovial tissue monocytes/macrophages in rheumatoid arthritis. *Arthritis Rheum* 44:1022–32.

Kavanaugh AF, Davis LS, Jain RI. 1996. A phase I/II open label study of the safety and efficacy of an anti-ICAM-1 (intercellular adhesion molecule-1; CD54) monoclonal antibody in early rheumatoid arthritis. *J Rheumatol* 23:1338–44.

Kotake S, Udagawa N, Hakoda M, et al. 2001. Activated human T cells directly induce osteoclastogenesis from human monocytes: Possible role of T cells in bone destruction in rheumatoid arthritis patients. *Arthritis Rheum* 44:1003–12.

Krzesicki RF, Fleming WE, Winterrowd GE, et al. 1991. T lymphocyte adhesion to human synovial fibroblasts. Role of cytokines and the interaction between intercellular adhesion molecule 1 and CD11a/CD18. *Arthritis Rheum* 34:1245–53.

Kyttaris VC, Juang YT, Tsokos GC. 2005 Immune cells and cytokines in systemic lupus erythematosus: An update. *Curr Opin Rheumatol* 17:518–22.

Langer HF, Chavakis T. 2009. Leukocyte-endothelial interactions in inflammation. *J Cell Mol Med* 13:1211–20.

Lawson C, Wolf S. 2009. ICAM-1 signaling in endothelial cells. *Pharmacol Rep* 61:22–32.

Leadbetter, EA, Rifkin IR, Hohlbaum AR, Beaudette BC, Shlomchik MJ, Marshak-Rothstein A. 2002. Chromatin-IgG complexes activate B cells by dual engagement of IgM and Toll-like receptors. *Nature* 416:603–7.

Lee CK, Lee EY, Chung SM, et al. 2004. Effects of disease-modifying antirheumatic drugs and antiinflammatory cytokines on human osteoclastogenesis through interaction with receptor activator of nuclear factor kappaB, osteoprotegerin, and receptor activator of nuclear factor kappaB ligand. *Arthritis Rheum* 50:3831–43.

Lee HY, Jeon HS, Song EK, et al. 2006. CD40 ligation of rheumatoid synovial fibroblasts regulates RANKL-mediated osteoclastogenesis: Evidence of NF-kappaB-dependent, CD40-mediated bone destruction in rheumatoid arthritis. *Arthritis Rheum* 54:1747–58.

Lefèvre S, Knedla A, Tennie C, et al. 2009. Synovial fibroblasts spread rheumatoid arthritis to unaffected joints. *Nat Med* 15:1414–20.

Ley K, Laudanna C, Cybulsky MI, et al. 2007. Getting to the site of inflammation: The leukocyte adhesion cascade updated. *Nat Rev Immunol* 7:678–89.

Ley K, Tedder TF. 1995. Leukocyte interactions with vascular endothelium: New insights into selectin-mediated attachment and rolling. *J Immunol* 155:525–8.

Liu K, Li QZ, Delgado-Vega AM, Abelson AK, Sánchez E, Kelly JA, Li L, et al. 2009. Kallikrein genes are associated with lupus and glomerular basement membrane-specific antibody-induced nephritis in mice and humans. *J Clin Invest* 119:911–23.

Luo B, Carman CV, Springer TA. 2007. Structural basis of integrin regulation and signaling. *Annu Rev Immunol* 25:619–47.

Mackay IR, Leskovsek NV, Rose NR. 2008. Cell damage and autoimmunity: A critical appraisal. *J Autoimmun* 30:5–11.

Maksymowych WP, Blackburn WD Jr, Tami JA, et al. 2002. A randomized, placebo controlled trial of an antisense oligodeoxynucleotide to intercellular adhesion molecule-1 in the treatment of severe rheumatoid arthritis. *J Rheumatol* 29:447–53.

Martel-Pelletier J, Fujimoto N, Obata K, et al. 1993. The imbalance between the synthesis level of metalloproteinases and TIMPs in osteoarthritic and rheumatoid arthritis cartilage can be enhanced by interleukin-1 [abstract]. *Arthritis Rheum* 36(suppl 9): S191.

Matsuoka N, Eguchi K, Kawakami A, et al. 1991. Phenotypic characteristics of T cells interacted with synovial cells. *J Rheumatol* 18:1137–42.

McLeod SJ, Shum AJ, Lee RL. 2004. The Rap GTPases regulate integrin-mediated adhesion, cell spreading, actin polymerization, and Pyk2 tyrosine phosphorylation in B lymphocytes. *J Biol Chem* 279:12009–19.

Means TK, Latz E, Hayashi F, Murali MR, Golenbock DT, Luster AD. 2005. Human lupus autoantibody-DNA complexes activate DCs through cooperation of CD32 and TLR9. *J. Clin Invest* 115:407–17.

Migliorini P, Pratesi F, Bongiorni F, Moscato S, Scavuzzo M, Bombardieri S. 2002. The targets of nephritogenic antibodies in systemic autoimmune disorders. *Autoimmun Rev* 1:168–73.

Miranda-Carús ME, Balsa A, Benito-Miguel M, et al. 2004. IL-15 and the initiation of cell contact-dependent synovial fibroblast-T lymphocyte cross-talk in rheumatoid arthritis: Effect of methotrexate. *J Immunol* 173:1463–76.

Muller WA. 2009. Mechanisms of transendothelial migration of leukocytes. *Circ Res* 105:223–30.

Muller WA, Weigl SA, Deng X. 1993. PECAM-1 is required for transendothelial migration of leukocytes. *J Exp Med* 178:449–60.

Müller-Ladner U, Neumann E. 2009 Synovial fibroblasts spread rheumatoid arthritis to unaffected joints. *Nat Med* 15:1414–20.

Murphy G, Caplice N, Molloy M. 2008. Fractalkine in rheumatoid arthritis: A review to date. *Rheumatology* 47:1446–51.

Nagase H. 1997. Activation mechanisms of matrix metalloproteinases. *Biol Chem* 378:151–60.

Nakatsuka K, Tanaka Y, Hubscher S, et al. 1997. Rheumatoid synovial fibroblasts are stimulated by the cellular adhesion to T cells through lymphocyte function associated antigen-1/intercellular adhesion molecule-1. *J Rheumatol* 24:458–64.

Nanki T, Shimaoka T, Kayashida K, et al. 2005. Pathogenic role of the CXCL16–CXCR6 pathway in rheumatoid arthritis. *Arthritis Rheum* 52:3004–14.

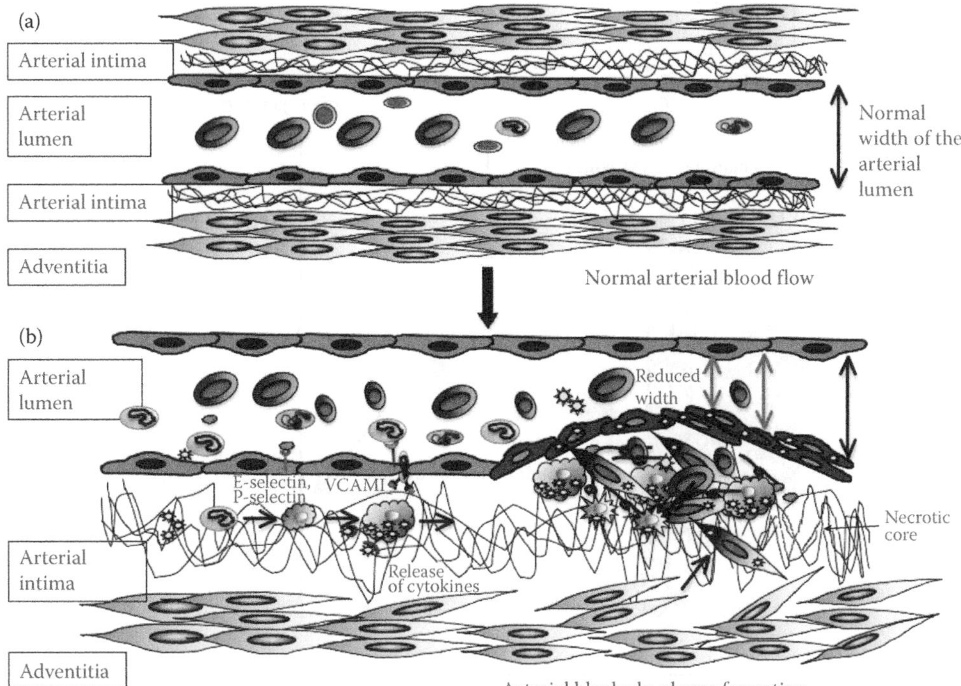

FIGURE 6.1 Schematics of normal artery and development of atherosclerosis. (a) Schematics of normal arterial sectional view, where blood flows without any obstruction. The lumen of the artery is lined by the endothelium followed by the arterial intima and the lining of vascular smooth muscle cells followed by adventitia. Normally a small number of circulating immune cells and progenitor cells and a large number of red blood cells along with white blood cells containing blood circulate within the lumen. (b) Atherosclerosis is an inflammatory disorder that develops within the arteries due to the injury caused to the endothelium in the presence of a high-cholesterol diet and other risk factors. Modified lipids such as oxidized low-density lipoprotein (ox-LDL) interact with the endothelium. The injured endothelium upregulates chemotactic proteins and adhesion molecules such as VCAM1, ICAM1, and selectins. These molecules attract the circulating leukocytes and platelets, which migrate into the arterial intima. In the arterial intima, ox-LDL is phagocytosed by macrophages that express scavenger receptor (SR) and convert into foam cells. The vSMCs also become pro-inflammatory and migrate to the intima, secrete extracellular matrix, and form lipid core that starts to obstruct the arterial blood flow. Pro-inflammatory endothelial cells, platelets, foam cells, and other immune cells (B, T, DC, and mast cells) also get trapped in the lipid-necrotic core forming a plaque, which if it ruptures, will block the blood supply, causing ischemia in adjacent tissues.

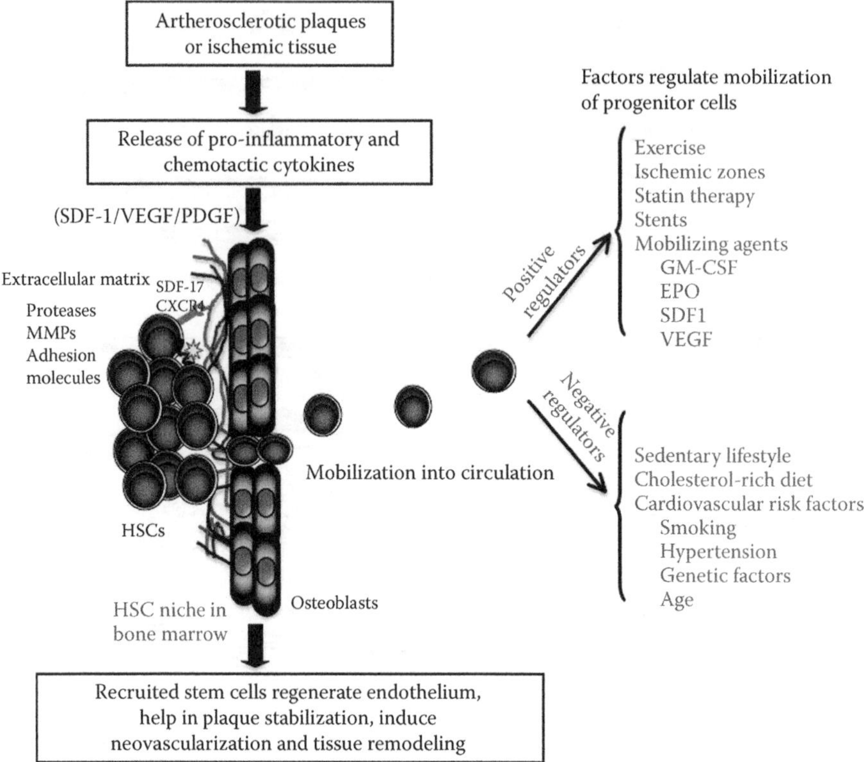

FIGURE 6.2 Mobilization and recruitment of progenitor cells. Injured vessel and ischemic zones release chemotactic factors (SDF-1, VEGF, PDGF, etc.) into the circulation. Release of mobilizing factors causes egress of progenitor cells (HSC, EPC, or vSMC) from bone marrow and recruits proteases such as MMP9, which mediates cleavage of c kit+ HSCs from the bone marrow. The positive regulators of mobilization of progenitor cells include ischemic zones, exercise, mobilizing agents, etc. Factors such as smoking, hypertension, lifestyle choices, and certain cardiovascular conditions cause decreased levels of circulating progenitor cells. The progenitor cells replace and regenerate injured endothelial cells and pro-inflammatory smooth muscle cells and may lead to intimal hyperplasia. In ischemic zones, progenitor cells help regenerate damaged tissues by inducing neovascularization and tissue remodeling.

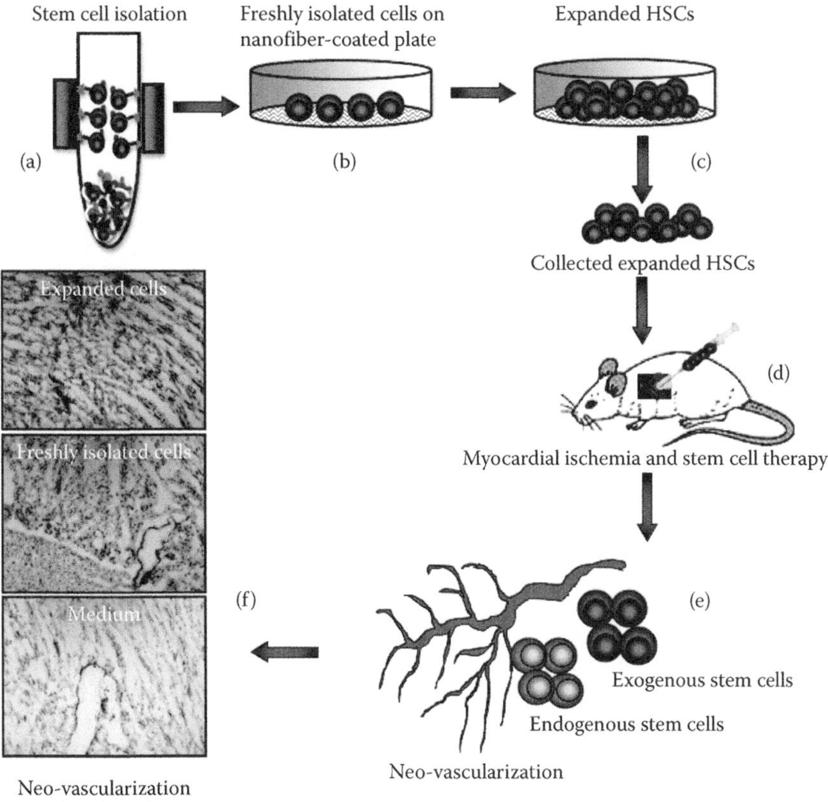

FIGURE 6.3 Stem cell therapy–mediated neovascularization in myocardial ischemia. (a) Schematics of autoMACS-mediated isolation of CD133+ cells. (b) Seeding of CD133+ isolated stem cells on the nanofiber (polyether sulfone, PES)-coated plates. (c) Expansion of stem cells over 10 days of culture in serum-free media supplemented with cytokines and growth factors. (d) Injection of either freshly isolated stem cells or nanofiber-expanded stem cells in immunocompromised rat model of myocardial ischemia via intra cardio-ventricular route. Media was used as control. (e) Besides the paracrine effect, exogenous and endogenous host stem cells take part in the neovascularization process. (f) After 4 weeks of therapy, the rat was sacrificed and cardiac tissues were stained for detection of blood vessels using alkaline phosphatase staining. In ischemic condition like myocardial ischemia, exogenous HSCs as well as circulating endogenous HSCs give rise to appropriate cell type, increase angiogenesis, and reduce fibrosis, resulting in recovery of cardiac function.

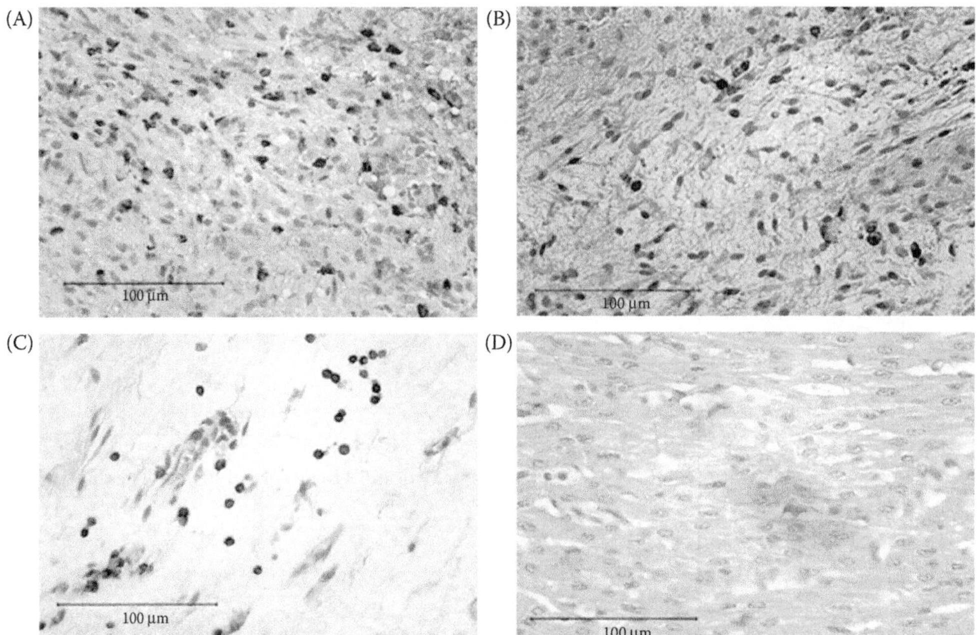

FIGURE 7.1 CD45 immunohistochemistry following myocardial infarction demonstrates markedly less cellular inflammatory response in fetal versus adult hearts. One week following myocardial infarction, the fetal heart (B) shows minimal numbers of inflammatory cells while the adult heart (A) demonstrates a significantly increased inflammatory infiltrate. One month following infarction, the inflammatory cell infiltrate in the fetal myocardium (D) has decreased, while it persists within the adult myocardium (C).

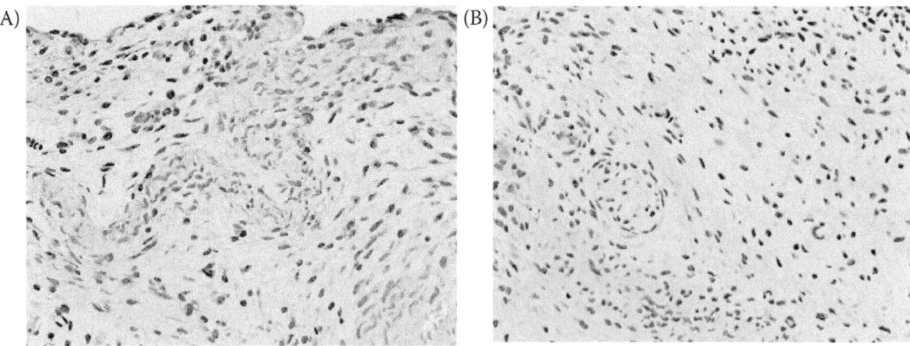

FIGURE 7.2 Immunohistochemical staining for CD45, the common leukocyte antigen, in 8mm (A) or 2mm (B) dermal wounds in fetal sheep 7 days after injury. Note: Increased wound size was associated with a dramatic increase in the number of inflammatory cells in the wound and subsequent scar formation.

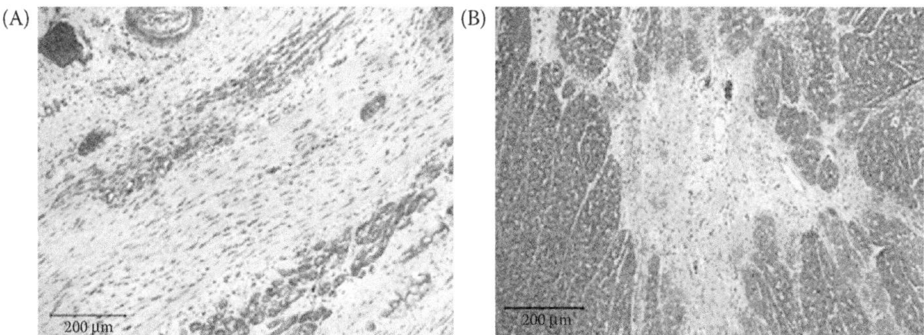

FIGURE 7.3 Trichrome staining of adult (A) and fetal (B) myocardium demonstrates increased collagen deposition and scar formation in the adult compared to the fetus 1 month following myocardial infarction.

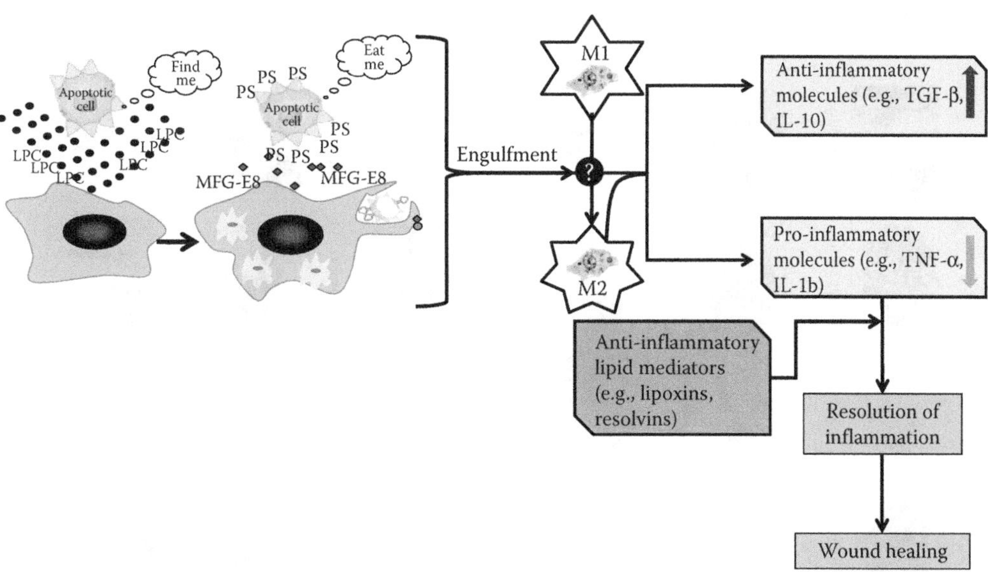

FIGURE 9.2 Lipid biosynthesis pathways that regulate inflammation and resolution. The pro- as well and anti-inflammatory lipid mediators regulate both initiation as well as the resolution phases of wound inflammation. These lipids are synthesized from polyunsaturated fatty acids, such as arachidonic acid (AA), eicosapentaenoic acid (EPA), and docosahexaenoic acid (DHA) by the enzymes cyclooxygenase (Cox) and/or lipoxygenase (LO). AA is converted to pro-inflammatory thromboxanes (TX), prostacyclin (PC), and prostaglandins (PG) via the cyclooxygenase (COX) pathway. In the LO pathway, hydroperoxyeicosatetraenoic acids (HPETEs) are produced that can be further enzymatically reduced to the hydroxylated form (HETE). LO enzymes are also involved in the production of anti-inflammatory lipoxins (LX), resolvins (RvEs and RvDs), and protectins (PD) from AA, EPA, and DHA. Lipoxins and resolvins are generated from AA and EPA/DHA by the enzymatic actions of both LO and acetylated COX-2.

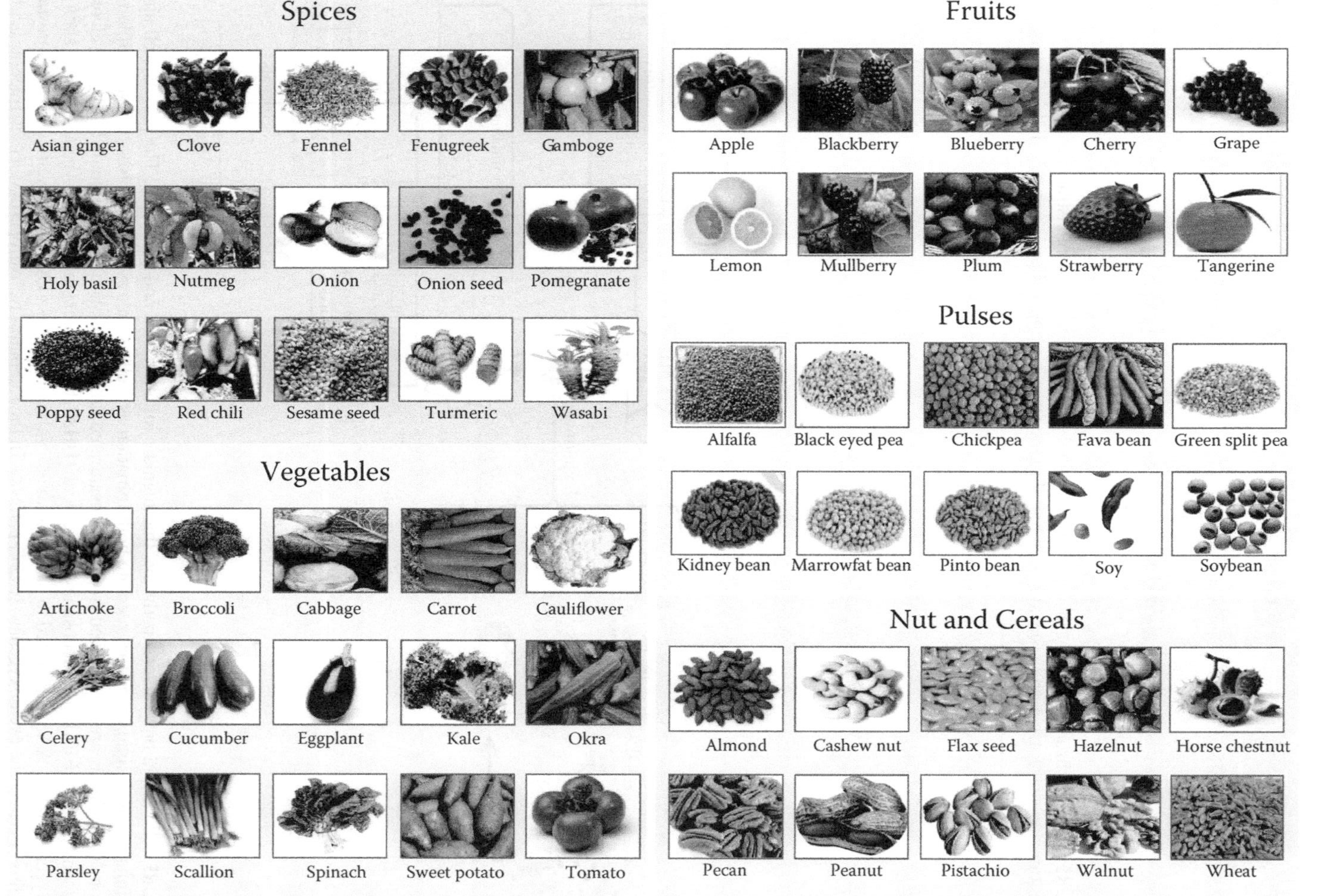

FIGURE 12.3 Common sources of plant-derived nutraceuticals. These sources include spices, vegetables, fruits, pulses, nuts, and cereals.

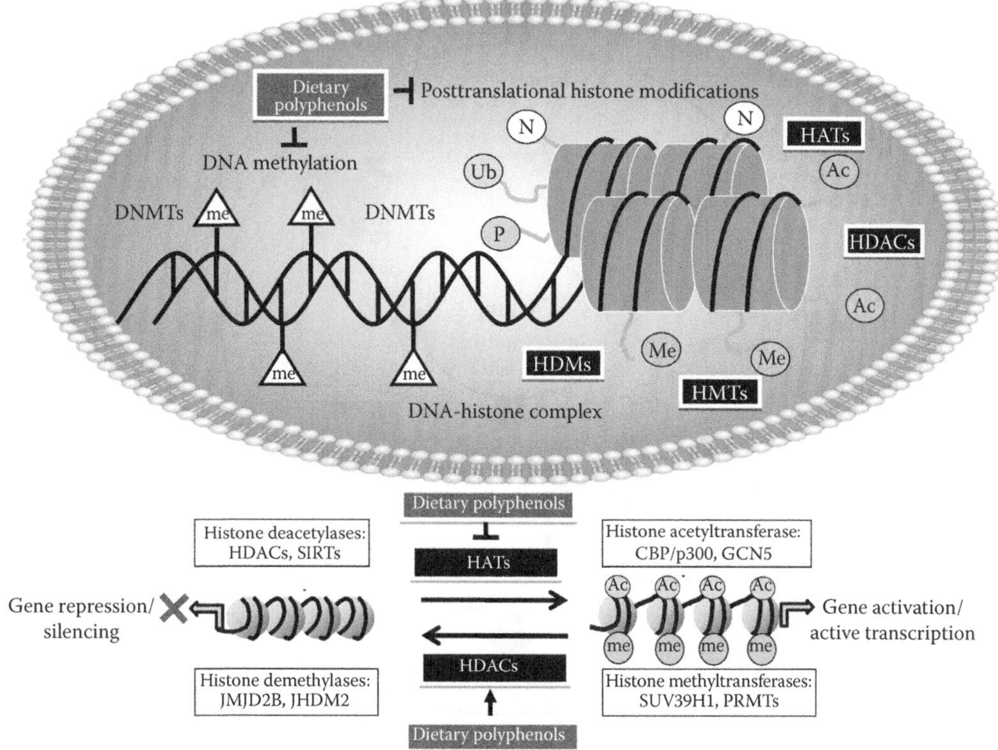

FIGURE 25.1 Regulatory role of dietary polyphenols in epigenetic modifications and gene expression. Epigenetic modifications by chromatin modification enzymes play a vital role in regulation of gene expression. Unwinding and rewinding of DNA is regulated by epigenetic alterations, such as histone acetylation/deacetylation and histone methylation/demethylation. This includes histone acetylation by histone acetyltransferases (HATs), histone deacetylation by histone deacetylases (HDACs), histone methylation by histone methyltransferases (HMTs), DNA methylation by DNA methyltransferases (DNMTs), histone demethylation by histone demethylases (HDMs), and histone phosphorylation and ubiquitination by kinases and ubiqutination enzymes, respectively. These epigenetic modifications result in conformation changes in the chromatin structure that can lead to alterations in DNA accessibility for transcription factors, coactivators, and polymerases, thereby resulting in either gene expression (transcriptional activation) or gene repression (silencing). Dietary polyphenols can affect these epigenetic chromatin modification enzymes specifically to modulate several cellular and molecular events. Ac, Acetylation; Me, Methylation; Ub, Ubiquitination; P, Phosphorylation.

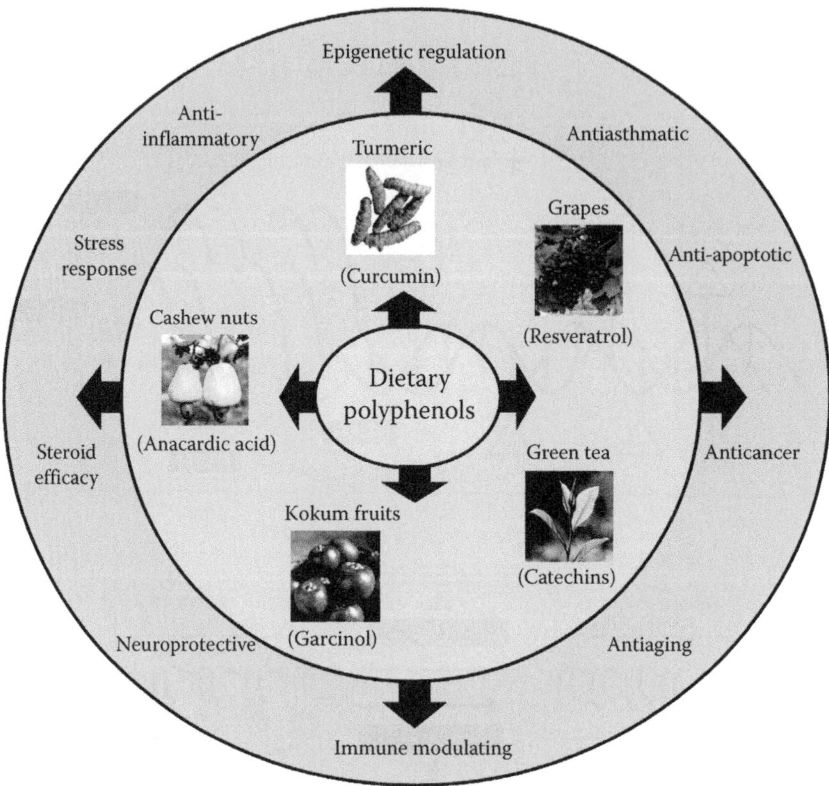

FIGURE 25.2 Dietary polyphenols from different plant sources and their molecular nutriepigenomics functions. Plant-derived polyphenols attenuate epigenetic alterations involved in several processes, such as anti-inflammatory, immune modulation, reversal of steroid resistance, as well as act as antiasthmatic, anticancer, and antiaging agents. The dietary polyphenols obtained from each of these plants are given in parenthesis.

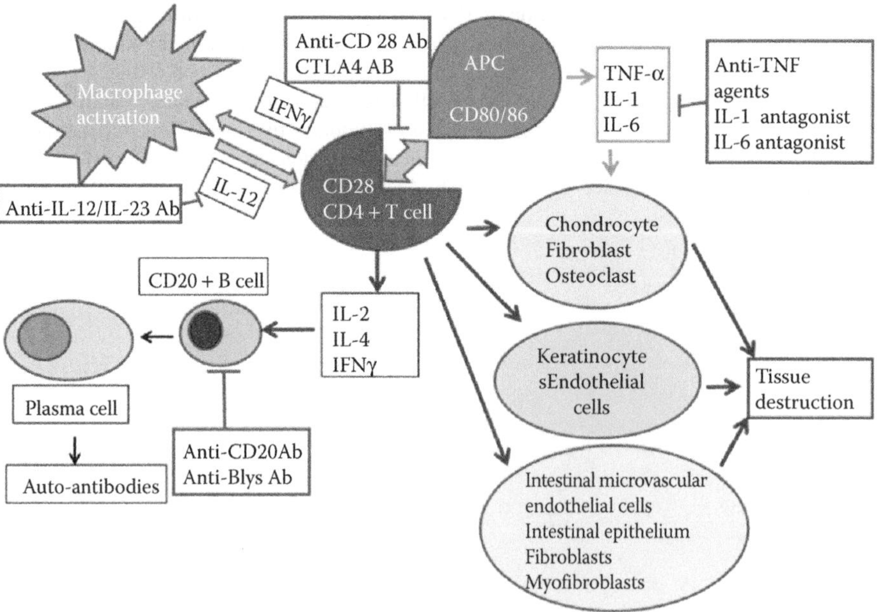

FIGURE 26.1 Different cell types associated with autoinflammation and tissue destruction and the potential targets for biologics.

Neumann E, Gay S, Müller-Ladner U. 2005. The RANK/RANKL/osteoprotegerin system in rheumatoid arthritis: New insights from animal models. *Arthritis Rheum* 52:2960–7.

Nolte D, Schmid P, Jager U, et al. 1994. Leukocyte rolling in venules of striated muscle and skin is mediated by P-selectin, not by L-selectin. *Am J Physiol* 267:H1637–42.

Palmer G, Gabay C, Imhof BA. 2006. Leukocyte migration to rheumatoid joints: Enzymes take over. *Arthritis Rheum* 54:2707–10.

Pap T, Muller-Ladner U, Gay RE. 2000. Fibroblast biology. Role of synovial fibroblasts in the pathogenesis of rheumatoid arthritis. *Arthritis Res* 2:361–7.

Patel DD, Zachariah JP, Whichard LP. 2001. CXCR3 and CCR5 ligands in rheumatoid arthritis synovium. *Clin Immunol* 98:39–45.

Perez DL, Maier H, Nieto E, et al. 2001. Chemokine expression precedes inflammatory cell infiltration and chemokine receptor and cytokine expression during the initiation of murine lupus nephritis. *J Am Soc Nephrol* 12:1369–82.

Phillipson M, Heit B, Colarusso P, et al. 2006. Intraluminal crawling of neutrophils to emigration sites: A molecularly distinct process from adhesion in the recruitment cascade. *J Exp Med* 203:2569–75.

Posthumus MD, Limburg PC, Westra J, et al. 2000. Serum matrix metalloproteinase 3 in early rheumatoid arthritis is correlated with disease activity and radiological progression. *J Rheumatol* 27:2761–68.

Rabquer BJ, Pakozdi A, Michel JE, et al. 2008. Junctional adhesion molecule C mediates leukocyte adhesion to rheumatoid arthritis synovium. *Arthritis Rheum* 58:3020–9.

Ronnblom L, Alm GV. 2003. Systemic lupus erythematosus and the type I interferon system. *Arthritis Res Ther* 5:68–75.

Rönnelid J, Ahlin E, Nilsson B, Nilsson-Ekdahl K, Mathsson L. 2008. Immune complex-mediated cytokine production is regulated by classical complement activation both in vivo and in vitro. *Adv Exp Med Biol* 632:187–201.

Sekine H, Kinser TT, Qiao F, Martinez E, Paulling E, Ruiz P, et al. 2011. The benefit of targeted and selective inhibition of the alternative complement pathway for modulating autoimmunity and renal disease in MRL/lpr mice. *Arthritis Rheum* 63:1076–85.

Sen M, Chamorro M, Reifert J, et al. 2001. Blockade of Wnt-5a/Frizzled-5 signaling inhibits rheumatoid synoviocytes activation. *Arthritis Rheum* 44:772–81.

Sen M, Lauterbach K, El-Gabalawy H, et al. 2000. Expression and function of wingless and frizzled homologs in rheumatoid arthritis. *Proc Natl Acad Sci USA* 97:2791–6.

Sen M, Reifert J, Lauterbach K, et al. 2002. Regulation of fibronectin and metalloproteinase expression by Wnt signaling in rheumatoid arthritis synoviocytes. *Arthritis Rheum.* 46:2867–77.

Sen M. 2005. Wnt signalling in rheumatoid arthritis. *Rheumatology* 44:708–13.

Shin MS, Lee N, Kang I. 2011. Effector T-cell subsets in systemic lupus erythematosus: Update focusing on Th17 cells. *Curr Opin Rheumatol.* Jun 30. [Epub ahead of print.]

Shoenfeld Y, Meroni PL, Cervera R.2008 Antiphospholipid syndrome dilemmas still to be solved: 2008 status. *Ann Rheum Dis* 67:438–42.

Simonet WS, Lacey DL, Dunstan CR, et al. 1997. Osteoprotegerin: A novel secreted protein involved in the regulation of bone density. *Cell* 89:309–19.

Smith CW. 2008. Adhesion molecules and receptors. *J Allergy Clin Immunol* 121:S375–9.

Springer TA. 1995. Traffic signals on endothelium for lymphocyte recirculation and leukocyte emigration. *Annu Rev Physiol* 57:827–72.

Sthoeger ZM, Sharabi A, Dayan M, Zinger H, Asher I, Sela U, et al. 2009. The tolerogenic peptide, hCDR1, down-regulates pathogenic cytokines and apoptosis and up-regulates immune-suppressive molecules and regulatory T cells in peripheral blood mononuclear cells of lupus patients. *Hum Immunol* Jan 28. [Epub ahead of print.]

Strieter RM, Polverini PJ, Kunkel SL, et al. 1995. The functional role of the ELR motif in CXC chemokine-mediated angiogenesis. *J Biol Chem* 270:27348–57.

Takayanagi H, Iizuka H, Juji T, et al. 2000. Involvement of receptor activator of nuclear factor κB ligand osteoclast differentiation factor in osteoclastogenesis from synoviocytes in rheumatoid arthritis. *Arthritis Rheum* 43:259–69.

Thomas W. 2006. For catch bonds, it all hinges on the interdomain region. *J Cell Biol* 174:911–3.

Tolboom TCA, Pieterman E, van der Laan WH, et al. 2002. Invasive properties of fibroblast-like synoviocytes: Correlation with growth characteristics and expression of MMP-1, MMP-3, and MMP-10. *Ann Rheum Dis* 61:975–80.

Toong C, Adelstein S, Phan TG. 2011 Clearing the complexity: Immune complexes and their treatment in lupus nephritis. *Int J Nephrol Renovascular Dis* 4:17–28.

Vega D, Maalouf NM, Sakhaee K. 2007. The role of receptor activator of nuclear factor-kappaB (RANK)/RANK ligand/osteoprotegerin: Clinical implications. *J Clin Endocrinol Metab* 92:4514–21.
von Hundelshausen P, Weber KS, Huo Y, et al. 2001. RANTES deposition by platelets triggers monocyte arrest on inflamed and atherosclerotic endothelium. *Circulation* 103:1772–7.
Wang S, Voisin MB, Larbi KY, et al. 2006. Venular basement membranes contain specific matrix protein low expression regions that act as exit points for emigrating neutrophils. *J Exp Med* 203:1519–32.
Waters ST, McDuffie M, Bagavant H, Deshmukh US, Gaskin F, Jiang C, et al. 2004. Breaking tolerance to double stranded DNA, nucleosome, and other nuclear antigens is not required for the pathogenesis of lupus glomerulonephritis. *J Exp Med* 199:255–64.
Williams EL, Gadola S, Edwards CJ. 2009. Anti-TNF-induced lupus. *Rheumatology (Oxford)* 48:716–20.
Xing Q, Wang B, Su H, Cui J, Li J. 2011. Elevated Th17 cells are accompanied by FoxP3+ Treg cells decrease in patients with lupus nephritis. *Rheumatol Int* Jan 18. [Epub ahead of print.]
Yamamura Y, Gupta R, Morita Y, et al. 2001. Effector function of resting T cells: Activation of synovial fibroblasts. *J Immunol* 166:2270–5.
Yung S, Tsang RC, Sun Y, Leung JK, Chan TM. 2005. Effect of human anti-DNA antibodies on proximal renal tubular epithelial cell cytokine expression: Implications on tubule-interstitial inflammation in lupus nephritis. *J Am Soc Nephrol* 16:3281–94.

17 Shear Stress and Vascular Inflammation

A Study in the Lung

John Noel and Shampa Chatterjee
University of Pennsylvania Medical Center
Philadelphia, Pennsylvania

CONTENTS

17.1 INTRODUCTION

As one of the human body's largest "organs," the endothelium (vascular and capillary system) plays a central role in the homeostasis of organ functions. It functions as a mechanical barrier, regulating fluid movement through the vasculature and transports nutrients to the different organs. The endothelial cells lining the vessel wall are the natural barrier that prevents microorganisms from invading tissues. The endothelial system can be damaged by inflammatory responses associated with infection, trauma, tissue breakdown, tumors, vascular anomalies, and toxicological or immunological responses and activation of the coagulation system. In addition, the endothelial cells are activated by a variety of stimuli. Among them is cyclical strain due to vessel wall distention by transmural pressure and shear stress, the frictional force generated by blood flow.

Shear stress arises from the friction between two virtual layers in a fluid. In vivo shear stress arises at the interface between the blood and endothelial layer where it leads to a deformation of endothelial cells. Long adaptation to different types of shear (laminar, disturbed, and oscillatory) affects the phenotype of endothelial cells. While laminar flow maintains endothelial integrity and function, non-laminar or irregular flow adversely modulates signaling and gene expression that contribute to vascular pathologies. The role of hemodynamic forces in endothelial dysfunction was first apparent when the earliest lesions of atherosclerosis were observed to develop preferentially at arterial branches and curvatures (or bends) [1]. Later studies showed that abnormal shear in these regions leads to local endothelial dysfunction via upregulation of pro-inflammatory genes and proteins. In addition, the endothelium in these regions is also susceptible to external inflammatory stimuli [2].

In the context of the lung, endothelial integrity and function are of paramount importance as the gas exchange occurs through the thin alveolar membrane or the alveolar-capillary barrier that is formed by Type 1 pneumocytes of the alveolar wall, the endothelial cells of the capillaries, and the basement membrane between the two cells. While physiological shear modulates the pulmonary vascular properties and function, chronic and abrupt changes in shear or stasis cause elevations in cytosolic Ca^{2+} [3] and activation of transcription factors [4] that regulate vascular inflammation via regulation of several inflammatory genes. All these changes eventually lead to pulmonary vascular remodeling that is often the key pathological determinant of pulmonary hypertension.

This chapter reviews the manner in which shear stress modulates inflammation in the endothelium. We focus on the molecular mechanisms that modulate inflammatory activation and discuss the resultant functional and structural vascular changes that lead to pulmonary disease.

17.2 HEMODYNAMIC FORCES AND THE QUIESCENT AND ACTIVATED ENDOTHELIUM

Shear stress is defined as the force that liquid flow exerts on the vessel wall. Shear is defined as force per unit area and has units, dynes/cm^2 or Newton/m^2 or Pascal (Pa). Laminar stress or steady shear stress arises from an orderly flow (i.e., constant velocity over time) of the fluid or blood that moves parallel to the vessel wall. Here shear is directly proportional to the flow rate. Laminar shear in a vessel is calculated by the following equation:

$$\text{Shear stress} = 4\ nQ/\pi r^2 l$$

where n = viscosity, Q = flow rate in mL/min, r = radius of the vessel, l = length of the vessel.

Shear stress may be either laminar or pulsatile or disturbed (turbulent or oscillatory). Regions of the arterial tree with uniform geometry are exposed to unidirectional, undisturbed laminar flow. Pulsatile flow is the intermittent flow of fluid or blood through a vessel; in general, blood flow in the arterial system is pulsatile because of the rhythmic property of the cardiac cycle. Disturbed flow occurs when the flow patterns within the blood vessel are random and unpredictable. It is either turbulent—in which case the flow forms vortices, eddies, and wakes—or oscillatory—where blood moves back and forth in responses to an oscillatory pressure gradient. Regions of the arterial tree that has arches and branches are exposed to disturbed flow (turbulent or oscillatory). Experimental measurements using different methods have shown that in humans the magnitude of shear stress ranges from 1 to 6 dynes/cm^2 in the venous system and from 10 to 70 dynes/cm^2 in arteries [5–7]. The magnitude of shear stress in the normal mouse aorta is an order of magnitude higher than in humans [8].

Endothelial cells are specifically equipped with a sensing mechanism to detect shear stress via specialized cellular mechanotransduction elements on the cytoskeleton. The cytoskeleton is coupled to the cellular membrane by elements that include integrins, cell–cell adhesion molecules, and surface expressed receptors. Shear-induced conformational changes in these complexes lead to activation of intracellular signaling molecules (such as GTPases, Ras, Rho, Rab, and Ran GTPases), which eventually leads to transcriptional activation of several target genes.

An endothelium at rest and exposed to laminar flow produces basal amounts of nitric oxide (NO) that contribute to the quiescence of the endothelium and help maintain vessel wall permeability and blood fluidity. While the former is maintained by intact endothelial barrier function [9, 10], the latter is controlled by endothelial cell dependent mechanisms that inhibit coagulation in the vascular system [11]. Activation of coagulation requires a phospholipid surface, enriched in phosphatidylserine, which is normally provided by activated platelets. Resting endothelial cells, by producing NO and prostaglandin I_2, inhibit platelet adhesion to the endothelium. Resting or quiescent endothelial cells also suppress adhesion molecules like E-selectin, vascular cell adhesion molecule (VCAM), intercellular adhesion molecule (ICAM), and leukocyte interactive protein P-selectin.

However, disturbances in blood flow give rise to turbulent or oscillatory flow, which causes reduction in NO production by endothelial cells and contributes to vasoconstriction. In addition, stasis or cessation of blood flow produces reactive oxygen species (ROS) [12, 13] that leads to endothelial activation. This implies a transcriptional increase of adhesion molecules (CAM) on the endothelial surface and the synthesis of cytokines, chemokines, and leukocyte adhesion molecule P-selectin. All these lead to localized recruitment and activation of circulating leukocytes and neutrophils to endothelial cells.

17.3 DISTURBED FLOW AND INFLAMMATORY RESPONSES OF ENDOTHELIAL CELLS

The previous section discussed that laminar blood flow acts on endothelial cells to generate molecules that promote a vasodilatory, anticoagulant, anti-inflammatory, and growth-inhibitory surface [14–18]. However, when blood flow is disturbed as is likely to occur at higher flow velocities, this creates sites of abnormally low and high shear stress. This is sensed by the endothelium and results in activation of a signaling cascade that leads to a procoagulant surface, production of inflammatory cytokines, and endothelial and smooth muscle cell proliferation, leading to thickening and remodeling of the vascular wall [18–22].

17.3.1 Effects of Laminar and Disturbed Flow on Endothelial Signaling and Function

Shear stress regulated gene expression occurs via activation of transcriptional factors such as nuclear factor kappa B (NF-kB), activator protein 1 (AP-1), kruppel-like factor 2 (KLF2), and nuclear respiratory factor 2 (Nrf2) [23–25]. Of these, NF-kB and AP-1 are well known to regulate vascular inflammation. Indeed, pro-inflammatory genes such as VCAM-1, E-selectin, and interleukin-8 (IL-8) require binding of both NF-kB and AP-1 for transcriptional activation. Overexpression of transcription factors KLF2 and Nrf2 also modulate pro-inflammatory signaling pathways.

Effect of shear on endothelial cells has been studied extensively in vitro. In cultured endothelial cells, onset of shear for short periods (5 min–1 h) activates NF-kB [26, 27] and triggers the expression of ICAM-1, VCAM-1, and E-selectin [28–30], while prolonged laminar shear (6–24 h) induces anti-inflammatory effects associated with alteration of NF-kB activation [31]. Laminar flow also upregulates the expression or activity of cytoprotective genes/proteins, including the complement-inhibitory proteins clusterin [32] and antioxidant enzymes superoxide dismutase (SOD), heme oxygenase-1 (HO-1) [33, 34], and NADPH quinone oxidoreductase 1 (NQO1) [34]. Overall the protective effects of laminar shear stress are mediated through the inhibition of cytokine-induced signaling and gene expression in endothelial cells [31, 35–38] and apoptosis in endothelium [39, 40].

Cells exposed to disturbed flow show high levels of activations of transcription factors such as NF-kB, Spl, Egr-1; adhesion molecules such as ICAM-1 VCAM-1, and E-selectin [41–44]; chemokines such as monocyte chemoattractant protein-1 (MCP-1); and matrix remodeling proteins such as matrix metalloproteinase-9 (MMP-9) [45–47]. In addition, other disruptors of endothelial function such as procoagulant molecules like tissue factor (TF) [48], vasoconstrictors such as endothelin-1 (ET-1) [49], and pro-oxidant sources such as bone morphogenic protein (BMP-4) [50] and gp91phox (a major component of NADPH oxidase complex) [51] are also upregulated by disturbed flow. Thus the upregulation of a large number of oxidant and pro-inflammatory genes occurs in endothelial cells in response to low or disturbed flow, and this is either absent or transient in cells subjected to steady or pulsatile laminar flow.

In general, endothelial signaling with disturbed flow leads to endothelial dysfunction that has the following manifestations:

- Increased expression of adhesion molecules ICAM, VCAM, E-selectin/P-selectin, and chemotactic molecules like MCP-1
- Increased recruitment of monocytes/macrophages/neutrophils into the vascular wall leading to intimal thickening and formation of plaques
- Increased permeability of the vasculature to macromolecules
- Irregular growth and proliferation of vascular cells
- Disturbed hemostatic balance, that is, increased expression of procoagulatory molecules such as tissue factor (TF), enhanced thrombin generation, and platelet aggregation

17.3.2 Effects of Disturbed Flow on Endothelial-Leukocyte Interactions

Attachment of circulating leukocytes to the endothelium, and their subsequent adhesion and transmigration through the endothelial monolayer, are the critical stages of endothelial dysfunction. This occurs through four stages: (1) selectin-dependent tethering and rolling of the leukocyte, (2) adhesion of the leukocyte via chemokines and adhesion molecules, (3) integrin dependent adhesion, and (4) transmigration across endothelium.

As noted in the earlier section, disturbed flow activates transcription factors NF-kB and AP-1, which drives expression of chemokines, cytokines, and adhesion molecules in endothelial cells. At the same time, shear induces rises in intracellular Ca^{2+} in these cells causing the formation of a Ca^{2+}–calmodulin complex that activates myosin light chain kinase (MLCK), which phosphorylates myosin light chain (MLC) [52]. Phosphorylated MLC initiates contraction of actin filaments that are attached to tight junction and adherens junction proteins, resulting in the opening of gaps between adjacent endothelial cells for transmigration of leukocytes. MLC also triggers the exocytosis of Weibel Palade bodies, bringing P-selectin to the luminal surface [10, 53]. At the same time, lysophosphatidylcholine, a by-product of arachidonic acid generation, is rapidly acetylated, generating an endothelial-cell derived acyl form of platelet-activating factor (PAF) [54]. P-selectin and PAF on the endothelial luminal plasma membrane cause the tethering of circulating neutrophils. This is followed by integrin activation that initiates neutrophil extravasation [55].

Leukocyte transmigration further amplifies endothelial inflammation and injury. After the initial leukocyte recruitment following pro-inflammatory responses, activated leukocytes attached to the endothelium produce cytokines, tumor necrosis factor α (TNF-α), and interleukin 1 (IL-1) [56]. TNF binds to its receptor (TNFR) and initiates (via TNFR associated death domain protein, TRADD) a signaling cascade that leads to new gene transcription mediated by NF-kB and AP-1 [57]. Thus transcription and translation of new proteins such as chemokine ligands CCL-2 occur. At this stage leukocyte recruitment is more effective. Overall, TNF and IL-1, often exacerbate endothelial cell injury by triggering endothelial cell death [56–58].

17.3.3 Effects of Disturbed Flow In Vivo

Disturbed flow occurs in the vascular system in the following regions: (1) in branch points of large arteries, (2) in regions where bypass grafts and stents have altered laminar flow, and (3) in regions where there is sudden cessation of flow and restart of the flow as would occur with ischemia/ischemia-reperfusion associated with thrombi or organ transplantation.

Several in vivo observations have shown a correlation between branch points and atherosclerotic lesions in the vascular tree in the aortic arch [59, 60], carotid bifurcation [61, 62], femoral artery [63], and branch points of the coronary artery [64]. This site selectivity is due to the modification of endothelial function by the pattern of disturbed blood flow or shear in these regions. In a mouse model, partial ligation of the carotid artery also showed inflammatory and atheroprone genes in regions of disturbed flow [65]. Ni et al. identified mechanosensitive genes in vivo and showed that (based on global gene expression profiles) disturbed flow for 12 h regulated genes involved in cell proliferation and growth; by 48 h genes that are involved in inflammation and immunologic disease such as inflammatory cytokines (CCL-11, CCL-4, CXCL-12, and CXCL-16), adhesion molecules (ICAM-1 and VCAM-1), and transcription factors (KLF2, KLF4) were induced [66]. Another in vivo study using endothelium isolated from disturbed flow regions that used microdissection and antisense RNA amplification methods demonstrated by DNA that disturbed flow regions exhibit a greater degree of heterogeneity in gene expression than those from laminar flow regions and that the heterogeneity may contribute to the initiation of atherosclerotic lesions [67].

Regions of disturbed flow in vivo were also found to have high transcriptional activity of NF-kB. Studies have shown activated NF-kB in the nucleus of endothelium from aortic arch in mice and pigs, suggesting that these regions may be primed for enhanced NF-kB activation [68, 69].

These findings have led to the concept that the disturbed flow pattern in branch points and curvatures causes the preferential localization of atherosclerotic lesions.

17.4 INFLAMMATION IN THE PULMONARY ENDOTHELIUM

The pulmonary vascular tree is the largest vascular bed in the human body, and the pulmonary endothelium coats this vascular system, forming a continuous, semipermeable barrier between blood and tissue. The pulmonary endothelium is responsible for maintaining adequate cardiovascular homeostasis and vascular tone. The cytoskeleton of the endothelium is connected to intercellular protein bridges called tight and adherens junctions and contact sites to the substratum called focal adhesions. Endothelial fluid permeability is controlled by these structures. Under conditions of inflammation, endothelial cells present chemotactic substances and adhesion molecules on their surface, along with production of oxidants either by the endothelial cells or by leukocytes or polymorphonuclear leukocytes (PMN) in the vicinity of these cells. Pro-inflammatory signals cause increased endothelial cell cytosolic Ca^{2+} concentration that initiates a signaling cascade (via activation of MLCK) that causes contraction of the tight and adherens junctions leading to disruption in the endothelial barrier [10, 52, 53]. In addition, Ca^{2+} triggers the expression of cell surface adhesion molecules, enabling docking of circulating PMNs to endothelial cells and their transmigration in the interstitium and air space. In the lung, these Ca^{2+} signals can be propagated along interconnected endothelial cells via gap junctions [70], making it more permissive or permeable to the spread of inflammation over large areas of the lung.

17.4.1 Effect on Neutrophil Recruitment and on Pulmonary Vascular Permeability

The pulmonary circulation has a higher concentration of neutrophils as compared to the systemic blood [71]. This is because neutrophils stick during transit through narrow pulmonary capillaries. Inflammatory stimuli increase the number of neutrophils in the pulmonary circulation; these neutrophils adhere to the endothelium and then transmigrate through the inter-endothelial gaps into the epithelial basement membrane and finally into the alveolar air space. Thus neutrophil recruitment by an activated endothelium is followed by increase in inter-endothelial space and thus increased pulmonary vascular permeability. Permeability leads to edema in the lung [72, 73] and results in loss of pulmonary function and can be a life-threatening condition.

How does the pulmonary microvascular endothelium maintain its barrier function? Transmembrane tight and adherens junctions are the main structures that maintain the intactness of the pulmonary endothelium. These junctions are balanced by contractile forces generated by actomyosin interaction that tend to pull the tightly connected cells apart. A balance between these adhesive and contractile forces maintains the endothelial monolayer in a semipermeable status. A disruption of the equilibrium results in endothelial barrier dysfunction and subsequent microvascular leakage. The physical process of "pulling" endothelial cells apart is a downstream event in multiple signaling pathways activated by inflammatory mediators.

Activated neutrophils disrupt the pulmonary endothelial barrier by generating oxidants (reactive oxygen species and free radicals) that cause tissue damage and injury [74, 75].

17.4.2 Interaction of Platelets with Endothelial Cells and Leukocytes

Under normal conditions, platelets circulate without interacting with quiescent endothelium. However, when the endothelium is activated due to inflammatory response, platelets bind to endothelial cells through P-selectin. Adherent platelets release oxidants and inflammatory mediators such as interleukin-1β (IL-1β), transforming growth factor-β (TGF-β), and platelet derived growth factor (PDGF) [76, 77]. Both PDGF and IL-1 play a role in recruitment of leukocytes and PMN and cause oxidative burst of neutrophils, thus further amplifying the injury.

17.4.3 Inflammation in Pulmonary Hypertension

Pulmonary hypertension (PH) is characterized by elevations in pulmonary artery pressure and pulmonary vascular resistance. The two most common features of PH are abnormal pulmonary vasoconstriction and alterations in pulmonary vascular structure due to remodeling of the blood vessel. The thickenings of the various layers that make the vascular wall arise from hypertrophy (cell growth) and hyperplasia (proliferation).

It is well accepted that several factors, such as mechanical stimuli (stretch, strain) [78], growth factors [79, 80], and inflammatory cytokines propel the remodeling of pulmonary arteries in pulmonary hypertension. That inflammation is an important factor for PH is now well recognized after the detection of inflammatory cells (macrophages, polymorphonuclear neutrophils, lymphocytes, mast cells) in the pulmonary vasculature in the vicinity of pulmonary vessels of patients with pulmonary hypertension (PPH) [81]. In addition, serum concentrations of the inflammatory cytokines, interleukin-1 and interleukin-6, were found to be increased with PH [82].

17.5 THE PULMONARY ENDOTHELIUM AS A THERAPEUTIC TARGET

Pulmonary vascular remodeling arises from endothelial dysfunction that is initiated either by altered shear as in the case of thrombi/occlusive pathologies or by inflammatory molecules. These pathologies cause abnormal vascular proliferation and growth of apoptosis-resistant endothelial cells leading to pulmonary vasoconstriction. Targeting the endothelium to restore its normal function can be achieved by multiple strategies:

- Blocking inflammation-associated signaling by treating with antibodies and inhibitors. Administration of antibody against αvβ5 was found to protect the lungs in vivo post inflammation [83].
- Targeting of genes or proteins with endothelial cell protective properties. Rat dermal fibroblasts transfected in vitro with the gene coding for angiopoietin-1 when injected intravenously to the intact animal protected the lung from subsequent inflammation injury because of the endothelial barrier sealing effect of angiopoietin-1 [84].
- Targeting oxidants produced post inflammation (by PMN and endothelium) in the lung. Antioxidant enzyme catalase conjugated to monoclonal antibodies against rat ACE; rat treated with this conjugate exhibited attenuation of lung injury post ischemia [85].
- Blocking MLC phosphorylation by pharmacological inhibition of myosin light chain kinase (MLCK) and Rho kinase to prevent stress fiber formation and endothelial barrier disruption [86].

- Endothelial progenitor cell (EPC) therapy for regeneration of the pulmonary vascular bed. EPCs are thought to have proliferative potential and may function to replace and/or restore damaged endothelial cells. Data from a prospective trial [87] reports that intravenously infused EPCs showed greater hemodynamic improvement (pulmonary arterial pressure, cardiac output, and pulmonary vascular resistance) in PH patients as compared to conventional therapy.

17.6 CONCLUSIONS

Shear stress arising from blood flow triggers mechanosensitive cell signaling events that maintain endothelial structure and function; however, regions of disturbed flow have altered signaling that propels an inflammatory phenotype in the endothelium. This leads to oxidant production, expression of adhesion molecules, cytokines, and chemokines that accelerate endothelial dysfunction. In the systemic vasculature, this can lead to atherosclerotic lesions, while in the pulmonary vessels, inflammation can lead to remodeling. There is compelling evidence linking dysfunctional pulmonary endothelial cells in the pathobiology of severe PH.

One of the major challenges for vascular and pulmonary scientists today is to develop an understanding of the role of inflammation in vascular remodeling and be able to dissect the role of transcription factors, adhesion molecules, cytokines, and chemokines, as well as the participation by the various players, the endothelium, leukocytes/PMN, and platelets and the various phases of their response. This will enable the development of a new therapeutic approach to inflammation that focuses on vascular endothelial cell responses.

TAKE-HOME MESSAGES

- Shear stress associated with laminar blood promotes a vasodilatory, anticoagulant, anti-inflammatory, and growth-inhibitory phenotype in endothelial cells.
- Regions of disturbed blood flow (arising at branch points in major blood vessels or with cessation and restart of flow) show preferential localization of atherosclerotic plaques or lesions.
- Mechanosensitive genes on the endothelium "sense" blood flow; disturbed flow for 12 h regulates genes involved in cell proliferation and growth. Longer periods of disturbed flow upregulate inflammatory gene expression.
- Inflammatory cytokines (CCL-11, CCL-4, CXCL-12, and CXCL-16), adhesion molecules (ICAM-1 and VCAM-1), and transcription factors (NF-kB, AP-1, KLF2, KLF4, Nrf2) that are induced with disturbed flow, drive an inflammatory phenotype.
- In the context of the pulmonary vasculature, an inflammatory phenotype causes platelets, leukocytes, and neutrophils to adhere to the endothelium, leading to oxidant injury and further release of inflammatory mediators.
- Inflammation in the pulmonary vascular tree is an important factor in driving pulmonary remodeling and eventually pulmonary hypertension.
- Blocking inflammation-associated signaling on the pulmonary and systemic endothelium can be a therapeutic strategy to combat endothelial dysfunction and vascular disease.

REFERENCES

1. Davies PF: Flow-mediated endothelial mechanotransduction, *Physiological Reviews* 1995, 75:519–560.
2. Sheikh S, Rainger GE, Gale Z, Rahman M, Nash GB: Exposure to fluid shear stress modulates the ability of endothelial cells to recruit neutrophils in response to tumor necrosis factor-alpha: A basis for local variations in vascular sensitivity to inflammation, *Blood* 2003, 102:2828–2834.

3. Brakemeier S, Eichler I, Hopp H, Kohler R, Hoyer J: Up-regulation of endothelial stretch-activated cation channels by fluid shear stress, *Cardiovasc Res* 2002, 53:209–218.
4. Sumpio BE, Chang R, Xu WJ, Wang XJ, Du W: Regulation of tPA in endothelial cells exposed to cyclic strain: Role of CRE, AP-2, and SSRE binding sites, *Am J Physiol Cell Physiol* 1997, 273:C1441–1448.
5. Chien S: Present state of blood rheology. In *Hemodilution: Theoretical Basis and Clinical Application.* Edited by Messmer K, Schmid-Schoenbein H. Basel, Switzerland: Karger, 1972.
6. Malek AM, Alper SL, Izumo S: Hemodynamic shear stress and its role in atherosclerosis, *JAMA* 1999, 282:2035–2042.
7. Nerem RM: Vascular fluid mechanics, the arterial wall, and atherosclerosis, *J Biomech Eng* 1992, 114:274–282.
8. Suo J, Ferrara DE, Sorescu D, Guldberg RE, Taylor WR, Giddens DP: Hemodynamic shear stresses in mouse aortas: Implications for atherogenesis, *Arterioscler Thromb Vasc Biol* 2007, 27:346–351.
9. Minshall RD, Malik AB: Transport across the endothelium: Regulation of endothelial permeability, *Handb Exp Pharmacol* 2006, 107–144.
10. Birch KA, Pober JS, Zavoico GB, Means AR, Ewenstein BM: Calcium/calmodulin transduces thrombin-stimulated secretion: Studies in intact and minimally permeabilized human umbilical vein endothelial cells, *J Cell Biol* 1992, 118:1501–1510.
11. Arnout J, Hoylaerts MF, Lijnen HR: Haemostasis, *Handb Exp Pharmacol* 2006, 176:1–41.
12. Chatterjee S, Chapman KE, Fisher AB: Lung ischemia: A model for endothelial mechanotransduction, *Cell Biochem Biophys* 2008, 52:125–138.
13. Chatterjee S, Fisher, AB: Shear stress, cell signaling and pulmonary vascular remodeling. In *Textbook of Pulmonary Vascular Disease.* Edited by JXJ Yuan, CA Hales, S Rich, SL Archer, JB West. New York: Springer Press, 2010, p. 787.
14. Davies PF: Flow-mediated endothelial mechanotransduction, *Physiol Rev* 1995, 75:519–560.
15. Uematsu M, Ohara Y, Navas JP, Nishida K, Murphy TJ, Alexander RW, et al.: Regulation of endothelial cell nitric oxide synthase mRNA expression by shear stress, *Am J Physiol Cell Physiol* 1995, 269:C1371–1378.
16. Diamond SL, Eskin SG, McIntire LV: Fluid flow stimulates tissue plasminogen activator secretion by cultured human endothelial cells, *Science* 1989, 243:1483–1485.
17. Malek A, Izumo S: Physiological fluid shear stress causes downregulation of endothelin-1 mRNA in bovine aortic endothelium, *Am J Physiol* 1992, 263:C389–396.
18. Nerem RM, Alexander RW, Chappell DC, Medford RM, Varner SE, Taylor WR: The study of the influence of flow on vascular endothelial biology, *Am J Med Sci* 1998, 316:169–175.
19. Gimbrone MA, Jr., Topper JN, Nagel T, Anderson KR, Garcia-Cardena G: Endothelial dysfunction, hemodynamic forces, and atherogenesis, *Ann N Y Acad Sci* 2000, 902:230–239; discussion 239–240.
20. Dirksen MT, van der Wal AC, van den Berg FM, van der Loos CM, Becker AE: Distribution of inflammatory cells in atherosclerotic plaques relates to the direction of flow, *Circulation* 1998, 98:2000–2003.
21. Tsao PS, Buitrago R, Chan JR, Cooke JP: Fluid flow inhibits endothelial adhesiveness. Nitric oxide and transcriptional regulation of VCAM-1, *Circulation* 1996, 94:1682–1689.
22. Ueba H, Kawakami M, Yaginuma T: Shear stress as an inhibitor of vascular smooth muscle cell proliferation. Role of transforming growth factor-beta 1 and tissue-type plasminogen activator, *Arterioscler Thromb Vasc Biol* 1997, 17:1512–1516.
23. Davies PF, Polacek DC, Shi C, Helmke BP: The convergence of haemodynamics, genomics, and endothelial structure in studies of the focal origin of atherosclerosis, *Biorheology* 2002, 39:299–306.
24. McCormick SM, Frye SR, Eskin SG, Teng CL, Lu CM, Russell CG, et al.: Microarray analysis of shear stressed endothelial cells, *Biorheology* 2003, 40:5–11.
25. Lehoux S, Tedgui A: Cellular mechanics and gene expression in blood vessels, *J Biomech* 2003, 36:631–643.
26. Lan Q, Mercurius KO, Davies PF: Stimulation of transcription factors NF kappa B and AP1 in endothelial cells subjected to shear stress, *Biochem Biophys Res Commun* 1994, 201:950–956.
27. Davis ME, Cai H, Drummond GR, Harrison DG: Shear stress regulates endothelial nitric oxide synthase expression through c-Src by divergent signaling pathways, *Circ Res* 2001, 89:1073–1080.
28. Morigi M, Micheletti G, Figliuzzi M, Imberti B, Karmali MA, Remuzzi A, et al.: Verotoxin-1 promotes leukocyte adhesion to cultured endothelial cells under physiologic flow conditions, *Blood* 1995, 86:4553–4558.
29. Nagel T, Resnick N, Atkinson WJ, Dewey CF, Jr., Gimbrone MA, Jr.: Shear stress selectively upregulates intercellular adhesion molecule-1 expression in cultured human vascular endothelial cells, *J Clin Invest* 1994, 94:885–891.

30. Gonzales RS, Wick TM: Hemodynamic modulation of monocytic cell adherence to vascular endothelium, *Ann Biomed Eng* 1996, 24:382–393.
31. Partridge J, Carlsen H, Enesa K, Chaudhury H, Zakkar M, Luong L, et al.: Laminar shear stress acts as a switch to regulate divergent functions of NF-kappaB in endothelial cells, *FASEB J* 2007, 21:3553–3561.
32. Urbich C, Fritzenwanger M, Zeiher AM, Dimmeler S: Laminar shear stress upregulates the complement-inhibitory protein clusterin: A novel potent defense mechanism against complement-induced endothelial cell activation, *Circulation* 2000, 101:352–355.
33. Chen XL, Varner SE, Rao AS, Grey JY, Thomas S, Cook CK, et al.: Laminar flow induction of antioxidant response element-mediated genes in endothelial cells. A novel anti-inflammatory mechanism, *J Biol Chem* 2003, 278:703–711.
34. Hosoya T, Maruyama A, Kang MI, Kawatani Y, Shibata T, Uchida K, et al.: Differential responses of the Nrf2-Keap1 system to laminar and oscillatory shear stresses in endothelial cells, *J Biol Chem* 2005, 280:27244–27250.
35. Berk BC: Atheroprotective signaling mechanisms activated by steady laminar flow in endothelial cells, *Circulation* 2008, 117:1082–1089.
36. Ni CW, Hsieh HJ, Chao YJ, Wang DL: Interleukin-6-induced JAK2/STAT3 signaling pathway in endothelial cells is suppressed by hemodynamic flow, *Am J Physiol Cell Physiol* 2004, 287:C771–780.
37. Tsai YC, Hsieh HJ, Liao F, Ni CW, Chao YJ, Hsieh CY, et al.: Laminar flow attenuates interferon-induced inflammatory responses in endothelial cells, *Cardiovasc Res* 2007, 74:497–505.
38. Pan S: Molecular mechanisms responsible for the atheroprotective effects of laminar shear stress, *Antioxid Redox Signal* 2009, 11:1669–1682.
39. Garin G, Abe J, Mohan A, Lu W, Yan C, Newby AC, et al.: Flow antagonizes TNF-alpha signaling in endothelial cells by inhibiting caspase-dependent PKC zeta processing, *Circ Res* 2007, 101:97–105.
40. Pi X, Yan C, Berk BC: Big mitogen-activated protein kinase (BMK1)/ERK5 protects endothelial cells from apoptosis, *Circ Res* 2004, 94:362–369.
41. Mohan S, Mohan N, Valente AJ, Sprague EA: Regulation of low shear flow-induced HAEC VCAM-1 expression and monocyte adhesion, *Am J Physiol Cell Physiol* 1999, 276:C1100–1107.
42. Yun S, Dardik A, Haga M, Yamashita A, Yamaguchi S, Koh Y, et al.: Transcription factor Sp1 phosphorylation induced by shear stress inhibits membrane type 1-matrix metalloproteinase expression in endothelium, *J Biol Chem* 2002, 277:34808–34814.
43. Dardik A, Chen L, Frattini J, Asada H, Aziz F, Kudo FA, et al.: Differential effects of orbital and laminar shear stress on endothelial cells, *J Vasc Surg* 2005, 41:869–880.
44. Chappell DC, Varner SE, Nerem RM, Medford RM, Alexander RW: Oscillatory shear stress stimulates adhesion molecule expression in cultured human endothelium, *Circ Res* 1998, 82:532–539.
45. Hsiai TK, Cho SK, Wong PK, Ing M, Salazar A, Sevanian A, et al.: Monocyte recruitment to endothelial cells in response to oscillatory shear stress, *FASEB J* 2003, 17:1648–1657.
46. Gambillara V, Montorzi G, Haziza-Pigeon C, Stergiopulos N, Silacci P: Arterial wall response to ex vivo exposure to oscillatory shear stress, *J Vasc Res* 2005, 42:535–544.
47. Magid R, Murphy TJ, Galis ZS: Expression of matrix metalloproteinase-9 in endothelial cells is differentially regulated by shear stress. Role of c-Myc, *J Biol Chem* 2003, 278:32994–32999.
48. Mazzolai L, Silacci P, Bouzourene K, Daniel F, Brunner H, Hayoz D: Tissue factor activity is upregulated in human endothelial cells exposed to oscillatory shear stress, *Thromb Haemost* 2002, 87:1062–1068.
49. Qiu Y, Tarbell JM: Interaction between wall shear stress and circumferential strain affects endothelial cell biochemical production, *J Vasc Res* 2000, 37:147–157.
50. Sorescu GP, Song H, Tressel SL, Hwang J, Dikalov S, Smith DA, et al.: Bone morphogenic protein 4 produced in endothelial cells by oscillatory shear stress induces monocyte adhesion by stimulating reactive oxygen species production from a nox1-based NADPH oxidase, *Circ Res* 2004, 95:773–779.
51. Hwang J, Saha A, Boo YC, Sorescu GP, McNally JS, Holland SM, et al.: Oscillatory shear stress stimulates endothelial production of O2- from p47phox-dependent NAD(P)H oxidases, leading to monocyte adhesion, *J Biol Chem* 2003, 278:47291–47298.
52. Stevens T, Garcia JG, Shasby DM, Bhattacharya J, Malik AB: Mechanisms regulating endothelial cell barrier function, *Am J Physiol Lung Cell Mol Physiol* 2000, 279:L419–422.
53. Birch KA, Ewenstein BM, Golan DE, Pober JS: Prolonged peak elevations in cytoplasmic free calcium ions, derived from intracellular stores, correlate with the extent of thrombin-stimulated exocytosis in single human umbilical vein endothelial cells, *J Cell Physiol* 1994, 160:545–554.
54. Prescott SM, Zimmerman GA, McIntyre TM: Human endothelial cells in culture produce platelet-activating factor (1-alkyl-2-acetyl-sn-glycero-3-phosphocholine) when stimulated with thrombin, *Proc Natl Acad Sci USA* 1984, 81:3534–3538.

55. Lorant DE, Patel KD, McIntyre TM, McEver RP, Prescott SM, Zimmerman GA: Coexpression of GMP-140 and PAF by endothelium stimulated by histamine or thrombin: A juxtacrine system for adhesion and activation of neutrophils, *J Cell Biol* 1991, 115:223–234.
56. Pober JS, Cotran RS: The role of endothelial cells in inflammation, *Transplantation* 1990, 50:537–544.
57. Pober JS: Tumor Necrosis Factor. In *Endothelial Biomedicine*. Edited by Aird WC. Cambridge: Cambridge University Press, 2007, p. 261.
58. Petrache I, Birukova A, Ramirez SI, Garcia JG, Verin AD: The role of the microtubules in tumor necrosis factor-alpha-induced endothelial cell permeability, *Am J Respir Cell Mol Biol* 2003, 28:574–581.
59. Caro CG, Fitz-Gerald JM, Schroter RC: Arterial wall shear and distribution of early atheroma in man, *Nature* 1969, 223:1159–1160.
60. Kilner PJ, Yang GZ, Mohiaddin RH, Firmin DN, Longmore DB: Helical and retrograde secondary flow patterns in the aortic arch studied by three-directional magnetic resonance velocity mapping, *Circulation* 1993, 88:2235–2247.
61. Carallo C, Irace C, Pujia A, De Franceschi MS, Crescenzo A, Motti C, et al.: Evaluation of common carotid hemodynamic forces. Relations with wall thickening, *Hypertension* 1999, 34:217–221.
62. Gnasso A, Irace C, Carallo C, De Franceschi MS, Motti C, Mattioli PL, et al.: In vivo association between low wall shear stress and plaque in subjects with asymmetrical carotid atherosclerosis, *Stroke* 1997, 28:993–998.
63. Moore JE, Jr., Xu C, Glagov S, Zarins CK, Ku DN: Fluid wall shear stress measurements in a model of the human abdominal aorta: Oscillatory behavior and relationship to atherosclerosis, *Atherosclerosis* 1994, 110:225–240.
64. Stone PH, Coskun AU, Kinlay S, Popma JJ, Sonka M, Wahle A, et al.: Regions of low endothelial shear stress are the sites where coronary plaque progresses and vascular remodelling occurs in humans: An in vivo serial study, *Eur Heart J* 2007, 28:705–710.
65. Nam D, Ni CW, Rezvan A, Suo J, Budzyn K, Llanos A, et al.: Partial carotid ligation is a model of acutely induced disturbed flow, leading to rapid endothelial dysfunction and atherosclerosis, *Am J Physiol Heart Circ Physiol* 2009, 297:H1535–1543.
66. Ni CW, Qiu H, Rezvan A, Kwon K, Nam D, Son DJ, et al.: Discovery of novel mechanosensitive genes in vivo using mouse carotid artery endothelium exposed to disturbed flow, *Blood* 2010, 116:e66–73.
67. Davies PF, Shi C, Depaola N, Helmke BP, Polacek DC: Hemodynamics and the focal origin of atherosclerosis: A spatial approach to endothelial structure, gene expression, and function, *Ann N Y Acad Sci* 2001, 947:7–16; discussion 16–17.
68. Passerini AG, Polacek DC, Shi C, Francesco NM, Manduchi E, Grant GR, et al.: Coexisting proinflammatory and antioxidative endothelial transcription profiles in a disturbed flow region of the adult porcine aorta, *Proc Natl Acad Sci USA* 2004, 101:2482–2487.
69. Parhar K, Ray A, Steinbrecher U, Nelson C, Salh B: The p38 mitogen-activated protein kinase regulates interleukin-1beta-induced IL-8 expression via an effect on the IL-8 promoter in intestinal epithelial cells, *Immunology* 2003, 108:502–512.
70. Parthasarathi K, Ichimura H, Monma E, Lindert J, Quadri S, Issekutz A, et al.: Connexin 43 mediates spread of Ca2+-dependent proinflammatory responses in lung capillaries, *J Clin Invest* 2006, 116:2193–2200.
71. Doerschuk CM, Mizgerd JP, Kubo H, Qin L, Kumasaka T: Adhesion molecules and cellular biomechanical changes in acute lung injury: Giles F. Filley Lecture, *Chest* 1999, 116:37S–43S.
72. Groeneveld AB: Vascular pharmacology of acute lung injury and acute respiratory distress syndrome, *Vascul Pharmacol* 2002, 39:247–256.
73. Bernard GR, Artigas A, Brigham KL, Carlet J, Falke K, Hudson L, et al.: The American-European Consensus Conference on ARDS. Definitions, mechanisms, relevant outcomes, and clinical trial coordination, *Am J Respir Crit Care Med* 1994, 149:818–824.
74. Flaishon R, Szold O, Weinbroum AA: Acute lung injury following pancreas ischaemia-reperfusion: Role of xanthine oxidase, *Eur J Clin Invest* 2006, 36:831–837.
75. Usatyuk PV, Parinandi NL, Natarajan V: Redox regulation of 4-hydroxy-2-nonenal-mediated endothelial barrier dysfunction by focal adhesion, adherens, and tight junction proteins, *J Biol Chem* 2006, 281:35554–35566.
76. Barry OP, Pratico D, Lawson JA, FitzGerald GA: Transcellular activation of platelets and endothelial cells by bioactive lipids in platelet microparticles, *J Clin Invest* 1997, 99:2118–2127.
77. Piccardoni P, Evangelista V, Piccoli A, de Gaetano G, Walz A, Cerletti C: Thrombin-activated human platelets release two NAP-2 variants that stimulate polymorphonuclear leukocytes, *Thromb Haemost* 1996, 76:780–785.

78. Kolpakov V, Rekhter MD, Gordon D, Wang WH, Kulik TJ: Effect of mechanical forces on growth and matrix protein synthesis in the in vitro pulmonary artery. Analysis of the role of individual cell types, *Circ Res* 1995, 77:823–831.
79. Dempsey EC, Stenmark KR, McMurtry IF, O'Brien RF, Voelkel NF, Badesch DB: Insulin-like growth factor I and protein kinase C activation stimulate pulmonary artery smooth muscle cell proliferation through separate but synergistic pathways, *J Cell Physiol* 1990, 144:159–165.
80. Tuder RM, Flook BE, Voelkel NF: Increased gene expression for VEGF and the VEGF receptors KDR/Flk and Flt in lungs exposed to acute or to chronic hypoxia. Modulation of gene expression by nitric oxide, *J Clin Invest* 1995, 95:1798–1807.
81. Tuder RM, Groves B, Badesch DB, Voelkel NF: Exuberant endothelial cell growth and elements of inflammation are present in plexiform lesions of pulmonary hypertension, *Am J Pathol* 1994, 144:275–285.
82. Humbert M, Monti G, Brenot F, Sitbon O, Portier A, Grangeot-Keros L, et al.: Increased interleukin-1 and interleukin-6 serum concentrations in severe primary pulmonary hypertension, *Am J Respir Crit Care Med* 1995, 151:1628–1631.
83. Su G, Hodnett M, Wu N, Atakilit A, Kosinski C, Godzich M, et al.: Integrin alphavbeta5 regulates lung vascular permeability and pulmonary endothelial barrier function, *Am J Respir Cell Mol Biol* 2007, 36:377–386.
84. McCarter SD, Mei SH, Lai PF, Zhang QW, Parker CH, Suen RS, et al.: Cell-based angiopoietin-1 gene therapy for acute lung injury, *Am J Respir Crit Care Med* 2007, 175:1014–1026.
85. Nowak K, Weih S, Metzger R, Albrecht RF, 2nd, Post S, Hohenberger P, et al.: Immunotargeting of catalase to lung endothelium via anti-angiotensin-converting enzyme antibodies attenuates ischemia-reperfusion injury of the lung in vivo, *Am J Physiol Lung Cell Mol Physiol* 2007, 293:L162–169.
86. Tinsley JH, Teasdale NR, Yuan SY: Myosin light chain phosphorylation and pulmonary endothelial cell hyperpermeability in burns, *Am J Physiol Lung Cell Mol Physiol* 2004, 286:L841–847.
87. Wang XX, Zhang FR, Shang YP, Zhu JH, Xie XD, Tao QM, et al.: Transplantation of autologous endothelial progenitor cells may be beneficial in patients with idiopathic pulmonary arterial hypertension: A pilot randomized controlled trial, *J Am Coll Cardiol* 2007, 49:1566–1571.

18 NGF and Its Receptor System in Inflammatory Diseases

Anupam Mitra, Smriti K. Raychaudhuri, and Siba P. Raychaudhuri
University of California, Davis
Davis, California

CONTENTS

18.1 NEUROGENIC INFLAMMATION

Neurogenic inflammation is a process in which inflammatory mediators, including neuropeptides, are released from afferent nerve terminals in the target tissue, resulting in vasodilatation, plasma extravasation, and hypersensitivity and inducing chemotaxis of neutrophils (Richardson and Vasko 2002; Boyle et al. 1985). Moreover, neuropeptides released at the site of inflammation also induce activation of immune cells like lymphocytes, mast cells, macrophages, and neutrophils resulting in activation of intracellular signaling cascade (Kulka et al. 2008; Susaki et al. 1996). In physiological and pathological conditions, neuropeptides known to influence immune reactions include substance P, neuropeptide Y (NPY), vasoactive intestinal peptide (VIP), nerve growth factor (NGF), and calcitonin gene-related peptide (CGRP) (Mosimann et al. 1993; Wagner et al. 1998). Neurogenic inflammation has been attributed to the pathogenesis of several inflammatory diseases such as psoriasis, asthma, atopic dermatitis, and inflammatory arthritis (Farber and Raychaudhuri 1999). Most extensive studies regarding the role of neuropeptides have been carried out in psoriasis. The role of neurogenic inflammation in the pathogenesis of psoriasis is substantiated by a number of observations such as exacerbations during periods of stress, marked proliferation of terminal cutaneous nerves, and upregulation of neuropeptides (SP, VIP, CGRP) in the psoriatic plaques compared to the normal skin and other inflammatory skin diseases (Farber and Raychaudhuri 1999; Raychaudhuri and Farber 1993). Moreover, therapeutic response to neuropeptide modulating agents such as capsaicin, somatostatin, and clearance of active plaques of psoriasis at the sites of anesthesia following traumatic denervation of cutaneous nerves provide convincing clinical evidence for a role of neurogenic inflammation in the pathogenesis of psoriasis (Farber and Raychaudhuri 1999; Raychaudhuri and Farber 1993; Bernstein et al. 1986).

18.2 NGF AND ITS RECEPTOR

Almost 50 years ago, a polypetide that induces neuronal growth was discovered and named as nerve growth factor (NGF; Levi-Montalcini 1952). Later on other molecules with similar structure and functions were identified, and together they form a family of polypeptide growth factors called the neurotrophins (NTs). Apart from NGF, this family is comprised of brain-derived neurotrophic factor (BDNF), neurotrophin 3 (NT-3), and NT-4/5. These are present in all tetrapods, except NT-4/5, which has not been found in birds (Hallböök 1999). Among all neurotrophins, NGF is the best characterized trophic polypeptide (Levi-Montalcini 1987). NGF is a glycoprotein of 118 amino acids, consisting of three subunits ($\alpha2$, $\beta2$, $\gamma2$). The gene of NGF is situated in the proximal short arm of chromosome 1. The α-NGF is inactive, the γ-NGF is a highly specific active protease that is able to process NGF precursor to its mature form, and the β-NGF is responsible for the NGF biological activity (Sofroniew et al. 2001). NGF is produced by astrocytes and oligodendrocytes in the nervous system, which in turn regulate the differentiation and survival of neurons mostly through the formation of paracrine circuits (Levi-Montalcini 1987). Apart from the nervous system, NGF is also produced and utilized by several non-nervous cell types like epithelial cells, keratinocytes, immune cells, and smooth muscle cells, which indicates the role of NGF outside the nervous system (Micera et al. 2003; Lambiase et al. 2004). NGF is synthesized as a precursor, pro-NGF in the tissues and undergoes post-translational processing by serine protease plasmin and matrix metalloproteinase 7 (MMP7) to generate and secrete the mature β-NGF (Micera et al. 2003; Lambiase et al. 2004). NGF is released from the sub-mandibular gland in mature form whereas in the other tissues like skin, hair follicles, thyroid gland, retina, colon, dorsal root ganglia, and prostate, NGF is mainly secreted as pro-NGF or a mixture of both forms (Sofroniew et al. 2001; Fahnestock et al. 2004). Both NGF and pro-NGF are biologically active, but pro-NGF is less potent in stimulating neurite outgrowth than NGF (Fahnestock et al. 2004). NGF produces its effect by interacting with the high affinity receptor TrkA rather than the low-affinity receptor $p75^{NGFR}$. The low-affinity receptor $p75^{NGFR}$ serves as a pan-neurotrophin receptor and belongs to the tumor necrosis factor receptor superfamily (Bothwell 1995). The functional receptor of NGF, TrkA, is present on cells of the nervous, immune, and endocrine systems, which further strengthen the belief that NGF participates in interactions among these three systems (Otten et al. 1994; Patterson and Childs 1994).

18.3 NGF AND ITS ROLE IN INFLAMMATORY CASCADE

18.3.1 NGF and Immunocompetent Cells

Despite being a neurotrophin, NGF regulates immune and inflammatory responses through direct and/or indirect effects on immune-competent cells (Thorpe et al. 1987; Table 18.1). Moreover, the presence of a highly dense array of sympathetic fibers in lymphoid tissues suggests a bridge between the nervous and immune systems, which probably translates neural messages into chemical signals that interact with specific cellular elements of the immune system (Felten et al. 1985). In transgenic mice overexpressing NGF, a hyper-sympathetic innervation of peripheral lymph nodes is seen (Carlson et al. 1995). In birds NGF and its functional receptor TrkA is strongly expressed in developing and adult lymphoid organs (Ciriaco et al. 1996), and also TrkA was found to be present on human thymic epithelial cells (Hannestad et al. 1997), which suggests the potential role of NGF in differentiation of immune cells. Moreover, it has been shown that the high affinity receptor TrkA is required for the development of the thymus organ for maturation of T cells (Garcia-Suarez et al. 2000).

18.3.1.1 Mast Cells

NGF can influence a variety of bone marrow–derived cells. Among the cells of immune lineage, mast cells were the first cells to be shown responsive to NGF both *in vivo* (Aloe et al. 1977) and *in vitro* (Böhm et al. 1986). Mast cells express only high affinity receptor TrkA, so NGF has an effect on mast cells but other neurotrophins do not have any effect on mast cells (Horigome et al. 1993). In neonatal rats, daily injection of NGF resulted in hyperplasia of connective tissue mast cells and mucosal mast cells (Aloe and Levi-Montalcini 1977; Marshall et al. 1990). Anti-NGF antibody decreased the number of rat peritoneal mast cells (Aloe 1988). Moreover, NGF acts as a cofactor with IL-3 in development of mast cells from human umbilical cord blood cells (Richard et al. 1992) as well as induces expression of mast cell surface markers in cell cultures (Welker et al. 2000). Human mast cell line, HMC-1, and cultured human mast cells produce active NGF and express functional TrkA (Tam et al. 1997). In the rat, it has been found that NGF promotes the survival of peritoneal mast cells by inhibiting apoptosis (Horigome et al. 1994). This pro-survival

TABLE 18.1
Role of NGF on Immunocompetent Cells

Cell Types	NGF Responses
Mast cells	Maturation[1, 2], survival[3], degranulation[4], proliferation[5, 6], modify expression of inflammatory cytokines[7]
Basophils	Differentiation[8], release of histamine[9, 10]
T lymphocytes	Proliferation[11, 12, 13], inhibition of apoptosis[14], transcriptional activation of *c-fos*[15], T cell dependent antibody synthesis[16], increase expression of IL-2 receptor[17, 18]
Memory B lymphocytes	Proliferation[19, 20], differentiation[12], survival of memory B cells[21]
Hemopoietic cells	Differentiation[17]
Neutrophils	Survival[22], superoxide production[22], chemotaxis[23, 24]
Monocytes/macrophages	Phagocytosis[25], survival[26, 27], increase in oxidative burst[28], increase in cathepsin S expression[29]

[1]Richard et al., 1992; [2]Welker et al., 2000; [3]Horigome et al., 1994; [4]Bruni et al., 1982; [5]Aloe and Levi-Montalcini, 1977; [6]Stead et al., 1987; [7]Marshall et al., 1999; [8, 9]Matsuda et al., 1988a; [10]Bischoff and Dahinden, 1992; [11]Raychaudhuri et al., 2011; [12]Otten et al., 1989; [13]Lambiase et al., 1997; [14]Raychaudhuri et al., 2011; [15]Vega et al., 2003; [16]Manning et al., 1985; [17]Matsuda et al., 1988; [18,19]Brodie and Gelfand, 1992; [20]Thorpe et al., 1987; [21]Torcia et al., 1996; [22]Kannan et al., 1991; [23]Gee et al., 1983; [24]Boyle et al., 1985; [25]Susaki et al., 1996; [26]la Sala et al., 2000; [27]Garaci et al., 1999; [28]Ehrhard et al., 1993a; [29]Liuzzo et al., 1999

effect of NGF is suggested to be mediated by increased expression and release of autocrine growth factors (Marshall et al. 1990). Moreover, it has been found that rat peritoneal mast cells and cells of the basophilic lineage can synthesize, store, and release biologically active NGF in physiologically relevant amounts (Leon et al. 1994), which indicates the possibility that NGF itself may be one of the autocrine agents inducing mast cell survival. At physiological condition, NGF is a poor secretagogue for rat peritoneal mast cells and presence of antigen or other secretagogue increases the secretagogue effect of NGF on these cells (Bruni et al. 1982). Thus NGF acts as an immunomodulator in the inflammatory response by regulating mediator release from mast cells. Moreover, locally produced NGF acts as a chemoattractant for other mast cells through mitogen-activated protein kinase (MAPK) and phosphatidylinositol 3-kinase signaling pathways resulting in mast cell accumulation in allergic and nonallergic inflammatory conditions (Sawada et al. 2000). In mast cells, NGF also modifies the expression of inflammatory cytokines (IL-6, TNF-α) by a prostanoid-dependent mechanism that establishes the influence of NGF on mast cells in local inflammatory response (Marshall et al. 1999).

18.3.1.2 Basophils

The functional receptor of NGF, TrkA, is expressed on human basophils (Bürgi et al. 1996). NGF promotes differentiation of basophils accompanied by release of histamine (Matsuda et al. 1988a). In human basophils also, NGF has a very similar sensitizing or priming effect on histamine release (Bischoff and Dahinden 1992). Subcutaneous injection of NGF produces plasma extravasation (Otten et al. 1984) and intravenous injection produces hypotension (Yan et al. 1991) by releasing histamine. NGF primes the basophils and enhances the release of mediators in the presence of different agonists (Bürgi et al. 1996). Thus, expression of NGF by mast cells constitutes an important link between mast cell activation and basophil function in late phase allergic reactions.

18.3.1.3 Eosinophils

NGF is stored and produced by eosinophils, and in disease conditions NGF activates the eosinophils to release specific inflammatory mediators like eosinophil peroxide, whereas it does not have any effect on the viability of eosinophils (Solomon et al. 1998). But according to another study, NGF increases the survival as well as the cytotoxic activity of human eosinophils (Hamada et al. 1996).

18.3.1.4 Neutrophils

Neutrophils play an important role in the early phase of inflammation and from several studies the effect of NGF on neutrophils has been well established. NGF at a concentration similar to that required for neurite outgrowth induces chemotaxis of human polymorphonuclear neutrophils *in vitro* (Gee et al. 1983) and also enhances the survival, phagocytosis, and superoxide production of murine neutrophils (Kannan et al. 1991). In *in vivo* study of mice, it has also been found that NGF acts as a chemotactic factor for neutrophils (Boyle et al. 1985).

18.3.1.5 Monocytes/Macrophages

The functional receptor of NGF, TrkA is present on human monocytes (Ehrhard et al. 1993a) and macrophages (Caroleo et al. 2001), but the expression of TrkA decreased during *in vitro* differentiation of monocytes to macrophages, suggesting a maturation-dependent TrkA expression (Ehrhard et al. 1993a). In murine study it has been found that NGF enhances the phagocytic property of monocytes/macrophages (Susaki et al. 1996). In humans, NGF increases the survival of monocytes by upregulating the anti-apoptotic Bcl-2 family proteins (la Sala et al. 2000). The interaction of NGF with its receptor TrkA on monocytes triggers oxidative burst, which is important for cytotoxic activity of monocytes (Ehrhard et al. 1993a). Moreover, NGF increases the expression of cathepsin S in macrophages, which is an important mediator of tissue destruction during inflammation (Liuzzo et al. 1999).

18.3.1.6 Lymphocytes

The role of NGF on lymphocytes was first described by Dean et al. (1987) on mouse spleen cells. There are reports that lymphocytes can synthesize and secrete NGF, thereby suggesting an autocrine/paracrine effect of NGF on the development and regulation of the immune system (Santambrogio et al. 1994). In human peripheral blood mononuclear cells, NGF increases the expression of IL-2 receptor, thereby inducing proliferation of these cells (Brodie and Gelfand 1992). *In vitro* NGF induces granulocyte differentation from hematopoietic stem cells (Matsuda et al. 1988). Also, NGF increases the expression of IL-2 receptors in human natural killer cells (NK), thereby contributing to the innate immune response (Thorpe et al. 1987). Regarding the expression of TrkA on lymphocytes, there are some conflicting results. In a study, Ehrhard et al. showed that TrkA is expressed only in mitogen activated mouse CD4+ T cell clones (1993). In contrast, Lambiase et al. demonstrated that stimulated as well as unstimulated T cells expressed TrkA (1997). Lambiase also showed that NGF is produced by both Th1 and Th2 cells, but upon mitogenic stimulation Th2 cells produce more NGF than Th1 cells (Lambiase et al. 1997). In a recent study, our group reported that TrkA expression on human T cells is upregulated by CD3/CD28 stimulation, which is consistent with findings of the Ehrhard group in murine T cells (Raychaudhuri et al. 2011). We also showed that NGF is mitogenic to T cells and is mediated through TrkA, which is consistent with other study of NGF-induced T cell proliferation (Otten et al. 1989). In activated human T cells, NGF induces the phosphorylation of AKT (p-AKT) and inhibits the TNF-α induced apoptosis of T cells (Raychaudhuri et al. 2011). NGF also increases the expression of IL-2 receptor on T cells (Thorpe et al. 1987), *c-fos* gene transcription (Vega et al. 2003), and T cell dependent antibody synthesis (Manning et al., 1985).

Similarly, NGF plays an important role in proliferation of B cells by increasing the expression of IL-2 receptor (Brodie and Gelfand 1992; Thorpe et al. 1987), differentiation of B cells into plasma cells (Otten et al. 1989) to secrete immunoglobulins (IgG, IgM, IgA; Kimata et al. 1991a; Kimata H et al. 1991b), and also to release specific cytokines like IL-4 and IL-5 (Otten et al. 1989; Lambiase et al. 2004). NGF also has an important role in induction and survival of specialized memory B cells through its receptor TrkA (Torcia et al. 1996).

18.3.1.7 Other cells

Apart from neuronal and immune cells, NGF is produced by structural cells like keratinocytes (Raychaudhuri et al. 1998) and fibroblasts of skin, lung, and joints (Micera et al. 2001; Manni et al. 2003; Raychaudhuri and Raychaudhuri 2009). NGF induces proliferation of keratinocytes (Pincelli et al. 1994) and increases expression of RANTES in keratinocytes (Raychaudhuri et al. 2000a), which plays an important role in recruiting inflammatory cells at the site of inflammation. In endothelial cells, NGF induces proliferation as well as increased expression of ICAM-1, thereby influencing angiogenesis and transmigration of leukocytes across the vessels (Raychaudhuri et al. 2001). NGF is produced by normal human synovial fibroblast and the functional receptor of NGF, TrkA, is expressed in these cells. The pro-inflammatory cytokines like IL-1β and TNF-α induce NGF synthesis in these cells and also upregulates the expression of TrkA (Manni et al. 2003; Raychaudhuri and Raychaudhuri 2009). There are some conflicting data from two studies regarding the mitogenic role of NGF on fibroblast like synoviocytes (FLS). In one study it has been reported that NGF does not induce proliferation of FLS (Manni et al. 2003), whereas a study by our group had demonstrated that NGF significantly induces proliferation of FLS (Raychaudhuri and Raychaudhuri 2009). This difference can be attributed to use of more physiological culture media in the latter study.

18.3.2 NGF and Cytokines

Cytokines play a pivotal role in maintaining the balance of the immune system. The pro-inflammatory cytokines like IL-1β and TNF-α markedly induce NGF synthesis from synovial fibroblasts (Raychaudhuri and Raychaudhuri 2009). In an *in vivo* study in marmosets, Villoslada et al. showed that NGF increases the anti-inflammatory cytokine IL-10 whereas it decreases the pro-inflammatory cytokine IFN-γ, thereby skewing the Th1/Th2 balance toward Th2 (2000). In rat pheochromocytoma cell line (PC12) it has been found that NGF induces the mRNA of IL-1-alpha (Alheim et al. 1991). In mouse macrophages NGF increases the secretion of TNF-α and K252a. TrkA inhibitor can inhibit the NGF induced TNF-α secretion (Barouch et al. 2001). Recently, Jiang et al. showed that NGF along with LPS significantly induces the secretion of pro-inflammatory cytokines like IL-1β, IL-6, and TNF-α in monocyte derived dendritic cell lines (2007).

18.4 ROLE OF NGF IN DIFFERENT INFLAMMATORY DISEASES

The role of NGF/TrkA has been established in different inflammatory diseases of neural and extra neural origin (Table 18.2).

18.4.1 Multiple Sclerosis

In patients with multiple sclerosis (MS), infiltration of brain parenchyma with immune cells like T cells and monocytes occurs, which leads to destruction of myelin (Hauser et al. 1986). These cells also have a role in experimental allergic encephalomyelitis (EAE; Sobel et al. 1984). Mast cells are also present in MS plaques (Olsson 1974). In the cerebrospinal fluid of MS patients, increased levels of NGF have been found (Bracci-Laudiero et al. 1992). NGF is produced from mast cells (Tam et al. 1997) and it induces proliferation of mast cells (Aloe and Levi-Montalcini 1977) and degranulation of mast cells (Bruni et al. 1982). Moreover, T cells and monocytes are responsive to NGF, resulting in induction of the inflammatory cascade (Raychaudhuri et al. 2011). These studies suggest that NGF plays an important role in the pathophysiology of MS. In contrast, Villoslada et al. showed in an EAE model of marmosets that NGF plays a protective role by downregulating the production of IFN-γ by infiltrating T cells and upregulates the production of IL-10 by glial cells (2000). Later on a protective role of NGF in EAE has been shown in mice (Parvaneh Tafreshi 2006). In a recent study it has been shown that in relapsing-remitting MS, mRNA expression of NGF is lower than healthy controls and it is mainly produced by T cells rather than monocytes, whereas in healthy controls

TABLE 18.2
Increased Concentration of NGF in Different Inflammatory Diseases

Inflammatory Diseases	Tissue
Multiple sclerosis	Cerebrospinal fluid[1]
Rheumatoid arthritis	Synovial fluid[2]
Psoriasis	Psoriasis plaques[3]
Psoriatic arthritis	Synovial fluid[4]
Osteoarthritis	Synovial fluid[5]
Systemic lupus erythematosus	Serum[6]
Vasculitic syndrome	Serum[7]
Allergic inflammatory disease of airway	Serum,[8] nasal lavage fluid[9]
Inflammatory bowel disease	Intestinal tissue[10]

[1]Bracci-Laudiero et al., 1992; [2]Raychaudhuri et al., 2011; [3]Fantini et al., 1995; [4]Raychaudhuri et al., 2009; [5]Aloe et al., 1992a; [6]Bracci-Laudiero et al., 1993; [7]Falcini et al., 1996; [8]Bonini et al., 1996; [9]Sanico et al., 2000; [10]Barada et al., 2007

monocytes produce more NGF than T cells (Urshansky et al. 2010). The exact role of NGF in MS is yet to be decided.

18.4.2 Rheumatoid Arthritis

Rheumatoid arthritis (RA) is characterized by increased pro-inflammatory cytokines and proliferation of fibroblast-like synoviocytes leading to formation of pannus, which leads to destruction of bone and cartilage. In a carrageenan induced arthritis (CIA) model, NGF has been found to be elevated compared to controls (Aloe et al. 1992b). In synovial fluid of RA patients, an elevated level of NGF was found compared to patients with noninflammatory arthritis like osteoarthritis (OA; Raychaudhuri et al. 2011). Contrarily, studies showed no difference in NGF level between RA and OA patients (Rihl et al., 2005) and six months of anti-TNF therapy did not alter serum NGF levels compared to its pre-treatment state (Del Porto et al., 2006). The mRNA expression of NGF and its receptor TrkA in synovial tissue was high in patients with RA compared to patients with OA (Barthel et al. 2009). A study by Raychaudhuri et al. showed that fibroblast-like synoviocytes (FLS) when stimulated with TNF-α/IL-1β produce more NGF and there is also upregulation of TrkA (Raychaudhuri et al. 2009). Moreover, NGF induced marked proliferation of FLS through its receptor TrkA in RA patients, resulting in formation of pannus, which is pathognomonic of RA (Raychaudhuri and Raychaudhuri 2009; Raychaudhuri et al. 2011). In synovial fluid of RA patients, activated T cells expressed TrkA, whereas in OA patients TrkA+ T cells were not detectable (Raychaudhuri et al. 2011). Treatment of RA patients with anti-TNF-α agents does not have significant impact on NGF, suggesting that TNF-α and NGF do not share the same regulatory pathway (Del Porto et al. 2006).

18.4.3 Psoriasis

Psoriasis is characterized by hyperkeratosis, parakeratosis, acanthosis, angiogenesis, neutrophilic microabscesses, and lymphomononuclear cell infiltrates (Farber and Raychaudhuri 1999). The role of neurogenic inflammation in the pathogenesis of psoriasis is substantiated by several observations: psoriasis exacerbates during stress (Farber et al. 1986), marked proliferation of terminal cutaneous nerves in psoriatic lesions, upregulation of neuropeptides like substance P, calcitonin gene-related peptide (Farber et al. 1990; Naukkarinen et al. 1989), and response of psoriatic plaques to neuropeptide-modulating agents such as capsaicin (Bernstein et al. 1986). NGF plays an important role in regulating neuropeptides like substance P and calcitonin gene-related peptide (Lindsay and Harmar 1989). NGF promotes regeneration of sensory nerve fibers in transplanted psoriatic skin compared to healthy skin (Raychaudhuri et al. 2008) and it has been reported that NGF is mitogenic to keratinocytes (Pincelli et al. 1994). NGF recruits mast cells and promotes their degranulation (Aloe and Levi-Montalcini 1977; Pearce and Thompson 1986), both of which are early events in a developing lesion of psoriasis. In addition, NGF activates T lymphocytes, recruits inflammatory cellular infiltrates (Thorpe et al. 1987; Lambiase et al. 1997), is mitogenic to endothelial cells, and induces ICAM on endothelial cells (Raychaudhuri et al. 2001), which altogether helps in recruitment of inflammatory cells at the site of lesion. Moreover, it has been shown that NGF is highly expressed in psoriatic patients (Fantini et al. 1995) and the receptor TrkA is also upregulated in terminal cutaneous nerves of psoriatic lesions (Raychaudhuri et al. 2000). In psoriatic patients it has been found that NGF is upregulated after the Koebner phenomenon, and the time required for the appearance of psoriatic features correlates with the peak level of NGF (Raychaudhuri et al. 2008). In a SCID mouse model, NGF-stimulated autologous peripheral blood mononuclear cells can induce features of psoriasis in transplanted grafts (Raychaudhuri et al. 2001a). Moreover, in a SCID psoriasis xenograft model, it has been shown that NGF-R modulating agents like K252a and NGF antibody improve psoriasis, which also establishes the role of NGF in psoriasis (Raychaudhuri et al. 2004).

18.4.4 Psoriatic Arthritis

In synovial fluid of psoriatic arthritis (PsA) patients, increased concentration of NGF was found from different studies and NGF induces marked proliferation of FLS (Raychaudhuri and Raychaudhuri 2009; Raychaudhuri et al. 2011), thereby contributing in the formation of pannus. The functional receptor of NGF, TrkA is present in FLS of PsA patients and upregulated when treated with TNF-α/IL-1β (Raychaudhuri and Raychaudhuri 2009). Moreover, this NGF induced FLS proliferation in PsA patients can be inhibited by NGF neutralizing monoclonal antibody and K252a, a TrkA receptor blocker, which in turn proves that proliferative effect of NGF is mediated through TrkA (Raychaudhuri et al. 2011). These studies suggest that NGF plays an important pathological role in PsA; however, more studies are required to confirm this.

18.4.5 Osteoarthritis

Osteoarthritis (OA) is characterized by disrupted tissue homeostasis of the articular cartilage and subchondral bone. In pathophysiology of OA, interactions between chondrocytes and the ECM play a major role. IL-1β and TNF-α are important pro-catabolic cytokines in OA (van de Loo et al. 1995). In synovial fluid of OA patients, NGF is marginally elevated compared to healthy controls (Aloe et al. 1992a) and chondrocytes secrete NGF and express TrkA (Iannone et al. 2002).

18.4.6 Progressive Systemic Sclerosis

Progressive systemic sclerosis (SSc) is characterized by activation of the immune system (White, 1994) with accumulation of extracellular matrix leading to fibrosis of skin (Strehlow and Korn 1998) and endothelial cell damage (Matucci-Cerinic et al. 1995). Mast cells (MC) increase in the early stage of disease but decline during disease progression (Hawkins et al. 1985). Fibroblasts release NGF, which induces proliferation, hypertrophy, and degranulation of MCs (Davies et al. 1987; Pearce and Thompson 1986). Mast cells can also synthesize and release NGF (Leon et al. 1994). A study by Tuveri and coworkers showed that in early stage of disease the densities of NGF positive cells and mast cells are more in SSc patients compared to healthy controls (Tuveri et al. 1993). In addition to skin, NGF also has an effect on pulmonary epithelial cells by increasing the permeability of pulmonary epithelial cells leading to pulmonary fibrosis (Piga et al. 2000).

18.4.7 Systemic Lupus Erythematosus

Systemic lupus erythematosus (SLE) is an immune-mediated disease characterized by B cell hyperactivity and dysregulated immune system with excess pathogenic subsets of auto-antibodies and immune complexes. It has been found that NGF is elevated in the serum of patients with SLE compared to healthy controls (Bracci-Laudiero et al. 1993) and NGF has a correlation with disease severity (Aalto et al. 2002).

18.4.8 Vasculitic Syndromes

Among different vasculitic syndromes, NGF is found to be elevated in Kawasaki disease (Falcini et al. 1996) and in Churg-Strauss syndrome (Yamamoto et al. 2001).

18.4.9 Allergic Inflammatory Disease of Airways

Allergic airway diseases are characterized by inflammation and hyperresponsiveness to different stimuli. It has been shown that hyperresponsiveness can be attributed to increased neural activity (Nockher et al. 2006; Sanico et al. 1998). Allergic bronchial asthma is characterized by airway inflammation, reversible airway obstruction in response to allergen inhalation, and enhanced bronchoconstrictor responses to unspecific stimuli (Braun et al. 2000). In mouse models of allergic inflammation and in NGF transgenic mice it has been shown that neurotrophins alter sensory innervations, enhance neuropeptide production, and induce airway hyperresponsiveness (Hoyle et al. 1998; Braun et al. 1998). In guinea pigs, exogenously administered NGF develops airway hyperresponsiveness to histamine (de Vries et al. 1999). Moreover, it has been found that subjects with allergic rhinitis exhibit hyperresponsiveness to nasal provocation with histamine (Mullins et al. 1989). In patients with allergic asthma, serum NGF level was found significantly elevated compared to nonallergic asthma and healthy controls (Bonini et al. 1996). Braun et al. showed that bronchial hyperreactivity of asthma is associated with an increase in NGF, which increases local Th2 responses resulting in increased production of IL-4, IL-5, IgG1, and IgE (1998).

In allergic rhinitis patients, the NGF level in nasal lavage fluid was found to be high compared to healthy controls (Sanico et al. 2000).

18.4.10 Inflammatory Bowel Disease

Inflammatory bowel disease (IBD) is comprised of two disease states: ulcerative colitis (UC) and Crohn's disease (CD). In different animal and human studies, involvement of the enteric nervous system during intestinal inflammation has been shown (Fiocchi 1997). In one study, Stead et al. showed that mast cells located in the lamina propria of gut are in close contact with the enteric nervous system, which showed a connection between the nervous and inflammatory systems (1989). In a rat model of colitis, NGF was found to be elevated in intestinal tissue compared to control rats (Barada et al. 2007). It has been found that in intestinal tissues of CD and UC patients, mRNA of NGF and TrkA are markedly increased compared to healthy controls (di Mola et al. 2000).

18.4.11 NGF/TrkA Interaction: Potential New Drug Target

The critical role of NGF and its receptor system in the pathophysiology of inflammation and inflammatory diseases has provided an attractive opportunity to develop a novel class of therapeutics for inflammatory diseases. The NGF/NGF-R interaction and the downstream signaling cascades offer putative drug targets (Figure 18.1). This can be achieved by NGF neutralizing agents, TrkA

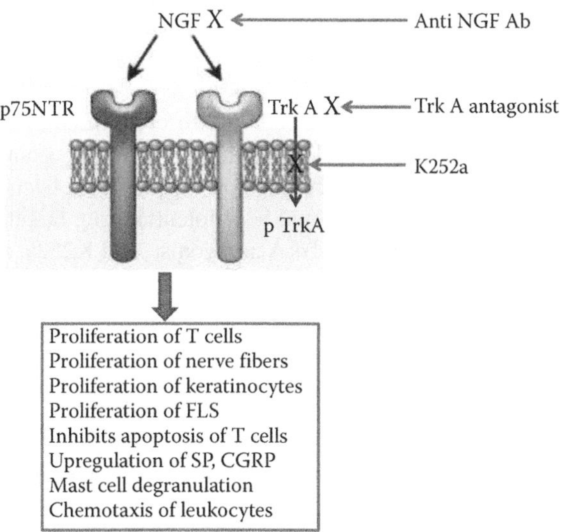

FIGURE 18.1 NGF/TrkA, a novel target for therapy of autoimmune diseases.

receptor antagonists, and by molecules that prevent activation of TrkA. Our group as well as others have demonstrated that psoriatic keratinocytes produce more functionally active NGF compared to normal individuals (Raychaudhuri et al. 2000; Fantini et al. 1995). The important role of NGF in the pathophysiology of psoriasis compelled our group to develop therapeutic strategies for psoriasis by manipulating NGF/TrkA interaction and its downstream signal transduction events. Our group had used a high-affinity NGF receptor (TrkA) inhibitor, K252a, and NGF-neutralizing antibody in SCID mouse xenograft model to substantiate the significance of NGF/NGF-R system in the pathogenesis of psoriasis (Raychaudhuri et al. 2004). We have demonstrated the efficacy of K252a and NGF-neutralizing antibody in psoriasis, evidenced by decreased thickness of the rete pegs, reduced infiltrates, and normalization of the stratum corneum, whereas the control group treated with normal saline did not improve (Raychaudhuri et al. 2004). This study provides direct evidence regarding the role of NGF/NGF-R in the pathology of human disease and therapeutic manipulation of NGF/TrkA interaction can be considered as a treatment of psoriasis.

Several investigators and pharmaceutical companies are currently in search of anti-NGF therapy for inflammatory diseases. A phase II clinical trial with topical K252a was recently completed for treating mild to moderate psoriasis vulgaris (U.S. NIH 2011a). Shelton et al. from Rinat Neuroscience Corporation (later acquired by Pfizer Inc.) have reported that treatment with anti-NGF antibody was efficacious for autoimmune arthritis of rats (2005). They found that treatment with anti-NGF caused a profound reversal of hyperalgesia within 24 hours of the infusion, and the efficacy was similar to daily doses of 3 mg/kg of indomethacin along with reversal of weight loss. These effects of pain relief and weight loss were achieved despite a lack of notable changes in the development or progression of the arthritis in this model. These results encouraged Pfizer to extend their study in chronic painful human diseases such as osteoarthritis (OA). In a phase III clinical trial, Tanezumab (RN624), a humanized anti-NGF antibody, was evaluated for safety and efficacy of subjects with knee or hip OA, but due to potential safety issues, FDA put a clinical hold on this study (U.S. NIH 2010). Another pharmaceutical company, Abbott, completed a phase I clinical trial in January 2011 with their anti NGF monoclonal antibody in OA of knee joints (U.S. NIH 2011b). Further studies are required to assess the efficacy and safety of anti-NGF monoclonal antibody in different inflammatory diseases.

18.5 CONCLUSIONS

Among the different neurotrophins, NGF and its functional receptor TrkA play an important role in maintaining the homeostasis of the immune system. NGF perpetuates the inflammatory process by influencing different steps of inflammation: (1) inducing proliferation of T cells/B cells/mast cells/nerve fibers/keratinocytes/FLS; (2) releasing histamine from mast cells; (3) promoting chemotaxis of leukocytes; and (4) inhibiting apoptosis of T cells. In different inflammatory diseases of neural as well as extra-neural origin, NGF remains elevated and plays an important role in disease pathophysiology. Among the inflammatory diseases, the role of NGF has been more substantiated in psoriasis. Interfering NGF/TrkA interaction could be a potential drug target. Clinical trials are ongoing with anti-NGF monoclonal antibodies, TrkA antagonist, and K252a, an inhibitor of tyrosine phosphorylation of TrkA in different disease conditions.

TAKE-HOME MESSAGES

- In addition to its biologic effect in the nervous system, NGF and its receptor system have a regulatory role in various other organs.
- NGF-induced signals are mediated by its high affinity receptor, tyrosine receptor kinase A (TrkA), and p75, the low affinity receptor.
- A growing number of studies on inflammatory diseases have demonstrated that the inflammatory state is characterized by upregulation of NGF synthesis.

- We have learned a new function of NGF: that NGF sensitizes nociceptive neurons and influences processing and transmission of pain signals.
- NGF influences an inflammatory reaction by regulating neuropeptides, angiogenesis, cell trafficking molecules, and T cell activation. All these functions of NGF are relevant in the maintenance or initiation of the critical biologic events in various inflammatory diseases.
- The recognition of a pathologic role of NGF and its receptor system has provided an attractive opportunity to develop a novel class of therapeutics for inflammatory diseases and chronic pain syndromes.
- Currently, anti-NGF pharmacologic agents are in the developing phase. It is likely that NGF-based therapy will be used in various disciplines of medicine, more so by the rheumatologists for the treatment of inflammatory diseases and chronic pain syndromes of musculoskeletal diseases.

REFERENCES

Aalto, K., Korhonen, L., Lahdenne, P., Pelkonen, P., Lindholm, D. 2002. Nerve growth factor in serum of children with systemic lupus erythematosus is correlated with disease activity. *Cytokine* 20:136–9.

Alheim, K., Andersson, C., Tingsborg, S., Ziolkowska, M., Schultzberg, M., Bartfai, T. 1991. Interleukin 1 expression is inducible by nerve growth factor in PC12 pheochromocytoma cells. *Proc Natl Acad Sci USA* 88:9302–6.

Aloe, L. and R. Levi-Montalcini. 1977. Mast cells increase in tissues of neonatal rats injected with nerve growth factor. *Brain Res* 133:358–66.

Aloe, L. 1988. The effect of nerve growth factor and its antibody on mast cells in vivo. *J Neuroimmunol* 18:1–12.

Aloe, L., Tuveri, M.A., Carcassi, U., Levi-Montalcini, R. 1992a. Nerve growth factor in the synovial fluid of patients with chronic arthritis. *Arthritis Rheum* 35:351–5.

Aloe, L., Tuveri, M.A., Levi-Montalcini, R. 1992b. Studies on carrageenan-induced arthritis in adult rats: Presence of nerve growth factor and role of sympathetic innervation. *Rheumatol Int* 12:213–6.

Barada, K.A., Mourad, F.H., Sawah, S.I., Khoury, C., et al. 2007. Up-regulation of nerve growth factor and interleukin-10 in inflamed and non-inflamed intestinal segments in rats with experimental colitis. *Cytokine* 37:236–45.

Barouch, R., Kazimirsky, G., Appel, E., Brodie, C. 2001. Nerve growth factor regulates TNF-alpha production in mouse macrophages via MAP kinase activation. *J Leukoc Biol* 69:1019–26.

Barthel, C., Yeremenko, N., Jacobs, R., et al. 2009. Nerve growth factor and receptor expression in rheumatoid arthritis and spondyloarthritis. *Arthritis Res Ther* 11:R82.

Bernstein, J.E., Parish, L.C., Rapaport, M., Rosenbaum, M.M., Roenigk, H.H., 1986. Effects of topically applied capsaicin on moderate and severe psoriasis vulgaris. *J Am Acad Dermatol* 15:504–7.

Bischoff, S.C. and C.A. Dahinden. 1992. Effect of nerve growth factor on the release of inflammatory mediators by mature human basophils. *Blood* 79:2662–9.

Böhm, A., Aloe, L., Levi-Montalcini, R. 1986. Nerve growth factor enhances precocious differentiation and numerical increase in mast cells in cultures of rat splenocytes. *Accad Naz Lincei* 80:1–6.

Bonini, S., Lambiase, A., Bonini, S., et al. 1996. Circulating nerve growth factor levels are increased in humans with allergic diseases and asthma. *Proc Natl Acad Sci USA* 93:10955–60.

Bothwell, M. 1995. Functional interactions of neurotrophins and neurotrophin receptors. *Annu Rev Neurosci* 18:223–53.

Boyle, M.D.P., Lawman, M.J.P., Gee, A.P., Young, M. 1985. Nerve growth factor: A chemotactic factor for polymorphonuclear leukocytes in vivo. *J Immunol* 134:564–8.

Bracci-Laudiero, L., Aloe, L., Levi-Montalcini, R., et al. 1992. Multiple sclerosis patients express increased levels of beta-nerve growth factor in cerebrospinal fluid. *Neurosci Lett* 147:9–12.

Bracci-Laudiero, L., Aloe, L., Levi-Montalcini, R., et al. 1993. Increased levels of NGF in sera of systemic lupus erythematosus patients. *Neuroreport* 4:563–5.

Braun, A., Appel, E., Baruch, R., et al. 1998. Role of nerve growth factor in a mouse model of allergic airway inflammation and asthma. *Eur J Immunol* 28:3240–51.

Braun, A., Lommatzsch, M., Renz, H., 2000. The role of neurotrophins in allergic bronchial asthma. *Clin Exp Allergy* 30:178–86.

Brodie, C. and E.W. Gelfand. 1992. Functional nerve growth factor receptors on human B lymphocytes: Interaction with IL 2. *J Immunol* 148:171–8.

Bruni, A., Bigon, E., Boarato, E., Mietto, L., Leon, A., Toffano, G. 1982. Interaction between nerve growth factor and lysophosphatidylserine on rat peritoneal mast cells. *FEBS Lett* 138:190–2.

Burgi, B., Otten, U. H., Ochensberger, B., et al. 1996. Basophil priming by neurotrophic factors. Activation through the trk receptor. *J Immunol* 157:5582–8.

Carlson, S.L., Albers, K.M., Beiting, D.J., Parish, M., Conner, J.M., Davis, B.M. 1995. NGF modulates sympathetic innervation of lymphoid tissues. *J Neurosci* 15:5892–9.

Caroleo, M.C., Costa, N., Bracci-Laudiero, L., Aloe, L. 2001. Human monocyte/macrophages activated by exposure to LPS overexpress NGF and NGF receptors. *J Neuroimmunol* 113:193–201.

Ciriaco, E., Dall'Aglio, C., Hannestad, J., et al. 1996. Localization of Trk neurotrophin receptor-like proteins in avian primary lymphoid organs (thymus and bursa of Fabricius). *J Neuroimmunol* 69:73–83.

Davies, A.M., Bandtlow, C., Heumann, R., Korsching, S., Rohrer, H., Thoenen, H. 1987. Timing and site of nerve growth factor synthesis in developing skin in relation to innervation and expression of the receptor. *Nature* 326:353–8.

de Vries, A., Dessing, M., Engels, F., Henricks, P., Nijkamp, F. 1999. Nerve growth factor induces a neurokinin-1 receptor-mediated airway hyperresponsiveness in guinea pigs. *Am J Respir Crit Care Med* 159:1541–4.

Dean, D.H., Hiramoto, R.N., Ghanta, V.K. 1987. Modulation of immune response. A possible role for murine salivary epidermal and nerve growth factors. *J Periodontol* 58:498–500.

Del Porto, F., Aloe, L., Lagana, B., Triaca, V., Nofroni, I., D'Amelio, R. 2006. Nerve growth factor and brain-derived neurotrophic factor levels in patients with rheumatic arthritis treated with TNF-alpha blockers. *Ann NY Acad Sci* 1069:438–43.

di Mola, F.F., Friess, H., Zhu, Z.W., et al. 2000. Nerve growth factor and Trk high affinity receptor (TrkA) gene expression in inflammatory bowel disease. *Gut* 46:670–9.

Ehrhard, P.B., Erb, P., Graumann, U., Otten, U. 1993. Expression of nerve growth factor and nerve growth factor receptor tyrosine kinase Trk in activated CD4-positive T-cell clones. *Proc Natl Acad Sci USA* 90:10984–8.

Ehrhard, P.B., Ganter, U., Bauer, J., Otten, U. 1993a Expression of functional *trk* protooncogene in human monocytes. *Proc Natl Acad Sci USA* 90:5423–7.

Fahnestock, M., Yu, G., Coughlin, M.D. 2004. ProNGF: A neurotrophic or an apoptotic molecule? *Prog Brain Res* 146:101–10.

Falcini, F., Matucci Cerinic, M., Ermini, M., et al. 1996. Nerve growth factor circulating levels are increased in Kawasaki disease: Correlation with disease activity and reduced angiotensin converting enzyme levels. *J Rheumatol* 23:1798–802.

Fantini, F., Magnoni, C., Brauci-Laudeis, L., Pincelli, C. 1995. Nerve growth factor is increased in psoriatic skin. *J Invest Dermatol* 105:854–5.

Farber, E.M., Nickoloff, B.J., Recht, B., Fraki, J.E. 1986. Stress, symmetry, and psoriasis: Possible role of neuropeptides. *J Am Acad Dermatol* 14:305–11.

Farber, E.M., Lanigan, S.W., Boer, J. 1990. The role of cutaneous sensory nerves in the maintenance of psoriasis. *Int J Dermatol* 6:418–20.

Farber, E.M., Raychaudhuri, S.P. 1999. Is psoriasis a neuroimmunologic disease? *Int J Dermatol* 38:12–5.

Felten, D.L., et al. 1985. Noradrenergic and peptidergic innervation of lymphoid tissue. *J Immunol* 135:755–65.

Fiocchi, C. 1997. Intestinal inflammation: A complex interplay of immune and nonimmune cell interaction. *Am J Physiol* 273:G769–75.

Garcia-Suarez, O., Germana, A., Hannestad, J., et al. 2000. TrkA is necessary for the normal development of the murine thymus. *J Neuroimmunol* 108:11–21.

Gee, A.P., Boyle, M.D., Munger, K.L., Lawman, M.J., Young, M. 1983. Nerve growth factor: Stimulation of polymorphonuclear leukocyte chemotaxis in vitro. *Proc Natl Acad Sci USA* 80:7215–8.

Hallböök, F. 1999. Evolution of the vertebrate neurotrophin and Trk receptor gene families. *Current Opinion Neurobiol* 9:616–21.

Hamada, A., Watanabe, N., Ohtomo, H., Matsuda H. 1996. Nerve growth factor enhances survival and cytotoxic activity of human eosinophils. *Br J Haematol* 93:299–302.

Hannestad, J., García-Suárez, O., Huerta, J.J., Esteban, I., Naves, F.J., Vega, J.A. 1997. TrkA neutrophin receptor protein in the rat and human thymus. *Anat Rec* 249:373–9.

Hauser, S.I.L., Bhan, A.K., Gilles, F., Kemp, M., Kerr, C., Weiner, H.L. 1986. Immunohistochemical analysis of the cellular infiltrate in multiple sclerosis lesions. *Ann Neurol* 19:578–87.

Hawkins, R.A., Claman, H.N., Clark, R.A.F., Steigerwald, J.C. 1985. Increased dermal mast cell populations in progressive systemic sclerosis: A link in chronic fibrosis? *Ann Inter Med* 102:182–6.

Horigome, K., Pryor, J.C., Bullock, E.D., Johnson, E.M.Jr. 1993. Mediator release from mast cells by nerve growth factor. Neurotrophin specificity and receptor mediation. *J Biol Chem* 268:14881–7.

Horigome, K., Bullock, E.D., Johnson, E.M.Jr. 1994. Effects of nerve growth factor on rat peritoneal mast cells. Survival promotion and immediate-early gene induction. *J Biol Chem* 269:2695–702.

Hoyle, G.W., Graham, R.M., Finkelstein, J.B., Nguyen, K.P.T., Gozal, D., Friedman, M. 1998. Hyperinnervation of the airways in transgenic mice overexpressing nerve growth factor. *Am J Respir Cell Mol Biol* 18:149–57.

Iannone, F., De Bari, C., Dell'Accio, F., et al. 2002. Increased expression of nerve growth factor (NGF) and high affinity NGF receptor (p140 TrkA) in human osteoarthritic chondrocytes. *Rheumatology* 41:1413–8.

Jiang, Y., Chen, G., Zhang, Y., Lu, L., Liu, S., Cao, X. 2007. Nerve growth factor promotes TLR4 signaling-induced maturation of human dendritic cells in vitro through inducible p75NTR 1. *J Immunol* 179:6297–304.

Kannan, Y.H., Ushuo, H., Koyama, H., et al. 1991. 2.5S Nerve growth factor enhances the survival, phagocytosis, and superoxide production of murine neutrophils. *Blood* 77:1320–5.

Kimata, H., Yoshida, A., Ishioka, C., Kusunoki, T., Hosoi, S., Mikawa, H. 1991a. Nerve growth factor specifically induces human IgG4 production. *Eur J Immunol* 21:137–41.

Kimata, H., Yoshida, A., Ishioka, C., Mikawa, H. 1991b. Stimulation of Ig production and growth of human lymphoblastoid B-cell lines by nerve growth factor. *Immunology* 72:451–2.

Kulka, M., Sheen, C.H., Tancowny, B.P., et al. 2008. Neuropeptides activate human mast cell degranulation and chemokine production. *Immunology* 123:398–410.

la Sala, A., Corinti, S., Federici, M., Saragovi, H.U., Girolomoni, G. 2000. Ligand activation of nerve growth factor receptor TrkA protects monocytes from apoptosis. *J Leukoc Biol* 68:104–10.

Lambiase, A., Bracci-Laudiero, L., Bonini, S., et al. 1997. Human CD4+ T cell clones produce and release nerve growth factor and express high affinity nerve growth factor receptors. *J Allergy Clin Immunol* 100:408–14.

Lambiase, A., Micera, A., Sgrulletta, R., Bonini, S., Bonini S. 2004. Nerve growth factor and the immune system: Old and new concepts in the cross-talk between immune and resident cells during pathophysiological conditions. *Curr Opin Allergy Clin Immunol* 4:425–30.

Leon, A., Buriani, A., Dal Toso, R., et al. 1994. Mast cells synthesize, store and release nerve growth factor. *Proc Natl Acad Sci USA* 91:3739–43.

Levi-Montalcini, R. 1952. Effects of mouse tumor transplantation on the nervous system. *Ann New York Acad Sci* 55:330–43.

Levi-Montalcini, R. 1987. The nerve growth factor 35 years later. *Science* 237:1154–62.

Lindsay, R.M. and A.J. Harmar. 1989. Nerve growth factor regulates expression of neuropeptides genes in adult sensory neurons. *Nature* 337:362–4.

Liuzzo, J.P., Petanceska, S.S., Devi, L.A. 1999. Neurotrophic factors regulate cathepsin S in macrophages and microglia: A role in the degradation of myelin basic protein and amyloid beta peptide. *Mol Med* 5:334–43.

Manni, L., Lundeberg, T., Fiorito, S., Bonini, S., Vigneti, E., Aloe, L. 2003. Nerve growth factor release by human synovial fibroblasts prior to and following exposure to tumor necrosis factor alpha, interleukin-1 beta and cholecystokinin-8: The possible role of NGF in the inflammatory response. *Clin Exp Rheumatol* 21:617–24.

Manning, P.T., Russell, J.H., Simmons, B., Johnson, E.M., Jr. 1985. Protection from guanethidine-induced neuronal destruction by nerve growth factor: Effect of NGF on immune function. *Brain Res* 340:61–9.

Marshall, J.S., Stead, R.H., McSharry, C., Nielsen, L., Bienenstock, J. 1990. The role of mast cell degranulation products in mast cell hyperplasia. I. Mechanism of action of nerve growth factor. *J Immunol* 144:1886–92.

Marshall, J.S., Gomi, K., Blennerhassett, M.G., Bienenstock, J. 1999. Nerve growth factor modifies the expression of inflammatory cytokines by mast cells via a prostanoid-dependent mechanism. *J Immunol* 162:4271–6.

Matsuda, H., Coughlin, M.D., Bienenstock, J., Denburg, J.A. 1988. Nerve growth factor promotes human hemopoietic colony growth and differentiation. *Proc Natl Acad Sci USA* 85:6508–12.

Matsuda, H., Switzer, J., Coughlin, M.D., Bienenstock, J., Denburg, J.A. 1988a. Human basophilic cell differentiation promoted by 2.5S nerve growth factor. *Int Arch Allergy Appl Immunol* 86:453–7.

Matucci-Cerinic, M., Kahaleh, B.M., LeRoy, E.C. 1995. The vascular involvement in systemic sclerosis. *Systemic Sclerosis*. Philadelphia (PA): Lea and Febiger, p. 513–74.

Micera, A., Vigneti, E., Pappo, O., et al. 2001. Nerve growth factor displays stimulatory effects on human skin and lung fibroblasts demonstrating a direct role for this factor in tissue repair. *Proc Natl Acad Sci USA* 98:6162–7.

Micera, A., Puxeddu, I., Aloe, L., Levi-Schaffer, F. 2003. New Insight on the involvement of nerve growth factor in allergic inflammation and fibrosis. *Cytokine Growth Factor Rev* 14:369–74.

Mosimann, B.L., White, M.V., Hohman, R.J., et al. 1993. Substance P, calcitonin gene-related peptide, and vasoactive intestinal peptide increase in nasal secretions after allergen challenge in atopic patients. *J Allergy Clin Immunol* 92:95–104.

Mullins, R.J., Olson, L.G., Sutherland, D.C. 1989. Nasal histamine challenges in symptomatic allergic rhinitis. *J Allergy Clin Immunol* 83:955–9.

Naukkarinen, A., Nickoloff, B.J., Farber, E.M. 1989. Quantification of cutaneous sensory nerves and their substance P content in psoriasis. *J Invest Dermatol* 92:126–9.

Nockher, W.A. and H. Renz. 2006. Neurotrophins and asthma: Novel insight into neuroimmune interaction. *J Allergy Clin Immunol* 117:67–71.

Olsson, Y. 1974. Mast cells in plaques of multiple sclerosis. *Acta Neurol Scand* 50:611–8.

Otten, U., Baumann, J.B., Girard, J. 1984. Nerve growth factor induces plasma extravasation in rat skin. *Eur J Pharmacol* 106:199–201.

Otten, U., Ehrhard, P., Peck, R. 1989. Nerve growth factor induces growth and differentiation of human B lymphocytes. *Proc Natl Acad Sci U S A* 86:10059–63.

Otten, U., Scully, J.L., Ehrhard, P.B., Gadient, R.A. 1994. Neurotrophins: Signals between the nervous and immune systems. *Prog Brain Res* 103:293–305.

Parvaneh Tafreshi, A. 2006. Nerve growth factor prevents demyelination, cell death and progression of the disease in experimental allergic encephalomyelitis. *Iran J Allergy Asthma Immunol* 5:177–81.

Patterson, J.C. and G.V. Childs. 1994. Nerve growth factor and its receptor in the anterior pituitary. *Endocrinology* 135:1689–96.

Pearce, F.L. and H.L. Thompson. 1986. Some characteristics of histamine secretion from rat peritoneal mast cells stimulated with nerve growth factor. *J Physiol* 372:379–93.

Piga, M., Passiu, G., Carta, P., et al. 2000. Increased pulmonary epithelial permeability in systemic sclerosis is associated with enhanced cutaneous nerve growth factor expression. *Eur J Int Med* 11:156–60.

Pincelli, C., Sevignani, C., Manfredini, R., et al. 1994. Expression and function of nerve growth factor and nerve growth factor receptor on cultured keratinocytes. *J Invest Dermatol* 103:13–18.

Raychaudhuri, S.P., Farber, E.M. 1993. Are sensory nerves essential for the development of psoriasis lesions? *J Am Acad Dermatol* 28:488–9.

Raychaudhuri, S.P., Jiang, W.Y., Farber, E.M. 1998. Psoriatic keratinocytes express high levels of nerve growth factor. *Acta Derm Venereol* 78:84–6.

Raychaudhuri, S.P., Jiang, W.Y., Smoller, B.R., Farber, E.M. 2000. Nerve growth factor and its receptor system in psoriasis. *Br J Dermatol* 143:198–200.

Raychaudhuri, S.P., Farber, E.M., Raychaudhuri, S.K. 2000a. Role of nerve growth factor in RANTES expression in keratinocytes. *Acta Derm Venereol* 80:247–50.

Raychaudhuri, S.K., Raychaudhuri, S.P., Weltman, H., Farber, E.M. 2001. Effect of nerve growth factor on endothelial cell biology: Proliferation and adherence molecule expression on human dermal microvascular endothelial cells. *Arch Dermatol Res* 293:291–5.

Raychaudhuri, S.P., Dutt, S., Raychaudhuri, S.K., Sanyal, M., Farber, E.M. 2001a. Severe combined immunodeficiency mouse-human skin chimeras: A unique animal model for the study of psoriasis and cutaneous inflammation. *Br J Dermatol* 144:931–9.

Raychaudhuri, S.P., Sanyal, M., Weltman, H., Raychaudhuri, S.K. 2004. K252a, a high affinity NGF receptor blocker improves psoriasis: An in vivo study using the SCID mouse-human skin model. *J Invest Dermatol* 122:812–9.

Raychaudhuri, S.P., Jiang, W.Y., Raychaudhuri, S.K. 2008. Revisiting the Koebner phenomenon: Role of NGF and its receptor system in the pathogenesis of psoriasis. *Am J Pathol* 172:961–71.

Raychaudhuri, S.P. and S.K. Raychaudhuri. 2009. The regulatory role of nerve growth factor and its receptor system in fibroblast-like synovial cells. *Scand J Rheumatol* 38:207–15.

Raychaudhuri, S.P., Raychaudhuri, S.K., Atkuri, K.R., Herzenberg, L.A., Herzenberg, L.A. 2011. Nerve growth factor: A key local regulator in the pathogenesis of inflammatory arthritis. *Arthritis Rheum* [Epub ahead of print].

Richard, A., McColl, S.R., Pelletier, G. 1992. Interleukin-4 and nerve growth factor can act as cofactors for interleukin-3-induced histamine production in human umbilical cord blood in serum-free culture. *Br J Haematol* 81:6–11.

Richardson, J.D. and M.R. Vasko. 2002. Cellular mechanisms of neurogenic inflammation. *J Pharmacol Exp Ther* 302:839–45.

Rihl, M., Kruithof, E., Barthel, C., et al. 2005. Involvement of neurotrophins and their receptors in spondyloarthritis synovitis: Relation to inflammation and response to treatment. *Ann Rheum Dis* 64:1542–9.

Sanico, A.M., Philip, G., Proud, D., Naclerio, R., Togias, A. 1998. Comparison of nasal mucosal responsiveness to neuronal stimulation in nonallergic and allergic rhinitis: Effects of capsaicin nasal challenge. *Clin Exp Allergy* 28:92–100.

Sanico, A.M., Stanisz, A.M., Gleeson, T.D., et al. 2000. Nerve growth factor expression and release in allergic inflammatory disease of the upper airways. *Am J Respir Crit Care Med* 161:1631–5.

Santambrogio, L., Benedetti, M., Chao, M.V., et al. 1994. Nerve growth factor production by lymphocytes. *J Immunol* 153:4488–95.

Sawada, J., Itakura, A., Tanaka, A., Furusaka, T., Matsuda, H. 2000. Nerve growth factor functions as a chemoattractant for mast cells through both mitogen-activated protein kinase and phosphatidylinositol 3-kinase signaling pathways. *Blood* 95:2052–8.

Shelton, D.L., Zeller, J., Ho, W.H., Pons, J., Rosenthal, A. 2005. Nerve growth factor mediates hyperalgesia and cachexia in auto-immune arthritis. *Pain* 116:8–16.

Sobel, R.A., Blanchette, B.W., Bhan, A.K., Colvin, R.B. 1984. The immunopathology of experimental allergic encephalomyelitis: Quantitative analysis of inflammatory cells in situ. *J Immunol* 132:2392–401.

Sofroniew, M.V., Howe, C.L., Mobley, W.C. 2001. Nerve growth factor signaling, neuroprotection, and neural repair. *Annu Rev Neurosci* 24:1217–81.

Solomon, A., Aloe, L., Pe'er, J., Frucht-Pery, J., Bonini, S., Bonini, S., et al. Nerve growth factor is preformed in and activates human peripheral blood eosinophils. *J Allergy Clin Immunol* 102:454–60.

Stead, R.H., Dixon, M.F., Bramwell, N.H., et al. 1989. Mast cells are closely apposed to nerves in the human gastrointestinal mucosa. *Gastroenterology* 97:575–85.

Strehlow, D. and J.H. Korn. 1998. Biology of the scleroderma fibroblast. *Curr Opin Rheumatol* 10:572–8.

Susaki, Y., Shimizu, S., Katakura, K., et al. 1996. Functional properties of murine macrophages promoted by nerve growth factor. *Blood* 88:4630–7.

Tam, S.Y., Tsai, M., Yamaguchi, M., et al. 1997. Expression of functional TrkA receptor tyrosine kinase in the HMC-1 human mast cell line and in human mast cells. *Blood* 90:1807–20.

Thorpe, L.W., Werrbach-Perez, K., Perez-Polo, J.R. 1987. Effects of nerve growth factor expression on interleukin-2 receptors on cultured human lymphocytes. *Ann NY Acad Sci USA* 496:310–1.

Torcia, M., Bracci-Laudiero, L., Lusibello, M., et al. 1996. Nerve growth factor is an autocrine survival factor for memory B lymphocytes. *Cell* 85:345–56.

Tuveri, M.A., Passiu, G., Mathieu, A., Aloe, L. 1993. Nerve growth factor and mast cell distribution in the skin of patients with systemic sclerosis. *Clin Exp Rheumatol* 11:319–22.

Urshansky, N., Mausner-Fainberg, K., Auriel, E., Regev, K., Farhum, F., Karni, A. 2010. Dysregulated neurotrophin mRNA production by immune cells of patients with relapsing remitting multiple sclerosis. *J Neurol Sci* 295:31–7.

U.S. National Institutes of Health. 2010. Tanezumab in osteoarthritis of the hip or knee. http://clinicaltrials.gov/ct2/show/NCT00985621?term=RN624%2C+OA&rank=5.

U.S. National Institutes of Health. 2011a. CT 327 in the treatment of psoriasis vulgaris. http://clinicaltrials.gov/ct2/show/NCT00995969?term=k252a%2C+psoriasis&rank=1.

U.S. National Institutes of Health. 2011b. Safety and tolerability of PG110 in patients with knee osteoarthritis pain. http://clinicaltrials.gov/ct2/show/NCT00941746?term=Anti+NGF&rank=3.

van de Loo, F.A., Joosten, L.A., van Lent, P.L., Arntz, O.J., van den Berg, W.B. 1995. Role of interleukin-1, tumor necrosis factor alpha, and interleukin-6 in cartilage proteoglycan metabolism and destruction. Effect of in situ blocking in murine antigen- and zymosan-induced arthritis. *Arthritis Rheum* 38:164–72.

Vega, J.A., Garcia-Suarez, O., Hannestad, J., Perez-Perez, M., Germana, A. 2003. Neurotrophins and the immune system. *J Anat* 203:1–19.

Villoslada, P., Hauser, S.L., Bartke, I., et al. 2000. Human nerve growth factor protects common marmosets against autoimmune encephalomyelitis by switching the balance of T helper cell type 1 and 2 cytokines within the central nervous system. *J Exp Med* 191:1799–806.

Wagner, U., Bredenbroker, D., Storm, B., et al. 1998. Effects of VIP and related peptides on airway mucus secretion from isolated rat trachea. *Peptides* 19:241–5.

Welker, P., Grabbe, J., Gibbs, B., Zuberbier, T., Henz, B.M. 2000. Nerve growth factor-β induces mast-cell marker expression during *in vitro* culture of human umbilical cord blood cells. *Immunology* 99:418–26.

White, B. 1994. Immune abnormalities in systemic sclerosis. *Clin Dermatol* 12:349–59.

Yamamoto, M., Ito, Y., Mitsuma, N., Li, M., Hattori, N., Sobue, G. 2001. Pathology-related differential expression regulation of NGF, GDNF, CNTF, and IL-6 mRNAs in human vasculitic neuropathy. *Muscle Nerve* 24:830–3.

Yan, Q., Settle, S.L., Wilkins, M.R. 1991. Hypotension induced by intravascular administration of nerve growth factor in the rat. *Clin Sci* 80:565–9.

Section III

Nutrition & Therapeutics for Inflammatory Diseases

19 Inflammation, Oxidative Stress, and Antioxidants

Naveen Kaushal, Vivek Narayan, Ujjawal H. Gandhi, Shakira M. Nelson, Anil Kumar Kotha, and K. Sandeep Prabhu
The Pennsylvania State University
University Park, Pennsylvania

CONTENTS

19.1 INTRODUCTION

Inflammation (*inflammare* (L.): to set on fire) is a fundamental physiological process and the body's protective reaction to injurious stimuli in the form of infection, trauma, pathogenic invasions, autoantibodies, ionizing radiation, and physical, chemical, or thermal stress. Broadly, there are two stages of inflammation, acute and chronic [1]. As part of an innate immune response, acute inflammation is beneficial for the host and is mediated through the activation of the immune system. On the other hand, chronic inflammation may predispose the host to various illnesses and forms the underlying basis for cancer, neurodegenerative diseases, HIV/AIDS, and gastrointestinal, cardiovascular, and autoimmune disorders [2] (Figure 19.1).

This understanding has led to the development of alternative therapies to mitigate inflammation that emphasize the identification of sources and mechanism(s) that regulate this complex process. There can be multiple sources of inflammation ranging from microbial to viral infections, obesity, autoimmune disorders, alcohol consumption, high calorie diet, and tobacco use [3]. These sources seem to be varied, but at the molecular level they converge into an intricate network of cellular signaling pathways that lead to the exacerbated production of inflammatory mediators, signifying oxidative stress.

Emerging evidence suggests that overproduction of reactive oxygen and nitrogen species (RONS), resulting in oxidative stress, plays a pivotal role in several disease states where inflammation serves

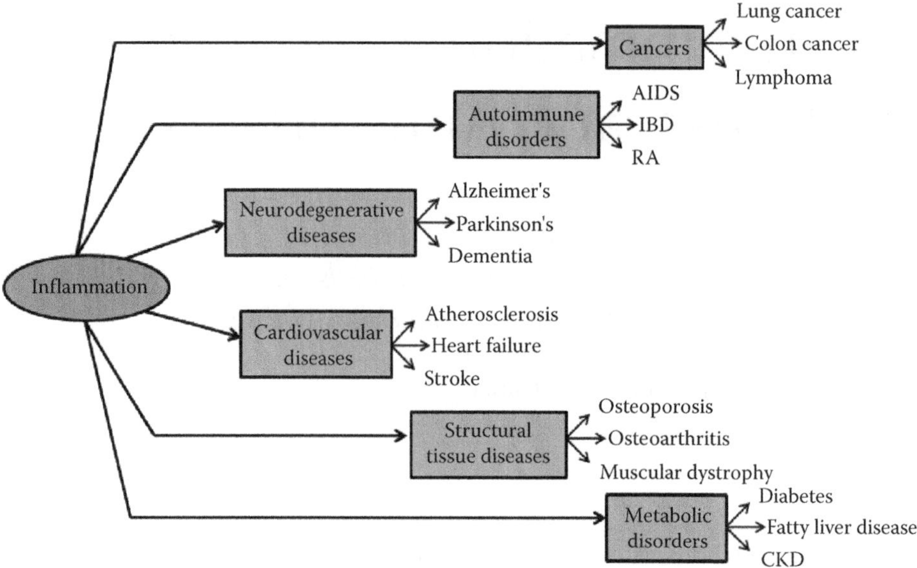

FIGURE 19.1 Schematic representation indicating diseases that have been linked to reactive oxygen species where inflammation plays a central role in pathogenesis. IBD, inflammatory bowel disease; CKD, chronic kidney disease.

as an underlying mechanism. RONS, particularly H_2O_2 and $ONOO^-$, have the ability to interact with many cellular lipids and proteins to elicit pathways of inflammation [4]. Apart from activating cytotoxic or cytocidal mechanisms, RONS can also modulate intracellular second messenger generating pathways leading to the activation of transcription factors of the nuclear factor-κB (NF-κB) family. Additionally, modulation of upstream kinase pathways involving STAT-3, p38, JNK, and Akt, contribute to shaping the inflammatory response [5] (Figure 19.2). Thus, knowledge of these

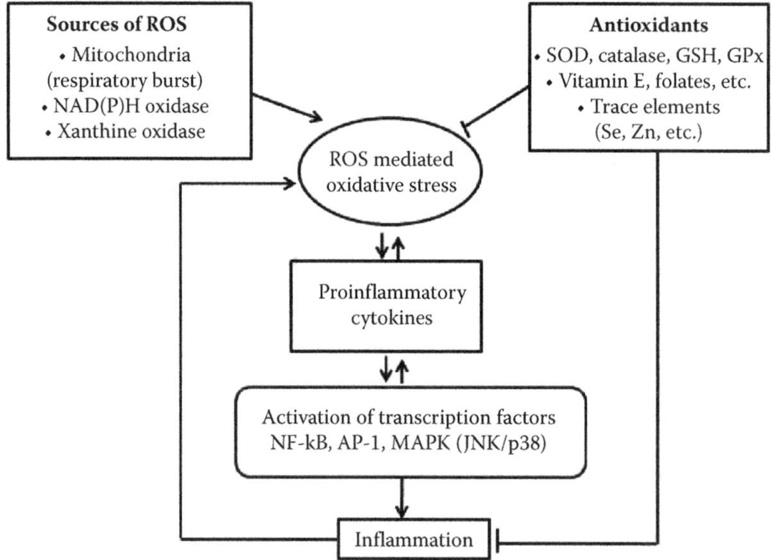

FIGURE 19.2 Schematic illustration of the imbalance between pro-oxidants (sources of ROS) and antioxidants leading to induction of ROS and inflammation.

mechanisms has elevated the importance of understanding both the molecular basis of inflammation and the regulatory systems.

Considering the critical role of oxidative stress in inflammation and pathogenesis, numerous preventive and therapeutic strategies have been developed, which incorporate antioxidants into the management of chronic diseases. At the primary level, biological systems have evolved a variety of antioxidant-based anti-inflammatory strategies to combat and help mitigate inflammatory stress. Such an elaborate antioxidant defense system comprises enzymes such as superoxide dismutase (SOD), catalase, glutathione peroxidase (GPx), and small peptides like reduced glutathione (GSH). Interestingly, many naturally occurring dietary supplements and nutrients have been shown to mitigate low-grade inflammation by specific mechanisms [6]. These antioxidants reduce inflammation by detoxification of reactive oxygen species (ROS), including H_2O_2, lipid and phospholipid hydroperoxides. Such changes in ROS impact gene expression signatures of key enzymes, such as the cyclooxygenases and lipoxygenases, which produce lipid mediators in the form of prostaglandins (PG), thromboxanes, prostacyclins, and oxidized fatty acids, respectively. Thus, knowledge of these mechanisms can not only lead to understanding the molecular basis of the control of inflammation, but can also greatly influence the development of treatment regimens with an emphasis on optimal antioxidant nutrition.

19.2 ROS MEDIATED OXIDATIVE STRESS AS A SOURCE OF INFLAMMATION

Free radicals, generated as a consequence of biochemical reactions, are regarded as the "necessary evil" that occupy the helm of a majority of all the pathological states. By definition, free radicals represent molecules containing one or more unpaired electrons that are formed due to alterations in redox homeostasis. An imbalance between the antioxidant defense system in cells and overwhelming levels of pro-oxidant species is thought to lead to the accumulation of free radicals, which is termed "oxidative stress." Oxidative stress results in the generation of ROS and reactive nitrogen species (RNS), which are together referred to as RONS. While these mediators are constantly generated under normal conditions as a consequence of aerobic metabolism, several enzymatic reactions such as autooxidation of catecholamine and the uncoupling of NO synthase (NOS) and xanthine oxidase (XO), NAD(P)H oxidase, and cytochrome P450 can lead to the production of RONS [7].

RONS are highly reactive products that exacerbate oxidative stress promoting inflammation. As a link between chronic inflammation and cancer, RONS activate intracellular signaling pathways leading to the transformation of normal cells to tumor cells [1]. Tumor promoters recruit inflammatory cells to stimulate and generate RONS, thereby causing a change in the tumor microenvironment to facilitate pathways that promote various stages of cancer development. Such a sustained inflammatory/oxidative environment, which constitutes a vicious circle, forms the underlying basis for tumorigenesis.

During inflammation, leukocytes are recruited to the site of damage, leading to increased release and accumulation of ROS at the site of damage due to "respiratory burst" [8]. H_2O_2 is a short-lived but highly toxic ROS that is also produced by other cellular components of relevance to inflammation. For example, both soluble immunoglobulins and the solubilized T-cell antigen receptor heterodimer can generate H_2O_2 [9]. It has been proposed that immunoglobulins use singlet oxygen ($^1O_2$) that oxidizes water to yield H_2O_2 [9]. Immunoglobulins, bound to the plasma membrane of B cells in association with accessory/costimulatory molecules to form the B cell receptor (BCR), serve as an important source of H_2O_2. However, plasma membrane NADPH oxidase, physically or functionally linked to the BCR, serves as the major source of H_2O_2 rather than the immunoglobulin molecule itself. The pivotal role of inducible NOS (iNOS) in inflammation is well known—it produces large amounts of NO and is expressed in immune cells, particularly macrophages, upon stimulation by a variety of mechanisms. However, endothelial NOS (eNOS) has recently been an object of similar consideration (Figure 19.3). Mechanical stimulation and inflammatory mediators upregulate

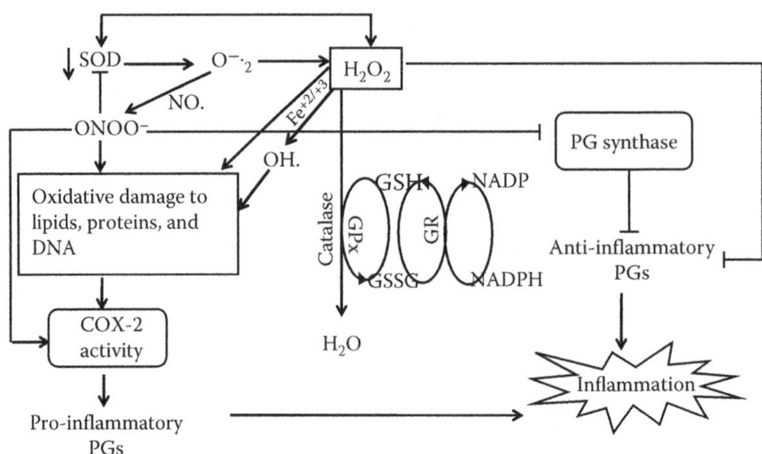

FIGURE 19.3 Schematic representation of the pivotal role of free radicals in inflammation. Upon stimulation of cells, free radicals cause oxidation of cellular lipids, proteins, and nucleic acids, thus inducing the inflammation by activating pro-inflammatory cytokine production.

eNOS [2]. Lack of eNOS-derived NO· was shown to upregulate the inflammatory pathway [2]. Thus, eNOS-derived NO· is a double-edged sword given its dual role in mitigating inflammation (as seen from its ability to decrease leukocyte endothelial cell interaction with inflammatory cells) and its ability to promote inflammation (in the form of perturbing plasma membrane permeability and thereby facilitating angiogenesis) [10].

Apart from H_2O_2 and NO·, inflammatory cells also produce oxidized lipid mediators that include metabolites of arachidonic acid, which along with cytokines and chemokines recruit inflammatory cells to the site of damage. The mobilization of arachidonic acid through the cytosolic phospholipase A2 (cPLA2) pathway produces oxidant species during the metabolic transformation by cyclooxygenase or lipoxygenase. Interestingly, this pathway might be responsible for ROS generation in cells stimulated via the TNF receptor I (TNFR1) or the FAS receptor, which further supports the notion that oxidative stress in the form of oxygen radicals serves as a critical player in TNF or FAS ligand (FASL)-dependent cell death [2].

One of the mechanisms of free radical–dependent exacerbation of inflammation originates from the ability of these highly reactive metabolites to oxidize cellular lipids, proteins, and nucleic acids [4]. In fact, ROS can react directly with biomolecules to cause cell damage and death. For example, superoxide and hydroxyl radicals initiate the process of autocatalytic lipid peroxidation. The products of oxidized lipids may themselves initiate further oxidative damage. Lipid peroxidation–induced inflammation not only impairs cellular homeostasis, but also plays an important role in the pathogenesis of chronic premalignant human diseases, where DNA damage is of common occurrence [11]. These pro-oxidants damage DNA by the generation of hydroxyl radicals (·OH) and singlet oxygen ($^1O_2$). The RONS produced by neutrophils and macrophages can contribute to mutagenesis of DNA by inducing strand breaks, oxidation of purine bases, and formation of DNA-protein cross-links. As a result of such effects on biomolecules, excessive ROS can mediate alterations in chromatin structure to modulate gene expression [12]. Along these lines, chronic inflammation increases the generation of highly reactive nitric oxide intermediates such as peroxynitrate, which can damage DNA by alkylation, deamination, oxidation, nitration of nucleobases, and single and double strand breaks [13]. Furthermore, DNA damage by itself can result in ROS generation, which can attack other targets that are vital for cell viability [14]. In addition, formation of DNA adducts with electrophiles generated from lipid peroxidation reactions also form an additional underlying molecular mechanism of disease pathogenesis. For example, reactive aldehydes like 4-hydroxy-2-nonenal (4-HNE), malondialdehyde, acrolein, and crotonaldehyde, which are products of lipid peroxidation, react directly with DNA bases or generate bifunctional intermediates that form exocyclic DNA adducts [15]. More importantly, guanine in DNA is the most targeted base that is oxidized by RONS to 8-nitroguanine that is generated at the site of injury by inflamed macrophages.

In addition to mitochondria, another source of oxidants is the endoplasmic reticulum (ER) where O_2^- is generated by leakage of electrons from NADPH cytochrome P450 reductase. Oxidative damage to resident proteins of the ER could also serve as one of the contributing factors in inflammation and ageing. The susceptibility of proteins to oxidative damage depends on oxidation-sensitive amino acid residues, presence of metal binding sites, subcellular localization, turnover, and structural perturbations. Modification of proteins by mediators of inflammation can lead to denaturation that can potentially render proteins nonfunctional. For example, carbonylation of lysine affects protein functions during signaling process. Carbonylated proteins that are not degraded form toxic aggregates that may affect cell viability. These modified proteins are often observed in patients with Parkinson's disease, Alzheimer's disease, and cancer. Similarly, alkylation of some proteins (p300/CBP) by fatty acid metabolites can target such proteins to ubiquitination followed by degradation by the proteosome. Apart from affecting protein function, mild oxidative stress can also impact mRNA translation, thus modulating protein synthesis [16].

The relationship of oxidative stress to inflammation has become synonymous with the fact that certain ROS function as messenger molecules to propagate inflammatory signals. The most important effects of oxidants on cellular signaling pathways, which are central to inflammation, include activation of MAPK/AP-1 and NF-κB pathways (Figure 19.2) [5]. ROS are implicated as second messengers involved in activation of redox sensitive transcription factor NF-κB [17]. NF-κB also controls the transcription of a number of pro-inflammatory cytokines, chemokines, enzymes of arachidonic acid cascade, and genes involved in cell transformation, proliferation, and angiogenesis [18]. In addition to NF-κB, ROS are also known to activate other transcription factors, such as signal transducer and activator of transcription 3 (STAT3), hypoxia-inducible factor (HIF)-1α, activator protein-1 (AP-1), nuclear factor of activated T cells (NFAT), and NF-E2 related factor-2 (Nrf-2), which also play a critical role in mediating cellular stress responses [1].

The induction of cyclooxygenase (COX)-2 and iNOS, aberrant expression of inflammatory cytokines (TNFα, IL-1β, IL-6), and chemokine receptor 4 (CXCR4), as well as alterations in the expression of specific miRNAs have also been reported to play a role in oxidative stress-induced inflammation [19].

While oxidants and free radicals activate a complex network of pathways leading to inflammation, a number of cellular defense mechanisms have evolved to mitigate this oxidative insult. Such an arsenal of strategies, endowed in most cells, is composed of the body's antioxidant defense system (ADS) constituting enzymes such as SOD, catalase, glutathione peroxidases (GPx), glutathione reductase, and glutathione-*S*-transferases (GSTs). In addition to this well-defined defense mechanism, certain metals, vitamins, and dietary supplements also are essential to maintain the levels of ROS and the homeostatic set points.

19.3 ANTIOXIDANTS AS ANTI-INFLAMMATORY AGENTS

While ROS play a significant role in host defenses, an overabundant production is far more detrimental, leading to the induction of stress and redox-sensitive signaling pathways, such as NF-κB [20]. As a compensatory mechanism, cells have evolved an elaborate "ADS" that is made up of antioxidant enzymes that metabolize ROS. As part of the ADS, catalase is identified as an efficient antioxidant enzyme and an important factor in mitigating inflammation. Catalase is a heme-containing protein found in the peroxisome of cells [21, 22]. Its main effects are seen with H_2O_2, via a two-step reduction, to water and molecular oxygen [22]. In specific cases, catalase protects the body against high levels of exogenous H_2O_2, which could lead to the induction of inflammatory genes from specific immune cells [21]. Additionally, catalase has a very high turnover rate, converting up to 60 million molecules of H_2O_2 per minute [22]. Studies have shown that a deficiency in catalase is a contributing factor to the development of diabetes [21].

Superoxide dismutases catalyze the dismutation of superoxide to H_2O_2 and, thus, couples enzymatically with catalase to reduce pro-oxidant superoxide to water and molecular oxygen. The mechanics of reduction mimics a "ping-pong" type mechanism, where a transition metal ion (such as Cu-Zn, Fe, Mn, Ni) is involved in a redox reaction in the active site [21, 22]. There are three major families of SOD, depending on the metal cofactor (Cu/Zn, Mn/Fe, and Ni). While each isoform has different functions and features, they all serve to reduce pro-oxidants. A wealth of literature supports the importance of SODs in mechanisms such as aging and neural tube development (dysraphic anomalies) to chemoprevention, suggesting the importance of ADS in normal physiology [21, 22].

Glutathione peroxidase exists in two forms, a selenium-dependent form (GPx) and a selenium-independent form (peroxidase activity associated with GST), with GPx being the most studied form [22]. As the name indicates, GPx provides protection against peroxides, creating water and an oxidized form of glutathione from hydrogen peroxide [22]. Studies have shown that uncontrolled production of peroxides, via lipid peroxidation and other methods, can lead to a dysregulation of inflammation and, subsequently, inflammatory diseases [23]. While GPx and catalase compete against each other for H_2O_2 as a substrate, the former enzyme acts on both lipid hydroperoxides in addition to H_2O_2, unlike catalase [22]. Being a heme protein, catalase also produces free radicals as by-products during the course of its catalytic cycle, which is not the case with GPx.

While catalase, SOD, and GPx are all antioxidant enzyme systems produced within the body, there are a number of small molecular antioxidants that are also found in foods. These antioxidants are in the forms of vitamins, minerals, and trace elements that complement the ADS.

Adequate intake of vitamins and minerals is required to maintain a homeostatic environment, where deficiencies can lead to underdeveloped or nonfunctional immune functions. As discussed before, the generation of ROS is part of the immune response. A decrease in the body's ability to respond to the increase in ROS results in escalating inflammation and vulnerability to oxidative stress [24]. Folate, vitamin E, vitamin B6, vitamin C, vitamin D, and vitamin A all play important roles in mitigating oxidative stress-dependent inflammation.

Used rapidly during an infection, vitamin C mitigates oxidative stress [24]. In fact, one of the main effects of vitamin C is maintaining a low oxidative state in cells during an inflammatory response, thus allowing the body to respond to an infection without the possible development of "inflammation chronicity." Attacking free oxygen radicals at a rate comparable to vitamin C, vitamin B6 has recently been identified as an efficient biological antioxidant [25]. Playing a large role in suppressing the progression of atherosclerosis and chronic inflammation, vitamin B6 is heavily involved in active-oxygen resistance [25]. In addition to these vitamins, folate has also been identified as an antioxidant. Attacking radicals different from vitamin C and B6, folate nonetheless reduces the presence of these radicals at rates similar to many antioxidants [25].

Like water-soluble vitamins, fat-soluble vitamins (vitamin E, vitamin D, and vitamin A) are also known to act as antioxidants. While both vitamin A and D have been studied and were found to play roles as antioxidants, it is vitamin E that has the greatest impact. Production of lipid peroxides alters the functions of cellular pathways and causes oxidative irreversible damage to membrane lipids [24, 26]. Vitamin E protects against this damage via its function as a free radical quencher. In asthma, ROS can induce bronchoconstriction to cause vascular leakage and pulmonary tissue damage [26]. Vitamin E has been shown to regulate these processes, protecting the lungs from ROS, decreasing inflammation, and preventing damage from occurring. In some cases, vitamin E has also been shown to optimize and enhance the immune response through its role as a free radical quencher [24].

Both water- and fat-soluble vitamins play a large role in combating the production of oxidative reagents, in conjunction with other antioxidants and trace elements. It is important to recognize that the combination of vitamin E, selenium (in the form of selenoprotein thioredoxin reductase), and dehydroascorbic acid functions as an efficient and regenerating antioxidant system to reduce free radicals in cells.

19.4 TRACE ELEMENTS AS ANTIOXIDANTS

Aside from the various antioxidants described earlier, some metal nutrients, like selenium (Se), zinc (Zn), copper (Cu), and iron (Fe), also facilitate protection from oxidative stress. They function as antioxidants primarily through their incorporation into proteins, either in the active site or as cofactors. In this section, we will describe the role of some of these trace nutrients as antioxidants and also explain their mode of action.

19.4.1 Selenium

The essential micronutrient Se has been attributed to a variety of beneficial physiologic properties because of its ability to be co-translationally incorporated into a class of cellular proteins, selenoproteins, in the form of the 21st amino acid, selenocysteine (Sec). About 30 different selenoproteins are known to be expressed in the mammalian system, which include the well-characterized GPx, TR, deiodinases, and others that are currently being studied, such as selenoproteins P, K, W, and R. Selenoenzymes of the glutathione peroxidase (GPx) and thioredoxin reductase (TR) families are involved in regeneration of antioxidant systems, maintenance of intracellular redox state and membrane integrity, as well as gene regulation by redox control of binding of transcription factors to DNA [27]. Epidemiological evidence shows that inadequate levels of Se and the associated production of RONS are linked with a higher incidence of cardiovascular diseases, progression of viral infections including HIV/AIDS, Alzheimer's disease, and infertility, where inflammation plays a significant role in pathogenesis. Anticarcinogenic properties of supra-nutritional doses of Se have been shown through mechanisms such as perturbation of tumor cell metabolism, induction of apoptosis, and inhibition of angiogenesis, which suggest a certain degree of overlap between the anti-inflammatory and antiproliferative effects of Se [28]. Also, supplementation of Se has been shown to improve the health status of patients suffering from inflammatory conditions like septic shock, autoimmune thyroiditis, pancreatitis, asthma, and rheumatoid arthritis.

Glutathione peroxidases, one of the most well-studied classes of selenoproteins, provide protection against oxidative damage by reducing hydrogen peroxide, organic hydroperoxides, and phospholipid hydroperoxides to their corresponding less-reactive alcohols. Five different isoenzymes of Sec-containing GPxs are known in humans (GPx1-4 and 6), each with specific substrate specificities and tissue-specific distribution [29]. The enzymatic reaction involves the selenocysteine in the active site and also two glutathione (GSH) molecules as cofactors. Studies have shown that GPx activity is greatly diminished in individuals deficient in Se, while Se supplementation of such individuals increases the activity.

Another family of selenoproteins, TRs, is also capable of alleviating oxidative stress by reducing hydrogen peroxide and lipid hydroperoxides, modulating the redox status of many proteins. The three isoenzymes of human TR (TR1-3) contain a selenocysteine residue in the C-terminus, which has been shown to be important for activity [29]. The antioxidant activity of TR occurs via a complex reaction cascade involving the transfer of electrons from NADPH to a disulfide in one subunit via FAD, to the selenylsulfide of the other subunit [30]. The reaction also involves thioredoxin (Trx) as a cofactor. TRs also provide protection from oxidative stress by maintaining Trx in a reduced state, as reduced Trx is a major electron donor in many redox reactions, restoring the enzymatic activity of oxidized peroxiredoxins, which are involved in the degradation of hydroperoxides and peroxynitrites [31].

Another ROS detoxifying selenoprotein is selenoprotein P (SePP), which is mainly secreted by the hepatocytes. Although its primary role is a transporter of Se to extrahepatic tissues (its C-terminus has up to 10 Sec residues), its antioxidant capacity has been demonstrated in rats, presumably owing to its N-terminus Sec [32]. SePP has been shown to reduce phospholipid hydroperoxides by using either GSH or Trx as a cosubstrate. SePP can also protect plasma proteins against oxidation and nitration by peroxynitrites, and also prevent the peroxidation of low-density lipoproteins (LDLs) [33].

Apart from selenoproteins, many organic selenocompounds have been discovered, which have ROS scavenging activity. Selenomethionine and ebselen (an Se-containing small molecule that functions as a GPx mimetic) have been reported to be able to scavenge $ONOO^-$ [34]. Selenocarbamate and selenourea compounds have been tested for superoxide scavenging activity, and the results suggest that these could be significant in the treatment of oxidative stress and superoxide radical-associated disorders [35].

19.4.1.1 Se and Its Role in Inflammation: Modulation of Lipid Metabolites as Key Regulators of Inflammation

Pathologically, inflammation is characterized by vascular events such as vasodilatation and exudation of plasma, and cellular events in the form of infiltration of the affected tissue by cells of the immune system such as neutrophils and macrophages, which are chiefly responsible for orchestrating the events of inflammation. Macrophages play a role in appropriately sustaining an inflammatory response and bringing a timely resolution. Thus, a tight regulation of macrophage function is critical for preventing the progression of a protective, physiological, acute inflammatory response to a destructive, pathological, chronic inflammatory state. Such chronic inflammation forms the basis of highly prevalent disease, including atherosclerosis, rheumatoid arthritis, inflammatory bowel disease, bronchial asthma, chronic pancreatitis, systemic vasculitides, periodontal disease, sarcoidosis, Alzheimer's disease, chronic glomerulonephritis, as well as a wide variety of malignancies.

Apart from undergoing the oxidative "respiratory" burst, "classical" macrophage activation includes the production of protein mediators like IL-1, IL-6, and TNF-α, as well as lipid mediators like the arachidonic acid (AA)-derived eicosanoids—PGE_2, PGD_2, thromboxane $(TX)A_2$, and 15d-PGJ_2. The vascular endothelial cells produce PGI_2 (prostacyclin), which plays a role in the vascular events on inflammation by counteracting the effects of TXA_2. The cell membrane-derived 20-carbon fatty acid, AA, is acted upon by COX-2, an enzyme that is rapidly induced by inflammatory stimuli. The resultant product, PGH_2, can be converted into a variety of different prostaglandins through reactions catalyzed by prostaglandin synthase enzymes viz. microsomal PGE synthase (mPGES-1), hematopoietic PGD synthase (H-PGDS), TXA synthase (TXAS), and prostacyclin synthase (PGIS) depending on the state of inflammation. Using a model of normal wound repair, Kapoor et al. have demonstrated a crucial role of AA-derived lipid mediators in the initiation and resolution of acute inflammation by shifting from pro-inflammatory PGE_2 to anti-inflammatory PGD_2 and its metabolite, 15d-PGJ_2 [36]. Extensive work on resolution of inflammation by Serhan et al. has revealed the presence of lipid mediators derived not only from AA but also from eicosapentaenoic acid (EPA) and docosahexaenoic acid (DHA), namely, resolvins and protectins, respectively, with high anti-inflammatory and pro-resolving properties [37]. Knowing that lipid mediators can play a dual role in inflammation, researchers have focused on studying the regulation of inflammation, including the discovery of compounds of dietary origin that exert control over inflammatory pathways.

Increasing interest in elucidating the anti-inflammatory mechanisms of Se has led to studies focusing on its role in the regulation of the AA-pathway metabolism that plays a major role in the events of inflammation as described earlier. It has become increasingly clear that Se could alter the metabolome given that many of the pathways of lipid oxidation (enzymatic and nonenzymatic) are sensitive to changes in the cellular redox status.

The links between lipid-derived free radicals and Se were elegantly elucidated by demonstrating that plasma GPx had fatty acid hydroperoxide-reducing properties, which prevented the deleterious inhibitory effect of such hydroperoxides on the protective AA pathway regulator prostacyclin synthase [38]. Separate studies on alveolar macrophages and lung neutrophils isolated from rats maintained on Se-deficient diets showed increased production of PGE_2 and TXA_2; the effects were attributed to changes in cellular hydroperoxide levels [39]. In a study involving an experimental myocardial ischemia model, rats fed with grains grown in Se-deficient soil showed higher activities of phospholipase A_2 (PLA_2), creatine kinase (CK), lactate dehydrogenase (LDH), and higher levels of AA, TXA_2, leukotriene C_4 (LTC_4), and lipid hydroperoxides. However, levels of PGI_2 were decreased 48 hours after ligation of the coronary vessels in rats. Supplementation with Se or vitamin E in the feed reversed the aforementioned effects, but the combination of Se and vitamin E gave the best results [40]. Using human placental explants, Eisenmann et al. demonstrated that different forms of Se variably affect the ratio of TXB_2/6-keto-$PGF_{1\alpha}$, which is a major determinant of preeclamptic pregnancies [41]. Furthermore, Cao et al. demonstrated that Se-deficient bovine mammary endothelial cells (BMEC) produced higher amounts of 15-hydroperoxyeicosatetraenoic acid (15-HPETE) and TXA_2, and lower amounts of PGI_2, which leads to vascular dysfunction [42]. These studies indicate that the AA pathway is redox sensitive and that Se can modulate the levels of downstream AA metabolites involved chiefly in the vascular events of inflammation.

An inverse co-relationship between the level of Se and that of cellular COX-2 and iNOS was demonstrated using a murine macrophage model attributing to an inverse causal regulation of the NF-κB family of transcription factors and cellular Se status. Vunta et al. showed that the anti-inflammatory effects of Se supplementation on macrophages were due to the increased production of the cyclopentenone eicosanoid mediator 15d-PGJ_2 [43]. Our studies further demonstrated that Se-dependent cellular production of 15d-PGJ_2 also inhibited the activation of NF-κB via the modulation of the redox status of a critical cysteine thiol of a pivotal upstream kinase, IKK-β. In addition, Se was demonstrated to activate the nuclear receptor and transcription factor, peroxisome proliferator-activated receptor (PPAR)γ, in a ligand-dependent manner that is known to transrepress many NF-κB target genes as well as activate a wide range of anti-inflammatory genes [43]. Recently, Gandhi et al. showed that bioavailable forms of Se that contribute to the formation of selenoproteins modulate the AA-pathway metabolism by shunting it from the production of pro-inflammatory mediators like PGE_2 and TXA_2 toward the production of anti-inflammatory downstream PGD_2 products, Δ^{12}-PGJ_2 and 15d-PGJ_2, as a result of differential regulation of the levels of various PG synthase enzymes [44] (Figure 19.4). Furthermore, Nelson et al. [45] have demonstrated that such a shunting in PG metabolism promoted a polarization of the "classically" (M1) activated macrophages (pro-inflammatory) toward the "alternative" (M2) activation state that favors resolution of inflammation (Figure 19.5). Taken together, these recent studies indicate a new mechanism in the control of inflammation, where dietary Se in the form of selenoproteins play a pivotal role. More studies are warranted to identify specific selenoproteins and address their role in mitigating inflammation and/or enhancing resolution of inflammation.

19.4.2 Zinc

Zinc (Zn) exerts its antioxidant properties via a group of low-molecular-weight metal binding proteins called metallothioneins (MT) apart from other Zn binding proteins. Chronic exposure to Zn induces the expression of MTs, which have the ability to scavenge ROS and protect tissues against ROS-induced oxidative damage [46]. MTs, which are Cys-rich proteins, act as thiol donors to protect against DNA damage induced by hydroxyl radicals [47]. MTs could also functionally substitute for Cu, Zn superoxide dismutase (SOD1) in $SOD1^{-/-}$ mice [48] in addition to their function as transporters of Zn, Cu and other physiological and xenobiotic heavy metals via the thiol group of their Cys residues. Upon reacting with ROS, the Cys residues in MTs get oxidized to cystine, which causes the release of the bound metals [49]. Zn released in this fashion can in turn induce the expression of MTs by binding to metal transcription factors, which then bind to metal response elements (MREs) in the MT genes [50]. Zinc has also been known to protect against iron (Fe^{2+})-induced

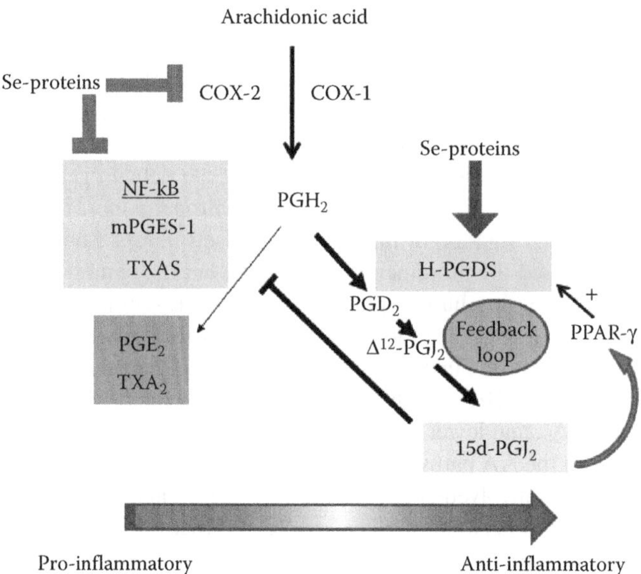

FIGURE 19.4 Schematic illustration of the shunting of arachidonic acid metabolism by selenoproteins in macrophages. Macrophages cultured in the presence of bioavailable selenium leads to the enhanced expression of H-PGDS and its product, PGD_2 and 15d-PGJ_2, while pro-inflammatory PGE_2 and TXA_2 that are products of m-PGES-1 and TXAS, respectively, are decreased. Inhibition of NF-κB and activation of PPAR-γ are two major pathways that are affected by selenoproteins in these cells to modulate pathways of shunting. Increased activation of PPAR-γ facilitates a positive feedback upregulation of H-PGDS, leading to increased levels of CyPGs. COX, cyclooxygenase; mPGES-1, microsomal PGE_2 synthase; H-PGDS, hematopoietic PGD_2 synthase; PPAR-γ, peroxisome proliferator activated receptor-γ.

lipid oxidation on membranes by acting synergistically with lipid- and water-soluble antioxidants like α-tocopherol and epicatechin [51]. It also has the capability to prevent the binding of highly oxidative metal ions like Fe^{2+} and Cu^+ to binding sites [51]. This prevents the local redox cycling of these metals, which can lead to oxidative damage. Another MT-independent antioxidant activity of Zn is to inhibit the generation of ROS (CYP4502E1) in liver in response to elevated alcohol levels [52]. Zn also contributes indirectly to oxidative stress protection by maintaining the conformation of SODs [53].

19.4.3 Copper

Copper (Cu) is essential in diet, as it is involved in the proper utilization of iron and is also involved in the production of important biomolecules such as SOD and cytochrome c oxidase [53]. Of the three SODs in humans, SOD1 (cytosolic) and SOD3 (extracellular), called Cu/Zn SOD, have Cu in their active sites as Cu^{2+} [54]. Cu/Zn SODs catalyze the dismutation of the superoxide radical (O_2^-) into molecular oxygen (O_2) and H_2O_2, which involves the alternate reduction and oxidation of the active site Cu. Apart from these enzymes, Cu is also important in the production of ceruloplasmin (Cp). Cp is a multi-copper oxidase (MCO) that displays antioxidant properties where it oxidizes free Fe^{2+} and Cu^+ to favor a "subdued" oxidizing environment in tissues [55].

19.5 SUMMARY AND FUTURE DIRECTIVES

Evidently, overproduction of RONS, resulting in oxidative stress, plays a prominent role in several disease states, including cardiovascular diseases, autoimmune disorders, arthritis, cancer, neurodegenerative diseases, and AIDS, where inflammation forms the underlying basis. The intervention

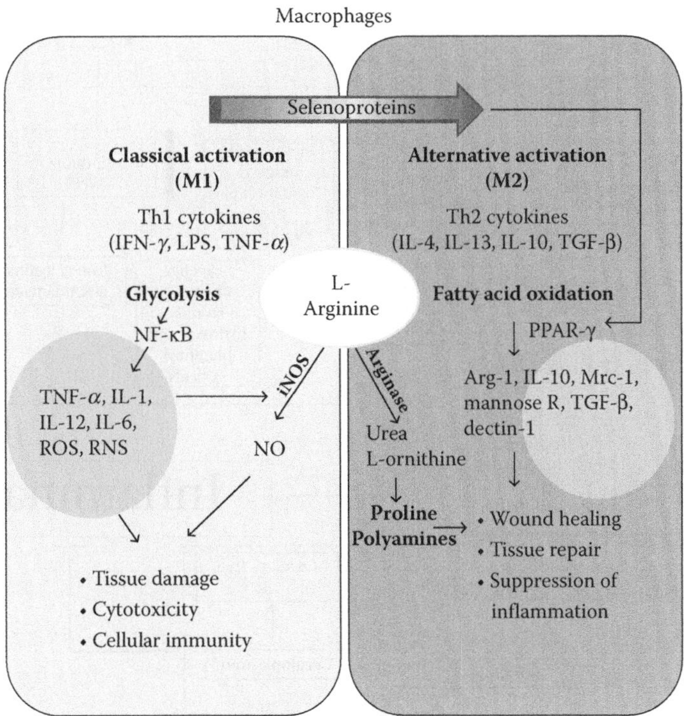

FIGURE 19.5 Implication of selenium-dependent eicosanoid shunting on pathways of anti-inflammation by macrophage phenotype switching. Selenoproteins are essential for the upregulation of cellular markers of M2 (anti-inflammatory) macrophages. Selenoproteins effectively mitigate RONS production, thus shunting of arachidonic acid metabolism toward CyPGs; changes in the gene expression within the pro-inflammatory (M1) macrophages facilitates their switching to anti-inflammatory (M2) macrophages to activate pro-resolution pathways.

studies based on reducing generation of ROS and RNS have been found to be beneficial in the treatment of inflammation. Studies strongly project the beneficial effects of Cu/Zn SOD, GPx, and other dietary antioxidants such as vitamin E and Se in the prevention and/or treatment of some of these diseases. These effects occur via the mitigation of inflammatory signaling pathways by endogenous and exogenous antioxidants as part of a complex signaling network that transduces signals and modulates the cellular intermediary metabolism (Figure 19.6).

However, lack of understanding of antioxidant therapies based on individual variations poses serious barriers to their introduction as standalone therapies in clinical medicine. These variations at the individual level makes it challenging to establish the decisive clinical biomarkers of inflammation. Furthermore, there exist no sensitive, facile, and accurate assays for oxidative stress that can predict the type of antioxidant supplementation that might be appropriate for a specific individual. The development of these treatment regimens have to rely on the understanding of multiple extracellular and intracellular pathways involved in ROS and RNS metabolism and their impact on the cellular signaling pathways. An important limitation to correlate changes in oxidant tone of cells or tissues is the ability to measure and quantitatively relate changes in oxidative stress-based biomarkers and antioxidant status of an individual to prevailing health conditions. Thus, there is an urgent need for a detailed longitudinal study to evaluate oxidative biomarkers along with traditional clinical end points. Research in this direction will usher in a new discipline of molecular medicine with an emphasis on oxidative stress and antioxidant therapeutics.

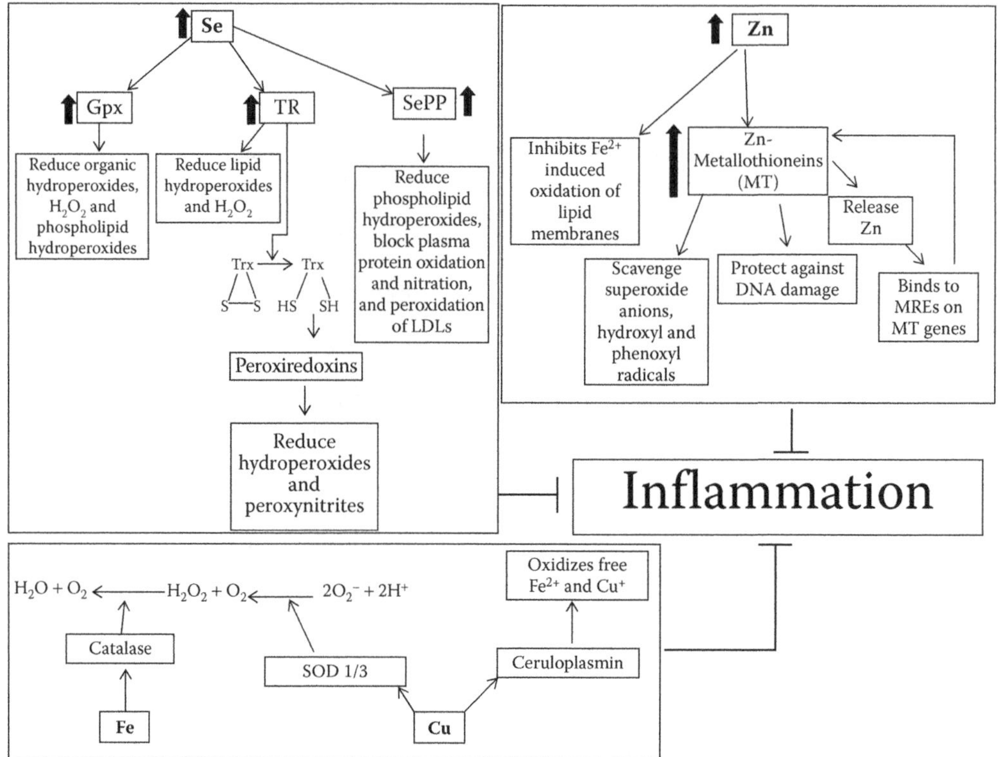

FIGURE 19.6 Scheme for the role of various trace elements such as Se, Zn, Fe, and Cu as antioxidants. These trace elements ablate inflammation primarily by mitigating the free radical–mediated oxidative stress as part of major antioxidant enzyme defense systems.

ACKNOWLEDGMENTS

Funding from the National Institutes of Health and American Institute for Cancer Research to KSP is gratefully acknowledged. We thank Dr. Tejo P. Nagaraja for his suggestions and editorial assistance.

TAKE-HOME MESSAGES

- Reactive oxygen and nitrogen species (RONS)-mediated oxidative stress occupies a central role in inflammation, which plays a prominent role in several disease states.
- Considering the critical role of oxidative stress in inflammation and pathogenesis, numerous preventive and therapeutic strategies have been developed that rationally incorporate antioxidants into the management of chronic diseases.
- Increased understanding of the mechanisms of inflammation has led us to think of alternative therapies to mitigate inflammation, which emphasizes the identification of sources and mechanism(s) that regulate this complex process.
- Apart from an elaborate antioxidant defense system in the body, various dietary antioxidants such as vitamins, folates, and some essential trace elements (micronutrients) also facilitate protection from oxidative stress.
- Lack of complete understanding of molecular mechanisms of action of antioxidants and a complex association of intraindividual variations pose serious barriers to the introduction of antioxidant therapies into clinical medicine.

- There is an urgent need to evaluate oxidative biomarkers along with traditional clinical end points to usher in a new discipline of antioxidant therapies.

REFERENCES

1. Reuter S, Gupta SC, Chaturvedi MM, Aggarwal BB: Oxidative stress, inflammation, and cancer: How are they linked? *Free Radic Biol Med* 2010, 49:1603–1616.
2. Di Virgilio F: New pathways for reactive oxygen species generation in inflammation and potential novel pharmacological targets, *Curr Pharm Des* 2004, 10:1647–1652.
3. Aggarwal BB, Vijayalekshmi RV, Sung B: Targeting inflammatory pathways for prevention and therapy of cancer: Short-term friend, long-term foe, *Clin Cancer Res* 2009, 15:425–430.
4. Zhou Q, Mrowietz U, Rostami-Yazdi M: Oxidative stress in the pathogenesis of psoriasis, *Free Radic Biol Med* 2009, 47:891–905.
5. Muller JM, Cahill MA, Rupec RA, Baeuerle PA, Nordheim A: Antioxidants as well as oxidants activate c-fos via Ras-dependent activation of extracellular-signal-regulated kinase 2 and Elk-1, *Eur J Biochem* 1997, 244:45–52.
6. Chen J, Berry MJ: Selenium and selenoproteins in the brain and brain diseases, *J Neurochem* 2003, 86:1–12.
7. Sawyer DB, Siwik DA, Xiao L, Pimentel DR, Singh K, Colucci WS: Role of oxidative stress in myocardial hypertrophy and failure, *J Mol Cell Cardiol* 2002, 34:379–388.
8. Hussain SP, Hofseth LJ, Harris CC: Radical causes of cancer, *Nat Rev Cancer* 2003, 3:276–285.
9. Reth M: Hydrogen peroxide as second messenger in lymphocyte activation, *Nat Immunol* 2002, 3:1129–1134.
10. Fukumura D, Gohongi T, Kadambi A, Izumi Y, Ang J, Yun CO, et al.: Predominant role of endothelial nitric oxide synthase in vascular endothelial growth factor-induced angiogenesis and vascular permeability, *Proc Natl Acad Sci USA* 2001, 98:2604–2609.
11. Rosin MP, Saad el Din Zaki S, Ward AJ, Anwar WA: Involvement of inflammatory reactions and elevated cell proliferation in the development of bladder cancer in schistosomiasis patients, *Mutat Res* 1994, 305:283–292.
12. Rahman I: Oxidative stress, chromatin remodeling and gene transcription in inflammation and chronic lung diseases, *J Biochem Mol Biol* 2003, 36:95–109.
13. Friedenson B: BRCA1 and BRCA2 pathways and the risk of cancers other than breast or ovarian, *Med Gen Med* 2005, 7:60.
14. Salmon TB, Evert BA, Song B, Doetsch PW: Biological consequences of oxidative stress-induced DNA damage in *Saccharomyces cerevisiae*, *Nucleic Acids Res* 2004, 32:3712–3723.
15. Wang G, Hong Y, Johnson MK, Maier RJ: Lipid peroxidation as a source of oxidative damage in *Helicobacter pylori*: Protective roles of peroxiredoxins, *Biochim Biophys Acta* 2006, 1760:1596–1603.
16. Avery SV: Molecular targets of oxidative stress, *Biochem J* 434:201–210.
17. Schulze-Osthoff K, Ferrari D, Los M, Wesselborg S, Peter ME: Apoptosis signaling by death receptors, *Eur J Biochem* 1998, 254:439–459.
18. Baldwin AS, Jr.: The NF-kappa B and I kappa B proteins: New discoveries and insights, *Annu Rev Immunol* 1996, 14:649–683.
19. Hussain SP, Harris CC: Inflammation and cancer: An ancient link with novel potentials, *Int J Cancer* 2007, 121:2373–2380.
20. Finkel T, Holbrook NJ: Oxidants, oxidative stress and the biology of ageing, *Nature* 2000, 408:239–247.
21. Goyal MM, Basak A: Human catalase: Looking for complete identity, *Protein Cell* 1:888–897.
22. Valko M, Rhodes CJ, Moncol J, Izakovic M, Mazur M: Free radicals, metals and antioxidants in oxidative stress-induced cancer, *Chem Biol Interact* 2006, 160:1–40.
23. Winrow VR, Winyard PG, Morris CJ, Blake DR: Free radicals in inflammation: Second messengers and mediators of tissue destruction, *Br Med Bull* 1993, 49:506–522.
24. Wintergerst ES, Maggini S, Hornig DH: Contribution of selected vitamins and trace elements to immune function, *Ann Nutr Metab* 2007, 51:301–323.
25. Endo N, Nishiyama K, Otsuka A, Kanouchi H, Taga M, Oka T: Antioxidant activity of vitamin B6 delays homocysteine-induced atherosclerosis in rats, *Br J Nutr* 2006, 95:1088–1093.
26. Riccioni G, Barbara M, Bucciarelli T, di Ilio C, D'Orazio N: Antioxidant vitamin supplementation in asthma, *Ann Clin Lab Sci* 2007, 37:96–101.
27. Rayman MP: The importance of selenium to human health, *Lancet* 2000, 356:233–241.

28. Jackson MI, Combs GF, Jr.: Selenium and anticarcinogenesis: Underlying mechanisms, *Curr Opin Clin Nutr Metab Care* 2008, 11:718–726.
29. Papp LV, Lu J, Holmgren A, Khanna KK: From selenium to selenoproteins: Synthesis, identity, and their role in human health, *Antioxid Redox Signal* 2007, 9:775–806.
30. Zhong L, Arner ES, Holmgren A: Structure and mechanism of mammalian thioredoxin reductase: The active site is a redox-active selenolthiol/selenenylsulfide formed from the conserved cysteine-selenocysteine sequence, *Proc Natl Acad Sci USA* 2000, 97:5854–5859.
31. Rhee SG, Chae HZ, Kim K: Peroxiredoxins: A historical overview and speculative preview of novel mechanisms and emerging concepts in cell signaling, *Free Radic Biol Med* 2005, 38:1543–1552.
32. Saito Y, Sato N, Hirashima M, Takebe G, Nagasawa S, Takahashi K: Domain structure of bi-functional selenoprotein P, *Biochem J* 2004, 381:841–846.
33. Traulsen H, Steinbrenner H, Buchczyk DP, Klotz LO, Sies H: Selenoprotein P protects low-density lipoprotein against oxidation, *Free Radic Res* 2004, 38:123–128.
34. Klotz LO, Sies H: Defenses against peroxynitrite: Selenocompounds and flavonoids, *Toxicol Lett* 2003, 140–141:125–132.
35. Takahashi H, Nishina A, Fukumoto RH, Kimura H, Koketsu M, Ishihara H: Selenoureas and thioureas are effective superoxide radical scavengers in vitro, *Life Sci* 2005, 76:2185–2192.
36. Kapoor M, Kojima F, Yang L, Crofford LJ: Sequential induction of pro- and anti-inflammatory prostaglandins and peroxisome proliferators-activated receptor-gamma during normal wound healing: A time course study, *Prostaglandins Leukot Essent Fatty Acids* 2007, 76:103–112.
37. Serhan CN, Chiang N, Van Dyke TE: Resolving inflammation: Dual anti-inflammatory and pro-resolution lipid mediators, *Nat Rev Immunol* 2008, 8:349–361.
38. Maddipati KR, Marnett LJ: Characterization of the major hydroperoxide-reducing activity of human plasma. Purification and properties of a selenium-dependent glutathione peroxidase, *J Biol Chem* 1987, 262:17398–17403.
39. Eskew ML, Zarkower A, Scheuchenzuber WJ, Hildenbrandt GR, Scholz RW, Reddy CC: Increased thromboxane A2 synthesis by rat lung neutrophils during selenium deficiency, *Prostaglandins* 1993, 46:319–329.
40. Liu W, Li G, Zhang X: Effects of selenium and vitamin E on arachidonic acid metabolism in experimental myocardial ischaemia, *Zhonghua Yu Fang Yi Xue Za Zhi* 1995, 29:279–282.
41. Eisenmann CJ, Miller RK: The effect of selenium compounds (selenite, selenate, ebselen) on the production of thromboxane and prostacyclin by the human term placenta in vitro, *Toxicol Appl Pharmacol* 1995, 135:18–24.
42. Cao YZ, Reddy CC, Sordillo LM: Altered eicosanoid biosynthesis in selenium-deficient endothelial cells, *Free Radic Biol Med* 2000, 28:381–389.
43. Vunta H, Davis F, Palempalli UD, Bhat D, Arner RJ, Thompson JT, et al.: The anti-inflammatory effects of selenium are mediated through 15-deoxy-Delta12,14-prostaglandin J2 in macrophages, *J Biol Chem* 2007, 282:17964–17973.
44. Gandhi UH, Kaushal N, Ravindra KC, Hegde S, Nelson SM, Narayan V, et al.: Selenoprotein-dependent upregulation of hematopoietic prostaglandin D2 synthase in macrophages is mediated through the activation of peroxisome proliferator-activated receptor (PPAR){gamma}, *J Biol Chem* 2011, 286:27471–27482.
45. Nelson SM, Lei X, Prabhu KS: Selenium levels affect the IL-4 induced expression of alternative activation markers in murine macrophages, *J Nutr* 2011, 141:1754–1761.
46. Formigari A, Irato P, Santon A: Zinc, antioxidant systems and metallothionein in metal mediated-apoptosis: Biochemical and cytochemical aspects, *Comp Biochem Physiol C Toxicol Pharmacol* 2007, 146:443–459.
47. Abel J, de Ruiter N: Inhibition of hydroxyl-radical-generated DNA degradation by metallothionein, *Toxicol Lett* 1989, 47:191–196.
48. Ghoshal K, Majumder S, Li Z, Bray TM, Jacob ST: Transcriptional induction of metallothionein-I and -II genes in the livers of Cu,Zn-superoxide dismutase knockout mice, *Biochem Biophys Res Commun* 1999, 264:735–742.
49. Kumari MV, Hiramatsu M, Ebadi M: Free radical scavenging actions of metallothionein isoforms I and II, *Free Radic Res* 1998, 29:93–101.
50. Pedersen MO, Jensen R, Pedersen DS, Skjolding AD, Hempel C, Maretty L, et al.: Metallothionein-I+II in neuroprotection, *Biofactors* 2009, 35:315–325.
51. Zago MP, Oteiza PI: The antioxidant properties of zinc: Interactions with iron and antioxidants, *Free Radic Biol Med* 2001, 31:266–274.

52. Zhou Z, Wang L, Song Z, Saari JT, McClain CJ, Kang YJ: Zinc supplementation prevents alcoholic liver injury in mice through attenuation of oxidative stress, *Am J Pathol* 2005, 166:1681–1690.
53. Johnson F, Giulivi C: Superoxide dismutases and their impact upon human health, *Mol Aspects Med* 2005, 26:340–352.
54. Antonyuk SV, Strange RW, Marklund SL, Hasnain SS: The structure of human extracellular copper-zinc superoxide dismutase at 1.7 A resolution: Insights into heparin and collagen binding, *J Mol Biol* 2009, 388:310–326.
55. Texel SJ, Xu X, Harris ZL: Ceruloplasmin in neurodegenerative diseases, *Biochem Soc Trans* 2008, 36:1277–1281.

20 Lipid Biomarkers of Inflammation

Ginger L. Milne
Vanderbilt University
Nashville, Tennessee

CONTENTS

20.1 INTRODUCTION

Eicosanoid is an umbrella term used to describe any oxygenated derivative of arachidonic acid (eicosatetraenoic acid, C20:4, ω-6). These molecules can be generated in vivo either enzymatically, through the action of cyclooxygenases (COX) and lipoxygenases (LO), or nonenzymatically via free radical–induced lipid peroxidation. Many eicosanoids are important mediators of human physiology and pathophysiology. This chapter focuses on the generation of eicosanoids in inflammation and the utility of measuring urinary metabolites as biomarkers to monitor formation of these molecules in humans.

20.2 PROSTAGLANDINS AND INFLAMMATION

Prostaglandins (PGs) are generated from arachidonic acid by the action of the COX enzymes. There are two COX enzymes—COX-1, frequently referred to as the constitutive COX, and COX-2, which is termed the inducible COX because its expression increases markedly in response to a variety of inflammatory stimuli. The parent prostaglandins (Figure 20.1)—PGE_2, PGD_2, PGI_2, $PGF_{2\alpha}$, and thromboxane A_2—are derived from the endoperoxide intermediate PGH_2 through the action of terminal prostaglandin synthases.

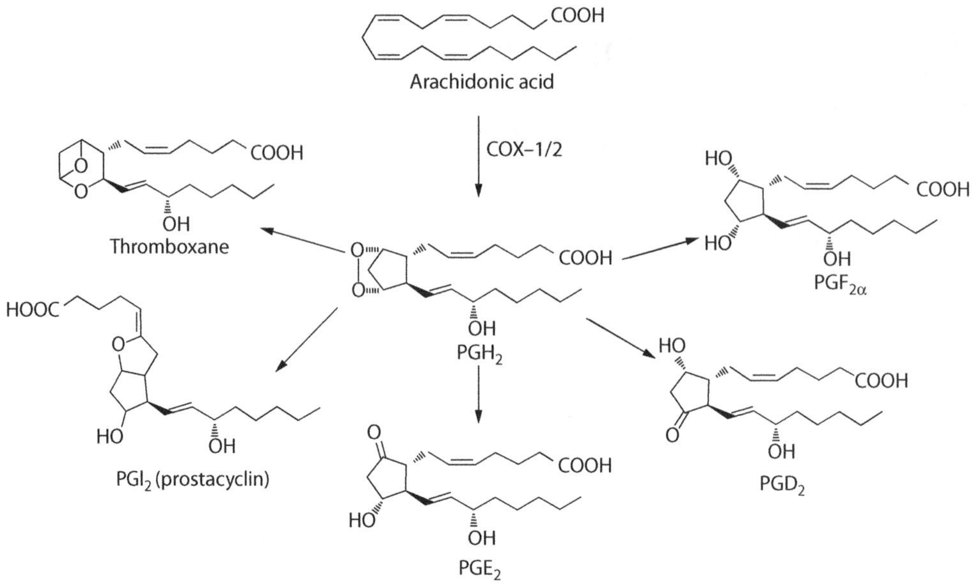

FIGURE 20.1 Five parent prostaglandins are generated via oxidation of arachidonic acid by the cyclooxygenase enzymes COX-1 and COX-2.

20.2.1 PGE_2

Of the parent PGs, considerable attention has focused on PGE_2. PGE_2 has been shown to exert a variety of physiological responses in cells and tissues. For example, PGE_2 modulates local vascular tone in various tissues, regulates sodium and water excretion by the kidney, and maintains normal gastric homeostasis. PGE_2 also plays an important role in inflammation; administration of PGE_2 can induce fever and increase hypersensitivity toward pain. Functional studies have also implicated PGE_2 in the pathophysiology of several inflammatory diseases, including arthritis and cardiovascular disease. There has also been significant interest in the role of PGE_2 in cancer, particularly in the areas of cellular growth and malignant transformation. PGE_2 was first shown to increase the proliferation, migration, and invasiveness of colorectal carcinoma cells [1]. A role for PGE_2 has now been implicated in the development and progression of non-small cell lung cancer (NSCLC) [2], head and neck cancers [3], breast cancer [4], and prostate cancer [5], among others.

PGE_2 is synthesized from the COX product PGH_2 through the action of isomerases termed PGE synthases. Three PGE synthases exist in human cells: two are microsomal and one is cytosolic. Importantly, similar to COX-2, microsomal PGE synthase-1 (mPGES-1) is inducible by proinflammatory stimuli and is considered to be a primary source of PGE_2 during inflammation [6]. The effects of PGE_2 are largely mediated by its interaction with four distinct PGE_2 receptors (EP1, EP2, EP3, and EP4) that are G-protein-coupled cell surface membrane receptors. COX inhibitors, including traditional nonsteroidal anti-inflammatory drugs (NSAIDs) such as aspirin and selective COX-2 inhibitors, and mPGES-1 inhibitors [7], currently in development, are utilized to block the synthesis and consequent deleterious effects of this molecule.

20.2.1.1 Quantification of PGE_2 Production In Vivo in Humans: Clinical Perspectives

PG production can be readily quantified in cell culture using a variety of approaches, including immunoassay and mass spectrometry (MS). However, quantification of endogenous PG production in animals and humans is significantly more challenging. Measurement of parent PGs in various biological fluids is often confounded by artifactual or ex vivo generation of these molecules during, for example, blood drawing, which results in the activation of platelets and other cells that are capable of synthesizing large amounts of eicosanoids. Systemic eicosanoid production, as measured

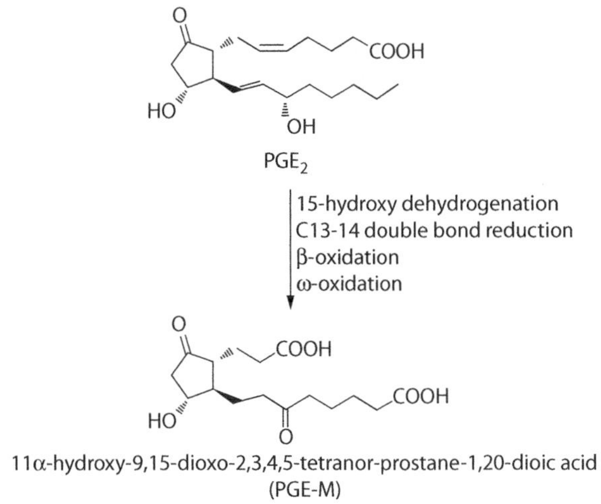

PGE_2

15-hydroxy dehydrogenation
C13-14 double bond reduction
β-oxidation
ω-oxidation

11α-hydroxy-9,15-dioxo-2,3,4,5-tetranor-prostane-1,20-dioic acid
(PGE-M)

FIGURE 20.2 Metabolism of prostaglandin E_2.

in the urine, is important for examining the consequences of various diseases and interventions on endogenous PG generation, but studies have shown, for example, that the primary source of parent PGE_2 in the urine is from production locally in the kidney and does not reflect systemic biosynthesis [8]. Thus, it is generally accepted that the most accurate index of endogenous eicosanoid production in humans is the measurement of excreted urinary metabolites [9].

In 2004, Murphey and colleagues reported the first facile and robust liquid chromatography-MS methodology to quantify the major urinary metabolite of PGE_2, 11α-hydroxy-9,15-dioxo-2,3,4,5-tetranor-prostane-1,20-dioic acid (PGE-M; Figure 20.2) [2]. The ability to simply and accurately quantify this metabolite has allowed for the direct quantification of PGE_2 in humans in a number of conditions and diseases. In particular, this biomarker, as described following, is proving to be especially useful in the study of cancer and related lifestyle factors. For example, increased urinary PGE-M levels are found in humans who smoke, with current and former smokers having higher levels than never smokers [10]. Additionally, PGE-M levels are increased in patients with colorectal cancer [11], head and neck squamous cell carcinoma [12], and NSCLC [2]; a major portion of this increase in endogenous PGE_2 is COX-2 derived.

Importantly, urinary PGE-M shows promise as a biomarker in the treatment and outcome prediction of cancer. For example, in a prospective study, Cai and colleagues determined that elevated levels of urinary PGE-M are associated with increased risk of colorectal cancer [13]. Also, in head and neck squamous cell carcinoma, PGE-M was prognostic for disease progression [12]. In patients who experienced disease progression, baseline levels of PGE-M were elevated compared to patients without disease progression. The authors thus postulated that perhaps individuals with high baseline PGE-M levels at diagnosis could benefit from treatment with a COX inhibitor. Finally, in perhaps the most clinically relevant setting studied to date, PGE-M was being used as a biomarker to select/predict patients who will positively respond to treatment with a COX-2 inhibitor, including celecoxib or apricoxib, in combination with chemotherapy, and other treatments, for recurrent NSCLC [14, 15]. In these studies, urinary PGE-M was measured in individuals at baseline and after treatment with a coxib for 5–7 days; patients with the largest decrease in PGE-M preliminarily have the most significant benefit from treatment with the COX-2 inhibitor. The utility of PGE-M as a biomarker in this setting is being confirmed in currently ongoing trials with both celecoxib and apricoxib.

Despite the benefits associated with the use of COX-2 inhibitors in the treatment of cancer, the efficacy of these molecules is limited due to severe cardiovascular side effects associated with these drugs. The reason for this cardiovascular toxicity is thought to, at least in part, result from a change in the prostacyclin-thromboxane ratio resulting from inhibition of COX-2 and not COX-1 [16, 17]. Recently, Chan and colleagues reported that plasma levels of high-sensitivity C-reactive protein (hsCRP) may be predictive of cardiovascular risk [18]. In the Adenoma Prevention with Celecoxib trial, patients with hsCRP levels > 3 mg/L were 2–3 times more likely to experience cardiovascular side effects during treatment with 200-mg-bid or 400-mg-bid celecoxib than patients with hsCRP levels ≤ 3 mg/L. Consequently, in patients with low hsCRP levels, the benefit from treatment with COX-2 may outweigh the cardiovascular risks. Having a biomarker such as hsCRP to potentially predict cardiovascular toxicity may again open the door for use of COX-2 inhibitors for treatment of inflammation, pain, and possibly cancer. Further, in response to the side effects associated with the COX-2 inhibitors, mPGES-1 inhibitors are in development for the treatment of inflammatory conditions [7]. Thus, the utility of urinary PGE-M as a biomarker of endogenous PGE_2 production and regulation is only likely to grow.

20.2.2 PGD_2

PGD_2 is synthesized much like PGE_2 from PGH_2 through the action of terminal PGD synthases. PGD_2 is a major product of mast cells released during anaphylaxis and is also produced by macrophages, dendritic cells, eosinophils, Th2 cells, and endothelial cells. Like PGE_2, the biological effects of PGD_2 are largely mediated by its interaction with G-protein-coupled cell surface membrane receptors. Two distinct receptors for PGD_2 exist—D-type prostanoid receptor-1 (DP1) and chemoattractant receptor-homologous molecule expressed on Th2 cells (CRTH2 or DP2). Through these receptors, PGD_2 affects smooth muscle contraction, sleep, and platelet aggregation. Additionally, this molecule is an important mediator of inflammation and the allergic response and is the major product of mast cells released during anaphylaxis. Interestingly, PGD_2 is also responsible for the flushing side effect in the skin observed after administration of pharmacologic doses of niacin for the purpose of lowering serum cholesterol [19].

20.2.2.1 Quantification of PGD_2 Production In Vivo in Humans: Clinical Perspectives

As discussed previously, the most accurate index of endogenous eicosanoid production in humans is the measurement of excreted urinary metabolites. The metabolic profile of PGD_2, however, is much more complex than that of PGE_2 or any of the other parent PGs. A summary of the metabolism of PGD_2 is shown in Figure 20.3. There are three major routes of metabolism of this molecule. In Route A in Figure 20.3, PGD_2 is converted to a series of metabolites with F-ring structures. Initially, PGD_2 is converted stereospecifically to $9\alpha,11\beta$-PGF_2 through the action of 11-ketoreductase. This molecule is subsequently converted to the terminal metabolite $9\alpha,11\beta$-dihydroxy-15-oxo-2,3,18,19-tetranorpost-5ene-1,20-dioic acid (PGD-M) [30]. This metabolite is formed in equilibrium with the shown cyclic structural isomer. Route B in Figure 20.3 shows that PGD_2 can be transformed through a mechanism similar to PGE_2 to yield a metabolite referred to as tetranor-PGDM [25]. Finally, as shown in Route C, PGD_2 can undergo dehydration at C-9 to yield the cyclopentenone molecule PGJ_2. PGJ_2 can then undergo rearrangement and subsequent dehydration at C-15 to yield 15-deoxy-$\Delta^{12,14}$-PGJ_2 (15-d-PGJ_2). Interestingly, 15-d-PGJ_2 itself has been shown to exert a variety of potent biological activities including activation of the anti-inflammatory peroxisome proliferator-activated receptor-gamma (PPARγ) [50].

Due to the complex nature of the metabolism of this molecule, the quantification of urinary metabolites of PGD_2 metabolites is not straightforward. PGD-M was initially reported in 1985 as the major urinary metabolite of PGD_2 [20]. The synthesis of this metabolite and an isotopically labeled internal standard was reported in 1988 [21]. Three years later, Morrow and colleagues reported a validated gas chromatography-mass spectrometric (GC/MS) assay for the quantification of this metabolite [22]. Normal levels of this metabolite in humans are defined to be 1.03 ± 0.36 ng/mg Cr. Quantification of PGD-M has been used to assess levels of PGD_2 production in a variety of clinical settings including mastocytosis and primary pulmonary hypertension [23, 24].

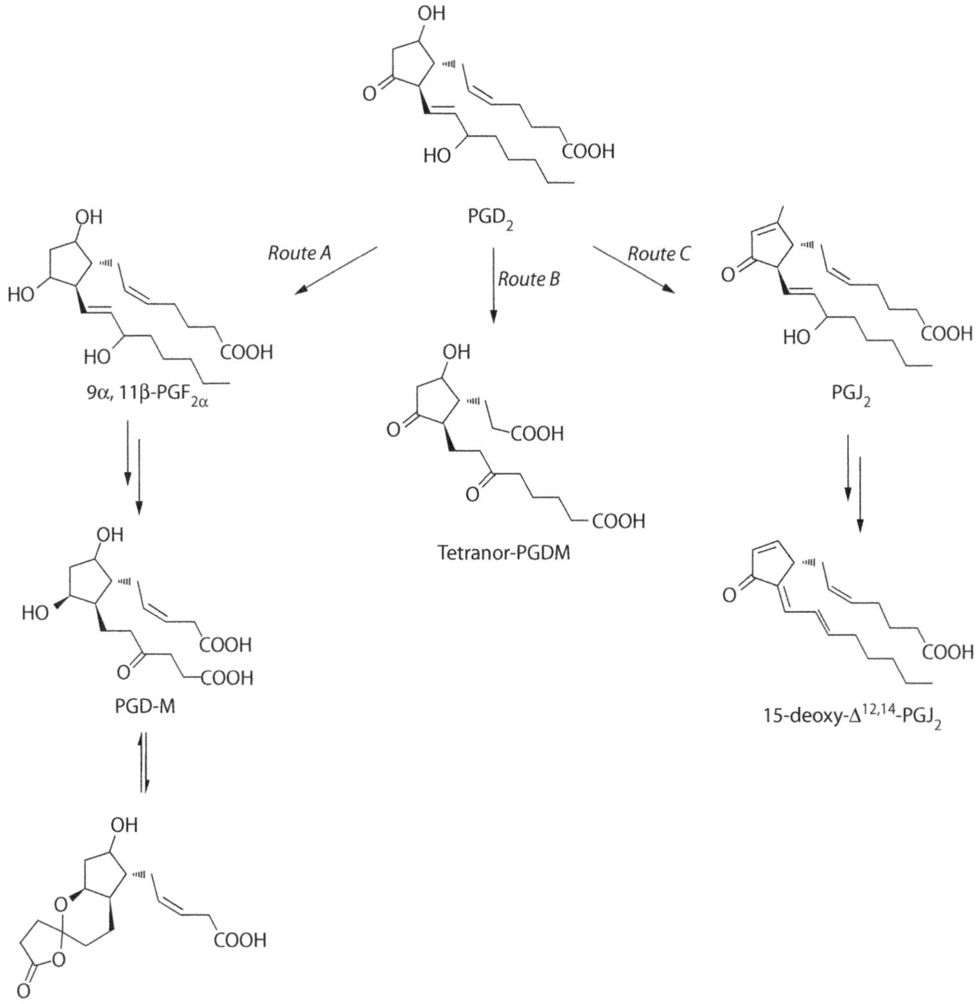

FIGURE 20.3 Metabolism of prostaglandin D_2.

The utility of measuring this metabolite, however, has been limited by the lack of a commercially available internal standard.

More recently, Song et al. reported tetranor-PGDM as a major urinary metabolite of PGD_2 in humans, formed in levels similar to PGD-M [25]. This metabolite is also the major urinary metabolite of PGD_2 detected in mouse urine. Deletion of lipocalin-type or hematopoietic PGD synthases in mice decreases levels of this metabolite. In humans, levels of tetranor-PGDM increased following niacin administration in concert with the flushing response in the skin, similarly to PGD-M [25]. Few other studies have yet quantified tetranor-PGDM in human disease. Recently, however, Zhang and colleagues reported an online solid phase extraction-LC/MS methodology to quantify both PGE-M and tetranor-PGDM in one analytical run [26]. Using this methodology, these authors quantified both metabolites in healthy human nonsmokers, healthy human smokers, and humans with chronic obstructive pulmonary disease (COPD). Both PGE-M and tetranor-PGDM were significantly elevated in persons with COPD compared to healthy nonsmokers. Levels of these compounds in healthy human smokers were comparable to levels in COPD. These two studies together suggest that tetranor-PGDM may prove to be a useful biomarker of endogenous PGD_2 production.

The quantification of PGJ_2 and 15-d-PGJ_2 in humans is not as straightforward as the quantification of the other metabolites of PGD_2 due to their chemical reactivity. In 2003, Bell-Parikh and colleagues reported that 15-d-PGJ_2, but not PGJ_2, is present in urine at concentrations in the very low picogram per mg creatinine range [27]. However, both PGJ_2 and 15-d-PGJ_2 possess electrophilic α,β-unsaturated carbonyl moieties that render these molecules capable of undergoing Michael addition reactions with endogenous nucleophiles including cysteine residues in proteins and small peptides, i.e., glutathione. To this end, work by Brunoldi et al. demonstrates that 15-d-PGJ_2 can be further metabolized via conjugation with glutathione in a human liver cell line (HepG2 cells), suggesting that the measurement of unmetabolized 15-d-PGJ_2 in urine might be an underrepresentation of actual endogenous levels of this compound [28]. Thus, further study of urinary metabolites of 15-d-PGJ_2 is warranted. Interestingly, Hardy and colleagues have shown that 15-d-PGJ_2 and isomeric compounds can be generated not only from COX-1/-2 but also from the free radical catalyzed peroxidation of arachidonic acid in settings of oxidative stress [29]. Thus, identification of a urinary biomarker of 15-d-PGJ_2 formation would be useful in identifying the source(s) of this active lipid mediator in humans.

As the metabolism of PGD_2 is incredibly complex and proceeds through three primary metabolic routes, it would be of interest to simultaneously measure metabolites from all routes in order to determine if differences in metabolism exist in different clinical settings.

20.3 LEUKOTRIENES AND INFLAMMATION

Leukotrienes (LTs) are formed from the oxidation of arachidonic acid via the enzyme 5-lipoxygenase (5-LO). This enzyme, in conjugation with 5-LO activating protein (FLAP), catalyzes the conversion of arachidonic acid to the unstable intermediate leukotriene A_4 (LTA_4), the 5,6-epoxide of arachidonic acid (Figure 20.4) [30]. To form LTA_4, arachidonic acid is initially oxidized to the unstable intermediate 5-hydroperoxyeicosatetraenoic acid (5-HPETE). 5-HPETE is converted to LTA_4 through the LTA_4 synthase activity of 5-LO. LTA_4 is then either a substrate for LTA4 hydrolase yielding LTB_4 or is conjugated with glutathione via the action of LTC4 synthase to yield LTC_4. LTC_4 can be metabolized through loss of glutamate and then glycine to yield LTD_4 and LTE_4, respectively. LTC_4, LTD_4, and LTE_4 are collectively referred to as cysteinyl LTs (cysLTs).

LTs exert their biological activities, like PGs, through G-protein-coupled cell surface membrane receptors. Two receptors bind LTB_4, BLT1 and BLT2, while two receptors bind LTC_4 and LTD_4, cystLT1 and cystLT2. Both LTB_4 and the cysteinyl LTs are potent mediators of inflammation. The cysLTs have specifically been shown to have profound effects on airway function and inflammation and play an important role in allergic asthma [31, 32]. LTB_4, a potent chemoattractant, is also important in pulmonary and cardiovascular inflammation. Because of their potent biological activities, a variety of anti-LT pharmacological agents, including 5-LO inhibitors, cysLT receptor antagonists, and most recently FLAP inhibitors, have been developed to treat allergic asthma and other pulmonary conditions.

20.3.1 Quantification of LT Production In Vivo in Humans: Clinical Perspectives

To assess in vivo 5-LO activity and LT production in humans, it is necessary to have an accurate, noninvasive methodology. LTB_4 is metabolized in vivo to yield a number of metabolites, including several glucuronic acid conjugates [33]. Recently, Mita and colleagues, for the first time, developed a methodology to quantify LTB_4 glucuronides in normal human urine by coupling HPLC with an enzyme immunoassay [34]. However, these authors were not able to detect significant increases in these metabolites in patients with mild asthma after allergen inhalation but did show increases in this metabolite in humans with aspirin-intolerant asthma after aspirin challenge.

In contrast to the complex metabolism of LTB_4, the cysLTs—LTC_4 and LTD_4—are metabolized to the end product of the 5-LO pathway LTE_4, which is excreted in the urine. Several methods have been developed to measure urinary LTE_4, including both radioimmunoassay and mass spectrometric

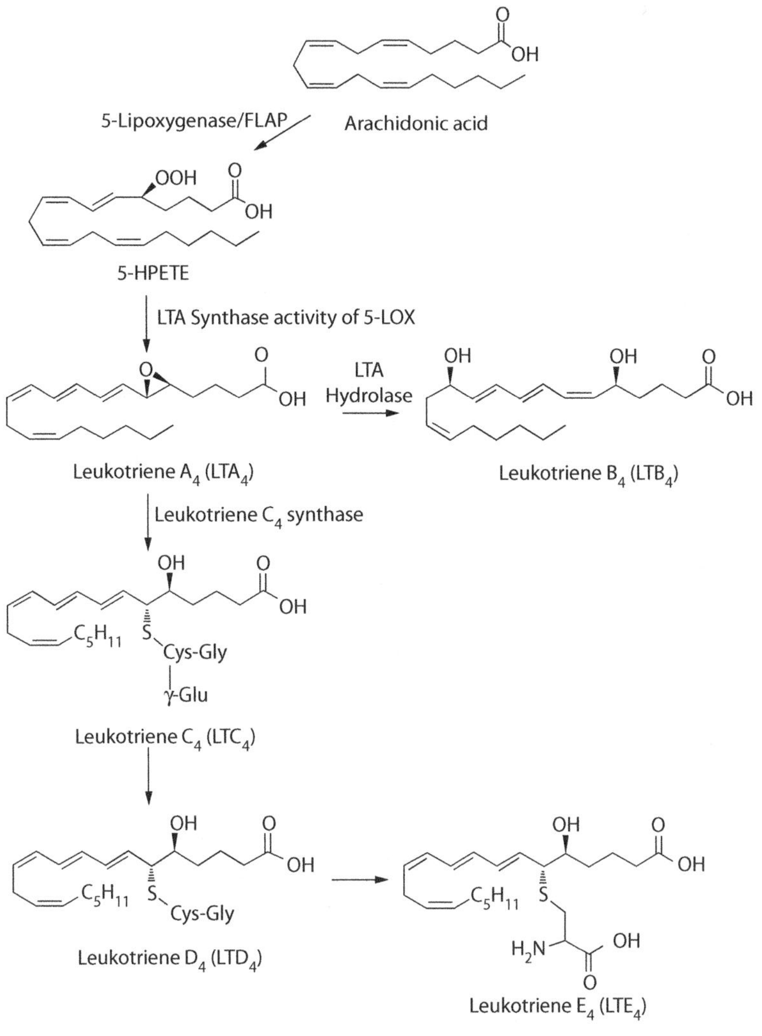

FIGURE 20.4 5-lipoxygenase in concert with the 5-lipoxygenase activating protein (FLAP) oxidized arachidonic acid to a series of compounds termed leukotrienes.

approaches. Radioimmunoassays, however, are inherently unreliable due to lack of specificity, and may require separate chromatographic purifications prior to quantification [35]. Urinary LTE_4 has been shown to be increased in a variety of inflammatory conditions and diseases, including asthma after challenge [36, 37], cardiac ischemia [38], atopic dermatitis [39], inflammatory bowel disease [40], and type 1 diabetes [41], among many others.

In 2009, Duffield-Lillico and colleagues determined that both urinary LTE_4 and PGE-M were increased in the urine of smokers, a population at risk for the development of inflammatory diseases such as cardiovascular disease and cancer [42]. Treatment of subjects with celecoxib (200 mg bid) for 5–7 days showed a significant decrease in PGE-M in all subjects, as expected. Subjects with the highest baseline levels of PGE-M showed the largest decrease in PGE-M after celecoxib, demonstrating increased COX-2 activity in this population. Interestingly, in this subject group, treatment with celecoxib led to an increase in urinary LTE_4, an effect not noted in individuals with low baseline PGE-M. Further study is needed to determine if this effect is a result of shunting of arachidonic acid from the COX pathway to the 5-LO pathway or if underlying signaling mechanisms are involved. However, this finding does highlight a potentially important interaction between the biosynthesis of PGs and LTs and underscores the importance of utilizing biomarkers to study the effect of therapeutics on the endogenous production of lipid mediators.

20.4 NONENZYMATIC LIPID PEROXIDATION: THE ISOPROSTANES

The isoprostanes (IsoPs) are a unique series of prostaglandin-like compounds formed in vivo via the nonenzymatic, free radical–catalyzed peroxidation of arachidonic acid. These compounds were first reported by Morrow and Roberts in 1990 [43]. Over the course of the past 20 years, many studies have defined the basic chemistry and biochemistry involved in the formation and metabolism of IsoPs. One particular class of IsoPs, F_2-IsoPs (mechanism of formation is shown in Figure 20.5), is known as the "gold standard" biomarker of endogenous lipid peroxidation resulting from oxidative stress. These molecules are excellent biomarkers as they are chemically stable and have been identified in all biological matrices analyzed. In fact, a multi-investigator study sponsored by the National Institute of Environmental Health Sciences (NIEHS), the Biomarkers of Oxidative Stress (BOSS) Study, found that quantification of plasma or urinary F_2-IsoPs by mass spectrometry was the most accurate method to assess endogenous oxidative stress [44].

20.4.1 Quantification of F_2-Isoprostanes in Humans: Clinical Perspectives

Normal levels of F_2-IsoPs in healthy humans have been defined. Normal levels in plasma are 35 ± 6 pg/mL while normal levels in urine are 1.6 ± 0.6 ng/mg creatinine [45]. Defining these

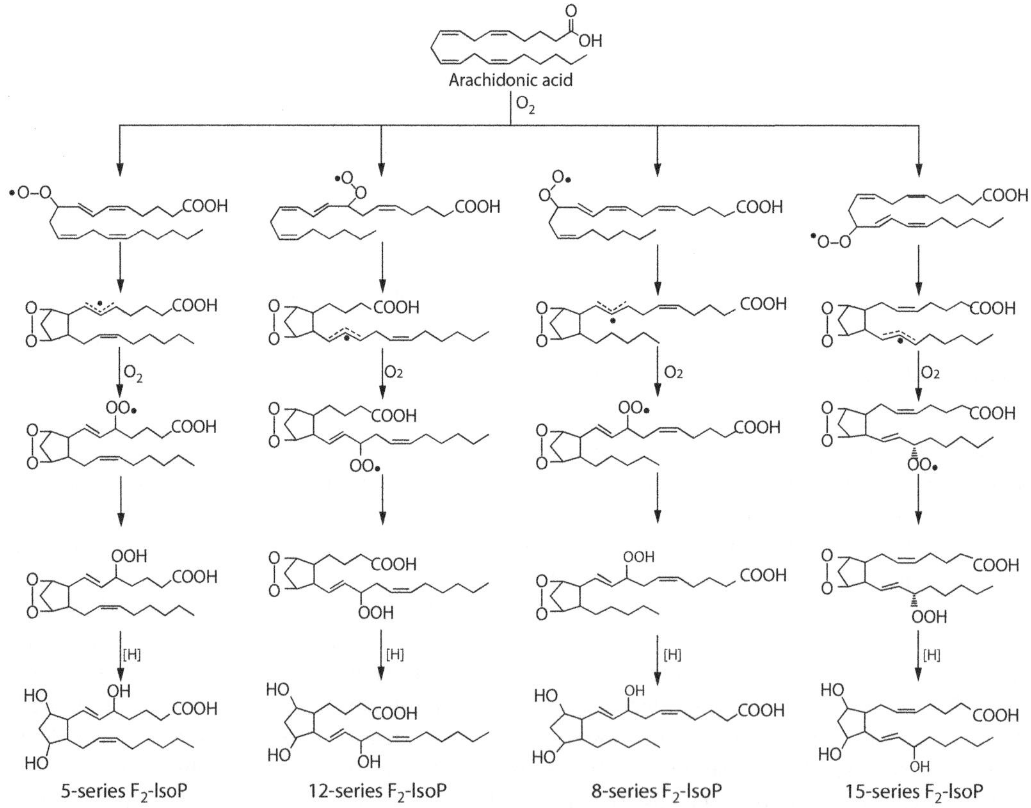

FIGURE 20.5 Arachidonic acid is nonenzymatically oxidized via the action of free radicals to yield a series of prostaglandin-like molecules termed isoprostanes. F_2-isoprostanes are chemically stable biomarkers of oxidative stress.

levels has allowed for assessment of oxidative stress and the effect of therapeutic interventions on oxidative stress in a variety of inflammatory diseases. For example, levels of F_2-IsoPs are increased in atherosclerosis and associated risk factors including smoking and obesity, ischemia/reperfusion injury, asthma, rheumatic diseases, certain types of cancers, and neurodegeneration, among others. Further, supplementation with antioxidants as well as lifestyle changes (i.e., cessation of cigarette smoking or weight loss) have been shown to decrease levels of F_2-IsoPs. (Please see these comprehensive reviews on IsoPs for specific references and further information [46, 47].) Consumption of marine fish oil, rich in eicosapentaenoic acid (EPA, C20:5, ω-3) and docosahexaenoic acid (DHA, C22:6, ω-3), also decreases levels of urinary and plasma F_2-IsoPs in vivo [48, 49]. Although there is no exact etiology or phenotype associated with oxidative stress and no specific drug or treatment to block this condition as one would block a specific enzyme or receptor, the F_2-IsoPs continue to be a useful biomarker and tool to study nonenzymatic lipid oxidation as a consequence of oxidative stress in research on the bench, in animal disease models, and in humans.

20.5 CONCLUSIONS AND FUTURE DIRECTIONS

Eicosanoids, including PGs, LTs, and IsoPs, are important lipid mediators in human physiology and pathophysiology, particularly inflammation. Development of robust, sensitive, and specific mass spectrometric methodologies to quantify urinary metabolites of these compounds has allowed for the accurate assessment of endogenous levels of these compounds in humans in both normal and disease settings and after therapeutic treatment. Current therapies and the ongoing development of PG synthase inhibitors, FLAP inhibitors, receptor antagonists, and novel antioxidant treatments will only increase the need and clinical utility of these biomarkers. Future studies, like the work of Duffield-Lillico and colleagues, should take care to quantify eicosanoids generated from the different enzymatic and nonenzymatic routes in order to assess important and potentially novel interactions in the biosynthesis of these lipid mediators.

TAKE-HOME MESSAGES

- Eicosanoids are oxidized derivatives of arachidonic acid.
- Prostaglandins (PGs) are products of the cyclooxygenase enzymes, COX-1 and COX-2.
- Leukotrienes (LTs) are products of the 5-lipoxygenase (5-LO).
- Isoprostanes (IsoP) are PG-like compounds generated nonenzymatically via free radical mechanisms.
- Measurement of urinary metabolites is the most accurate way to quantify endogenous eicosanoid production.
- PGE_2 is the most abundant PG produced in humans and is increased particularly in cancer.
- The urinary metabolite of PGE_2 (PGE-M) is a useful biomarker to predict cancer patients who might benefit from treatment with a COX-2 inhibitor.
- PGD_2 is increased during anaphylaxis and the allergic response. Metabolism of PGD_2 is more complex than PGE_2 and is limited by the lack of available internal standards.
- Measurement of urinary LTE_4 has implicated upregulation of 5-LO in a number of inflammatory diseases. Better biomarkers of LTB_4 production are needed.
- F_2-IsoPs are robust, chemically stable markers of lipid peroxidation and oxidative stress.

REFERENCES

1. Sheng, H.; Shao, J.; Washington, M. K.; DuBois, R. N. 2001. Prostaglandin E2 increases growth and motility of colorectal carcinoma cells. *J Biol Chem* 276:18075.

2. Murphey, L. J.; Williams, M. K.; Sanchez, S. C.; Byrne, L. M.; Csiki, I.; Oates, J. A.; et al. 2004. Quantification of the major urinary metabolite of PGE2 by a liquid chromatographic/mass spectrometric assay: Determination of cyclooxygenase-specific PGE2 synthesis in healthy humans and those with lung cancer. *Anal Biochem* 334:266.
3. Camacho, M.; Leon, X.; Fernandez-Figueras, M. T.; Quer, M.; Vila, L. 2008. Prostaglandin E(2) pathway in head and neck squamous cell carcinoma. *Head Neck* 30:1175.
4. Subbaramaiah, K.; Hudis, C.; Chang, S. H.; Hla, T.; Dannenberg, A. J. 2008. EP2 and EP4 receptors regulate aromatase expression in human adipocytes and breast cancer cells. Evidence of a BRCA1 and p300 exchange. *J Biol Chem* 283:3433.
5. Jain, S.; Chakraborty, G.; Raja, R.; Kale, S.; Kundu, G. C. 2008. Prostaglandin E2 regulates tumor angiogenesis in prostate cancer. *Cancer Res* 68:7750.
6. Kudo, I.; Murakami, M. 2005. Prostaglandin E synthase, a terminal enzyme for prostaglandin E2 biosynthesis. *J Biochem Mol Biol* 38:633.
7. Iyer, J. P.; Srivastava, P. K.; Dev, R.; Dastidar, S. G.; Ray, A. 2009. Prostaglandin E(2) synthase inhibition as a therapeutic target. *Expert Opin Ther Targets* 13:849.
8. Frolich, J. C.; Wilson, T. W.; Sweetman, B. J.; Smigel, M.; Nies, A. S.; Carr, K.; et al. 1975. Urinary prostaglandins. Identification and origin. *J Clin Invest* 55:763.
9. Catella, F.; Nowak, J.; Fitzgerald, G. A. 1986. Measurement of renal and non-renal eicosanoid synthesis. *Am J Med* 81:23.
10. Gross, N. D.; Boyle, J. O.; Morrow, J. D.; Williams, M. K.; Moskowitz, C. S.; Subbaramaiah, K.; et al. 2005. Levels of prostaglandin E metabolite, the major urinary metabolite of prostaglandin E2, are increased in smokers. *Clin Cancer Res* 11:6087.
11. Johnson, J. C.; Schmidt, C. R.; Shrubsole, M. J.; Billheimer, D. D.; Joshi, P. R.; Morrow, J. D.; et al. 2006. Urine PGE-M: A metabolite of prostaglandin E2 as a potential biomarker of advanced colorectal neoplasia. *Clin Gastroenterol Hepatol* 4:1358.
12. Kekatpure, V. D.; Boyle, J. O.; Zhou, X. K.; Duffield-Lillico, A. J.; Gross, N. D.; Lee, N. Y.; et al. 2009. Elevated levels of urinary prostaglandin E metabolite indicate a poor prognosis in ever smoker head and neck squamous cell carcinoma patients. *Cancer Prev Res* 2:957.
13. Cai, Q.; Gao, Y. T.; Chow, W. H.; Shu, X. O.; Yang, G.; Ji, B. T.; et al. 2006. Prospective study of urinary prostaglandin E2 metabolite and colorectal cancer risk. *J Clin Oncol* 24:5010.
14. Reckamp, K.; Gitlitz, B.; Chen, L. C.; Patel, R.; Milne, G.; Syto, M.; et al. Biomarker-based phase I dose-escalation, pharmacokinetic, and pharmacodynamic study of oral apricoxib in combination with erlotinib in advanced nonsmall cell lung cancer. *Cancer* 117:809.
15. Csiki, I.; Morrow, J. D.; Sandler, A.; Shyr, Y.; Oates, J.; Williams, M. K.; et al. 2005. Targeting cyclooxygenase-2 in recurrent non-small cell lung cancer: A phase II trial of celecoxib and docetaxel. *Clin Cancer Res* 11:6634.
16. Fitzgerald, G. A. 2004. Coxibs and cardiovascular disease. *N Engl J Med* 351:1709.
17. Funk, C. D.; FitzGerald, G. A. 2007. COX-2 inhibitors and cardiovascular risk. *J Cardiovasc Pharmacol* 50:470.
18. Chan, A. T.; Sima, C. S.; Zauber, A. G.; Ridker, P. M.; Hawk, E. T.; Bertagnolli, M. M. C-reactive protein and risk of colorectal adenoma according to celecoxib treatment. *Cancer Prev Res (Phila)* 4:1172.
19. Morrow, J. D.; Awad, J. A.; Oates, J. A.; Roberts, L. J., 2nd. 1992. Identification of skin as a major site of prostaglandin D2 release following oral administration of niacin in humans. *J Invest Dermatol* 98:812.
20. Liston, T. E.; Roberts, L. J., 2nd. 1985. Metabolic fate of radiolabeled prostaglandin D2 in a normal human male volunteer. *J Biol Chem* 260:13172.
21. Prakash, C.; Roberts, L. J., 2nd; Saleh, S.; Taber, D. F.; Blair, I. A. 1987. Synthesis of putative prostaglandin D2 metabolites. *Adv Prostaglandin Thromboxane Leukot Res* 17B:781.
22. Morrow, J. D.; Prakash, C.; Duckworth, T. A.; Zackert, W. E.; Blair, I. A.; Oates, J. A.; et al. 1991. A stable isotope dilution mass spectrometric assay for the major urinary metabolite of PGD2. *Adv Prostaglandin Thromboxane Leukot Res* 21A:315.
23. Awad, J. A.; Morrow, J. D.; Roberts, L. J., 2nd. 1994. Detection of the major urinary metabolite of prostaglandin D2 in the circulation: Demonstration of elevated levels in patients with disorders of systemic mast cell activation. *J Allergy Clin Immunol* 93:817.
24. Robbins, I. M.; Barst, R. J.; Rubin, L. J.; Gaine, S. P.; Price, P. V.; Morrow, J. D.; et al. 2001. Increased levels of prostaglandin D(2) suggest macrophage activation in patients with primary pulmonary hypertension. *Chest* 120:1639.

25. Song, W. L.; Wang, M.; Ricciotti, E.; Fries, S.; Yu, Y.; Grosser, T.; et al. 2008. Tetranor PGDM, an abundant urinary metabolite reflects biosynthesis of prostaglandin D2 in mice and humans. *J Biol Chem* 283:1179.
26. Zhang, Y.; Zhang, G.; Clarke, P. A.; Huang, J. T.; Takahashi, E.; Muirhead, D.; et al. Simultaneous and high-throughput quantitation of urinary tetranor PGDM and tetranor PGEM by online SPE-LC-MS/MS as inflammatory biomarkers. *J Mass Spectrom* 46:705.
27. Bell-Parikh, L. C.; Ide, T.; Lawson, J. A.; McNamara, P.; Reilly, M.; FitzGerald, G. A. 2003. Biosynthesis of 15-deoxy-delta12,14-PGJ2 and the ligation of PPARgamma. *J Clin Invest* 112:945.
28. Brunoldi, E. M.; Zanoni, G.; Vidari, G.; Sasi, S.; Freeman, M. L.; Milne, G. L.; et al. 2007. Cyclopentenone prostaglandin, 15-deoxy-Delta12,14-PGJ2, is metabolized by HepG2 cells via conjugation with glutathione. *Chem Res Toxicol* 20:1528.
29. Hardy, K. D.; Cox, B. E.; Milne, G. L.; Yin, H.; Roberts, L. J., 2nd. Nonenzymatic free radical-catalyzed generation of 15-deoxy-Delta(12,14)-prostaglandin J-like compounds (deoxy-J-isoprostanes) in vivo. *J Lipid Res* 52:113.
30. Silverman, E. S.; Drazen, J. M. 1999. The biology of 5-lipoxygenase: function, structure, and regulatory mechanisms. *Proc Assoc Am Physicians* 111:525.
31. Busse, W. W. 1998. Leukotrienes and inflammation. *Am J Respir Crit Care Med* 157:S210.
32. Lewis, R. A.; Austen, K. F.; Soberman, R. J. 1990. Leukotrienes and other products of the 5-lipoxygenase pathway. Biochemistry and relation to pathobiology in human diseases. *N Engl J Med* 323:645.
33. Berry, K. A.; Borgeat, P.; Gosselin, J.; Flamand, L.; Murphy, R. C. 2003. Urinary metabolites of leukotriene B4 in the human subject. *J Biol Chem* 278:24449.
34. Mita, H.; Turikisawa, N.; Yamada, T.; Taniguchi, M. 2007. Quantification of leukotriene B4 glucuronide in human urine. *Prostaglandins Other Lipid Mediat* 83:42.
35. Rabinovitch, N. 2007. Urinary leukotriene e(4). *Immunol Allergy Clin North Am* 27:651.
36. Bochenek, G.; Nizankowska, E.; Gielicz, A.; Swierczynska, M.; Szczeklik, A. 2004. Plasma 9alpha,11beta-PGF2, a PGD2 metabolite, as a sensitive marker of mast cell activation by allergen in bronchial asthma. *Thorax* 59:459.
37. Kikawa, Y.; Miyanomae, T.; Inoue, Y.; Saito, M.; Nakai, A.; Shigematsu, Y.; et al. 1992. Urinary leukotriene E4 after exercise challenge in children with asthma. *J Allergy Clin Immunol* 89:1111.
38. Carry, M.; Korley, V.; Willerson, J. T.; Weigelt, L.; Ford-Hutchinson, A. W.; Tagari, P. 1992. Increased urinary leukotriene excretion in patients with cardiac ischemia. In vivo evidence for 5-lipoxygenase activation. *Circulation* 85:230.
39. Hishinuma, T.; Suzuki, N.; Aiba, S.; Tagami, H.; Mizugaki, M. 2001. Increased urinary leukotriene E4 excretion in patients with atopic dermatitis. *Br J Dermatol* 144:19.
40. Stanke-Labesque, F.; Pofelski, J.; Moreau-Gaudry, A.; Bessard, G.; Bonaz, B. 2008. Urinary leukotriene E4 excretion: A biomarker of inflammatory bowel disease activity. *Inflamm Bowel Dis* 14:769.
41. Hardy, G.; Boizel, R.; Bessard, J.; Cracowski, J. L.; Bessard, G.; Halimi, S.; et al. 2005. Urinary leukotriene E4 excretion is increased in type 1 diabetic patients: A quantification by liquid chromatography-tandem mass spectrometry. *Prostaglandins Other Lipid Mediat* 78:291.
42. Duffield-Lillico, A. J.; Boyle, J. O.; Zhou, X. K.; Ghosh, A.; Butala, G. S.; Subbaramaiah, K.; et al. 2009. Levels of prostaglandin E metabolite and leukotriene E(4) are increased in the urine of smokers: Evidence that celecoxib shunts arachidonic acid into the 5-lipoxygenase pathway. *Cancer Prev Res (Phila)* 2:322.
43. Morrow, J. D.; Hill, K. E.; Burk, R. F.; Nammour, T. M.; Badr, K. F.; Roberts, L. J., 2nd. 1990. A series of prostaglandin F2-like compounds are produced in vivo in humans by a non-cyclooxygenase, free radical-catalyzed mechanism. *Proc Natl Acad Sci USA* 87:9383.
44. Kadiiska, M. B.; Gladen, B. C.; Baird, D. D.; Germolec, D.; Graham, L. B.; Parker, C. E.; et al. 2005. Biomarkers of oxidative stress study II: Are oxidation products of lipids, proteins, and DNA markers of CCl4 poisoning? *Free Radic Biol Med* 38:698.
45. Morrow, J. D.; Roberts, L. J., 2nd. 1999. Mass spectrometric quantification of F2-isoprostanes in biological fluids and tissues as measure of oxidant stress. *Methods Enzymol* 300:3.
46. Milne, G. L.; Sanchez, S. C.; Musiek, E. S.; Morrow, J. D. 2007. Quantification of F2-isoprostanes as a biomarker of oxidative stress. *Nat Protoc* 2:221.
47. Milne, G. L.; Yin, H.; Hardy, K. D.; Davies, S. S.; Roberts, L. J. 2011. Isoprostane generation and function. *Chem Rev* 10:5973.
48. Gao, L.; Yin, H.; Milne, G. L.; Porter, N. A.; Morrow, J. D. 2006. Formation of F-ring isoprostane-like compounds (F3-isoprostanes) in vivo from eicosapentaenoic acid. *J Biol Chem* 281:14092.

49. Mas, E.; Woodman, R. J.; Burke, V.; Puddey, I. B.; Beilin, L. J.; Durand, T.; et al. The omega-3 fatty acids EPA and DHA decrease plasma F(2)-isoprostanes: Results from two placebo-controlled interventions. *Free Radic Res* 44:983.
50. Surh, Y. J.; Na, H. K.; Park, J. M; Lee, H. N.; Kim, W.; Yoon, I. S.; Kim, D. D. 2011. 15-Deoxy-$\Delta^{12,14}$-prostaglandin J_2, an electrophilic lipid mediator of anti-inflammatory and pro-resolving signaling. *Biochem Pharmacol* 82:1335.

21 Physical Activity and Inflammation

An Overview

Edite Teixeira de Lemos[*]
Coimbra University
Coimbra, Portugal

ESAV, Polytechnic Institute of Viseu
Viseu, Portugal

Flávio Reis[*]
Coimbra University
Coimbra, Portugal

CONTENTS

21.1 THE STATE OF LOW-GRADE INFLAMMATION—HEALTHY SIGNIFICANCE

21.1.1 The Inflammatory Response and Its Mechanisms

The inflammatory process can be initiated through a variety of mechanisms, which include the introduction of pathogens as well as challenges to the system through chemical, thermal, and

[*] Equally contributed to the chapter.

mechanical stresses. Regardless of the inciting factors, the events accompanying inflammation are somewhat consistent. The local response to infections or tissue injury involves the production of cytokines that are released at the site of inflammation. Cytokines are small polypeptides, which were originally discovered to have immunoregulatory roles (Majno and Joris 2004). Some of these cytokines facilitate the influx of lymphocytes, neutrophils, monocytes, and other cells. The local inflammatory response is accompanied by a systemic response known as the acute-phase response (Table 21.1). This response includes the production of a large number of hepatocyte-derived acute phase proteins, such as C-reactive protein (CRP), and can be mimicked by the injection of the cytokines tumor necrosis factor alpha (TNF-α), interleukin (IL)-1β, or IL-6 into laboratory animals or humans (Majno and Joris 2004; Pedersen 2006; Edwards et al. 2007). The initial cytokines in the cytokine cascade are TNF-α, IL-1, IL-6, IL-1 receptor antagonist (IL-1ra), and soluble TNF-α receptors (sTNF-R). IL-1ra inhibits IL-1 signal transduction and sTNF-R represents the naturally occurring inhibitors of TNF-α (Majno and Joris 2004; Luster et al. 2005). In response to an acute infection or trauma, the cytokines and inhibitors may increase several-fold and decrease when the infection or trauma is healed. Additionally, these acute phase reactants have considerable effects on the metabolism during acute illness, leading to hyperglycemia, insulin resistance, and increased glucogenesis (Jenkins and Ross 1999). Elevation in these markers also increases proteolysis (Hasselgren 1999), bone resorption (Smith et al. 2002), and dyslipidemia (Marik 2006), in addition to upregulating other members of the inflammatory cascade, each of which has its own downstream biologic effects (Table 21.1).

TABLE 21.1
Different Cytokine/Substances Mediating the Inflammatory Process, Cells That Produce Them, and the Effects That They Have on the Metabolism

Cytokine	Producing Cell	Action
Interleukin-1	Macrophages	Stimulation of various cells, e.g., T cells, acts to initiate inflammation, induces hypothalamus to increase body temperature
Interleukin-2	T cells	Causes proliferation of activated T and B cells, induces antibody synthesis
Interleukin-3	T cells	Induces growth and differentiation of immune cells in bone marrow
Interleukin-4	T cells	Promotes B cell growth and differentiation
Interleukin-5	T cells	Induces differentiation of B cells and activates some macrophages
Interleukin-6	T cells, macrophages	Co-stimulator of T cells, induces growth in B cells
Interleukin-10	T cells	Activates B cells and inhibits macrophage function
Interleukin-12	Macrophages	Activates T cells and NK cells
Interleukin-13	T cells	Induces proliferation of B cells and differentiation of T cells
Gamma-interferon	T cells, NK cells	Activates macrophages
Tumor necrosis factor	Macrophages	Causes activation of some macrophages. Induces inflammation and fever. Induces catabolism of muscle and fat, thus leading to cachexia (bodily wasting)
Transforming growth factor	T cells, macrophages	Inhibits T cell growth and macrophage activation
Lymphotoxin	T cells	Similar to TNF, activates macrophages
Histamine	Mast cells	Not actually a cytokine, but an important chemical mediator that induces blood vessel dilation and increases cell wall permeability

21.1.2 Inflammation as a Cause of Chronic Disease

Inflammation is a key function in the process by which the body responds to an injury or an infection, and the acute phase of inflammation normally leads to recovery from infection to healing, and a return to normal values within a few days. However, if the response is not properly phased, the process can develop into a chronic low-grade inflammatory state that may trigger different diseases under pathological conditions (Hotamisligil 2006; Kahn et al. 2006).

Several parameters of the inflammatory reaction can be measured in plasma. Chronic low-grade systemic inflammation has been characterized by a two- to three-fold elevation in the systemic concentrations of pro-inflammatory and anti-inflammatory cytokines, naturally occurring cytokine antagonists, and the acute phase reactant CRP (Zeyda et al. 2007). In the latter case, the stimuli for the cytokine production are not known, but it is assumed that the origin of TNF-α in chronic low-grade systemic inflammation is mainly the adipose tissue (Gil et al. 2007; Bulló et al. 2003). The inflammatory markers that have been shown to be associated with obesity or the metabolic syndrome include acute phase proteins, pro-inflammatory cytokines, adhesion molecules, and adipokines (proteins secreted by adipose tissue). Plasma concentrations of IL-6 (Bennet at al. 2003) and TNF-α have been shown to predict the risk of myocardial infarction in several studies (Reilly et al. 2007), and CRP has emerged as a particularly stronger independent risk factor for cardiovascular disease (CVD) than the low-density lipoprotein cholesterol level (Hansson 2005; Pearson et al. 2003; Ridker et al. 2001 and 2003). In addition to CVD, increased risk for several other diseases seems to be associated with elevated CRP levels. Most studies use CRP as the only marker of inflammation; however, choosing a wider spectrum of inflammatory markers can give us a better picture of the specific mechanisms involved. Other commonly used acute-phase reactants include complement factors C3 and C4, serum amyloid A, and ceruloplasmin. Pro-inflammatory cytokines, such as IL-6, IL-1β, and TNF-α, are often included in the panel of inflammatory markers and have been associated with obesity and components of the metabolic syndrome. The endothelial expression of vascular adhesion molecules (VCAM), intracellular adhesion molecule (ICAM), and E-selectin are used as markers of the infiltration of inflammatory cells in the arterial wall. It now seems clear that circulating levels of inflammatory mediators are often strongly correlated with one another as a result of their tightly regulated production. Moreover, levels of inflammatory mediators are correlated with other risk factors in chronic morbidity, including levels of fibrinogen, albumin, cholesterol, arterial blood pressure, and body mass index (BMI), among others.

Despite the fact that the changes in acute-phase reactants are much smaller than those in acute infections, the chronicity of low-grade inflammation is strongly associated with increasing age, lifestyle factors (such as smoking and obesity), together with increased risk of CVD and type 2 diabetes (T2DM). In spite of the different backgrounds and symptoms related to the various diseases, all systemic chronic inflammation shares common characteristics, including elevated circulating levels of cytokines TNF-α and IL-6 under basal or resting conditions (Higashimoto et al. 2008; Mohamed-Ali et al. 2001; White et al. 2006). Indeed, the source of these inflammatory markers (e.g., TNF-α) depends on the disease. T2DM is associated with an overproduction of TNF-α by the adipocyte, whereas in diseases with an autoimmune component, macrophages and T cells are the main source (Abbas 2007).

Mounting evidence suggests that TNF-α plays a direct role in the metabolic syndrome (Pedersen and Febbraio 2008), linking insulin resistance to vascular disease (Plomgaard et al. 2005). In CVDs, activated immune cells also play a major role, particularly in the etiology of atherosclerosis (Matter and Handschin 2007). Several downstream mediators and signaling pathways seem to provide the cross-talk between inflammatory and metabolic signaling (Plomgaard et al. 2005). With regard to IL-6, its role in insulin resistance is highly controversial. Nevertheless, when diabetic patients were given rhIL-6-infusion, plasma concentrations of insulin declined to levels comparable with that in age- and BMI-matched healthy controls, indicating that IL-6 can enhance insulin sensitivity. Pedersen and Febbraio (2008) reviewed a number of studies indicating that IL-6 enhances lipolysis, as well as fat oxidation, via an activation of AMPK without causing hypertriacylglycerolemia. Given the different biological profiles of TNF-α and IL-6, and given that TNF-α may trigger an IL-6 release, one theory holds that it is TNF-α derived from adipose tissue that is actually the major "driver" behind inflammation-induced insulin resistance and atherosclerosis.

The chronic inflammatory state of hyperinsulinemia with elevated levels of insulin and free IGF-1 promote proliferation of colon cells and lead to a survival benefit of transformed cells, ultimately resulting in colorectal cancer (Berster and Goke 2008). In addition, a number of neurodegenerative diseases are linked to a local inflammatory response in the brain (neuroinflammation), such as IL-1β, and TNF-related apoptosis-inducing ligand (TRAIL) and other cytokines have been postulated to be involved in the etiology of Alzheimer's disease (Zipp and Aktas 2006). Moreover, in addition to the neuroinflammation found in many neurodegenerative disorders, systemic inflammation may further exacerbate the progression of neurodegeneration (Perry et al. 2007).

Obesity is also strongly associated with enhanced circulating TNF-α levels and visceral fat, which has been correlated with endothelial dysfunction and with C-reactive protein levels, indicating an inflammatory component to the adverse effects of visceral adipose tissue (VAT). Adipose tissue from obese individuals shows accumulation of macrophages, which provide the major cellular source of a concomitant enhanced local expression of the TNF-α protein (Preis et al. 2010). Rather, local TNF-α may stimulate production of IL-6 and subsequent mediators in the inflammatory cascade. Moreover, tumor initiation, promotion, and progression are stimulated by systemic elevation of pro-inflammatory cytokines (Lin and Karin 2007).

Pedersen (2007) analyzed all the diseases focused herein in a "diseasome of physical inactivity" (p. 413) considering that despite the fact that all these diseases had highly different phenotypical presentations, they shared important pathogenetic mechanisms. According to his view, chronic systemic inflammation is associated with physical inactivity independent of obesity, further hypothesizing that physical inactivity leads to the accumulation of visceral fat and, consequently, to the activation of a network of inflammatory pathways, which promote development of insulin resistance, atherosclerosis, neurodegeneration, and tumor growth and, thereby, the development of the diseases belonging to the "diseasome of physical inactivity." This hypothesis was further corroborated by Laye et al. (2007) and by Olsen et al. (2008) both in animal models and humans with physical inactivity. Thus, inactivity and visceral fat seem to precede systemic inflammation and chronic diseases.

21.2 THE ANTI-INFLAMMATORY NATURE OF PHYSICAL ACTIVITY

Physical activity, inflammation, and immunity are tightly linked in an interesting and complex way (Febbraio 2007). Regular and moderate physical exercise reduces systemic inflammation, but the mediators of this beneficial effect remain to be fully elucidated. However, several candidate mechanisms have been identified. Exercise increases the release of epinephrine, cortisol, growth hormone, prolactin, and other factors that have immunomodulatory effects (Nieman 2003). Furthermore, exercise results in decreased expression of Toll-like receptor on monocytes, suggesting an involvement in whole body inflammation. Other mediators and mechanisms have also been postulated, including factors derived from the muscle and from the adipose tissue, as well as from the innate immune response.

21.2.1 Physical Activity and Myokines

The recent discovery of "myokines," cytokines produced and secreted from skeletal muscle, analogous to "adipokines" made from fat tissue, shed light on this bivalent association between exercise and inflammation (Febbraio 2007). The first myokine described was interleukin-6 (IL-6); similar factors synthesized and secreted upon contraction of muscle fibers include IL-8 and IL-15 (Pedersen 2011). In addition to these muscle-derived cytokines, increased IL-1 receptor antagonist (IL-1ra),

IL-10 and TNF-α are found in the circulation after exercise. The cytokine response to exercise differs from that elicited by severe infections. Classical pro-inflammatory cytokines, TNF-α and IL-1β, in general do not increase with exercise, indicating that the cytokine cascade induced by exercise is markedly different from the cytokine cascade induced by infections. However, systemic elevation of TNF-α is restricted to physical activity of extremely high intensity and, therefore, could be responsible for the elevated inflammatory state upon prolonged and intense exercise. Such myokines may exert a direct effect on fat metabolism and thereby result in indirect anti-inflammatory effects. Moreover, myokines may exert direct anti-inflammatory effects or stimulate the production of anti-inflammatory components. It is suggested that contracting skeletal muscles release myokines, which work in a hormone-like fashion, exerting specific endocrine effects on visceral fat and other ectopic fat deposits. Other myokines work locally within the muscle via paracrine mechanisms, exerting their effects on signaling pathways involved in fat oxidation.

There is an ongoing debate regarding the anti- or pro-inflammatory effects of IL-6. IL-6 secreted by myocytes appears to be anti-inflammatory, as opposed to IL-6 secreted chronically by adipose tissue. When IL-6 is secreted by muscle, it has been shown that it increases anti-inflammatory cytokines such as IL-10 and IL-1ra (Brandt and Pedersen 2010) and inhibits IL-1β and TNF-α release with exercise (Mathur and Pedersen 2008). All these studies support the anti-inflammatory role of IL-6 secreted by myocytes in the response to exercise (Figure 21.1). The Pedersen group also presented evidence that IL-6 is produced by working muscle itself in the absence of injury or signs of inflammation. They hold that IL-6 acts as an anti-inflammatory cytokine, since IL-6 exerts inhibitory effects on TNF-α and IL-1 production (e.g., by LPS-stimulated monocytes). In addition, IL-6 also stimulates the production of IL-1ra and IL-10, and it is one of the primary inducers of CRP, which can also have anti-inflammatory properties, all of this supporting the anti-inflammatory effects of exercise. Moreover, it is been considered that IL-6 may act as an energy sensor and its release by the contraction of skeletal muscle may be attenuated or even totally inhibited by glucose ingestion during exercise (Pedersen and Febraio 2008).

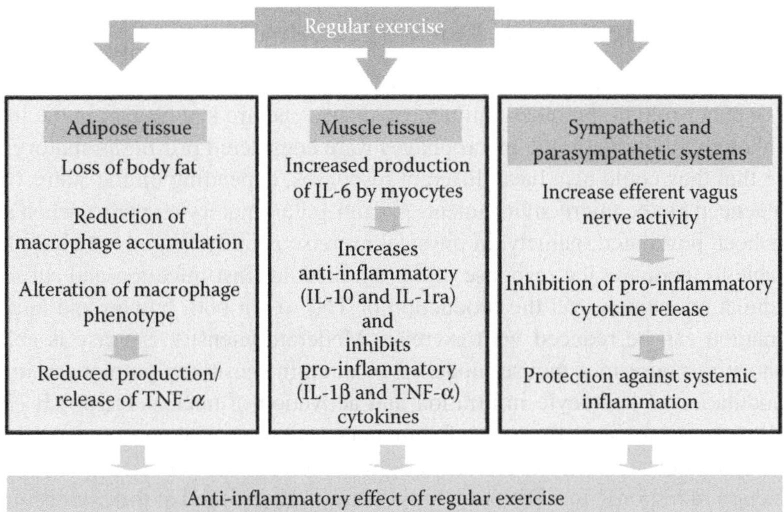

FIGURE 21.1 Anti-inflammatory nature of regular exercise. Several potential mechanisms for the anti-inflammatory effect of exercise have been postulated, which includes an adipose tissue-innate immune system axis, in which the exercise is able to reduce body fat; the accumulation of macrophages in the adipose tissue, further altering the macrophage phenotype; an increased muscle production of IL-6, which then promotes anti-inflammatory actions; and a "cholinergic anti-inflammatory pathway hypothesis," which claims that exercise promotes an imbalance between the sympathetic and parasympathetic nervous systems, which is responsible for protection against systemic inflammation.

IL-15 has recently been discovered as another anabolic myokine, with putative involvement in muscle fat cross-talk (Nielsen et al. 2007). Some works supported the idea that IL-15 secretion from muscle tissue may modulate visceral fat mass specifically *via* an endocrine mechanism (Quinn 2008; Nielsen et al. 2008). Riechman et al. (2004) reported an increase in IL-15 after an acute resistance training bout, thus suggesting that IL-15 may accumulate within the muscle as a consequence of regular training. Meanwhile, and according to the works of Nielsen and Pedersen (2007), IL-15 did not change after 2.5 h treadmill running. However, it seems that the role of muscle contraction on IL-15 regulation is not clearly defined.

The plasma concentrations of IL-8, another myokine, seem to be increased in response to exhaustive exercise involving eccentric muscle contraction (running); nevertheless, no changes were observed in relation to concentric exercise. According to Akerstrom et al. (2005), high local IL-8 expression occurs in contracting muscle but with only a small and transient net release; this may indicate that muscle-derived IL-8 acts locally and exerts its effect in an autocrine or paracrine fashion (Akerstrom et al. 2005). IL-8 signaling promotes angiogenic responses in endothelial cells, increases proliferation and survival of endothelial and cancer cells, and potentiates the migration of cancer cells, endothelial cells, and infiltrating neutrophils at the tumor site. Accordingly, IL-8 expression correlates with the angiogenesis, tumorigenicity, and metastasis of tumors in numerous xenograft and orthotopic *in vivo* models (Waugh and Wilson 2008).

In a recent revision about metabolic responses and environmental adaptations of muscle and their myokines, Pedersen (2011) focuses on the effects not only of the myokines but also of other factors released by exercise, such as brain-derived neurotrophic factor (which presented a role in neurobiology and metabolism, both central and from the skeletal muscle) and fibroblast growth factor-21 (expressed in human skeletal muscle in response to insulin stimulation). Cytokine production can be affected by other physiological factors affected by exercise, such as stress hormones, acidosis, oxidative stress and heat, among others (Radom-Aizik et al. 2007). In addition, cytokine response may vary by the type of exercise, intensity, duration, and recovery between bouts and training status (Miles 2008).

21.2.2 The Adipose Tissue—Innate Immune System Hypothesis

The activated macrophages, through the release of a milieu of chemokines, cytokines, and other proteins, play a major role in the innate immune response and are key players in the inflammatory responses. Although traditionally the macrophages were considered pro-inflammatory secretors, it is now known that they could also have different responses, depending on the status of activation, which are influenced by the microenvironment. An anti-inflammatory response, when alternatively activated, has been postulated, namely, in physical exercise activity (Woods et al. 2009). Exercise seems to be able to modulate the response of an organism against infection and subsequent presence of inflammation, namely, via the production of TNF-α. In both human and animal models, acute inflammation can be reduced with exercise. Moderate-intensity exercise is able to reduce the inflammatory response in a mouse model of ovalbumin-sensitized pulmonary inflammation, viewed by modulation of leukocyte infiltration and activation of nuclear factor kB (Pastva et al. 2005). According to the authors, the anti-inflammatory action was due to exercise-released glucocorticoids. Using rats exhaustively exercised, Bagby et al. (1994) found that there was a reduced TNF-α production in response to LPS for up to six hours, suggesting that the exercise-induced suppression of macrophage function was due to increased corticosterone.

A similar anti-inflammatory effect was reported by Starkie et al. (2003) in a study in which young males at rest were submitted to an LPS challenge, exercised exhaustively, or administered IL-6. The results obtained showed that after the LPS challenge, both the exercise and the IL-6 groups showed decreased plasma TNF-α levels when compared with rest subjects. The authors suggested that the exercise induces the production of IL-6, which acts as an anti-inflammatory cytokine by suppressing the production of TNF-α (Figure 21.1). The anti-inflammatory nature of exercise in chronic inflammation and its association with the macrophage/innate immune response was revised by Woods et al. (2009) in both cross-sectional and longitudinal studies. Almost all the longitudinal studies analyzed reported an anti-inflammatory effect of exercise training in subjects with chronic diseases, including heart disease, metabolic syndrome, and overweight individuals. Furthermore, the duration of the training seems to influence the anti-inflammatory effect, with shorter studies demonstrating less activity.

Although the mechanisms responsible for the anti-inflammatory effect of regular exercise remain to be fully elucidated, several potential mechanisms have been postulated. One of them includes the existence of an adipose tissue–innate immune system axis that plays a central role. According to this possibility, exercise is able to reduce body fat, the accumulation of macrophages in the adipose tissue, further altering the macrophage phenotype in the adipose tissue (Woods et al. 2009). Furthermore, as previously commented, exercise is able to induce muscle production of IL-6, or, as suggested by Pavlov and Tracey (2005) in their "cholinergic anti-inflammatory pathway hypothesis" (p. 5559) to alter the sympathetic and parasympathetic nervous systems equilibrium (Figure 21.1). These mechanisms together might underlie the anti-inflammatory effect of exercise and are the basis of the adipocyte-macrophage-innate immune response hypothesis. Innate immunity is a central player of the inflammatory process. The better definition of the precise effect of regular exercise on the innate immunity, namely, on the adipocyte tissue and macrophages, will be decisive for clarifying the anti-inflammatory nature of exercise and the underlying mechanisms.

21.2.3 Type and Intensity of Exercise versus Anti- or Pro-Inflammatory Response

Physical exercise has been recommended as a therapeutic measure in preventive medicine, particularly for the complications of cardiovascular (CV) and metabolic diseases (Bassuk and Manson 2005; Moore 2004). Regular exercise/training improves the lipid profile by positive effects on circulating lipid and lipoprotein content, lowers blood pressure (BP), increases insulin sensitivity, and ameliorates arterial and heart functions and inflammation. Moreover, it adjusts body weight by reduction of body fat and has been particularly recommended, with proven benefits, for diabetes and metabolic syndrome patients (Tanasescu et al. 2003). However, high-intensity exercise or the lack of adaptation of individuals might be harmful, in a dual effect known as "exercise paradox." The final effect seems to depend, thus, on the intensity and work performance during training, as well as on the type of exercise protocol practiced, as was previously suggested (Selamoglu et al. 2000; Miyazaki et al. 2001; Kemi et al. 2005; Baptista et al. 2008).

Observational data from large-population cohort studies consistently show an association between physical activity and inflammation. The effects of training depend on the intensity of exercise training status, age, and involvement of disorders. Beavers et al. (2010) reviewed the findings from 26 observational studies examining the association between markers of systemic inflammation and aerobic exercise. Several interesting reviews, both epidemiological studies and clinical trials, addressing the influence of physical activity and fitness on low-grade inflammation on the general adult population have been published (Autenrieth et al. 2009), also in athletes (Tomaszewski et al. 2003), and to a lesser extent in children and adolescents (Zahner et al. 2006; Wamberg et al. 2007). Although CRP is by far the most commonly used, a wide range of inflammatory markers have been measured and assessed against physical activity; these include fibrinogen, cytokines, and leukocytes.

Our group found that lean nondiabetic ZDF rats submitted to a swimming protocol (1 h/day and 3 days/week) during 12 weeks do not present statistically significant modifications of IL-6 and TNF-α levels. Nevertheless, serum levels of CRP exhibit a significant decrease when compared to their age-matched sedentary group (Teixeira de Lemos et al. 2009). Our results confirmed that exercise training can attenuate or blunt the response of a single bout of exercise. Thus, while physical exercise could present anti-inflammatory effects in regular and moderate conditions, strenuous exercise affects resting levels of inflammatory markers during and after a period of intensive training.

Although it has been generally accepted that some of the beneficial effects of exercise involve the stimulation of the innate and/or inflammatory response, preventing the organism against infection, today many researchers support the concept that the beneficial effects of exercise are mainly mediated by its anti-inflammatory properties. This fact becomes important in order to use exercise as a "therapeutic help" in people with inflammatory disorders. However, the existence of a pro-inflammatory effect of exercise has been increasingly discussed. Therefore, in contrast to the reduction of chronic inflammation by regular, moderate exercise, prolonged, high-intensity training results in increased systemic inflammation and elevated risk of infection. In fact, subsequent to this type of exercise, athletes exhibit a transient immunodepression (Gleeson et al. 2004).

21.3 EXERCISE-INDUCED INFLAMMATION

21.3.1 Exercise-Induced Muscle Damage and Inflammation

Exhaustive exercise causes substantial tissue injury, which might activate the inflammation-induced cytokine cascade, analogous to what occurs in sepsis. Biphasic kinetics of IL-6 might putatively explain the dual effect of exercise, as recently commented by several authors. In a systematic review based on the effect of exercise on CRP and on other inflammatory markers, Kasapis and Thompson (2005) concluded that exercise produces a short-term inflammatory response, whereas studies based on cross-sectional comparisons and longitudinal exercise training demonstrated a long-term anti-inflammatory effect, also suggesting that anti-inflammatory responses may contribute to the beneficial effects of habitual physical activity. The authors emphasize that this anti-inflammatory response of habitual exercise is mainly highlighted by changes in the circulating concentration of CRP, supporting the idea that habitual exercise training reduces CRP levels by altering the inflammatory process (Kasapis and Thompson 2005; Beavers et al. 2010).

Other authors also related that during and immediately after acute prolonged and very demanding exercise there is a dramatic increase in leukocyte cell count and serum IL-6 levels, which, in the hours following exercise, is accompanied by an increase in the anti-inflammatory cytokine IL-10 (Philippou et al. 2009; Fallon et al. 2001). Furthermore, acute prolonged exercise has been shown to increase the expression of the oxidant-responsive anti-inflammatory gene HO-1 in lymphocytes (Urakawa et al. 2003). Our group, using 20-week-old lean ZDF rats forced to swim until exhaustion and then immediately sacrificed, recently observed a statistically significant increased TNF-α (27.8%) and CRP (5%) when compared to non-exercised rats (unpublished data). The studies from Giraldo et al. (2008) in healthy people have shown that the circulating levels of pro-inflammatory cytokines IL-1β and IFN-γ increase after a single bout of either moderate or intense exercise, and peritoneal macrophages are also stimulated after acute exercise and training in rodents (Ortega 2003).

It is now accepted that during acute exercise muscles release IL-6 and plasma concentrations of CRP, and both pro- and anti-inflammatory cytokines (TNF-α, IL-1, IL-1ra, IL-10, and sTNF-r) increase to various magnitudes as well as leukocyte subsets, such as neutrophils, lymphocytes (including their subsets T, B, and NK cells), and monocytes. Following the cessation of intense exercise, neutrophils and monocytes continue to increase into the recovery period. During this period, other leukocyte subsets decrease in number, while plasma concentrations of the above-mentioned cytokines stay elevated for some more hours. Strenuous and eccentric exercise seems to exert the most prominent changes in immune parameters (Malm 2004; Pedersen and Hoffman 2000). It is therefore clear that acute bouts of exercise exert various effects on the immune system and are typically transient in nature. The extent to which these changes occur in patients with a chronic inflammatory disease is important to address in order to ensure that exercise is performed in a safe manner where inflammation is not being further amplified.

21.3.2 The Paradox of IL-6 in Physical Activity and Inflammation

Cytokines are released not only from mononuclear cells but also from muscle cells. Starkie et al. (2003) showed that physical exercise directly inhibits endotoxin-induced TNF-α production in humans, most likely through IL-6 release from exercising muscle (Starkie et al. 2003). Typically, IL-6 is the first cytokine present in circulation after exercise practice, followed by an increase in IL-1ra and IL-10 (Pedersen and Febbraio 2008). The ubiquitous role of IL-6 and the hypothesis of an exercise-induced anti-inflammatory IL-6 release were already postulated (Pedersen 2007; Fisman and Tenenbaum 2010). Therefore, IL-6, a multifactorial cytokine, regulates cellular and humoral responses and plays a pivotal role in inflammation, being associated with several pathological conditions, emerging as an independent early predictor for T2DM and as a marker of low-grade inflammation (Pedersen 2007; Fisman and Tenenbaum 2010). However, even more interesting concerning IL-6, as Fisman and Tenenbaum (2010) recently commented, is the putative beneficial effects played as an anti-inflammatory factor, which is particularly evident in insulin sensitivity during exercise. Therefore, a marked increase in circulating levels of IL-6 after exercise without muscle damage has been a remarkably consistent finding. The magnitude by which plasma IL-6 increases is related to exercise duration, intensity of effort, muscle mass involved in the mechanical work, and endurance capacity (Febbraio and Pedersen 2002). The release by muscle of a humoral factor into the circulation after exercise improves insulin sensitivity, most probably through AMPK (Fisman and Tenenbaum 2010). IL-6 has been indicated as the strongest candidate for the humoral factor released after exercise, working in a hormone-like fashion, in which it is released by the muscle, now viewed as an endocrine organ, for influencing other organs (Fisman and Tenenbaum 2010). Although this hypothesis deserves further clarification, the role of IL-6 as both the "good" and the "bad," depending on the circumstances, as commented by Fisman and Tenenbaum (2010), opens new windows on the way interleukins act, and in particular concerning the effects of exercise in insulin resistance and diabetes. In this anti-inflammatory environment, IL-6 inhibits TNF-α production, which was confirmed by studies in animals (Matthys et al. 2002). Furthermore, exercise also suppresses secretion of TNF-α by pathways independent of IL-6, as shown by the results obtained with knockout mice for IL-6 submitted to exercise (Keller et al. 2004). The anti-inflammatory nature of regular exercise (training) has been associated to a reduced CVD, particularly due to the training-evoked increased expression of antioxidant and anti-inflammatory mediators in the vascular wall, which could directly inhibit atherosclerosis development (Wilund 2007).

21.4 PHYSICAL ACTIVITY FOR THE TREATMENT OF LOW-GRADE INFLAMMATION IN CARDIOMETABOLIC DISORDERS: FOCUS ON TYPE 2 DIABETES MELLITUS

Physical inactivity is considered one of the major risk factors that promotes the development and progression of CVD. Sedentary lifestyles increase all causes of mortality, double the risk of CVD, diabetes, and obesity, and substantially increase the risk of colon cancer, high blood pressure, osteoporosis, depression, and anxiety. Empirical evidence now links low-grade inflammation with disorders of several body systems and tissues, including the circulatory (atherosclerosis, heart failure), endocrine (insulin resistance, metabolic syndrome, type 1 and type 2 diabetes, obesity), skeletal (sarcopenia, arthritis, osteoporosis), and pulmonary (chronic obstructive pulmonary disease) and neurological (dementia, depression) systems (Niklas and Brinkley 2009). Furthermore, age is a contributing factor to an elevated, yet subclinical, state of inflammation. Cross-sectional data show that circulating concentrations of cytokines and acute phase reactants, as well as cellular production of cytokines, are, on average, two- to four-fold higher in older persons (Niklas and Brinkley 2009). Since people are living longer, the public health burden of aging-related disability, as well as costs with medical care, has become a critical concern. One of the most concerning disorders, because it is associated with several serious complications, is type 2 diabetes mellitus. Physical exercise has been recommended as a therapeutic measure in preventive medicine, particularly for the complications of cardiovascular (CV) and metabolic diseases, such as T2DM (Bassuk and Manson 2005; Moore 2004). Regular exercise/training improves the lipid profile by positive effects on circulating lipid and lipoprotein content, lowers blood pressure (BP), increases insulin sensitivity, ameliorates arterial and heart functions, and reduces inflammation, and has been particularly recommended, with proven benefits, for diabetes and metabolic syndrome patients (Tanasescu et al. 2003). This section is an overview of some of the key aspects related to physical (in)activity, inflammation, and T2DM, focusing on both animal and human studies.

21.4.1 Effects of Physical Exercise in Type 2 Diabetes

The previous data highlighted the idea that the beneficial effects of exercise seem to be related to its ability to decrease inflammatory cytokine levels and/or increase anti-inflammatory ones, which might also be true for pathological conditions, such as T2DM.

In diabetic rats, training was able to prevent the increase of pro-inflammatory cytokines and CRP. Martin-Cordero et al. (2009) found that obese Zucker rats, a model of metabolic syndrome, presents impairment of pro-inflammatory cytokine release by macrophages (TNF-α, IL-6, IL-1β and IFN-γ), an effect that was improved by habitual physical activity (Martin-Cordero et al. 2009; Martin-Cordero et al. 2011). Previously, our group (Teixeira de Lemos et al. 2007b and 2009) clearly demonstrated the anti-inflammatory capacity of swimming exercise training in the ZDF rat, a model of obese T2DM. Furthermore, an increment of serum adiponectin in trained obese diabetic ZDF (fa/fa) rats to levels near those found in the control lean rats was also found by us (Teixeira de Lemos et al. 2009). The anti-inflammatory effects of adiponectin have been associated with an improvement in the cardiometabolic profile, which might be due, at least in part, to regulatory actions on other factors, including TNF-α, IL-6, and CRP levels (Ravin et al. 2005), which was also demonstrated in our study using the ZDF rat submitted to swimming regular exercise training (Teixeira de Lemos et al. 2007b). Considering that all measurements were performed 48 hours after the last training session, the results may suggest an extension of the anti-inflammatory effect obtained by a single bout of exercise.

Pancreatic islets from type 2 diabetic patients present amyloid deposits, fibrosis, and increased cell death, which are associated with the inflammatory response (Hull et al. 2004). T2DM is also characterized by hyperglycemia, dyslipidemia, increased circulating inflammatory factors, and cellular stress, which are critical in precipitating islet inflammation in vivo. The impact of islet-derived inflammatory factors and islet inflammation on β-cell function and mass may be both beneficial and/or deleterious. Depending on their roles in regulating pancreatic β-cells function, some cytokines are protective while others can be detrimental. Actually, chronic exposure of islets to IL-1β, IFN-γ, TNF-α, and resistin inhibits insulin secretion and induces β-cell apoptosis. Other cytokines, such as adiponectin and visfatin, exert protective effects on pancreatic β-cell function. In addition to circulating cytokines, islets also produce a variety of cytokines in response to physiologic and pathologic stimuli, and these locally produced cytokines play important roles in regulation of pancreatic β-cell function as well (Donath et al. 2010). To maintain the normal pancreatic β-cell function, the deleterious and protective cytokines need to be balanced. The abnormal control of cytokine profile in islets and in plasma is associated with pancreatic β-cell dysfunction and type 2 diabetes (Donath et al. 2010). All the emerging evidence reinforces the paradigm that islet inflammation is involved in the regulation of β-cell function and survival in T2DM. Few studies have previously reported the putative beneficial effects of regular exercise practice (training) on the pancreas, *per se*. Studies on distinct rat strains, including Otsuka Long Evans Tokushima Fatty (OLETF), Goto-Kakizaki (GK), Zucker Fatty (ZF), ZDF, and F344 rats, have shown improvements in whole-body insulin sensitivity and preservation of β-cell mass with exercise training (Minato et al. 2002; Shima et al. 1997). Insulin sensitivity improvements by exercise may confer an indirect beneficial effect on β cells by decreasing insulin demand and minimizing β-cell exhaustion; at the same time, minimizing hyperglycemia mediates the loss of β-cell function (Dela et al. 2004), but a direct effect on pancreatic function could not be excluded. Although almost all the studies have demonstrated β-cell mass preservation with exercise training, none of them focuses on inflammation. The recognition that islet inflammation is a key factor in TD2M pathogenesis has highlighted the concern regarding the protection of pancreatic islets and endocrine function. Thus, restoring the normal cytokine profile in endocrine pancreas and plasma may hold great promise for more efficient β-cell dysfunction treatment and T2DM management. Teixeira de Lemos et al. (2009) demonstrated, using the aforementioned animal model of obese T2DM, the ZDF rat, that exercise swimming training was able to prevent accumulation of pro-inflammatory cytokines (IL-6 and TNF-α) on the endocrine pancreas. A decrease in pancreas immunostaining of both cytokines was observed, suggesting a protective effect of regular physical exercise against local inflammation. Figure 21.2 summarizes some of the anti-inflammatory properties of regular exercise, involving direct effects on adipose tissue and on the innate immune system, as well as indirect contributions of the liver and pancreas beta cell.

In humans data from small intervention studies suggest a beneficial effect of regular exercise on inflammation. For example, in a "before-after" trial of a 6-mo individualized exercise intervention in which 43 participants exercised for an average of 2.5 h/wk, a 35% decrease in CRP was observed while mononuclear cell production of atherogenic cytokines fell by 58%, whereas the production of atheroprotective cytokines rose by 36% (Smith et al. 1999). Observational data from populations with metabolic syndrome consistently reveal an association between physical activity and inflammation. In the ATTICA study, physically active individuals with metabolic syndrome, compared to sedentary persons, had 30% lower IL-6, 15% lower TNFα, 19% lower serum amyloid A (SAA), and 15% lower white blood cell (WBC) counts (Pitsavos et al. 2005). Despite the limitations observed in interventional studies that may contribute to some disparities in the results, regular/chronic exercise lead to lower basal levels of circulating inflammatory markers (Kadoglou et al. 2007; Balducci et al. 2010). As was observed with other cardiometabolic risk factors, physical

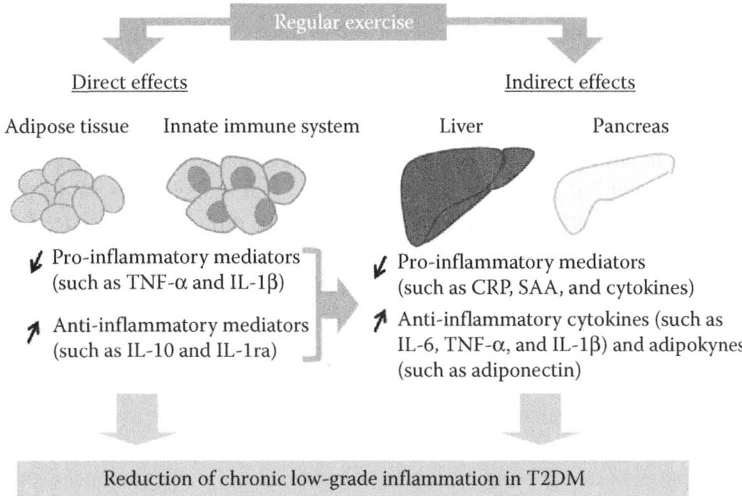

FIGURE 21.2 Direct and indirect anti-inflammatory actions of regular exercise in T2DM. Regular exercise reduces adipose tissue and innate immune system pro-inflammatory (due to reduction of pro-inflammatory mediators, including TNF-α and IL-1β) and reinforces anti-inflammatory (due to increment of anti-inflammatory mediators, including IL-10 and IL-1 receptor antagonist, IL-1ra) state. These effects promote a positive impact on the liver, viewed by reduced production of pro-inflammatory mediators, such as C-reactive protein (CRP), serum amyloid protein A (SAA), and cytokines. Furthermore, the anti-inflammatory state is also promoted in the pancreas beta cell. In connection, these direct and indirect mechanisms contribute to the reduction of chronic low-grade inflammation found in T2DM conditions.

activity–induced improvements in inflammatory status may be exclusive to individuals with high levels of inflammatory markers at baseline (Lakka et al. 2005). In all, the available evidence would support the contention that regular exercise, of sufficient intensity, appears to be anti-inflammatory in nature (Das 2004).

Nevertheless, results considering the reduction of inflammation by chronic exercise should consider the dependency of weight loss; the sex and age; the intensity and frequency of exercise; as well as concomitant medication. We must consider that to be beneficial exercise must improve fitness and muscle strength without exacerbating underlying inflammation associated with the T2D. Thus understanding the balance between safe and effective exercise is necessary to provide an evidence-based approach to exercise prescription for individuals with a chronic inflammatory disease.

21.4.2 Beyond the Anti-Inflammatory Effect of Physical Activity in T2DM: Pleiotropic Effects of Regular Exercise

Although this chapter is focused on the effects of physical activity (exercise) on inflammation, we should mentioned that there's more to physical activity (exercise) than a beneficial anti-inflammatory action that should be used as a therapeutic tool in cardiometabolic disorders, including in T2DM. The studies performed by our group (Teixeira de Lemos et al. 2007a, 2007b, and 2009), using an animal model of obese T2DM, clearly demonstrated that the practice of a regular and moderate intensity swimming protocol (training), was able to promote a decrease of the pro-inflammatory profile and increase in anti-inflammatory markers. However, beyond the beneficial anti-inflammatory action, we found other important effects. Although unable to fully reverse pancreatic lesions, exercise has prevented and/or delayed the worsening of diabetic dysmetabolism. The obtained results highlighted the pleiotropic effect of exercise training, viewed by several properties, including improvement of arterial vascular compliance and blood pressure; correction of dysglycemia and dyslipidemia; increase of antioxidant defenses, thus promoting a reduction of oxidative aggression; and reduced pancreatic dysfunction in Langerhans islets, responsible for the cell failure and appearance of relative insulin deficiency with insulin resistance, a feature of advanced stages of T2DM. The cardiometabolic protective role of exercise training in T2DM becomes clearer when considering the pleiotropic actions encountered by our group, which were corroborated by other studies in animal models, as well as in humans, as previously discussed in detail, together with other important action that undoubtedly contributes to prevent or attenuate diabetes evolution and its complications, which includes accentuation of the reduced myocardial β-adrenergic responsiveness in diabetic rats stems principally from the reduction in β2-adrenoceptor expression, which might have protective action (Lahaye et al. 2010); decreases in resting systolic blood pressure and 24-hour proteinuria in obese diabetic patients with chronic kidney disease (CKD), which is in favor of reduced cardiovascular complications in these patients (Leehey et al. 2009); reduction in plasma endothelin 1 (ET-1) and NO content, together with beneficial effects on anthropometric measurements and plasma oxidant stress markers, suggesting an improvement of endothelial dysfunction in patients with IGT (Kasımay et al. 2010); and improvement of TNF-α and IL-6 release impairment by noninfiltrated peritoneal macrophages in a rat model of obese metabolic syndrome (Martin-Cordero et al. 2009 and 2011).

Type 2 diabetic patients usually need more than an antidiabetic drug, namely for the correction of other risk factors encountered in T2DM patients, including other drugs: lipid-lowering drugs (statins, ezetimibe, fibrates, or combinations); antihypertensive drugs (ACE inhibitors, angiotensin II receptor antagonists [ARAs], beta blockers, diuretics, calcium entry blockers); antiplatelet drugs (acetylsalicylic acid [ASA], clopidogel, triflusal, or associations).

The similarities between the effects of chronic exercise and a putative antidiabetic polypill deserve to be highlighted, with the additional advantage that exercise, when practiced regularly and under moderate intensity (training), does not cause relevant side effects and presents a greater metabolic effectiveness if compared with a hypothetical antidiabetic polypill.

However, we must recognize that further research is needed in humans to establish the preferred type, duration, and intensity of training that should be practiced in order to maximize the benefits of exercise for different subgroups of T2DM patients, including their anti-inflammatory properties now reviewed.

21.5 CONCLUSIONS AND FUTURE DIRECTIONS

Current research seems to indicate that regular exercise protects against diseases associated with low-grade systemic inflammation. Several potential mechanisms for the anti-inflammatory effect of exercise have been postulated, which include an adipose tissue-innate immune system axis, in which the exercise is able to reduce body fat, the accumulation of macrophages in the adipose tissue, alterations of macrophage phenotype, as well as effects on muscles, namely, the release of cytokines/myokines that putatively mediate the health beneficial effects of exercise. These actions, most probably acting in concert, might play a crucial role in the protection against chronic diseases, including cardiovascular and metabolic conditions, such as type 2 diabetes mellitus, and cancer, among others. In particular, the long-term effect of exercise may to some extent be ascribed to the anti-inflammatory response elicited by an acute bout of exercise, which is partly mediated by muscle-derived IL-6 and other anti-inflammatory interleukins. Nevertheless, when the usual balance between the pro- and anti-inflammatory cytokines is upset by an acute exhaustive bout of exercise, an aggravation of inflammatory state occurs. The deleterious consequences involve a pathological combination of exercise stimulation of immune signals. In this context, we may hypothesize that acute exercise may be considered an immunization and should act like a vaccination stimulating immunity without causing disease. The possibility exists that, with regular exercise, the anti-inflammatory effects of an acute bout of exercise will protect against chronic systemic low-grade inflammation and, thereby, offer protection against insulin resistance and the development of T2DM. The results obtained by our group in an animal model of T2D (male *Zucker Diabetic Fatty* rats) seems to corroborate this hypothesis and even suggest a positive influence of exercise training in the inflammation responsible for pancreatic islet degeneration.

To fully understand the mechanism underlying the protective, anti-inflammatory, effects of exercise, we need to focus on the nature of exercise that more effectively could alleviate the effects of chronic inflammation in several diseases. In addition, regular/chronic exercise can afford pleiotropic cardioprotective actions by preventing/alleviating several other aspects underlying the pathological evolution of T2DM. Thus, the correct use of regular exercise could promote identical or even greater metabolic effectiveness if compared with a hypothetical antidiabetic polypill. However, we must recognize that further research is needed in humans to establish the preferred type, duration, and intensity of training that should be practiced in order to maximize the benefits of exercise not only in prevention but also in treatment in T2D patients.

TAKE-HOME MESSAGES

- A sedentary lifestyle, or physical inactivity, can be considered a strong predictor of chronic disease development and a contributor for earlier morbidity and mortality.
- A state of low-grade chronic inflammation is associated with several disorders, including those of circulatory (atherosclerosis, heart failure), endocrine/metabolic (insulin resistance, metabolic syndrome, type 1 and type 2 diabetes, obesity), skeletal (sarcopenia, arthritis, osteoporosis), pulmonary (chronic obstructive pulmonary disease), and neurological (dementia, depression) nature.
- Due to its inflammatory activity, regular exercise offers protection against chronic diseases and, thus, may be viewed as a valid nonpharmacological therapeutic option.
- The mechanisms underlying the anti-inflammatory nature of regular exercise seem to involve direct influences on adipose tissue and on the innate immune system, as well as

indirect effects on liver, and includes increased production of anti-inflammatory mediators and reduction of release of pro-inflammatory cytokines.

- Although acute and strenuous exercise can cause elevated systemic inflammation, regular exercise (training) appears to reduce levels of inflammatory markers.
- In T2DM patients, regular exercise induces favorable changes on the metabolic profile, including a notorious anti-inflammatory action, thus contributing to clinical benefits by reducing cardiovascular and metabolic risk.
- In order to obtain the maximum beneficial effects, exercise programs should be tailored to individual needs on the basis of factors such as age, gender, and health status.

REFERENCES

Abbas AK, Lichtman AH, and Pillai S. 2007. *Cellular and Molecular Immunology*, 6th ed. Philadelphia, PA: WB Saunders.

Akerstrom T, Steensberg A, Keller P, Keller C, Penkowa M, Pedersen BK. 2005. Exercise induces interleukin-8 expression in human skeletal muscle. *J Physiol.* 563(Pt 2):507–16.

Autenrieth C, Schneider A, Döring A, Meisinger C, Herder C, Koenig W, et al. 2009. Association between different domains of physical activity and markers of inflammation. *Med Sci Sports Exerc.* 41(9):1706–13.

Bagby GJ, Sawaya DE, Crouch LD, Shepherd RE.1994. Prior exercise suppresses the plasma tumor necrosis factor response to bacterial lipopolysaccharide. *J Appl Physiol.* 77(3):1542–7.

Balducci S, Zanuso S, Nicolucci A, De Feo P, Cavallo S, Cardelli P, et al. 2010. Italian Diabetes Exercise Study (IDES) Investigators Effect of an intensive exercise intervention strategy on modifiable cardiovascular risk factors in subjects with type 2 diabetes mellitus: A randomized controlled trial: The Italian Diabetes and Exercise Study (IDES). *Arch Intern Med.* 170(20):1794–803.

Baptista S, Piloto N, Reis F, Teixeira-de-Lemos E, Garrido AP, Dias A, et al. 2008. Treadmill running and swimming imposes distinct cardiovascular physiological adaptations in the rat: Focus on serotonergic and sympathetic nervous systems modulation. *Acta Physiol Hung.* 95(4):365–81.

Bassuk SS, Manson JE. 2005. Epidemiological evidence for the role of physical activity in reducing risk of type 2 diabetes and cardiovascular disease. *J Appl Physiol.* 99:1193–1204.

Beavers KM, Brinkley TE, Nicklas BJ. 2010. Effect of exercise training on chronic inflammation., *Clin Chim Acta.* 411(11-12):785–93.

Bennet AM, Prince JA, Fei GZ, Lyrenäs L, Huang Y, Wiman B, et al. 2003. Interleukin-6 serum levels and genotypes influence the risk for myocardial infarction. *Atherosclerosis* 171(2):359–67.

Berster JM, Göke B. 2008. Type 2 diabetes mellitus as risk factor for colorectal cancer. *Arch Physiol Biochem.* 114(1):84–98.

Brandt C, Pedersen BK. 2010. The role of exercise-induced myokines in muscle homeostasis and the defense against chronic diseases. *J Biomed Biotechnol.* 2010:520258.

Bulló M, García-Lorda P, Megias I, Salas-Salvadó J. 2003. Systemic inflammation, adipose tissue tumor necrosis factor, and leptin expression. *Obesity Research.* 11(4):525–31.

Das UN. Anti-inflammatory nature of exercise. 2004. *Nutrition.* 20:323–36.

Dela F, von Linstow ME, Mikines KJ, Galbo H. 2004. Physical training may enhance beta-cell function in type 2 diabetes. *Am J Physiol Endocrinol Metab.* 287(5):E1024–31.

Donath MY, Böni-Schnetzler M, Ellingsgaard H, Halban PA, Ehses JA. 2010. Cytokine production by islets in health and diabetes: Cellular origin, regulation and function. *Trends in Endocrinol Metab.* 21(5):261–7.

Edwards KM, Burns VE, Carroll D, Drayson M, Ring C. 2007. The acute stress-induced immunoenhancement hypothesis. *Exercise and Sport Sciences Reviews.* 35(3):150–5.

Fallon KE, Fallon SK, Boston T. 2001. The acute phase response and exercise: Court and field sports. *Br J Sports Med.* 35, 170–3.

Febbraio MA, Pedersen BK. 2002. Muscle-derived interleukin-6: Mechanisms for activation and possible biological roles. *FASEB J.* 16(11):1335–47.

Febbraio MA. 2007. Exercise and inflammation. *J Appl Physiol.* 103(1):376–7.

Fisman EZ, Tenenbaum A. 2010. The ubiquitous interleukin-6: A time for reappraisal. *Cardiovasc Diabetol.* 9:62.

Gil A, María Aguilera C, Gil-Campos M, Cañete R. 2007. Altered signalling and gene expression associated with the immune system and the inflammatory response in obesity. *Br J Nutr.* 98(supplement 1):S121–6.

Giraldo E, Hinchado MD, García JJ, Ortera E. 2008. Influence of gender and oral contraceptives intake on innate and inflammatory response. Role of neuroendocrine factors. *Mol Cell Biochem.* 313:147–53.

Gleeson M, Nieman DC, Pedersen BK. 2004. Exercise, nutrition and immune function. *J Sports Sci.* 22(1):115–25.

Hansson GK. 2005. Inflammation, atherosclerosis, and coronary artery disease. *N Engl J Med.* 352:1685–95.

Hasselgren PO. 1999. Role of ubiquitin-proteasome pathway in sepsis-induced muscle catabolism. *Mol Biol Rep.* 26(1-2):71–6.

Higashimoto Y, Yamagata Y, Taya S, Iwata T, Okada M, Ishiguchi T, et al. 2008. Systemic inflammation in chronic obstructive pulmonary disease and asthma: Similarities and differences. *Respirology* 13:128–33.

Hotamisligil, GS. 2006. Inflammation and metabolic disorders. *Nature* 444:860–7.

Hull RL, Westermark GT, Westermark P, Kahn SE. 2004. Islet amyloid: A critical entity in the pathogenesis of type 2 diabetes. *J Clin Endocrinol Metab.* 89:3629–43.

Jenkins RC, Ross RJM. 1999. *The endocrine response to acute illness.* Front Horm Res. Basel, Karger. vol 24, p. 13.

Kadoglou NP, Iliadis F, Angelopoulou N, Perrea D, Ampatzidis G, Liapis CD, et al. 2007. The anti-inflammatory effects of exercise training in patients with type 2 diabetes mellitus. *Eur J Cardiovasc Prev Rehabil.* 14:837–43.

Kahn SE, Hull RL, Utzschneider KM. 2006. Mechanisms linking obesity to insulin resistance and type 2 diabetes. *Nature.* 444, 840–6.

Kasapis C, Thompson PD. 2005. The effects of physical activity on serum C-reactive protein and inflammatory markers: A systematic review. *J Am Coll Cardiol.* 45:1563–9.

Kasımay O, Ergen N, Bilsel S, Kaçar O, Deyneli O, Gogas D, et al. 2010. Diet-supported aerobic exercise reduces blood endothelin-1 and nitric oxide levels in individuals with impaired glucose tolerance. *J Clin Lipidol.* 4(5):427–34.

Keller C, Keller P, Giralt M, Hidalgo J, Pedersen BK. 2004. Exercise normalises overexpression of TNF-alpha in knockout mice. *Biochem Biophys Res Commun.* 321(1):179–182.

Kemi OJ, Haram PM, Loennechen JP, Osnes JB, Skomedal T, Wisloff U, et al. 2005. Moderate *vs.* high exercise intensity: Differential effects on aerobic fitness, cardiomyocyte contractility, and endothelial function. *Cardiovasc Res.* 67:161–72.

Lahaye SD, Gratas-Delamarche A, Malardé L, Vincent S, Zguira MS, Morel SL, et al. 2010. Intense exercise training induces adaptation in expression and responsiveness of cardiac b-adrenoceptors in diabetic rats. *Cardiovasc Diabetol.* 9:72.

Lakka TA, Lakka HM, Rankinen T, Leon AS, Rao DC, Skinner JS, et al. 2005. Effect of exercise training on plasma levels of C-reactive protein in healthy adults: The HERITAGE Family Study. *Eur Heart J* 26:2018–25.

Laye MJ, Thyfault JP, Stump CS, Booth FW. 2007. Inactivity induces increases in abdominal fat. *J Appl Physiol.* 102:1341–7.

Leehey DJ, Moinuddin I, Bast JP, Qureshi S, Jelinek CS, Cooper C, et al. 2009. Aerobic exercise in obese diabetic patients with chronic kidney disease: A randomized and controlled pilot study. *Cardiovasc Diabetol.* 8:62.

Lin WW, Karin M. 2007. A cytokine-mediated link between innate immunity, inflammation, and cancer. *J Clin Invest.* 117(5):1175–83.

Luster AD, Alon R, von Andrian UH. 2005. Immune cell migration in inflammation: Present and future therapeutic targets. *Nat Immunol.* 6:1182–90.

Majno G, Joris I. 2004. *Cells, Tissues, and Disease: Principles of General Pathology.* 2nd ed. Oxford, UK: Oxford University Press.

Malm C. 2004. Exercise immunology: The current state of man and mouse. *Sports Med.* 34(9):555–66.

Marik PE. 2006. Dyslipidemia in the critically ill. *Crit Care Clin.* 22(1):151–9.

Martin-Cordero L, Garcia JJ, Giraldo E, De la Fuente M, Manso R, Ortega E. 2009. Influence of exercise on the circulating levels and macrophage production of IL-1beta and IFN gamma affected by metabolic syndrome: An obese Zucker rat experimental animal model. *Eur J Appl Physiol.* 107(5):535–43.

Martín-Cordero L, García JJ, Hinchado MD, Bote E, Manso R, Ortega E: Habitual physical exercise improves macrophage IL-6 and TNF-α deregulated release in the obese Zucker rat model of the metabolic syndrome. *Neuroimmunomodulation* 18(2):123–30.

Mathur N, Pedersen BK. 2008. Exercise as a mean to control low-grade systemic inflammation. *Mediators Inflamm.* 109502: Epub.

Matter CM, Handschin C. 2007. RANTES (regulated on activation, normal T cell expressed and secreted), inflammation, obesity, and the metabolic syndrome. *Circulation.* 115(8):946–8.

Matthys P, Mitera T, Heremans H, Van Damme J, Billiau A. 2002. Anti-gamma interferon and anti-interleukin-6 antibodies affect staphylococcal enterotoxin B-induced weight loss, hypoglycemia, and cytokine release in D- physique anxiety in older adults: Fitness and efficacy influences. *Aging Ment Health*. 6(3):222–30.

Miles MP. 2008. How do we solve the puzzle of unintended consequences of inflammation? Systematically. *J Appl Physiol*. 105: 1023–5.

Minato K, Shiroya Y, Nakae Y, Kondo T. 2002. The effect of chronic exercise on the rat pancreas. *Int J Pancreatol*. 27(2):151–6.

Miyazaki H, Oh-Ishi S, Ookawara T, Kizaki T, Toshinai K, Ha S, et al. 2001. Strenuous endurance training in humans reduces oxidative stress following exhausting exercise. *Eur J Appl Physiol*. 84:1–6.

Mohamed-Ali V, Armstrong L, Clarke D, Bolton CH, Pinkney JH. 2001. Evidence for the regulation of levels of plasma adhesion molecules by proinflammatory cytokines and their soluble receptors in type 1 diabetes. *J Intern Med*. 250: 415–21.

Moore GE. 2004. The role of exercise prescription in chronic disease. *J Sports Med*. 38:6–7.

Nicklas BJ, Brinkley TE. 2009. Exercise training as a treatment for chronic inflammation in the elderly. *Exerc Sport Sci Rev*. 37(4):165–70.

Nielsen AR, Hojman P, Erikstrup C, Fischer CP, Plomgaard P, Mounier R, et al. 2008. Association between interleukin-15 and obesity: Interleukin-15 as a potential regulator of fat mass. *J Clin Endocrinol Metab*. 93(11):4486–93.

Nielsen AR, Mounier R, Plomgaard P, Mortensen OH, Penkowa M, Speerschneider T, et al. 2007. Expression of interleukin-15 in human skeletal muscle effect of exercise and muscle fibre type composition. *J Physiol*. 584(Pt 1):305–12.

Nielsen AR, Pedersen BK. 2007. The biological roles of exercise induced cytokines: IL-6, IL-8, and IL-15. *Appl Physiol Nutr Metab*. 32:833–9.

Nieman DC. 2003. Current perspective on exercise immunology. *Curr Sports Med Rep*. 2(5):239–42.

Olsen RH, Krogh-Madsen R, Thomsen C, Booth FW, Pedersen BK. 2008. Metabolic responses to reduced daily steps in healthy nonexercising men. *JAMA*. 299:1261–63.

Ortega E. 2003. Neuroendocrine mediators in the modulation of phagocytosis by exercise: Physiological implications. *Exerc Immunol Rev*. 9:70–94.

Pastva A, Estell K, Schoeb TR, Schwiebert LM. 2005. RU486 blocks the anti-inflammatory effects of exercise in a murine model of allergen-induced pulmonary inflammation. *Brain Behav Immun*. 19(5):413–22.

Pavlov VA, Tracey KJ. 2005. The cholinergic anti-inflammatory pathway. *Brain Behav Immun*. 19:493–99.

Pearson, TA, Mensah, GA, Alexander, RW, et al. 2003. Markers of inflammation and cardiovascular disease: Application to clinical and public health practice: A statement for healthcare professionals from the Centers for Disease Control and Prevention and the American Heart Association. *Circulation*. 107:499–511.

Pedersen BK, Hoffman-Goetz L. 2000. Exercise and the immune system: Regulation, integration, and adaptation. *Physiol Rev*. 80:1055–81.

Pedersen BK, Febbraio MA. 2008. Muscle as an endocrine organ: Focus on muscle-derived interleukin-6. *Physiol Rev*. 88(4):1379–406.

Pedersen BK. 2011. Muscles and their myokines. *J Exp Biol*. 214(Pt 2):337–46.

Pedersen BK. 2006. The anti-inflammatory effect of exercise: Its role in diabetes and cardiovascular disease control. *Essays in Biochemistry*. 42:105–17.

Pedersen BK. 2009. The diseasome of physical inactivity—and the role of myokines in muscle—fat cross talk. *J Physiol*. 587:5559–68.

Pedersen BK. 2007. IL-6 signalling in exercise and disease. *Biochemical Society Transactions* 35:1295–97.

Pedersen BK. 2011. Exercise-induced myokines and their role in chronic diseases. *Brain Behav Immun*. 25(5):811–6.

Perry VH, Cunningham C, Holmes C. 2007. Systemic infections and inflammation affect chronic neurodegeneration. *Nat Rev Immunol*. 7(2):161–7.

Philippou A, Bogdanis G, Maridaki M, Halapas A, Sourla A, Koutsilieris M. 2009. Systemic cytokine response following exercise-induced muscle damage in humans. *Clin Chem Lab Med*. 47:777–82.

Pitsavos C, Panagiotakos DB, Chrysohoou C, Kavouras S, Stefanadis C. 2005. The associations between physical activity, inflammation, and coagulation markers, in people with metabolic syndrome: The ATTICA study. *Eur J Cardiovasc Prev Rehabil*. 12:151–8.

Plomgaard P, Penkowa M, Pedersen BK. 2005. Fiber type specific expression of TNF-alpha, IL-6 and IL-18 in human skeletal muscles. *Exerc Immunol Rev*. 11:53–6.

Preis SR, Massaro JM, Robins SJ, Hoffmann U, Vasan RS, Irlbeck T, et al. 2010. Abdominal subcutaneous and visceral adipose tissue and insulin resistance in the Framingham heart study. *Obesity (Silver Spring).* 18(11):2191–8.

Quinn LS. 2008. Interleukin-15: A muscle-derived cytokine regulating fat-to-lean body composition. *J Anim Sci.* 86(14 Suppl):E75–E83.

Quinn LS, Anderson BG, Strait-Bodey L, Wolden-Hanson T. 2010. Serum and muscle interleukin-15 levels decrease in aging mice: Correlation with declines in soluble interleukin-15 receptor alpha expression. *Exp Gerontol.* 45(2):106–12.

Radom-Aizik S, Leu SY, Cooper DM, Zaldivar F Jr. 2007. Serum from exercising humans suppresses t-cell cytokine production. *Cytokine.* 40:75–81.

Ravin KR, Kamari Y, Navni I, Grossman E, Sharabi Y. 2005. Adiponectin: Linking the metabolic syndrome to its cardiovascular consequences. *Expert Rev Cardiovasc Ther.* 3:465–71.

Reilly MP, Rohatgi A, McMahon K, Wolfe ML, Pinto SC, Rhodes T, Girman C, et al. 2007. Plasma cytokines, metabolic syndrome, and atherosclerosis in humans. *Journal of Investigative Medicine.* 55(1):26–35.

Ridker PM, Buring JE, Cook NR, et al. 2003. C-reactive protein, the metabolic syndrome, and risk of incident cardiovascular events: An 8-year follow-up of 14,719 initially healthy American women. *Circulation* 107:391–7.

Ridker PM, Stampfer MJ, Rifai N. 2001. Novel risk factors for systemic atherosclerosis: A comparison of C-reactive protein, fibrinogen, homocysteine, lipoprotein(a), and standard cholesterol screening as predictors of peripheral arterial disease. *JAMA.* 285:2481–5.

Riechman SE, Balasekaran G, Roth SM, Ferrell RE. 2004. Association of interleukin-15 protein and interleukin-15 receptor genetic variation with resistance exercise training responses. *J Appl Physiol.* 97:2214–9.

Selamoglu S, Turgay F, Kayatekin BM, Gunenc S, Yslegen C. 2000. Aerobic and anaerobic training effects on the antioxidant enzymes of the blood. *Acta Physiol Hung.* 87:267–73.

Shima K, Zhu M, Noma Y, Mizuno A, Murakami T, Sano T, et al. 1997. Exercise training in Otsuka Long-Evans Tokushima Fatty rat, a model of spontaneous non-insulin-dependent diabetes mellitus: Effects on the B-cell mass, insulin content and fibrosis in the pancreas. *Diabetes Res Clin Pract.* 35(1):11–19.

Smith JK, Dykes R, Douglas JE, Krishnaswamy G, Berk S. 1999. Long-term exercise and atherogenic activity of blood mononuclear cells in persons at risk of developing ischemic heart disease. *JAMA* 281:1722–27.

Smith LM, Cuthbertson B, Harvie J, Webster N, Robins S, Ralston SH. 2002. Increased bone resorption in the critically ill: Association with sepsis and increased nitric oxide production. *Crit Care Med.* 4:837–40.

Starkie R, Ostrowski SR, Jauffred S, Febbraio M, Pedersen BK. 2003. Exercise and IL-6 infusion inhibit endotoxin-induced TNF-alpha production in humans. *FASEB J.* 17:884–6.

Tanasescu M, Leitzmann MF, Rimm EB, Hu FB. 2003. Physical activity in relation to cardiovascular disease and total mortality among men with type 2 diabetes. *Circulation* 7:2435–9.

Teixeira de Lemos E, Reis F, Baptista S, Garrido AP, Pinto R, Sepodes B, et al. 2007a. Efeitos do exercício físico aeróbio no perfil metabólico e oxidativo de ratos diabéticos tipo 2. *Bull SPHM.* (1):16–28.

Teixeira de Lemos E, Reis F, Baptista S, Pinto R, Sepodes B, Vala H, et al. 2007b. Exercise training is associated with improved levels of C-reactive protein and adiponectin in ZDF (type 2) diabetic rats. *Med Sci Monit.* 13(8):BR168–174.

Teixeira de Lemos E, Reis F, Baptista S, Pinto R, Sepodes B, Vala H, et al. 2009. Exercise training decreases proinflammatory profile in Zucker diabetic (type 2) fatty rats. *Nutrition.* 25(3):330–9.

Tomaszewski M, Charchar FJ, Przybycin M, Crawford L, Wallace AM, Gosek K, et al. 2003. Strikingly low circulating CRP concentrations in ultramarathon runners independent of markers of adiposity: How low can you go? *Arterioscler Thromb Vasc Biol.* 23:1640–4.

Urakawa H, Katsuki A, Sumida Y, Gabazza EC, Murashima S, Morioka K, et al. 2003. Oxidative stress is associated with adiposity and insulin resistance in men. *J Clin Endocrinol Metab.* 88(10):4673–6.

Wärnberg J, Nova E, Romeo J, Moreno LA, Sjöström M, Marcos A. 2007. Lifestyle-related determinants of inflammation in adolescence. *Br J Nutr.* 98:S116–S120.

Waugh DJJ, Wilson C. 2008. The interleukin-8 pathway in cancer. *Clin Cancer Res.* 14: 6735–41.

White LJ, Castellano V, McCoy SC. 2006. Cytokine responses to resistance training in people with multiple sclerosis. *J Sports Sci.* 24: 911–14.

Wilund KR. 2007. Is the anti-inflammatory effect of regular exercise responsible for reduced cardiovascular disease? *Clin Sci (Lond).* 112(11):543–55.

Woods JA, Vieira VJ, Keylock KT. 2009. Exercise, inflammation, and innate immunity. *Immunol Allergy Clin North Am.* 29(2):381–93.

Zahner L, Puder JJ, Roth R, Schmid M, Guldimann R, Pühse U, et al. 2006. A school-based physical activity program to improve health and fitness in children aged 6–13 years ("Kinder-Sportstudie KISS"): Study design of a randomized controlled trial [ISRCTN15360785]. *BMC Public Health* 6:147.

Zeyda M, Farmer D, Todoric J, Aszmann O, Speiser M, Györi G, et al. 2007. Human adipose tissue macrophages are of an anti-inflammatory phenotype but capable of excessive pro-inflammatory mediator production. *International Journal of Obesity* 31(9):1420–8.

Zipp F, Aktas O. 2006. The brain as a target of inflammation: Common pathways link inflammatory and neurodegenerative diseases. *Trends Neurosci.* 29(9):518–27.

22 Omega-6 and Omega-3 Polyunsaturated Fatty Acids and Inflammatory Processes

Philip C. Calder
University of Southampton
Southampton, United Kingdom

CONTENTS

22.1 OMEGA-6 AND OMEGA-3 POLYUNSATURATED FATTY ACIDS—NAMING, BIOSYNTHESIS, SOURCES, AND INTAKES

This chapter will consider two families of polyunsaturated fatty acids (PUFAs), the omega-6 (ω-6) family and the omega-3 (ω-3) family. Omega-6 and ω-3 PUFAs each have a characteristic, distinguishing structural feature, separate dietary sources, and distinct biological properties. PUFAs are fatty acids that have two or more double bonds within the fatty acyl hydrocarbon chain. They form a smaller part of the human diet than saturated or monounsaturated fatty acids. The terms "ω-6" and "ω-3" refer to the characteristic, distinguishing structural feature of these fatty acids: the number (i.e., 6 or 3) indicates the carbon atom in the hydrocarbon chain on which the first double bond is found if the terminal methyl carbon is defined as carbon number one. The simplest ω-6 PUFA is linoleic acid, an 18-carbon fatty acid with two double bonds in the hydrocarbon chain, which is described as 18:2ω-6. The simplest ω-3 PUFA is α-linolenic acid, an 18-carbon fatty acid with three double bonds in the hydrocarbon chain, described as 18:3ω-3. Neither linoleic acid nor α-linolenic acid can be synthesized in animals. However, they are synthesized in plants. Furthermore, animals are not able to interconvert ω-6 and ω-3 PUFAs, but plants can do this. In animals, linoleic and α-linolenic acids are able to be converted to other fatty acids (Figure 22.1). During this conversion the position of the methyl-terminal double bond is retained so that linoleic acid is a precursor to other ω-6 PUFAs, and α-linolenic acid is a precursor to other ω-3 PUFAs (Figure 22.1). The conversion of linoleic and α-linolenic acids to these other fatty acids occurs through the insertion of additional double bonds into the hydrocarbon chain in a process called unsaturation and by elongation of the hydrocarbon chain. By this pathway, linoleic acid can be converted to arachidonic acid (20:4ω-6; ARA; Figure 22.1) and α-linolenic acid can be converted to eicosapentaenoic acid (20:5ω-3; EPA; Figure 22.1). Both arachidonic acid and EPA can be further metabolized, EPA ultimately giving rise to docosahexaenoic acid (22:6ω-3; DHA; Figure 22.1).

Since linoleic acid is synthesized in plants, it is not surprising that it is found in significant quantities in seeds and nuts, in many commonly used vegetable oils (e.g., corn, sunflower, and soybean oils), and in products made from vegetable oils like margarines. α-linolenic acid is found in green plant tissues, in some common vegetable oils (e.g., soybean and rapeseed oils),

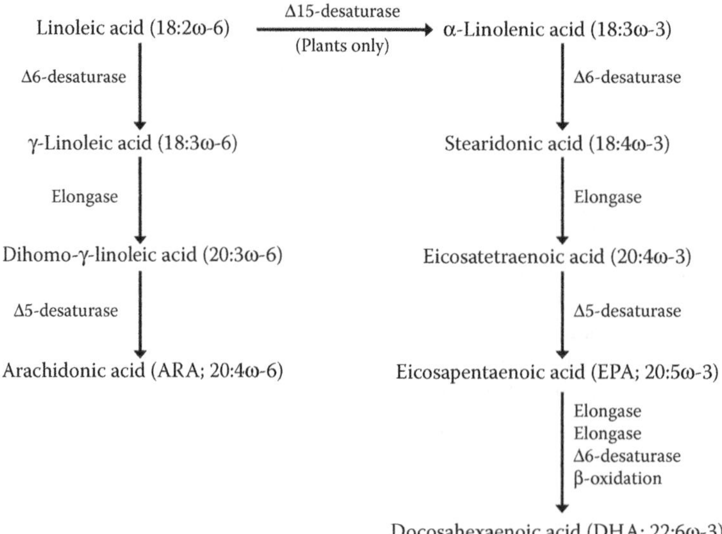

FIGURE 22.1 The biosynthesis of polyunsaturated fatty acids. ARA, arachidonic acid; DHA, docosahexaenoic acid; EPA, eicosapentaenoic acid.

in some nuts (e.g., walnuts), and in flaxseed (also known as linseed) and flaxseed oil. In most Western diets linoleic and α-linolenic acids together make up more than 95%, and often as much as 98%, of dietary PUFA, other PUFAs usually being consumed in low amounts. In most diets linoleic acid intake is in considerable excess of α-linolenic acid intake. Typical Western adult intakes of linoleic acid and α-linolenic acid would be approximately 10 and 1 g/day. Dietary intakes of ARA, EPA and DHA are much lower than those of linoleic and α-linolenic acids. ARA is found in meat, eggs, and offal (e.g., liver) and typically several hundreds of mg/day are consumed. EPA and DHA are found in seafood, especially in what are referred to as "fatty" or "oily" fish; these are fish like tuna, salmon, mackerel, herring, and sardines. They are also found in eggs and meat but at low levels. One oily fish meal could provide 1 to 3.5 g of n-3 PUFAs depending upon the type of fish. A white fish meal (e.g., cod) would provide about one tenth of this amount of EPA and DHA. Because most people do not eat oily fish often or at all, typical intakes of EPA and DHA are in the tens to hundreds of mg/day (British Nutrition Foundation 1999; Meyer et al. 2003). Fish oil supplements contain EPA and DHA. In a standard supplement EPA and DHA make up about 30% of the fatty acids present, so in a 1-g capsule there will be about 300 mg of EPA+DHA.

22.2 OMEGA-6 AND OMEGA-6 POLYUNSATURATED FATTY ACIDS IN CELLS INVOLVED IN INFLAMMATION

PUFAs are key structural and functional components of the phospholipids in cell membranes. Phospholipids of macrophages, neutrophils, and lymphocytes from rats or mice fed on standard laboratory chow usually have a high ARA content, with ARA contributing as much as 15% to 20% of the fatty acids present; furthermore, they usually contain very little EPA or DHA (Calder et al. 1990, 1994). However, these compositions can be changed by certain dietary manipulations. For example, if rats are fed a diet containing ARA, their lymphocytes have a higher ARA content (Peterson et al. 1998). If rats or mice are fed a diet containing fish oil, their macrophages, neutrophils, and lymphocytes become enriched in EPA and DHA (Peterson et al. 1998; Wallace et al. 2001). The incorporation of EPA and DHA into inflammatory cell phospholipids with fish oil feeding is mirrored by a reduction in the amount of ARA present (Peterson et al. 1998; Wallace et al. 2001).

Similar observations are made in humans. The total phospholipids of monocytes, neutrophils, and lymphocytes taken from the blood of humans consuming typical Western diets contains about 10% to 20% of fatty acids as ARA, with about 0.5%–1% EPA and about 1.5%–3% DHA (Lee et al. 1985; Endres et al. 1989; Sperling et al. 1993; Caughey et al. 1996; Yaqoob et al. 2000; Healy et al. 2000). There are, however, differences in fatty acid contents between the different phospholipid classes (Sperling et al. 1993). As seen in animal studies, the fatty acid composition of human inflammatory cells can be modified as a result of changed intakes of ARA, EPA, and DHA. Increased intake of ARA by elderly human subjects resulted in a higher proportion of ARA in blood mononuclear cells, a mixture of lymphocytes and monocytes (Thies et al. 2001). A number of studies have demonstrated that increased intake of EPA+DHA, usually from fish oil supplements, results in a higher content of EPA and DHA in mononuclear cells and neutrophils (Lee et al. 1985; Endres et al. 1989; Sperling et al. 1993; Caughey et al. 1996; Yaqoob et al. 2000; Healy et al. 2000; Thies et al. 2001; Kew et al. 2003, 2004; Rees et al. 2006). The increased content of EPA and DHA is associated with a decreased content of ω-6 PUFAs, especially ARA. Time-course studies suggest that the net incorporation of EPA and DHA into human inflammatory cells begins within days and reaches its peak within a few weeks (Yaqoob et al. 2000; Healy et al. 2000; Thies et al. 2001; Kew et al. 2004; Rees et al. 2006; Faber et al. 2011). Figure 22.2 shows data from two studies using almost the same dose of EPA and DHA to demonstrate the time course of incorporation into human mononuclear cells. Studies that have used multiple doses of fish oil show that the incorporation of EPA and DHA occurs in a dose-response manner (Healy et al. 2000; Rees et al. 2006; Calder 2008a).

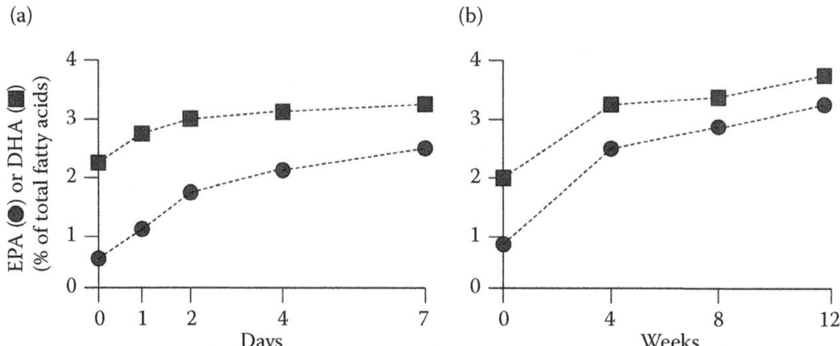

FIGURE 22.2 Time course of incorporation of EPA and DHA into blood inflammatory cells in humans. (a) Healthy male volunteers consumed 2.4 g EPA plus 1.3 g DHA per day for a week, and blood samples were collected at 0, 1, 2, 4, and 7 days. Fatty acid composition of white blood cells was measured. Data are mean of 12 subjects and are taken from Faber et al. (2011). (b) Healthy male and female volunteers consumed 2.1 g EPA plus 1.1 g DHA per day for 12 weeks, and blood samples were collected at 0, 4, 8, and 12 weeks. Fatty acid composition of blood mononuclear cells was measured. Data are mean of 8 subjects and are taken from Yaqoob et al. (2000).

22.3 ARACHIDONIC ACID IS A PRECURSOR OF EICOSANOID MEDIATORS INVOLVED IN INFLAMMATION

Eicosanoids are synthesised from 20-carbon PUFAs. They include prostaglandins (PGs), thromboxanes, and leukotrienes (LTs), and they are key mediators and regulators of inflammation (Lewis et al. 1990; Tilley et al. 2001). Because of its high content in membrane phospholipids, ARA is the usual precursor for eicosanoid synthesis (Figure 22.3). ARA is released from the phospholipids through the action of phospholipase A_2 enzymes, which are activated by inflammatory stimuli. The free ARA then acts as a substrate for cyclooxygenase (COX) or lipoxygenase (LOX) enzymes. COX enzymes lead to PGs and thromboxanes while LOX enzymes lead to LTs. The eicosanoid mediators have a variety of roles in inflammation. For example, PGE_2 has a number of pro-inflammatory effects including increasing vascular permeability, vasodilation, blood flow, and local pyrexia, and potentiation of pain caused by other agents. It also promotes the production of some matrix metalloproteinases that can cause local tissue damage and destruction (e.g., to bone in rheumatoid arthritis [RA]). A number of pharmaceutical agents used as anti-inflammatories target the COX pathway. LTB_4 increases vascular permeability, enhances local blood flow, is a potent chemotactic agent for leukocytes, induces release of lysosomal enzymes, and enhances release of reactive oxygen species and inflammatory cytokines.

Expression of COX is increased in the synovium of patients with RA (Sano et al. 1992; Feldmann and Maini 1999) and in joint tissues in rat models of arthritis (Sano et al. 1992). PGE_2, LTB_4, and 5-hydroxyeicosatetraenoic acid are found in the synovial fluid of patients with active RA (Sperling 1995). Infiltrating leukocytes such as neutrophils, monocytes, and synoviocytes are important sources of eicosanoids in RA (Sperling 1995). The efficacy of nonsteroidal anti-inflammatory drugs (NSAIDs), which act to inhibit COX activity, in RA indicates the importance of this pathway in the pathophysiology of the disease.

Induction of colitis in laboratory animals results in the appearance of inflammatory eicosanoids such as PGE_2 and LTB_4 in the colonic mucosa (Hirata et al. 2001; Nieto et al. 2002). In human inflammatory bowel disease (IBD) the intestinal mucosa contains elevated levels of inflammatory eicosanoids such as LTB_4 (Sharon and Stenson 1984). An LT biosynthesis inhibitor had a protective effect on mucosal injury in acetic acid–induced rat colitis (Empey et al. 1991). This effect was associated with decreased colonic LTB_4 concentrations. However, the role of PGE_2 in IBD is less certain.

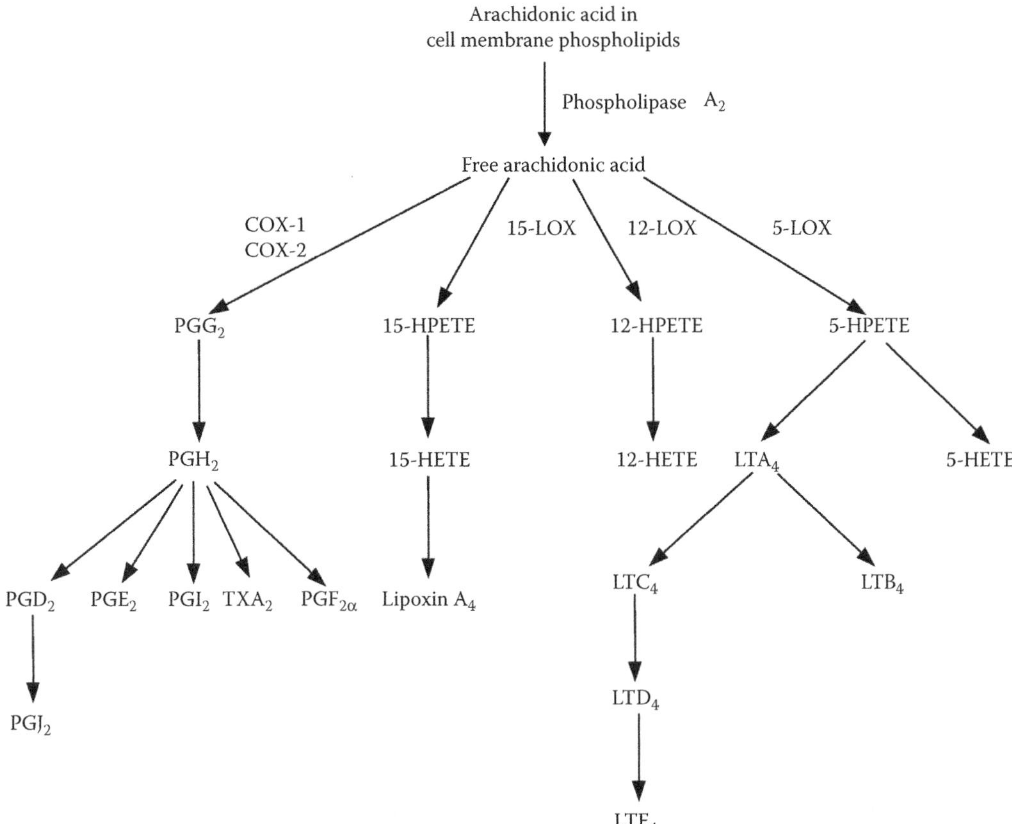

FIGURE 22.3 Outline of the pathway of eicosanoid synthesis from arachidonic acid. COX, cyclooxygenase; HETE, hydroxyeicosatetraenoic acid; HPETE, hydroperoxyeicosatetraenoic acid; LOX, lipoxygenase; LT, leukotriene; PG, prostaglandin; TX, thromboxane. (Reproduced from Calder, P.C., *Am. J. Clin. Nutr.* 83, 2006. With permission.)

Although it is generally considered that eicosanoids derived from ARA are pro-inflammatory in nature, this is an oversimplification (Calder 2009). For example, PGE_2 inhibits the production of two of the classic pro-inflammatory cytokines, tumor necrosis factor (TNF)-α and interleukin (IL)-1, by monocytes and macrophages (Calder 2009). Furthermore, in recent years it has been discovered that PGE_2 inhibits 5-LOX and so decreases production of the inflammatory 4-series LTs (Levy et al. 2001), and induces 15-LOX so promoting the formation of lipoxins (Levy et al. 2001; Vachier et al. 2002) that have been found to have anti-inflammatory effects (Gewirtz et al. 2002). These findings show that PGE_2 possesses both pro- and anti-inflammatory actions. There may be a temporal change in the role of PGE_2 with an initial pro-inflammatory effect followed by a role in resolution of inflammation through inhibition of 5-LOX and induction of lipoxin production.

22.4 OMEGA-3 FATTY ACIDS AND LIPID MEDIATORS

22.4.1 Omega-3 Fatty Acids Decrease Production of Eicosanoids from Arachidonic Acid

Animal studies have shown that production of ARA-derived eicosanoids like PGE_2 is decreased by EPA or DHA feeding (Chapkin et al. 1991; Yaqoob and Calder 1995; Peterson et al. 1998). Numerous studies with healthy human volunteers have described decreased production of PGE_2

and 4 series-LTs by inflammatory cells following use of fish oil supplements for a period of weeks to months (Lee et al. 1985; Endres et al. 1989; Meydani et al. 1991; Sperling et al. 1993; von Schacky et al. 1993; Caughey et al. 1996). Similar effects are seen in patients with chronic inflammatory diseases. For example, fish oil supplements decreased LTB_4 production by neutrophils (Kremer et al. 1995, 1987; Cleland et al. 1988) and monocytes (Cleland et al. 1988; Tullekan et al. 1990) and decreased PGE_2 production by mononuclear cells (Cleland et al. 2006) in patients with RA. In patients with IBD, fish oil decreased LTB_4 production by neutrophils (McCall et al. 1989; Hawthorne et al. 1992; Shimizu et al. 2003) and colonic mucosa (Stenson et al. 1992; Shimizu et al. 2003), decreased PGE_2 and thromboxane B_2 production by colonic mucosa (McCall et al. 1989), and decreased production of PGE_2 by blood mononuclear cells (Trebble et al. 2004).

The studies in humans demonstrating that oral ω-3 PUFAs decrease production of ARA-derived eicosanoids have usually used fairly high intakes of the ω-3 PUFAs, most often several grams per day. A dose-response study in healthy volunteers reported that an EPA intake of 1.35 g/day for 3 months was not sufficient to influence ex vivo PGE_2 production by endotoxin-stimulated mononuclear cells, whereas an EPA intake of 2.7 g/day did significantly decrease PGE_2 production (Rees et al. 2006).

22.4.2 EPA Gives Rise to Alternative Eicosanoids

EPA is also a substrate for the COX and LOX enzymes that produce eicosanoids, but the mediators produced have a different structure from those made from ARA. Increased generation of 5-series LTs has been demonstrated using macrophages from fish oil–fed mice (Chapkin et al. 1991) and neutrophils from humans taking fish oil supplements for several weeks (Lee et al. 1985; Endres et al. 1989; Sperling et al. 1993). Transgenic ("*fat*-1") mice bearing the *C. elegans* "ω-3 desaturase" gene and so able to convert ω-6 to ω-3 PUFAs, resulting in greatly elevated ω-3 PUFA content in their tissues, were shown to generate large amounts of PGE_3 within colonic tissue after chemical induction of colonic inflammation (Hudert et al. 2006). The functional significance of generation of eicosanoids from EPA is that EPA-derived mediators are often much less biologically active than those produced from ARA (Goldman et al. 1983; Lee et al. 1984; Bagga et al. 2003; Wada et al. 2007). Thus, EPA results in decreased production of potent eicosanoids from ARA and increased production of weak eicosanoids.

22.4.3 EPA and DHA Give Rise to Anti-Inflammatory and Inflammation Resolving Mediators Called Resolvins and Protectins

In the last 10 years or so new families of ω-3 PUFA-derived lipid mediators have been discovered. These include the resolvins (E-series from EPA and D-series from DHA) and protectins; the latter are referred to as neuroprotectins when generated within neural tissue. The synthesis of resolvins and protectins involves the COX and LOX pathways, with different epimers being produced in the presence and absence of aspirin (Serhan et al. 2000a, b). Resolvin synthesis is increased by feeding fish oil–rich diets to laboratory rodents (Hong et al. 2003) and was shown to occur in *fat*-1 mice in which colitis had been induced (Hudert et al. 2006). The biological effects of resolvins and protectins have been examined extensively in cell culture and animal models of inflammation. These models have shown them to be anti-inflammatory and inflammation resolving (Serhan et al. 2002). For example, resolvin E1, resolvin D1, and protectin D1 all inhibited transendothelial migration of neutrophils, so preventing the infiltration of neutrophils into sites of inflammation; resolvin D1 inhibited IL-1β production; and protectin D1 inhibited TNF-α and IL-1β production (Serhan et al. 2000a, b, 2002, 2008). Resolvin E1 protected against chemically induced colitis in mice, and this was associated with decreased infiltration of granulocytes into colonic tissue and decreased expression of several inflammatory genes (TNF-α, IL-12p40, COX-2, inducible nitric oxide synthase) within that tissue (Arita et al. 2005).

22.5 OMEGA-3 FATTY ACIDS DECREASE NFκB-MEDIATED INFLAMMATORY SIGNALING

22.5.1 The NFκB System

A key transcription factor involved in inflammatory responses is nuclear factor kappa B (NFκB), which is the principal transcription factor involved in upregulation of inflammatory cytokine, adhesion molecule, and COX-2 genes (Kumar et al. 2004; Sigal 2006). NFκB is activated as a result of a signaling cascade triggered by extracellular inflammatory stimuli and involving phosphorylation of an inhibitory subunit (inhibitory subunit of NFκB [IκB]), which then allows translocation of the remaining NFκB dimer to the nucleus (Perkins 2007). Thus, expression of inflammatory genes is upregulated. NFκB is a recognized target for controlling inflammation.

22.5.2 EPA and DHA Inhibit NFκB Activation and Induction of NFκB Targets

EPA and DHA inhibited endotoxin-stimulated production of IL-6 and IL-8 by cultured human endothelial cells (de Caterina et al. 1994; Khalfoun et al. 1997) and EPA or fish oil inhibited endotoxin-induced TNF-α production by cultured monocytes (Lo et al. 1999; Babcock et al. 2002; Novak et al. 2003; Zhao et al. 2004). EPA or fish oil decreased endotoxin-induced activation of NFκB in human monocytes (Lo et al. 1999; Novak et al. 2003; Zhao et al. 2004) and this was associated with decreased IκB phosphorylation (Novak et al. 2003; Zhao et al. 2004). These observations suggest direct effects of ω-3 PUFAs on inflammatory gene expression via inhibition of activation of the transcription factor NFκB. Animal feeding studies with fish oil support the observations made in cell culture with respect to the effects of ω-3 PUFAs on NFκB activation and inflammatory cytokine production. For example, compared with feeding corn oil, fish oil lowered NFκB activation in endotoxin-activated murine spleen lymphocytes (Xi et al. 2001). Feeding fish oil to mice decreased ex vivo production of TNF-α, IL-1β, and IL-6 by endotoxin-stimulated macrophages (Billiar et al. 1988; Renier et al. 1993; Yaqoob and Calder 1995) and decreased circulating TNF-α, IL-1β, and IL-6 concentrations in mice injected with endotoxin (Sadeghi et al. 1999). Several studies providing fish oil supplements to healthy human volunteers have reported decreased production of TNF-α, IL-1β, and IL-6 by endotoxin-stimulated monocytes or mononuclear cells (Endres et al. 1989; Meydani et al. 1991; Caughey et al. 1996; Trebble et al. 2003), although not all studies confirm this effect (Calder 2006). In patients with RA, fish oil supplements resulted in decreased IL-1 production by monocytes (Kremer et al. 1990), decreased plasma IL-1β concentrations (Esperson et al. 1992), and decreased serum TNF-α concentrations (Kolahi et al. 2010).

22.5.3 EPA and DHA May Promote an Anti-Inflammatory Interaction between PPAR-γ and NFκB

A second transcription factor involved in the regulation of inflammation is peroxisome proliferator activated receptor (PPAR)-γ, which is thought to act in an anti-inflammatory manner (Szanto and Nagy 2008). PPAR-γ knockdown mice show enhanced susceptibility to chemically induced colitis (Desreumaux et al. 2001), and PPAR-γ agonists reduce colitis in murine models (Su et al. 1999; Desreumaux et al. 2001). Thus, upregulation of PPAR-γ is a likely target for controlling inflammation (Dubuquoy et al. 2006). While PPAR-γ directly regulates inflammatory gene expression, it also interferes with the translocation of NFκB to the nucleus (van den Berghe et al. 2003). PPAR-γ may be activated by ω-3 PUFAs and by various eicosanoids (Forman et al. 1995, 1997; Kleiwer et al. 1995, 1997; Gottlicher et al. 1992). In dendritic cells, DHA induced PPAR-γ (Kong et al. 2010) and a number of known PPAR-γ target genes (Zapata-Gonzalez et al. 2008). These effects were linked to decreased production of the inflammatory cytokines TNF-α and IL-6 upon endotoxin stimulation (Kong et al. 2010). Thus, activation of PPAR-γ may be one of the anti-inflammatory mechanisms of action of ω-3 PUFAs, and this may link to the inhibition of NFκB activation described previously.

22.5.4 EPA and DHA May Act through a Cell Surface Receptor That Inhibits NFκB Activation

Some of the G-protein coupled cell membrane receptors can bind certain fatty acids. For example, GPR40 and GPR120 are both able to bind long chain fatty acids, and both these GPRs activate intracellular signaling pathways. GPR120 is highly expressed on inflammatory macrophages but GPR40 is not (Oh et al. 2010). An agonist of GPR120 inhibited the macrophage response to endotoxin, an effect that involved maintenance of cytosolic IκB and a decrease in production of TNF-α and IL-6 (Oh et al. 2010), effects that are similar to those of EPA and DHA. These observations indicate that GPR120 is involved in anti-inflammatory signaling. GPR120-mediated gene activation was enhanced by EPA and DHA and the ability of DHA to inhibit responsiveness of macrophages to endotoxin was abolished in GPR120 knockdown cells. These findings indicate that the inhibitory effect of DHA (and probably also EPA) on NFκB occurs via GPR120. Thus, there appear to be at least two mechanisms by which ω-3 PUFAs inhibit NFκB activation, one involving GPR120 and the other involving PPAR-γ, although these may be linked.

22.6 EFFECTS OF EPA AND DHA ON T CELLS

Cell culture studies have demonstrated effects of various fatty acids, including ω-3 PUFAs, on T cell functional responses and signaling, that these effects are dose-dependent (Calder et al. 1991), and that they relate to changes in the fatty acid composition of lymphocyte phospholipids (Calder et al. 1994). In cell cultures, both EPA and DHA inhibit T cell proliferation (Calder et al. 1991, 1994; Calder and Newsholme 1992a, b) and the production of IL-2 (Calder and Newsholme 1992a, b; Wallace et al. 2001). Animal feeding studies with fairly high amounts of fish oil, or of EPA or DHA, have also reported reduced T cell proliferative responses (Yaqoob et al. 1994; Jolly et al. 1997; Wallace et al. 2001). Mechanisms believed to be involved include alterations in the physical state of the plasma membrane (Calder et al. 1994), modification of the profile of eicosanoid mediators, which in turn influence T cell function (Calder et al. 1992; Miles et al. 2003), and direct effects on transcription factor activation (Miles and Calder 1998). In recent years there has been some focus on whether ω-3 PUFAs influence T cell functions via effects on lipid rafts (Stulnig et al. 1998, 2001; Stulnig and Zeyda 2004; Zeyda et al. 2006), since lipid rafts appear to be intimately involved in T lymphocyte responses to activation (Katagiri et al. 2001; Razzaq et al. 2004; Harder 2004). Certainly increased exposure of T cells to ω-3 PUFAs either in cell culture or in animal feeding studies affects the chemical structure of rafts in T cells and alters raft functioning (Fan et al. 2003, 2004; Zeyda et al. 2002, 2003). Despite the consistent picture of the effect of ω-3 PUFAs on T cells that emerges from in vitro and animal studies and the increasing understanding of the possible mechanism involved, human data are inconsistent; although some studies show that increased intake of ω-3 PUFAs from fish oil decreases human T cell proliferation (Meydani et al. 1991; Thies et al. 2001) and IL-2 production (Meydani et al. 1991), several other studies show no effect. One reason for this may be an insufficient dose of n-3 PUFAs provided in some studies, but that does not explain all the inconsistency.

22.7 OMEGA-3 FATTY ACIDS AS A THERAPEUTIC OPTION FOR CHRONIC INFLAMMATION

22.7.1 General Comments

The generally opposing roles of ω-6 and ω-3 PUFAs in shaping and regulating inflammatory processes and responses suggest that the balance of these fatty acids might be important in determining the development and severity of inflammatory diseases. For example, a high intake of ω-6 PUFAs, especially ARA, could contribute to inflammatory processes and so could predispose to or exacerbate inflammatory diseases. Conversely, the recognition that the ω-3 PUFAs have anti-inflammatory actions suggests that increasing their intake by patients with inflammatory diseases, for example, through dietary supplementation, may be of clinical benefit. In this section studies of ω-3 PUFAs in two chronic inflammatory conditions, RA and IBD, will be discussed.

22.7.2 Rheumatoid Arthritis

22.7.2.1 Introduction

RA is a chronic inflammatory autoimmune disease of the joints and bones (Firestein 2003). Joint inflammation is manifested by swelling, pain, functional impairment, morning stiffness, osteoporosis, and muscle wasting. Erosion of bone occurs commonly in the joints of the hands and feet. The joint lesions are characterized by infiltration of inflammatory cells and contain high concentrations of the chemical mediators they produce (Feldmann and Maini 1999). One pharmaceutical treatment for the inflammation involved in RA involves the use of NSAIDs, which target the metabolism of the ω-6 PUFA ARA via COX enzymes, suggesting a key involvement of these eicosanoid mediators in the pathology of RA. As indicated before, ω-3 PUFAs from oily fish and fish oils target ARA availability and metabolism and also influence several other inflammatory responses that are involved in RA. Cleland et al. (2006) found that RA patients using fish oil supplements were more likely to reduce the use of NSAIDs and to be in remission than those patients who did not use fish oil.

22.7.2.2 Omega-3 PUFAs and Animal Models of RA

An early study compared the effects of vegetable and fish oils in a mouse model involving induction of arthritis with collagen (Leslie et al. 1985). It was found that fish oil delayed the onset of arthritis (mean 34 days vs. 25 days) and reduced its incidence (69% vs. 93%) and severity (mean peak severity score 6.7 vs. 9.8; Leslie et al. 1985). Both EPA and DHA suppressed arthritis induced in rats by streptococcal cell walls with EPA being more effective (Volker et al. 2000). A recent study of collagen-induced arthritis in the susceptible DBA/1 mouse strain showed that dietary ω-3 PUFAs slowed the onset of arthritis, decreased its severity, reduced paw swelling, and improved knee joint pathology (Ierna et al. 2010).

22.7.2.3 Trials of ω-3 PUFAs in RA

A number of randomized, placebo-controlled, double-blind studies of the clinical outcomes of fish oil in RA have been reported; the characteristics and findings of these trials have been summarized elsewhere (Calder 2008b) and are summarized in Table 22.1. The dose of ω-3 PUFAs used in these trials has typically been high, between 1.6 and 7.1 g/day and averaging about 3.5 g/day (Table 22.1). This dose would be difficult to achieve through the diet, but can be achieved through the use of

TABLE 22.1
Summary of Randomized Controlled Trials (RCTs) of ω-3 PUFAs in Rheumatoid Arthritis and Inflammatory Bowel Disease

	Rheumatoid Arthritis	Inflammatory Bowel Disease
Number of RCTs	19	14
Years published	1985–2008	1989–2008
Dose of EPA+DHA used (g/day)	1.6 to 7.1	2.7 to 5.6
Duration (weeks)	12 to 52	12 to 104
Number of RCTs reporting a positive effect of ω-3 PUFAs	17	7

Note: Further details can be found in Calder 2008b, c.

supplements or liquid oil. Almost all of the trials showed some benefit of fish oil, with many trials reporting a number of positive outcomes (Table 22.1). Benefits have included reduced duration of morning stiffness, reduced number of tender or swollen joints, reduced joint pain, reduced time to fatigue, increased grip strength, and decreased use of NSAIDs (Table 22.1). The study of Adam et al. (2003) demonstrated that clinical benefits were greater if the intake of ARA was decreased at the same time as intake of ω-3 PUFAs was increased.

22.7.2.4 Meta-Analyses of Trials of ω-3 PUFAs in RA

Meta-analyses bring together data from several studies and reanalyze the combined data set. A meta-analysis of ω-3 PUFAs in RA that included data from nine trials published between 1985 and 1992 inclusive and from one unpublished trial concluded that dietary fish oil supplementation for three months significantly reduced tender joint count (mean difference –2.9; $P = 0.001$) and morning stiffness (mean difference –25.9 minutes; $P = 0.01$) (Fortin et al. 1995). A more recent meta-analysis that included data from trials published between 1985 and 2002 was conducted (MacLean et al. 2004), although this included one study of flaxseed oil, one study that did not use a control for fish oil, and one study in which transdermal administration of ω-3 PUFAs by ultrasound, rather than the oral route, was used. This meta-analysis concluded that fish oil supplementation has no effect on patient report of pain, swollen joint count, disease activity, or patient's global assessment. However, this conclusion may be flawed, because of the inappropriate manner in which studies were combined and because of a poor understanding of the study designs used. For example, the meta-analysis fails to recognize that patients' ability to reduce the need for using NSAIDs or their ability to be withdrawn from NSAID use, as was done in some designs, must indicate a reduction in pain with ω-3 PUFA use. This meta-analysis does state that in a qualitative analysis of seven studies that assessed the effect of ω-3 fatty acids on anti-inflammatory drug or corticosteroid requirement, six demonstrated reduced requirement for these drugs and concluded that ω-3 fatty acids may reduce requirements for corticosteroids. The effect of ω-3 PUFAs on tender joint count was not assessed by this meta-analysis, which simply restated the findings of the earlier meta-analysis (Fortin et al. 1995) that ω-3 fatty acids reduce tender joint counts. The most recent meta-analysis of ω-3 PUFAs and clinical outcomes in RA was published in 2007 (Goldberg and Katz 2007); this included data from 17 trials, including one trial in RA with flaxseed oil and two trials of fish oil not in RA patients, but which reported joint pain. Data on six outcomes were analyzed. This analysis indicated that fish oil reduces patient assessed joint pain (26% reduction; $P = 0.03$), duration of morning stiffness (57% reduction; $P = 0.003$), number of painful and/or tender joints (71% reduction; $P = 0.003$), and consumption of NSAIDs (60% reduction; $P = 0.01$). This meta-analysis provides fairly strong evidence of the robustness of the efficacy of ω-3 PUFAs in RA.

22.7.3 Inflammatory Bowel Diseases

22.7.3.1 Introduction

Ulcerative colitis (UC) and Crohn's disease (CD) are the two main forms of IBD. CD can affect any part of the gastrointestinal tract, while UC primarily affects the colon (Shanahan 2002; Farrell and Peppercorn 2002). Inflammatory bowel diseases are multifactorial conditions involving both genetic and environmental components, with the final outcome being driven by an aberrant immune response to normal commensal microbiota in individuals who have a weakened gut epithelial barrier (Macfarlane et al. 2004). In both forms of IBD, there are large infiltrates of inflammatory cells in the inflamed tissue with activated NFκB and increased inflammatory cytokines and other mediators.

22.7.3.2 Omega-3 PUFAs and Animal Models of IBD

Animal studies report that dietary fish oil decreases chemically induced colonic damage and inflammation compared with an ω-6 PUFA-rich diet (see Calder 2008c). The effects on disease severity were, in all cases, associated with a reduction in production of ARA-derived eicosanoids. A more recent study investigated dextran sodium sulfate–induced colitis in *fat-1* mice (Hudert et al. 2006). The mice show much less colonic damage and inflammation than wild type mice, and this was associated with a marked change in the pattern of inflammatory mediators present in colonic tissue. A study in IL-10 knockout mice that spontaneously develop colitis demonstrated significantly reduced colonic inflammation of the mice that were fed fish oil compared with ω-6 PUFA-rich corn oil (Chapkin et al. 2007).

22.7.3.3 Trials of ω-3 PUFAs in IBD

Omega-3 PUFAs are incorporated into gut mucosal tissue of patients with IBD who supplement their diet with fish oil and this is associated with reduced inflammation (Calder 2008c). A number of randomized, placebo-controlled, double-blind studies of fish oil in IBD have been reported; the characteristics and findings of these trials have been summarized elsewhere (Calder 2008c) and are summarized in Table 22.1. The dose of ω-3 PUFAs used in these trials has typically been high, between 2.7 and 5.6 g/day and averaging about 4.5 g/day. Some of these trials indicate benefits of fish oil, which include improved clinical score, improved gut mucosal histology, improved sigmoidoscopic score, lower rate of relapse, and decreased use of corticosteroids (Table 22.1). However, a number of trials do not report such benefits (Table 22.1). Although one study with an enterically coated fish oil showed a significantly lower rate of relapse over 12 months in patients with CD (Belluzzi et al. 1996), two recent trials with a similar design and fish oil preparation and using a similar dose of ω-3 PUFAs could not replicate this finding (Feagan et al. 2008).

22.7.3.4 Meta-Analyses of Trials of ω-3 PUFAs in RA

A meta-analysis identified 13 studies of fish oil supplementation in IBD (i.e., both UC and CD) reporting outcomes related to clinical score, sigmoidoscope score, gut mucosal histology score, induced remission, and relapse, but concluded that there were sufficient data to perform meta-analysis only for relapse and only for UC (MacLean et al. 2004). The pooled risk of relapse with ω-3 PUFAs relative to placebo from these studies was 1.13 (95% CI: 0.91, 1.57), and it was concluded that ω-3 fatty acids have no effect on relative risk of relapse in ulcerative colitis and that there was a statistically non-significant reduction in requirement for corticosteroids for ω-3 fatty acids relative to placebo in two studies. In the past few years, a series of meta-analyses on this subject have been published (Turner et al. 2007, 2009; De Ley et al. 2007). One of these considered maintenance of remission in CD and included six studies, including one published in abstract form only and one in pediatric patients (Turner et al. 2009). The authors identified a marginal significant benefit of ω-3 PUFAs for maintaining remission (relative risk 0.77; 95% CI: 0.61, 0.98), but because of heterogeneity among studies and a lack of effect in the two largest studies included, the overall conclusion was that ω-3 PUFAs are probably ineffective for maintenance of remission in Crohn's Disease. The other two meta-analyses considered maintenance of remission in UC (Turner et al. 2007; De Ley et al. 2007). The first of these included three studies and concluded that ω-3 PUFAs are not effective in maintaining remission in ulcerative colitis (Turner et al. 2007); relative risk was 1.02 (95% CI: 0.51, 2.03). The second of these included six studies, including two published only as abstracts, and concluded that there is insufficient information to reach a conclusion (De Ley et al. 2007). Thus, despite some favorable studies (Table 22.1), the overall view at the moment must be that there is only weak evidence that ω-3 PUFAs have clinical benefits in human IBD.

22.8 CONCLUSIONS

Fatty acids can influence inflammation through a variety of mechanisms, including acting via cell surface and intracellular receptors/sensors that control inflammatory cell signaling and gene expression patterns. Some effects of fatty acids on inflammatory cells appear to be mediated by, or at least are associated with, changes in fatty acid composition of cell membranes. Changes in fatty acid composition can modify membrane fluidity, lipid raft formation, cell signaling leading to altered gene expression, and the pattern of lipid and peptide mediator production. Cells involved in the inflammatory response are typically rich in the ω-6 fatty acid ARA, but the contents of ARA and of the ω-3 fatty acids EPA and DHA can be altered through oral administration of EPA and DHA. Eicosanoids produced from ARA have roles in inflammation, although these are not always pro-inflammatory. EPA also gives rise to eicosanoids, and these may have differing properties from those of ARA-derived analogs. EPA and DHA give rise to resolvins, and DHA to protectins, which are anti-inflammatory and inflammation resolving. Increased membrane content of EPA and DHA (and decreased ARA content) results in a changed pattern of production of eicosanoids and resolvins. Thus, fatty acid exposure and the fatty acid composition of human inflammatory cells influences their function; the contents of ARA, EPA, and DHA appear to be especially important. As a result of their anti-inflammatory actions, ω-3 PUFAs may have therapeutic efficacy in inflammatory diseases. Work with animal models of RA has demonstrated efficacy of fish oil. There have been a number of clinical trials of fish oil in patients with RA. Most of these trials report clinical improvements (e.g., improved patient assessed pain, decreased morning stiffness, fewer painful or tender joints, decreased use of NSAIDs), and when the trials have been pooled in meta-analyses, statistically significant clinical benefit has emerged. Thus, evidence for clinical efficacy of ω-3 PUFAs in RA is robust. Dietary fish oil has beneficial effects in animal models of colitis induced by chemicals or by knockout of the IL-10 gene. EPA and DHA are incorporated into the gut mucosal tissue of patients with IBD who supplement their diet with fish oil, and there are reports that this is associated with anti-inflammatory effects. A number of clinical trails of fish oil in both UC and CD have been reported. Although some of these trials indicate benefits, which include improved clinical score, improved gut mucosal histology, improved sigmoidoscopic score, lower rate of relapse, and decreased use of corticosteroids, a number of trials do not report such benefits. Meta-analyses conclude that there is currently no clear evidence of efficacy of ω-3 fatty acids in human IBD. It is not clear why the anti-inflammatory effects observed after fish oil treatment do not translate into clinical improvements in IBD, and there is a need to understand this issue further.

TAKE-HOME MESSAGES

- Cells involved in the inflammatory response are typically rich in the ω-6 fatty acid arachidonic acid, which is the precursor of inflammatory eicosanoids.
- Omega-3 fatty acids from fish oil are readily incorporated into inflammatory cells.
- The fatty acid composition of inflammatory cells affects membrane fluidity, lipid raft formation, cell signaling processes leading to altered gene expression, and the pattern of lipid and peptide mediators produced.
- Fatty acids can also act through cell surface and cytosolic receptors that influence inflammatory signaling.
- Eicosapentaenoic acid (EPA) derived eicosanoids are often weaker in activity than those produced from arachidonic acid.
- EPA and docosahexaenoic acid (DHA) give rise to resolvins, and DHA to protectins, which are anti-inflammatory and inflammation resolving.
- Fatty acid exposure and the fatty acid composition of human inflammatory cells influences their function: the contents of arachidonic acid, EPA, and DHA appear to be especially important.
- As a result of their anti-inflammatory actions, ω-3 fatty acids may have therapeutic efficacy in patients with inflammatory diseases.
- Evidence from studies in rheumatoid arthritis indicates clinical benefit from ω-3 fatty acids.
- Evidence from studies in inflammatory bowel disease is inconsistent and allows no clear conclusion about clinical benefit to be made.

REFERENCES

Adam, O., Beringer, C., Kless, T., et al. 2003. Antiinflammatory effects of a low arachidonic acid diet and fish oil in patients with rheumatoid arthritis. *Rheumatol. Int.* 23:27–36.

Arita, M., Yoshida, M., Hong, S., et al. 2005. Resolvin E1, an endogenous lipid mediator derived from omega-3 eicosapentaenoic acid, protects against 2,4,6-trinitrobenzene sulfonic acid-induced colitis. *PNAS* 102:7621–6.

Babcock, T.A., Novak, T., Ong, E., et al. 2002. Modulation of lipopolysaccharide-stimulated macrophage tumor necrosis factor-α production by ω-3 fatty acid is associated with differential cyclooxygenase-2 protein expression and is independent of interleukin-10. *J. Surg. Res.* 107:135–9.

Bagga, D., Wang, L., Farias-Eisner, R., Glaspy, J.A., and Reddy, S.T. 2003. Differential effects of prostaglandin derived from ω-6 and ω-3 polyunsaturated fatty acids on COX-2 expression and IL-6 secretion. *PNAS* 100:1751–6.

Belluzzi, A., Brignola, C., Campieri, M., Pera, A., Boschi, S., and Miglioli, M. 1996. Effect of an enteric-coated fish-oil preparation on relapses in Crohn's disease. *N. Engl. J. Med.* 334:1557–60.

Billiar, T., Bankey, P., Svingen, B., et al. 1988. Fatty acid uptake and Kupffer cell function: Fish oil alters eicosanoid and monokine production to endotoxin stimulation. *Surgery* 104:343–9.

British Nutrition Foundation. 1999. Briefing Paper: n-3 Fatty acids and health. London: British Nutrition Foundation.

Calder, P.C. 2006. N-3 polyunsaturated fatty acids, inflammation, and inflammatory diseases. *Am. J. Clin. Nutr.* 83:1505S–19S.

Calder, P.C. 2008a. The relationship between the fatty acid composition of immune cells and their function. *Prostagland. Leukotr. Essent. Fatty Acids* 79:101–8.

Calder, P.C. 2008b. PUFA, inflammatory processes and rheumatoid arthritis. *Proc. Nutr. Soc.* 67:409–18.

Calder, P.C. 2008c. Polyunsaturated fatty acids, inflammatory processes and inflammatory bowel diseases. *Mol. Nutr. Food Res. 52*:885–97.

Calder, P.C., 2009. Polyunsaturated fatty acids and inflammatory processes: New twists in an old tale. *Biochimie* 91:791–5.

Calder, P.C., Bevan, S.J., and Newsholme, E.A. 1992. The inhibition of T-lymphocyte proliferation by fatty acids is via an eicosanoid-independent mechanism. *Immunol.* 75:108–15.

Calder, P.C., Bond, J.A., Bevan, S.J., Hunt, S.V., and Newsholme, E.A. 1991. Effect of fatty acids on the proliferation of concanavalin A-stimulated rat lymph node lymphocytes. *Int. J. Biochem.* 23:579–88.

Calder, P.C., Bond, J.A., Harvey, D.J., Gordon, S., and Newsholme, E.A. 1990. Uptake and incorporation of saturated and unsaturated fatty acids into macrophage lipids and their effect upon macrophage adhesion and phagocytosis. *Biochem. J.* 269:807–14.

Calder, P.C. and Newsholme, E.A. 1992a. Polyunsaturated fatty acids suppress human peripheral blood lymphocyte proliferation and interleukin-2 production. *Clin. Sci.* 82:695–700.

Calder, P.C. and Newsholme, E.A. 1992b. Unsaturated fatty acids suppress interleukin-2 production and transferrin receptor expression by concanavalin A-stimulated rat lymphocytes. *Mediat. Inflamm.* 1:107–15.

Calder, P.C., Yaqoob, P., Harvey, D.J., Watts, A., and Newsholme, E.A. 1994. The incorporation of fatty acids by lymphocytes and the effect on fatty acid composition and membrane fluidity. *Biochem. J.* 300:509–18.

Caughey, G.E., Mantzioris, E., Gibson, R.A., Cleland, L.G., and James, M.J. 1996. The effect on human tumor necrosis factor α and interleukin 1β production of diets enriched in n-3 fatty acids from vegetable oil or fish oil. *Am. J. Clin. Nutr.* 63:116–22.

Chapkin, R.S., Akoh, C.C., and Miller, C.C. 1991. Influence of dietary n-3 fatty acids on macrophage glycerophospholipid molecular species and peptidoleukotriene synthesis. *J. Lipid Res.* 32:1205–13.

Chapkin, R.S., Davidson, L.A., Ly, L., Weeks, B.R., Lupton, J.R., and McMurray, D.N. 2007. Immunomodulatory effects of (n-3) fatty acids: Putative link to inflammation and colon cancer. *J. Nutr.* 137:200S–4S.

Cleland, L.G., Caughey, G.E., James, M.J., and Proudman, S.M. 2006. Reduction of cardiovascular risk factors with longterm fish oil treatment in early rheumatoid arthritis. *J. Rheumatol.* 33:1973–9.

Cleland, L.G., French, J.K., Betts, W.H., Murphy, G.A., and Elliot, M.J. 1988. Clinical and biochemical effects of dietary fish oil supplements in rheumatoid arthritis. *J. Rheumat.* 15:1471–5.

De Caterina, R., Cybulsky, M.I., Clinton, S.K., Gimbrone, M.A., and Libby, P. 1994. The omega-3 fatty acid docosahexaenoate reduces cytokine-induced expression of proatherogenic and proinflammatory proteins in human endothelial cells. *Arterioscler. Thromb.* 14:1829–36.

De Ley, M., de Vos, R., Hommes, D.W., and Stokkers, P. 2007. Fish oil for induction of remission in ulcerative colitis. *Cochrane Database Systemat. Rev.* 17:CD005986.

Desreumaux, P., Dubuquoy, L., Nutten, S., et al. 2001. Attenuation of colon inflammation through activators of the retinoid X receptor (RXR)/peroxisome proliferator-activated receptor gamma (PPARgamma) heterodimer. A basis for new therapeutic strategies. *J. Exp. Med.* 193:827–38.

Dubuquoy, L., Rousseaux, C., Thuru, X., et al. 2006. PPARgamma as a new therapeutic target in inflammatory bowel diseases. *Gut* 55:1341–9.

Empey, L.R., Jewell, L.D., Garg, M.L., Thomson, A.B., Clandinin, M.T., and Fedorak, R.N. 1991. Indomethacin worsens and a leukotriene biosynthesis inhibitor accelerates mucosal healing in rat colitis. *Can. J. Physiol. Pharmacol.* 69:480–7.

Endres, S., Ghorbani, R., Kelley, V.E., et al. 1989. The effect of dietary supplementation with n-3 polyunsaturated fatty acids on the synthesis of interleukin-1 and tumor necrosis factor by mononuclear cells. *N. Engl. J. Med.* 320:265–71.

Esperson, G.T., Grunnet, N., Lervang, H.H., et al. (1992). Decreased interleukin-1 beta levels in plasma from rheumatoid arthritis patients after dietary supplementation with n-3 polyunsaturated fatty acids. *Clin. Rheumatol.* 11:393–5.

Faber, J., Berkhout, M., Vos, A.P., et al. 2011. Supplementation with a fish oil-enriched, high-protein medical food leads to rapid incorporation of EPA into white blood cells and modulates immune responses within one week in healthy men and women. *J. Nutr.* 141:964–70.

Fan, Y.Y., Ly, L.H., Barhoumi, R., McMurray, D.N., and Chapkin, R.S. 2004. Dietary docosahexaenoic acid suppresses T cell protein kinase Cθ lipid raft recruitment and IL-2 production. *J. Immunol.* 173:6151–60.

Fan, Y.Y., McMurray, D.N., Ly, L.H., and Chapkin, R.S. 2003. Dietary n-3 polyunsaturated fatty acids remodel mouse T-cell lipid rafts. *J. Nutr.* 133:1913–20.

Farrell, R.J. and Peppercorn, M.A. 2002. Ulcerative colitis. *Lancet* 359:331–40.

Feagan, B.G., Sandborn, W.J., Mittmann, U., et al. 2008. Omega-3 free fatty acids for the maintenance of remission in Crohn disease: The EPIC Randomized Controlled Trials. *JAMA* 299:1690–7.

Feldmann, M. and Maini, R.N. 1999. The role of cytokines in the pathogenesis of rheumatoid arthritis. *Rheumatol.* 38 (Suppl. 2):3–7.

Firestein, G.S. 2003. Evolving concepts of rheumatoid arthritis. *Nature* 423:356–61.

Forman, B.M., Chen, J., and Evans, R.M. 1997. Hypolipidemic dreugs, polyunsaturated fatty acids, and eicosanoids are ligands for peroxisome proliferator-activated receptors α and δ. *PNAS* 94:4312–7.

Forman, B.M., Tontonoz, P., Chen, J., Brun, R.P., Spiegelman, B.M., and Evans. R.M. 1995. 15-Deoxy-D12,14 prostaglandin J_2 is a ligand for adipocyte determination factor PPARγ. *Cell* 83:803–12.

Fortin, P.R., Lew, R.A., Liang, M.H., et al. 1995. Validation of a metaanalysis: The effects of fish oil in rheumatoid arthritis. *J. Clin. Epidemiol.* 48:1379–90.

Gewirtz, A.T., Collier-Hyams, L.S., Young, A.N., et al. 2002. Lipoxin A4 analogs attenuate induction of intestinal epithelial proinflammatory gene expression and reduce the severity of dextran sodium sulfate-induced colitis. *J. Immunol.* 168:5260–7.

Goldberg, R.J. and Katz, J. 2007. A meta-analysis of the analgesic effects of omega-3 polyunsaturated fatty acid supplementation for inflammatory joint pain. *Pain* 129:210–33.

Goldman, D.W., Pickett, W.C., and Goetzl, E.J. 1983. Human neutrophil chemotactic and degranulating activities of leukotriene B_5 (LTB_5) derived from eicosapentaenoic acid. *Biochem. Biophys. Res. Commun.* 117:282–8.

Gottlicher, M., Widmaek, E., Li, Q., and Gustafsson, J-A. 1992. Fatty acids activate a chimera of the clofibric acid-activated receptor and the glucocorticoid receptor. *PNAS* 89:4653–7.

Harder, T. 2004. Lipid raft domains and protein networks in T-cell receptor signal transduction. *Curr. Opin. Immunol.* 16:353–9.

Hawthorne, A.B., Daneshmend, T.K., Hawkey, C.J., et al. 1992. Treatment of ulcerative colitis with fish oil supplementation: A prospective 12 month randomised controlled trial. *Gut* 33:922–8.

Healy, D.A., Wallace, F.A., Miles, E.A., Calder, P.C., and Newsholme, P. 2000. The effect of low to moderate amounts of dietary fish oil on neutrophil lipid composition and function. *Lipids* 35:763–8.

Hirata, I., Murano, M., Nitta, M., et al. 2001. Estimation of mucosal inflammatory mediators in rat DSS-induced colitis—possible role of PGE_2 in protection against mucosal damage. *Digestion* 63 (Suppl. 1):73–80.

Hong, S., Gronert, K., Devchand, P., Moussignac, R-L., and Serhan, C.N. 2003. Novel docosatrienes and 17S-resolvins generated from docosahexaenoic acid in murine brain, human blood and glial cells: Autocoids in anti-inflammation. *J. Biol. Chem.* 278:14677–87.

Hudert, C.A., Weylandt, K.H., et al. 2006. Transgenic mice rich in endogenous omega-3 fatty acids are protected from colitis. *PNAS* 103:11276–81.

Ierna, M., Kerr, A., Scales, H., Berge, K., and Griinari, M. 2010. Supplementation of diet with krill oil protects against experimental rheumatoid arthritis. *BMC Musculoskel. Disord.* 11:136.

Jolly, C.A., Jiang, Y.H., Chapkin, R.S., and McMurray, D.N. 1997. Dietary (n-3) polyunsaturated fatty acids suppress murine lymphoproliferation, interleukin-2 secretion, and the formation of diacylglycerol and ceramide. *J. Nutr.* 127:37–43.

Katagiri, Y.U., Kiyokawa, N., and Fujimoto, J. 2001. A role for lipid rafts in immune cell signaling. *Microbiol. Immunol.* 45:1–8.

Kew, S., Mesa, M.D., Tricon, S., et al. 2004. Effects of oils rich in eicosapentaenoic and docosahexaenoic acids on immune cell composition and function in healthy humans. *Am. J. Clin. Nutr.* 79:674–81.

Kew, S., Wells, S., Thies, F., et al. 2003. The effect of eicosapentaenoic acid on rat lymphocyte proliferation depends upon its position in dietary triacylglycerols. *J. Nutr.* 133:4230–8.

Khalfoun, B., Thibault, F., Watier, H., et al. 1997. Docosahexaenoic and eicosapentaenoic acids inhibit in vitro human endothelial cell production of interleukin-6. *Adv. Exp. Biol. Med.* 400:589–97.

Kleiwer, S.A., Lenhard, J.M., Willson, T.M., Patel, I., Morris, D.C., and Lehman, J.M. 1995. A prostaglandin J2 metabolite binds peroxisome proliferator-activated receptor γ and promotes adipocyte differentiation. *Cell* 83:813–9.

Kliewer, S.A., Sundseth, S.S., Jones, S.A., et al. 1997. Fatty acids and eicosanoids regulate gene expression through direct interactions with peroxisome proliferator-activated receptors α and γ. *PNAS* 94:4318–23.

Kolahi, S., Ghorbanihaghjo, A., Alizadeh, S., et al. 2010. Fish oil supplementation decreases serum soluble receptor activator of nuclear factor-kappa B ligand/osteoprotegerin ratio in female patients with rheumatoid arthritis *Clin. Biochem.* 43:576–80.

Kong, W., Yen, J.H., Vassiliou, E., Adhikary, S., Toscano, M.G., and Ganea, D. 2010. Docosahexaenoic acid prevents dendritic cell maturation and in vitro and in vivo expression of the IL-12 cytokine family. *Lipids Health Dis.* 9:12.

Kremer, J.M., Jubiz, W., Michalek, A., et al. 1987. Fish-oil supplementation in active rheumatoid arthritis. *Ann. Int. Med.* 106:497–503.

Kremer, J.M., Lawrence, D.A., Jubiz, W., DiGiacomo, R., Rynes, R., Bartholomew, L.E., and Sherman, M. 1990. Dietary fish oil and olive oil supplementation in patients with rheumatoid arthritis: Clinical and immunologic effects. *Arthritis Rheum.* 33:810–20.

Kremer, J.M., Lawrence, D.A., Petrillo, G.F., et al. 1995. Effects of high-dose fish oil on rheumatoid arthritis after stopping nonsteroidal anti-inflammatory drugs: Clinical and immune correlates. *Arth. Rheum.* 38:1107–14.

Kumar, A., Takada, Y., Boriek, A.M., and Aggarwal, B.B. 2004. Nuclear factor-kappaB: Its role in health and disease. *J. Mol. Med.* 82:434–48.

Lee, T.H., Hoover, R.L., Williams, J.D., et al. 1985. Effects of dietary enrichment with eicosapentaenoic acid and docosahexaenoic acid on in vitro neutrophil and monocyte leukotriene generation and neutrophil function. *N. Engl. J. Med.* 312:1217–24.

Lee, T.H., Mencia-Huerta, J.M., Shih, C., et al. 1984. Characterization and biologic properties of 5,12-dihydroxy derivatives of eicosapentaenoic acid, including leukotriene-B5 and the double lipoxygenase product. *J. Biol. Chem.* 259:2383–9.

Leslie, C.A., Gonnerman, W.A., Ullman, M.D., et al. 1985. Dietary fish oil modulates macrophage fatty acids and decreases arthritis susceptibility in mice. *J. Exp. Med.* 162:1336–9.

Levy, B.D., Clish, C.B., Schmidt, B., Gronert, K., and Serhan, C.N. 2001. Lipid mediator class switching during acute inflammation: Signals in resolution. *Nature Immunol.* 2:612–9.

Lewis, R.A., Austen, K.F., and Soberman, R.J. 1990. Leukotrienes and other products of the 5-lipoxygenase pathway: Biochemistry and relation to pathobiology in human diseases. *N. Engl. J. Med.* 323:645–55.

Lo, C.J., Chiu, K.C., Fu, M., Lo, R., and Helton, S. 1999. Fish oil decreases macrophage tumor necrosis factor gene transcription by altering the NF kappa B activity. *J. Surg. Res.* 82:216–21.

Macfarlane, S., Furrie, E., Kennedy, A., Cummings, J.H., and Macfarlane, G.T. 2004. Mucosal bacteria in ulcerative colitis. *Brit. J. Nutr.* 93:S67–72.

MacLean, C.H., Mojica, W.A., Morton, S.C., et al. 2004. Effects of omega-3 fatty acids on inflammatory bowel disease, rheumatoid arthritis, renal disease, systemic lupus erythematosus, and osteoporosis, Evidence Report/Technical Assessment no. 89. AHRQ Publication no. 04-E012-2. Rockville, MD: Agency for Healthcare Research and Quality; available at http://www.ahrq.gov/clinic/tp/o3lipidtp.htm

McCall, T.B., O'Leary, D., Bloomfield, J., and O'Morain, C.A. 1989. Therapeutic potential of fish oil in the treatment of ulcerative colitis. *Aliment. Pharmacol. Therapeut.* 3:415–24.

Meydani, S.N., Endres, S., Woods, M.M., et al. 1991. Gorbach. Oral (n-3) fatty acid supplementation suppresses cytokine production and lymphocyte proliferation: Comparison between young and older women. *J. Nutr.* 121:547–55.

Meyer, B.J., Mann, N.J., Lewis, J.L., Milligan, G.C., Sinclair, A.J., and Howe, P.R. 2003. Dietary intakes and food sources of omega-6 and omega-3 polyunsaturated fatty acids. *Lipids* 38:391–8.

Miles, E.A., Aston, L., and Calder, P.C. 2003. In vitro effects of eicosanoids derived from different 20-carbon fatty acids on T helper type 1 and T helper type 2 cytokine production in human whole-blood cultures. *Clin. Exp. Allergy* 33:624–32.

Miles, E.A. and Calder, P.C. 1998. Modulation of immune function by dietary fatty acids. *Proc. Nutr. Soc.* 57:277–92.

Nieto, N., Torres, M.I., Rios, A., and Gil, A. 2002. Dietary polyunsaturated fatty acids improve histological and biochemical alterations in rats with experimental ulcerative colitis. *J. Nutr.* 132:11–9.

Novak, T.E., Babcock, T.A., Jho, D.H., Helton, W.S., and Espat, N.J. 2003. NF-kappa B inhibition by omega-3 fatty acids modulates LPS-stimulated macrophage TNF-alpha transcription. *Am. J. Physiol.* 284:L84–9.

Oh, D.Y., Talukdar, S., Bae, E.J., et al. 2010. GPR120 is an omega-3 fatty acid receptor mediating potent anti-inflammatory and insulin-sensitizing effects. *Cell* 142:687–98.

Perkins, N.D. 2007. Integrating cell-signalling pathways with NF-kappaB and IKK function. *Nature Rev. Mol. Cell Biol.* 8:49–62.

Peterson, L.D., Jeffery, N.M., Thies, F., et al. 1998. Eicosapentaenoic and docosahexaenoic acids alter rat spleen leukocyte fatty acid composition and prostaglandin E_2 production but have different effects on lymphocyte functions and cell-mediated immunity. *Lipids* 33:171–80.

Razzaq, T.M., Ozegbe, P., Jury, E.C., et al. (2004). Regulation of T-cell receptor signaling by membrane microdomains. *Immunol.* 113:413–26.

Rees, D., Miles, E.A., Banerjee, T., et al. 2006. Dose-related effects of eicosapentaenoic acid on innate immune function in healthy humans: A comparison of young and older men. *Am. J. Clin. Nutr.* 83:331–42.

Renier, G., Skamene, E., de Sanctis, J., and Radzioch, D. 1993. Dietary n-3 polyunsaturated fatty acids prevent the development of atherosclerotic lesions in mice: Modulation of macrophage secretory activities. *Arterioscler. Thomb.* 13:1515–24.

Sadeghi, S., Wallace, F.A., and Calder, P.C. 1999. Dietary lipids modify the cytokine response to bacterial lipopolysaccharide in mice. *Immunol.* 96:404–10.

Sano, H., Hla, T., Maier, J.A.M., et al. 1992. In vivo cyclooxygenase expression in synovial tissues of patients with rheumatoid arthritis and osteoarthritis and rats with adjuvant and streptococcal cell wall arthritis. *J. Clin. Invest.* 89:97–108.

Serhan, C.N., Chiang, N., and van Dyke, T.E. 2008. Resolving inflammation: Dual anti-inflammatory and pro-resolution lipid mediators. *Nature Rev. Immunol.* 8:349–61.

Serhan, C.N., Clish, C.B., Brannon, J., et al. 2000a. Anti-inflammatory lipid signals generated from dietary n-3 fatty acids via cyclooxygenase-2 and transcellular processing: A novel mechanism for NSAID and n-3 PUFA therapeutic actions. *J. Physiol. Pharmacol.* 4:643–54.

Serhan, C.N., Clish, C.B., Brannon, J., et al. 2000b. Novel functional sets of lipid-derived mediators with anti-inflammatory actions generated from omega-3 fatty acids via cyclooxygenase 2-nonsteroidal antiinflammatory drugs and transcellular processing. *J. Exp. Med.* 192:1197–1204.

Serhan, C.N., Hong, S., Gronert, K., et al. 2002. Resolvins: A family of bioactive products of omega-3 fatty acid transformation circuits initiated by aspirin treatment that counter pro-inflammation signals. *J. Exp. Med.* 196:1025–37.

Shanahan, F. 2002. Crohn's disease. *Lancet* 359:62–9.

Sharon, P. and Stenson, W.F. 1984. Enhanced synthesis of leukotriene B_4 by colonic mucosa in inflammatory bowel disease. *Gastroenterol.* 86:453–60.

Shimizu, T., Fujii, T., Suzuki, R., Igarashi, J., Ohtsuka, Y., Nagata, S., et al., 2003. Effects of highly purified eicosapentaenoic acid on erythrocyte fatty acid composition and leukocyte and colonic mucosa leukotriene B4 production in children with ulcerative colitis. *Pediat. Gastroenterol. Nutr.* 37:581–5.

Sigal, L.H. 2006. Basic science for the clinician 39: NF-kappaB-function, activation, control, and consequences. *J. Clin. Rheumatol.* 12:207–11.

Sperling, R.I. 1995. Eicosanoids in rheumatoid arthritis. *Rheum. Dis. Clin. N. Am.* 21:741–58.

Sperling, R.I., Benincaso, A.I., Knoell, C.T., et al. 1993. Dietary ω-3 polyunsaturated fatty acids inhibit phosphoinositide formation and chemotaxis in neutrophils. *J. Clin. Invest.* 91:651–60.

Stenson, W.F., Cort, D., Rodgers, J., et al. 1992. Dietary supplementation with fish oil in ulcerative colitis. *Ann. Intern. Med.* 116:609–14.

Stulnig, T., Berger, M., Sigmund, T., et al. 1998. Polyunsaturated fatty acids inhibit T cell signal transduction by modification of detergent-soluble membrane domains. *J. Cell Biol.* 143:637–44.

Stulnig, T.M., Huber, J., Leitinger, N., et al. 2001. Polyunsaturated eicosapentaenoic acid displaces proteins from membrane rafts by altering raft lipid composition. *J. Biol. Chem.* 276:37335–40.

Stulnig, T.M. and Zeyda, M. 2004. Immunomodulation by polyunsaturated fatty acids: Impact on T-cell signaling. *Lipids* 39:1171–5.

Su, G.G., Wen, X., Bailey, S.T., et al. 1999. A novel therapy for colitis utilizing PPARgamma ligands to inhibit the epithelial inflammatory response. *J. Clin. Invest.* 104:383–9.

Szanto, A. and Nagy, L. 2008. The many faces of PPARgamma: Anti-inflammatory by any means? *Immunobiol.* 213:789–803.

Thies, F., Nebe-von-Caron, G., Powell, J.R., et al. 2001. Dietary supplementation with γ-linolenic acid or fish oil decreases T lymphocyte proliferation in healthy older humans. *J. Nutr.* 131:1918–27.

Tilley, S.L., Coffman, T.M., and Koller, B.H. 2001. Mixed messages: Modulation of inflammation and immune responses by prostaglandins and thromboxanes. *J. Clin. Invest.* 108:15–23.

Trebble, T., Arden, N.K., Stroud, M.A., et al. 2003. Inhibition of tumour necrosis factor-α and interleukin-6 production by mononuclear cells following dietary fish-oil supplementation in healthy men and response to antioxidant co-supplementation. *Brit. J. Nutr.* 90:405–12.

Trebble, T.M., Arden, N.K., Wootton, S.A., et al. 2004. Fish oil and antioxidants alter the composition and function of circulating mononuclear cells in Crohn's disease. *Am. J. Clin. Nutr.* 80:1137–44.

Tullekan, J.E., Limburg, P.C., Muskiet, F.A.J., and van Rijswijk, M.H. 1990. Vitamin E status during dietary fish oil supplementation in rheumatoid arthritis. *Arth. Rheum.* 33:1416–9.

Turner, D., Steinhart, T.H., and Griffiths, A.M. 2007. Omega 3 fatty acids (fish oil) for maintenance of remission in ulcerative colitis. *Cochrane Database Systemat. Rev.* 18:CD006443.

Turner, D., Zlotkin, S.H., Shah, P.S., and Griffiths, A.M. 2009. Omega 3 fatty acids (fish oil) for maintenance of remission in Crohn's disease. *Cochrane Database Systemat. Rev.* 21:CD006320.

Vachier, I., Chanez, P., Bonnans, C., Godard, P., Bousquet, J., and Chavis, C. 2002. Endogenous anti-inflammatory mediators from arachidonate in human neutrophils. *Biochem. Biophys. Res. Commun.* 290:219–24.

Van den Berghe, W., Vermeulen, L., Delerive, P., et al. 2003. A paradigm for gene regulation: Inflammation, NF-kappaB and PPAR. *Adv. Exp. Med. Biol.* 544:181–96.

Volker, D.H., FitzGerald, P.E.B., and Garg, M.L. 2000. The eicosapentaenoic to docosahexaenoic acid ratio of diets affects the pathogenesis of arthritis in Lew/SSN rats. *J. Nutr.* 130:559–65.

Von Schacky, C., Kiefl, R., Jendraschak, E., and Kaminski, W.E. 1993. N-3 fatty acids and cysteinyl-leukotriene formation in humans in vitro, ex vivo and in vivo. *J. Lab. Clin. Med.* 121:302–9.

Wada, M., DeLong, C.J., Hong, Y.H., et al. 2007. Enzymes and receptors of prostaglandin pathways with arachidonic acid-derived versus eicosapentaenoic acid-derived substrates and products. *J. Biol. Chem.* 282:22254–66.

Wallace, F.A., Miles, E.A., Evans, C., et al. 2001. Dietary fatty acids influence the production of Th1- but not Th2-type cytokines. *J. Leuk. Biol.* 69:449–57.

Xi, S., Cohen, D., Barve, S., and Chen, L.H. 2001. Fish oil suppressed cytokines and nuclear factor kappaB induced by murine AIDS virus infection. *Nutr. Res.* 21:865–78.

Yaqoob, P. and Calder, P.C. 1995. Effects of dietary lipid manipulation upon inflammatory mediator production by murine macrophages. *Cell. Immunol.* 163:120–8.

Yaqoob, P., Newsholme, E.A., and Calder, P.C. 1994. The effect of dietary lipid manipulation on rat lymphocyte subsets and proliferation. *Immunol.* 82:603–10.

Yaqoob, P., Pala, H.S., Cortina-Borja, M., Newsholme, E.A., and Calder, P.C. 2000. Encapsulated fish oil enriched in α-tocopherol alters plasma phospholipid and mononuclear cell fatty acid compositions but not mononuclear cell functions. *Eur. J. Clin. Invest.* 30:260–74.

Zapata-Gonzalez, F., Rueda, F., Petriz, J., et al. 2008. Human dendritic cell activities are modulated by the omega-3 fatty acid, docosahexaenoic acid, mainly through PPAR(gamma): RXR heterodimers: Comparison with other polyunsaturated fatty acids. *J. Leuk. Biol.* 84:1172–82.

Zeyda, M., Staffler, G., Horejsi, V., and Waldhausl, W. 2002. LAT displacement from lipid rafts as a molecular mechanism for the inhibition of T cell signalling by polyunsaturated fatty acids. *J. Biol. Chem.* 277:28418–23.

Zeyda, M. and Stulnig, T.M. 2006. Lipid Rafts & Co.: An integrated model of membrane organization in T cell activation. *Prog. Lipid Res.* 45:187–202.

Zeyda, M., Szekeres, A.B., Saemann, M.D., et al. 2003. Suppression of T cell signaling by polyunsaturated fatty acids: Selectivity in inhibition of mitogen-activated protein kinase and nuclear factor activation. *J. Immunol.* 170:6033–9.

Zhao, Y., Joshi-Barve, S., Barve, S., and Chen, L.H. 2004. Eicosapentaenoic acid prevents LPS-induced TNF-alpha expression by preventing NF-kappaB activation. *J. Am. Coll. Nutr.* 23:71–8.

23 Anti-Inflammatory Phytochemicals, Obesity, and Diabetes

An Overview

Srujana Rayalam, MaryAnne Della-Fera, and Clifton A. Baile
University of Georgia
Athens, Georgia

CONTENTS

23.1 INTRODUCTION

Adipose tissue produces proteins that are classical mediators of the inflammatory response. Pro-inflammatory cytokines like tumor necrosis factor alpha (TNF-α), plasminogen activator inhibitor-1, interleukin-1β (IL-1β), C-reactive protein, and IL-6 are secreted by adipocytes, resulting in the enhanced systemic levels of these cytokines in obese subjects [1]. These observations provided the first link between metabolic conditions such as obesity, insulin resistance, and diabetes. Obesity and diabetes have since then been considered inflammatory conditions and the pathogenesis of inflammation in obesity and diabetes is an emerging field of research.

23.2 OBESITY AND DIABETES AS MAJOR PUBLIC HEALTH ISSUES

Changes in the human diet and lifestyle have surpassed the ability of the human genome to make adjustments to a changing environment, and as a result billions of people worldwide are now overweight. The World Health Organization declared obesity as a global epidemic in 2003, and the latest data from the National Health and Nutrition Examination Survey (NHANES) indicated that in the United States by 2008 the overall prevalence of obesity was 32.2% for adult men and 35.5% for adult women [2]. Further, it is predicted that by 2040, 100% of adults will be either overweight or obese [3]. Because of the increasing prevalence of obesity and overweight, the economic impact has become significant. Since 1995, the direct costs have steadily risen with an estimated cost of overweight and obesity being $167 billion in 2006 [4], and, in addition, Medicare, Medicaid, and private insurers increased spending due to obesity from 6.5% in 1998 to 9.1% in 2006.

One of the major consequences of obesity apart from coronary heart disease is non-insulin-dependent diabetes mellitus (NIDDM) and metabolic syndrome. The prevalence of diabetes worldwide is also increasing rapidly in association with the increase in obesity. With approximately 30% of obese adolescents having metabolic syndrome, the personal and financial cost of associated complications of diabetes could become overwhelming [5]. According to National Institute of Diabetes and Digestive and Kidney Diseases (NIDDK), diabetes is the leading cause of kidney failure, heart disease, and stroke and is the seventh leading cause of death in the United States with about 8.3% of the population being affected. Estimated total costs associated with diabetes in 2007 were $174 billion with the medical expenses for people with diabetes being two-fold higher than for people without diabetes.

23.3 ADIPOSE TISSUE AS A SECRETORY ORGAN

Obesity is often associated with an increase in adipose tissue mass, and the overall regulation of adipose tissue mass involves complex interactions between endocrine, paracrine, and autocrine systems. Recent developments in understanding the pathophysiology of adipose tissue indicate that adipose tissue is no longer considered merely an energy storage unit and is recognized as an endocrine organ that is biologically active.

There are two types of adipose tissues: white adipose tissue (WAT) and brown adipose tissue (BAT). Scientists have extensively studied WAT but less has been known about BAT in humans until the past decade. The major difference between WAT and BAT is that WAT mainly acts as an energy storage tissue while BAT is a thermogenic tissue. To facilitate thermogenesis, the number and size of mitochondria are much greater in BAT than in WAT. BAT is present in concentrated masses in humans during infancy but gets dispersed to various parts of the body during adulthood, although it may still play some important roles. WAT secretes a variety of peptides known as adipokines, and most of these adipokines are pro-inflammatory in nature. A number of these adipokines and receptors contribute to adipocyte growth and development.

Until the late 1990s, it was believed that adipocyte number does not change over one's lifetime. However, the finding that adipocytes can be both gained and lost opened new doors in understanding adipocyte biochemistry and has led to the proposition of a life cycle for adipocytes. Studies on the life cycle for adipocytes have shown that the molecular changes that occur during various stages of adipocyte life cycle can be controlled [6] by dietary factors, resulting in a net effect on adipose tissue mass. Adipocytes are derived from mesenchymal stem cells (MSCs), which are also precursor cells for osteoblasts. The life cycle of the adipocyte includes a growth phase followed by growth arrest, clonal expansion, and a complex sequence of changes in gene expression leading to storage of lipids and finally cell death [7]. Appropriate environmental and gene expression cues will facilitate the conversion of preadipocytes to mature adipocytes. Further, a coordinated interplay between adipokines, transcription factors, hormones, and dietary factors influence the adipogenesis and lipogenesis processes. Modulation of these factors will result in either hyperplasia or hypertrophy of adipocytes and storage or mobilization of lipid. No attempt will be made to discuss all the factors affecting adipogenesis but only major factors that significantly affect the adipocyte life cycle, and, consequently, adipose tissue mass will be discussed briefly in this chapter.

Molecular control of adipogenesis is initiated by activation of a series of transcription factors. These transcription factors ultimately upregulate the genes necessary for adipocyte function and lipogenesis such as lipoprotein lipase, glucose transporter-4 (GLUT4), and acetyl co-A carboxylase. The major transcription factors involved in the differentiation of adipocytes are peroxisome proliferator-activated receptor gamma (PPARγ), the CCAAT/enhancer binding protein (C/EBP) family, and the steroid regulatory element binding protein 1 (SREBP1). PPARγ is considered a master regulator of adipogenesis, and both PPAR and C/EBP families must function cooperatively to transactivate adipocyte genes to bring about adipocyte differentiation [7]. Hormones like insulin, leptin, glucocorticoids, and insulin-like growth factor-1 (IGF-1) also significantly influence adiposity. Leptin is produced by adipocytes and is involved in regulating feed intake and appetite. Insulin promotes adipocyte differentiation by upregulating GLUT4 and activating lipogenic enzymes. Glucocorticoids and IGF-1 are necessary to trigger the differentiation program in preadipocytes [7].

23.4 RELATIONSHIP BETWEEN OBESITY, DIABETES, AND INFLAMMATION

Obesity is associated with low-grade inflammation of adipose tissue resulting in activation of the immune system, which subsequently leads to insulin resistance and diabetes. Adipose tissue as discussed earlier is not a simple energy storage organ but exerts important immune functions that are achieved primarily through the release of adipocytokines like leptin, resistin, plasminogen activator inhibitor type-1 (PAI-1), and adiponectin, as well as TNF, IL-6, and IL-1. These cytokines are critically involved in chronic inflammation under obesity conditions.

TNFα is generally considered a major mediator of inflammation leading to obesity and insulin resistance. Genetically obese mice are known to overexpress TNFα, and in human adipocytes, TNFα expression is elevated in obese subjects and is decreased after weight loss. Also, mice lacking TNFα are protected from obesity and insulin resistance, indicating that TNFα is directly linked to obesity; therefore, targeting this pathway may have huge potential in prevention and therapy of metabolic diseases. Like TNFα, NF-kB, a transcription factor regulating the expression of a majority of genes that control inflammation, is also evidenced to be closely linked to obesity and insulin resistance. TNFα itself is a potent activator of NF-kB, and, in turn, expression of TNFα is regulated by NF-kB. Moreover, antidiabetic drugs upregulate an inhibitor of NF-kB to suppress NF-kB activation, indicating that NF-kB could be a target in diabetes. Additionally, adiponectin and insulin also inhibit NF-kB activation.

Toll-like receptors (TLRs), a family of pattern-recognition receptors that play a critical role in the innate immune system by activating pro-inflammatory signaling pathways in response to microorganisms, are also considered to play a critical role in the development of insulin resistance in obese people. Recent studies indicate that TLR4-deficient mice are protected against the development of diet-induced obesity and inflammation through NF-kB activation and insulin resistance [8]. NF-kB in turn regulates PAI-1, IL-6, and several other pro-inflammatory signaling molecules. Thus, these molecules play an important role in the obesity-associated low-grade inflammation, recently termed as metaflammation, and could therefore be another important target for obesity and diabetes management.

It is now understood that under obese conditions WAT is infiltrated with macrophages, and these locally present macrophages are responsible for the major part of the locally produced pro-inflammatory cytokines. High glucose intake may result in reactive oxygen species production by these infiltrated macrophages leading to activation of NF-kB. TNFα also plays a major role in the pathophysiology of insulin resistance and diabetes by phosphorylating insulin receptor substrate-1 protein, thereby preventing its interaction with insulin receptor beta subunit and inhibiting the insulin signaling pathway. Likewise, the receptors for IL-6 and many of the other interleukins interact with the insulin signaling pathway, leading to an impaired biological effect of insulin. Thus, with obesity a chronic increase in circulating cytokine levels is seen, which contributes to insulin resistance (reviewed in [9] and [10]).

23.5 COMPLICATIONS OF OBESITY AND DIABETES

Obesity and osteoporosis share several features including a common progenitor cell, the mesenchymal stem cell. It was once believed that a positive correlation exists between body mass index and bone mass, leading to the suggestion that fat mass may have beneficial effects on bone [11]. However, more recent studies reveal that adiposity does not protect against decreases in bone mass, and may in fact contribute to reduced bone mass [12]. For example, overweight children were reported to have lower lumbar spine and total body bone mineral content and bone area relative to weight, compared to children with normal body mass index-for-age percentiles [12]. Furthermore, adipocytes in bone marrow secrete IL-6, which can promote osteoclastogenesis and inhibit osteoblast activity in culture [13], and once a certain level of bone marrow adiposity is reached, conditions will promote further adipogenesis at the expense of osteogenesis. Additional evidence from environmental factors and medical interventions also support an inverse correlation between fat mass and bone mass.

On the other hand, hyperglycemia and hyperlipidemia under diabetic conditions are associated with oxidative stress. Increased levels of fatty acids and modified lipoproteins induce inflammatory immune responses and oxidative stress reactions leading to generation of free radicals that account for the cardiovascular complications and mortality of obesity and type 2 diabetes [14]. Thus, controlling inflammation could have multiple beneficial effects in the prevention and treatment of obesity, diabetes, osteoporosis, and other related diseases.

23.6 PHYTOCHEMICALS IN OBESITY AND DIABETES

Phytochemicals and plant extracts have numerous beneficial effects on insulin sensitivity and other parameters related to metabolic health, as evidenced both in rodents and human clinical trials [15, 16]. Among phytochemicals, polyphenols in fruits, vegetables, berries, and beverages modify imbalanced lipid and glucose homeostasis, thereby reducing the risk of the metabolic syndrome and type 2 diabetes complications. A vegetarian diet comprised of fruits and vegetables is reported to be associated with a more favorable profile of metabolic risk factors and a lower risk of metabolic syndrome [17]. Research reports that people adhering to a vegetarian diet are at a 56% lower risk for developing the metabolic syndrome than nonvegetarians, and that triglycerides, blood glucose, waist circumference, and body mass index are all significantly lower in vegetarians than in nonvegetarians [17].

Although the specific mechanisms by which the majority of polyphenols act is not clear, it is understood that most of the phytochemicals act on multiple targets and have numerous signaling pathways, making them ideal for multifocal signal modulation therapy in metabolic syndrome and cancer. Since inflammation is an underlying factor in both obesity and diabetes, anti-inflammatory phytochemicals have been investigated extensively as a treatment for obesity and diabetes.

In vitro experiments frequently employ 3T3-L1 adipocytes to examine the anti-inflammatory effects of phytochemicals in adipocytes. This cell line is considered a well-established cell model for studying adipocyte differentiation *in vitro*. Upon stimulation with pro-adipogenic agents like insulin and glucocorticoids, 3T3-L1 fibroblasts that are under growth arrest undergo one or two rounds of cell division, which is accompanied by induction of C/EBPβ and C/EBPδ followed by an increased expression of C/EBPα and PPARγ. As the adipocytes mature, they become spherical and start accumulating fat droplets. Mature adipocytes will later undergo either apoptosis or lipolysis followed by apoptosis. Several phytochemicals have the potential to reduce adipogenesis in vitro (reviewed in [18]). Among the phytochemicals studied, curcumin, resveratrol, epigallocatechin-3-gallate (EGCG), and genistein received considerable attention.

23.6.1 Curcumin

Curcumin is a yellow pigment derived from the spice turmeric, which is very popular in Asia. In 3T3-L1 cells, curcumin suppressed adipocyte differentiation and induced apoptosis [19]. Curcumin is considered the best example for multifocal signal modulation therapy as it modulates numerous targets that have been linked to obesity and insulin resistance. Curcumin interacts with a wide range of cells including adipocytes, pancreatic cells, hepatocytes, myocytes, and even macrophages. The major anti-inflammatory effects of curcumin include downregulation of TNFα and NF-kB in various tissues and cell lines including adipocytes. Curcumin suppresses activation of NF-kB by first inhibiting the degradation of IkBα, an inhibitory protein that keeps NF-kB in an inactivated state in the cytosol. Second, curcumin also inhibits the activation of IkB kinase, an enzyme that phosphorylates IkBα and causes degradation by the proteosome, which will eventually result in the activation of NF-kB. Third, curcumin suppresses NF-kB activation through induction of adiponectin, a protein made almost exclusively by adipose tissue. Other pro-inflammatory cytokines like IL-1, IL-6, and monocyte chemotactic protein-1 (MCP-1) are suppressed by curcumin in numerous cell lines [10].

In adipocytes, curcumin decreases adipocyte differentiation by inhibiting the Wnt/β-catenin signaling pathway. Curcumin also inhibited the expression of adipocyte-specific transcription factors—C/EBPs, PPARγ, SREBP-1c—and fatty acid synthase in adipocytes. AMP-activated protein kinase (AMPK), which regulates several intracellular processes including glucose uptake and beta oxidation of fatty acids, is activated by curcumin in adipocytes, leading to fatty acid oxidation and downregulation of PPAR-γ, thus inhibiting adipocyte differentiation. Inhibition of adipocyte differentiation by curcumin is also accompanied by inhibition of phosphorylation of extracellular signal-related protein kinase (ERK) 1/2, c-Jun NH2 terminal kinase (JNK), and p38. Since inhibitors of JNK block TNF-alpha upregulation, it is possible that the anti-inflammatory effects of curcumin in adipocytes are mediated via inhibition of JNK signaling [10].

Direct effects of curcumin on pancreatic beta cells contribute to its antidiabetic effects. Curcumin causes enhanced insulin release from the beta cells and also decreases beta cell volume. In hepatic cells, curcumin increases AMPK activation and decreases gluconeogenic genes. AMPK-mediated suppression of gluconeogenesis may also be a potential mechanism mediating glucose-lowering effects of curcumin [19]. Curcumin also decreased glucose-6-phosphatase activity in hepatocytes, contributing to a reduction of hepatic glucose production. Finally, curcumin also acts as a free radical scavenger as evidenced by its protective effects on pancreatic islets from streptozotocin-induced oxidative stress.

23.6.1.1 Rodent Studies

When curcumin was administered intraperitoneally, the progression of streptozotocin-induced diabetes was prevented, accompanied by suppression of inflammatory cytokines like TNFα and IL-1 in serum. Curcumin also lowered relative liver weight and reduced serum and liver cholesterol levels in diabetic rats. In hamsters fed a high-fat and high-cholesterol diet, curcumin lowered free fatty acid and triglyceride levels in serum and also suppressed insulin resistance. Even in leptin-deficient ob/ob mice, curcumin ameliorated diabetes.

It is evident that infiltration of WAT with macrophages leads to increased production of inflammatory cytokines. Curcumin decreases macrophage infiltration of WAT and increases adipose tissue adiponectin, which in turn inhibits NF-kB activation. Similar to its effects in *in vitro* studies, curcumin supplementation lowered the production of TNFα, IL-6, and MCP-1 in rodents.

23.6.1.2 Human Clinical Trials

Curcumin dates back to 3000 B.C.E. and has been quoted in Ayurveda for several medical conditions including obesity. Although the anti-inflammatory, antiobesity, and antidiabetic effects of curcumin are demonstrated both *in vitro* and *in vivo* using rodent models, so far there are only pilot studies undertaken with human subjects to examine its effects on obesity and diabetes. These studies confirm that curcumin lowers blood sugar levels in diabetic patients. In healthy human volunteers, curcumin reduced serum lipid peroxides and serum cholesterol.

A search on ClinicalTrials.gov currently reveals about sixty human trials using curcumin, of which three trials are investigating curcumin therapy in diabetic patients while the majority are investigating its chemopreventive or chemotherapeutic potential in cancer. Scientists believe that the preliminary data from rodents and pilot studies using human subjects hold promise, and interest in curcumin as a therapeutic agent for obesity and diabetes continues to grow [20].

23.6.2 Resveratrol

Among all the phytochemicals, resveratrol has received enormous attention lately. Resveratrol belongs to a class of polyphenolic compounds called stilbenes and was initially known just as a phytoalexin produced in high amounts in the skin of grapes in response to infection or mechanical injury. Resveratrol subsequently became popular when it was found in red wine. An observation that high red wine consumption by the French is the primary reason for their low incidence of coronary heart disease, despite having a diet relatively rich in saturated fats, has been called the French Paradox. Some have attributed the beneficial effects of red wine to resveratrol. However, there is no evidence that resveratrol has cardioprotective effects in humans, particularly with the amounts present in one to two glasses of red wine. In fact, resveratrol can be found in fresh grape skins at higher concentrations of 50–100 mg per gram, compared to 0.2 to 7.7 mg/L of red wine [21].

There are reports of anti-inflammatory, antiproliferative, apoptotic, and antioxidant effects of resveratrol in several cell lines, including tumor cell lines. In adipocytes, resveratrol has been shown to have specific effects, particularly the ones that contribute to its anti-inflammatory effects. In addition to decreasing the synthesis of lipids, resveratrol reduced the levels of pro-inflammatory mediators, TNFα, IL-6, and COX-2 in murine adipocytes [22, 23]. Also, the TNFα-induced secretion and mRNA expression of plasminogen activator inhibitor-1, IL-6, and adiponectin are reduced by resveratrol [24]. Like curcumin, resveratrol is also a potent inhibitor of NF-kB activation, indirectly influencing adipocyte differentiation [25].

Reactive oxygen species belong to the family of free radicals, which are highly reactive oxidizing agents. The presence of unpaired electrons ready to be shared by other molecules makes free radicals very reactive. Polyphenols like resveratrol are potent ROS scavengers reducing the oxidative stress, which contributes to the suppression of inflammation. Particularly in tumor cells, resveratrol's antioxidant effects suppress tumorigenesis. Further, resveratrol exhibits a protective effect against lipid peroxidation in cell membranes and DNA damage caused by ROS. The antioxidant activity of resveratrol in adipocytes caused a decrease in membrane potential, resulting in an increased number of apoptotic cells associated with an increase in caspase-3 activity and downregulation of Bcl2 proteins [26].

SIRT-1, AMPK, PGC1-α, and SREBP-1c were reported as targets for resveratrol in various studies. In adipocytes, a decrease in adipogenesis was associated with a decrease in adipocyte-specific transcription factors PPARγ, C/EBPα, and SREBP-1c. Activation of SIRT1 by resveratrol results in mobilization of lipids from adipocytes in vitro. Resveratrol also increased cAMP levels in various cell lines, which is crucial for lipolysis. In human adipocytes, resveratrol increased basal and insulin-stimulated glucose uptake. It should be noted that the anti-adipogenic effects of resveratrol are also influenced by its effects on mitochondrial biogenesis. Genes that modulate mitochondrial function such as uncoupling protein 1, mitofusin, and PGC-1α are upregulated by resveratrol. Finally, resveratrol also promotes osteoblast differentiation and increases alkaline phosphatase activity in vitro (reviewed in [27]).

23.6.2.1 Rodent Studies

A majority of animal studies with resveratrol are focused on prevention of cancer. There are comparatively few rodent studies examining the antiobesity and antidiabetic effects of resveratrol. Supplementation of 400 mg/kg/day of resveratrol with high-fat diets to mice increased their resistance to obesity by causing a decrease in total body fat content without altering food intake [28]. Increased energy expenditure as evident with the upregulation of mitochondrial genes like UCP-1 and PGC-1α with resveratrol treatment likely contributed to the resistance to weight gain. In contrast, a low dose of 22.4 mg/kg/day of resveratrol with a high-fat diet did not alter the weight gain but improved insulin sensitivity and increased the survival rate in mice [29]. Interconnected SIRT1 and AMPK pathways are identified as key targets of resveratrol in these studies [30]. Activation of SIRT1 and AMPK leads to upregulation of several downstream factors like PGC-1α, FOXO, and SREB-1c, contributing to reduced lipogenesis, increased energy expenditure and resistance to weight gain. Recently, intracerebroventricular infusion of resveratrol was reported to normalize hyperglycemia in diet-induced obese and diabetic mice [31].

23.6.2.2 Human Clinical Trials

Although there are no human clinical trials with resveratrol exploring the beneficial effects on obesity and diabetes, currently available studies have described various aspects of resveratrol's safety and bioavailability, indicating that resveratrol is well tolerated without any toxic side effects but has poor bioavailability [32]. Resveratrol from grape juice is reported to be absorbed in biologically active quantities and amounts that are likely to reduce the risk of atherosclerosis [33]. According to ClinicalTrials.gov, there are about 35 trials either completed or currently ongoing with resveratrol in humans. Seven or eight of these trials are investigating the beneficial effects of resveratrol under conditions of obesity, diabetes, impaired glucose tolerance, and insulin resistance. From these studies, both *in vitro* and *in vivo*, it is clear that resveratrol holds great potential in the prevention and therapy of obesity, diabetes, and other complications of metabolic syndrome.

23.6.3 Epigallocatechin-3-Gallate

Green tea is a widely consumed beverage, and the body of evidence regarding the health-promoting effects of green tea has grown considerably since 2000. Green tea catechins are polyphenolic flavonoids, and health benefits of green tea are primarily associated with epigallocatechin-3-gallate (EGCG). There are several *in vitro* and *in vivo* reports that EGCG lowers the incidence of cancers. Identification of the EGCG receptor 67-kDa laminin receptor (LR) has helped scientists explain the numerous biological effects of EGCG. Initially, 67-kDa LR was found in cancer cells but was later found to be present in several isoforms even in normal cells. EGCG receptor associates with lipid rafts of other types of receptors, and lipid rafts are known to contain specific kinases that enable generation of second messengers by phosphorylating proteins, thus explaining the marked effects of EGCG on kinase activity and the subsequent selective phosphorylation of downstream proteins. In addition to decreasing the activity of 67-kDa LR receptor, EGCG also decreases the activity of IGF-1, FGF, insulin, TLR-4, estrogen, and VEGF receptors [34]. The anti-inflammatory effects of EGCG *in vitro* were mainly attributed to its inhibitory effects on NF-kB activation as well as suppression of other pro-survival pathways such as AKT, mTOR, and MAPKs [35].

Increase in adipose tissue mass associated with obesity is a result of both increased adipocyte size and increased adipocyte number. EGCG acts at both levels by decreasing adipocyte size by promoting lipolysis and decreasing adipocyte number by either inducing adipocyte apoptosis or by inhibiting adipocyte differentiation. These effects of EGCG are mediated by acting specifically on the MAPK family, which is an essential part of the signal transduction machinery in transmitting signals from cell surface receptors and exogenous stimulants. In particular, EGCG acts on ERK 1/2 and not other MAPKs by decreasing its phosphorylation in adipocytes, suggesting that the antimitogenic effect of EGCG on 3T3-L1 preadipocytes is dependent on the ERK MAPK. EGCG also activates AMPK and downregulates lipogenic enzymes such as ACC, FAS, and GPDH and transcription factors such as PPARγ and C/EBPα, resulting in decreased adipogenesis and enhanced lipolysis. It is interesting to note that the effects of green tea on thermogenesis are mediated through an interaction between EGCG and caffeine. EGCG and caffeine synergistically interact with norepinephrine to stimulate the thermogenesis of brown adipose tissue in rodents.

23.6.3.1 Rodent Studies

Dietary supplementation of EGCG to diet-induced obese rats decreased body weight and adipose tissue mass. In a similar study, intraperitoneal administration of EGCG to female Sprague-Dawley rats for 7 days resulted in rats losing 29% of their body weight relative to control weight. In support of these antiobesity effects of EGCG, other *in vivo* data have shown that EGCG or EGCG-containing green tea extract reduces food uptake, lipid absorption, and serum triglycerides and cholesterol, as well as stimulates energy expenditure and fat oxidation. Like resveratrol, EGCG also activated AMPK, which is now considered a novel target for the treatment of obesity and type 2 diabetes. Thus, dietary supplementation with EGCG contributes to the prevention and treatment of both obesity and type 2 diabetes.

23.6.3.2 Human Clinical Trials

Intervention studies in humans have demonstrated that green tea containing EGCG lowered body weight and body fat. However, these results have not always been repeatable. Although studies have not provided clear-cut evidence for a link between tea consumption and body weight, tea intake has been associated with decreased serum levels of total cholesterol and LDL cholesterol. Likewise, green tea consumption was also associated with an increased proportion of HDL cholesterol. More recently, consuming green tea extract containing 25% EGCG for 3 months was reported to reduce body weight and waist circumference in obese patients. However, the concentration of EGCG in human plasma has been reported to reach no higher than 1 μmol/L, raising concerns about the amount of EGCG to be consumed to attain physiological levels. Interestingly, EGCG-containing green tea extracts that contain caffeine are more potent than caffeine alone at stimulating 24-h energy expenditure and fat oxidation in humans. Scientists believe that long-term but not short-term consumption of green tea containing EGCG may have beneficial effects of reducing body weight and body fat [34].

23.6.4 Genistein

Genistein, a phytoestrogen, is an isoflavone primarily found in soy, which possesses a high affinity for estrogen receptor β (ERβ). Due to its possible influence on the physiology of the mammalian reproductive tract, there is an increasing interest in genistein. Structurally, genistein, like other flavonoids, has two benzene rings linked together with a heterocyclic pyran or pyrone ring and has potent antioxidant and anti-inflammatory properties. The anti-inflammatory effects of genistein are also mediated in part by its inhibitory effects on NF-kB activity. Further, genistein decreased the activity of inducible nitric oxide synthase (iNOS), which is responsible for prolonged production of larger amounts of nitric oxide (NO). In general, NO production is increased in inflammation and has pro-inflammatory effects. Genistein also decreases signal transducer and activator of transcription 1 (STAT-1), which is a transcription factor for iNOS resulting in decreased NO production in response to inflammatory stimuli [36].

In vitro, genistein decreased lipid accumulation and the expression of a wide variety of adipocyte specific genes including PPARγ, C/EBPα, aP2, LPL, and FAS in murine and human adipocytes [18]. Lipogenesis is also suppressed and lipolysis is enhanced by genistein in rat and human adipocytes. A decrease in lipid synthesis and promotion of lipolysis in adipocytes with genistein is also associated with an inhibition of adipocyte differentiation and induction of adipocyte apoptosis, resulting in a decrease of both adipocyte number and adipocyte size (reviewed in [18]). Additionally, genistein directly modulates beta-cell function by activation of the cAMP/PKA-dependent ERK1/2 signaling pathway [37], suggesting that genistein might be a useful dietary supplement to control insulin-independent diabetes mellitus.

23.6.4.1 Rodent Studies

In rodents, high doses of genistein decreased body weight and adipose tissue mass. It should be noted that the decrease in adipose tissue mass was in part due to induction of apoptosis in adipocytes. In ovariectomized female mice, genistein at 1500 mg/kg reduced food intake and body weight, while body composition was not significantly affected. Weights of parametrial and inguinal fat depots were decreased, and decrease in fat depot weights correlated with an increase in apoptosis compared to control. However, lower doses of genistein were either ineffective in females or increased fat mass in males [38, 39].

Genistein possesses the capacity to reduce hyperglycemia via minimization of islet cell loss in a dosage-dependent manner after the onset of diabetes [40]. In streptozotocin diabetic mice, administration of 6 mg/kg of genistein for 4 weeks reverted the pro-inflammatory cytokine and reactive oxygen species overproduction and restored the nerve growth factor content in the diabetic sciatic nerve, improved the antioxidant enzymes activities, and restored vascular deficits associated with diabetes [41].

23.6.4.2 Human Clinical Trials

Healthy, normal-weight postmenopausal women did not experience improvement in metabolic parameters when given high-dose isoflavones despite an increase in serum adiponectin levels [42]. Soy isoflavone supplements containing genistein, diadzein, and other flavonoids moderately decreased the bone resorption marker deoxypyridoline but did not affect the bone formation markers bone alkaline phosphatase and osteocalcin in menopausal women [43]. Thus, clinical studies in humans have overall failed to detect a decrease in body weight perhaps due to differences in bacterial flora in the gut resulting in different metabolites and varying levels of absorption. Scientists believe that better results could be achieved with higher doses of genistein [39]. On the other hand, circulating concentrations of genistein are markedly higher in Japanese men and women than those reported for a UK population due to a higher soy intake in the Japanese population, and this may explain, in part, the lower incidence of metabolic syndrome in Asian populations compared with Western populations [44].

23.7 PERSPECTIVES

Genomics and proteomics have led to considerable progress in the life sciences, but the questions regarding the role of herbs and natural products remain unanswered. The lack of concrete results in clinical studies using anti-inflammatory phytochemicals for treating metabolic diseases can to some extent be explained by the difficulty in translating doses from *in vitro* studies to *in vivo* situations. Furthermore, phytochemicals may be metabolized differently across species and in different population groups. Data on the optimal concentration required to produce a specific effect *in vivo* need to be generated, and in order to determine whether ingestion of phytochemicals will result in biological effect, better intake data are required on how much the human population is exposed to on a daily basis through fruits, vegetables, and other food supplements.

On the contrary, targeting several signaling pathways simultaneously using multiple phytochemicals at much lower doses to achieve synergistic effects, called *multifocal signal modulation therapy* for metabolic disorders, is gaining considerable attention. Studies indicate that phytochemicals like resveratrol, genistein, and quercetin are more potent in combination and synergistically inhibiting adipogenesis *in vitro* than either compound alone [45, 46]. Likewise, several other combinations of phytochemicals have been investigated both *in vitro* and in rodent models with significant success in decreasing body weight gain and decreasing bone loss. Phytochemicals used in these studies may act by blocking one or more targets in signal transduction pathways, by potentiating effects of other phytochemicals, or by increasing the bioavailability or half-life of other chemicals [47, 48].

23.8 CONCLUSIONS AND FUTURE DIRECTIONS

The majority of *in vitro* rodent and human studies suggests that anti-inflammatory phytochemicals like curcumin, resveratrol, genistein, and EGCG favorably modulated the levels of various biomarkers linked with insulin resistance and obesity (Figure 23.1). However, more research, particularly in humans, is needed to strengthen the link between phytochemicals and metabolic disorders like obesity and insulin resistance and also to establish the effective dose required to modulate these metabolic responses. The clinical studies performed so far with phytochemicals for metabolic disorders are limited by their small sample size and lack of long-term follow-up.

Obesity and related metabolic disorders are epidemic in nature, and according to the Centers for Disease Control and Prevention, three out of four Americans are overweight. It seems absurd to suggest that 75% of the U.S. population be on weight-reduction medications. Also, the treatment strategy for the prevention of obesity and type 2 diabetes is difficult since the process of the metabolic deregulation takes years to manifest in clinical disease. Therefore, lifestyle changes accompanied by the intake of dietary supplements that reduce overall inflammation accompanied by reduction of weight gain could be a plausible recommendation. As in this quote from Hippocrates, "Let food be thy medicine and medicine be thy food," consumption of fruits and vegetables rich in phytochemicals could ameliorate obesity-induced inflammatory responses and pathologies by suppressing the inflammatory signals. Further, phytochemicals have the potential to act on multiple targets, and this approach needs to be further explored for the basis of synthesizing new and more powerful drugs that could help prevent compensatory mechanisms in obesity and type 2 diabetes.

TAKE-HOME MESSAGES

- Obesity and metabolic syndrome are major public health crises not only in the United States but also globally.
- Obesity is associated with low-grade chronic inflammation.

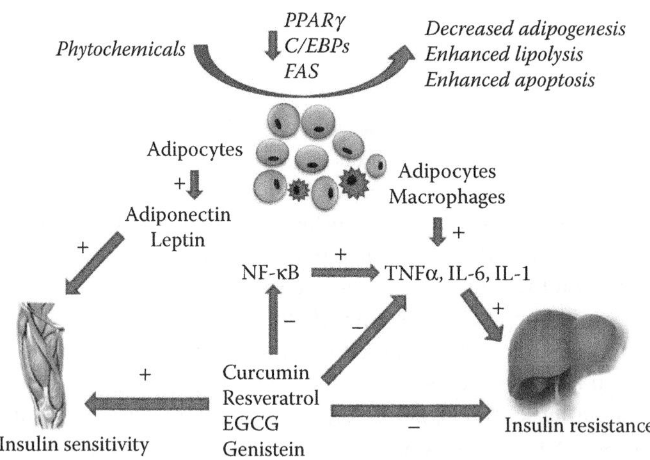

FIGURE 23.1 An overview of antiobesity and antidiabetic effects of phytochemicals. Phytochemicals inhibit NF-kB activation, leading to a decrease of pro-inflammatory cytokines and indirectly decreasing insulin resistance. Phytochemicals also directly act on adipocytes to decrease adipogenesis and promote lipolysis. PPARγ: peroxisome proliferator-activated receptor gamma; C/EBP: CCAAT/enhancer binding protein; FAS: fatty acid synthase; TNFα: tumor necrosis factor alpha; IL: interleukin; NF-kB: nuclear factor kappa B.

- Adipose tissue is not a simple energy storage site but is a source of hormones, growth factors, cytokines, and signaling molecules that regulate body metabolism.
- Major pro-inflammatory cytokines, for example, tumor necrosis factor and interleukins 1 and 6, are secreted by adipose tissue.
- Phytochemicals suppress the pro-inflammatory transcription factor nuclear factor-kappa B, leading to the downregulation of the adipocytokines tumor necrosis factor and interleukin-6 and the upregulation of antiobesity factors such as adiponectin.
- Adipocyte-specific transcription factors such as peroxisome proliferator activated receptor γ and the CCAAT/enhancer binding protein family and lipogenesis-related genes such as fatty acid synthase are downregulated; and lipolysis-associated genes such as hormone sensitive lipase and lipoprotein lipase are upregulated by most anti-inflammatory phytochemicals.
- *In vitro* doses of phytochemicals are difficult to translate into *in vivo* treatments owing to differences and variability in bioavailability, absorption, and metabolism.
- Phytochemicals act by blocking one or more targets in signal transduction pathways, making them ideal for multifocal signal modulation therapies to avoid compensatory mechanisms associated with obesity and related diseases.
- Using combinations of natural products holds greater promise for the prevention and treatment of obesity and type 2 diabetes through targeted decreases in adipogenesis and enhanced lipolysis in adipocytes and hepatocytes.
- More human studies are needed to establish the therapeutic potential of anti-inflammatory phytochemicals for chronic diseases associated with aging such as obesity and insulin resistance.

REFERENCES

1. Fain, J. N., Madan, A. K., Hiler, M. L., Cheema, P., Bahouth, S. W., Comparison of the release of adipokines by adipose tissue, adipose tissue matrix, and adipocytes from visceral and subcutaneous abdominal adipose tissues of obese humans. *Endocrinology* 2004, *145*, 2273–2282.
2. Flegal, K. M., Carroll, M. D., Ogden, C. L., Curtin, L. R., Prevalence and trends in obesity among US adults, 1999–2008. *JAMA* 2010, *303*, 235–241.
3. Wang, Y., Beydoun, M. A., Liang, L., Caballero, B., Kumanyika, S. K., Will all Americans become overweight or obese? Estimating the progression and cost of the US obesity epidemic. *Obesity (Silver Spring)* 2008, *16*, 2323–2330.
4. Cawley, J., Meyerhoefer, C., The medical care costs of obesity: An instrumental variables approach. *National Bureau of Economic Research Working Paper* 2010.
5. Biro, F. M., Wien, M., Childhood obesity and adult morbidities. *Am J Clin Nutr* 2010, *91*, 1499S–1505S.
6. Rayalam, S., Della-Fera, M. A., Baile, C. A., Phytochemicals and regulation of the adipocyte life cycle. *J Nutr Biochem* 2008, *19*, 717–726.
7. Gregoire, F. M., Adipocyte differentiation: From fibroblast to endocrine cell. *Exp Biol Med (Maywood)* 2001, *226*, 997–1002.
8. Tsukumo, D. M., Carvalho-Filho, M. A., Carvalheira, J. B., Prada, P. O., et al., Loss-of-function mutation in Toll-like receptor 4 prevents diet-induced obesity and insulin resistance. *Diabetes* 2007, *56*, 1986–1998.
9. Bastard, J. P., Maachi, M., Lagathu, C., Kim, M. J., et al., Recent advances in the relationship between obesity, inflammation, and insulin resistance. *Eur Cytokine Netw* 2006, *17*, 4–12.
10. Aggarwal, B. B., Targeting inflammation-induced obesity and metabolic diseases by curcumin and other nutraceuticals. *Annu Rev Nutr* 2010, *30*, 173–199.
11. Felson, D. T., Zhang, Y., Hannan, M. T., Anderson, J. J., Effects of weight and body mass index on bone mineral density in men and women: The Framingham study. *J Bone Miner Res* 1993, *8*, 567–573.
12. Janicka, A., Wren, T. A., Sanchez, M. M., Dorey, F., et al., Fat mass is not beneficial to bone in adolescents and young adults. *J Clin Endocrinol Metab* 2007, *92*, 143–147.
13. Maurin, A. C., Chavassieux, P. M., Frappart, L., Delmas, P. D., et al., Influence of mature adipocytes on osteoblast proliferation in human primary cocultures. *Bone* 2000, *26*, 485–489.

14. Pickup, J. C., Inflammation and activated innate immunity in the pathogenesis of type 2 diabetes. *Diabetes Care* 2004, *27*, 813–823.
15. Dembinska-Kiec, A., Mykkanen, O., Kiec-Wilk, B., Mykkanen, H., Antioxidant phytochemicals against type 2 diabetes. *Br J Nutr* 2008, *99 E Suppl 1*, ES109–117.
16. Minich, D. M., Bland, J. S., Dietary management of the metabolic syndrome beyond macronutrients. *Nutr Rev* 2008, *66*, 429–444.
17. Rizzo, N. S., Sabate, J., Jaceldo-Siegl, K., Fraser, G. E., Vegetarian dietary patterns are associated with a lower risk of metabolic syndrome: The adventist health study 2. *Diabetes Care* 2011, *34*, 1225–1227.
18. Andersen, C., Rayalam, S., Della-Fera, M. A., Baile, C. A., Phytochemicals and adipogenesis. *Biofactors* 2010, *36*, 415–422.
19. Ejaz, A., Wu, D., Kwan, P., Meydani, M., Curcumin inhibits adipogenesis in 3T3-L1 adipocytes and angiogenesis and obesity in C57/BL mice. *J Nutr* 2009, *139*, 919–925.
20. Epstein, J., Sanderson, I. R., Macdonald, T. T., Curcumin as a therapeutic agent: The evidence from in vitro, animal and human studies. *Br J Nutr* 2010, *103*, 1545–1557.
21. Aggarwal, B. B., Bhardwaj, A., Aggarwal, R. S., Seeram, N. P., et al., Role of resveratrol in prevention and therapy of cancer: Preclinical and clinical studies. *Anticancer Res* 2004, *24*, 2783–2840.
22. Gonzales, A. M., Orlando, R. A., Curcumin and resveratrol inhibit nuclear factor-kappaB-mediated cytokine expression in adipocytes. *Nutr Metab (Lond)* 2008, *5*, 17.
23. Picard, F., Kurtev, M., Chung, N., Topark-Ngarm, A., et al., Sirt1 promotes fat mobilization in white adipocytes by repressing PPAR-gamma. *Nature* 2004, *429*, 771–776.
24. Ahn, J., Lee, H., Kim, S., Ha, T., Resveratrol inhibits TNF-alpha-induced changes of adipokines in 3T3-L1 adipocytes. *Biochem Biophys Res Commun* 2007, *364*, 972–977.
25. Kundu, J. K., Surh, Y. J., Molecular basis of chemoprevention by resveratrol: NF-kappaB and AP-1 as potential targets. *Mutat Res* 2004, *555*, 65–80.
26. Hsu, C. L., Yen, G. C., Induction of cell apoptosis in 3T3-L1 pre-adipocytes by flavonoids is associated with their antioxidant activity. *Mol Nutr Food Res* 2006.
27. Rayalam, S., Della-Fera, M. A., Baile, C. A., Synergism between resveratrol and other phytochemicals: Implications for obesity and osteoporosis. *Mol Nutr Food Res* 2011.
28. Lagouge, M., Argmann, C., Gerhart-Hines, Z., Meziane, H., et al., Resveratrol improves mitochondrial function and protects against metabolic disease by activating SIRT1 and PGC-1alpha. *Cell* 2006, *127*, 1109–1122.
29. Baur, J. A., Pearson, K. J., Price, N. L., Jamieson, H. A., et al., Resveratrol improves health and survival of mice on a high-calorie diet. *Nature* 2006, *444*, 337–342.
30. Um, J. H., Park, S. J., Kang, H., Yang, S., et al., AMP-activated protein kinase-deficient mice are resistant to the metabolic effects of resveratrol. *Diabetes* 2010, *59*, 554–563.
31. Ramadori, G., Gautron, L., Fujikawa, T., Vianna, C. R., et al., Central administration of resveratrol improves diet-induced diabetes. *Endocrinology* 2009, *150*, 5326–5333.
32. Smoliga, J. M., Baur, J. A., Hausenblas, H. A., Resveratrol and health—A comprehensive review of human clinical trials. *Mol Nutr Food Res* 2011.
33. Pace-Asciak, C. R., Rounova, O., Hahn, S. E., Diamandis, E. P., Goldberg, D. M., Wines and grape juices as modulators of platelet aggregation in healthy human subjects. *Clin Chim Acta* 1996, *246*, 163–182.
34. Moon, H. S., Lee, H. G., Choi, Y. J., Kim, T. G., Cho, C. S., Proposed mechanisms of (-)-epigallocatechin-3-gallate for anti-obesity. *Chem Biol Interact* 2007, *167*, 85–98.
35. Syed, D. N., Afaq, F., Kweon, M. H., Hadi, N., et al., Green tea polyphenol EGCG suppresses cigarette smoke condensate-induced NF-kappaB activation in normal human bronchial epithelial cells. *Oncogene* 2007, *26*, 673–682.
36. Hamalainen, M., Nieminen, R., Vuorela, P., Heinonen, M., Moilanen, E., Anti-inflammatory effects of flavonoids: Genistein, kaempferol, quercetin, and daidzein inhibit STAT-1 and NF-kappaB activations, whereas flavone, isorhamnetin, naringenin, and pelargonidin inhibit only NF-kappaB activation along with their inhibitory effect on iNOS expression and NO production in activated macrophages. *Mediators Inflamm* 2007, *2007*, 45673.
37. Fu, Z., Zhang, W., Zhen, W., Lum, H., et al., Genistein induces pancreatic beta-cell proliferation through activation of multiple signaling pathways and prevents insulin-deficient diabetes in mice. *Endocrinology* 2010, *151*, 3026–3037.
38. Kim, H. K., Nelson-Dooley, C., Della-Fera, M. A., Yang, J. Y., et al., Genistein decreases food intake, body weight, and fat pad weight and causes adipose tissue apoptosis in ovariectomized female mice. *J Nutr* 2006, *136*, 409–414.

39. Penza, M., Montani, C., Romani, A., Vignolini, P., et al., Genistein affects adipose tissue deposition in a dose-dependent and gender-specific manner. *Endocrinology* 2006, *147*, 5740–5751.
40. Yang, W., Wang, S., Li, L., Liang, Z., Wang, L., Genistein reduces hyperglycemia and islet cell loss in a high-dosage manner in rats with alloxan-induced pancreatic damage. *Pancreas* 2011, *40*, 396–402.
41. Valsecchi, A. E., Franchi, S., Panerai, A. E., Rossi, A., et al., The soy isoflavone genistein reverses oxidative and inflammatory state, neuropathic pain, neurotrophic and vasculature deficits in diabetes mouse model. *Eur J Pharmacol* 2011, *650*, 694–702.
42. Charles, C., Yuskavage, J., Carlson, O., John, M., et al., Effects of high-dose isoflavones on metabolic and inflammatory markers in healthy postmenopausal women. *Menopause* 2009, *16*, 395–400.
43. Taku, K., Melby, M. K., Kurzer, M. S., Mizuno, S., et al., Effects of soy isoflavone supplements on bone turnover markers in menopausal women: Systematic review and meta-analysis of randomized controlled trials. *Bone* 2010, *47*, 413–423.
44. Morton, M. S., Arisaka, O., Miyake, N., Morgan, L. D., Evans, B. A., Phytoestrogen concentrations in serum from Japanese men and women over forty years of age. *J Nutr* 2002, *132*, 3168–3171.
45. Rayalam, S., Della-Fera, M. A., Yang, J. Y., Park, H. J., et al., Resveratrol potentiates genistein's antiadipogenic and proapoptotic effects in 3T3-L1 adipocytes. *J Nutr* 2007, *137*, 2668–2673.
46. Park, H. J., Yang, J. Y., Ambati, S., Della-Fera, M. A., et al., Combined effects of genistein, quercetin, and resveratrol in human and 3T3-L1 adipocytes. *J Med Food* 2008, *11*, 773–783.
47. Rayalam, S., Yang, J. Y., Della-Fera, M. A., Baile, C. A., Novel molecular targets for prevention of obesity and osteoporosis. *J Nutr Biochem* 2011, Epub.
48. Lai, C. Y., Yang, J. Y., Rayalam, S., Della-Fera, M. A., et al., Preventing bone loss and weight gain with combinations of vitamin D and phytochemicals. *J Med Food* 2011, *14*, 1352–1362.

24 Anti-Inflammatory Nutraceuticals and Herbal Medicines for the Management of Metabolic Syndrome

George Q. Li, Ka H. Wong, Antony Kam, Xian Zhou, Eshaifol A. Omar, Ali Alqahtani, and Kong M. Li
University of Sydney
New South Wales, Australia

Valentina Razmovski-Naumovski and Kelvin Chan
University of Sydney
New South Wales, Australia

University of Western Sydney
New South Wales, Australia

CONTENTS

24.1 INTRODUCTION

Metabolic syndrome, previously known as syndrome X and insulin resistance syndrome, refers to a constellation of disorders including central obesity, insulin resistance, glucose intolerance, dyslipidemia, and hypertension. These risk factors have been linked to an increased probability of cardiovascular diseases and type 2 diabetes mellitus. The World Health Organization (WHO) estimates that by 2015, approximately 2.3 billion adults worldwide will be overweight, and more than 700 million adults will be obese (WHO 2006). Additionally, the global prevalence of type 2 diabetes will rise to 336 million by 2030 (International Diabetes Foundation 2009).

Among the complex interactions between genetic, metabolic, and environmental factors that are associated with the present prevalence of obesity and metabolic syndrome, diet is considered as the major culprit (Babio et al. 2009a). The increasing incidence of metabolic syndrome relates to the adoption of modern lifestyles, which consist of high-fat, high-carbohydrate, and low-macronutrient foods, long-term and continuous stress, and disruption of chronobiology. Recent meta-analysis and clinical trials illustrate that the Mediterranean diet (low carbohydrate, high in polyunsaturated fats) is associated with a significantly lower prevalence of metabolic syndrome and its associated cardiovascular risk (Babio et al. 2009b; Esposito et al. 2004; Kastorini et al. 2011; Kastorini and Panagiotakos 2010; Panagiotakos and Polychronopoulos 2005).

Emerging scientific evidence has demonstrated that low-grade chronic inflammation is one of the pivotal mechanisms in the induction of metabolic syndrome and its associated pathophysiological consequences (Gustafson 2010; Konner and Bruning 2010). Although there is a suite of pharmacological interventions, the management of metabolic syndrome and its complications remains unsatisfactory.

In the past few decades, traditional Chinese medicine (TCM) has been increasingly used in the treatment of metabolic syndrome, and a growing body of evidence supports the usage of medicinal herbs, foods, and spices to alleviate the development of metabolic syndrome and its associated complications (Omar et al. 2010; Xie and Du 2010; Yin et al. 2008). A literature search was performed on Web of Science, PubMed, and Medline using *inflammation*, *metabolic syndrome*, *herb*, *nutraceutic*, and *natural product* as the key words. Over thirty herbs and nutraceuticals have been comprehensively studied and show promising clinical applications for hyperglycemia, anti-inflammation, antioxidation, and lipid lowering. These common herbs and nutraceuticals include aloe (*Aloe vera*), *Andrographis paniculata*, *Astragalus membranaceus*, baical skullcap (*Scutellaria baicalensis*), bilberry (*Vaccinium myrtillus*), bitter melon (*Momordica charantia*), cinnamon (*Cinnamomum zeylanicum*), clove (*Syzygium aromaticum*), evening primrose oil (*Oenothera biennis*), fenugreek (*Trigonella foenum graecum*), fish oil, flaxseed oil (*Linum usitatissimum*), garlic (*Allium sativum*), ginger (*Zingiber officinale*), ginkgo (*Ginkgo biloba*), ginseng (*Panax ginseng*), goldenseal (*Hydrastis canadensis*), grape seed (*Vitis vinifera*), green tea (*Camellia sinensis*), *Gymnema montanum*, hawthorn (*Crataegus monogyna*), honey, licorice (*Glycyrrhiza glabra*), oats (*Avena sativa*), olive (*Olea europaea*), psyllium (*Plantago ovata*), turmeric (*Curcuma longa*), yam (*Dioscorea opposita*), *Withania somnifera*, wolfberry (*Lycium barbarum*), fish oil, flaxseed oil (*Linum usitatissimum*), onion (*Allium cepa*), Gotu kola (*Centella asiatica*), pomegranate (*Punica granatum*), kudzu (*Pueraria lobata*), and propolis. Among the list, three categories of nutraceuticals are most prominent for tackling inflammation in metabolic syndrome: TCM heat clearing herbs (such as baical skullcap), foods (such as green tea), and fatty acids (such as evening primrose oil and fish oil), and they will be further evaluated in this chapter (Table 24.1).

24.2 TREATMENT OF METABOLIC SYNDROME IN TRADITIONAL CHINESE MEDICINE

In TCM, metabolic syndrome, including obesity, prediabetes, hypertension, and hyperlipidemia, is closely related to "Phlegm" (fat issue, sputum, coagulated blood, gull stone, greasy and thick

TABLE 24.1
Herbal Medicines with Scientific Evidence for the Management of Metabolic Syndrome

Common Name	Dose	Methods	Pharmacological Actions	References
Fish oil			Decreased production of prostaglandin E2, thromboxane B2, leukotriene B4 and E4, 5-hydroxyeicosatetraenoic acid.	(Caughey et al. 1996; Calder 2006)
Flaxseed oil	0.4 g/day	LDL receptor-deficient mouse	Effectively inhibited the expression of inflammatory markers, including IL-6, mac-3, and VCAM-1 in aortic atherosclerotic tissue.	(Dupasquier et al. 2007)
Pomegranate	50 mL/day	Double blinded, randomized, placebo-controlled trial	No significant effect on diabetic parameters, including fasting blood sugar, HbA1c levels, and serum insulin levels, but significantly lowered serum C-peptide.	(Rock et al. 2008; Rosenblat, Hayek, and Aviram 2006)
	40 g/day	Type 2 diabetic patients with hyperlipidemia	Reduction in serum total cholesterol and LDL levels, total cholesterol:HDL and LDL:HDL ratios.	(Esmaillzadeh et al. 2004, 2006)
	100 and 300 mg/kg/day	Diabetic Wistar rats	Reduced serum glucose, urine glucose and urinary protein levels, serum angiotensin converting enzyme (ACE) levels.	(Mohan, Waghulde, and Kasture 2010)
Green tea		Cross-sectional study	Significantly reduced the plasma level of total cholesterol, LDL, VLDL, and triglyceride levels, and increased HDL.	(Sasazuki et al. 2000)
	50 μL of 0.75% green tea polyphenols	SD rats, adipocytes	Improved insulin sensitivity by significantly increasing basal and insulin-stimulated glucose uptake.	(Wu et al. 2004)
	0.5 g/day of lyophilized green tea powder	SD rats	Reduced fasting serum triglyceride and free fatty acid levels.	(Wu et al. 2004)
Baical skullcap		Streptozotocin-induced diabetic rats	Decreased blood glucose levels, increased body weight, and reversed increase of sucrase activity in small intestine of rats.	(Gu et al. 2009)
		Rat mesenteric arteries	Relaxed rat mesenteric arteries via large-conductance Ca^{2+}-activated K^+ (BKCa) channel activation and voltage-dependent Ca^{2+} channel (VDCC) inhibition by endothelium-independent mechanisms.	(Lin et al. 2010)
		Macrophage foam cells	Promoted cholesterol efflux by increasing protein phosphatase 2B-dependent dephosphorylation at ATP-binding cassette transporter-A1, leading to reduced cholesterol accumulation.	(Chen et al. 2011)
Goldthread		L6 muscle cells model	Inhibited fatty acid uptake by reducing peroxisome proliferator-activated receptor-α and fatty acid transferase/CD36 expressions, and through AMP-activated protein kinase (AMPK) and p38 mitogen activated protein kinase (MAPK) pathway.	(Chen et al. 2009; Cheng et al. 2006)

(*continued*)

TABLE 24.1
Herbal Medicines with Scientific Evidence for the Management of Metabolic Syndrome (Continued)

Common Name	Dose	Methods	Pharmacological Actions	References
Goldthread		Type 2 diabetes mellitus Wistar rats	Decreased fasting blood glucose and triglyceride levels, and improved endothelium-dependent vasorelaxation in aorta through enhanced NO bioavailability by upregulating endothelial nitric oxide synthase (eNOS) expression and downregulating expression of NADPH oxidase.	(Wang et al. 2009)

tongue coating) accumulated in the body (Zhang and Chen 2004). There are complete theories on the pathology and the management of each condition of metabolic syndrome.

24.2.1 Obesity

Similar to Western medicine, TCM considers obesity as the overconsumption of heavy and greasy food. In TCM, fat or adipose tissue is mostly due to "Phlegm" (fat issue) and "Dampness" (related to heaviness and turbidness) evils, which means the "Spleen" (referring to the digestive and metabolic system) fails to process and move waste fluids and foods; they consolidate and transform into "Phlegm." TCM takes a holistic approach to obesity by focusing on the underlying changes in the body. The principle of the treatment is to strengthen and balance the whole body by focusing on the "Spleen," and efficiently dissolve and discharge turbidity and "Phlegm" from the body (Li and Wu 2011).

24.2.2 Prediabetes

According to the Chinese medicine textbook *Huang Di Nei Jing* (*The Yellow Emperor's Canon of Medicine*, 25–225 C.E.), obesity causes cardiovascular complications. Obesity is accompanied by anxiety, gradual damage of the "Heart" (which controls blood and vessels, facial complexion, and the mind), "Spleen," "Liver" (mainly for storing blood), and "Kidney" (which stores the "Essence" and is essential in the growth and development of the human body), leading to disharmony of Yin (negative and inactive sides of all phenomena) and Yang (positive, active sides of all phenomena; Zhang and Chen 2004). As a result, obesity causes excessive Yang energy and Yin energy disorder, and then consumes Yin fluids of the body to its depletion (Covington 2001). The "Dryness-Heat" in the body (related to inflammatory conditions in orthodox medicine) leads to prediabetes. In treatment, the insufficiency of Yin will disappear only when excessive Yang is corrected (Li et al. 2004). Therefore, the TCM treatment of prediabetes focuses on replenishing Yin (promoting fluid production in the body) and evacuating "Fire" (Heat) from the body (Li et al. 2004; Liu and Tang 2000).

24.2.3 Hypertension

TCM treats the symptoms and resulting medical problems of hypertension as insufficient blood flow to the brain and other organs. The symptoms present themselves in the form of dizziness, headache, fatigue, shortness of breath, chest pains, irregular heartbeat, and vision problems (Ping 1997). The pathology is divided into four categories: disorder of the "Liver" causing hyperactivity, headache, and dizziness; deficiency of "Blood" and "Qi" (the energy of the body, the meridians, the food, and the universe), which fail to nourish the brain and lead to dizziness; deficiency in the "Kidneys," which causes insufficient blood flow to the brain; and the interior retention of "Phlegm-Damp" (obesity; Zhang and Chen 2004). The treatment of hypertension includes promotion of bowel movements (remove stagnation), promotion of urination (relieving water retention within vessels), promotion of circulation, and nourishing the "Liver," "Lung," and "Kidney" organs.

24.2.4 Hyperlipidemia

Hyperlipidemia is diagnosed as the accumulation of "Damp" and "Phlegm" (in the form of blood lipids) in the blood vessels (Zhang et al. 1990). The pathology is closely related to the disorder of the "Liver," "Spleen," and "Kidney" in obesity. Due to the weakness of the "Spleen" and "Kidney," incorrect transferral of fluids occurs and results in "Interior Phlegm," which causes elevated blood lipids. TCM treatment for lowering lipids focuses on nourishing Yin, strengthening "Kidney" and "Spleen," clearing "Fire" (Heat) and discharging "Phlegm."

Therefore, based on the disorder differentiation of metabolic syndrome, the treatment principles in TCM are to remove the "Phlegm" and "Dampness," to replenish Yin and clear the excessive heat from the body. Herbs and spices used to remove the "Phlegm" and "Dampness" include green tea (*Camellia sinensis*), hawthorn (*Crataegus monogyna*), turmeric (*Curcuma longa*), cinnamon (*Cinnamomum zeylanicum*), and clove (*Syzygium aromaticum*). The heat clearing natural products include goldthread *(Coptis chinensis)*, bitter melon (*Momordica charantia)*, baical skullcap (*Scutellaria baicalensis*), rhubarb (*Rheum officinalis*), and propolis.

In summary, TCM has a unique theory on the pathology and the management of metabolic syndrome. However, scientific evidence is required to evaluate the efficacy, safety, and quality of these nutraceuticals, herbal medicines, spices, and teas.

24.3 HEAT CLEARING CHINESE HERBS FOR METABOLIC SYNDROME

In TCM, herbs classified as "Heat Clearing Herbs" have the primary function of clearing internal Heat and expelling Dampness, cooling Blood, and eliminating Toxins from the human body. Traditionally, these herbs have been used effectively as antipyretic, antimicrobial, and, to a certain extent, as antitumor hemostatics for bleeding conditions and sedative agents (Joseph and Jin 2005). Likewise, modern science has revealed that patients with a "Heat Pattern" demonstrated higher levels of neuroendocrine activation, thermogenesis, and metabolism when compared with normal people. These changes could be found in patients with, for example, hypertension, diabetes, thyroidism, or schizophrenia (Jiang 2005). In TCM, baical skullcap, goldthread, and *Phellodendron amurense*, collectively known as the "San Huang" (the three yellow) are popular heat clearing herbs. Used in combination, they synergize the therapeutic actions of one another.

24.3.1 Baical Skullcap

In TCM, baical skullcap clears Heat, particularly in the Lungs, and Damp-Heat in the Large Intestine. It is traditionally used to treat acute upper and lower tract infections such as acute bronchitis, pneumonia, acute tonsillitis, and inflammatory conditions of the intestines such as enteritis, colitis, Crohn's disease, and intestinal parasitic diseases (Joseph and Jin 2005). Current pharmacological research has confirmed the strong anti-inflammatory properties of baical skullcap through its inhibition of particular inflammatory cascades, including lipopolysaccharide (LPS)-induced nitric oxide, inducible nitric oxide synthase (iNOS), cytokines (e.g., interleukin (IL)-1B, IL-2, IL-6, and tumor necrosis factor (TNF)-α), chemokines, and vascular endothelial growth factor (VEGF) production in macrophages, nuclear factor-kappa B (NF-κB), cyclooxygenase (COX)-2, and mitogen-activated protein kinase (MAPK) signalling pathways (Kim et al. 2009; Yoon et al. 2009; Kim et al. 2006). Thus, this herb has been given extensive attention by researchers for the treatment of metabolic syndrome.

Four-week treatment with baicalin (the main active compound of baical skullcap) decreased blood glucose levels, reversed the decreased body weight, and reversed the increase of sucrase activity in the small intestine of streptozotocin-induced diabetic rats. It is believed that baicalin's antidiabetic effect is partly via the inhibition of disaccharidases in the small intestine (Gu et al. 2009).

It has been demonstrated that baicalin could be a potential agent for the management of cardiovascular disorders. In another study, baicalin relaxed rat mesenteric arteries via large-conductance Ca^{2+}-activated K^+ (BKCa) channel activation, and voltage-dependent Ca^{2+} channel (VDCC) inhibition by endothelium-independent mechanisms (Lin et al. 2010).

An antiobesity herbal formula, which includes baical skullcap, inhibits fat accumulation in 3T3-L1 adipocytes and in high-fat-diet–induced obese mice through the modulation of the adipogenesis pathway (Lee, Kang, and Yoon 2010). Wogonin, another active compound in baical skullcap extract, promoted cholesterol efflux by increasing protein phosphatase 2B-dependent dephosphorylation at the ATP-binding cassette transporter-A1, thus leading to reduced cholesterol accumulation in macrophage foam cells (Chen et al. 2011). In a different study performed on type 2 diabetic Goto-Kakizaki rats, baicalin demonstrated a potent antioxidant property by reducing hyperglycemia-induced oxidative stress through the increased expression of antioxidant enzyme activities. It was also associated with anti-hypertriglyceridemic, as well as anti-hypercholesterolemic, effects, when compared to metformin (Waisundara et al. 2011).

24.3.2 Goldthread

In contrast, goldthread enters the "'Heart and Stomach Meridians." In TCM, it is used as a remedy for restlessness, insomnia, irritability, thirst, and a bitter taste in the mouth, hypochondriac pain and distention, acid regurgitation, poor appetite, and nausea symptoms originating from the "Stomach" or "Spleen" (Joseph and Jin 2005). Goldthread and its most prominent compound, berberine, have been shown to play a vital role in anti-inflammatory, immunomodulating, and antioxidant processes (Yan et al. 2011; Zhang et al. 2011; Hsiang et al. 2005; Song, Chen, and Zhu 1992; Kim et al. 2010; Ko et al. 2007). These multitargeted actions may indicate the beneficial properties of goldthread for the treatment of metabolic syndrome. Based on an *in vivo* study, it has been suggested that a Chinese herbal formula containing goldthread may be a new oral agent for treating the metabolic syndrome and preventing type 2 diabetes (Tan et al. 2011).

Berberine improved the free fatty acid-induced insulin resistance in an L6 muscle cell model through inhibiting fatty acid uptake, partly by reducing peroxisome proliferator-activated receptor-α (PPAR-α) and fatty acid transferase/CD36 expressions (Chen et al. 2009), as well as through the AMP-activated protein kinase (AMPK) and p38 mitogen-activated protein kinase (MAPK) pathway, which may account for the antihyperglycemic effects of this compound (Cheng et al. 2006). Furthermore, Kong et al. (2009) reported that berberine increased insulin receptor mRNA and protein expression in a dose- and time-dependent manner in cultured human liver cells and L6 rat skeletal muscle cells, through a protein kinase C-dependent activation of its promoter. In animal models, treatment of type 2 diabetes mellitus rats with berberine also lowered fasting blood glucose and fasting serum insulin, increased insulin sensitivity, and elevated insulin receptor mRNA, as well as PKC activity, in the liver (Kong et al. 2009; Wang et al. 2011).

In an *in vivo* study investigating the dysfunction of aortas in type 2 diabetes mellitus Wistar rats, berberine not only significantly decreased fasting blood glucose and triglyceride levels, but also improved endothelium-dependent vasorelaxation in aorta through enhanced NO bioavailability by upregulating endothelial nitric oxide synthase (eNOS) expression and downregulating expression of NADPH oxidase (Wang et al. 2009). This finding may support the use of goldthread in the management of high blood pressure resulting from endothelial dysfunction.

Following treatment with berberine, the mRNA and protein expression levels of specific genes for adipocyte differentiation, GATA-2 and GATA-3, were elevated and accompanied by inhibited adipogenesis (Hu and Davies 2009). It was demonstrated that diabetic rats treated with berberine

(100 mg/kg) had lower plasma triglyceride levels compared with the controls (Wang et al. 2009). In a current comprehensive review, the effects and the underlying mechanisms of berberine on carbohydrate and lipid metabolism in relation to endothelial function and the cardiovascular system has been described (Affuso et al. 2010).

24.4 ANTI-INFLAMMATORY FOOD AND TEA

24.4.1 POMEGRANATE

Punica granatum Linn., commonly known as pomegranate (Figure 24.1), belongs to the Punicacae family. It is a well-recognized fruit and beverage with high antioxidant content. A recent review highlighted the anti-inflammatory activities of pomegranate in the treatment of cancer (Lansky and Newman 2007). Thus, pomegranate shows great potential for treating the conditions of metabolic syndrome.

Four weeks of oral consumption of pomegranate juice (100 and 300 mg/kg/day) significantly reduced the serum glucose, urine glucose, and urinary protein levels in diabetic Wistar rats with or without hypertension (Mohan et al. 2010). Pomegranate juice consumption has also shown positive antihypertensive effects in several studies. The consumption of pomegranate juice orally for 4 weeks (100 and 300 mg/kg/day) significantly lowered the serum angiotensin converting enzyme (ACE) levels in both diabetic and diabetic hypertensive rats, compared to controls (Mohan et al. 2010). In addition, pomegranate juice caused a significant reduction in the mean levels and changes in arterial blood pressure in response to several catecholamine drugs (i.e., adrenaline, noradrenaline, phenylalanine, angiotensin II, and 5-hydroxytryptamine) in angiotensin II-induced hypertensive diabetic rats (Mohan et al. 2010).

Aviram and Dornfeld (2001) conducted two clinical studies to evaluate the effectiveness of pomegranate juice on blood pressure. The study found that 50 mL of pomegranate juice per day for 2 weeks reduced systolic blood pressure in hypertensive patients by 5%, together with an inhibition of ACE activities by 36% (Aviram and Dornfeld 2001). Another clinical study that involved 19 patients with asymptomatic carotid artery stenosis revealed that pomegranate juice resulted in a reduction in systolic blood pressure by 21% after 1 year of consumption. However, the blood pressure was not further reduced with 3-year consumption (Aviram et al. 2004). The consumption of concentrated pomegranate juice (40 g/day) in 22 type 2 diabetic patients with hyperlipidemia showed a significant reduction in serum total cholesterol and low-density lipoprotein (LDL) levels, as well as total cholesterol: high-density lipoprotein (HDL) and LDL:HDL ratios (Esmaillzadeh et al. 2004, 2006). Two

FIGURE 24.1 Pomegranate plant and fruit.

clinical trials that involved 20 and 30 human subjects, respectively, demonstrated that the consumption of pomegranate juice did not significantly affect diabetic parameters, including fasting blood sugar, HbA1c levels, and serum insulin levels, but significantly lowered serum C-peptide (a cleavage product of pro-insulin) by 23% in diabetic patients (Rock et al. 2008; Rosenblat et al. 2006).

24.4.2 Green Tea

Biological and clinical evidence has illustrated the role of green tea in the management of metabolic syndrome. A recent review has highlighted its anti-inflammatory activities (Tipoe et al. 2007).

It has been reported that the oral consumption of green tea for 12 weeks reduced blood glucose and insulin levels in rats (Wu et al. 2004). Similarly, 4-week oral consumption of green tea significantly reduced serum glucose level in high-cholesterol-fed rats (Ramadan et al. 2009). Green tea improved insulin sensitivity, as reflected by a significant increase in basal and insulin-stimulated glucose uptake by adipocytes (Wu et al. 2004).

Green tea extracts significantly attenuated the increase in blood pressure caused by angiotensin-II in rats (Antonello et al. 2007). Green tea supplementation for 12 weeks reduced fasting serum triglycerides and free fatty acid levels in rats (Wu et al. 2004). In high-cholesterol-fed rats, 4 weeks of oral consumption of green tea attenuated the total lipid, triglyceride, and phospholipid levels when compared to the control (Ramadan et al. 2009). It has been shown that oral administration of green tea extract for 4 weeks significantly reduced the body weight and liver-to-body weight ratio in high-cholesterol-diet-fed rats (Ramadan et al. 2009).

A single consumption of green tea extract reduced carbohydrate absorption by 25% and attenuated the plasma glucose levels following an oral glucose tolerance test in humans (Zhong et al. 2006; Tsuneki et al. 2004). Chronic consumption of green tea for 30 days also significantly reduced plasma glucose and insulin levels in type 2 diabetic patients (Hosoda et al. 2003). A cohort study that involved 600 subjects reported that the consumption of tea (>120 mL/day) for 1 year was associated with a 46% reduction in the risk of developing hypertension (Yang et al. 2004). Two cross-sectional studies demonstrated that the daily consumption of green tea significantly reduced the plasma level of total cholesterol and triglyceride levels (Imai and Nakachi 1995; Sasazuki et al. 2000). The plasma level of HDL was also significantly increased, together with a reduction in LDL and very low density lipoprotein (VLDL) levels (Sasazuki et al. 2000). Several clinical studies reported that the consumption of green tea for 12 weeks was associated with a significant reduction in body weight, body fat mass, and BMI in overweight subjects (Nagao et al. 2007; Nagao et al. 2005).

24.5 ANTI-INFLAMMATORY EFFECTS OF POLYUNSATURATED FATTY ACIDS IN THE MANAGEMENT OF METABOLIC SYNDROME

It is well known that the harmful impact of high-fat diets relates to the amount and the type of dietary fatty acids. While a high intake of saturated and *trans* fatty acids has been associated with an increased incidence of metabolic syndrome and its cardiovascular complications, the consumption of polyunsaturated fatty acids (PUFAs) has been associated with lower cardiovascular risk. More scientific evidence has emerged on the beneficial role of PUFAs, both vegetable n-6 PUFAs and marine n-3 PUFAs (e.g., fish oil), including eicosapentaenoic acid (EPA, 20:5, n-3) and docosahexaenoic acid (DHA, 22:6, n-3), in cardiovascular and metabolic protection. Clinical evidence suggests that 1.2–2 g/day of the polyunsaturated fatty acids (PUFAs) was associated with a significant reduction in blood pressure, improvements in vascular function, and reduction in inflammation in patients with symptoms of metabolic syndrome (Dangardt et al. 2010; Cicero et al. 2010). Alpha-linolenic acid (ALA, 18:3, n-3), a plant-derived n-3 PUFA, and its biosynthetic form, stearidonic acid, represent a relatively high proportion of the total fatty acids in some vegetable oils, such as flaxseed oil, soybean, evening primrose oil, perilla, canola, grape seed, and walnut.

The observed beneficial effects on cardiovascular complications are mainly correlated with the effect of PUFAs in the reduction of plasma triacylglycerols. However, recent studies suggest that the anti-inflammatory potential of PUFAs could play a pivotal role in the prevention and treatment of metabolic syndrome. In obese diabetic mice, it has been shown that n-3 PUFA could mediate a beneficial effect on the development of diabetes and metabolic complications by preventing adipose tissue inflammation induced by a high-fat diet, and may ameliorate insulin sensitivity (Todoric et al. 2006). In the same study, n-3 PUFAs also changed the inflammatory gene expression associated with macrophage infiltration. Furthermore, it has been reported that n-3 PUFAs significantly decreased circulating inflammatory markers in elderly patients with chronic heart failure (Zhao et al. 2009).

A variety of molecular mechanisms underlying the anti-inflammatory effects of n-3 PUFAs have been described. Fish oil supplementation of the human diet has been shown to result in the decreased production of prostaglandin E2, thromboxane B2 (Caughey et al. 1996), leukotriene B4, leukotriene E4, 5-hydroxyeicosatetraenoic acid, and inflammatory cells (Calder 2006). The high intake of long-chain n-3 PUFAs also results in an increased proportion of these fatty acids in inflammatory cell phospholipids (Yaqoob et al. 2000). At sufficiently high intake, n-3 PUFAs decrease the production of inflammatory eicosanoids, pro-inflammatory cytokines, and reactive oxygen species, and the expression of adhesion molecules. Long-chain n-3 PUFAs act both directly (e.g., by replacing arachidonic acid as an eicosanoid substrate and inhibiting arachidonic acid metabolism) and indirectly (e.g., by altering the expression of inflammatory genes via transcription factor activation; Calder 2006). In addition, a recent *in vivo* study has shown that stearidonic and eicosapentaenoic acids inhibited obesity-associated increases of IL-6 expression in genetically obese *ob/ob* mice (Hsueh et al. 2011). Furthermore, it was demonstrated that ALA significantly reduced tumor necrosis factor-α (TNF-α) and interleukin-6 (IL-6) concentrations as well as superoxide production and malonaldialdehyde formation in ischemic hearts of diabetic rats (Xie et al. 2011).

The beneficial effect of flaxseed is likely due to the omega-3 fatty acid content of its oil fraction, which is also one of the richest sources of ALA (57%). Flaxseed and its constituents may have protective and/or therapeutic effects against cardiovascular and metabolic complications through numerous mechanisms, including antioxidant effect, and reduction in serum total cholesterol, LDL-C, inflammatory markers, platelet aggregation, and expression of adhesion molecules. Dietary flaxseed (0.4 g/day) has been shown to effectively inhibit the expression of inflammatory markers, including IL-6, Mac-3, and VCAM-1, in aortic atherosclerotic tissue in LDL receptor-deficient mice after 24 weeks of diet intervention (Dupasquier et al. 2007). In addition, a study on a metabolic syndrome animal model reported that flax oil feeding is associated with a reduction of hepatic lipid accumulation through the activation of PPAR-γ (Chechi et al. 2010).

24.6 DISCUSSION AND SUMMARY

Metabolic syndrome is a chronic disease that will target all age groups due to the modern sedentary and convenience lifestyle. It is a collection of dysfunctions including glucose intolerance, central obesity, dysliplidemia, and hypertension, which have been associated with chronic low-grade inflammation. Many herbs and nutraceuticals are used in the management of this complex disease due to their multitargeted actions. Heat clearing herbs in TCM, teas, and fatty acids are well known for their anti-inflammatory properties and have been shown clinically to be effective agents against the symptoms of metabolic syndrome. Several reviews have demonstrated the strong linkage between low-grade chronic inflammation and metabolic syndrome. To the best of our knowledge, however, there is a lack of literature revealing the potential clinical mechanism of action. Therefore, this chapter gives us a platform to evaluate the use of herbal medicines and nutraceuticals for the treatment of metabolic syndrome in targeting inflammation. The literature on selected nutraceuticals and herbs indicates a strong link between the anti-inflammatory mechanism and their effects in metabolic syndrome (Figure 24.2).

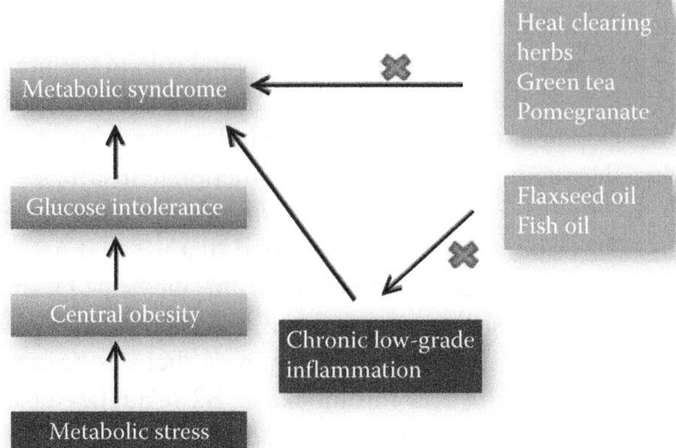

FIGURE 24.2 Anti-inflammatory pathway contributing to the effects of nutraceuticals in metabolic syndrome. Metabolic syndrome is caused by various factors such as metabolic stress and chronic low-grade inflammation, which can be ameliorated by the aforementioned nutraceuticals and herbal medicines including green tea, pomegranate, flaxseed oil, fish oil, and heat clearing herbs such as goldthread and baical skullcap.

The three categories discussed here have been used worldwide for thousands of years. Documented traditional use, safety, and the multiconstituent profile are the perceived advantages over pharmaceuticals with narrow therapeutic windows.

Due to the complexity of chemical components in nutraceuticals and herbal medicines, a comprehensive platform is required to define the nature of the chemical mixture that is used for biological testing and clinical studies (Li, et al. 2010a, 2010b; Razmovski-Naumovski et al. 2010). To meet the requirements of evidence-based medicine, and to prove their safety, quality, and efficacy, high-quality preclinical and clinical studies are warranted. In conclusion, nutraceuticals and herbal medicines, particularly those with anti-inflammatory activity, offer considerable scope in the discovery and for the development of novel therapeutic agents for the prevention and treatment of metabolic syndrome and associated cardiovascular complications.

TAKE-HOME MESSAGES

- Low-grade chronic inflammation is one of the crucial mechanisms in the induction of metabolic syndrome and its associated pathophysiological complications.
- A growing body of evidence supports the usage of medicinal herbs and nutraceuticals to alleviate the development of metabolic syndrome.
- In TCM, metabolic syndrome refers to "Phlegm" and "Dampness" (fat issue, sputum, coagulated blood, gull stone, greasy and thick tongue coating) accumulated in the body.
- The treatment principles in TCM are to remove the "Phlegm" and "Dampness," to replenish Yin and clear the excessive Heat from the body.
- Heat clearing herbs such as baical skullcap and goldthread are used clinically for metabolic syndrome, and have been shown to have anti-inflammatory activities in preclinical studies.
- Both preclinical and clinical studies on pomegranate and green tea have indicated both are beneficial in metabolic syndrome.
- Polyunsaturated fatty acids in vegetable oils, such as flaxseed, evening primrose oil, and fish oil are now well recognized to be beneficial in metabolic syndrome and cardiovascular diseases.
- Specific studies on herbal medicines to show their inhibition effects on the expression of inflammatory markers and underlining mechanisms are still required.

- There is a lack of large sample size, multicenter, and multinational clinical trial in investigating the efficacy of herbal medicine on metabolic syndrome.
- To meet the requirements of evidence-based medicine, high-quality preclinical and clinical studies are warranted.

ACKNOWLEDGMENTS

This project is supported by the International Science Linkages established under the Australian Government's innovation statement, "Backing Australia's Ability," and the National Institute of Complementary Medicine, Australia.

REFERENCES

Affuso, F., V. Mercurio, V. Fazio, and S. Fazio. 2010. Cardiovascular and metabolic effects of berberine. *World J Cardiol* 2 (4):71–7.

Antonello, M., D. Montemurro, M. Bolognesi, M. Di Pascoli, A. Piva, F. Grego, D. Sticchi, L. Giuliani, S. Garbisa, and G. P. Rossi. 2007. Prevention of hypertension, cardiovascular damage and endothelial dysfunction with green tea extracts. *Am J Hypertens* 20 (12):1321–8.

Aviram, M., and L. Dornfeld. 2001. Pomegranate juice consumption inhibits serum angiotensin converting enzyme activity and reduces systolic blood pressure. *Atherosclerosis* 158 (1):195–8.

Aviram, M., M. Rosenblat, D. Gaitini, S. Nitecki, A. Hoffman, L. Dornfeld, et al. 2004. Pomegranate juice consumption for 3 years by patients with carotid artery stenosis reduces common carotid intima-media thickness, blood pressure and LDL oxidation. *Clinical Nutrition* 23 (3):423–33.

Babio, N., M. Bullo, J. Basora, M. A. Martinez-Gonzalez, J. Fernandez-Ballart, F. Marquez-Sandoval, et al. 2009a. Adherence to the Mediterranean diet and risk of metabolic syndrome and its components. *Nutr Metabol & Cardiovas Dis* 19 (8):563–70.

Babio, N., M. Bullo, and J. Salas-Salvado. 2009b. Mediterranean diet and metabolic syndrome: The evidence. *Public Health Nutrition* 12 (9A):1607–17.

Calder, P. C. 2006. n-3 polyunsaturated fatty acids, inflammation, and inflammatory diseases. *Am J Clin Nutr* 83 (6 Suppl):1505S–1519S.

Caughey, G. E., E. Mantzioris, R. A. Gibson, L. G. Cleland, and M. J. James. 1996. The effect on human tumor necrosis factor alpha and interleukin 1 beta production of diets enriched in n-3 fatty acids from vegetable oil or fish oil. *Am J Clin Nutr* 63 (1):116–22.

Chechi, K., N. Yasui, K. Ikeda, Y. Yamori, and K. Cheema S. 2010. Flax oil-mediated activation of PPAR-gamma correlates with reduction of hepatic lipid accumulation in obese spontaneously hypertensive/NDmcr-cp rats, a model of the metabolic syndrome. *Br J Nutr* 104 (9):1313–21.

Chen, C. Y., S. K. Shyue, L. C. Ching, K. H. Su, Y. L. Wu, Y. R. Kou, et al. 2011. Wogonin promotes cholesterol efflux by increasing protein phosphatase 2B-dependent dephosphorylation at ATP-binding cassette transporter-A1 in macrophages. *J Nutr Biochem* 22 (11):1015–21.

Chen, Y., Y. Li, Y. Wang, Y. Wen, and C. Sun. 2009. Berberine improves free-fatty-acid-induced insulin resistance in L6 myotubes through inhibiting peroxisome proliferator-activated receptor [gamma] and fatty acid transferase expressions. *Metabolism* 58 (12):1694–1702.

Cheng, Z., T. Pang, M. Gu, A. H. Gao, C. M. Xie, J. Y. Li, et al. 2006. Berberine-stimulated glucose uptake in L6 myotubes involves both AMPK and p38 MAPK. *Biochimica et Biophysica Acta (BBA)* 1760 (11):1682–1689.

Cicero, A. F., G. Derosa, V. Di Gregori, M. Bove, A. V. Gaddi, and C. Borghi. 2010. Omega 3 polyunsaturated fatty acids supplementation and blood pressure levels in hypertriglyceridemic patients with untreated normal-high blood pressure and with or without metabolic syndrome: A retrospective study. *Clin Exp Hypertens* 32 (2):137–44.

Covington, M. B. 2001. Traditional Chinese medicine in the treatment of diabetes. *Diabetes Spectrum* 14:154–159.

Dangardt, F., W. Osika, Y. Chen, U. Nilsson, L. M. Gan, E. Gronowitz, et al. 2010. Omega-3 fatty acid supplementation improves vascular function and reduces inflammation in obese adolescents. *Atherosclerosis* 212 (2):580–5.

Dupasquier, C. M., E. Dibrov, A. L. Kneesh, P. K. Cheung, K. G. Lee, H. K. Alexander, et al. 2007. Dietary flaxseed inhibits atherosclerosis in the LDL receptor-deficient mouse in part through antiproliferative and anti-inflammatory actions. *Am J Physiol Heart Circ Physiol* 293 (4):H2394–402.

Esmaillzadeh, A., F. Tahbaz, I. Gaieni, H. Alavi-Majd, and L. Azadbakht. 2004. Concentrated pomegranate juice improves lipid profiles in diabetic patients with hyperlipidemia. *J Medicinal Food* 7 (3):305–8.

Esmaillzadeh, A., F. Tahbaz, I. Gaieni, H. Alavi-Majd, and L. Azadbakht. 2006. Cholesterol-lowering effect of concentrated pomegranate juice consumption in type II diabetic patients with hyperlipidemia. *Internat J Vitamin & Nutr Res* 76 (3):147–51.

Esposito, K., R. Marfella, M. Ciotola, C. Di Palo, F. Giugliano, G. Giugliano, et al. 2004. Effect of a mediterranean-style diet on endothelial dysfunction and markers of vascular inflammation in the metabolic syndrome: A randomized trial. *JAMA* 292 (12):1440–6.

Gu, S. J., L. Liu, Y. W. Liu, X. Y. Jing, X. D. Liu, and L. Xie. 2009. Inhibitory effect of baicalin on disaccharidase in small intestine. *Chinese J Nat Med* 7 (2):129–133.

Gustafson, B. 2010. Adipose tissue, inflammation and atherosclerosis. *J Atherosclerosis & Thrombosis* 17 (4):332–41.

Hosoda, K., M. F. Wang, M. L. Liao, C. K. Chuang, M. Iha, B. Clevidence, et al. 2003. Antihyperglycemic effect of oolong tea in type 2 diabetes. *Diabetes Care* 26 (6):1714–8.

Hsiang, C. Y., S. L. Wu, S. E. Cheng, and T. Y. Ho. 2005. Acetaldehyde-induced interleukin-1beta and tumor necrosis factor-alpha production is inhibited by berberine through nuclear factor-kappaB signaling pathway in HepG2 cells. *J Biomed Sci* 12 (5):791–801.

Hsueh, H. W., Z. Zhou, J. Whelan, K. G. Allen, N. Moustaid-Moussa, H. Kim, and K. J. Claycombe. 2011. Stearidonic and eicosapentaenoic acids inhibit interleukin-6 expression in ob/ob mouse adipose stem cells via Toll-like receptor-2-mediated pathways. *J Nutr* 141 (7):1260–6.

Hu, Y., and G. E. Davies. 2009. Berberine increases expression of GATA-2 and GATA-3 during inhibition of adipocyte differentiation. *Phytomedicine* 16 (9):864–873.

Imai, K., and K. Nakachi. 1995. Cross sectional study of effects of drinking green tea on cardiovascular and liver diseases. *BMJ* 310 (6981):693–6.

International Diabetes Foundation. Diabetes and impaired glucose tolerance: Global burden. International Diabetes Federation Altas. 2009. http://www.diabetesatlas.org/content/diabetes-and-impaired-glucose-tolerance

Jiang, W. Y. 2005. Therapeutic wisdom in traditional Chinese medicine: A perspective from modern science. *Trends Pharmaco Sci* 26 (11):558–63.

Joseph, H., and Y. Jin. 2005. *The healing power of Chinese herbs and medicinal recipes*. New York, NY: Haworth Press Inc.

Kastorini, C. M., H. J. Milionis, K. Esposito, D. Giugliano, J. A. Goudevenos, and D. B. Panagiotakos. 2011. The effect of Mediterranean diet on metabolic syndrome and its components: A meta-analysis of 50 studies and 534,906 individuals. *J Amer Coll Cardiol* 57 (11):1299–313.

Kastorini, C. M., and D. B. Panagiotakos. 2010. The role of the Mediterranean diet on the development of the metabolic syndrome. *Frontiers in Bioscience* 2:1320–33.

Kim, D. H., H. K. Kim, S. Park, J. Y. Kim, Y. Zou, K. H. Cho, et al. 2006. Short-term feeding of baicalin inhibits age-associated NF-kappaB activation. *Mech Ageing Dev* 127 (9):719–25.

Kim, E. H., B. Shim, S. Kang, G. Jeong, J. S. Lee, Y. B. Yu, et al. 2009. Anti-inflammatory effects of *Scutellaria baicalensis* extract via suppression of immune modulators and MAP kinase signaling molecules. *J Ethnopharmacol* 126 (2):320–31.

Kim, J. M., H. A. Jung, J. S. Choi, B. S. Min, and N. G. Lee. 2010. Comparative analysis of the anti-inflammatory activity of Huang-lian extracts in lipopolysaccharide-stimulated RAW264.7 murine macrophage-like cells using oligonucleotide microarrays. *Arch Pharm Res* 33 (8):1149–57.

Ko, Y. J., J. S. Lee, B. C. Park, H. M. Shin, and J. A. Kim. 2007. Inhibitory effects of Zoagumhwan water extract and berberine on angiotensin II-induced monocyte chemoattractant protein (MCP)-1 expression and monocyte adhesion to endothelial cells. *Vascul Pharmacol* 47 (2-3):189–96.

Kong, W. J., H. Zhang, D. Q. Song, R. Xue, W. Zhao, J. Wei, et al. 2009. Berberine reduces insulin resistance through protein kinase C-dependent up-regulation of insulin receptor expression. *Metabolism* 58 (1):109–19.

Konner, A. C., and J. C. Bruning. 2010. Toll-like receptors: Linking inflammation to metabolism. *Trends in Endocrinology & Metabolism* 22 (1):16–23.

Lansky, E. P., and R. A. Newman. 2007. *Punica granatum* (pomegranate) and its potential for prevention and treatment of inflammation and cancer. *J Ethnopharmacology* 109 (2):177–206.

Lee, H., R. Kang, and Y. Yoon. 2010. SH21B, an anti-obesity herbal composition, inhibits fat accumulation in 3T3-L1 adipocytes and high fat diet-induced obese mice through the modulation of the adipogenesis pathway. *Journal of Ethnopharmacology* 127 (3):709–17.

Li, G. Q., V. Razmovski-Naumovski, B. Kimble, V. L. Qiao, W. Tongkao-on, B. L. Lin, et al. 2010a. Quality control methods for herbal medicines. In *Comprehensive Bioactive Natural Products – Quality Control & Standardization*, edited by V. K. Gupta, A. K. Verma, and S. Kaul. Houston, TX: Studium Press.

Li, G. Q., V. Razmovski-Naumovski, E. Omar, A. W. Teoh, M-K. Song, W. Tongkao-on, et al. 2010b. Evaluation of biological activity in quality control of herbal medicines. In *Comprehensive Bioactive Natural Products – Quality Control & Standardization*, edited by V. K. Gupta, A. K. Verma, and S. Kaul. Houston, TX: Studium Press.

Li, S. W., and J. Wu. 2011. Clinical observation on simple obesity of spleen deficiency and dampness excess treated by hour-prescription of point. *Zhongguo Zhen Jiu* 31 (2): 125–8.

Li, W. L., H. C. Zheng, J. Bukuru, and N. D. Kimpe. 2004. Natural medicines used in the traditional Chinese medical system for therapy of diabetes mellitus. *J Ethnopharmacology* 92:1–21.

Lin, Y. L., Z. K. Dai, R. J. Lin, K. S. Chu, I. J. Chen, J. R. Wu, et al. 2010. Baicalin, a flavonoid from *Scutellaria baicalensis Georgi*, activates large-conductance Ca2+-activated K+ channels via cyclic nucleotide-dependent protein kinases in mesenteric artery. *Phytomedicine* 17 (10):760–70.

Liu, T. H., and D. Y. Tang. 2000. Consideration and measurement for diabetes therapy in traditional Chinese medical system. *Chinese Journal of Information on Traditional Chinese Medicine* 7:8–9.

Mohan, M., H. Waghulde, and S. Kasture. 2010. Effect of pomegranate juice on angiotensin II-induced hypertension in diabetic Wistar rats. *Phytotherapy Research* 24 (S2):S196–S203.

Nagao, T., T. Hase, and I. Tokimitsu. 2007. A green tea extract high in catechins reduces body fat and cardiovascular risks in humans. *Obesity (Silver Spring)* 15 (6):1473–83.

Nagao, T., Y. Komine, S. Soga, S. Meguro, T. Hase, Y. Tanaka, and I. Tokimitsu. 2005. Ingestion of a tea rich in catechins leads to a reduction in body fat and malondialdehyde-modified LDL in men. *Am J Clin Nutr* 81 (1):122–9.

Omar, E. A., A. Kam, A. Alqahtani, K. M. Li, V. Razmovski-Naumovski, S. Nammi, et al. 2010. Herbal medicines and nutraceuticals for diabetic vascular complications: Mechanisms of action and bioactive phytochemicals. *Current Pharmaceutical Design* 16 (34):3776–807.

Panagiotakos, D. B., and E. Polychronopoulos. 2005. The role of Mediterranean diet in the epidemiology of metabolic syndrome: Converting epidemiology to clinical practice. *Lipids in Health & Disease* 4:7.

Ping, C., ed. 1997. *Concepts and Theories of Traditional Chinese Medicine*. Edited by L. Yubin and L. Chengcai. Vol. 2. Tokyo: Science Press.

Ramadan, G., N. M. El-Beih, and E. A. Abd El-Ghffar. 2009. Modulatory effects of black v. green tea aqueous extract on hyperglycaemia, hyperlipidaemia and liver dysfunction in diabetic and obese rat models. *Br J Nutr* 102 (11):1611–9.

Razmovski-Naumovski, V., W. Tongkao-on, B. Kimble, V. L. Qiao, B-L. Lin, K. M. Li, et al. 2010. Multiple chromatographic and chemometric methods for quality standardisation of Chinese herbal medicines. *World Sci & Tech* 12 (1):99–106.

Rock, W., M. Rosenblat, R. Miller-Lotan, A. P. Levy, M. Elias, and M. Aviram. 2008. Consumption of wonderful variety pomegranate juice and extract by diabetic patients increases paraoxonase 1 association with high-density lipoprotein and stimulates its catalytic activities. *J Agri & Food Chem* 56 (18):8704–8713.

Rosenblat, M., T. Hayek, and M. Aviram. 2006. Anti-oxidative effects of pomegranate juice (PJ) consumption by diabetic patients on serum and on macrophages. *Atherosclerosis* 187 (2):363–371.

Sasazuki, S., H. Kodama, K. Yoshimasu, Y. Liu, M. Washio, K. Tanaka, et al. 2000. Relation between green tea consumption and the severity of coronary atherosclerosis among Japanese men and women. *Ann Epidemiol* 10 (6):401–8.

Song, L. C., K. Z. Chen, and J. Y. Zhu. 1992. [The effect of Coptis chinensis on lipid peroxidation and antioxidases activity in rats]. *Zhongguo Zhong Xi Yi Jie He Za Zhi* 12 (7):421–3, 390.

Tan, Y., M. A. Kamal, Z. Z. Wang, W. Xiao, J. P. Seale, and X. Qu. 2011. Chinese herbal extracts (SK0506) as a potential candidate for the therapy of the metabolic syndrome. *Clin Sci (Lond)* 120 (7):297–305.

Tipoe, G. L., T. M. Leung, M. W. Hung, and M. L. Fung. 2007. Green tea polyphenols as an anti-oxidant and anti-inflammatory agent for cardiovascular protection. *Cardiovasc Hematol Disord Drug Targets* 7 (2):135–44.

Todoric, J., M. Loffler, J. Huber, M. Bilban, M. Reimers, A. Kadl, et al. 2006. Adipose tissue inflammation induced by high-fat diet in obese diabetic mice is prevented by n-3 polyunsaturated fatty acids. *Diabetologia* 49 (9):2109–19.

Tsuneki, H., M. Ishizuka, M. Terasawa, J. B. Wu, T. Sasaoka, and I. Kimura. 2004. Effect of green tea on blood glucose levels and serum proteomic patterns in diabetic (db/db) mice and on glucose metabolism in healthy humans. *BMC Pharmacol* 4:18.

Waisundara, V. Y., S. Y. Siu, A. Hsu, D. Huang, and B. K. H. Tan. 2011. Baicalin upregulates the genetic expression of antioxidant enzymes in type-2 diabetic Goto-Kakizaki rats. *Life Sciences* 88 (23-24):1016–1025.

Wang, C., J. Li, X. Lv, M. Zhang, Y. Song, L. Chen, and Y. Liu. 2009. Ameliorative effect of berberine on endothelial dysfunction in diabetic rats induced by high-fat diet and streptozotocin. *Euro J Pharmacol* 620 (1-3):131–137.

Wang, Y., T. Campbell, B. Perry, C. Beaurepaire, and L. Qin. 2011. Hypoglycemic and insulin-sensitizing effects of berberine in high-fat diet- and streptozotocin-induced diabetic rats. *Metabolism* 60 (2):298–305.

World Health Organisation. Obesity and overweight. Fact sheet number 311. 2006. http://www.who.int/mediacentre/factsheets/fs311/en/index.html

Wu, L. Y., C. C. Juan, L. T. Ho, Y. P. Hsu, and L. S. Hwang. 2004. Effect of green tea supplementation on insulin sensitivity in Sprague-Dawley rats. *J Agric Food Chem* 52 (3):643–8.

Xie, N., W. Zhang, J. Li, H. Liang, H. Zhou, W. Duan, et al. 2011. Alpha-linolenic acid intake attenuates myocardial ischemia/reperfusion injury through anti-inflammatory and anti-oxidative stress effects in diabetic but not normal rats. *Arch Med Res* 42 (3):171–81.

Xie, W., and L. Du. 2010. Diabetes is an inflammatory disease: Evidence from traditional Chinese medicines. *Diabetes, Obesity & Metabolism* 13 (4):289–301.

Yan, H., X. Sun, S. Sun, S. Wang, J. Zhang, R. Wang, et al. 2011. Anti-ultraviolet radiation effects of *Coptis chinensis* and *Phellodendron amurense* glycans by immunomodulating and inhibiting oxidative injury. *Int J Biological Macromol* 48 (5):720–25.

Yang, Y. C., F. H. Lu, J. S. Wu, C. H. Wu, and C. J. Chang. 2004. The protective effect of habitual tea consumption on hypertension. *Arch Intern Med* 164 (14):1534–40.

Yaqoob, P., H. S. Pala, M. Cortina-Borja, E. A. Newsholme, and P. C. Calder. 2000. Encapsulated fish oil enriched in alpha-tocopherol alters plasma phospholipid and mononuclear cell fatty acid compositions but not mononuclear cell functions. *Eur J Clin Invest* 30 (3):260–74.

Yin, J., H. Zhang, and J. Ye. 2008. Traditional Chinese medicine in treatment of metabolic syndrome. *Endocr, Metabo & Immu Disord Drug Targets* 8 (2):99–111.

Yoon, S. B., Y. J. Lee, S. K. Park, H. C. Kim, H. Bae, H. M. Kim, et al. 2009. Anti-inflammatory effects of *Scutellaria baicalensis* water extract on LPS-activated RAW 264.7 macrophages. *J Ethnopharmacol* 125 (2):286–90.

Zhang, E., J. Zhang, & Z. Liu. 1990. *Clinic of Traditional Chinese Medicine (1)* Edited by Zhang, E. Shanghai, House of Shanghai University of Traditional Chinese Medicine.

Zhang, J. C., and K. Y. Chen. 2004. Metabolic syndrome and the treatment by combining of Western and Chinese medicines. *Journal of Zhong Xi Jie He* 24 (11):1029–32.

Zhang, Q., X. L. Piao, X. S. Piao, T. Lu, D. Wang, and S. W. Kim. 2011. Preventive effect of *Coptis chinensis* and berberine on intestinal injury in rats challenged with lipopolysaccharides. *Food and Chemical Toxicology* 49 (1):61–9.

Zhao, Y. T., L. Shao, L. L. Teng, B. Hu, Y. Luo, X. Yu, et al. 2009. Effects of n-3 polyunsaturated fatty acid therapy on plasma inflammatory markers and N-terminal pro-brain natriuretic peptide in elderly patients with chronic heart failure. *J Int Med Res* 37 (6):1831–41.

Zhong, L., J. K. Furne, and M. D. Levitt. 2006. An extract of black, green, and mulberry teas causes malabsorption of carbohydrate but not of triacylglycerol in healthy volunteers. *Am J Clin Nutr* 84 (3):551–5.

25 Epigenetics and Nutriepigenomics in Chronic Inflammatory Lung Diseases

Nutritional and Therapeutic Interventions

Saravanan Rajendrasozhan
University of Hail
Hail, Saudi Arabia

Isaac K. Sundar and Irfan Rahman
University of Rochester Medical Center
Rochester, New York

CONTENTS

25.1 LUNG INFLAMMATION AND PATHOGENESIS OF CHRONIC LUNG DISEASES (ASTHMA AND CHRONIC OBSTRUCTIVE PULMONARY DISEASE)

Chronic obstructive pulmonary disease (COPD) and asthma are the most prevalent lung diseases characterized by progressive/persistent lung inflammation. The inflammatory pattern is distinct in each disease; however, oxidants play a potential role in increasing the pro-inflammatory gene expression and development of the disease. Oxidants generated from inhaled noxious agents, such as cigarette smoke, allergens, or environmental pollutants, disrupt the oxidant-antioxidant balance, which results in oxidative stress. Increased level of oxidative stress/damage and decreased antioxidant defense are found in the lungs of asthmatics and patients with COPD. Oxidizing environments induce activation of transcription factors, which regulate inflammatory molecules, such as nuclear factor-kappaB (NF-κB) and activator protein-1 (AP-1) via several redox sensitive kinases. Activated transcription factors are translocated to the nucleus and bind to specific promoter regions of DNA to induce transcription of pro-inflammatory genes (Rajendrasozhan et al. 2008a).

Posttranslational modifications of transcription factors, such as phosphorylation and acetylation, can facilitate their DNA binding ability. As the DNA is present in the form of chromatin (DNA-histone protein complex) in eukaryotic cells, the chromatin structure is critical for the access of transcription factors to DNA. Tight coiling of DNA around core histones (H2A, H2B, H3, and H4) renders it inaccessible for binding of transcription factors to DNA, which results in inhibition of gene transcription (gene silencing). Uncoiling stretches the DNA into a linear structure and exposes the gene promoters for access of transcription factors. This leads to the formation of transcription co-activator assembly and gene transcription. Thus, regulation of coiling and uncoiling of DNA plays a pivotal role in gene expression (Figure 25.1). This chapter focuses on epigenetic changes that modulate chromatin structure and inflammatory conditions in the lungs (Rajendrasozhan et al. 2008a; Rajendrasozhan et al. 2009). It also focuses on how natural dietary products influence these phenomena that are defined as nutriepigenomics under a broad category of nutrigenomics.

25.2 EPIGENETIC/EPIGENOMIC CHANGES

Potentially reversible modifications in gene expression caused by heritable changes in chromatin structure that occur without any changes in genetic code are termed as epigenetics. Epigenetic events can alter the chromatin structure by modifying DNA structure (via DNA methylation) or highly conserved core histone proteins (via posttranslational modifications such as acetylation, methylation, phosphorylation, ubiquitination, SUMOylation, and poly-ADP-ribosylation). Histone acetylation and histone/DNA methylation are relatively well-studied epigenetic modifications in chronic inflammatory lung diseases (Rajendrasozhan et al. 2009).

Acetylation/deacetylation of histone tails at the ε-amino groups of lysine residues is directly linked to transcriptional activity. Acetylation neutralizes the basic charge of the histone tails, and thereby reduces the histone-DNA as well as histone-histone interactions. Moreover, acetylated histones (H3 and H4) form a molecular tag for the recruitment of chromatin remodeling enzymes, such as CREB-binding protein (CBP)/p300, Brg1, and Brg3, which allows chromatin unwinding and the recruitment of other transcription factors. Thus, acetylation of histones, by histone acetyltransferases (HATs), facilitates gene expression. Deacetylation of histones, by histone deacetylases (HDACs), suppresses gene expression by facilitating DNA rewinding (tight coiling of DNA on histones).

Methylation of histones, by histone methyltransferases (HMTs), provides a binding site or interacting domain for other regulatory proteins to be recruited on promoters. Interestingly, depending

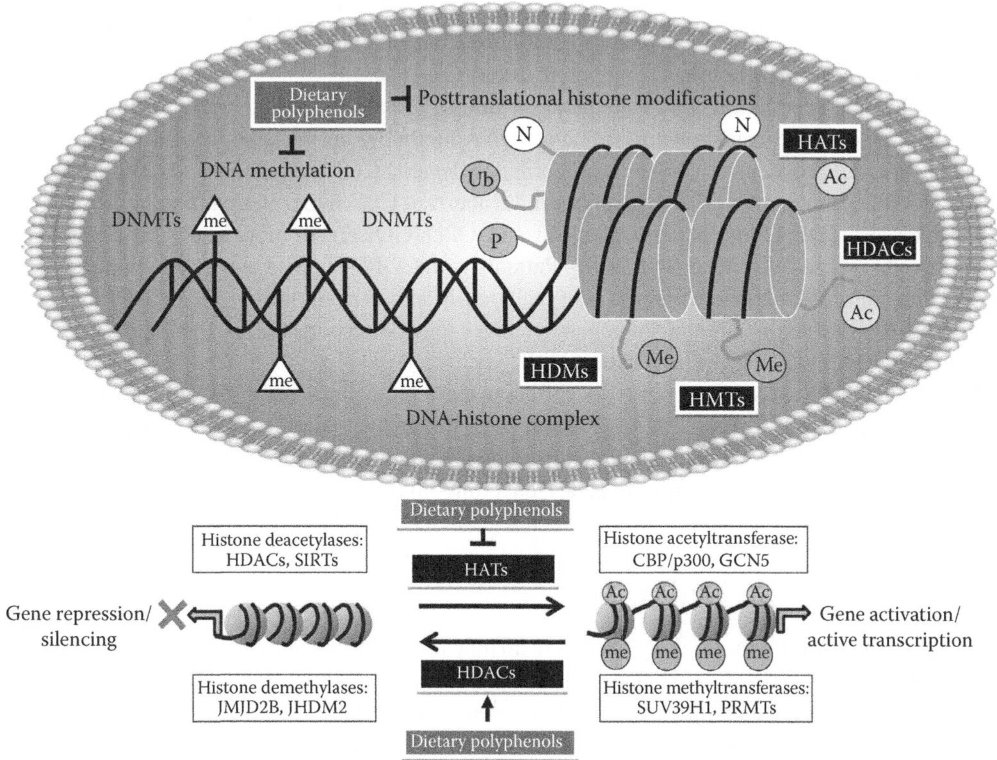

FIGURE 25.1 (see color insert) Regulatory role of dietary polyphenols in epigenetic modifications and gene expression. Epigenetic modifications by chromatin modification enzymes play a vital role in regulation of gene expression. Unwinding and rewinding of DNA is regulated by epigenetic alterations, such as histone acetylation/deacetylation and histone methylation/demethylation. This includes histone acetylation by histone acetyltransferases (HATs), histone deacetylation by histone deacetylases (HDACs), histone methylation by histone methyltransferases (HMTs), DNA methylation by DNA methyltransferases (DNMTs), histone demethylation by histone demethylases (HDMs), and histone phosphorylation and ubiquitination by kinases and ubiqutination enzymes, respectively. These epigenetic modifications result in conformation changes in the chromatin structure that can lead to alterations in DNA accessibility for transcription factors, coactivators, and polymerases, thereby resulting in either gene expression (transcriptional activation) or gene repression (silencing). Dietary polyphenols can affect these epigenetic chromatin modification enzymes specifically to modulate several cellular and molecular events. Ac, Acetylation; Me, Methylation; Ub, Ubiquitination; P, Phosphorylation.

on the methylated lysine residues, the histone methylation can signal either gene expression or repression (Gosden and Feinberg 2007). Methylation of DNA in CpG islands is also an epigenetic heritable event regulated by DNA methyltransferases. DNA methylation represses gene expression by the recruitment of methyl-CpG-binding protein MeCP2, which in turn recruits HDACs (Rajendrasozhan et al. 2009).

25.2.1 Epigenetic Changes in COPD

COPD is characterized by airflow limitation in the lungs that is not fully reversible and associated with progressive inflammatory condition. The molecular mechanism of COPD pathogenesis is considered to be mediated by epigenetic chromatin changes. Lung tissues of current smokers and ex-smokers with COPD showed an increased acetylation in histones H4 and H3, respectively, as compared to nonsmokers. Increased H4 acetylation is correlated with an increased expression of inflammatory protein (p40 subunit of interleukin-12) (Szulakowski et al. 2006). Cigarette smoke exposure, a major etiological factor for COPD, induces histone H3 phosphorylation (H3S10) and acetylation (H3K9 and H4K12), which are associated with increased pro-inflammatory cytokine/chemokine release in mouse lungs (Yang et al. 2008).

Unlike histone acetylation, DNA methylation in COPD is complicated and poorly understood. Cigarette smoking has been shown to influence DNA methylation. Variable DNA methylation patterns may be useful to identify smokers susceptible to develop COPD (DeMeo et al. 2009). An international COPD genetic network study reported the aberrant CpG hypermethylation in patients with COPD. Moreover, DNA from patients with stage 4 COPD (GOLD) are more likely to be methylated as compared to stage 2 (DeMeo et al. 2010). Patients with COPD showed elevated hypomethylation of CpG sites in tumor suppressor genes that include p16 and Wnt inhibitory factor 1 (Suzuki et al. 2010). These heritable/imprintable epigenetic changes may be responsible for the increased risk for patients with COPD to develop lung cancer.

The inflammatory process is markedly different between asthma and COPD, but clinical signs overlap in many cases. Asthma and COPD can be distinguished based on their DNA methylation profiles in bronchial epithelial cells obtained from brushings of small airways. Both diseases showed distinct DNA methylation profiles in genes encoding matrix metalloproteinases (MMPs), pro-inflammatory cytokines, and chemokines, which regulate inflammatory processes (Vucic et al. 2010).

25.2.2 Epigenetic Changes in Asthma

Bronchial asthma is characterized by reversible airflow obstruction due to airway inflammation and hyperresponsiveness. Aberrant epigenetic regulation, such as histone acetylation and DNA methylation, are involved in the development and progression of asthma. Asthmatic lungs show hypomethylation of genes involved in lymphocyte proliferation, leukocyte activation, cytokine biosynthesis, cytokine secretion, immune responses, inflammation and immunoglobulin binding, and hypermethylation of genes involved in ectoderm development, hemostasis, wound healing, calcium ion binding, and oxidoreductase activity (Cheong et al. 2011). Exposure to high levels of airborne pollutants could cause hypermethylation of the forkhead box transcription factor 3 (FOXO3) locus, which results in impaired regulatory T-cell function and increased asthma risk and morbidity (Nadeau et al. 2010).

Asthma is the most common chronic disease in both children and adults. Interestingly, several groups have shown that maternal dietary supplementation of methyl donors (diets rich in folate, choline, betaine) or exposure to airborne pollutants can have a profound impact on the phenotype of the offspring through epigenetic reprogramming (Hollingsworth et al. 2008; Perera et al. 2009). Transplacental exposure to polycyclic aromatic hydrocarbons (abundant in high-traffic areas) induces DNA methylation in acyl-CoA synthetase long-chain family member 3 gene that might be mechanistically related to childhood asthma or airway inflammation (Perera et al. 2009). Prenatal dietary exposure to high-methyl donor diet, as mentioned earlier, has been shown to induce airway hyperresponsiveness in mice due to hypermethylation of various promoters (Hollingsworth et al. 2008). In addition, histone acetylation is significantly elevated in asthmatic lungs (Ito et al. 2002), which plays a pivotal role in sustained pro-inflammatory gene expression as seen in asthmatics.

25.3 MOLECULAR EPIGENOMIC TARGETS FOR THERAPEUTIC INTERVENTIONS

Epigenetic changes in chronic lung inflammatory diseases can be maintained through cell division and inherited across generations, although there is no change in genetic code. The epigenetic changes and chromatin modifications are reversible and can be regulated by specific enzymes/proteins involved in the process. Emerging epigenetic paradigm in chronic and allergic airways diseases, such as asthma and COPD, include epigenetic marks that affect gene expression in the lung, which will provide novel insight regarding these complex epigenetic mechanisms and pathways that harness the dynamic biology of epigenetic chromatin remodeling in the lung. Understanding these epigenetic mechanisms/pathways may lead to the development of novel diagnostic and epigenetics-based therapeutic approaches to treat and manage chronic lung diseases (Martino and Prescott 2011; Yang and Schwartz 2011).

25.3.1 Histone Acetyltransferases

Histone acetyltransferases (HATs) catalyze the acetylation of lysine residues in histone amino-terminal domains, which has been positively linked to transcriptional activation. It also acetylates a wide range of non-histone proteins, including transcription factors, in which the effect of acetylation depends on the functional domain. Several HATs or proteins with intrinsic HAT activity have been identified. CBP and p300 are ubiquitously expressed HATs with high structural homology. CBP and p300 are of particular interest as they are the key co-activators in regulating pro-inflammatory gene transcription. The co-activator function of CBP/p300 is regulated via phosphorylation by p38 mitogen-activated protein kinase, which is activated in oxidative stress conditions.

The activities of HATs are shown to be affected in inflammatory conditions. Elevated HAT activity is found in asthmatic airways as compared to those in normal subjects (Ito et al. 2002). Interestingly, lung tissues (bronchial biopsies) of asthmatics showed increased HAT activity, without any changes in protein levels or site of expression of CBP/p300 and p300/CBP-associated factor (PCAF) (Ito et al. 2002). In contrast with asthmatics, lungs of patients with COPD showed no significant change in HAT activity (Ito et al. 2005). These observations suggest that increased histone acetylation and pro-inflammatory gene expression seen in asthma and COPD is mediated by different molecular mechanisms.

Corticosteroids reduces the nuclear HAT activity in patients with severe asthma; however, they are not effective in reducing pro-inflammatory response and improving lung function (Hew et al. 2006). The corticosteroid insensitivity to inhibit inflammatory response is due to the reduction in HDAC2 activity. This observation suggests the use of a therapeutic drug with a multiple mode of action or a combination of multiple drugs (HAT inhibitors and HDAC activators) may be important in treatment of COPD and asthma.

25.3.2 Histone Deacetylases

Histone deacetylases (HDACs) catalyze the removal of the acetyl group from lysine residues of the core histone tails and non-histone proteins, such as transcription factors (e.g., NF-κB and AP-1). Counter-regulatory to acetylation, HDAC-mediated deacetylation is linked to transcriptional repression. To date, 18 different HDACs have been identified in humans and mice. The role of HDAC2 and sirtuin 1 (SIRT1) is more extensively studied in lung inflammatory processes than other HDACs.

HDAC2 is a class I HDAC and is primarily localized in the nucleus to regulate gene expression. Monocyte-derived macrophages, the main orchestrators of chronic inflammatory response, show a decreased HDAC2 level in smokers and patients with COPD as compared to healthy non-smokers (Mercado et al. 2011). Peripheral lung tissues of patients with COPD showed decreased levels of both mRNA and protein expression of HDAC2, which subsequently leads to a reduction in HDAC2 activity. The extent of reduction is associated with increased histone H4 acetylation and pro-inflammatory gene transcription (interleukin-8 mRNA) in different stages of COPD severity (GOLD stages 1 to 4) (Ito et al. 2005).

Asthmatic lungs show a distinct pattern of HDAC level/activity depending on disease severity, cell type and heterogeneity of the disease phenotype. Reduction in HDAC2 activity was observed in peripheral monocytes derived from patients with severe asthma as compared to non-severe asthma (Hew et al. 2006). Interestingly, mild intermittent asthmatics showed reduction in HDAC2 activity and HDAC2 protein expression in alveolar macrophages and bronchial biopsies compared to normal subjects (Ito et al. 2002; Cosio et al. 2004). In contrast, no significant reduction in HDAC2 activity was observed in lungs of mild asthmatics (Ito et al. 2005). Decreased HDAC2 activity is directly related to increased acetylation of histone proteins and transcription factors, such as NF-κB, which lead to increased pro-inflammatory gene expression. The reduction in HDAC2 is induced by cigarette smoke/oxidant-mediated phosphorylation of serine/threonine residues by a protein kinase CK2-mediated mechanism, which resulted in ubiquitin-proteasome-dependent HDAC2 degradation (Adenuga et al. 2009).

The reduction in HDAC2 is directly correlated with corticosteroid insensitivity seen in severe asthmatics and patients with COPD because corticosteroids recruit HDAC2 to switch off pro-inflammatory gene expression. Reduced HDAC2 activity in COPD is correlated with acetylation of nuclear factor erythroid-related factor 2 (Nrf2, which regulates antioxidant gene expression) and the subsequent loss of Nrf2 stability (Mercado et al. 2011). Interestingly, Nrf2 deficiency leads to reduced HDAC2 levels and deacetylase activity in lungs. The dysregulation of HDAC2-Nrf2 balance results in impaired antioxidant defenses and steroid insensitivity that worsen the inflammatory condition (Adenuga et al. 2010). This suggests the use of HDAC2 as a potential target to increase corticosteroid sensitivity and manage chronic inflammatory lung diseases.

SIRT1 is a metabolic NAD^+-dependent class III HDAC. It regulates a range of biological processes including stress resistance, inflammation, apoptosis, autophagy, and aging by deacetylating histone H3K56, NF-κB, AP-1, p53, FOXO3, and poly(ADP-ribose)-polymerase-1 (PARP-1). Smokers and patients with COPD show a decreased nuclear SIRT1 level in peripheral lungs as compared to nonsmokers (Rajendrasozhan et al. 2008b). Exposure to cigarette smoke, oxidants, or aldehydes induces posttranslational modifications, such as formation of nitrotyrosine, aldehyde carbonyl adducts, and alkylation of SIRT1, which results in ubiquitin-proteasome-dependent degradation and decreased enzymatic activity (Rajendrasozhan et al. 2008b; Caito et al. 2010). Loss of SIRT1 is associated with increased acetylation (K310) of the RelA/p65 subunit of NF-κB and subsequently increased inflammation (Rajendrasozhan et al. 2008b). SIRT1 deficiency also increased the autophagy of lung cells via the PARP-1-dependent mechanism, which is involved in the pathogenesis of cigarette smoke-mediated age-related diseases, such as emphysema and COPD (Hwang et al. 2010). An anti-inflammatory effect of SIRT1 overexpression against cigarette smoke exposure is shown (Rajendrasozhan et al. 2008b), suggests the potential use of SIRT1 activators to treat chronic inflammatory lung diseases.

25.3.3 Histone Methyltransferases

Histone methyltransferases (HMTs) are enzymes that cause histone methylation either at lysine (lysine methyltransferase) or arginine (arginine methyltransferase) residues. HMTs are deregulated in several chronic lung diseases, thereby affecting global methylation. Histones H3 (H3K4, H3K9, and H3K27) and H4 (H4K20) are frequently and preferentially methylated as mono-, di-, or tri-methylated histone H3 and histone H4 (Seligson et al. 2009). Interestingly, histone methylation is associated with both gene activation and repression, depending on the lysine residue that is methylated in the histones (H3 and H4). For instance, methylation at H3K4, H3K36, and H3K79 is linked to gene activation, whereas H3K9, H3K27, and H3K20 methylation is associated with gene repression (Gosden and Feinberg 2007). However, it remains to be known whether these modifications occur on various pro- and anti-inflammatory, pro- and anti-apoptotic, protease and anti-protease, antimicrobial and antioxidant genes in response to cigarette smoke in lungs of patients with asthma and COPD. Furthermore, it is not known whether dietary natural products would have any influence on these modifications that would either occur *in utero* or later in adult life.

25.3.4 Histone Demethylases

Histone demethylases (HDMs) catalyze the removal of the methyl group from lysine or arginine residues of histones. The HDMs are classified into two types, that is, lysine-specific demethylase 1 (LSD1) and Jumonji C (JmjC) domain family proteins, which are involved in regulation of gene expression. Identification of several HDMs suggests their dynamic role in regulation of biological processes including histone methylation (Lim et al. 2010). The role of HDMs in asthma and COPD is not known. Depletion of H3K4 demethylase JMJD1A sufficiently reduces tumor growth *in vivo*, demonstrating its role in regulating histone methylation in response to hypoxia. Under hypoxic conditions, the increased levels of H3K4me3 in human alveolar A549 and bronchial Beas-2B epithelial cell lines occur because of inhibition of the demethylation process, particularly demethylase (JARID1A). Knockdown of JARID1A induces H3K4me3 at the promoters of heme oxygenase-1 (HMOX1) and decay-accelerating factor (DAF) genes in Beas-2B. Therefore, hypoxia (which occurs in the lungs of patients with COPD and cancer) targets JARID1A demethylase, which induces H3K4me3 both globally and at gene-specific promoters (Zhou et al. 2010). This has implications in epigenetic targeting of cigarette smoke-induced COPD and lung cancer.

25.3.5 NF-κB

NF-κB is a pleiotropic transcription factor involved in the regulation of inflammatory and immune responses. Five family members of NF-κB exist in mammals: NF-κB1 (p50 and the precursor p105), NF-κB2 (p52 and the precursor p100), RelA (p65), RelB, and c-Rel that exist in cells as hetero- or homodimers. NF-κB regulates the expression of cytokines, chemokines, and inducible enzymes that are involved in the inflammatory processes. NF-κB is activated via phospho-acetylation and causes a pro-inflammatory response in lungs of asthmatics and patients with COPD (Edwards et al. 2009). The RelA/p65-p50 heterodimer, the most abundant form of NF-κB, is released from the inhibitor kappaB (IκB) by the IκB kinase (IKK)-mediated mechanism. Subsequently, RelA/p65 is phosphorylated at S276 and S311 by protein kinase A (PKA), mitogen- and stress-activated protein kinases 1 and 2 (MSK1 and MSK2), and protein kinase C zeta (PKCζ). This phosphorylation facilitates its interaction with p300/CBP, which acetylates RelA/p65 at K218, K211, and K310. Acetylation of RelA/p65 at K218/K221 increases the DNA binding, while acetylation at K310 increases the transactivation potential. However, post-translational modifications of chromatin are critical for the binding of RelA/p65 with DNA. Unwinding of DNA, primarily by acetylation of histones, increases the accessibility of NF-κB to pro-inflammatory gene promoters.

Impaired HAT-HDAC balance mediates histone acetylation and DNA unwinding. In addition, NF-κB plays a pivotal role in histone acetylation as it recruits transcription co-activators, including p300/CBP, to the chromatin. All four histone proteins (H2A, H2B, H3, and H4) can be acetylated by p300/CBP in the nucleosomes in concert with PCAF. NF-κB also has intrinsic HAT activity. Recruitment of NF-κB in the chromatin is accompanied by the ordered elevation of histone H4 and H3 acetylation. Moreover, NF-κB-induced acetylation occurs preferentially on histone H4 at K8 and K12 at NF-κB-responsive regulatory elements, which leads to transcriptional activation. Thus, pro-inflammatory gene expression requires coordination of transcription factors, co-activators, and chromatin remodeling. Because of the central role of NF-κB in inflammatory conditions, it is now considered as the primary target in therapeutic intervention of chronic inflammatory lung diseases.

25.3.6 MicroRNAs

MicroRNAs (miRNAs) are small noncoding RNAs that regulate mRNA stability and/or translation, which play an important role in a variety of biological processes ranging from developmental patterning and stress responses to epigenetic inheritance (Oglesby et al. 2010). MicroRNAs bind to target sites in the 3' UTR of mRNAs, causing posttranscriptional repression or degradation of mRNAs. Schembri et al. (2009) found that microRNA-218 modulates smoking-induced gene expression changes in human airway epithelium, and plays a role in the host response to environmental exposures and in pathogenesis of smoking-related lung disease. However, miRNA expression is not altered in response to anti-inflammatory corticosteroid (budesonide) treatment in mild asthmatics (Schembri et al. 2009). Furthermore, certain microRNAs (e.g., microRNA-18 and microRNA-124a) downregulate glucocorticoid receptor levels (Uchida et al. 2008; Kotani et al. 2009), which implicates the involvement of microRNA signaling in glucocorticoid responsiveness. However, there is no information available regarding the involvement of specific microRNA in steroid sensitivity. Future studies on severe asthmatics with steroid resistance may reveal changes in miRNA expression (Schembri et al. 2009; Williams et al. 2009).

Recent studies have also demonstrated that miRNA expression levels are downregulated in lungs of mice and rats exposed to cigarette smoke, which correlates with the miRNA expression changes observed in human airway epithelium (Izzotti et al. 2009; Izzotti et al. 2010). Pottelberge et al. (2011) identified 8 miRNAs, including let-7c and miR-125b, which are expressed lower in sputum cells of patients with COPD who are current smokers compared to never smokers without airflow limitations. TNFRII was identified as a predicted target of let-7c, which is shown to be elevated in the sputum of current smokers with COPD (Pottelberge et al. 2011). Thus, it is possible that miRNA alterations have potential clinical and biological relevance in phenotypic characterization and disease progression of smoking-related lung diseases and possibly in asthma (Oglesby et al. 2010; Pottelberge et al. 2011).

25.4 NUTRIEPIGENOMICS

Food and food components (phytochemicals) are of considerable interest in the management of various chronic lung diseases because of their presumed safety and traditional use. Natural product–containing compounds—such as curcumin, resveratrol, catechins, garcinol, and anacardic acid—regulate epigenetic programming termed nutriepigenomics (Figure 25.2). Dietary intake of

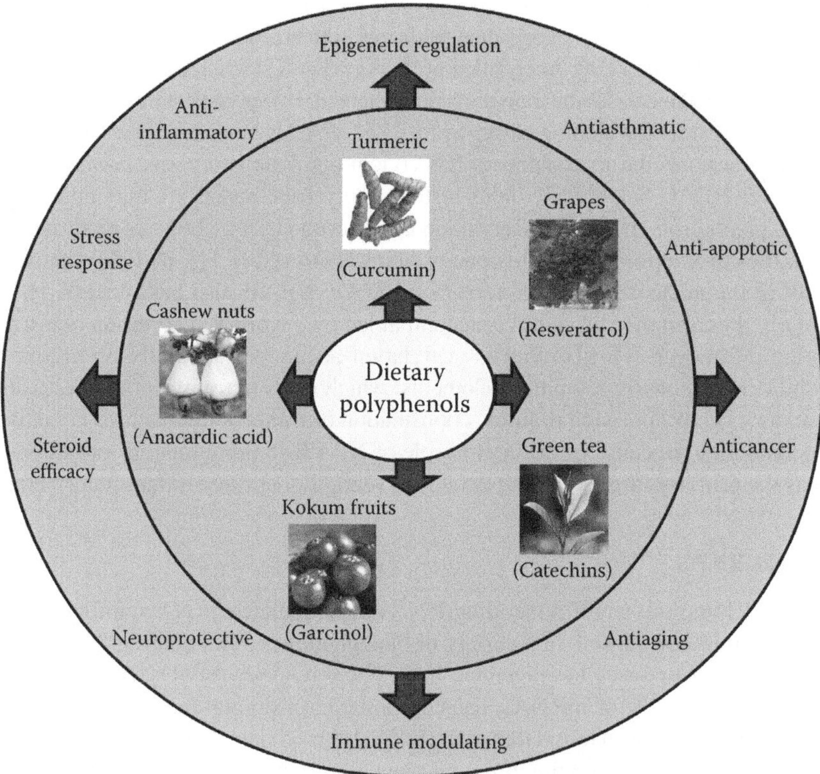

FIGURE 25.2 (see color insert) Dietary polyphenols from different plant sources and their molecular nutriepigenomics functions. Plant-derived polyphenols attenuate epigenetic alterations involved in several cellular processes, such as anti-inflammatory, immune modulation, reversal of steroid resistance, as well as act as antiasthmatic, anticancer, and antiaging agents. The dietary polyphenols obtained from each of these plants are given in parentheses.

fruits and vitamin E is correlated with decreased severity of COPD and mortality (Walda et al. 2002). Consumption of tea, apples, and other food products that contain catechin, flavonol, and flavon (polyphenols, 58 mg/day) has been reported to improve pulmonary function and reduce COPD symptoms, such as chronic cough and breathlessness (Tabak et al. 2001). Intake of polyphenols, such as quercetin, naringenin, and hesperetin is negatively correlated with the incidence of asthma (Knekt et al. 2002). These studies indicate that natural products have significant therapeutic potential in the management of chronic inflammatory diseases, including COPD and asthma (Mathers et al. 2010; vel Szic et al. 2010). Further research on anti-inflammatory effects of phytochemicals, especially on gene expression/chromatin remodeling, may also provide potential therapeutic strategies for the treatment and management of lung inflammation in chronic lung diseases.

25.4.1 Curcumin

Curcumin (diferuloylmethane) is the major constituent in the rhizomes of *Curcuma longa*, which is known as turmeric and Indian saffron. It is used as a spice or coloring agent in food preparations and has a long history of use to treat a wide variety of inflammatory conditions. Accumulating experimental evidence suggests that curcumin is a highly pleiotropic molecule capable of interacting with numerous molecular targets involved in inflammation, such as transcription factors, inflammatory cytokines, protein kinases, and other enzymes involved in epigenetic regulation. While the mechanism of action of curcumin is complicated and well studied, curcumin-mediated epigenetic regulation via its anti-inflammatory activity is less studied.

Curcumin regulates the epigenetic silencing of pro-inflammatory genes by shifting the histone acetylation-deacetylation balance toward the deacetylation side. It is a potent p300/CBP-specific HAT inhibitor, whereas it does not affect the activities of PCAF and methyltransferase (Balasubramanyam et al. 2004). Curcumin-mediated HAT inhibition is closely correlated with its ability to act as a Michael reaction acceptor due to the presence of alpha, beta-unsaturated carbonyl groups in the side chain (Marcu et al. 2006). Presumably, curcumin binds not only to the active site, but to some other specific site in p300/CBP that leads to conformational changes, resulting in a decreased binding efficiency of both histones and acetyl CoA to the enzyme (Balasubramanyam et al. 2004). It can effectively repress the acetylation of histone/non-histone proteins and HAT-dependent chromatin transcription (Balasubramanyam et al. 2004).

In addition to the inhibition of p300/CBP, curcumin can also induce HDAC2 levels, which is associated with decreased NF-κB-DNA binding and inhibition of pro-inflammatory cytokine release (Yun et al. 2011). This is supported by the finding that cigarette smoke/oxidant-mediated reduction in HDAC2 protein level and activity was attenuated by curcumin treatment in monocytes and macrophages (Meja et al. 2008). Curcumin-mediated restoration of HDAC2 is associated with increased corticosteroid sensitivity in response to cigarette smoke exposure because corticosteroid-mediated repression of pro-inflammatory gene expression that occurs via HDAC2 recruitment (Meja et al. 2008).

Curcumin has the ability to regulate microRNA expression, thereby impacting signaling pathway and cellular function (Li et al. 2010). However, it is unknown whether the regulation of microRNA by curcumin contributes to its ability to reverse steroid resistance. Curcumin also regulates the transcription factor activity to mitigate pro-inflammatory gene expression. It inhibits cigarette smoke–induced activation of NF-κB through inhibition of IKK, which is involved in degradation of inhibitory protein κB. By inhibiting the activation, curcumin indirectly prevents the NF-κB-DNA binding, co-activator assembly on chromatin, and thereby prevents the NF-κB-mediated epigenetic changes and pro-inflammatory gene expression. Curcumin has been shown to inhibit CS-induced lung inflammatory cell influx and COPD-like airway inflammation in mice (Moghaddam et al. 2009; Suzuki et al. 2009).

Antioxidant activity of curcumin plays a complementary role to regulate epigenetic modifications by reducing the oxidative modifications on epigenetic enzymes. Curcumin exhibits potent antioxidant activity by quenching free radicals and activating Nrf2-mediated expression of antioxidant genes, such as heme oxygenase-1, glutathione peroxidase, modulatory subunit of glutamylcysteine ligase, and NAD(P)H quinone oxidoreductase 1 (Goel and Aggarwal 2010). The phenolic hydroxyl groups are important for antioxidant activities of curcumin. Oral administration of curcumin induced the expression of Nrf2-regulated antioxidants and attenuated CS-induced pulmonary inflammation and emphysema in mice (Suzuki et al. 2009).

Human clinical trials indicated that curcumin is safe even at a dose of 12 g/day. Few clinical studies have shown that oral intake of curcumin has a protective effect in inflammatory diseases (White and Judkins 2011). However, the major obstacle for the use of curcumin as a therapeutic drug is its low bioavailability. Poor absorption due to its hydrophobic nature, rapid metabolism, such as conjugation/glucuronidation, and rapid systemic clearance contribute to the low plasma and tissue levels of curcumin upon administration. Multiple approaches are being sought to increase the bioavailability of curcumin that include coadministration with an adjuvant (like piperine) to delay metabolic clearance and improve formulations for curcumin delivery, such as liposome, nanoparticles, and phospholipid complex. Enhanced bioavailability of curcumin, and clinical trials in patients with asthma or COPD, are likely to address its potential therapeutic benefit for the intervention of lung inflammation in chronic lung diseases.

25.4.2 Resveratrol

Resveratrol (a phytoalexin), found in grapes, wine, peanuts, and soy, possesses diverse pharmacologic properties, such as anticancer, anti-inflammatory, neuroprotective, and antiaging. The beneficial effects of resveratrol are mediated primarily by SIRT1 activation. Resveratrol increases SIRT1 activity by lowering the Km value for both acetylated substrate and NAD^+ (Howitz et al. 2003).

The anti-inflammatory activity of resveratrol has been reported against cigarette smoke exposure in cells in culture. Resveratrol inhibits NF-κB activation and expression of pro-inflammatory cytokines (Yang et al. 2007; Csiszar et al. 2008), via SIRT1-mediated deacetylation of RelA/p65 subunit of NF-κB at K310, which results in transcriptional repression of pro-inflammatory genes (Yang et al. 2007). Interestingly, the anti-inflammatory activity of resveratrol was abolished by knockdown of SIRT1, suggesting that resveratrol exerts its activity through a SIRT1-dependent mechanism (Csiszar et al. 2008). Resveratrol also induces the association between SIRT1 and p300, which leads to inhibition of NF-κB acetylation and subsequent nuclear translocation (Shakibaei et al. 2011). Radical-scavenging and Nrf2-activating activities of resveratrol are also important, because resveratrol-induced downregulation of pro-inflammatory cytokines in airway inflammation, which is shown to be mediated by NF-κB-independent mechanism (Kode et al. 2008).

Resveratrol-induced activation of SIRT1 is associated with decreased expression of MMP-9 and autophagy, which play a pivotal role in COPD pathogenesis (Hwang et al. 2010; Lee and Kim 2011). Resveratrol mitigates the release of steroid-resistant inflammatory cytokines and MMP-9 from alveolar macrophage isolated from patients with COPD (Knobloch et al. 2011). The anti-inflammatory activity of resveratrol is more potent than corticosteroids, and therefore resveratrol could be an alternative to corticosteroids in COPD therapy (Knobloch et al. 2011). Resveratrol has also been shown to possess anti-inflammatory and anti-asthmatic effects in a mouse model of allergic asthma (Lee et al. 2009).

About 75% of orally administered resveratrol is absorbed in humans; however, the bioavailability is considerably low because of extensive metabolism in the intestine and liver. Variable approaches are being used to overcome these limitations, which include the use of methylated derivatives, micronized resveratrol, and Tween-80 as an adjuvant. The aforementioned observations showed that resveratrol studies warrant further research for the development of a potential therapeutic agent involving resveratrol in management of asthma and COPD.

25.4.3 Catechins

Catechins are the major component of green tea extract. There are several isomers of catechins, which include catechin, catechin gallate (CG), gallocatechin, gallocatechin gallate (GCG), epicatechin, epicatechin gallate (ECG), epigallocatechin, and epigallocatechin gallate (EGCg). Dietary intake of catechin has been reported to independently improve the lung function (as measured by FEV_1) and reduce the symptoms of COPD, such as chronic cough, breathlessness, and chronic phlegm (Tabak et al. 2001). The potential use of catechins as a therapeutic agent has raised considerable interest due to its mechanism of action. EGCG is the most effective catechin and specific global inhibitor of HAT activity.

EGCG inhibits p300-mediated acetylation of NF-κB and thereby abrogates acetylation-dependent activation and nuclear translocation of NF-κB. It also inhibits the binding of p300 to the pro-inflammatory gene promoter with an increased recruitment of HDAC3, which results in downregulated expression of NF-κB-related inflammatory proteins (Choi et al. 2009). EGCG showed a potent anti-inflammatory activity against cigarette smoke exposure by inhibiting NF-κB activation, which is independent of its antioxidant activity (Yang et al. 2001). EGCG showed poor oral bioavailability due to its limited absorption and intestinal metabolism. Peracetate protected EGCG, nano-lipidic EGCG, or coadministration with piperine are useful approaches to increase its oral bioavailability.

25.4.4 Garcinol

Garcinol (polyisoprenylated benzophenone derivative) is derived from kokum (*Garcinia indica*) fruit. It is a potent nonspecific inhibitor of CBP/p300 and PCAF HATs. Garcinol inhibits HAT-mediated histone acetylation *in vivo* by acting as a competitive inhibitor for core histones and an uncompetitive type of inhibitor for acetyl-CoA (Balasubramanyam et al. 2004). Phenolic catechol and β-diketone groups in garcinol are important in inhibiting acetyltransferase activity through binding with histone and acetyl-CoA binding sites of HAT (Arif et al. 2009). It is reported to repress chromatin transcription and downregulate global gene expression (Balasubramanyam et al. 2004). Garcinol pretreatment downregulated the expression of pro-inflammatory mediators in response to cigarette smoke exposure. The anti-inflammatory activity of garcinol is mediated by inhibition of HAT activity and presumably the disruption of interaction between HAT and NF-κB (Yang et al. 2009).

25.4.5 Anacardic Acid

Anacardic acid (6-pentadecyl salicylic acid) is derived from cashew nuts and many other medicinal plants, such as *Amphipterygium adstringens* and *Ozoroa insignis*. It inhibits the HAT activity of CBP/p300 and PCAF (Sung et al. 2008). Anacardic acid interferes with the NF-κB pathway by inhibiting NF-κB activation and acetylation, which results in downregulated expression of pro-inflammatory gene products. Knockdown of the p300 histone acetyltransferase gene by RNA interference abrogated the anti-inflammatory effect of anacardic acid, suggesting that anacardic acid inhibits NF-κB-mediated gene expression by a HAT-dependent mechanism (Sung et al. 2008). Further research is needed to determine the therapeutic potential of anacardic acid against asthma and COPD.

The elucidation of potential therapeutic agents based on epigenetic modifications from the vast range of phytochemicals remains a major challenge in the management of chronic lung diseases.

25.5 PERSPECTIVES

Epigenetic/epigenomic mechanisms of disease pathogenesis have an impact in both basic sciences and clinical areas of research.

25.5.1 Clinical

- Epigenetic/epigenomic changes could be used as diagnostic tools for asthma and COPD and their disease severity.
- Epigenetic/epigenomic therapy as nutriepigenomics/nutrigenomics, using phytochemicals/natural products, may provide comparatively nontoxic and less-expensive alternatives to current therapy for chronic inflammatory lung diseases. Furthermore, the use of phytochemicals (especially HDAC2 and SIRT1 activators) in conjunction with conventional medications may be useful to improve the therapeutic potential in managing chronic lung diseases.

25.5.2 Basic Sciences

- Understanding the mechanism of phytochemical-mediated epigenetic modifications is not only useful for evaluation of novel compounds for inflammatory processes but also has an effect in chronic inflammatory diseases.
- Optimization of natural products to improve absorption, bioavailability, and biological action by preparing analogues, derivatives, nanoparticles, or coadministration with adjuvant will be a useful approach to develop effective pharmacological agents.
- Structural research on anti-inflammatory phytochemicals will provide effective synthetic chemicals in the management of asthma and COPD.

25.6 CONCLUSIONS AND FUTURE DIRECTIONS

In this chapter, we have discussed the epigenetic alterations in asthma and COPD, as well as their role in pro-inflammatory gene expression. However, further epigenome-wide research is needed on histone acetylation and methylation patterns and DNA methylation to understand the distinct pathological differences between asthma and COPD. Natural products showed a potential regulatory role in epigenetic modifications as nutriepigenomic/nutrigenomic agents. Within the anti-inflammatory phytochemicals discussed, curcumin and resveratrol appear to be promising because of their multiple modes of action, such as their ability to cause HDAC2 activation/HAT inhibition and NF-κB inactivation. However, larger-scale double-blind clinical trials are required to establish the role of those natural products in management of asthmatics and patients with COPD. Further research is needed to establish whether nutritional intake of natural products (as a food/dietary supplements) can result in epigenetic alterations and anti-inflammatory effect against asthma and COPD as nutriepigenomics/nutrigenomics therapies.

TAKE-HOME MESSAGES

- Cigarette smoke and environmental pollutants are known to impose oxidative burden in lungs, and thereby activate redox-sensitive transcription factors, which regulate the expression of genes encoding pro-inflammatory cytokines/chemokines.
- Epigenetic changes, such as posttranslational modifications of histone proteins and DNA methylation, regulate chromatin (DNA-protein complex) structure and thereby modulate pro-inflammatory gene expression.
- Epigenetic mechanisms underpin the pathogenesis of chronic inflammatory lung diseases, such as asthma and COPD.
- Environmental factors influence on pathogenesis and progression of asthma and COPD, at least in part, is due to epigenetic modifications.
- Histone acetyltransferases, histone deacetylases, histone methyltransferases, histone demethylases, nuclear factor-kappaB, and microRNAs are potential molecular targets by nutraceuticals for epigenetic modifications.

- Epigenetic therapy based on dietary agents may be useful to switch off the expression of pro-inflammatory genes and hence manage chronic inflammatory lung diseases.
- Natural products, such as curcumin, resveratrol, catechins, garcinol, and anacardic acid regulate epigenetic programming, known as nutriepigenomics/nutrigenomics, in the lung in response to cigarette smoke/oxidant exposure, and thereby play an anti-inflammatory role in chronic lung inflammation (Table 25.1).
- Understanding the epigenetic regulation will be helpful in elucidation of novel and specific nutriepigenomics therapy in the management of asthma and COPD.

TABLE 25.1
Molecular Targets of Phytochemicals and Their Potential Cellular and Functional Roles in Epigenetic Regulation and Inflammation

Phytochemicals	Molecular Targets	Cellular and Functional Roles
Curcumin	HATs (CBP/p300) at H3K9, H4K8 and H4K12, HDAC2, RelA/p65 at K218, K221 and K310, NF-κB-dependent pro-inflammatory genes, miRNAs, and Nrf2-mediated antioxidant genes	Regulates epigenetic silencing of pro-inflammatory genes. Represses acetylation of histones and non-histone proteins. Induces HDAC2 expression and decreases NF-κB-DNA binding and pro-inflammatory gene expression. Restores HDAC2, thereby increases corticosteroid sensitivity in response to CS. Regulates miRNA expression, thus impacting signaling pathways and their cellular functions. Induces expression of Nrf2-regulated antioxidant genes such as HO-1, GPx, GCL, and NQO1.
Resveratrol	SIRT1, HAT, H4K8 and H4K12, NF-κB, RelA/p65 at K310, Nrf2, and MMP9	SIRT1 activation decreases MMP-9 and autophagy proteins. Anti-inflammatory activity against CS *in vitro*. Demonstrates anti-inflammatory and anti-asthmatic effects in experimental model of allergic asthma in mice. Inhibits NF-κB activation and expression of pro-inflammatory genes. Mitigates the release of steroid-resistant inflammatory cytokines.
Catechins (EGCG)	HATs, NF-κB and NF-κB-dependent pro-inflammatory genes, and HDAC3	Improves lung function and reduces symptoms of COPD. Global inhibitor of HAT activity. Inhibits p300-mediated acetylation of NF-κB. Increases HDAC3 recruitment, downregulates expression of NF-κB-dependent pro-inflammatory genes. Anti-inflammatory activity against CS exposure by inhibiting NF-κB activation.
Garcinol	HATs, and NF-κB	Nonspecific inhibitor of CBP/p300 and PCAF (inhibits HAT activity). Represses chromatin transcription and downregulates global gene expression. Anti-inflammatory activity due to inhibition of HAT activity and disruption of interaction between HATs and NF-κB.
Anacardic acid	HATs, and NF-κB	Inhibits HAT activity of CBP/p300 and PCAF. Blocks NF-κB activation and acetylation thereby downregulates pro-inflammatory gene expression.

CBP, CREB-binding protein; PCAF, p300/CBP-associated factor; CS, Cigarette smoke; GCL, Glutamyl-cysteine ligase; GPx, Glutathione peroxidase; HATs, Histone acetyltransferases; HDAC2, Histone deacetylase 2; HMT, Histone methyltransferase; HDM, Histone demethylase; NF-κB, Nuclear factor-κB; HO-1, Hemeoxygenase-1; MMP-9, Matrix metalloproteinase-9; miRNAs, MicroRNAs; NF-κB, Nuclear factor-kappa B; NQO1, NAD(P)H quinone oxidoreductase 1; SIRT1, Sirtuin 1.

ACKNOWLEDGMENTS

Supported by the NIH 1R01HL085613, 1R01HL097751, 1R01HL092842, and NIEHS Environmental Health Sciences Center Grant ES-01247.

REFERENCES

Adenuga, D., S. Caito, H. Yao, I. K. Sundar, J. W. Hwang, S. Chung, and I. Rahman. 2010. Nrf2 deficiency influences susceptibility to steroid resistance via HDAC2 reduction. *Biochem Biophys Res Commun* 403: 452–6.

Adenuga, D., H. Yao, T. H. March, J. Seagrave, and I. Rahman. 2009. Histone deacetylase 2 is phosphorylated, ubiquitinated, and degraded by cigarette smoke. *Am J Respir Cell Mol Biol* 40: 464–73.

Arif, M., S. K. Pradhan, G. R. Thanuja, B. M. Vedamurthy, S. Agrawal, D. Dasgupta, et al. 2009. Mechanism of p300 specific histone acetyltransferase inhibition by small molecules. *J Med Chem* 52: 267–77.

Balasubramanyam, K., M. Altaf, R. A. Varier, V. Swaminathan, A. Ravindran, P. P. Sadhale, et al. 2004. Polyisoprenylated benzophenone, garcinol, a natural histone acetyltransferase inhibitor, represses chromatin transcription and alters global gene expression. *J Biol Chem* 279: 33716–26.

Balasubramanyam, K., R. A. Varier, M. Altaf, V. Swaminathan, N. B. Siddappa, U. Ranga, et al. 2004. Curcumin, a novel p300/CREB-binding protein-specific inhibitor of acetyltransferase, represses the acetylation of histone/nonhistone proteins and histone acetyltransferase-dependent chromatin transcription. *J Biol Chem* 279: 51163–71.

Caito, S., S. Rajendrasozhan, S. Cook, S. Chung, H. Yao, A. E. Friedman, et al. 2010. SIRT1 is a redox-sensitive deacetylase that is post-translationally modified by oxidants and carbonyl stress. *FASEB J* 24: 3145–59.

Cheong, H. S., S. M. Park, M. O. Kim, J. S. Park, J. Y. Lee, J. Y. Byun, et al. 2011. Genome-wide methylation profile of nasal polyps: relation to aspirin hypersensitivity in asthmatics. *Allergy* 66: 637–44.

Choi, K. C., M. G. Jung, Y. H. Lee, J. C. Yoon, S. H. Kwon, H. B. Kang, et al. 2009. Epigallocatechin-3-gallate, a histone acetyltransferase inhibitor, inhibits EBV-induced B lymphocyte transformation via suppression of RelA acetylation. *Cancer Res* 69: 583–92.

Cosio, B. G., B. Mann, K. Ito, E. Jazrawi, P. J. Barnes, K. F. Chung, et al. 2004. Histone acetylase and deacetylase activity in alveolar macrophages and blood mononocytes in asthma. *Am J Respir Crit Care Med* 170: 141–7.

Csiszar, A., N. Labinskyy, A. Podlutsky, P. M. Kaminski, M. S. Wolin, C. Zhang, et al. 2008. Vasoprotective effects of resveratrol and SIRT1: Attenuation of cigarette smoke-induced oxidative stress and proinflammatory phenotypic alterations. *Am J Physiol Heart Circ Physiol* 294: H2721–35.

DeMeo, D. L., N. Boutaoui, B. Klanderman, A. Baccarelli, E. K. Silverman, and M. A. Boston. 2009. Variable DNA methylation patterns may identify smokers susceptible to develop COPD. *Am J Respir Crit Care Med* 179: A3981.

DeMeo, D., N. Boutaoui, B. J. Klanderman, D. Lomas, S. I. Lomas, A. Agusti, et al. 2010. DNA methylation varies by GOLD stage in the international COPD genetics network. *Am J Respir Crit Care Med* 181: A2027.

Edwards, M. R., N. W. Bartlett, D. Clarke, M. Birrell, M. Belvisi, and S. L. Johnston. 2009. Targeting the NF-kappaB pathway in asthma and chronic obstructive pulmonary disease. *Pharmacol Ther* 121: 1–13.

Goel, A. and B. B. Aggarwal. 2010. Curcumin, the golden spice from Indian saffron, is a chemosensitizer and radiosensitizer for tumors and chemoprotector and radioprotector for normal organs. *Nutr Cancer* 62: 919–30.

Gosden, R. G. and A. P. Feinberg. 2007. Genetics and epigenetics—nature's pen-and-pencil set. *N Engl J Med* 356: 731–3.

Hew, M., P. Bhavsar, A. Torrego, S. Meah, N. Khorasani, P. J. Barnes, et al. 2006. Relative corticosteroid insensitivity of peripheral blood mononuclear cells in severe asthma. *Am J Respir Crit Care Med* 174: 134–41.

Hollingsworth, J. W., S. Maruoka, K. Boon, S. Garantziotis, Z. Li, J. Tomfohr, N. Bailey, et al. 2008. In utero supplementation with methyl donors enhances allergic airway disease in mice. *J Clin Invest* 118: 3462–9.

Howitz, K. T., K. J. Bitterman, H. Y. Cohen, D. W. Lamming, S. Lavu, J. G. Wood, et al. 2003. Small molecule activators of sirtuins extend *Saccharomyces cerevisiae* lifespan. *Nature* 425: 191–6.

Hwang, J. W., S. Chung, I. K. Sundar, H. Yao, G. Arunachalam, M. W. McBurney, et al. 2010. Cigarette smoke-induced autophagy is regulated by SIRT1-PARP-1-dependent mechanism: Implication in pathogenesis of COPD. *Arch Biochem Biophys* 500: 203–9.

Ito, K., G. Caramori, S. Lim, T. Oates, K. F. Chung, P. J. Barnes, et al. 2002. Expression and activity of histone deacetylases in human asthmatic airways. *Am J Respir Crit Care Med* 166: 392–6.

Ito, K., M. Ito, W. M. Elliott, B. Cosio, G. Caramori, O. M. Kon, et al. 2005. Decreased histone deacetylase activity in chronic obstructive pulmonary disease. *N Engl J Med* 352: 1967–76.

Izzotti, A., G. A. Calin, P. Arrigo, V. E. Steele, C. M. Croce, and S. De Flora. 2009. Downregulation of microRNA expression in the lungs of rats exposed to cigarette smoke. *FASEB J* 23: 806–12.

Izzotti, A., P. Larghero, M. Longobardi, C. Cartiglia, A. Camoirano, V. E. Steele, et al. 2010. Dose-responsiveness and persistence of microRNA expression alterations induced by cigarette smoke in mouse lung. *Mutat Res*. PMID: 21185844.

Knekt, P., J. Kumpulainen, R. Jarvinen, H. Rissanen, M. Heliovaara, A. Reunanen, et al. 2002. Flavonoid intake and risk of chronic diseases. *Am J Clin Nutr* 76: 560–8.

Knobloch, J., H. Hag, D. Jungck, K. Urban, and A. Koch. 2011. Resveratrol impairs the release of steroid-resistant cytokines from bacterial endotoxin-exposed alveolar macrophages in chronic obstructive pulmonary disease. *Basic Clin Pharmacol Toxicol* 109: 138–43.

Kode, A., S. Rajendrasozhan, S. Caito, S. R. Yang, I. L. Megson, and I. Rahman. 2008. Resveratrol induces glutathione synthesis by activation of Nrf2 and protects against cigarette smoke-mediated oxidative stress in human lung epithelial cells. *Am J Physiol Lung Cell Mol Physiol* 294: L478–88.

Kotani, A., D. Ha, J. Hsieh, P. K. Rao, D. Schotte, M. L. den Boer, et al. 2009. miR-128b is a potent glucocorticoid sensitizer in MLL-AF4 acute lymphocytic leukemia cells and exerts cooperative effects with miR-221. *Blood* 114: 4169–78.

Lee, M., S. Kim, O. K. Kwon, S. R. Oh, H. K. Lee, and K. Ahn. 2009. Anti-inflammatory and anti-asthmatic effects of resveratrol, a polyphenolic stilbene, in a mouse model of allergic asthma. *Int Immunopharmacol* 9: 418–24.

Lee, S. J. and M. M. Kim. 2011. Resveratrol with antioxidant activity inhibits matrix metalloproteinase via modulation of SIRT1 in human fibrosarcoma cells. *Life Sci* 88: 465–72.

Li, Y., D. Kong, Z. Wang, and F. H. Sarkar. 2010. Regulation of microRNAs by natural agents: an emerging field in chemoprevention and chemotherapy research. *Pharm Res* 27: 1027–41.

Lim, S., E. Metzger, R. Schule, J. Kirfel, and R. Buettner. 2010. Epigenetic regulation of cancer growth by histone demethylases. *Int J Cancer* 127: 1991–8.

Marcu, M. G., Y. J. Jung, S. Lee, E. J. Chung, M. J. Lee, J. Trepel, et al. 2006. Curcumin is an inhibitor of p300 histone acetylatransferase. *Med Chem* 2: 169–74.

Martino, D. and S. Prescott. 2011. Epigenetics and prenatal influences on asthma and allergic airways disease. *Chest* 139: 640–7.

Mathers, J. C., G. Strathdee, and C. L. Relton. 2010. Induction of epigenetic alterations by dietary and other environmental factors. *Adv Genet* 71: 3–39.

Meja, K. K., S. Rajendrasozhan, D. Adenuga, S. K. Biswas, I. K. Sundar, G. Spooner, et al. 2008. Curcumin restores corticosteroid function in monocytes exposed to oxidants by maintaining HDAC2. *Am J Respir Cell Mol Biol* 39: 312–23.

Mercado, N., R. Thimmulappa, C. M. Thomas, P. S. Fenwick, K. K. Chana, L. E. Donnelly, et al. 2011. Decreased histone deacetylase 2 impairs Nrf2 activation by oxidative stress. *Biochem Biophys Res Commun* 406: 292–8.

Moghaddam, S. J., P. Barta, S. G. Mirabolfathinejad, Z. Ammar-Aouchiche, N. T. Garza, T. T. Vo, et al. 2009. Curcumin inhibits COPD-like airway inflammation and lung cancer progression in mice. *Carcinogenesis* 30: 1949–56.

Nadeau, K., C. McDonald-Hyman, E. M. Noth, B. Pratt, S. K. Hammond, J. Balmes, et al. 2010. Ambient air pollution impairs regulatory T-cell function in asthma. *J Allergy Clin Immunol* 126: 845–52e10.

Oglesby, I. K., N. G. McElvaney, and C. M. Greene. 2010. MicroRNAs in inflammatory lung disease—master regulators or target practice? *Respir Res* 11: 148.

Perera, F., W. Y. Tang, J. Herbstman, D. Tang, L. Levin, R. Miller, et al. 2009. Relation of DNA methylation of 5'-CpG island of ACSL3 to transplacental exposure to airborne polycyclic aromatic hydrocarbons and childhood asthma. *PLoS One* 4: e4488.

Pottelberge, G. R., P. Mestdagh, K. R. Bracke, O. Thas, Y. M. Durme, G. F. Joos, et al. 2011. MicroRNA expression in induced sputum of smokers and patients with chronic obstructive pulmonary disease. *Am J Respir Crit Care Med* 183: 898–906.

Rajendrasozhan, S., S. R. Yang, I. Edirisinghe, H. Yao, D. Adenuga, and I. Rahman. 2008a. Deacetylases and NF-kappaB in redox regulation of cigarette smoke-induced lung inflammation: Epigenetics in pathogenesis of COPD. *Antioxid Redox Signal* 10: 799–811.

Rajendrasozhan, S., S. R. Yang, V. L. Kinnula, and I. Rahman. 2008b. SIRT1, an antiinflammatory and antiaging protein, is decreased in lungs of patients with chronic obstructive pulmonary disease. *Am J Respir Crit Care Med* 177: 861–70.

Rajendrasozhan, S., H. Yao, and I. Rahman. 2009. Current perspectives on role of chromatin modifications and deacetylases in lung inflammation in COPD. *COPD* 6: 291–7.

Schembri, F., S. Sridhar, C. Perdomo, A. M. Gustafson, X. Zhang, A. Ergun, et al. 2009. MicroRNAs as modulators of smoking-induced gene expression changes in human airway epithelium. *Proc Natl Acad Sci USA* 106: 2319–24.

Seligson, D. B., S. Horvath, M. A. McBrian, V. Mah, H. Yu, S. Tze, et al. 2009. Global levels of histone modifications predict prognosis in different cancers. *Am J Pathol* 174: 1619–28.

Shakibaei, M., C. Buhrmann, and A. Mobasheri. 2011. Resveratrol-mediated SIRT-1 interactions with p300 modulate receptor activator of NF-kappaB ligand (RANKL) activation of NF-kappaB signaling and inhibit osteoclastogenesis in bone-derived cells. *J Biol Chem* 286: 11492–505.

Sung, B., M. K. Pandey, K. S. Ahn, T. Yi, M. M. Chaturvedi, M. Liu, et al. 2008. Anacardic acid (6-nonadecyl salicylic acid), an inhibitor of histone acetyltransferase, suppresses expression of nuclear factor-kappaB-regulated gene products involved in cell survival, proliferation, invasion, and inflammation through inhibition of the inhibitory subunit of nuclear factor-kappaBalpha kinase, leading to potentiation of apoptosis. *Blood* 111: 4880–91.

Suzuki, M., T. Betsuyaku, Y. Ito, K. Nagai, N. Odajima, C. Moriyama, et al. 2009. Curcumin attenuates elastase- and cigarette smoke-induced pulmonary emphysema in mice. *Am J Physiol Lung Cell Mol Physiol* 296: L614–23.

Suzuki, M., H. Wada, M. Yoshino, L. Tian, H. Shigematsu, H. Suzuki, et al. 2010. Molecular characterization of chronic obstructive pulmonary disease-related non-small cell lung cancer through aberrant methylation and alterations of EGFR signaling. *Ann Surg Oncol* 17: 878–88.

Szulakowski, P., A. J. Crowther, L. A. Jimenez, K. Donaldson, R. Mayer, T. B. Leonard, et al. 2006. The effect of smoking on the transcriptional regulation of lung inflammation in patients with chronic obstructive pulmonary disease. *Am J Respir Crit Care Med* 174: 41–50.

Tabak, C., I. C. Arts, H. A. Smit, D. Heederik, and D. Kromhout. 2001. Chronic obstructive pulmonary disease and intake of catechins, flavonols, and flavones: The MORGEN Study. *Am J Respir Crit Care Med* 164: 61–4.

Uchida, S., A. Nishida, K. Hara, T. Kamemoto, M. Suetsugi, M. Fujimoto, et al. 2008. Characterization of the vulnerability to repeated stress in Fischer 344 rats: Possible involvement of microRNA-mediated down-regulation of the glucocorticoid receptor. *Eur J Neurosci* 27: 2250–61.

vel Szic, K. S., M. N. Ndlovu, G. Haegeman, and W. Vanden Berghe. 2010. Nature or nurture: Let food be your epigenetic medicine in chronic inflammatory disorders. *Biochem Pharmacol* 80: 1816–32.

Vucic, E. A., I. M. Wilson, B. P. Coe, D. D. Sin, H. Coxson, W. L. Lam, et al. 2010. Global DNA methylation analysis of bronchial epithelial cells from patients with asthma, COPD with and without lung cancer. *Am J Respir Crit Care Med* 181: A6772.

Walda, I. C., C. Tabak, H. A. Smit, L. Rasanen, F. Fidanza, A. Menotti, et al. 2002. Diet and 20-year chronic obstructive pulmonary disease mortality in middle-aged men from three European countries. *Eur J Clin Nutr* 56: 638–43.

White, B. and D. Z. Judkins. 2011. Clinical inquiry. Does turmeric relieve inflammatory conditions? *J Fam Pract* 60: 155–6.

Williams, A. E., H. Larner-Svensson, M. M. Perry, G. A. Campbell, S. E. Herrick, I. M. Adcock, et al. 2009. MicroRNA expression profiling in mild asthmatic human airways and effect of corticosteroid therapy. *PLoS One* 4: e5889.

Yang, C. M., I. T. Lee, C. C. Lin, Y. L. Yang, S. F. Luo, Y. R. Kou, et al. 2009. Cigarette smoke extract induces COX-2 expression via a PKCalpha/c-Src/EGFR, PDGFR/PI3K/Akt/NF-kappaB pathway and p300 in tracheal smooth muscle cells. *Am J Physiol Lung Cell Mol Physiol* 297: L892–902.

Yang, F., H. S. Oz, S. Barve, W. J. de Villiers, C. J. McClain, and G. W. Varilek. 2001. The green tea polyphenol (-)-epigallocatechin-3-gallate blocks nuclear factor-kappa B activation by inhibiting I kappa B kinase activity in the intestinal epithelial cell line IEC-6. *Mol Pharmacol* 60: 528–33.

Yang, I. V. and D. A. Schwartz. 2011. Epigenetic control of gene expression in the lung. *Am J Respir Crit Care Med* 183: 1295–301.

Yang, S. R., S. Valvo, H. Yao, A. Kode, S. Rajendrasozhan, I. Edirisinghe, et al. 2008. IKK alpha causes chromatin modification on pro-inflammatory genes by cigarette smoke in mouse lung. *Am J Respir Cell Mol Biol* 38: 689–98.

Yang, S. R., J. Wright, M. Bauter, K. Seweryniak, A. Kode, and I. Rahman. 2007. Sirtuin regulates cigarette smoke-induced proinflammatory mediator release via RelA/p65 NF-kappaB in macrophages in vitro and in rat lungs in vivo: Implications for chronic inflammation and aging. *Am J Physiol Lung Cell Mol Physiol* 292: L567–76.

Yun, J. M., I. Jialal, and S. Devaraj. 2011. Epigenetic regulation of high glucose-induced proinflammatory cytokine production in monocytes by curcumin. *J Nutr Biochem* 22: 450–8.

Zhou, X., H. Sun, H. Chen, J. Zavadil, T. Kluz, A. Arita, et al. 2010. Hypoxia induces trimethylated H3 lysine 4 by inhibition of JARID1A demethylase. *Cancer Res* 70: 4214–21.

26 Biologics
Molecular Medicine from Bench to Bedside

Ananya Datta Mitra
University of California-Davis
Davis, California

Debasis Bagchi
University of Houston
Houston, Texas

Siba P. Raychaudhuri and Smriti K. Raychaudhuri
University of California-Davis
Davis, California

CONTENTS

26.1 INTRODUCTION

Owing to the recent advancement of science and technology, the use of biological agents has opened a novel, effective, and relatively target-specific approach to the treatment of autoimmune diseases. These biological agents, commonly called biologics, are therapeutic serum, toxin, antitoxin, vaccine, virus, blood, blood component or derivative, allergenic product, or analogous product, or derivatives used for the prevention, treatment, or cure of human diseases. In this chapter, we discuss the biologics used in the treatment of autoimmune diseases.

Biologics are proteins of high molecular weight, designed by genetic engineering, with inherent immunologic activity targeting specifically different components of the immune and inflammatory cascade and leading to the modulation of the disease. These agents are expensive and associated with immediate and delayed hypersensitivity reactions and immunodeficiency and immune imbalance categories of adverse reactions (Borchers et al. 2011), requiring continuous laboratory monitoring during therapy. In spite of their cost, side effects, and difficulty in administration, patients have accepted them as a new hope for their unfulfilled wishes in the treatment of autoimmune diseases.

26.2 INTERPLAY OF THE IMMUNE SYSTEM IN THE DEVELOPMENT OF AUTOIMMUNE DISEASE

Autoimmune diseases are characterized by the presence of autoantibodies and inflammation, including mononuclear phagocytes, autoreactive T lymphocytes, and plasma cells. Autoimmune diseases can be classified as organ-specific or non-organ-specific depending on whether the autoimmune response is directed against a particular tissue like the synovium in rheumatoid arthritis or against widespread antigens such as cell nuclear antigens in lupus.

Pathogenesis of autoimmune disease includes a complex interplay of immune cells. The major infiltrating cells include macrophages, neutrophils, self-reactive CD4+T helper cells and CD8+cytolytic T cells, with natural killer cells, mast cells, and dendritic cells. These immune cells destroy the tissue architecture by damaging the cells or by releasing cytotoxic cytokines, prostaglandins, reactive nitrogen, or oxygen intermediates. Tissue macrophages and monocytes play the role of antigen-presenting cells to begin an autoimmune response, or as effector cells to establish an immune response that has already been initiated.

26.2.1 T Cells in Autoimmunity

A defect in thymic selection may result in clonal expansion of auto-reactive CD4 and CD8 T cells in autoimmune diseases. Although both CD4 and CD8 T cells take part in autoimmune processes, due to molecular complexities, the exact role of CD8 cells can only be elucidated in animal models of type 1 diabetes, experimental autoimmune encephalomyelitis (EAE), and rheumatoid arthritis (Najafian et al. 2003; Davila et al. 2005). Naïve CD4 T cells are progenitors of at least four distinct types of effector/regulatory cells: Th1, Th2, Th17, and Treg cells. Local cytokine environment orchestrates the CD4 T cell subset differentiation to different effector lineages. In the presence of IL-12, naïve T cells differentiate to Th1 cells, which activate macrophages and cytotoxic T cells and mount a cell-mediated immune response in rheumatoid arthritis, psoriasis, psoriatic arthritis, acute allograft rejection, graft-versus-host disease, and many other autoimmune diseases. They produce pro-inflammatory cytokines TNF-α, IFN-γ, IL-2, and IL-18. Similarly, in the presence of IL-23 and TGF-β and IL-6 naïve T cells, they differentiate to Th17 cells and induce their signature cytokines IL-17A, IL-17F, IL-22, and IL-26, which are also pro-inflammatory in nature. Th17 cells constitute a division of the adaptive immune system that clears extracellular pathogens mounting a massive inflammatory response. Animal studies reveal the role of Th17 cells and Th17-associated cytokines in infectious diseases, autoimmune conditions, transplantation reactions, allergy, and tumor biology. Experimental evidence also suggests the role of Th17 cells in human psoriasis (Krueger et al.

2007), rheumatoid arthritis (Kirkham et al. 2006), multiple sclerosis (Matusevicius et al. 1999), inflammatory bowel disease (Duerr et al. 2006), asthma (Barczyk et al. 2003), and some bacterial and fungal infections. In psoriasis, T cells extracted from psoriatic skin lesions showed predominantly a Th17 phenotype (Pene et al. 2008), which is in line with the observation that CCL20/CCR6 signaling is an important pathway for chemoattraction of inflammatory cells to epithelial tissues. Similarly, in rheumatoid arthritis patients, cytokine expression of IL-17 was prognostic of joint destruction (Kirkham et al. 2006). IL17 induces a plethora of pro-inflammatory cytokines, chemokines, and proteases in the chondrocytes and osteoblasts. Furthermore, RANKL expression on the surface of Th17 cells induces osteoclastogenesis (Kotake et al. 1999; Miranda-Carus et al. 2006), promoting cartilage and bone destruction/resorption (Koenders et al. 2006). It is quite intriguing to find that an immunosuppressive cytokine is inducing a highly inflammatory T helper cell line. In the molecular level, TGF-β induced differentiation of Tregs and Th17 cells is necessary to maintain a balance between tolerance and autoimmunity. Normally, TGF-β induces Tregs, which are a regulatory subset of the naïve T cell, taking part in maintaining self-tolerance. In the presence of pathological triggers, immune cells start producing IL-6, which prevent the development of the Tregs and induce the development of Th17 cells leading to a blast of immune response. Experiments with the rheumatic synovium showed that Tregs are present in abundance in the synovial fluid (Lawson et al. 2006) but are incapable of suppressing an immune response. In contrast, if the naïve T helper cell encounters IL-4, it will differentiate into a Th2 cell, which secretes anti-inflammatory cytokines IL-4, IL-5, IL-9, IL-10, and IL-13. Th2 cells also stimulate B cells to make antibodies that bind to mast cells, basophils, and eosinophils.

26.2.2 B Cells in Autoimmunity

B lymphocytes express cell-surface immunoglobulin (Ig) receptors skilled to recognize antigens (Ags). The major role of these cells is to generate antibody-secreting plasma cells and memory B cells in response to specific Ags. Secreted antibodies are involved in humoral immunity, which takes an active part in neutralizing toxins and pathogens and also in assisting their elimination by activating several effector mechanisms, such as phagocytosis or complement system proteins. B cells can also act as antigen presenting cells because they exclusively arrest and internalize Ag to an endocytic vacuole optimized for peptide loading onto class II MHC through their receptors. The MHC-peptide complex is transported to the cell surface and is recognized by the T-cell receptor of the CD4 T cells. Genetic perturbations that affect B-cell activation thresholds (e.g., CD19, CD22, lyn, SHP2) or that regulate apoptosis (e.g., bcl-2, bcl-x, Fas/FasL) dictate the escapade of an autoreactive B cell, its clonal expansion, and the induction of an autoimmune process (Maniati et al. 2008). These autoreactive B cells secrete autoantibodies directed to self-peptides or cell surface receptors. In patients with systemic lupus erythematosus (SLE) and rheumatoid arthritis (RA), autoantibodies form immune complexes with circulating auto-Ags, which accumulate in the kidneys and joint synovium and activate complement cascade, resulting in local inflammation and tissue destruction. Although commonly regarded as a T cell–dependent disease, the therapeutic success of monoclonal antibody (mAb) against the B-cell surface molecule CD20 in RA shifted focus on the role of B cells in autoimmune diseases.

26.2.3 Cytokines in Autoimmunity

Cytokines are small cell-signaling protein molecules produced throughout the body by a wide variety of cells. They mainly act as chemical messengers that recruit inflammatory cells (T cells, neutrophils) to the site of inflammation. Normally, the pro-inflammatory cytokines (such as TNF-alpha, interleukin (IL) IL-1, IL-6, IL-8, IL-12, IL-17) as well as the anti-inflammatory cytokines (such as IL-4, IL-10, IL-11, IL-13) walk hand in hand during inflammation. The role of these pro-inflammatory cytokines is well established in respect to tissue destruction in autoimmune diseases. TNF-α, the signature pro-inflammatory cytokine of Th1 cells, induces apoptotic cell death and inflammation, and inhibits tumorigenesis and viral replication. TNF-α stimulates the release of the inflammatory cytokines (IL-1beta, IL-6, IL-8, GM-CSF), upregulates a series of critical chemokines (MCP-1, MIP-2, RANTES, MIP-1α), and activates endothelial adhesion molecules (ICAM-1, VCAM-1, E-selectin). These functions of TNF-α regulate initiation and perpetuation of important events associated with an inflammatory reaction. Levels of TNF-α are elevated both locally and systemically in autoimmune disorders such as RA, ankylosing spondylitis, psoriasis, and Crohn's disease, suggesting that higher levels of TNF-α may directly contribute to tissue damage (Braun et al. 1995; Feldman et al. 1996). Similarly, IL-1 is involved in the pathogenesis of autoimmune arthritis by activating T cells and stimulating matrix metalloproteinases (MMP) from fibroblasts and chondrocytes. Studies of arthritis in animals have strongly implicated interleukin-1 in joint damage. IL-6 stimulates both B and T cell functions as well as the production of acute phase reactants (Van Snick 1990). IL-17 induces pro-inflammatory cytokines, cyclooxygenase-2 in chondrocytes, and osteoclast differentiation factor expression in osteoblasts, thus causing development of autoimmune arthritis. Similarly, the anti-inflammatory cytokines produced by the Th2 cells inhibit proliferation and activation of Th1 lymphocytes, activated macrophages, and dendritic cells, which in turn decreases the production of pro-inflammatory cytokines like IL-1, TNF-α, IL-6, and IL-8, and inhibits tissue damage. Furthermore, B cells also produce a vast array of cytokines and chemokines, including IL-1, IL-4, IL-6, IL-8, and IL-7, granulocyte colony-stimulating factor (CSF), granulocyte macrophage-CSF, IL-10 IL-12, TNF-α, lymphotoxin, TGF-β, bone morphogenic protein-6/7, vascular endothelial growth factor-A, macrophage inflammatory protein-1α/β, IL-16, and CXCL13, which take part in inflammation and activation of T cells.

26.3 TARGETS FOR BIOLOGICS

Important components of the pathological apparatus that are being targeted for therapy include cytokines, chemokines responsible for chemotaxis of lymphocytes to target tissues, enzymes responsible for extracellular matrix and vascular penetration by immune cells, pathogenic cell population (T and B cells), cell bound receptors and receptors on immune cells (e.g., the T cell receptor), immunoglobulins, and complement components (Figure 26.1).

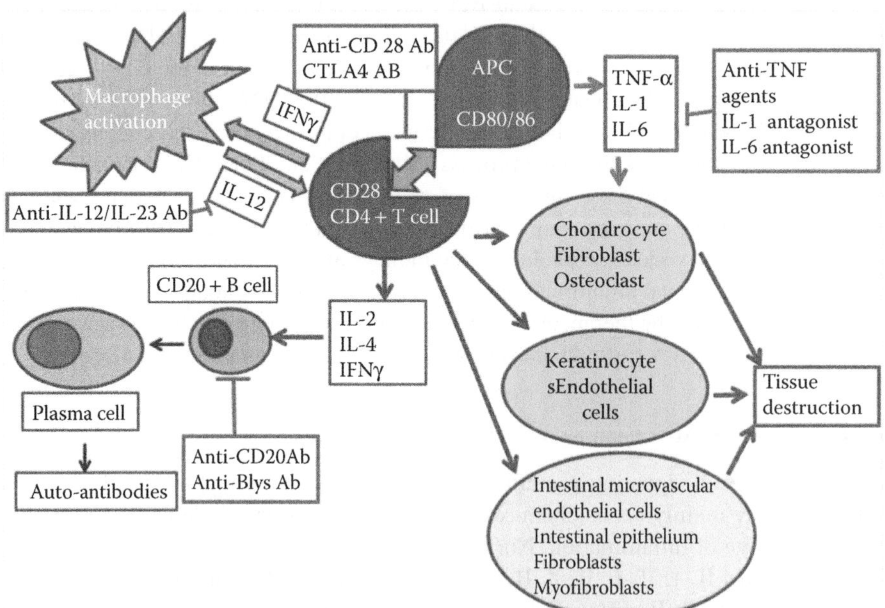

FIGURE 26.1 (see color insert) Different cell types associated with autoinflammation and tissue destruction and the potential targets for biologics.

26.3.1 Anti-Cytokine Therapy

Anti-cytokines are molecules that modulate functions of cytokines either by interfering with their binding to membrane-associated cytokine receptors or by blocking different steps of their synthesis and maturation (Tartour et al. 1994).

26.3.1.1 Targeting Tumor Necrosis Factor-α

TNF-α is involved in systemic inflammation and is a member of a group of cytokines that stimulate the acute phase reaction. Human TNF-α is translated as a transmembrane precursor protein molecule. It is then cleaved in the extracellular domain by matrix metalloproteinases to release mature soluble TNF-α. Both the transmembrane precursor and the cleaved soluble form of TNF are biologically active and require trimerization of three TNF monomers to form trimeric TNF for biological activity. This trimeric TNF binds to one of two types of receptors: TNFR1 or TNFR2, also known as p55 and p75, respectively. Most tissues express TNF-R1 constitutively, which can be fully activated by both the membrane-bound and soluble trimeric forms of TNF, while TNF-R2 is only found in immune cells and responds to the membrane-bound form of the TNF homotrimer. This binding causes a conformational change in the receptor, leading to dissociation of the inhibitory protein SODD from the intracellular death domain, thus enabling the adaptor protein TRADD to bind to the death domain, serving as a platform for intracellular signal transduction and activation of NF-κB and MAPK. These transcription factors then translocate to the nucleus and mediate the transcription of a vast array of proteins involved in cell survival and proliferation, inflammatory response, and anti-apoptotic factors. Thus, TNF inhibitors offer a targeted strategy inhibiting these critical steps in the inflammatory cascade. There has been a major clinical breakthrough with the use of TNF-α blocking biologics. TNF-α blockers are being used in a number of immunological diseases like psoriasis, psoriatic arthritis, Crohn's disease, RA, and ankylosing spondylitis (Table 26.1). As of 2010, millions of patients have been treated with TNF-α blockers for the treatment of inflammatory diseases (Taylor et al. 2006; Finckh et al. 2006; Gall and Kalb 2008; Gordon et al. 2006).

The Food and Drug Administration (FDA) has thus far approved five TNF-α inhibitors for the treatment of a variety of inflammatory conditions (Table 26.2). Extensive information has been gathered on etanercept, infliximab, and adalimumab over the years and has been fully illustrated in a series of publications (Raychaudhuri and Raychaudhuri 2011). In this chapter we are mainly focusing on the newer agents like golimumab and certolizumab.

Golimumab (a human monoclonal anti-TNF-α antibody) and certolizumab (a pegylated Fab fragment of humanized monoclonal TNF-α antibody) are the two latest additions to the anti-TNF regimen. Both of these medicines were approved by the FDA in the early part of 2009 for rheumatoid arthritis (RA) and are likely to be used for various autoimmune diseases.

Golimumab (Simponi®), marketed by Centocor, received FDA approval in April 2009 for the treatment of moderate to severe rheumatoid arthritis, active psoriatic arthritis, and ankylosing

TABLE 26.1
Food and Drug Administration (FDA)-Approved Clinical Indications for the Use of TNF-α Inhibitors

Indications	Medical Conditions
Rheumatologic	Moderately to severely active rheumatoid arthritis (RA)
	Moderately to severely active polyarticular juvenile idiopathic arthritis (JIA) in children ages 2 years and older
	Ankylosing spondylitis
	Psoriatic arthritis
Gastrointestinal	Moderate to severe Crohn's disease (including fistulating Crohn's disease)
Dermatological	Chronic moderate to severe plaque psoriasis in adults ages 18 years and older

TABLE 26.2
FDA-Approved TNF Inhibitors for Clinical Use

TNF-α Inhibitor	Description	Half-Life	Frequency of Administration	Indications
Infliximab (Remicade®)	Chimeric mouse–human anti-TNF α monoclonal antibody (mAb); risk of immunogenicity	8–10 days	Intravenous infusion with first 3 doses within 10 weeks and thereafter once every 8 weeks	Crohn's disease, ulcerative colitis, ankylosing spondylitis, psoriatic arthritis, severe plaque psoriasis and first-line therapy with Mtx in moderate-severe RA
Adalimumab (Humira®)	Fully humanized anti-TNF-α mAb; low risk of immunogenicity	14 days	Subcutaneous injections (SQ) once every 2 weeks	Psoriasis, psoriatic arthritis, rheumatoid arthritis, ankylosing spondylitis and Crohn's disease
Etanercept (Enbrel®)	Soluble p75 TNF-α receptor fusion protein formed by fusing two soluble TNFR2 molecules and Fc fragment; inhibits TNF-α with LT-α	4–6 days	SQ once or twice per week	Plaque psoriasis, psoriatic arthritis, ankylosing spondylitis, rheumatoid arthritis, and juvenile rheumatoid arthritis
Golimumab (Simponi®)	Fully humanized anti-TNF-α mAb; low risk of immunogenicity	14 days	SQ once in a month	Moderate to severe rheumatoid arthritis, active psoriatic arthritis, and ankylosing spondylitis
Certolizumab pegol (Cimzia®)	Pegylated Fab fragment of humanized monoclonal TNF-α antibody, no chance of ADCC due to absence of Fc fragment	14 days	SQ once in a month	Crohn's disease and treatment of moderate to severe rheumatoid arthritis

spondylitis. It is a fully human anti-TNFα monoclonal antibody. It is developed by genetic engineering of transgenic mice with the insertion of human antibody genes into its genome, which suppresses mice antibody genes and can produce human anti-TNF antibodies when immunized with a target antigen (Ishida et al. 2002; Shealy et al. 2007). Golimumab forms high-affinity, stable complexes with human TNF. It acts by neutralizing both circulating and membrane-bound forms of human TNF (Shealy et al. 2007; FDA 2009b). It is given once a month subcutaneously. In several multicenter, randomized, double-blind, controlled trials, the efficacy and safety of golimumab in RA have been evaluated. Studies done are RA-1 study, known as GO-AFTER; RA-2, known as GO-FORWARD; and RA-3 (Melmed et al. 2008; McCluggage et al. 2009). Study RA-1 evaluated 461 patients who were previously treated (at least 8 to 12 weeks prior to administration of the study agent) with one or more doses of a TNF-blocker without any serious adverse reaction. Study RA-2 evaluated 444 patients who had active RA despite a stable dose of minimum 15 mg/week of MTX and who had never been treated with a biologic TNF-blocker. Study RA-3 evaluated 637 patients with active RA who were MTX-naïve and had never been exposed to a TNF-blocker. The primary endpoint in study RA-1 and study RA-2 was the percentage of patients achieving an ACR 20 (ACR: American College of Rheumatology) response at week 14, and the primary endpoint in study RA-3 was the percentage of patients achieving an ACR 50 response at week 24 (FDA 2009b; Emery et al. 2009; Keystone et al. 2009). Golimumab was found to be more effective than the placebo and methotrexate (MTX) in these trials. The combination therapy of golimumab and MTX achieved higher percentage ACR responses at week 14 (studies RA-1 and RA-2) and week 24 (studies RA-1, RA-2, and RA-3) versus patients treated with the MTX alone (FDA 2009b). However, there was no clear evidence of improved ACR response with the higher golimumab dose group (100 mg) compared to the lower golimumab dose group (50 mg).

Safety and efficacy studies have been carried out in RA, psoriatic arthritis, psoriasis, and ankylosing spondylitis. A majority of the studies have been done in RA patients, including early onset untreated patients. In a recent phase III, multicenter, randomized, double-blind, placebo-controlled study, it has also been reported that in RA golimumab can be considered as first-line therapy for early onset rheumatoid arthritis (Emery et al. 2009). Similarly, the safety and efficacy of golimumab have also been evaluated in a multicenter, randomized, double-blind, placebo-controlled trial in psoriatic arthritis (PsA). This study was done in 405 adult patients with moderate to severe forms of active PsA (≥ 3 swollen joints and ≥ 3 tender joints; Shealy et al. 2007; Kavanaugh et al. 2009). Patients having PsA with a median duration of 5.1 years and with a psoriatic skin lesion of at least 2 cm in diameter were randomly assigned to placebo, golimumab 50 mg, or golimumab 100 mg given subcutaneously every 4 weeks. Although patients included may have received methotrexate/oral corticosteroid/NSAID, they were naïve to TNF-blocker. The primary endpoint was the percentage of patients achieving ACR 20 response at week 14. At week 14 ACR20 responses occurred in 9%, 51%, and 45% of the three groups, respectively, and at least 75% improvement in the Psoriasis Area and Severity Index (PASI) scores occurred in 3%, 40%, and 58%, respectively. There was no clear evidence of improved ACR response with the higher golimumab dose group (100 mg) compared to the lower golimumab dose group (50 mg). In another randomized, double-blind, placebo-controlled, phase III trial, golimumab has been found to be effective in 356 adult patients with active ankylosing spondylitis (FDA 2009b; Inman et al. 2008).

Certolizumab pegol, also known as Cimzia®, was approved in April 2008 for the treatment of Crohn's disease, and in 2009 for the treatment of moderate to severe rheumatoid arthritis. It is comprised of the pegylated Fab fragment of a humanized monoclonal antibody directed against TNF. It does not contain an Fc portion, and therefore does not induce complement activation, antibody-dependent cellular cytotoxicity, or apoptosis (Melmed et al. 2008). The pegylation of the antibody delays elimination and extends the half-life of certolizumab. As a result, the medication may be administered monthly subcutaneously. In RA, efficacy and safety data of certolizumab are available from a series of studies denoted as RAPID 1, RAPID 2, and FAST4WARD. Certolizumab has been compared to placebo in 1,821 patients with moderate to severe forms of active RA in these multicenter, double-blind, randomized controlled trials. Outcomes of efficacy were determined by percentage of patients achieving ACR 20 response at week 24. In the FAST4WARD trial, Cimzia 400 mg demonstrated a significant therapeutic response in terms of ACR20, ACR50, and ACR70 as compared to placebo in RA patients who failed at least one prior disease-modifying antirheumatic drug (DMARD). Patient-reported outcomes and physical function were also better in the certolizumab arm (Fleischmann et al. 2009). Patients were randomized into two treatment groups: Cimzia 400 mg (n = 111) every 4 weeks from baseline to week 20; placebo (n = 109) every 4 weeks from baseline to week 20. ACR 20 response rate at week 24 was defined as the primary endpoint; ACR 50 and ACR 70 were considered secondary endpoints of this study. RAPID 1 compared the combination of MTX and certolizumab to MTX monotherapy in TNF-inhibitor-naïve patients with active, uncontrolled RA despite treatment with MTX monotherapy. Primary efficacy outcomes included ACR 20 response rate at week 24 and total modified Sharp score at week 52. At week 24, higher numbers of patients in the combination treatment arms achieved the primary endpoint of ACR 20 response. At week 52, modified total Sharp score in patients who received combination treatment compared to MTX monotherapy indicated less bone erosion and joint-space narrowing. RAPID 2 compared the combination of certolizumab and MTX with MTX monotherapy in patients with active RA whose symptoms were inadequately controlled with ≥6 months of treatment with MTX monotherapy. Patients (n = 619) were randomized 2:2:1 to subcutaneous certolizumab pegol (liquid formulation) 400 mg at weeks 0, 2, and 4, followed by 200 mg or 400 mg plus MTX, or placebo plus MTX, every 2 weeks for 24 weeks. The results showed that significantly more patients in the certolizumab pegol 200 mg and 400 mg groups achieved an ACR 20 response versus placebo. Certolizumab pegol 200 and 400 mg also significantly inhibited radiographic progression. Most adverse events were mild or moderate, with low incidence of withdrawals due to adverse events. Five patients developed tuberculosis (Smolen, Landewé, et al. 2009). Randomized, double-blind, placebo-controlled trials (also known as PRECiSE1 and PRECiSE2) to evaluate the efficacy and safety of certolizumab pegol in the treatment of Crohn's disease demonstrated that in moderate to severe Crohn's disease, induction and maintenance therapy with certolizumab pegol was associated with a modest improvement in response rates, as compared with placebo (Sandborn et al. 2007; Scheiber et al. 2007; Allez et al. 2009).

On the basis of experience gained in cytokine modulation therapy of chronic inflammatory diseases such as rheumatoid arthritis, psoriasis, psoriatic arthritis, ankylosing spondylitis, and inflammatory bowel disease, the application of TNF-inhibitors shifted our vision from nonspecific immunomodulatory agents to a novel, targeted therapeutic option for distinct chronic inflammatory diseases. Studies have also demonstrated the efficacy of anti-TNF therapy in inflammatory conditions like Behçet's disease, pyoderma gangrenosum, cutaneous Crohn's disease, and subcorneal pustular dermatitis (Jacobi et al. 2006).

26.3.1.2 Adverse Effects Associated with Anti-TNF Agents

As TNF-α is responsible for transcription of a vast array of proteins involved in cell survival and proliferation, inflammatory response, and anti-apoptotic factors, it plays a critical role in host defenses against bacterial infections, particularly against mycobacterial infections. Blocking of TNF-α will counter these critical processes required for immuno-surveillance of pathologic microbial agents. Therefore, it is not surprising that all anti-TNF-α agents have been associated with a variety of serious and "routine" opportunistic infections because they suppress the inflammatory response (Raychaudhuri et al. 2009). Both clinical trials and post-marketing surveillance show evidence of multiple adverse effects due to TNF inhibition (Day 2002). Side effects of anti-TNF therapy are mentioned in Table 26.3. Although TNF-α blockers are generally well tolerated, physicians need to be extremely cautious about the potential of serious side effects of anti-TNF drugs and should review the indications/contraindications of anti-TNF agents in every patient. Before the use of TNF-α blockers, detailed history of heart failure, chronic liver disease, neurological disorders, and neoplasia is mandatory. TNF-α blockers are contraindicated in heart failure of NYHA III or IV class and have to be used with caution in patients with chronic liver disease, neurological disorders, or history of

TABLE 26.3
Adverse Effects Associated with Anti-TNF Therapy

Infections (Listing et al. 2005)	Noninfectious Side Effects
1. Upper and lower respiratory tract infections including pneumonia (Wolfe et al. 2006)	10. Risk of malignancies including lymphoma/leukemia (Tubach et al. 2005)
2. Pulmonary TB (Askling 2009; Dixon et al. 2008; Favalli et al. 2009)	11. Congestive heart failure
3. Viral infections	12. Induction of autoimmunity and autoimmune diseases (e.g., SLE)
4. Skin and subcutaneous tissue infections (Dixon et al. 2006; Favalli et al. 2009)	13. Aplastic anemia
5. Meningitis	14. Intestinal perforation
6. Bone and joint infections	15. Diabetes mellitus (etanercept)
7. Gastrointestinal and urogenital tract infections	16. Vasculitis
8. Sepsis	17. Infusion and injection site reactions
9. Opportunistic infections (Tubach et al. 2005):	
a) Atypical mycobacterium	
b) Histoplasmosis	
c) Listeriosis	
d) Candidiasis	
e) Aspergillosis	
f) *Pneumocystis carinii*	

TABLE 26.4
Suggested Screening Tests for Specific Infections in TNF-α Inhibitor Recipients (Crum et al. 2005; Winthrop 2006)

Infection	Recommended Screening
Tuberculosis (TB)	1. Chest radiograph and PPD at baseline and PPD every 12 months 2. Screen patients for risk factors for *Mycobacterium tuberculosis* 3. Risk factors comprise of birth or residence in a TB endemic country, or history of residing in a mass setting, a positive PPD, substance abuse, exposure to TB patients in health care facility and chest radiographic findings consistent with prior tuberculosis 4. Diagnosis and treatment of latent tuberculosis infection (LTBI) 5. Assume TB as a potential cause of febrile or respiratory infection in all patients receiving TNF-blocking agents
Histoplasmosis	1. Chest radiograph and urine histoplasmin antigen testing at baseline and every 3–4 months for patients who live or have lived in endemic areas 2. Histoplasmin skin tests at baseline However, neither skin nor serologic testing for histoplasmosis is recommended in HIV-infected persons.
Coccidioidomycosis	1. Chest radiograph and serologic testing with IgM and IgG tests at baseline and follow-up testing every 3–4 months for patients who live or have lived in endemic areas 2. Not recommended for HIV-infected persons
Listeria	1. Patient education regarding food preparation and safety 2. To stay away from unpasteurized milk products, undercooked meat, and ready-to-eat foods

malignancy, especially lymphoma. Moderate and severe infections require appropriate treatment before biologic therapy is started because active and quiescent infections may worsen during therapy. The existence of any contraindications to the use of these agents (Table 26.4) needs to be screened before the commencement of therapy. The development of anti-TNF agent has been a major success in the management of patients with autoimmune diseases. Although there are risks associated with anti-TNF therapy, still the role of these agents cannot be underestimated in patient care.

26.3.1.3 Targeting IL-1

IL-1 superfamily includes IL-1α, IL-1β, and the IL-1 receptor antagonist (IL-1RA). The IL-1RA blocks the role of IL-1α and IL-1β in immune activation by competing with them for their receptor. IL-1α and IL-1β are produced by macrophages, monocytes, and dendritic cells. Although IL-1 blockers showed efficacy in preclinical animal models of RA, they are not as effective as TNF-α in humans. Approaches to IL-1 inhibition include IL-1RA gene therapy, IL-1 trap, and anakinra. Among these IL-1 inhibitors, anakinra (Kineret®) is only available for treatment. Anakinra (recombinant form of human IL-1 receptor antagonist) is FDA approved for the treatment of moderate to severe refractory rheumatoid arthritis with MTX (Donahue et al. 2008). It is administered subcutaneously once a day. It suppresses the inflammation and cartilage degradation associated with rheumatoid arthritis and has been reported as safe and well tolerated for up to 3 years of continuous use in patients with the disease. Headache, nausea, flu-like symptoms, and local injection site reactions are common. Severe adverse events include infections (mainly pneumonia and upper respiratory tract infections), sinusitis, and diarrhea. Sepsis and opportunistic infections, including mycobacterial infection and histoplasmosis, have occasionally been reported (Fleischmann et al. 2006; Donahue et al. 2008; Mertens and Singh 2009; Weisman 2002).

Although both TNF-α and IL-1 represent a group of acute phase reactants responsible for inflammation, a combination of anti-TNF and anti-IL-1 therapy could be potentially dangerous. Trials involving the combination of etanercept and anakinra in patients of rheumatoid arthritis with MTX failure showed no added benefit but increased risks of serious infection, injection site reaction, and neutropenia (Genovese et al. 2004). In comparison to anti-TNF agents, IL-1 inhibitors could not make a significant impact on the treatment of rheumatic disease.

26.3.1.4 Targeting IL-6

IL-6 plays essential roles not only in the immune response but also in hematopoiesis and the central nervous system. Deregulated production of IL-6 has been found in chronic inflammatory autoimmune diseases such as RA, systemic onset juvenile idiopathic arthritis (soJIA), Crohn's disease, and SLE. IL-6 accounts for the symptoms of these diseases, and disease activity can be correlated with the serum levels of IL-6. IL-6 binds to both soluble and membrane-bound receptors and leads to the transduction of intracellular signals through the interaction of this complex with gp130, mediating gene activation, and a wide range of biologic activities (Hirano 1998). IL-6 has the ability to activate T cells, B cells, macrophages, and osteoclasts, and is a pivotal mediator of the hepatic acute-phase response. Toclizumab (Actemra®) is a humanized anti IL-6 receptor antibody, which suppresses IL-6 signaling mediated by both membranous and soluble IL-6 receptor. It was tested in patients with RA and showed reduced disease activity and dose-dependent improvement in ACR 20 (Nishimoto et al. 2004). The FDA approved toclizumab in January 2011 for treatment of moderate to severe rheumatoid arthritis in adults who have failed other approved therapies. Main side effects include elevated liver enzymes, elevated low-density lipoprotein (LDL), hypertension, and gastrointestinal perforations.

26.3.1.5 Other Interleukin Antagonists

A human monoclonal Ig G1 kappa anti-IL-15 Ab, AMG714, is currently under phase I and phase II trials for the treatment of rheumatoid arthritis (McInnes et al. 2004). Reports suggest reduction of colitis severity with IL-18 antibody in animal models (Siegmund et al. 2001). Clinical trials of a human anti-IL-18 antibody or IL-18 binding protein are expected (Holmes et al. 2000). Among these therapies, the most promising approach is to target IL-23, which is necessary for differentiation and survival of Th17. Mice deficient in IL-23 are found to be resistant to experimental autoimmune encephalitis, collagen induced arthritis (CIA), and inflammatory bowel disease (Cua et al. 2003; Murphy et al. 2003; Hue et al. 2006). ROR gamma transcription factor and IL-17A and IL-17F are expressed by Th17 cells. IL-17 plays an important role in the induction of TNF alpha and IL-6, growth factor (GM-CSF and G-CSF), and chemokines CXCL8, CXCL1, and CXCL10. Th17 blockade has been shown to be effective in a number of animal models of disease including CIA (Nakee et al. 2003; Lubberts et al. 2001; Lubberts et al. 2004) and thus is a potential target for psoriasis and RA. Moreover, IL-23 induces production of IL-22 in the Th-17 cells. IL-22 and its receptor IL-22R1 are expressed in synovial tissues in RA. The pro-inflammatory role of IL-22 is evident by an increase in MCP-1 expression and proliferation of fibroblast in vitro. IL-22, IL-17, and Th17 cells are an important area of research in the development of therapy for psoriasis. Targeting the shared IL-12/IL-23 p40 subunit as an effective treatment for psoriasis is the area of current interest. Ustekinumab (Stelara, Centocor), a human monoclonal antibody against IL-12/IL-23, is approved in Canada, Europe, and the United States to treat moderate to severe plaque psoriasis. It is now being tested for psoriatic arthritis. In phase III placebo controlled trials, ustekinumab for a 12-week period showed a greater number of patients achieving PASI 75 than in the placebo. The dosing of ustekinumab is more spaced out than previous biologics with subcutaneous injections given at week 0, week 4, and then at 12-week intervals, making treatment more convenient. A major limitation of ustekinumab is the development of neutralizing antibodies, resulting in partial response (defined as achievement of PASI 50 but not PASI 75 by 28 weeks) in patients.

Another IL-12/IL-23 targeted antibody is briakinumab (ABT-874, Abbott), which has just completed phase III trials for treatment of plaque psoriasis. A recent study found that etanercept rapidly downregulates expression of Th17-associated cytokines like IL-17, IL-22, IL-6, IL-1β, and IL-23 as well as CCL20 (Zaba et al. 2007).

A recent report (Genovese et al. 2010) showed that LY2439821 (Eli Lilly), a humanized anti-IL-17 monoclonal antibody, along with oral DMARDs improves signs and symptoms of RA with no major adverse events. Percentages of ACR20, ACR50, and ACR70 responses were higher in the LY2439821-treated group than placebo-controlled groups at multiple time points. Patients also showed improvement in 28-joint disease activity score (DAS28) as early as week 1 of study and lasted for at least 16 weeks (8 weeks after the last dose of drug). Apart from a few cases of mild to moderate leukopenia and neutropenia, no major adverse events were reported, and these events did not increase with escalation of drug doses. Phase II clinical trials are being conducted with LY2439821 in moderate to severe psoriasis.

A recent study (Hueber et al. 2010) showed the efficacy of human anti-IL-17A monoclonal antibody, AIN457 (Novartis), in psoriasis, RA, and uveitis. In that study, 140 patients were enrolled (60, AIN457 treated, and 44, placebo treated). Single administration of AIN457 (3 mg/kg) caused significant reduction in PASI score and investigator global assessment score (IGA) in the psoriasis trial. In immunohistochemical assessment, treatment with AIN457 showed a significant decrease in the area occupied by IL-17A+ and CD3+ T cells at week 4. Reverse transcription polymerase chain reaction (RT-PCR) and microarray analysis of gene expression for different pro-inflammatory cytokines apart from TNF-α showed downregulation with AIN457 treatment. In general, AIN457 treatment for 12 weeks caused a faster and greater relief than placebo at all time points. In the RA trial, the DAS28, C reactive protein (CRP) values decreased with AIN457 treatment. The ACR 20 rate was higher with AIN457 treatment than with placebo at week 6. A phase II clinical trial with AIN457 is now going on in patients with psoriatic arthritis.

26.3.2 Targeting T Cells

In the absence of antigenic stimulus, naïve T cells remain in a dormant state as they circulate through secondary lymphoid tissues, the blood and lymph. T cell activation by antigen-presenting cells (APC) is dependent on two distinct signals. Signal one is mediated through the T cell receptor (TCR), which forms a trimolecular complex with the antigenic peptide and MHC class II molecule from the antigen-presenting cell. Full activation of the T cell requires another signal, known as the signal two, which results from the interaction of costimulatory receptors CD28 and cytotoxic T lymphocyte-associated antigen-4 (CTLA4) (CD152) on the T cell surface, with their respective ligands CD80 (B7-1) and CD86 (B7-2) on the antigen-presenting cells. Binding of CD28 with CD80/86 leads to phosphorylation of a number of intracellular signaling pathways that lead to the recruitment of transcription factors (e.g., NFAT, AP-1, and NF-κB) critical to the transcription of the *IL2* gene. This IL-2 is a multifunctional cytokine which acts via the IL-2 receptor resulting in stimulation of growth of activated T cells via paracrine and autocrine signaling (Malek and Castro 2010). It can promote Th1 and Th2 cell differentiation; it is required for the generation of CD8+ T cell memory responses and plays a crucial role in the maintenance and homeostasis of Treg cells in the periphery. In contrast, CTLA4-Ig binds to B7-1 and B7-2 molecules on APCs and blocks the CD28 mediated costimulatory signal for T cell activation. Therefore, the B7 family of molecules on APCs regulates T cell activation by delivering stimulatory signals through CD28 and inhibitory signals through CD152. There are other ligands like the CD40L (CD154) present on the T cell which bind CD40 on the APC surface. This results in activation of the APCs and induction of *potent microbicidal substances* in them, including reactive oxygen species and nitric oxide, leading to the destruction of ingested microbe. The binding of CD40L to the CD40 receptor on B cells, along with IL-4 secretion from the T cells causes resting B cell activation. As a result, the B cell can undergo division, antibody isotype switching, and differentiation to antibody producing plasma cells.

This unique mechanism of T cell activation has provided several target molecules for therapeutic manipulation of immune responses. These include:

1. Targeting the T lymphocyte costimulatory systems such as CD28, CTLA4 (an inhibitory receptor), and CD80 or CD86 on antigen presenting cells or to target the CD40/CD40 ligand system. Costimulatory signal blockade with CTLA4-Ig inhibited graft rejection and induced long-term tolerance in mice (Lenschow et al. 1992), which led to successful clinical trials with CTLA4-Ig in psoriasis and rheumatoid arthritis (Abrams et al. 1999; Kremer et al. 2003). Although phase I studies with intravenous infusion of abatacept (Orencia®) in patients with moderate to severe psoriasis were successful, but due to lack of additional efficacy and safety data, it is thus far not approved for psoriasis. In a multicenter open-labeled trial (Abrams et al. 1999), it showed that after 26 weeks of treatment, 46% of the treated patients achieved greater than 50% improvement in disease activity, compared with baseline values as compared to only 4% of the control patients. Abatacept was approved in the United States and Europe in 2005 for treatment of RA in adult patients with an inadequate response to DMARDs or TNF inhibitors. In January 2010, it was approved in Europe for moderate to severe active polyarticular juvenile idiopathic arthritis in patients 6 years of age and older. It is marketed by Bristol-Myers Squibb. It is self-administered subcutaneously. Structurally, ABT consists of the extracellular domain of human cytotoxic T-lymphocyte–associated antigen 4 (CTLA4) linked to the modified Fc portion of human immunoglobulin G1. ABT selectively modulates the CD80/CD86:CD28 costimulatory signal required for full T cell activation, downregulating subsequent immune-effector mechanisms (e.g., production of pro-inflammatory cytokines, autoantibodies). The modified Fc portion of ABT is not active; thus, ABT is not associated with adverse events (AEs) resulting from either complement- or antibody-dependent cell-mediated cytotoxicity. Abatacept should not be used concurrently with TNF inhibitors or with the interleukin-1 receptor antagonist because of significant immunosuppression leading to severe infections. Live vaccines should not be given concurrently or within 3 months of stopping abatacept.

Recently, we demonstrated that a monoclonal anti-CD28 antibody (FR255734) prepared by Fujisawa Pharmaceutical Co., Ltd. (now Astellas Pharmaceuticals Inc., Tokyo,

Japan) effectively inhibits cell activation by blocking CD28/B7 co-stimulatory interactions (Raychaudhuri , Kundu-Raychaudhuri, et al. 2008). FR255734 is a humanized IgG2κ anti-human CD28 antibody having complementary determining regions of the mouse anti-human monoclonal antibody TN228 (developed by immunizing BALB/c mice with human CD28-transfected mouse fibroblast L cells and fusing immune splenocytes with P3 U1 myeloma cells) and the Fc domain of human IgG2M3 with two amino acid mutations into human γ2 chain to abolish binding of the antibody to FcγR. In vitro studies with activated human T lymphocytes purified from human peripheral blood mononuclear cells found that FR255734 inhibits proliferation of human T cells stimulated with anti-CD3 and P815/human CD80+ cells as well as pro-inflammatory cytokine production in a concentration-dependent manner. We also used the severe combined immunodeficiency (SCID) mouse–psoriasis xenograft model to evaluate the therapeutic efficacy of this compound in vivo, where we noticed significant improvement in epidermal thickness and reduction in inflammatory infiltrates in the FR255734-treated group as compared to negative controls. These results definitely form a platform for the development of novel therapeutic agents, like FR255734, targeting the interaction of CD28 and B7 costimulatory molecules of activated T cells, which play an active role in psoriasis, rheumatoid arthritis, and multiple sclerosis.

2. Therapies can be developed exploiting T cell receptor (TCR) and MHC-antigen interactions. The first approach is to identify and target the TCR of the pathogenic T cells with monoclonal antibodies. Another strategy will be to block T cell recognition at the level of the antigen-presenting cell by using an altered peptide sequence mimicking the antigenic peptide responsible for disease initiation. This approach remains ineffective because of the lack of information regarding the specific antigen(s) recognized by pathogenic T cells.
3. Targeting the effector memory T cells (TEM): It is well recognized that naïve and memory cells have diverse capacities to travel in lymphoid and nonlymphoid tissues. Recent evidence indicates that memory CD4+ and CD8+ T cells comprise at least two functionally distinct subsets: (1) nonpolarized "central memory" T cells (TCM), which express the chemokine receptor CCR7 and CD62 ligand and home to the T cell areas of secondary lymphoid organs; and (2) polarized "effector memory" T cells (TEM), which have lost the expression of CCR7 and have acquired the capacity to migrate to nonlymphoid tissues. It has been reported that most T cells in psoriatic lesions are of the TEM phenotype. LFA-3/IgG1 fusion protein (alefacept) preferentially targets TEM cells and has been used for treatment of psoriasis with partial success (Krueger and Ellis 2003). The exact mechanism of action is very intricate, involving inhibition of activation of effector T cells by blocking the costimulatory molecule LFA-3/CD2 interaction and also inducing apoptosis of memory-effector T lymphocytes. The activated T cells would stimulate proliferation of keratinocytes, resulting in the typical psoriatic symptoms. Therefore, it leads to clinical improvement of moderate to severe psoriasis by blunting these reactions. The FDA approved alefacept for treatment of adult patients with moderate to severe chronic plaque psoriasis who receive systemic therapy or phototherapy. Administered as 15 mg intramuscular injections weekly for 12 weeks, it is required to check CD4 cell counts every week or every other week while on therapy, and the dose should be stopped if the count is less than 250/μL and should be discontinued if the CD4 count remains below 250/μL for 1 month. Due to its effect on the CD4 cell counts, it is contraindicated in patients infected with HIV.
4. K+ channels in the immune system: K+ channels regulate membrane potential and Ca2+ signaling in both excitable and nonexcitable cells The voltage-gated Kv1.3 expression is increased four- to five-fold in activated CD4+ and CD8+ memory T cells, whereas human naïve or TCM cells upregulate the calcium-activated KCa3.1 channel to regulate membrane potential and Ca2+ signaling in the activated state. Binding of APC with the TCR triggers a Ca2+ influx through voltage-independent Ca2+ channels, leading to an increase in cytosolic Ca2+ resulting in translocation of NFAT to the nucleus ultimately resulting in

cytokine secretion and T cell proliferation. The membrane potential of this Ca2+ influx is maintained by K+ efflux through Kv1.3 and/or KCa3.1. So targeting KCa3.1 for naïve T cell-mediated acute immune reactions and Kv1.3 for memory T cell-induced chronic immune reactions offer potential for novel therapeutic molecules. As KV1.3 channels are expressed predominantly in the memory T cell, Kv1.3 blockers may constitute a promising new drug candidate for the treatment of TEM-cell-mediated inflammatory skin diseases like allergic contact dermatitis and psoriasis (Azam et al. 2007). Gilhar et al. (2011) found that there was suppression of development of psoriatic features in three of six human grafts in the SCID mice model by ShK, a known Kv1.3 channel blocker. It also reduced T cell infiltrates and cytokine (IFN-γ and TNF-α) production. However, there was incomplete recovery of the psoriaform grafts with ShK treatment. This may be due to production of TNF-α by keratinocytes, Langerhans cells, and mast cells, which most likely are not affected by ShK. Still, this drug, due to its varied mechanism of action from other immunotherapies, has enormous potential to be a highly effective anti-psoriatic agent with fewer side effects in humans.

26.3.3 Targeting B Cells in the Treatment of Systemic and Cutaneous Autoimmune Diseases

Preclinical studies with autoimmune-prone MRL-lpr/lpr mice lacking B cells failed to develop autoimmune kidney destruction, vasculitis, or autoantibodies (Chaudhari et al. 2001), but MRL-lpr/lpr mice with B cells not capable of producing antibodies still develop autoimmune disease. B cells have a double role in the pathogenesis of autoimmune diseases, including presentation of antigen to the T cells and antibody production. So targeting the B cells will be a novel approach that will not only remove the B cell antigen presenting cells (APCs) but also will get rid of the autoantibody producing B cells. Keeping this in mind, various markers of B cells such as CD20, CD22, and B cell growth factors like Blys and APRIL have been targeted to treat autoimmune diseases.

26.3.3.1 Anti CD20 Therapies

- Rituximab: Rituximab is a chimeric monoclonal IgG1 antibody directed against the B-lymphocyte surface antigen, CD20. Its exact mechanism of action remains unclear, believed to induce CD20-positive B-cell lysis by several mechanisms including complement-dependent toxicity, antibody-dependent cell cytotoxicity, and induction of apoptosis. RTX has been FDA approved for the treatment of patients with RA who have had an inadequate response to anti-TNF-α therapy. It is given as an intravenous infusion 500 to 1000 mg every 2 weeks for two doses for RA. Rituximab has a half-life of approximately 21 days and can deplete peripheral B cells for up to 9 months or longer after a single course of therapy. Shown to improve the signs and symptoms of disease, functional status, quality of life, and retard radiographic progression of disease in patients with RA who had previously failed to respond to methotrexate or TNF inhibitors. Although rituximab can be used alone or in combination with DMARDs, combination therapy yielded better clinical outcome. Moreover, patients seropositive for rheumatoid factor (RF) seemed to have a greater clinical response than seronegative patients (Quartuccio et al. 2009). Rituximab has also shown promising results in the treatment of other autoimmune diseases such as SLE, primary Sjögren syndrome, idiopathic thrombocytopenic purpura, chronic inflammatory demyelinating polyneuropathy, multiple sclerosis, and vasculitis (e.g., Wegener granulomatosis, relapsing polychondritis).
- Frequent infusion reactions and potentially neutralizing antichimeric antibody associated with rituximab have prompted development of several humanized and fully human anti-CD20 mAbs. Two humanized anti-CD20 monoclonal antibodies that target CD20 in RA are in phases 2 and 3 of clinical development: ocrelizumab (Genentech, South San Francisco, California) and ofatumumab (Genmab, Copenhagen, Denmark). Both ocrelizumab and

ofatumumab are given intravenously and have shown promising results among patients with RA who have failed to respond to at least one DMARD and/or TNF inhibitors.

- TRU-015: CD20-directed small modular immunopharmaceutical protein (SMIP): These are dimeric, single-chain polypeptides approximately one-third to one-half the size of mAbs. Although SMIPs can induce antibody-dependent cell cytotoxicity and apoptosis, by design, they have attenuated complement activation and subsequent lower levels of complement-dependent toxicity. TRU-015 is a SMIP that binds to CD20 on pre-B to mature B cells and causes dose-dependent B-cell depletion. Earlier studies have been promising with favorable clinical response and safety profile among patients with RA (Burge et al. 2008; Stromatt et al. 2009). Studies are ongoing to assess the long-term efficacy and safety in the treatment of RA and lymphoma.

26.3.3.2 Targeting B-Cell Activating Factor and a Proliferation-Inducing Ligand

B cell–activating factor (BAFF) or B-lymphocyte stimulator (BLyS) and a proliferation-inducing ligand (APRIL) are members of TNF cytokine family and are essential for survival and growth of B cells. Belimumab is a recombinant fully human anti-BLyS monoclonal antibody (LymphoStat-B) that binds to soluble BAFF and prevents its binding to BAFF-R. In SLE, in phase I, II, and III trials showed stabilization of disease activity with fewer side effects. In RA, in a phase II trial, although belimumab decreased the levels of B cells and RF, it failed to improve the overall RA disease activity.

Atacicept is a recombinant fusion protein comprised of a portion of the transmembrane activator and calcium-modulator and an immunoglobulin chain (TACI-Ig or atacicept) fused with the Fc fragment of human IgG1 and is designed to have a broader effect on different stages of B-cell differentiation. Atacicept targets molecules on the B cell surface that promote B cell survival (BLyS and APRIL). Several early studies in patients with RA and SLE have shown that atacicept can reduce peripheral blood B cells and serum immunoglobulin without associated toxicity (Tak et al. 2008). Phase II and III clinical trials are currently under way to assess the efficacy and safety of atacicept in the treatment of SLE, RA, and relapsing multiple sclerosis.

26.3.4 Angiogenesis Factor

Vascular endothelial growth factor (VEGF), a potent endothelial cell mitogen and vascular permeability factor, may play a pivotal role in psoriasis and RA, by supplying oxygen and nutrients necessary for cell metabolism and division, as well as by bringing in leukocytes and signaling mediators such as cytokines, chemoattractants, and growth factors. Within the synovium, resident macrophages and fibroblast-like synovial cells have been found to express VEGF protein, while mRNA encoding for both VEGF receptors (Flt-1 and KDR) has been detected on microvascular endothelial cells. Elevated levels of VEGF have been detected in the synovial fluid and serum from RA patients when compared to healthy controls, and serum VEGF levels have been shown to correlate with disease severity (Miotia et al. 2000; Afuwape et al. 2003). Inhibition of VEGF in vivo is accompanied by side effects, such as impaired wound healing, hemorrhage, and gastrointestinal perforation.

Placental growth factor (PlGF), like VEGF, binds to VEGF-R1 (and soluble VEGF-R1), but, in comparison to VEGF, PlGF does not bind VEGF-R2 (Autiero et al. 2003; Tjwa et al. 2003). PlGF in preclinical studies may play a more pronounced role in pathological angiogenesis, shown by impaired tumor growth and vascularization without accompanying major vascular abnormalities (Oosthuyse et al. 2001). Furthermore, PlGF is expressed in synovial fluid, making it a potentially important therapeutic target (Bottomley et al. 2000).

26.3.5 Drugs That Inhibit Leukocyte Adhesion

Inflammation can be downregulated by blocking leukocyte migration. Intercellular adhesion molecule-1 (ICAM-1) is a transmembrane glycoprotein that facilitates propagation of inflammatory cells. Communication between T cells and mononuclear phagocytes is mediated by their interaction with lymphocyte function-associated antigen 1 (CD11a) and ICAM-1 (CD54).

Efalizumab: Multicenter randomized, controlled trials have shown that efalizumab (Raptiva®), a humanized monoclonal antibody to CD11a, has benefit in the treatment of psoriasis (46), but in April 2009, it was voluntarily withdrawn from the market due to the risks of progressive multifocal leukoencephalopathy (FDA 2009a).

23.3.6 Targeting Nerve Growth Factor

The recognition that nerve growth factor (NGF) and its receptor system (NGF-R) has a critical role in pathogenesis of inflammation, inflammatory disease, and pain mechanisms has provided an astonishing and attractive opportunity to develop a novel class of therapeutics for inflammatory diseases and chronic pain syndromes. Raychaudhuri et al. (2008) demonstrated that increased levels of NGF in transplanted psoriatic plaques led to marked proliferation of murine cutaneous nerve fibers with upregulation of neuropeptides in transplanted plaques as compared to transplanted normal human skin on SCID mouse. They have shown further that K252a, a high-affinity receptor inhibitor and neutralizing NGF antibody, is therapeutically effective in psoriasis (Raychaudhuri, Jiang, et al. 2008). Interestingly, K252a is now in phase II clinical trial as a novel topical treatment for psoriasis. In 2005, Shelton et al. reported that treatment with anti-NGF antibody is efficacious for autoimmune arthritis of rats (Shelton et al. 2005), it reversed weight loss and decreased joint pain, and moderately decreased inflammatory cytokine levels. These results encouraged Lane et al. to do a phase I study in osteoarthritis (OA; Lane et al. 2005) with RN624 (anti-NGF), where RN624 improved pain and function in subjects with moderate knee OA.

26.4 CONCLUSIONS

Development of biologics for autoimmune diseases has revolutionized the treatment of chronic autoimmune diseases like RA, psoriasis, and many others. These agents have transformed our perspective toward the modern-day therapeutic armamentarium using highly “targeted immunotherapy” focusing specific steps in the immune and pro-inflammatory response in autoimmune diseases. The development of TNF-α blocker biologics for the treatment of psoriasis, psoriatic arthritis, RA, Crohn’s, and ankylosing spondylitis is a major breakthrough. Similarly, B cell-depletion therapy has provided a very effective therapeutic option for critical patients of lupus, pemphigus, rheumatoid arthritis, and ANCA-associated vasculitis.

A major challenge in biologic therapy is the development of serious side effects, its cost of therapy as well as for side effects, and the development of autoantibodies decreasing its long-term efficacy. Clinical familiarity of biological therapies is presently an ongoing process and still the long-term safety is uncertain. The future aim will be to improve the safety and efficacy of immunotherapy from the current state of establishing remission to eventually cure of the diseased state without increasing adverse effects and costs of treatment.

TAKE-HOME MESSAGES

- The development of TNF-α blockers for the treatment of psoriasis, psoriatic arthritis, RA, Crohn’s, and ankylosing spondylitis is a major breakthrough.
- Anti-TNF agents have considerable advantages over the existing immunomodulators. They are designed to target a very specific component of the immune-mediated inflammatory cascades and may have lower risks of systemic side effects.
- B cell-depletion therapy has provided a very effective therapeutic option for critical patients of lupus, pemphigus, rheumatoid arthritis, and ANCA-associated vasculitis.

- PML (progressive multifocal leukoencephalopathy) is a severe emerging infection in immunocompromised patients. Unexpected cases of PML have been reported in patients who received the immune modulatory monoclonal antibodies rituximab, efalizumab, and natalizumab.
- IL-23 induces production of IL-17 and IL-22 in the Th17 cells. IL-23, IL-22, IL-17, and Th17 cells are an important area of research for novel therapies of autoimmune diseases.
- High cost and the potential for serious side effects of biologics are social and clinical challenges to the current generation of physicians.

REFERENCES

Abrams, J.R., Lebwohl, M.G., Guzzo, C.A., Jegasothy, B.V., Goldfarb, M.T., Goffe, B.S., et al. 1999. CTLA4Ig-mediated blockade of T-cell costimulation in patients with psoriasis vulgaris. *J Clin Invest* 103:1243–1252.

Afuwape, A.O., Feldmann, M., Paleolog, E.M. 2003. Adenoviral delivery of soluble VEGF receptor 1 (sFlt-1) abrogates disease activity in murine collagen-induced arthritis. *Gene Ther* 10:1950–1960.

Allez, M., Vermeire, S., Mozziconacci, N., Michetti, P., Laharie, D., Louis, E., et al. 2009. Efficacy and safety of a third anti-TNF monoclonal antibody in Crohn's disease after failure of two other anti-TNF. *Aliment Pharmacol Ther* [Epub ahead of print].

Askling, J. 2009. Risk for tuberculosis following treatment of rheumatoid arthritis with anti-TNF therapy: the Swedish experience 1998–2008. *Ann Rheum Dis* 68(suppl3):422.

Autiero, M., Luttun, A., Tjwa, M., Carmeliet, P. 2003. Placental growth factor and its receptor, vascular endothelial growth factor receptor-1: Novel targets for stimulation of ischemic tissue revascularization and inhibition of angiogenic and inflammatory disorders. *J Thromb Haemost* 1:1356–1370.

Azam, P., Sankaranarayanan, A., Homerick, D., Griffey, S., Wulff, H. 2007. Targeting effector memory T cells with the small molecule Kv1.3 blocker PAP-1 suppresses allergic contact dermatitis. *J Invest Dermatol* 127:1419–1429.

Barczyk, A., Pierzchala, W., Sozanska, E. 2003. Interleukin-17 in sputum correlates with airway hyperresponsiveness to methacholine. *Respir Med* 97:726–733.

Borchers, A.T., Leibushor, N., Cheema, G.S., Naguwa, S.M., Gershwin, M.E. 2011. Immune-mediated adverse effects of biologicals used in the treatment of rheumatic diseases. *J Autoimmun* doi:10.1016/j.jaut.2011.08.002 [Epub ahead of print].

Bottomley, M.J., Webb, N.J., Watson, C.J., Holt, L., Bukhari, M., Denton, J., et al. 2000. Placenta growth factor (PlGF) induces vascular endothelial growth factor (VEGF) secretion from mononuclear cells and is co-expressed with VEGF in synovial fluid. *Clin Exp Immunol* 119:182–188.

Braun, J., Bollow, M., Neure, L., Seipelt, E., Seyrekbasan, F., Herbst, H., et al. 1995. Use of immunohistologic and in situ hybridization techniques in the examination of sacroiliac joint biopsy specimens from patients with ankylosing spondylitis. *Arthritis Rheum* 38:499–505.

Burge, D.J., Bookbinder, S.A., Kivitz, A., Fleischmann, R.M., Shu, C., Bannink, J. 2008. Pharmacokinetic and pharmacodynamic properties of TRU-015, a CE20-directed small modular immunopharmaceutical protein therapeutic, in patients with rheumatoid arthritis: A phase I, open-label, dose-escalation clinical study. *Clin Ther* 30:1806–1816.

Chaudhari, U., Romano, P., Mulcahy, L.D., Dooley, L.T., Baker, D.G., Gottlieb, A.B. 2001. Efficacy and safety of infliximab monotherapy for plaque-type psoriasis: A randomised trial. *Lancet* 357(9271):1842–1847.

Crum, N.F., Lederman, E.R., Wallace, M.R. 2005. Infections associated with tumor necrosis factor-alpha antagonists. *Medicine* (Baltimore) 84:291–302.

Cua, D.J., Sherlock, J., Chen, Y., Murphy, C.A., Joyce, B., Seymour, B., et al. 2003. Interleukin-23 rather than interleukin-12 is the critical cytokine for autoimmune inflammation of the brain. *Nature* 421:744–748.

Davila, E., Kang, Y.M., Park, Y.W., Sawai, H., He, X., Pryshchep, S., et al. 2005. Cell-based immunotherapy with suppressor CD8+ T cells in rheumatoid arthritis. *J Immunol* 174(11):7292–301.

Day, R. 2002. Adverse reactions to TNF-alpha inhibitors in rheumatoid arthritis. *Lancet* 359:540–541.

Dixon, W.G., Hyrich, K.L., Watson, K.D., Lunt, M. 2008. Drug-specific risk of tuberculosis in patients with rheumatoid arthritis treated with anti-TNF therapy: Results from the BSR Biologics Register (BSRBR). American College of Rheumatology 2008 Annual Scientific Meeting: October 24–29, 2008; San Francisco, California. *Arthritis Rheum Abstract* THU0134.

Dixon, W.G., Watson, K., Lunt, M., Hyrich, K.L., Silman, A.J., Symmons, D.P. 2006. Rates of serious infection, including site-specific and bacterial intracellular infection, in rheumatoid arthritis patients receiving anti-tumor necrosis factor therapy: Results from the British Society for Rheumatology Biologics Register. *Arthritis Rheum* 54:2368–2376.

Donahue, K.E., Gartlehner, G., Jonas, D.E., Lux, L.J., Thieda, P., Jonas, B.L., et al. 2008. Systemic review: Comparative effectiveness and harms of disease-modifying medications for rheumatoid arthritis. *Ann Intern Med* 148:124–134.

Duerr, R.H., Taylor, K.D., Brant, S.R., Rioux, J.D., Silverberg, M.S., Daly, M.J., et al. 2006. A genome-wide association study identifies IL23R as an inflammatory bowel disease gene. *Science* 314:1461–1463

Emery, P., Fleischmann, R.M., Moreland, L.W., Hsia, E.C., Strusberg, I., Durez, P., et al. 2009. Golimumab, a human anti-tumor necrosis factor alpha monoclonal antibody, injected subcutaneously every four weeks in methotrexate-naive patients with active rheumatoid arthritis: Twenty-four-week results of a phase III, multicenter, randomized, double-blind, placebo-controlled study of golimumab before methotrexate as first-line therapy for early-onset rheumatoid arthritis. *Arthritis Rheum* 60(8):2272–2283.

Favalli, E.G., Desiati, F., Atzeni, F., Sarzi-Puttini, P., Caporali, R., Pallavicini, F.B., et al. 2009. Serious infections during anti-TNFalpha treatment in rheumatoid arthritis patients. *Autoimmun Rev* 8:266–273.

Feldmann, M., Brennan, F.M., Maini, R.N. Role of cytokines in rheumatoid arthritis. 1996. *Annu Rev Immunol* 14:397–440.

Finckh, A., Simard, J.F., Duryea, J., Liang, M.H., Huang, J., Daneel, S., et al. 2006.The effectiveness of anti-tumor necrosis factor therapy in preventing progressive radiographic joint damage in rheumatoid arthritis: A population-based study. *Arthritis Rheum* 54:54–59.

Fleischmann, R., Vencovsky, J., van Vollenhoven, R.F., Borenstein, D., Box, J., Coteur, G., et al. 2009. Efficacy and safety of certolizumab pegol monotherapy every 4 weeks in patients with rheumatoid arthritis failing previous disease-modifying antirheumatic therapy: The FAST4WARD study. *Ann Rheum Dis* 68:805–811.

Fleischmann, R.M., Tesser, J., Schiff, M.H., Schechtman, J., Burmester, G.R., Bennett, R., et al. 2006. Safety of extended treatment with anakinra in patients with rheumatoid arthritis. *Ann Rheum Dis* 65:1006–1012.

Food and Drug Administration. 2009a. US Food and Drug Administration FDA public health advisory: Updated safety information about Raptiva (efalizumab) [online]. http://www.fda.gov/Drugs/DrugSafety/PostmarketDrugSafetyInformationforPatientsandProviders/DrugSafetyInformationforHeathcareProfessionals/PublicHealthAdvisories/ucm110605.htm.

Food and Drug Administration. 2009b. FDA labelling information. FDA Web site: http://www.accessdata.fda.gov/drugsatfda_docs/label/2009/125289s000lbl.pdf.

Gall, J.S. and Kalb, R.E. 2008. Infliximab for the treatment of plaque psoriasis. *Biologics* 2(1):115–124.

Genovese, M.C., Cohen, S., Moreland, L., Lium, D., Robbins, S., Newmark, R., et al. 2004. Combination therapy with etanercept and anakinra in the treatment of patients with rheumatoid arthritis who have been treated unsuccessfully with methotrexate. *Arthritis Rheumat* 50:1412–1419.

Genovese, M.C., Van den Bosch, F., Roberson S.A., Bojin, S., Biagini, I.M., Ryan, P., et al. 2010. LY2439821, a humanized anti-intereukin-17 monoclonal antibody in the treatment of patients with rheumatoid arthritis. *Arthritis & Rheumatism* 62:929–939.

Gilhar, A., Bergman, R., Assay, B., Ullmann, Y., Etzioni, A. 2011. The beneficial effect of blocking Kv1.3 in the psoriasiform SCID mouse model. *J Invest Dermatol* 131(1):118–124.

Gordon, K.B., Langley, R.G., Leonardi, C., Toth, D., Menter, M.A., Kang, S., et al. 2006. Clinical response to adalimumab treatment in patients with moderate to severe psoriasis: Double-blind, randomized controlled trial and open-label extension study. *J Am Acad Dermatol* 55:598.

Hirano, T. 1998. Interleukin 6 and its receptor: Ten years later. *Int Rev Immunol* 16:249.

Holmes, S., Abrahamson, J.A., Al-Mahdi, N., Abdel-Meguid, S.S., Ho, Y.S. 2000. Characterization of the in vitro and in vivo activity of monoclonal antibodies to human IL-18. *Hybridoma* 19:363.

Hue, S., Ahern, P., Buonocore, S., Kullberg, M.C., Cua, D.J., McKenzie, B.S., et al. 2006. Interleukin-23 drives innate and T cell-mediated intestinal inflammation. *J Exp Med* 203:2473–2483.

Hueber, W., Patel, D.D., Dryja, T., Wright, A.M., Koroleva, I., Bruin, G., Antoni, C., Draelos, Z., Gold, M.H.; Psoriasis Study Group, Durez, P., Tak, P.P., Gomez-Reino, J.J.; Rheumatoid Arthritis Study Group, Foster, C.S., Kim, R.Y., Samson, C.M., Falk, N.S., Chu, D.S., Callanan, D., Nguyen, Q.D.; Uveitis Study Group, Rose, K., Haider, A., Di Padova, F. 2010. Effects of AIN457, a fully human antibody to interleukin-17A, on psoriasis, rheumatoid arthritis and uveitis. *Sci Transl Med* 2:52ra72.

Inman, R.D., Davis, J.C. Jr., Heijde, D., Diekman, L., Sieper, J., Kim, S.I., et al. 2008. Efficacy and safety of golimumab in patients with ankylosing spondylitis: Results of a randomized, double-blind, placebo-controlled, phase III trial. *Arthritis Rheum* 58:3402–3412.

Ishida, I., Tomizuka, K., Yoshida, H., Tahara, T., Takahashi, N., Ohguma, A., et al. 2002. Production of human monoclonal and polyclonal antibodies in TransChromo animals. *Cloning Stem Cells* 4(1):91–102.

Jacobi, A., Mahler, V., Schuler, G., Hertl, M. 2006. Treatment of inflammatory dermatoses by tumour necrosis factor antagonists. *J Eur Acad Dermatol Venereol* 20(10):1171–1187.

Kavanaugh, A., McInnes, I., Mease, P., Krueger, G.G., Gladman, D., Gomez-Reino, J., et al. 2009. Golimumab, a new human tumor necrosis factor α antibody, administered every four weeks as a subcutaneous injection in psoriatic arthritis: Twenty-four-week efficacy and safety results of a randomized, placebo-controlled study. *Arthritis Rheum* 60:976–986.

Keystone, E.C., Genovese, M.C., Klareskog, L., Hsia, E.C., Hall, S.T., Miranda, P.C., et al. 2009. Golimumab, a human antibody to tumour necrosis factor α given by monthly subcutaneous injections, in active rheumatoid arthritis despite methotrexate therapy: The GO-FORWARD study. *Ann Rheum Dis* 68:789–796.

Kirkham, B.W., Lassere, M.N., Edmonds, J.P., Juhasz, K.M., Bird, P.A., Lee, C.S., et al. 2006. Synovial membrane cytokine expression is predictive of joint damage progression in rheumatoid arthritis: A two-year prospective study (the DAMAGE study cohort). *Arthritis Rheum* 54:1122–1131.

Koenders, M.I., Lubberts, E., van de Loo, F.A., Oppers-Walgreen, B., Van Den Bersselaar, L., Helsen, M.M., et al. 2006. Interleukin-17 acts independently of TNF-α under arthritic conditions. *J Immunol* 176:6262–6269.

Kotake, S., Udagawa, N., Takahashi, N., Matsuzaki, K., Itoh, K., Ishiyama, S., et al. 1999. IL-17 in synovial fluids from patients with rheumatoid arthritis is a potent stimulator of osteoclastogenesis. *J Clin Invest* 103:1345–1352.

Kremer, J.M., Westhovens, R., Leon, M., Di Giorgio, E., Alten, R., Steinfeld, S., et al. 2003. Treatment of rheumatoid arthritis by selective inhibition of T-cell activation with fusion protein CTLA4Ig. *N Engl J Med* 349:1907–1915.

Krueger, G.G., Ellis, C.N. 2003. Alefacept therapy produces remission for patients with chronic plaque psoriasis. *Br J Dermatol* 148(4):784–788.

Krueger, G.G., Langley, R.G., Leonardi, C., Yeilding, N., Guzzo, C., Wang, Y., et al. 2007. A human interleukin-12/23 monoclonal antibody for the treatment of psoriasis. *N Engl J Med* 356:580–592.

Lane, N., Webster, L., Lu, S., Gray, M., Hefti, F., Walicke, P. 2005. RN624 (Anti-NGF) improves pain and function in subjects with moderate knee osteoarthritis: A phase I study. *Arthritis Rheum* 52(Suppl.):9.

Lawson, C.A., Brown, A.K., Bejarano, V., Douglas, S.H., Burgoyne, C.H., Greenstein, A.S., et al. 2006. Early rheumatoid arthritis is associated with a deficit in the CD4+CD25high regulatory T cell population in peripheral blood. *Rheumatology* (Oxford) 45(10):1210–1217.

Lenschow, D.J., Zeng, Y., Thistlethwaite, J.R., Montag, A., Brady, W., Gibson, M.G., et al.1992. Long-term survival of xenogeneic pancreatic islet grafts induced by CTLA4Ig. *Science* 257:789–792.

Listing, J., Strangfeld, A., Kary, S., Rau, R., von Hinueber, U., Stoyanova-Scholz, M., et al. 2005. Infections in patients with rheumatoid arthritis treated with biologic agents. *Arthritis Rheum* 52:3403–412.

Lubberts, E., Joosten, L.A., Oppers, B., van den Bersselaar, L., Coenende Roo, C.J., Kolls, J.K., et al. 2001. IL-1-independent role of IL-17 in synovial inflammation and joint destruction during collagen-induced arthritis. *J Immunol* 167:1004–1013.

Lubberts, E., Koenders, M.I., Oppers-Walgreen, B., van den Bresselaar, L., Coenen-de Roo, C.J., Joosten, L.A., et al. 2004. Treatment with a neutralizing anti-murine interleukin-17 antibody after the onset of collagen-induced arthritis reduces joint inflammation, cartilage destruction and bone erosion. *Arthritis Rheum* 50:650–659.

Malek, T.R., Castro, I. 2010. Interleukin-2 receptor signaling: At the interface between tolerance and immunity. *Immunity* 33(2):153–165.

Maniati, E., Potter, P., Rogers, N.J., Morley, B.J. 2008. Control of apoptosis in autoimmunity. *J Pathol* 214(2):190–198.

Matusevicius, D., Kivisakk, P., He, B., Kostulas, N., Ozenci, V., Fredrikson, S., et al. 1999. Interleukin-17 mRNA expression in blood and CSF mononuclear cells is augmented in multiple sclerosis. *Mult Scler* 5(2):101–104.

McCluggage, L.K., Scholtz, J.M. 2010. Golimumab: A tumor necrosis factor alpha inhibitor for the treatment of rheumatoid arthritis. *Ann Pharmacother* 44(1):135–144.

McInnes, I., Martin, R., Zimmerman-Gorska, I., Nayiager, S., Sun, G., Patel, A., et al. 2004. Safety and efficacy of a human monoclonal antibody to IL-15 (AMG 714) in patients with rheumatoid arthritis (RA): Results from a multicenter, randomized, double-blind, placebo-controlled trial. *Arthritis Rheumat* 50 (Suppl 9):Abstract 527.

Melmed, G.Y., Targan, S.R., Yasothan, U., Hanicq, D., Kirkpatrick, P. 2008. Certolizumab pegol. *Nat Rev Drug Discov* 7(8):641–642.

Mertens, M., Singh, J.A. 2009. Anakinra for rheumatoid arthritis. *Cochrane Database Syst Rev* 1:CD005121.

Miotla, J., Maciewicz, R., Kendrew, J., Feldmann, M., Paleolog, E. 2000. Treatment with soluble VEGF receptor reduces disease severity in murine collagen-induced arthritis. *Lab Invest* 80:1195–1205.

Miranda-Carus, M.E., Benito-Miguel, M., Balsa, A., Cobo-Ibanez, T., Perez de Ayala, C., Pascual-Salcedo, D., et al. 2006. Peripheral blood T lymphocytes from patients with early rheumatoid arthritis express RANKL and interleukin-15 on the cell surface and promote osteoclastogenesis in autologous monocytes. *Arthritis Rheum* 54:1151–1164.

Murphy, C.A., Langrish, C.L., Chen, Y., Blumenschein, W., McClanahan, T., Kastelein, R.A., et al. 2003. Divergent pro- and antiinflammatory roles for IL-23 and IL-12 in joint autoimmune inflammation. *J Exp Med* 198:1951–1957.

Najafian, N., Chitnis, T., Salama, A.D., Zhu, B., Benou, C., Yuan, X., et al. 2003. Regulatory functions of CD8+CD28- T cells in an autoimmune disease model. *J Clin Invest* 112(7):1037–1048.

Nakee, S., Nambu, A., Sudo, K., Iwakura, Y. 2003. Suppression of immune induction of collagen-induced arthritis in IL-17 deficient mice. *J Immunol* 171:6173–6177.

Nishimoto, N., Yoshizaki, K., Miyasaka, N., Yamamoto, K., Kawai, S., Takeuchi, T., et al. 2004. Treatment of rheumatoid arthritis with humanized anti-interleukin-6 receptor antibody: A multicenter, double-blind, placebo-controlled trial. *Arthritis Rheumat* 50:1761–1769.

Oosthuyse, B., Moons, L., Storkebaum, E., Beck, H., Nuyens, D., Brusselmans, K., et al. 2001. Deletion of the hypoxia-response element in the vascular endothelial growth factor promoter causes motor neuron degeneration. *Nat Genet* 28:131–138.

Pene, J., Chevalier, S., Preisser, L., Venereau, E., Guilleux, M.H., Ghannam, S., et al. 2008. Chronically inflamed human tissues are infiltrated by highly differentiated Th17 lymphocytes. *J Immunol* 180:7423–7430.

Quartuccio, L., Fabris, M., Salvin, S., Atzeni, F., Saracco, M., Benucci, M., et al. 2009. Rheumatoid factor positivity rather than anti-CCP positivity, a lower disability and a lower number of anti-TNF agents failed are associated with response to rituximab in rheumatoid arthritis. *Rheumatology* 48:1557–1559.

Raychaudhuri, S. P., Jiang, WY, Raychaudhuri, SK. 2008. Revisiting the Koebner phenomenon: Role of NGF and its receptor system in the pathogenesis of psoriasis. *Am J Pathol* 172(4):961–971.

Raychaudhuri, S.P. and Raychaudhuri, S.K. 2011. Biologics: Target-specific treatment of systemic and cutaneous autoimmune diseases. In: *Arthritis Pathophysiology, Prevention and Therapeutics*, ed. Bagchi, D., Moriyama, H., and Raychaudhuri, S.P., 147–163. Boca Raton: Taylor & Francis, CRC Press.

Raychaudhuri, S.P., Kundu-Raychaudhuri, S., Tamura, K., Masunaga, T., Kubo, K., Hanaoka, K., et al. 2008. FR255734, a humanized, Fc-silent, anti-CD28 antibody, improves psoriasis in the SCID mouse-psoriasis xenograft model. *J Invest Dermatol* 128:1969–1976.

Raychaudhuri, S.P., Nguyen, C.T., Raychaudhuri, S.K., Gershwin, M.E. 2009. Incidence and nature of infectious disease in patients treated with anti-TNF agents. *Autoimmun Rev* 9(2):67–81.

Sandborn, W.J., Feagan, B.G., Stoinov, S., Honiball, P.J., Rutgeerts. P., Mason, D., et al. and the PRECiSE 1 Study Investigators. 2007. Certolizumab pegol for the treatment of Crohn's disease. *N Engl J Med* 357:228–238.

Schreiber, S., Khaliq-Kareemi, M., Lawrance, I.C., Thomsen, O.Ø., Hanauer, S.B., McColm, J., et al. and the PRECiSE 2 Study Investigators. 2007. Maintenance therapy with certolizumab pegol for Crohn's disease. *N Engl J Med* 357:239–250.

Shealy, D., Cai, A., Lacy, E., Nesspor, T., Staquet, K., Johns, L., et al. 2007. Characterization of golimumab (CNTO 148), a novel fully human monoclonal antibody specific for TNFalpha (EULAR abstract THU0088). *Ann Rheum Dis* 66(Suppl II):151.

Shelton, D.L., Zeller, J., Ho, W.H., Pons, J., Rosenthal, A. 2005. Nerve growth factor mediates hyperalgesia and cachexia in auto-immune arthritis. *Pain* 116(1-2):8–16.

Siegmund, B., Fantuzzi, G., Rieder, F., Gamboni-Robertson, F., Lehr, H.A., Hartmann, G., et al. 2001. Neutralization of interleukin-18 reduces severity in murine colitis and intestinal IFN-gamma and TNF-alpha production. *Am J Physiol Regul Integr Comp Physiol* 281:R1264.

Smolen, J., Landewé, R.B., Mease, P., Brzezicki, J., Mason, D., Luijtens, K., et al. 2009. Efficacy and safety of certolizumab pegol plus methotrexate in active rheumatoid arthritis: The RAPID 2 study. A randomised controlled trial. *Ann Rheum Dis* 68:797–804.

Smolen, J.S., Kay, J., Doyle, M.K., Landewé, R., Matteson, E.L., Wollenhaupt, J., et al. 2009. Golimumab in patients with active rheumatoid arthritis after treatment with tumour necrosis factor α inhibitors (GO-AFTER study): A multicentre, randomised, double-blind, placebo-controlled, phase III trial. *Lancet* 374:210–221.

Stromatt, S., Chopiak, V., Dvoretskiy, L., Koshukova, G., Nasonov, E., Povoroznyuk, V., et al. 2009. Sustained safety and efficacy of TRU-015 with continued retreatment of rheumatoid arthritis subjects following a phase 2B study [abstract 403]. *Arthritis Rheum* 60(suppl 10):S148–149.

Tak, P.P., Thurlings, R.M., Rossier, C., Nestorov, I., Dimic, A., Mircetic, V., et al. 2008. Atacicept in patients with rheumatoid arthritis: Results of a multicenter, phase Ib, double blind, placebo-controlled, dose-escalating, single- and repeated-dose study. *Arthritis Rheum* 58:61–72.

Tartour, E., Lee, R.S., Fridman, W.H. 1994. Anti-cytokines: Promising tools for diagnosis and immunotherapy. *Biomed Pharmacother* 48(10):417–424.

Taylor, P.C., Steuer, A., Gruber, J., McClinton, C., Cosgrove, D.O., Blomley, M.J., et al. 2006. Ultrasonographic and radiographic results from a two-year controlled trial of immediate or one-year-delayed addition of infliximab to ongoing methotrexate therapy in patients with erosive early rheumatoid arthritis. *Arthritis Rheum* 54:47–53.

Tjwa, M., Luttun, A., Autiero, M., Carmeliet, P. 2003. VEGF and PlGF: Two pleiotropic growth factors with distinct roles in development and homeostasis. *Cell Tissue Res* 314:5–14.

Tubach, F., Salmon-Ceron, D., Ravaud, P., Mariette, X. 2005.The RATIO observatory: French registry of opportunistic infections, severe bacterial infections, and lymphomas complicating anti-TNFalpha therapy. *Joint Bone Spine* 72:456–460.

Van Snick, J. Interleukin-6: An overview. 1990. *Annu Rev Immunol* 8:253–278.

Weisman, M.H. 2002. What are the risks of biological therapy in RA? An update on safety. *J Rheumatol* 65(suppl):33–38.

Winthrop, K.L. 2006. Risk and prevention of tuberculosis and other serious opportunistic infections associated with the inhibition of tumor necrosis factor. *Nature Clinical Practice Rheumatology* 2:602–610.

Wolfe, F., Caplan, L., Michaud, K. 2006. Treatment for rheumatoid arthritis and the risk of hospitalization for pneumonia: Associations with prednisone, disease-modifying antirheumatic drugs, and anti-tumor necrosis factor therapy. *Arthritis Rheum* 54:628–634.

Zaba, L.C., Cardinale, I., Gilleaudeau, P., Sullivan-Whalen, M., Suárez-Fariñas, M., Fuentes-Duculan, J., et al. 2007. Amelioration of epidermal hyperplasia by TNF inhibition is associated with reduced Th17 responses. *J Exp Med* 204:3183–3194.

27 Disease Modifying Anti-Rheumatic Drugs

Nigil Haroon and Vinod Chandran
University of Toronto
Toronto, Ontario, Canada

CONTENTS

27.1 INTRODUCTION

A disease modifying antirheumatic drug (DMARD) is defined by Outcome Measures in Rheumatology (OMERACT) as a drug that retards the progression of structural joint damage. They are immunosuppressive drugs that have the capacity to not only modify the immune response in the body but also to modify the course of the rheumatic disease. The term is generally used with reference to the treatment of rheumatoid arthritis (RA), and so traditional DMARDs may not have "disease modifying" ability in conditions like ankylosing spondylitis (AS). DMARDs have an onset of action that is typically slow, and it can be weeks to months before clinical effect is evident.

The classic DMARDs are methotrexate (MTX), sulfasalazine (SSZ), hydroxychloroquine (HCQ), and leflunomide (LEF). Cyclosporine (Cys) and azathioprine (AZA) are also considered DMARDS. Although steroids and biological response modifiers have disease modifying ability, only traditional DMARDs are discussed in this chapter.

27.2 METHOTREXATE

Methotrexate (MTX) is considered by most rheumatologists to be the pivotal DMARD in the management of RA. MTX was primarily used as an antineoplastic agent, and at much lower doses it is used as an immunosuppressive for the management of rheumatic diseases. It has been successfully used to treat some nonrheumatic conditions as well, including psoriasis and bronchial asthma.

27.2.1 Mechanism of Action

MTX is a well-known folate antagonist. Folic acid plays a cardinal role in 1-carbon metabolism. The transfer of 1-carbon units is critical to the synthesis of DNA, RNA, membrane lipids and neurotransmitters. Based on this known link, it was proposed that MTX affects cell survival by decreasing the availability of folate.

Methylene tetrahydrofolate (MTHF) is critical in the synthesis of thymidilate (dTMP) by transferring a methyl group to uridylate (dUMP). This reaction (Figure 27.1) is catalyzed by the enzyme thymidilate synthase (TS) and in the process MTHF is converted to dihydrofolate (DHF). DHF is converted to tetrahydrofolate (THF) by dihydrofolate reductase (DHFR). MTX inhibits this enzyme, thereby depleting the 1-carbon supplier (MTHF) required for dTMP synthesis. By depleting the cell of its sole source of dTMP, cell death is triggered.

Adenosine accumulation is also considered to contribute to the immunosuppressive effect of MTX. MTX can inhibit the enzyme amino imidazole carboxamide ribonucleotide transformylase (AICAR-T), which converts AICAR to formyl-AICAR. AICAR accumulation leads to the inhibition of adenosine deaminase (ADA), an enzyme that metabolizes adenosine and deoxy-adenosine to inosine. The accumulation of adenosine has been shown to have anti-inflammatory effects on immune cells including inhibiting the release of pro-inflammatory cytokines [1, 2].

In addition to the aforementioned mechanisms, MTX is also known to inhibit other enzymes including TS and amidophosphoribosyltransferase. There may be other anti-inflammatory mechanisms of MTX that are yet to be discovered. MTX is effective in a weekly dose even though the serum half-life is very short. This is possible because MTX is converted to polyglutamates after entering cells. MTX polyglutamates have anti-inflammatory properties like MTX and are retained in the cell increasing the half-life.

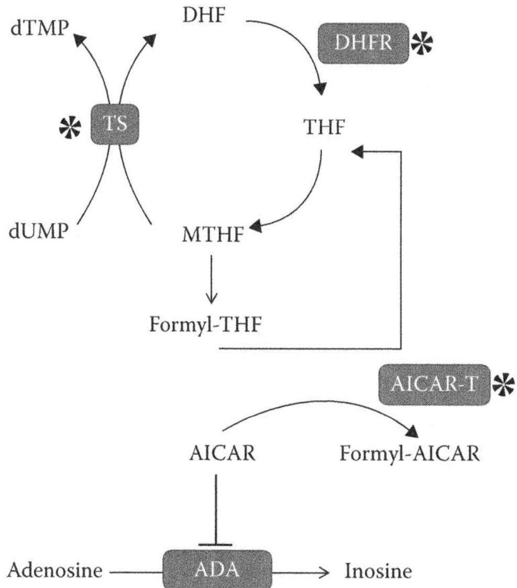

FIGURE 27.1 The 1-carbon cycle and the role of methotrexate (enzymes marked with a star are inhibited by MTX).

27.2.2 Dose

MTX is used in a dose range of 7.5 to 25 mg weekly as a single dose. The minimum clinically effective dose in RA is considered to be 7.5 mg. An increase in dose to between 12.5 and 20 mg weekly has been shown to increase the efficacy, while toxicity increases at doses above 25 mg weekly [3]. The toxicity is also higher if given on a split-daily basis than as a single weekly dose. As the half-life of MTX polyglutamate is 3 days, a twice-weekly dose was tried with negative results [4]. Subcutaneous dosing of MTX can increase the bioavailability of the drug and has been shown to have better efficacy [5].

Folic acid (FA) should be taken along with MTX to reduce the adverse effects. The dose of FA to be taken and the schedule is still debated but a minimum of 5 mg weekly is recommended by the 3e-initiative (Table 27.1). The two commonly followed FA dosing regimens are 5 mg twice weekly or 1 mg daily for 5 days in a week. Folic acid decreases the incidence of oral mucosal and gastrointestinal side effects by 79% [6]. It has also been shown that FA taken along with MTX does not affect its clinical efficacy. Folinic acid may decrease the clinical efficacy of MTX without any additional benefit over folic acid and so is recommended only to reverse toxicity [7]. Folinic acid can bypass the DHFR step and directly replenish the cellular level of tetrahydrofolates, thus being very effective in MTX toxicity.

27.2.3 Use in Rheumatic Diseases

MTX is effective in RA, including early arthritis, and has been shown to reduce the progression of erosions. MTX is also used for treating peripheral arthritis in patients with spondyloarthritis although the evidence to support this is thin. Other indications include psoriatic arthritis (PsA), chronic sarcoidosis, resistant arthritis in SLE and scleroderma, limited non-life threatening granulomatosis with polyangiitis, and chronic or recurrent iritis.

The long-term safety of MTX has been well established and is considered an effective and relatively safe drug in the management of RA [8]. MTX is known to reduce progression of erosions in RA [9]. In case of suboptimal response, a combination with other DMARDs may be effective [10]. However, a recent meta-analysis showed that combining conventional DMARDs was beneficial

TABLE 27.1
Multinational Evidence-Based Recommendations from the 3e Initiative for the Use of MTX in Rheumatic Disorders

Recommendations	Grade
In DMARD-naive patients, the efficacy/toxicity favors MTX monotherapy over combination with other conventional DMARDs; MTX should be considered as the anchor for combination therapy when MTX alone does not achieve disease control.	A
Taking at least 5 mg folic acid per week with MTX is strongly recommended.	A
Oral MTX should be started at 10–15 mg/week, with escalation of 5 mg every 2–4 weeks up to 20–30 mg/week, depending on clinical response and tolerability; parenteral administration should be considered in the case of inadequate clinical response or intolerance.	B
MTX, as a steroid-sparing agent, is recommended in giant-cell arteritis and polymyalgia rheumatica and can be considered in patients with systemic lupus erythematosus or (juvenile) dermatomyositis.	B
MTX can be safely continued in the perioperative period in RA patients undergoing elective orthopaedic surgery.	B
Based on its acceptable safety profile, MTX is appropriate for long-term use.	B
MTX should not be used for at least 3 months before planned pregnancy for men and women and should not be used during pregnancy or breast feeding.	C
When starting MTX or increasing the dose, ALT with or without AST, creatinine and CBC should be performed every 1–1.5 months until a stable dose is reached, and every 1–3 months thereafter; clinical assessment for side effects and risk factors should be performed at each visit.	C
The work-up for patients starting MTX should include assessment of risk factors for toxicity (including alcohol intake), patient education, AST, ALT, albumin, CBC, creatinine, chest x-ray (obtained within the previous year); consider serology for HIV, hepatitis B/C, blood fasting glucose, lipid profile, and pregnancy test.	C
MTX should be stopped if there is an increase in ALT/AST > 3X the normal upper limit and restarted at a lower dose following normalization. If the ALT/AST levels are persistently elevated, the dose should be adjusted. Diagnostic procedures should be considered in case of persistently elevated ALT/AST more than three times the normal upper limit after discontinuation.	C

Source: Visser, K., Katchamart, W., Loza, E., Martinez-Lopez, J. A., Salliot, C., Trudeau, J., et al. *Annals of the Rheumatic Diseases,* 68, 7, 2009. With permission.

only in reducing pain and functional capacity as measured by HAQ [11]. Smoking, female gender, younger age, and longer disease duration are predictors of poor response to MTX [12].

27.2.4 Adverse Effects of Methotrexate

Gastrointestinal adverse effects are common in patients on low-dose MTX therapy. The common gastrointestinal adverse effects seen are oral ulcers, nausea, vomiting, anorexia, abdominal pain, diarrhea, and elevated liver enzymes. Gastrointestinal side effects can be effectively reduced by folic acid supplementation. In those patients without a response to folic acid supplementation, either a reduced dose or change to subcutaneous route of administration can help. Alcohol ingestion should be avoided in patients on MTX as the risk of adverse effects, especially hepatotoxicity, is increased.

Hematological adverse effects, though rare, can be life threatening in patients on low dose MTX. The prevalence of pancytopenia is estimated to be 1% in patients on MTX and is usually reversible [13]. Bone marrow toxicity increases with renal failure, older age, alcohol use, folic acid deficiency, and combination with other marrow-toxic drugs. The use of MTX on a split multiple dose regimen is more toxic than a single weekly dose. This can be explained by the fact that MTX acts only during the DNA synthesis phase (S-phase) of cells, and the longer the marrow is exposed to the drug, the greater the number of cells in S-phase is affected. Hence, by decreasing the marrow exposure to once weekly, the chance of marrow suppression is significantly decreased.

Pulmonary toxicity is a well-known complication of MTX. The best characterized type of pulmonary toxicity is acute interstitial pneumonitis. As the name suggests, the symptoms start acutely and unpredictably with shortness of breath and cough. There are associated systemic symptoms like fever, and investigation shows hypoxemia reduced diffusion capacity of the lung and bilateral fluffy interstitial radiological opacities. Histopathology shows interstitial pneumonitis with lymphocytic infiltration and occasionally granulomas. Other pulmonary adverse effects include interstitial fibrosis, pulmonary nodulosis, pleurisy, and pulmonary edema.

Apart from these well-defined adverse effects, rarer adverse effects seen with MTX include MTX-induced accelerate subcutaneous nodulosis (MIAN), hyperhomocysteinemia, osteopathy, and headache. MTX is FDA category X and should not be used during pregnancy. MTX can result in fetal death and congenital abnormalities.

27.3 LEFLUNOMIDE

Leflunomide (LEF) was first introduced for the management of RA in 1998 and has since become an important DMARD in the management of RA with efficacy similar to that of MTX.

27.3.1 Mechanism of Action

LEF is a prodrug that is metabolized in the gut wall and the liver to the active metabolite "teriflunomide" (A77 1726). The precise mechanism of action of LEF is not known. In therapeutic doses A77 1726 selectively inhibits de novo pyrimidine synthesis by inhibiting the enzyme dihydroorotate dehydrogenase (DHODH) [14]. When lymphocyte pools proliferate they require an almost eightfold expansion of their pyrimidine pool, which is achieved by activation of both de novo and salvage pathways of pyrimidine synthesis. By inhibiting the rate-limiting step in de novo pyrimidine synthesis, lymphocyte proliferation is restricted because a sufficient pool of pyrimidines required for DNA synthesis is unavailable. Other modes of action include suppression of interleukin 1 (IL-1) and tumor necrosis factor α (TNFα) secretion upon T lymphocyte/monocyte contact activation, as well as suppression of activation of NFκB. A77 1726 may also reduce cell-cell contact activation during inflammation by downregulating glycosylation of adhesion molecules [14].

27.3.2 Dose

LEF is available in oral tablets at doses of 10 mg, 20 mg, and 100 mg. The standard recommendation is to start therapy with a loading dose of 100 mg daily for 3 days and then switch to a maintenance dose of 20 mg daily. However, many clinicians avoid the loading dose due to higher gastrointestinal toxicity.

Food does not interfere with drug absorption. Circulating A77 1726 is bound by >99% to plasma proteins and has low apparent volume of distribution. The half-life of A77 1726 is approximately 2 weeks, and steady state is reached in 7 weeks after daily dosing. In healthy subjects, 90% of LEF is excreted by 28 days but some may be present for much longer periods. The apparent half-life of A77 1726 can be reduced to 1 to 2 days by the oral administration of cholestyramine, 8 g three times daily. Activated charcoal, 50 g every 6 hours, can reduce plasma levels by 50% within 24 hours.

27.3.3 Use in Rheumatic Diseases

27.3.3.1 Rheumatoid Arthritis

LEF was demonstrated to be superior to placebo in improving signs and symptoms in a 6-month randomized dose-ranging controlled trial. Mladenovic et al. demonstrated that compared to placebo, LEF in doses of 10 mg and 25 mg was safe and efficacious, whereas 5 mg was not [15].

Subsequently, two pivotal trials compared LEF to MTX and SSZ. Strand et al. compared LEF to MTX and placebo [16]. To patients in the LEF arm, treatment protocol included a loading dose of LEF (100 mg for 3 days) followed by 20 mg daily. Patients in the MTX arm received the drug initially at a dose of 7.5 mg daily. If active disease was still present at week 6, the dose of MTX was increased to 15 mg over weeks 7 through 9 and continued thereafter. All patients received folate supplementation. LEF was found to be superior to placebo and the response rates similar to that obtained with MTX. Radiographs demonstrated less damage progression with LEF and MTX therapy than with placebo. Both LEF and MTX treatment improved measures of physical function and health-related quality of life [16]. In a 24-week randomized controlled trial (RCT), Smolen et al. compared LEF (100 mg daily on days 1–3, then 20 mg daily) to sulfasalazine (0.5 g daily, titrated progressively to 2.0 g daily at week 4) and placebo [17]. LEF was found to be more effective than placebo and showed similar efficacy to sulfasalazine. Radiographic damage progression was also significantly slower with LEF and sulfasalazine than with placebo [17]. The two RCTs thus established the role of LEF in the management of RA.

27.3.3.2 Psoriatic Arthritis

The efficacy of LEF in PsA was evaluated in one RCT that compared LEF to placebo [18]. In this study LEF was found to be superior to placebo for the treatment of arthritis as well as psoriasis. Unfortunately, radiographs were not included in the study; therefore, the effect of LEF on progression of joint damage was not studied.

27.3.3.3 Other Rheumatic Diseases

In AS, LEF was found to be effective in treating peripheral arthritis but not axial arthritis in a small open label study with 20 patients [19]. In systemic lupus erythematosus, in a small double-blind RCT in which 12 patients were randomized to either LEF or placebo, LEF was more effective than placebo in treating patients with mild to moderate disease activity and was safe and well tolerated [20]. Tam et al. (2006) also found LEF to be effective in the treatment of lupus nephritis refractory or intolerant to traditional immunosuppressive therapy in an open label trial [21]. LEF at a dose of 30 mg daily was shown to be useful in maintaining remission after induction of remission with cyclophosphamide in patients with granulomatosis with polyangiitis [22].

27.3.4 Adverse Effects of Leflunomide

In the major clinical trials withdrawal due to side effects were significantly higher in LEF-treated patients compared to placebo as well as MTX, but similar to sulfasalazine. Diarrhea is the most common side effect that limits treatment; the incidence is lower if a loading dose is not used. Diarrhea usually responds to reducing the dose, but may require drug discontinuation. Liver toxicity is another major side effect; therefore, liver function tests need to be monitored. Hypertension occurs more frequently in LEF-treated patients compared to placebo-treated patients. Pneumonitis and peripheral neuropathy have been reported. Skin rash, sometimes severe, most commonly occurring between months 2 and 5, may require drug discontinuation. Other reported side effects include cytopenia, nausea, and vomiting, abdominal pain, and weight loss. In case of significant side effects, the active metabolite of LEF may be washed out by the oral administration of cholestyramine, 8 g three times daily for 11 days.

LEF is rated FDA pregnancy category X since animal studies have shown substantial teratogenic effects with small doses. Therefore, LEF should not be prescribed for women who are not practicing reliable birth control methods. It is also recommended that nursing mothers not receive LEF. If a woman who has previously received LEF desires to become pregnant, A771726 levels should be measured. Active elimination of LEF from the body using one or more courses of cholestyramine 8 g three times daily for 11 days should be considered for levels greater than 0.02 mg/L. Before attempting pregnancy, levels less than 0.02 mg/L should be confirmed on two separate occasions, at least 14 days apart, and women should wait an additional three full menstrual cycles.

LEF is contraindicated in patients with impaired liver function, severe renal impairment, cytopenia, severe immunodeficiency, severe hypoproteinemia, or known hypersensitivity to the drug. Patients taking LEF should have a baseline complete blood count and liver and kidney function tests and chronic hepatitis serology. The complete blood count and liver tests should be repeated monthly, with dosage adjustments for abnormal test results. While patients are on treatment with LEF, live vaccinations should be avoided.

27.4 SULFASALAZINE

Sulfasalazine (SSZ) was originally known as sulfasalazopyridine and was synthesized as a combination of mesalazine or 5-aminosalicylic acid (5-ASA) and sulfapyridine. This was possibly synthesized in this combination because of the theory that RA is an inflammatory disease caused by an infection, hence the combination of an anti-inflammatory substance with an antibacterial agent. Less than 30% of SSZ is absorbed in the small intestines, and the rest is broken down in the large intestine into the individual components sulfapyridine and mesalazine. While sulfapyridine is entirely absorbed, less than 30% of mesalazine is.

Initially, SSZ was thought to act by antibacterial effects on the host, especially in the gut. This hypothesis has long been discarded, and SSZ is now considered to be a good immunosuppressive drug.

27.4.1 Mechanism of Action

SSZ has more immunosuppressive activity than the individual components. The exact mechanism of action of SSZ is controversial. It is known to suppress neutrophil activities including chemotaxis, degranulation, and proteolytic enzyme production [23]. It is also known to suppress activation of T, B, and NK cells, with decrease in pro-inflammatory cytokine and immunoglobulin production. SSZ can directly inhibit proliferation of endothelial cells and fibroblasts, as well as suppress osteoclasts and angiogenesis. These effects lead to the disease-modifying effect on rheumatoid synovitis. In addition, SSZ is known to inhibit prostaglandin E2 synthetase, the arachidonic acid cascade, as well as the extracellular release of phospholipase A2 [24]. Like MTX, SSZ is known to inhibit AICAR-T, resulting in the accumulation of adenosine [25].

27.4.2 Dose

Sulfasalazine is given at doses ranging from 2.0 to 3.0 g/day. It is generally started at a lower dose and increased on a weekly basis. This is known to improve the tolerability of the drug, especially to the gastrointestinal toxicity.

27.4.3 Use in Rheumatic Diseases

SSZ is a good DMARD for RA. A meta-analysis of placebo-controlled trials enrolling a total of 335 patients showed a significant improvement in joint counts and erythrocyte sedimentation rate (ESR) [26]. A meta-analysis comparing the efficacy of DMARDs in RA showed similar efficacy of SSZ with MTX, antimalarials, and injectable gold [27]. SSZ has been reported to have similar efficacy to LEF [17]. SSZ may have a quicker onset of action compared to HCQ [28].

Apart from RA, SSZ has been used successfully in PsA, reactive arthritis (ReA), inflammatory bowel disease associated arthritis, AS, and juvenile idiopathic arthritis (JIA). A meta-analysis of SSZ in AS showed benefits in improving ESR and the severity of back stiffness [29]. In one of the trials included in the meta-analysis, patients had relatively shorter disease duration and more peripheral arthritis [30]. This was the only study in which there was improvement in back pain and metrology.

This led to the conclusion that SSZ may be effective in early AS and in those with peripheral arthritis. The benefit of SSZ in peripheral arthritis has been established in another trial enrolling patients with AS, ReA, and PsA [31]. Clinical benefit of SSZ was seen only in patients with peripheral arthritis.

In a Cochrane review including 13 RCTs, SSZ was found to be significantly better than placebo in the treatment of PsA [32]. SSZ is used in ReA and it has been found in a trial to start working as early as after 4 weeks of initiation [33]. The trial showed significantly higher response rates to SSZ compared to placebo.

27.4.4 Adverse Effects

The most common adverse effects seen with SSZ are gastrointestinal, including nausea, vomiting, dyspepsia, and anorexia. Other common adverse effects include headache and skin rash. Bone marrow toxicity is rare, and oligospermia leading to male infertility is reversible. The adverse effects are more common during the early weeks to months of treatment and relatively rarer later. Enteric-coated tablets reduce the rate of achieving peak plasma concentrations and result in lower adverse effects. Starting therapy at a lower dose of 500 mg per day and increasing slowly has resulted in fewer failures due to intolerance.

27.5 ANTIMALARIALS

The 4-aminoquinolines chloroquine (CHQ) and HCQ are the most commonly used antimalarials. HCQ differs from CHQ by a hydroxyl group and is less toxic. The remarkable feature of antimalarials is the low incidence of adverse effects. In lupus it is considered a universal treatment that should not be stopped if possible. They have not only therapeutic but also preventive effects in lupus. In RA, HCQ can be used in combination with other DMARDS, resulting in higher efficacy. In combination with MTX in RA, they are known to prevent or improve MIAN [34].

27.5.1 Mechanism of Action

The exact mechanism of action of antimalarials is not known. They are thought to interfere with lysosomal acidification, leading to effects on antigen presentation [35]. Lysosomal acidification can also lead to inhibition of Toll-like receptors (TLR) [35]. They affect phagocytosis, suppress inflammatory cytokine production, and dampen the calcium signaling by T and B cell receptors [36–38]. Antimalarials bind and stabilize the DNA. They can decrease the matrix metalloproteinase MMP9 and increase tissue inhibitor of metalloproteinase (TIMP), thereby favorably affecting the balance of metalloproteinases [39].

Recently a unique function of HCQ to help prevent thrombosis was identified. Annexin A5, a natural anticoagulant, binds to phospholipids, preventing the activation of coagulation factors. In antiphospholipid antibody syndrome (APLA), the antibodies inhibit the action of annexin A5, which can be normalized with HCQ [40].

27.5.2 Dose

Chloroquine is used at a dose of 500 mg and HCQ at a dose of 400–600 mg per day. Antimalarials may be given as a single dose at night to reduce nausea, dose-related transient vision problems, and possibly photosensitivity.

27.5.3 Use in Rheumatic Diseases

The classic rheumatic indications for using HCQ are RA and SLE. Other indications include myositis, Sjögren's syndrome, palindromic rheumatism, and sarcoidosis. It is now being used in APLA to prevent thrombosis.

Although HCQ has a slower onset of action, it has been found to be as effective as MTX and SSZ in a meta-analysis [27].

27.5.4 Adverse Effects

The adverse effects of antimalarials are rare but can be dramatic and sometimes irreversible, especially if late to recognize. The adverse effects include retinopathy, myopathy, cardiomyopathy, tinnitus, hearing difficulty, ataxia, dizziness, nightmares, and psychotic behavior.

More common and less severe adverse effects include abdominal pain, nausea, vomiting, anorexia, and diarrhea. Skin pigmentation can be a major deterrent for young patients despite the tremendous benefits of treatment with HCQ. Rarely, hemolysis can occur with glucose-6-phosphate dehydrogenase (G6PD) deficiency.

27.6 AZATHIOPRINE

Since its first use as a "disease modifying" agent in the early 1960s, AZA has been commonly used in the treatment of chronic inflammatory diseases such as inflammatory bowel disease, systemic lupus erythematosus, vasculitis, and RA.

27.6.1 Mechanism of Action

AZA is a long-lived prodrug of 6-mercaptopurine (6-MP), a purine analogue that acts as an antagonist to the endogenous purines that are essential for synthesis of DNA, RNA, and some coenzymes. It exerts its cytotoxic effects after activation by the enzyme hypoxanthine-guanine phosphoribosyl transferase (HGPRT). Both in vitro and in vivo, AZA is metabolized to 6-MP through reduction by glutathione and other sulfhydryl-containing compounds and then enzymatically converted into 6-thiouric acid, 6-methyl-MP, and 6-thioguanine (6-TG). AZA then becomes incorporated into replicating DNA and also blocks the de novo pathway of purine synthesis [41]. However, blockade of DNA replication does not fully explain all of the laboratory and clinical findings of AZA-induced immunosuppression. Tiede et al. [42] recently showed that AZA and its metabolites induce apoptosis of T cells. The induction of apoptosis requires co-stimulation with CD28 and is mediated by specific blockade of Rac1 activation through binding of AZA-generated 6-thioguanine triphosphate (6-Thio-GTP) to Rac1 instead of GTP. The activation of Rac1 target genes such as mitogen-activated protein kinase, NF-κB, and bcl-x(L) is suppressed by AZA, leading to a mitochondrial pathway of apoptosis. AZA thus converts a co-stimulatory signal into an apoptotic signal by modulating Rac1 activity [42].

27.6.2 Dose

The drug is available as 50 mg tablets. The dose of AZA ranges from 1–2.5 mg/kg. It is usually started at a dose of 50 mg daily, and the dose titrated to 2 mg/kg with close monitoring of complete blood count.

The oral bioavailability of AZA ranges from 27% to 83% and is inversely proportional to the dose. After absorption, the prodrug AZA undergoes approximately 90% conversion to 6-MP by nonenzymatic attack by sulfhydryl-containing compounds such as glutathione or cysteine. 6-MP is a substrate for the enzyme thiopurine methyl transferase (TPMT) that catabolizes 6-MP to 6-methyl-MP. 6-MP is converted to thioxanthine and thiouric acid by xanthine oxidase. When the xanthine oxidase inhibitor allopurinol is administered with AZA, plasma 6-MP concentrations are five-fold higher, and the toxicity of 6-MP is enhanced [43]. Determining a patient's TPMT genotype or phenotype, a priori, can identify the rare patient (1 in 300) who should not receive a thiopurine because of little or no TPMT enzyme activity that can lead to severe neutropenia on exposure to

AZA. Patients with intermediate TPMT levels can safely receive AZA at lower doses and can undergo dose escalation safely under close monitoring [43].

27.6.3 Use in Rheumatic Diseases

Although AZA is an effective immunosuppressant and is commonly used in the management of severe manifestations of systemic lupus erythematosus and systemic vasculitides, it has a minor role in the management of inflammatory arthritides. At least 6 RCTs have evaluated the efficacy of AZA in RA. Compared to placebo, there is some evidence of efficacy on swollen joint count [44]. However, MTX has been shown to be superior to AZA, and therefore, with the availability of MTX and a number of agents with superior efficacy, AZA is seldom used in the management of RA but is an option if other drugs are contraindicated, failed, or are unavailable.

Although there are no randomized trials, AZA has been used in patients with refractory PsA. Data from an observational cohort showed that there was no statistically significant difference in the reduction in number of actively inflamed joints between AZA-treated patients and controls treated with other agents. Some patients had a significant improvement in psoriasis [45].

AZA is used in the management of severe manifestations of SLE, especially proliferative nephritis. In the original studies conducted by the National Institutes of Health (NIH), AZA was not inferior to cyclophosphamide as an induction treatment for lupus nephritis. More recent studies show that AZA and cyclophosphamide suppress the clinical and histologic activity of lupus nephritis to a similar extent, but cyclophosphamide may be more effective in preventing relapse and progression. For other manifestations of SLE, including cutaneous disease, AZA is widely used as a corticosteroid-sparing agent. AZA in combination with corticosteroids is useful in the treatment of a number of other autoimmune diseases, including inflammatory myositis, Behçet's disease, and systemic vasculitis.

27.6.4 Adverse Effects

Common side effects include nausea, vomiting, and abdominal pain. The most serious early toxicity is neutropenia. Therefore, complete blood counts need to be monitored closely after initiating therapy. Consideration should be given to undergoing TPMT testing to identify the rare patient (1 in 300) who should not receive AZA because of little or no TPMT enzyme activity. However, TPMT testing cannot substitute for complete blood count monitoring in patients receiving AZA. Other side effects include increased risk of infection and drug-induced hepatitis (rare). Long-term therapy may increase risk of malignancies.

AZA is rated FDA pregnancy category D. Whenever possible, use in pregnancy should be avoided. However, AZA may be continued or used to treat severe disease manifestations after weighing risk and benefits. Use in nursing mothers is not recommended.

AZA is contraindicated in patients with absent TPMT activity but may be given in lower doses in patients with intermediate activity with careful dose adjustment. Since one of the pathways for inactivation of AZA is inhibited by allopurinol, the two drugs should not be given together. In the rare clinical situation when the two drugs have to be given, the dose of AZA should be reduced to one-quarter.

27.7 CYCLOSPORIN A

Cyclosporin A (CsA) was first used in the management of autoimmune disease in the late 1980s. It is now used only in refractory disease given the significant risk of toxicity on long-term therapy.

27.7.1 Mechanism of Action

Cyclosporine is a lipophilic endecapeptide first isolated from the fungus *Tolypocladium inflatum*. Cyclosporine impairs production of interleukin-2 and other cytokines, reducing lymphocyte proliferation. Cyclosporine complexes with a cytosolic-binding protein (immunophilin) called cyclophilin. The CsA-cyclophilin complex binds to and inhibits calcineurin, a serine/threonine phosphatase. Inhibition of calcineurin phosphatase activity prevents the translocation of cytosolic nuclear factor of activated T cells (NFAT) to the nucleus. Translocation of NFAT is required for the transcription of genes for cytokines such as interleukin-2 and for T cell activation. Inhibition of this translocation thus prevents IL-2 production and T cell activation.

27.7.2 Dose

CsA is available in 25 mg and 100 mg capsules. It is usually started at a dose of 2.5 mg/kg/day, administered in divided doses. Clinical response is seen only after a period of 4 to 8 weeks and is maximal after 12 weeks or more of treatment. To improve efficacy, the dosage can be increased by 0.5 mg/kg/day at 4- to 8-week intervals to a maximal dosage of 4 mg/kg/day. If there is no clinical response in 6 months, CsA should be discontinued.

The newer microemulsion-based Neoral formulation has better bioavailability and less inter-subject and intra-subject variability than the older oil-based Sandimmune formulation. The bioavailability is about 30%. CsA is lipophilic and widely distributed in body tissues. Metabolism of CsA depends on P-glycoprotein (Pgp), a drug efflux pump that pumps CsA out of cells. Pgp is expressed on intestinal epithelial cells and in the liver. CsA is metabolized by the CYP3A enzyme system, which is also active in the liver and intestinal epithelium. Pgp, by limiting drug uptake, and CYP3A4, by facilitating drug metabolism in the gut and liver, act to limit the bioavailability of CsA [46]. CsA elimination is not altered in renal failure; however, because of its nephrotoxicity, cyclosporine is avoided in patients with impaired renal function. Liver disease impairs the excretion of cyclosporine metabolites.

27.7.3 Use in Rheumatic Diseases

In RA, CsA is effective as a single agent and in combination with MTX or hydroxychloroquine. When compared to placebo, CsA improves swollen joint count, pain, and disability, as well as ACR20 response criteria. The combination of CsA + MTX or LEF was more efficacious on swollen joint count and ACR50 and ACR70 response than cyclosporine monotherapy, but not on pain, disability, radiographic damage, or ACR [44].

CsA is effective in controlling psoriasis. A three-arm RCT comparing CsA 3 mg/kg/day added to standard therapy, SSZ 2 g/day added to standard therapy, and standard therapy alone showed that CsA was well tolerated and was more efficacious that standard therapy and SSZ [47]. In the most recently published RCT, CsA was compared to placebo as an add-on treatment in patients with PsA demonstrating an incomplete response to MTX monotherapy. There was significant improvement at 12 months in the swollen joint count, C-reactive protein, PASI, and synovitis detected by high-resolution ultrasound. There was no improvement in the Health Assessment Questionnaire or pain scores [48]. Thus, CsA has a role in the management of PsA either on its own or as an add-on treatment to MTX. However, it is not well tolerated. Its effect on joint damage has not been assessed.

In SLE, small uncontrolled studies have shown that CsA improves disease activity, has a steroid-sparing effect, and improves proteinuria, thrombocytopenia, and leukopenia. CsA has also been used in the maintenance phase of ANCA associated vasculitis and for macrophage-activation syndrome and Behçet's disease.

27.7.4 Adverse Effects

Short-term treatment with CsA is usually well tolerated. Common side effects include gastrointestinal upset, hypertrichosis, gingival hyperplasia, tremor, paresthesias, breast tenderness, hyperkalemia, hypomagnesemia, and hyperuricemia. Important side effects, especially with long-term treatment, include hypertension and renal dysfunction. In transplant recipients, CsA use has been associated with an increased risk of skin cancer and lymphoma, but data from patients treated for inflammatory arthritides are limited.

Cyclosporine is rated FDA category C, and its use in pregnancy is not recommended unless the potential benefit exceeds the potential risk to the fetus. Breastfeeding should be avoided when a patient is on treatment with CsA.

CsA has a number of significant drug interactions because of the influence of Pgp and CYP3A4 enzyme activity on its metabolism. Drugs that inhibit CYP3A4 and Pgp—such as erythromycin, azole antifungal drugs, and some calcium channel antagonists—may result in a two- to five-fold increase in CsA concentrations, resulting in increased toxicity. Hepatic enzyme inducers such as rifampicin, phenytoin, and phenobarbitone decrease CsA concentration. Aminoglycosides, quinolone antibiotics, amphotericin B, NSAIDs, and ACE inhibitors may increase renal toxicity. CsA increases the risk of myotoxicity of statins.

Prior to starting CsA, renal function and blood pressure should be confirmed to be normal. Liver enzymes, serum potassium, magnesium, uric acid, and lipid levels should also be monitored prior to and during therapy with CsA. Live vaccines should be avoided.

27.8 OTHER DMARDs

Intramuscular gold therapy is beneficial for RA and was shown to improve swollen joint count, pain, and disability when compared to placebo. There is no clear evidence for better efficacy of MTX (at tested doses) versus intramuscular gold [44]. Early use of gold salt injections may retard progression of joint erosions. Although not shown to protect from progression of joint damage, gold has been used in the treatment of PsA, with intramuscular gold being more efficacious [49]. Despite the known benefits, gold salts are being used less by rheumatologists because of the need for meticulous monitoring for serious toxicity (e.g., cytopenias, proteinuria) and the costs of administration and monitoring. Oral gold therapy with auranofin has different and less severe toxicity than intramuscular gold. Cytopenia and proteinuria do not occur, but diarrhea is not uncommon. Auranofin is less efficacious than MTX, injectable gold, or SSZ.

D-penicillamine was used in the management of RA. However, that drug has shown no significant advantage versus placebo and also has serious adverse effects including aplastic anemia, agranulocytosis, thrombocytopenia, Goodpasture's syndrome, myasthenia gravis, and membranous nephritis, and is therefore best avoided.

The broad spectrum tetracycline antibiotic minocycline that also inhibits matrix metalloproteinases and decreases synovial T cell proliferation and cytokine production was shown to have some efficacy on swollen joint count but not joint damage or function in RA. There is limited data comparing this drug to other standard drugs. Side effects include feelings of light-headedness, vertigo, and liver toxicity. Drug-induced lupus is an important side effect with chronic therapy.

With the advent of newer, more effective therapy, and with better understanding of the use of older DMARDs, these drugs are rarely, if ever, used.

27.9 CONCLUSIONS

A number of DMARDs are currently used in the management of peripheral inflammatory arthritis, with MTX, LEF, SSZ, and HCQ the most commonly studied and used by practitioners. These drugs have a major role as the first-line agent in the management of early disease (especially RA) and also in combination with newer biologic agents. Some of these drugs are also effective in the management of peripheral spondyloarthritis. They are, however, not effective in the management of axial spondyloarthritis.

TAKE-HOME MESSAGES

- Disease modifying antirheumatic drugs (DMARD) are antirheumatic agents referring to their ability to modify the course of disease, as opposed to simply treating symptoms such as inflammation and pain.
- Agents in this group include azathioprine, cyclosporine, gold salts, hydroxychloroquine, leflunomide, methotrexate, D-penicillamine, and sulfasalazine.
- MTX is considered by most rheumatologists to be the pivotal DMARD in the management of RA.
- MTX was primarily used as an antineoplastic agent, and at much lower doses it is used as an immunosuppressive for the management of rheumatic diseases.
- Sulfasalazine and leflunomide are also very effective DMARDs for rheumatoid arthritis.
- In addition to treating the disease process, DMARDs also help reduce use of corticosteroids and thus reduce the toxicities of corticosteroids.

REFERENCES

1. Bouma MG, Stad RK, van den Wildenberg FA, Buurman WA. Differential regulatory effects of adenosine on cytokine release by activated human monocytes. *Journal of Immunology* 1994 Nov 1; 153(9):4159–4168.
2. Cronstein BN, Naime D, Firestein G. The antiinflammatory effects of an adenosine kinase inhibitor are mediated by adenosine. *Arthritis and Rheumatism* 1995 Aug; 38(8):1040–1045.
3. Furst D, Koehnke R, Burmeister L, Kohler J, Cargill I. Increasing methotrexate effect with increasing dose in the treatment of resistant rheumatoid-arthritis. *Journal of Rheumatology* 1989 Mar; 16(3):313–320.
4. Pandya S, Aggarwal A, Misra R. Methotrexate twice weekly vs once weekly in rheumatoid arthritis: A pilot double-blind, controlled study. *Rheumatology International* 2002 May; 22(1):1–4.
5. Braun J, Kaestner P, Flaxenberg P, Waehrisch J, Hanke P, Demary W, et al. Comparison of the clinical efficacy and safety of subcutaneous versus oral administration of methotrexate in patients with active rheumatoid arthritis. *Arthritis and Rheumatism* 2008 Jan; 58(1):73–81.
6. Ortiz Z, Shea B, Suarez Almazor M, Moher D, Wells G, Tugwell P. Folic acid and folinic acid for reducing side effects in patients receiving methotrexate for rheumatoid arthritis. Cochrane database of systematic reviews (Online) 2000; (2)(2):CD000951.
7. Joyce DA, Will RK, Hoffman DM, Laing B, Blackbourn SJ. Exacerbation of rheumatoid arthritis in patients treated with methotrexate after administration of folinic acid. *Annals of the Rheumatic Diseases* 1991 Dec; 50(12):913–914.
8. Salliot C, van der Heijde D. Long-term safety of methotrexate monotherapy in patients with rheumatoid arthritis: A systematic literature research. *Annals of the Rheumatic Diseases* 2009 Jul; 68(7):1100–1104.
9. Rich E, Moreland LW, Alarcon GS. Paucity of radiographic progression in rheumatoid arthritis treated with methotrexate as the first disease modifying antirheumatic drug. *Journal of Rheumatology* 1999 Feb; 26(2):259–261.
10. Capell HA, Madhok R, Porter DR, Munro RA, McInnes IB, Hunter JA, et al. Combination therapy with sulfasalazine and methotrexate is more effective than either drug alone in patients with rheumatoid arthritis with a suboptimal response to sulfasalazine: Results from the double-blind placebo-controlled MASCOT study. *Annals of the Rheumatic Diseases* 2007 Feb; 66(2):235–241.
11. Katchamart W, Trudeau J, Phumethum V, Bombardier C. Methotrexate monotherapy versus methotrexate combination therapy with non-biologic disease modifying anti-rheumatic drugs for rheumatoid arthritis. *Cochrane Database of Systematic Reviews* (Online) 2010 Apr 14; (4)(4):CD008495.
12. Saevarsdottir S, Wallin H, Seddighzadeh M, Ernestam S, Geborek P, Petersson IF, et al. Predictors of response to methotrexate in early DMARD naive rheumatoid arthritis: Results from the initial open-label phase of the SWEFOT trial. *Annals of the Rheumatic Diseases* 2011 Mar; 70(3):469–475.

13. Gutierrez-Urena S, Molina JF, Garcia CO, Cuellar ML, Espinoza LR. Pancytopenia secondary to methotrexate therapy in rheumatoid arthritis. *Arthritis and Rheumatism* 1996 Feb; 39(2):272–276.
14. Breedveld FC, Dayer JM. Leflunomide: Mode of action in the treatment of rheumatoid arthritis. *Annals of the Rheumatic Diseases* 2000 Nov; 59(11):841–849.
15. Mladenovic V, Domljan Z, Rozman B, Jajic I, Mihajlovic D, Dordevic J, et al. Safety and effectiveness of leflunomide in the treatment of patients with active rheumatoid arthritis. Results of a randomized, placebo-controlled, phase II study. *Arthritis and Rheumatism* 1995 Nov; 38(11):1595–1603.
16. Strand V, Cohen S, Schiff M, Weaver A, Fleischmann R, Cannon G, et al. Treatment of active rheumatoid arthritis with leflunomide compared with placebo and methotrexate. Leflunomide Rheumatoid Arthritis Investigators Group. *Archives of Internal Medicine* 1999 Nov 22; 159(21):2542–2550.
17. Smolen JS, Kalden JR, Scott DL, Rozman B, Kvien TK, Larsen A, et al. Efficacy and safety of leflunomide compared with placebo and sulphasalazine in active rheumatoid arthritis: A double-blind, randomised, multicentre trial. European Leflunomide Study Group. *Lancet* 1999 Jan 23; 353(9149):259–266.
18. Kaltwasser JP, Nash P, Gladman D, Rosen CF, Behrens F, Jones P, et al. Efficacy and safety of leflunomide in the treatment of psoriatic arthritis and psoriasis: A multinational, double-blind, randomized, placebo-controlled clinical trial. *Arthritis and Rheumatism* 2004 Jun; 50(6):1939–1950.
19. Haibel H, Rudwaleit M, Braun J, Sieper J. Six months open label trial of leflunomide in active ankylosing spondylitis. *Annals of the Rheumatic Diseases* 2005 Jan; 64(1):124–126.
20. Tam LS, Li EK, Wong CK, Lam CW, Szeto CC. Double-blind, randomized, placebo-controlled pilot study of leflunomide in systemic lupus erythematosus. *Lupus* 2004; 13(8):601–604.
21. Tam LS, Li EK, Wong CK, Lam CW, Li WC, Szeto CC. Safety and efficacy of leflunomide in the treatment of lupus nephritis refractory or intolerant to traditional immunosuppressive therapy: An open label trial. *Annals of the Rheumatic Diseases* 2006 Mar; 65(3):417–418.
22. Metzler C, Miehle N, Manger K, Iking-Konert C, de Groot K, Hellmich B, et al. Elevated relapse rate under oral methotrexate versus leflunomide for maintenance of remission in Wegener's granulomatosis. *Rheumatology* (Oxford, England) 2007 Jul; 46(7):1087–1091.
23. Molin L, Stendahl O. The effect of sulfasalazine and its active components on human polymorphonuclear leukocyte function in relation to ulcerative colitis. *Acta Medica Scandinavica* 1979; 206(6):451–457.
24. Plosker GL, Croom KF. Sulfasalazine: A review of its use in the management of rheumatoid arthritis. *Drugs* 2005; 65(13):1825–1849.
25. Gadangi P, Longaker M, Naime D, Levin RI, Recht PA, Montesinos MC, et al. The anti-inflammatory mechanism of sulfasalazine is related to adenosine release at inflamed sites. *Journal of Immunology* 1996 Mar 1; 156(5):1937–1941.
26. Suarez-Almazor ME, Belseck E, Shea B, Wells G, Tugwell P. Sulfasalazine for rheumatoid arthritis. Cochrane database of systematic reviews (Online) 2000; (2)(2):CD000958.
27. Felson DT, Anderson JJ, Meenan RF. The comparative efficacy and toxicity of second-line drugs in rheumatoid arthritis. Results of two metaanalyses. *Arthritis and Rheumatism* 1990 Oct; 33(10):1449–1461.
28. Nuver-Zwart IH, van Riel PL, van de Putte LB, Gribnau FW. A double blind comparative study of sulphasalazine and hydroxychloroquine in rheumatoid arthritis: Evidence of an earlier effect of sulphasalazine. *Annals of the Rheumatic Diseases* 1989 May; 48(5):389–395.
29. Chen J, Liu C. Is sulfasalazine effective in ankylosing spondylitis? A systematic review of randomized controlled trials. *Journal of Rheumatology* 2006 Apr; 33(4):722–731.
30. Nissila M, Lehtinen K, Leirisalo-Repo M, Luukkainen R, Mutru O, Yli-Kerttula U. Sulfasalazine in the treatment of ankylosing spondylitis. A twenty-six-week, placebo-controlled clinical trial. *Arthritis and Rheumatism* 1988 Sep; 31(9):1111–1116.
31. Clegg DO, Reda DJ, Abdellatif M. Comparison of sulfasalazine and placebo for the treatment of axial and peripheral articular manifestations of the seronegative spondylarthropathies: A Department of Veterans Affairs cooperative study. *Arthritis and Rheumatism* 1999 Nov; 42(11):2325–2329.
32. Jones G, Crotty M, Brooks P. Interventions for psoriatic arthritis. Cochrane database of systematic reviews (Online) 2000; (3)(3):CD000212.
33. Clegg DO, Reda DJ, Weisman MH, Cush JJ, Vasey FB, Schumacher HR,Jr, et al. Comparison of sulfasalazine and placebo in the treatment of reactive arthritis (Reiter's syndrome). A Department of Veterans Affairs Cooperative Study. *Arthritis and Rheumatism* 1996 Dec; 39(12):2021–2027.
34. Combe B, Guttierrez M, Anaya JM, Sany J. Possible efficacy of hydroxychloroquine on accelerated nodulosis during methotrexate therapy for rheumatoid arthritis. *Journal of Rheumatology* 1993 Apr; 20(4):755–756.
35. Kalia S, Dutz JP. New concepts in antimalarial use and mode of action in dermatology. *Dermatologic Therapy* 2007 Jul-Aug; 20(4):160–174.

36. Nowell J, Quaranta V. Chloroquine affects biosynthesis of Ia molecules by inhibiting dissociation of invariant (gamma) chains from alpha-beta dimers in B cells. *Journal of Experimental Medicine* 1985 Oct 1; 162(4):1371–1376.
37. Sperber K, Quraishi H, Kalb TH, Panja A, Stecher V, Mayer L. Selective regulation of cytokine secretion by hydroxychloroquine: Inhibition of interleukin 1 alpha (IL-1-alpha) and IL-6 in human monocytes and T cells. *Journal of Rheumatology* 1993 May; 20(5):803–808.
38. Wozniacka A, Lesiak A, Narbutt J, McCauliffe DP, Sysa-Jedrzejowska A. Chloroquine treatment influences proinflammatory cytokine levels in systemic lupus erythematosus patients. *Lupus* 2006; 15(5):268–275.
39. Lesiak A, Narbutt J, Sysa-Jedrzejowska A, Lukamowicz J, McCauliffe DP, Wozniacka A. Effect of chloroquine phosphate treatment on serum MMP-9 and TIMP-1 levels in patients with systemic lupus erythematosus. *Lupus* 2010 May; 19(6):683–688.
40. Rand JH, Wu XX, Quinn AS, Ashton AW, Chen PP, Hathcock JJ, et al. Hydroxychloroquine protects the annexin A5 anticoagulant shield from disruption by antiphospholipid antibodies: Evidence for a novel effect for an old antimalarial drug. *Blood* 2010 Mar 18; 115(11):2292–2299.
41. Aarbakke J, Janka-Schaub G, Elion GB. Thiopurine biology and pharmacology. *Trends in Pharmacological Sciences* 1997 Jan; 18(1):3–7.
42. Tiede I, Fritz G, Strand S, Poppe D, Dvorsky R, Strand D, et al. CD28-dependent Rac1 activation is the molecular target of azathioprine in primary human CD4+ T lymphocytes. *Journal of Clinical Investigation* 2003 Apr; 111(8):1133–1145.
43. Sahasranaman S, Howard D, Roy S. Clinical pharmacology and pharmacogenetics of thiopurines. *European Journal of Clinical Pharmacology* 2008 Aug; 64(8):753–767.
44. Gaujoux-Viala C, Smolen JS, Landewe R, Dougados M, Kvien TK, Mola EM, et al. Current evidence for the management of rheumatoid arthritis with synthetic disease-modifying antirheumatic drugs: A systematic literature review informing the EULAR recommendations for the management of rheumatoid arthritis. *Annals of the Rheumatic Diseases* 2010 Jun; 69(6):1004–1009.
45. Lee JC, Gladman DD, Schentag CT, Cook RJ. The long-term use of azathioprine in patients with psoriatic arthritis. *Journal of Clinical Rheumatology: Practical Reports on Rheumatic & Musculoskeletal Diseases* 2001 Jun; 7(3):160–165.
46. Lown KS, Mayo RR, Leichtman AB, Hsiao HL, Turgeon DK, Schmiedlin-Ren P, et al. Role of intestinal P-glycoprotein (mdr1) in interpatient variation in the oral bioavailability of cyclosporine. *Clinical Pharmacology and Therapeutics* 1997 Sep; 62(3):248–260.
47. Salvarani C, Macchioni P, Olivieri I, Marchesoni A, Cutolo M, Ferraccioli G, et al. A comparison of cyclosporine, sulfasalazine, and symptomatic therapy in the treatment of psoriatic arthritis. *Journal of Rheumatology* 2001 Oct; 28(10):2274–2282.
48. Fraser AD, van Kuijk AW, Westhovens R, Karim Z, Wakefield R, Gerards AH, et al. A randomised, double blind, placebo controlled, multicentre trial of combination therapy with methotrexate plus ciclosporin in patients with active psoriatic arthritis. *Annals of the Rheumatic Diseases* 2005 Jun; 64(6):859–864.
49. Carette S, Calin A, McCafferty JP, Wallin BA. A double-blind placebo-controlled study of auranofin in patients with psoriatic arthritis. *Arthritis and Rheumatism* 1989 Feb; 32(2):158–165.
50. Visser K, Katchamart W, Loza E, Martinez-Lopez JA, Salliot C, Trudeau J, et al. Multinational evidence-based recommendations for the use of methotrexate in rheumatic disorders with a focus on rheumatoid arthritis: Integrating systematic literature research and expert opinion of a broad international panel of rheumatologists in the 3E Initiative. *Annals of the Rheumatic Diseases* 2009 Jul; 68(7):1086–1093.

28 Nonsteroidal Anti-Inflammatory Drugs

Alakendu Ghosh and Pradyot Sinhamahapatra
Institute of Postgraduate Medical Education and Research
Kolkata, India

CONTENTS

28.1 INTRODUCTION

Inflammation is a host defense mechanism initiated by foreign molecules, generating a cascade of mediators including prostaglandins leading to tissue damage as a bystander effect in an attempt by the body to get rid of the foreign antigen. Inflammation needs to be controlled when it is inappropriate, aberrant, or sustained or leads to significant tissue damage. There are different mechanisms to halt the inflammatory cascade, one of which targets the key pathway of production of key mediators of inflammation—the cyclooxygenase pathway.

Nonsteroidal anti-inflammatory drugs (NSAIDs) are a heterogeneous group of chemical compounds that share the common property of having anti-inflammatory, analgesic, and antipyretic actions. These drugs are mostly organic acids.

28.2 HISTORY

The search for remedies for pain relief is old. In the past, willow (*Willow* spp.) and other salicylate-rich plants were serendipitously discovered to be the source of herbal pain relief medications. In 1853, Charles Frédéric Gerhardt produced acetyl salicylic acid for the first time by reacting acetyl chloride with sodium salicylate. By the end of the 19th century, the then dye firm Bayer commercially produced acetylsalicylic acid and branded it as aspirin. Today, aspirin and acetylsalicylic acid are used synonymously as a generic name. Acetaminophen (paracetamol) and ibuprofen were subsequently discovered. The mechanism of action of aspirin was subsequently elucidated by John Vane, who received a Nobel Prize for his work (Vane 1971; Vane 1996).

28.3 MECHANISM OF ACTION

All drugs categorized under NSAIDs inhibit cyclooxygenase (COX), the enzyme responsible for conversion of arachidonic acid (derived from cell membrane phospholipid by phospholipase A2) to prostaglandins. All the agents are competitive and reversible antagonists of COX except aspirin, which acetylates the enzyme and thus is irreversible. This is important because the antiplatelet activity (inactivated for the platelet's life) persists longer.

The enzyme COX is now known to be present in two isoforms (Hawkey 1999): COX-1 and COX-2. COX-1 is constitutive ("housekeeping") and ubiquitous—present in gastrointestinal tract, kidneys, and platelets—and it performs a homeostatic function during physiological situations. Thus, inhibition of COX-1 results in the gastrointestinal side effects on one hand and useful antiplatelet activities on the other. COX-2 is induced during inflammation and facilitates inflammatory response. Thus, inhibition of COX-2 results in the anti-inflammatory, antipyretic, and analgesic action (Vane 1971, 1996; Figure 28.1). There is another type of this enzyme, COX-3, which is a splice variant of COX-1, the role of which is less clear (Timothy 2002). It has been proposed that acetaminophen's actions could be mediated by its ability to inhibit COX-3 in brain (Botting 2005).

There are some other less-characterized mechanisms of action of these drugs as well. These drugs downregulate interleukin-1 (IL-1) production, decrease production of free radicals, interfere with intracellular calcium-mediated events, and inhibit chemotaxis (Daniel 2009). Further, NSAIDs decrease the sensitivity of vessels to histamine and bradykinin.

While the agents that inhibit both the cyclooxygenase enzymes are labelled as "nonselective" or "traditional" NSAIDs, those that preferentially block cyclooxygenase-2 are called "coxibs" (Lipsky 1998). The drugs, however, vary in their relative selectivity for COX-1 and COX-2 in vitro (FitzGerald 2001; Grosser 2006). The COX-2 selective drugs were thought to be safer in respect to the gastrointestinal toxicity, but they are more likely to increase thrombotic cardiovascular events because of their inability to inhibit platelets (which only express COX-1; FitzGerald 2003).

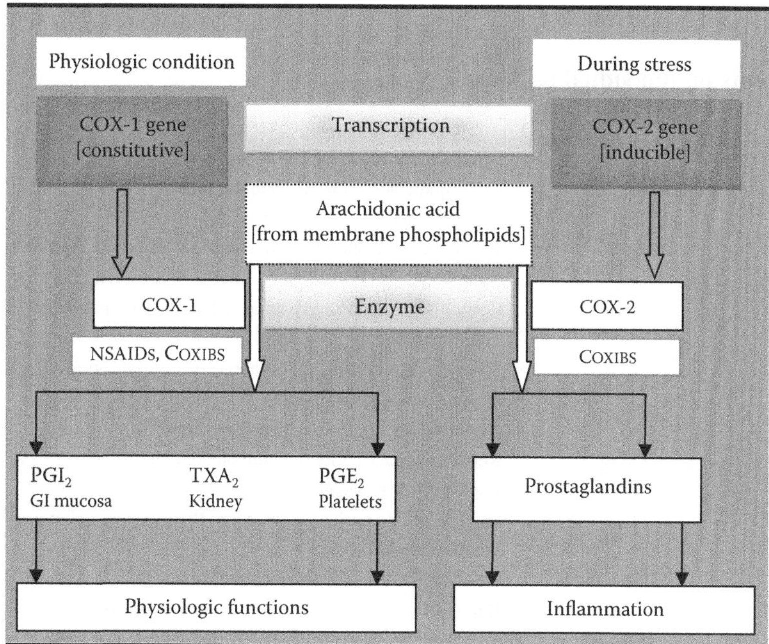

FIGURE 28.1 Mechanisms of action of NSAIDs.

The mechanism of action of acetaminophen is less clear. It is believed to act by inhibiting cyclo-oxygenase in the brain, thereby explaining its superior antipyretic action and less demonstrable anti-inflammatory action. For this reason, it is often not classified under NSAIDs.

28.4 CLASSIFICATION

The NSAIDs are weak acids. The different chemical classes of the drugs are salicylates (aspirin [acetylsalicylic acid]); propionic acid derivatives (ibuprofen, naproxen, ketoprofen, fenoprofen, flurbiprofen, oxaprozin); acetic acid derivatives (indomethacin, etodolac, ketorolac, diclofenac, nabumetone, sulindac); enolic acid (oxicam) derivatives (piroxicam, meloxicam, lornoxicam); fenamic acid derivatives (mefenamic acid); sulfonanilides (nimesulide); selective COX-2 inhibitors or coxibs (celecoxib, etoricoxib). This classification has little clinical value except that the coxibs are a distinct class. Acetaminophen (paracetamol), although often discussed within NSAIDs and sharing some properties with them, is a different molecule with limited anti-inflammatory property.

28.5 PHARMACOKINETICS

- Bioavailability: The drugs are rapidly absorbed and the oral bioavailability of most of the agents is high with peak concentrations occurring within 1 to 4 hours.
- Protein binding: The drugs are highly protein-bound (average 95% to 99%).
- Half-lives: The agents differ widely in their duration of action and half-lives (Table 28.1).
- Metabolism and clearance: The drugs are metabolized in the liver to inactive metabolites and are then excreted. The excretion of salicylate (aspirin) occurs with first-order kinetics with a half-life between 0.5 and 19 hours, depending on the dose of aspirin administered.

28.6 PHARMACODYNAMICS

Most of the NSAIDs are weak organic acids. Nabumetone is an exception; it is the only non-acid NSAID in current use. The NSAIDs act by inhibiting the COX pathway as illustrated

TABLE 28.1
Brief Comments on Individual NSAIDs

Drug Name	Half-Life (hours)	Notes
Aspirin	4 to 6	Decrease dose in renal failure and hepatic disease.
Celecoxib	11	Contraindicated with sulfonamide allergy. Less gastrointestinal toxicity.
Diclofenac	2	Incidence of increased transaminase levels higher than with other NSAIDs.
Etodolac	6 to 7	Relatively less GI toxicity.
Etoricoxib	22	Contraindicated in severe renal or liver disease patients. Caution in mild to moderate disease.
Flurbiprofen	3 to 4	Rarely causes neurotoxicity (ataxia, myoclonus, tremor, rigidity).
Ibuprofen	2	Reports of septic meningitis.
Indomethacin	2 to 13	Approved for treatment of patent ductus.
Ketorolac	4 to 6	Renal toxicities are more common than others. Used mainly as analgesic.
Nabumetone	24	Food increases peak concentration. Rarely causes phototoxicity, pseudoporphyria.
Naproxen	12 to 15	Does not increase cardiovascular risk. Rarely causes allergic vasculitis, pseudoporphyria.
Oxaprozin	49 to 60	Mild uricosuric. Decrease dose in renal failure and low body weight.
Piroxicam	3 to 86	Decrease dose in hepatic disease.
Sulindac	16	Used in polyposis.
Tolmetin	1 to 1.5	Rarely causes thrombotic thrombocytopenic purpura (TTP). Ineffective in gout.

Note: Drugs are listed alphabetically.

before. There are subtle differences among the different agents. A single dose of aspirin suppresses thromboxane production in platelets for several days. This alters the thromboxane versus prostacyclin (TXA_2-PGI_2) balance toward PGI_2, which leads to inhibition of platelets and vasodilation; thus, it is an antithrombotic (David 2005). Acetaminophen has very little anti-inflammatory effect because it inhibits COX-2 only weakly. It is not technically considered to be an NSAID.

Regarding efficacy, coxibs and other NSAIDs are similar at comparable doses (FitzGerald 2001). However, there are differences in the response of individual persons to the particular drugs. For this reason, a trial with a different agent is necessary if the patient fails to respond to one NSAID.

28.7 CLINICAL USES OF NSAIDs

NSAIDs are used (1) for symptomatic relief of acute musculoskeletal pain, (2) to treat fever and malaise and headache, (3) to treat inflammatory arthritides like rheumatoid arthritis, and (4) as an antiplatelet agent to treat ischemic heart disease and ischemic stroke (aspirin only), other conditions like systemic mastocytosis, Bartter's syndrome, and so forth.

28.7.1 Osteoarthritis

The LOGICA study demonstrated that most patients with osteoarthritis (OA) requiring NSAIDs for pain control showed a high prevalence of gastrointestinal (GI) and cardiovascular (CV) risk factors.

Over half of the patients were at either high GI or CV risk, or both, such that the prescription of OA treatments should be very carefully considered (Lanas 2010).

28.7.2 Other Inflammatory Arthropathies

NSAIDs are used for anti-inflammatory and analgesic properties in rheumatoid arthritis, spondyloarthropathy, juvenile idiopathic arthritis, and so forth. They are mostly used short term. In ankylosing spondylitis, NSAIDs given for prolonged periods of up to a year have led to improvement in spinal mobility and acute-phase reactants, and continuous therapy retards radiographic progression (Dougados 1999; Wanders 2005).

28.7.3 Other Roles of NSAIDs

NSAIDs have been noted to prevent the development of certain diseases and experimentally tried in such situations as cancer chemoprevention and Alzheimer's disease (Thun 2002; McGeer 2000).

28.8 HOW TO GUIDE THERAPY

In a recent study of prescription patterns in osteoarthritis patients, it was observed that prescription of NSAIDs was not in accordance with current recommendations made by regulatory agencies (Lanas 2011). NSAIDs are prescribed in two fashions: short duration in local injury or inflammation for pain relief and also as antipyretic in fever and for a relatively longer duration in some situations demanding anti-inflammatory as well as analgesic properties such as rheumatic conditions (e.g., rheumatoid arthritis, juvenile arthritis, ankylosing spondylitis).

The clinician must ask the patient for any history of allergy to such drugs in the past, exacerbation of bronchial asthma, history suggestive of peptic ulcer disease or gastrointestinal bleeding, and any significant cardiovascular, renal, or hepatic disease. Before starting therapy for longer duration, it is prudent to check baseline hemoglobin, blood counts (total and differential, platelets), liver function tests, serum creatinine, and serum electrolytes (Table 28.2). For subjects who have any of the aforementioned issues, the therapy has to be tailored accordingly, if not contraindicated. The situations are discussed in detail in Section 28.13.

Regarding monitoring of long-term NSAID therapy, the patient needs to be clinically assessed periodically during follow-up. The follow-up protocol depends on the baseline risk profile. Serum creatinine, electrolytes, blood counts, and liver function tests need to be repeated at appropriate intervals.

TABLE 28.2
Checklist for Therapy with NSAIDs

Check	
Initiation	Allergy
	Asthma
	Peptic ulcer, GI bleeding
	CAD, heart failure
	Renal dysfunction
	Full blood counts
	Creatinine, electrolytes
	Liver function tests
Monitoring	Full blood counts
	Creatinine, electrolytes
	Liver function tests

TABLE 28.3
Adverse Reactions to NSAIDs

System	Toxicities
Mucocutaneous	Allergy, urticarial, angioedema, toxic epidermal necrolysis, photosensitivity
Gastrointestinal	Anorexia, nausea, vomiting, diarrhea, peptic ulcer, gastrointestinal bleeding
Hepatic	Elevated transaminases, cholestasis, acute liver injury
Renal	Acute interstitial nephritis, papillary necrosis, renal tubular acidosis, sodium retention, hyperkalemia, renal failure
Cardiovascular	Hypertension, ischemic heart disease, myocardial infarction, worsening of heart failure
Pulmonary	Exacerbation of allergic asthma (aspirin sensitive), vasomotor rhinitis
Hematologic	Cytopenias, anemia (from gastrointestinal bleeding)
Neurologic	Confusion, drowsiness, dizziness, aseptic meningitis, seizure
Others	Reye's syndrome

28.9 ADVERSE REACTIONS AND TOXICITIES OF NSAIDs

The drugs share some common adverse effects. They are summarized in Table 28.3.

28.10 CHOICE OF NSAIDs

The different NSAIDs are comparable in terms of efficacy; however, there may be considerable interindividual variation in response to particular molecules. The factors that have bearing while choosing a particular NSAID are individual response, toxicity, half-life, cost, and physician or patient preferences. Important issues regarding particular agents are briefly summarized in the following (Table 28.1).

28.11 PRECAUTIONS FOR USE OF NSAIDs

- Gastrointestinal disease: Discussed in Section 28.13
- Cardiovascular disease: Discussed in Section 28.13
- Renal disease: Discussed in Section 28.13
- Concomitant medications: Discussed in Section 28.12

28.11.1 Anemia

The risk of NSAID-induced gastrointestinal blood loss is increased in the presence of severe anemia. Anemia, especially if iron deficiency, even in absence of overt blood loss in a patient taking NSAIDs, requires investigation to rule out GI blood loss.

28.11.2 Hypersensitivity Reactions: Asthma

The coexistence of hypersensitivity to aspirin (and to other NSAIDs) with upper airway (rhinosinusitis/nasal polyps) and lower airway (asthma) disease is referred to as aspirin triad, asthma triad, or Samter's syndrome (Samter 1968). This condition is now termed as aspirin-exacerbated respiratory disease (AERD) because it is not drug hypersensitivity but exacerbation of the underlying chronic inflammatory respiratory disease by aspirin or other NSAIDs (Berges-Gimeno 2002). It is proposed to be due to deprivation of PGE2 by NSAIDs, which leads to activation of inflammatory pathways, including increased leukotriene synthesis. Management would be avoidance of NSAIDs and use of alternative drugs like acetaminophen and possibly COX-2 inhibitors (Celik 2005).

28.11.3 Delayed Hypersensitivity

Cutaneous symptoms are the most frequent manifestations of delayed hypersensitivity to NSAIDs. Prompt discontinuation is recommended to decrease the mortality. Symptomatic treatment involves systemic corticosteroids and antihistamines (Kowalski 2011). Patients with SJS/TEN should be treated in intensive care units offering typical management as for burns. There is no specific pharmacological treatment, and use of corticosteroids, plasmapheresis, intravenous immunoglobulins, or immunosuppressive drugs is still controversial (French 2006).

28.11.4 Pregnancy

NSAIDs are not recommended during pregnancy, particularly during the third trimester. They may cause premature closure of the fetal ductus arteriosus and renal dysfunction in the fetus. Additionally, they are linked with premature birth (Østensen 2004). In contrast, acetaminophen is regarded as relatively safe and well tolerated in pregnancy (Graham 2005). However, aspirin is used in pregnant women with antiphospholipid antibodies (Cervera 2004).

28.11.5 Elderly

The elderly are at higher risk of adverse reactions to NSAIDs. Furthermore, the aged person might be taking other drugs for comorbid conditions, increasing the chance of drug interactions. If at all required, NSAIDs should be used at the lowest possible dose for the shortest period. It is prudent to use acetaminophen in such conditions (Zachary 2010).

28.11.6 Surgery

Surgical or wound site bleeding can occur theoretically due to inhibition of platelet TXA2 production. NSAIDs should be discontinued for a period equal to five times their half-life prior to surgery to ensure hemostasis. Aspirin should be discontinued 1–2 weeks before surgery because its effect on platelets is irreversible.

28.12 DRUG INTERACTIONS

Recognizing drug interaction and its avoidance is of paramount importance for any clinician. Many new drugs are introduced each year, and new interactions are recognized with use. There is software to detect potential interactions. However, they tend to flag each interaction and prescription becomes difficult. The important (i.e., clinically relevant) interactions are to be taken seriously. NSAIDs often need to be co-prescribed with other drugs. For example, a patient with rheumatoid arthritis who is on therapy with methotrexate might require the NSAID for control of pain from the disease itself. On the other hand, a patient with diabetes mellitus who is being treated with sulfonylurea drugs may take an NSAID for another reason. The clinician needs to be very careful in such situations and patients need to be explained what could happen so as to avoid inadvertent drug interactions. The different categories of possible drug interactions follow (Table 28.4). The list is only illustrative and not exhaustive. Physicians are advised to follow a software-based drug interaction checker (such as PDR Network, http://www.pdr.net/drugpages; British National Formulary, http://bnf.org/bnf/index.htm; Medscape, http://reference.medscape.com) and standard references including product information brochures.

TABLE 28.4
List of Common Drug Interactions

Drug Category	Drugs	Effect	Mechanisms
Antidiabetic drugs	Oral hypoglycemic drugs (e.g., glibenclamide, glipizide, glimipiride)	Increased risk of hypoglycemia	Competition for plasma protein binding, inhibition of metabolism
Anticoagulant drugs	Anticoagulants (e.g., warfarin)	Increased risk of bleeding	Competition for plasma protein binding, inhibition of metabolism
Antihypertensive drugs	All antihypertensives (e.g., angiotensin converting enzyme inhibitor, β-blocker)	Blunted control of hypertension	Sodium retention by NSAIDs
	Angiotensin converting enzyme inhibitor, angiotensin receptor blockers	Renal toxicity	Blocking renal prostaglandin mediated vasodilation maintaining blood flow
Diuretics	All diuretics	Renal toxicity	Hypovolemia
	K^+ sparing diuretics	Hyperkalemia	Blocking aldosterone axis
Antimicrobial drugs	Quinolones	Seizure	Competition for plasma protein binding
	Aminoglycosides	Increased blood level of the drug	Decreased renal excretion of the drug
Antiepileptic drugs	Phenytoin, valproate	Increased blood level of the drug	Competition for plasma protein binding, inhibition of metabolism
Drugs used in psychiatry	Lithium	Increased blood level of the drug	Decreased renal excretion of the drug
	Alcohol	Increased risk of gastrointestinal bleeding	Additive effect
Immunosuppressive drugs	Corticosteroids, cyclosporine	Increased nephrotoxicity	Additive effect
	Methotrexate	Increased blood level of the drug	Decreased renal excretion of the drug
Cardiovascular drugs	Digoxin	Increased blood level of the drug	Decreased renal excretion of the drug

28.13 SPECIAL SITUATIONS AND TOXICITIES

The three most important issues while prescribing NSAIDs are gastrointestinal, cardiovascular, and renal toxicities. In the perspective of an average primary care grouping of 100,000 (Bandolier 2000), in the population with 3,500 over-65s taking NSAIDs, there would be 18, 10, and 22 hospital admissions every year for upper gastrointestinal bleeding, acute renal failure, and congestive heart failure, respectively.

The key issues are described below.

28.13.1 NSAID-Induced Gastrointestinal Toxicity

Since COX-1 pathway–derived prostaglandins are responsible for regulating gastric acid secretion (PGE_2 and PGI_2 reduces acid secretion) and maintenance of mucosal defense and cytoprotection (PGE_2 stimulates epithelial mucus production, bicarbonate secretion; while PGE_2 and PGI_2 maintains mucosal blood flow by vasodilation and epithelial proliferation) in the GI tract, blocking them with NSAIDs would be expected to cause ulcerations by altering these (Scheiman 1996; Wolfe 1999). All NSAIDs are associated with GI toxicity, including prophylactic low-dose aspirin (Lanas 2007; Sostres 2011). The coxibs can also cause asymptomatic GI ulcers and are found in around 30% of patients receiving long-term NSAIDs (Bombardier 2000). The prevalence of asymptomatic ulcers with low-dose prophylactic aspirin is 7.3% at 3 months (Wilcox 1994). *H. pylori* infection increases the risk of these complications and eradication of the infection would be advantageous in reducing the risk (Aalykke 1999; Laine 2002; Ghosh 2005). Even the COX-2 inhibitors are also associated with a definite GI toxicity, although lower than nonselective NSAIDs. In the CLASS study, the GI toxicity with celecoxib was lower than with diclofenac, however, this difference was not observed at one year (Silverstein 2000).

The risk factors for gastrointestinal toxicity are history of peptic ulcer or gastrointestinal bleeding, use of concomitant medications (glucocorticoid, anticoagulants like warfarin), use of higher doses of NSAIDs, use of NSAIDs for long duration, presence of other medical problems (diabetes, hypertension, cardiovascular disease, renal insufficiency, hepatic impairment), and older age (age >60 years; Wolfe 1999). The risk would be highest if the patient has a history of recent complicated peptic ulcer or at least two risk factors (history of uncomplicated ulcer; use of high dose of NSAID; concomitant use aspirin [including low dose], anticoagulant or corticosteroids; age older than 65 years) (Lanza 2009). Using COX-2 with PPI has the lowest rebleeding risk when NSAIDs are used in such patients (Chan 2007). A number of large studies (OMNIUM—omeprazole vs. misoprostol; ASTRONAUT—ranitidine vs. omeprazole; MUCOSA—Misoprostol Ulcer Complications Outcomes Study Assessment) and a recent meta-analysis (Rostom 2007) demonstrate that misoprostol, PPIs, and double doses (not standard dose) of H2RAs are effective at reducing the risk of endoscopic gastric and duodenal NSAID-induced ulcers. Standard doses of H2RAs are ineffective in this regard. Regarding prophylaxis, the only agent studied in outcome trial is misoprostol, and it was shown to reduce the risk of NSAID-induced ulcer complications. However, the poor tolerance limits the use of the drug. There had been interest in using a lower dose of misoprostol (400 μg compared to 800 μg) to avoid the intolerance, mostly diarrhea. Both doses are associated with diarrhea, and the effectiveness of misprostol at the lower dose is not documented. Thus, possibly the use of low-dose misoprostol to prevent NSAID-related ulcers is not justified. Regarding use of COX-2 selective blockers, the Celecoxib Long-Term Arthritis Safety Study (CLASS) had demonstrated superior tolerance to celecoxib compared to others (Silverstein 2000). In the MEDAL, VIGOR, and other studies, COX-2 inhibitors are demonstrated to have lower GI toxicity (Cannon 2006; Bombardier 2000). The best way to prevent the gastrointestinal toxicities of NSAIDs is to avoid the use of nonselective NSAIDs with or without cotherapy with mucosal protective agent or proton pump inhibitors. An alternative would be to use acetaminophen, which has very little gastric adverse effects (Cryer 1998).

28.13.2 Hepatotoxicity

NSAIDs are common causes of Drug-Induced Liver Injury (DILI). The incidence is low, however, ranging from 0.29/100,000 to 9/100,000. NSAIDs can cause a wide spectrum of liver damage ranging from asymptomatic, transient, hyper-transaminasemia to fulminant hepatic failure. Presence of jaundice is prognostically bad, with 25% developing severe disease (Bessone 2010). Ibuprofen has the highest liver safety profile among NSAIDs and showed no severe liver injury in larger studies. Uncommonly, piroxicam may cause severe hepatocellular damage. Nimesulide was removed from the market in several countries due to severe liver damage; in others it is recommended for use in adults for short term. Coxib-induced liver injury is an uncommon event, occurring in 1 per 100,000 exposed persons (Boelsterli 2002).

28.13.3 NSAID-Induced Cardiovascular Toxicity

Endothelial cell COX-2 produces PGI_2, which leads to dilation of vascular smooth muscle and blocks platelet aggregation. On the other hand, platelets, expressing only COX-1, produce TXA_2, which leads to platelet aggregation. In stress, COX-2 expression overwhelms COX-1. Thus, blocking COX-2 leads to the adverse cardiac events as may be expected (Cheng 2006). The three landmark trials—Celecoxib Long-Term Arthritis Safety Study (CLASS), Vioxx Outcomes Research Study (VIGOR), and Adenomatous Polyp Prevention On Vioxx (APPROVe) (Silverstein 2000; Bombardier 2000; Bresalier 2005)—pointed out that the adverse cardiovascular "class effect" was an inherent property of all coxibs. The conclusions of a recent meta-analysis (31 trials in 116,429 patients with more than 115,000 patient years of follow-up) of randomized trials broadly confirm previous research showing an increased risk of cardiovascular events with many NSAIDs. Compared with placebo, rofecoxib (withdrawn from the market) was associated with the highest risk of myocardial infarction (rate ratio 2.12). Ibuprofen was associated with the highest risk of stroke (3.36, 1.00 to 11.6), followed by diclofenac (2.86, 1.09 to 8.36). Etoricoxib (4.07, 1.23 to 15.7) and diclofenac (3.98, 1.48 to 12.7) were associated with the highest risk (approximately four-fold) of cardiovascular death (Trelle 2011). The risk is present for both traditional NSAIDs (risk average 1.19; diclofenac 1.38; Singh 2006) and COX-2 selective NSAIDs. Naproxen is associated with the lowest CV risk overall. Interpretation of the results for ibuprofen is hampered by the absence of stratification by dose, which other work suggests is important. The cardiovascular safety issues led to the withdrawal of a number of coxibs like rofecoxib. Presently, there is concrete evidence of cardiovascular risk with most NSAIDs, and this should guide clinicians to modify their prescription habit.

The Framingham risk factors like blood pressure, glycemic status, lipid profile, smoking status, and family history of ischemic heart disease should be noted. The cardiovascular risk assessment can objectively be done using online calculators such as that from the National Cholesterol Education Initiative (http://hp2010.nhlbihin.net/atpiii/calculator.asp); therapy may be planned accordingly. For patients with high cardiovascular risk, naproxen is preferred, if there is low GI risk. These patients are also supposed to be on low-dose prophylactic aspirin, which increases their GI risk. Thus, addition of a PPI would be prudent (Rostom 2009). In patients with both GI and CV risk, it is advisable to avoid any NSAID. The American Heart Association had issued a guideline for use of NSAIDs in patients with CV risk (Antman 2007).

28.13.3.1 Patient on Low-Dose Aspirin

Patients on low-dose aspirin for cardiac protection present a complex situation. NSAIDs diminish the protective effect of aspirin (Catella-Lawson 2001), probably competing with aspirin to the acetylation site of platelet COX-1, but they are weak in this regard (as they are a reversible inhibitor as compared to aspirin). Clopidogrel may be used as a rational alternative to low-dose aspirin for the patient with significant cardiovascular risk factors who requires chronic NSAIDs. The coxibs do not interfere with the inhibitory effects of low-dose aspirin on platelets, but they increase the cardiovascular risk as explained previously, and there is no evidence that low-dose aspirin provides consistent protection against this increased risk (FitzGerald 2001; Grosser 2006).

Further, NSAIDs can cause volume overload and a modest elevation in blood pressure, can blunt the therapeutic effect of antihypertensive drugs, and can worsen heart failure. Etoricoxib may be associated with more frequent problems with blood pressure control. An approximately 5 mmHg rise in supine blood pressure may be expected with NSAIDs (Johnson 2003). This should also be taken into consideration when prescribing NSAIDs in patients with cardiovascular disease.

28.13.4 NSAID-Induced Renal Toxicity

Renal prostaglandins are important in maintaining the renal perfusion in individuals with parenchymal renal disease and when the person is dehydrated or the circulating volume is decreased (e.g., heart failure). In these states, the renin-angiotensin-aldosterone axis is stimulated, leading to vasoconstriction. Prostaglandins compensate this by dilating the vessels. Thus, NSAIDs alter the intrarenal hemodynamics (reduce renal blood flow) by blocking the COX enzyme, leading to a fall in glomerular filtration and consequent sodium and water retention (Whelton 2000).

NSAIDs can cause sodium retention and edema, hyperkalemia, acute renal failure, and rarely proteinuria, interstitial nephritis, renal papillary necrosis/chronic interstitial nephritis (De Broe

TABLE 28.5
Treatment Options with NSAIDs Based on Risk Stratification

Risk Group	Treatment Options
Low GI and low CV risk	NSAIDs
Low GI and high CV risk	Naproxen + possibly PPI + clopidogrel[a] (PPI is added as the GI toxicity of the combination is as much as adding low-dose aspirin)
High GI and low CV risk	NSAIDs + PPI
	Coxibs ± PPI (added depending upon risk)
High GI and high CV risk	Avoid NSAIDs, use acetaminophen

[a] Effectiveness of low-dose aspirin may be reduced with NSAIDs.
PPI, proton pump inhibitors, such as omeprazole (Grosser 2006; Wolfe 1999; Bijlsma 2010).

2005). There are also reports of subclinical nephrotoxicity (Calvo-Alen 1994). Hyperkalemia is an unusual complication occurring mainly when used with potassium sparing diuretics, ACE inhibitor, or in myeloma or heart failure. Renal papillary necrosis/chronic interstitial nephritis occur with massive dosage of NSAIDs. Acute interstitial nephritis may develop at any time during treatment, but typically occurs months after the therapy. Although it generally resolves upon discontinuation of therapy, it is a significant cause of acute renal failure.

The risk factors for predisposing NSAID-induced renal failure are advanced age; preexisting intrinsic renal disease; volume depleted states (dehydration); use of diuretics or angiotensin-converting enzyme inhibitor; and, comorbid conditions like cirrhosis and heart failure. NSAIDs (including the coxibs) should be prescribed (if it is a compelling situation) with extreme caution when one or more of the risk factors for NSAID-induced hemodynamically mediated renal failure is present and should not be used when the creatinine clearance is less than 30 mL/min.

28.13.5 Topical NSAIDs

The clinical knowledge may be summarized as in Table 28.5.

28.14 TREATMENT OPTIONS FOR PATIENTS WITH GASTROINTESTINAL AND CARDIOVASCULAR RISK

Topical preparations (gel, ointment, etc.) of NSAIDs are in use for localized musculoskeletal pain and inflammation like sprains. There is clinical evidence of efficacy of topical NSAIDs in strains and sprains and in arthritic conditions (a systematic review of 86 randomized controlled trials involving 10,160 patients; Moore 1998). The drugs found to have efficacy are ketoprofen (NNT 2.6), felbinac (3.0), ibuprofen (3.5), and piroxicam (4.2). In a more recent Cochrane review (Massey 2010), topical diclofenac, ibuprofen, ketoprofen, and piroxicam were of similar efficacy, but indomethacin and benzydamine were not significantly better than placebo. Local skin reactions were generally mild and transient and did not differ from placebo. Plasma concentrations are low after topical application (usually less than 5%), and there are very few systemic adverse events. Data were inadequate to reliably compare individual topical NSAIDs with each other or the same oral NSAID. Topical NSAIDs are effective in both acute (NNT 3.9 [3.4–4.4]) and chronic (NNT 3.1 [2.7–3.8]) conditions. Topical NSAIDs are safer than oral NSAIDs, especially with less severe gastrointestinal toxicity. However, a substantial proportion of older adults report systemic adverse reactions with topical agents (Makris et al. 2010). Topical preparations should not be used if there are broken skin, ulcerations, and so forth, for obvious reasons.

TABLE 28.6
Instructions to Patients

- Do not take the drug unless prescribed.
- Do not take this medicine if you have ever suffered allergic or other adverse reactions to aspirin or any analgesic (NSAIDs).
- Consult with your physician if you have a history of gastrointestinal bleeding or ulcer before taking NSAIDs.
- Tell your physician if you are pregnant, intend to become pregnant, or are breastfeeding as the drugs might harm your baby.
- Tell your physician if you are taking any other medicines for other reasons so that he or she knows if the drugs are compatible.
- Inform your physician if you have any bleeding or blood clotting problems.
- Some NSAIDs can cause drowsiness; thus take precautions for performing jobs requiring alertness like driving/using machinery.
- Remember, NSAIDs can also affect other medical conditions like hypertension, heart problems, kidney problems, asthma, and others.
- Inform your physician if you regularly consume alcohol.

TAKE-HOME MESSAGES

- NSAIDs are a heterogeneous group of chemical compounds that share the common property of having anti-inflammatory, analgesic, and antipyretic action.
- Most of the NSAIDs are weak organic acids. The NSAIDs act by inhibiting the COX pathway as illustrated previously.
- The three most important issues while prescribing NSAIDs are gastrointestinal, cardiovascular, and renal toxicities.
- NSAIDs are common causes of drug-induced liver injury (DILI). The incidence is, however, low.
- NSAIDs should be prescribed with extreme caution in the following conditions: advanced age; preexisting intrinsic renal disease; volume-depleted states (dehydration); use of diuretics or angiotensin-converting enzyme inhibitor; comorbid conditions like cirrhosis and heart failure.
- Instruction to patients: Patients being prescribed NSAIDs need to be informed about the drug and necessary precautions (Table 28.6).

WEB SITES

PDR Network, http://www.pdr.net/drugpages/concisemonographlist.aspx
British National Formulary, http://bnf.org/bnf/index.htm
Medscape, http://reference.medscape.com/drug-interactionchecker

GENERAL RESOURCES

Burke A, Smyth E, FitzGerald GA. Analgesic-antipyretic agents; pharmacotherapy of gout. In Brunton LL (ed.), *Goodman & Gilman's The Pharmacological Basis of Therapeutics*. New York: McGraw-Hill, 2006.

Imboden JB. Nonsteroidal anti-inflammatory drugs. In Imboden JB, Hellmann DB, Stone JH (eds.), *Current Diagnosis and Treatment in Rheumatology*. New York: McGraw Hill, 2007.

REFERENCES

Aalykke C, Lauritsen JM, Hallas J, Reinholdt S, Krogfelt K, Lauritsen K. *Helicobacter pylori* and risk of ulcer bleeding among users of nonsteroidal anti-inflammatory drugs: A case-control study. *Gastroenterology* 1999; 116(6):1305–1309.

Antman EM, Bennett JS, Daugherty A, Furberg C, Roberts H, Taubert KA, American Heart Association. Use of nonsteroidal antiinflammatory drugs: An update for clinicians: A scientific statement from the American Heart Association. *Circulation* 2007; 115(12):1634–1642.

Bandolier. More on NSAID adverse effects. 2000. Available at http://www.jr2.ox.ac.uk/bandolier/band79/b79-6.html#Heading8.

Berges-Gimeno MP, Simon RA, Stevenson DD. The natural history and clinical characteristics of aspirin-exacerbated respiratory disease. *Ann Allergy Asthma Immunol.* 2002; 89:474–478.

Bessone F. Non-steroidal anti-inflammatory drugs: What is the actual risk of liver damage? *World J Gastroenterol* 2010; 16(45):5651–5661.

Bijlsma JWJ. Patient benefit risk in arthritis: A rheumatologist's perspective. *Rheumatology* 2010; 49:ii11–ii17.

Boelsterli UA. Mechanisms of NSAID-induced hepatotoxicity: Focus on nimesulide. *Drug Saf* 2002; 25: 633–648.

Bombardier C, Laine L, Reicin A, et al. Comparison of upper gastrointestinal toxicity of rofecoxib and naproxen in patients with rheumatoid arthritis. VIGOR Study Group. *N Engl J Med* 2000; 343:1520–1528.

Botting R, Ayoub SS. COX-3 and the mechanism of action of paracetamol/acetaminophen. *Prostaglandins Leukot Essent Fatty Acids* 2005; 72:85–87.

Bresalier R, Sandler R, Quan H, et al. Adenomatous polyp prevention on Vioxx: Cardiovascular events associated with rofecoxib in a colorectal adenoma chemoprevention trial. APPROVe Trial Investigators. *N Engl J Med* 2005; 352:1092–1102.

Calvo-Alen J, Angeles De Cos M, Rodriguez-Valverde V, Escalladv R, Florez J, Arias M. Subclinical renal toxicity. *J Rheumatol* 1994; 214:1742–1747.

Cannon CP, Curtis SP, Bolognese JA, et al. Clinical trial design and patient demographics of the Multinational Etoricoxib and Diclofenac Arthritis Long-term (MEDAL) study program: Cardiovascular outcomes with etoricoxib versus diclofenac in patients with osteoarthritis and rheumatoid arthritis. *Am Heart J* 2006; 152:237–245.

Catella-Lawson F, Reilly MP, Kapoor SC, et al. Cyclooxygenase inhibitors and the antiplatelet effects of aspirin. *N Engl J Med* 2001; 345:1809–1817.

Celik G, Pasaoglu G, Bavbek S, Abadoglu O, Dursun B, Mungan D, et al. Tolerability of selective cyclooxygenase inhibitor, celecoxib, in patients with analgesic intolerance. *J Asthma* 2005; 42:127–131.

Cervera R, Balasch J. The management of pregnant patients with antiphospholipid syndrome. *Lupus* 2004; 13(9):683–687.

Chan FK, Wong VW, Suen BY, Wu JC, Ching JY, Hung LC, et al. Combination of a cyclo-oxygenase-2 inhibitor and a proton-pump inhibitor for prevention of recurrent ulcer bleeding in patients at very high risk: A double-blind, randomised trial. *Lancet* 2007; 369:1621–1626.

Cheng Y, Wang M, Yu Y, et al. Cyclooxygenases, microsomal prostaglandin E synthase-1, and cardiovascular function. *J Clin Invest* 2006; 116:1391–1399.

Cryer B, Kimmer MB. Gastrointestinal side effects of nonsteroidal anti-inflammatory drugs. *Am J Med* 1998; 105:20S–30S.

De Broe ME. Drug-induced nephropathies. In Davison AMA, Cameron JS, Grunfeld JP et al. (eds.) *Oxford Textbook of Nephrology,* 2581–2598. Oxford: Oxford University Press. 2005

Dougados M, Gueguen A, Nakache JP, et al. Ankylosing spondylitis: What is the optimum duration of a clinical study? A one year versus 6 weeks non-steroidal anti-inflammatory drug trial. *Rheumatology* 1999; 38:235–244.

Dudzinski DM, Serhan CN. Pharmacology of eicosanoids. In David E Golan (ed.) *The Pathologic Basis of Drug Therapy.* Baltimore: Lippincott Williams and Wilkins. 2005; p. 627–643.

Einhorn T. Do inhibitors of cyclooxygenase-2 impair bone healing? *J Bone Miner Res* 2002; 17:977.

FitzGerald GA. COX-2 and beyond: Approaches to prostaglandin inhibition in human disease. *Nat Rev Drug Discov.* 2003 Nov; 2(11):879–890.

FitzGerald GA, Patrono C. The coxibs, selective inhibitors of cyclooxygenase-2. *N Engl J Med* 2001; 345(6):433–442.

French LE, Trent JT, Kerdel FA. Use of intravenous immunoglobulin in toxic epidermal necrolysis and Stevens-Johnson syndrome: Our current understanding. *Int Immunopharmacol* 2006; 6:543–549.

Furst D, Ulrich RN, Varkey-Altamirano C. Nonsteroidal antiinflammatory drugs, disease modifying antirheumatic drugs, nonopioid analgesics, and drugs used in gout. In: Bertram G Katzung (ed.) *Basic and Clinical Pharmacology*. New Delhi: Tata McGraw-Hill Education Private Limited. 2009: p. 621–629.

Ghosh B, Ghosh A. Influence of *H. pylori* and omeprazole on celecoxib induced gastroduodenal mucosal injury. *Ind J Gastroenterol* 2005; 24:177.

Graham G, Scott K, Day R. Tolerability of paracetamol. *Drug Saf* 2005; 28(3):227–240.

Grosser T, Fries S, FitzGerald GA. Biological basis for the cardiovascular consequences of COX-2 inhibition: Therapeutic challenges and opportunities. *J Clin Invest.* 2006; 116(1):4–15.

Hawkey CJ. COX-2 inhibitors. *Lancet* 1999; 353:307–314.

Johnson DL, Hisel TM, Phillips BB. Effect of cyclooxygenase-2 inhibitors on blood pressure. *Ann Pharmacother* 2003; 37:442–446.

Kowalski ML, Makowska JS, Blanca M, Bavbek S, Bochenek G, Bousquet J, et al. Hypersensitivity to nonsteroidal anti-inflammatory drugs (NSAIDs)—classification, diagnosis and management: Review of the EAACI/ENDA and GA2LEN/HANNA. *Allergy* 2011; 66:818–829.

Laine L. Review article: The effect of *Helicobacter pylori* infection on nonsteroidal anti-inflammatory drug-induced upper gastrointestinal tract injury. *Aliment Pharmacol Ther* 2002; 16(Suppl 1):34–39.

Lanas A, Garcia-Tell G, Armada B, Oteo-Alvaro A. Prescription patterns and appropriateness of NSAID therapy according to gastrointestinal risk and cardiovascular history in patients with diagnoses of osteoarthritis. *BMC Medicine* 2011; 9:38. doi:10.1186/1741-7015-9-38.

Lanas A, Scheiman J. Low-dose aspirin and upper gastrointestinal damage: Epidemiology, prevention and treatment. *Curr Med Res Opin.* 2007; 23(1):163–173.

Lanas A, Tornero J, Zamorano JL. Assessment of gastrointestinal and cardiovascular risk in patients with osteoarthritis who require NSAIDs: The LOGICA study. *Ann Rheum Dis.* 2010; 69(8):1453–1458.

Lanza FL, Chan FK, Quigley EM. Practice Parameters Committee of the American College of Gastroenterology. Guidelines for prevention of NSAID-related ulcer complications. *Am J Gastroenterol* 2009; 104(3):728–738.

Lipsky LP, Abramson, SB, Crofford L, Dubois RN, Simon LS, van de Putte LB. The classification of cyclooxygenase inhibitors. *Journal of Rheumatology* 1998; 25:2298–2303.

Makris UE, Kohler MJ, Fraenkel L. Adverse effects (AEs) of topical NSAIDs in older adults with osteoarthritis (OA): A systematic review of the literature. *J Rheumatol.* 2010; 37(6):1236–1243.

Marcum ZA, Hanlon JT. Recognizing the risks of chronic nonsteroidal anti-inflammatory drug use in older adults. *Ann Longterm Care* 2010; 18(9):24–27.

Massey T, Derry S, Moore RA, McQuay HJ. Topical NSAIDs for acute pain in adults. *Cochrane Database Syst Rev.* 2010; June 16(6).

McGeer PL. Cyclooxygenase-2 inhibitors: Rationale and therapeutic potential for Alzheimer's disease. *Drugs Aging* 2000; 17:1–11.

Moore RA, Tramèr MR, Carroll D, et al. Quantitative systematic review of topically-applied non-steroidal anti-inflammatory drugs. *BMJ* 1998; 316:333–338.

Østensen ME, Skomsvoll JF. Anti-inflammatory pharmacotherapy during pregnancy. *Expert Opinion on Pharmacotherapy* Mar. 2004; 5(3):571–580.

Rostom A, Dube C, Wells GA, Tugwell P, Welch V, Jolicoeur E, McGowan J, Lanas A. Prevention of NSAID-induced gastroduodenal ulcers (Review). *The Cochrane Library* 2007. http://www.med.upenn.edu/gastro/documents/CochranedatabaseNSAIDSpud.pdf.

Rostom A, Moayyedi P, Hunt R, for the Canadian Association of Gastroenterology Consensus group. Canadian consensus guidelines on long-term nonsteroidal anti-inflammatory drug therapy and the need for gastroprotection: Benefits versus risks. *Alimentary Pharmacology and Therapeutics* 2009; 29(5):481–496.

Samter M, Beers RF Jr. Intolerance to aspirin. Clinical studies and consideration of its pathogenesis. *Ann Intern Med* 1968; 68:975–983.

Scheiman JM. NSAIDs, gastrointestinal injury, and cytoprotection. *Gastroenterol Clin North Am* 1996; 25:279–298.

Silverstein FE, Faich G, Goldstein JL, et al. Gastrointestinal toxicity with celecoxib vs nonsteroidal anti-inflammatory drugs for osteoarthritis and rheumatoid arthritis. The CLASS study: A randomized controlled trial. Celecoxib Long-term Arthritis Safety Study. *JAMA* 2000; 284:1247–1255.

Singh G, Wu O, Langhorne P, et al. Risk of acute myocardial infarction with non-selective non-steroidal anti-inflammatory drugs: A meta-analysis. *Arthritis Res Ther* 2006; 8:R153.

Sostres C, Lanas A. Gastrointestinal effects of aspirin. *Nature Reviews Gastroenterology and Hepatology* 8, 385–394 (July 2011) | doi:10.1038/nrgastro.2011.97.

Thun MJ, Henley SJ, Patrono C. Nonsteroidal anti-inflammatory drugs as anticancer agents: Mechanistic, pharmacologic, and clinical issues. *J Natl Cancer Inst* 2002; 94:252–266.

Trelle S, Reichenbach S, Wandel S, Hildebrand P, Tschannen B, Villiger P, et al. Cardiovascular safety of non-steroidal anti-inflammatory drugs: Network meta-analysis. *BMJ* 2011; 342:c7086doi:10.1136/bmj.c7086.

Vane JR. Inhibition of prostaglandin synthesis as a mechanism of action for the aspirin-like drugs. *Nature* 1971; 231:232–235.

Vane JR. Introduction: Mechanism of action of NSAIDs. *British Journal of Rheumatology* 1996; 35(Suppl l):l–3.

Vestergaard P, Rejnmark L, Mosekilde L. Fracture risk associated with use of nonsteroidal anti-inflammatory drugs, acetylsalicylic acid, and acetaminophen and the effects of rheumatoid arthritis and osteoarthritis. *Calcif Tissue Int* 2006; 79:84–94.

Wanders A, van der Heijde D, Landewe R, et al. Nonsteroidal anti-inflammatory drugs reduce radiographic progression in patients with ankylosing spondylitis: A randomized controlled trial. *Arthritis Rheum* 2005; 52:1756–1765.

Warner TD, Mitchell JA. Cyclooxygenase-3 (COX-3): Filling in the gaps toward a COX continuum? *Proc Natl Acad Sciences* 2002; 99:13371–13373.

Whelton, A. Renal and related cardiovascular effects of conventional and COX-2 specific NSAIDs and non-NSAID analgesics. *American Journal of Therapeutics* 2000; 7:63–74.

Wilcox CM, Shalek KA, Cotsonis G. Striking prevalence of over-the-counter nonsteroidal anti-inflammatory drug use in patients with upper gastrointestinal hemorrhage. *Arch Intern Med* 1994; 154:42–46.

Wolfe MM, Lichtenstein DR, Singh G. Gastrointestinal toxicity of nonsteroidal antiinflammatory drugs. *N Engl J Med* 1999; 340:188.

Index

A

D

E

F

G

H

J

K

L

M

N

O

P

R

S

T

U

V

W

Z